The Structural Basis of Biological Energy Generation

This illustration is based on a scanned image of a leaf. The left side of the illustration shows the original leaf, followed, to the right, by a black and white representation. Consecutive strips to the right show the same leaf at increasing resolution, thereby revealing increased structural detail. The illustration exemplifies how scientists can explore living systems at different scales that reveal different information and insights, which are represented by different colors.

Advances in Photosynthesis and Respiration
Including Bioenergy and Related Processes

VOLUME 39

The book series Advances in Photosynthesis and Respiration *Including Bioenergy and Related Processes* provides a comprehensive and state-of-the-art account of research in photosynthesis, respiration and related processes. Virtually all life on our planet Earth ultimately depends on photosynthetic energy capture and conversion to energy-rich organic molecules. These are used for food, fuel, and fiber. Photosynthesis is the source of almost all bioenergy on Earth. The fuel and energy uses of photosynthesized products and processes have become an important area of study, and competition between food and fuel has led to resurgence in photosynthesis research. This series of books spans topics from physics to agronomy and medicine; from femtosecond processes through season-long production to evolutionary changes over the course of the history of the Earth; from the photophysics of light absorption, excitation energy transfer in the antenna to the reaction centers, where the highly-efficient primary conversion of light energy to charge separation occurs, through the electrochemistry of intermediate electron transfer, to the physiology of whole organisms and ecosystems; and from X-ray crystallography of proteins to the morphology of organelles and intact organisms. In addition to photosynthesis in natural systems, genetic engineering of photosynthesis and artificial photosynthesis is included in this series. The goal of the series is to offer beginning researchers, advanced undergraduate students, graduate students, and even research specialists, a comprehensive, up-to-date picture of the remarkable advances across the full scope of research on photosynthesis and related energy processes. The purpose of this series is to improve understanding of photosynthesis and respiration at many levels both to improve basic understanding of these important processes and to enhance our ability to use photosynthesis for the improvement of the human condition.

For further volumes:
www.springer.com/series/5599

The Structural Basis of Biological Energy Generation

Edited by

Martin F. Hohmann-Marriott
Norwegian University of Science and Technology
Trondheim
Norway

Springer

Editor
Martin F. Hohmann-Marriott
Department of Biotechnology
Norwegian University of Science
and Technology, NTNU
Trondheim, Norway
martin.hohmann-marriott@ntnu.no

ISSN 1572-0233 ISSN 2215-0102 (electronic)
ISBN 978-94-017-8741-3 ISBN 978-94-017-8742-0 (eBook)
DOI 10.1007/978-94-017-8742-0
Springer Dordrecht Heidelberg New York London

Library of Congress Control Number: 2014939010

Printed on acid-free paper

Springer is part of Springer Science+Business Media (www.springer.com)

From the Series Editors

Advances in Photosynthesis and Respiration Including Bioenergy and Related Processes

Volume 39: The Structural Basis of Biological Energy Generation

We are delighted to announce the publication of Volume *39* in this series. This is the fifth volume with the new cover and enhanced web presence. The series publisher, Springer, now makes the table of contents of all of the volumes freely available online. Links to each volume are given below. Readers may also see that this volume and the past few volumes have had significantly more color figures, and the color figures are now better integrated into the chapters, instead of being collected in one section of the book. This improvement was possible because of changes in how the books are produced. Another change is that references to chapters in books are now tracked by bibliographic services. This will help authors provide evidence of the importance of their work. We hope that these updates will maintain the importance of these edited volumes in the dissemination of the science of photosynthesis and bioenergy.

We are fortunate to have Martin F. Hohmann-Marriott, of Norway, to take the lead of editing a unique volume on *The Structural Basis of Biological Energy Generation.* It is first of its kind in the field of photosynthesis. Martin is an authority on bioenergetics of photosynthesis, and on the evolution of photosynthetic organisms. He has expertise on a variety of photosynthetic organisms including green sulfur bacteria, cyanobacteria, green algae and many others.

The Book

As stated in the Preface by Martin, this book provides an "overview of the structural foundation for bioenergetics in bacteria, algae and plants from the molecular to the organism level". It deals with how organisms channel energy into generating what we call "*the stuff of life*" giving rise to the living world around us. We learn about the details of the mechanisms that organisms employ to capture light energy, transport electrons and protons, and ultimately fix carbon. Biological energy generation also requires accessing the energy stored by photosynthetic reactions, and so mitochondria are covered in several chapters as well. The organisms covered in this book illustrate the range of mechanisms of biological energy generation rather than the range of organisms most familiar to people. This emphasis is a better representation of the diversity of biological energy generation. We are sure that this book will have a great impact in the field of photosynthesis for a long time.

Authors

The current book contains 24 chapters written by 49 authors from 15 countries. We thank all the authors for their valuable contribution to this book; their names (arranged alphabetically) are: N. **A**dir (Israel; Chap. 4); M. Asao (USA; Chap. 13); M. **B**arták (Czech Republic; Chap. 20); F. Baymann (France; Chap. 8); M. Benchimol (Brazil; Chap. 22); M. Berney (New Zealand; Chap. 15); E.J. Boekema (The Netherlands; Chap. 12); B. Böttcher (United Kingdom; Chap. 6); Z.L. Bouzon (Brazil; Chap. 16); H.-P. Braun (Germany; Chap. 12); L.S. Brown (Canada; Chap. 1); C. Büchel (Germany; Chap. 2); S.J. Butcher (Finland; Chap. 5); A.M. **C**ollins (USA; Chap. 13); G.M. Cook (New Zealand; Chap. 15); V. **D**askalakis (Cyprus; Chap. 10); L. David (Israel; Chap. 4); A. Dikiy (Norway; Chap. 11); O. Dobrovolska (Norway; Chap. 11); N.V. Dudkina (United Kingdom; Chap. 12); D. **G**argano (Norway; Chap. 23); M.L. Genova (Italy; Chap. 21); H.L. Gorton (USA; Chap. 19); P. Gräber (Germany; Chap. 6); K. Gundermann (Germany; Chap. 2); E. Hoffmann (Germany; Chap. 3); M.S. Kimber (Canada; Chap. 7); P.G. **K**roth (Germany; Chap. 18); J. **M**aple (Norway; Chap. 23); A. Marx (Israel; Chap. 4); Y. Matsuda (Japan; Chap. 18); A.E. McDonald (Canada; Chap. 9); J.A. Mears (USA; Chap. 24); S.G. Møller (USA; Chap. 23); T. **P**olívka (Czech Republic; Chap. 3); J. Pšenčík (Czech Republic; Chap. 5); W.M. **S**attley (USA; Chap. 13); E.C. Schmidt (Brazil; Chap. 16); M. Schmidt (Germany; Chap. 17); E. Shumilina (Norway; Chap. 11); K.-H. **T**ang (USA; Chap. 13); F. ten Brink (France; Chap. 8); C.S. Ting (USA; Chap. 14); R. Tuma (United Kingdom; Chap. 5); G.C. **V**anlerberghe (Canada; Chap. 9); C. Varotsis (Cyprus; Chap. 10); T.C. Vogelmann (USA; Chap. 19); C. **W**ilhelm (Germany; Chap. 17); and C.S. **Z**itta (Brazil; Chap. 16).

Our Books: Now 38 Volumes

We list below information on the 38 volumes that have been published thus far (see http://www.springer.com/series/5599 for the series web site). We are pleased to note that Springer, our publisher, is now producing complete *Tables of Contents* of these books. Electronic access to individual chapters depends on subscription (ask your librarian) but Springer provides free downloadable front matter as well as indexes. The available web sites of the books in the Series are listed below.

- **Volume 38 (2014): Microbial BioEnergy: Hydrogen Production,** edited by Davide Zannoni and Roberto De Phillipis, both from Italy. Fifteen chapters, XXXV + 366 pp, Hardcover, ISBN: 978-94-017-8553-2 (HB); ISBN 978-94-017-8554-9 (e-book) [http://www.springer.com/life+sciences/plant+sciences/book/978-94-017-8553-2]
- **Volume 37 (2014): Photosynthesis in Bryophytes and Early Land Plants,** edited by David T. Hanson, and Steven K. Rice, both from USA. Eighteen chapters, XXVII + 343 pp, Hardcover, ISBN: 978-94-007-6987-8 (HB); ISBN 978-94-007-6988-5 (e-book) [http://www.springer.com/life+sciences/plant+sciences/book/978-94-007-6987-8]
- **Volume 36 (2013): Plastid Development in Leaves during Growth and Senescence**, edited by Basanti Biswal, Karin Krupinska and Udaya Biswal, from India and Germany. Twenty-eight chapters, 837 pp, Hardcover, ISBN: 978-94-007-5723-3 (HB); ISBN 978-94-007-5724-0 (e-book) [http://www.springer.com/life+sciences/plant+sciences/book/978-94-007-5723-3]
- **Volume 35 (2012): Genomics of Chloroplasts and Mitochondria**, edited by Ralph Bock and Volker Knoop, from Germany. Nineteen chapters, 475 pp, Hardcover, ISBN: 978-94-007-2919-3 (HB) ISBN 978-94-007-2920-9 (e-book) [http://www.springer.com/life+sciences/plant+sciences/book/978-94-007-2919-3]
- **Volume 34 (2012): Photosynthesis – Plastid Biology, Energy Conversion and Carbon Assimilation**, edited by Julian Eaton-Rye, Baishnab C. Tripathy, and Thomas D. Sharkey, from New Zealand, India, and USA. Thirty-three chapters, 854 pp, Hardcover, ISBN: 978-94-007-1578-3 (HB) ISBN 978-94-007-1579-0 (e-book) [http://www.springer.com/life+sciences/plant+sciences/book/978-94-007-1578-3/]
- **Volume 33 (2012): Functional Genomics and Evolution of Photosynthetic Systems**, edited by Robert L. Burnap and Willem F.J. Vermaas, from USA. Fifteen chapters, 428 pp, ISBN:

978-94-007-1532-5 [http://www.springer.com/life+sciences/book/978-94-007-1532-5/]

- **Volume 32 (2011): C4 Photosynthesis and Related CO2 Concentrating Mechanisms,** edited by Agepati S. Raghavendra and Rowan Sage, from India and Canada. Nineteen chapters, 425 pp, Hardcover, ISBN: 978-90-481-9406-3 [http://www.springer.com/life+sciences/plant+sciences/book/978-90-481-9406-3]
- **Volume 31 (2010): The Chloroplast: Basics and Applications,** edited by Constantin Rebeiz (USA), Christoph Benning (USA), Hans J. Bohnert (USA), Henry Daniell (USA), J. Kenneth Hoober (USA), Hartmut K. Lichtenthaler (Germany), Archie R. Portis (USA), and Baishnab C. Tripathy (India). Twenty-five chapters, 451 pp, Hardcover, ISBN: 978-90-481-8530-6 [http://www.springer.com/life+sciences/plant+sciences/book/978-90-481-8530-6]
- **Volume 30 (2009): Lipids in Photosynthesis: Essential and Regulatory Functions,** edited by Hajime Wada and Norio Murata, both from Japan. Twenty chapters, 506 pp, Hardcover, ISBN: 978-90-481-2862-4; e-book, ISBN: 978-90-481-2863-1 [http://www.springer.com/life+sciences/plant+sciences/book/978-90-481-2862-4]
- **Volume 29 (2009): Photosynthesis in Silico: Understanding Complexity from Molecules,** edited by Agu Laisk, Ladislav Nedbal, and Govindjee, from Estonia, The Czech Republic, and USA. Twenty chapters, 525 pp, Hardcover, ISBN: 978-1-4020-9236-7 [http://www.springer.com/life+sciences/plant+sciences/book/978-1-4020-9236-7]
- **Volume 28 (2009): The Purple Phototrophic Bacteria,** edited by C. Neil Hunter, Fevzi Daldal, Marion C. Thurnauer and J. Thomas Beatty, from UK, USA and Canada. Forty-eight chapters, 1053 pp, Hardcover, ISBN: 978-1-4020-8814-8 [http://www.springer.com/life+sciences/plant+sciences/book/978-1-4020-8814-8]
- **Volume 27 (2008): Sulfur Metabolism in Phototrophic Organisms**, edited by Christiane Dahl, Rüdiger Hell, David Knaff and Thomas Leustek, from Germany and USA. Twenty-four chapters, 551 pp, Hardcover, ISBN: 978-4020-6862-1 [http://www.springer.com/life+sciences/plant+sciences/book/978-1-4020-6862-1]
- **Volume 26 (2008): Biophysical Techniques Photosynthesis**, Volume II, edited by Thijs Aartsma and Jörg Matysik, both from The Netherlands. Twenty-four chapters, 548 pp, Hardcover, ISBN: 978-1-4020-8249-8 [http://www.springer.com/life+sciences/plant+sciences/book/978-1-4020-8249-8]
- **Volume 25 (2006): Chlorophylls and Bacteriochlorophylls: Biochemistry, Biophysics, Functions and Applications,** edited by Bernhard Grimm, Robert J. Porra, Wolfhart Rüdiger, and Hugo Scheer, from Germany and Australia. Thirty-seven chapters, 603 pp, Hardcover, ISBN: 978-1-40204515-8 [http://www.springer.com/life+sciences/plant+sciences/book/978-1-4020-4515-8]
- **Volume 24 (2006): Photosystem I: The Light-Driven Plastocyanin: Ferredoxin Oxidoreductase,** edited by John H. Golbeck, from USA. Forty chapters, 716 pp, Hardcover, ISBN: 978-1-40204255-3 [http://www.springer.com/life+sciences/plant+sciences/book/978-1-4020-4255-3]
- **Volume 23 (2006): The Structure and Function of Plastids**, edited by Robert R. Wise and J. Kenneth Hoober, from USA. Twenty-seven chapters, 575 pp, Softcover, ISBN: 978-1-4020-6570–6; Hardcover, ISBN: 978-1-4020-4060-3 [http://www.springer.com/life+sciences/plant+sciences/book/978-1-4020-4060-3]
- **Volume 22 (2005): Photosystem II: The Light-Driven Water: Plastoquinone Oxidoreductase**, edited by Thomas J. Wydrzynski and Kimiyuki Satoh, from Australia and Japan. Thirty-four chapters, 786 pp, Hardcover, ISBN: 978-1-4020-4249-2 [http://www.springer.com/life+sciences/plant+sciences/book/978-1-4020-4249-2]
- **Volume 21 (2005): Photoprotection, Photoinhibition, Gene Regulation, and Environment,** edited by Barbara Demmig-Adams, William W. Adams III and Autar K. Mattoo, from USA. Twenty-one chapters, 380 pp, Hardcover, ISBN: 978-14020-3564-7 [http://www.springer.com/life+sciences/plant+sciences/book/978-1-4020-3564-7]

- **Volume 20 (2006): Discoveries in Photosynthesis**, edited by Govindjee, J. Thomas Beatty, Howard Gest and John F. Allen, from USA, Canada and UK. One hundred and eleven chapters, 1304 pp, Hardcover, ISBN: 978-1-4020-3323-0 [http://www.springer.com/life+sciences/plant+sciences/book/978-1-4020-3323-0]
- **Volume 19 (2004): Chlorophyll *a* Fluorescence: A Signature of Photosynthesis**, edited by George C. Papageorgiou and Govindjee, from Greece and USA. Thirty-one chapters, 820 pp, Hardcover, ISBN: 978-1-4020-3217-2 [http://www.springer.com/life+sciences/biochemistry+%26+biophysics/book/978-1-4020-3217-2]
- **Volume 18 (2005): Plant Respiration: From Cell to Ecosystem**, edited by Hans Lambers and Miquel Ribas-Carbo, from Australia and Spain. Thirteen chapters, 250 pp, Hardcover, ISBN: 978-14020-3588-3 [http://www.springer.com/life+sciences/plant+sciences/book/978-1-4020-3588-3]
- **Volume 17 (2004): Plant Mitochondria: From Genome to Function**, edited by David Day, A. Harvey Millar and James Whelan, from Australia. Fourteen chapters, 325 pp, Hardcover, ISBN: 978-1-4020-2399-6 [http://www.springerlink.com/content/978-1-7923-2399-6]
- **Volume 16 (2004): Respiration in Archaea and Bacteria: Diversity of Prokaryotic Respiratory Systems,** edited by Davide Zannoni, from Italy. Thirteen chapters, 310 pp, Hardcover, ISBN: 978-14020-2002-5 [http://www.springer.com/life+sciences/plant+sciences/book/978-1-4020-2002-5]
- **Volume 15 (2004): Respiration in Archaea and Bacteria: Diversity of Prokaryotic Electron Transport Carriers**, edited by Davide Zannoni, from Italy. Thirteen chapters, 350 pp, Hardcover, ISBN: 978-1-4020-2001-8 [http://www.springer.com/life+sciences/biochemistry+%26+biophysics/book/978-1-4020-2001-8]
- **Volume 14 (2004): Photosynthesis in Algae**, edited by Anthony W. Larkum, Susan Douglas and John A. Raven, from Australia, Canada and UK. Nineteen chapters, 500 pp, Hardcover, ISBN: 978-0-7923-6333-0 [http://link.springer.com/book/10.1007/978-94-007-1038-2/page/1]
- **Volume 13 (2003): Light-Harvesting Antennas in Photosynthesis**, edited by Beverley R. Green and William W. Parson, from Canada and USA. Seventeen chapters, 544 pp, Hardcover, ISBN:978-07923-6335-4 [http://www.springer.com/life+sciences/plant+sciences/book/978-0-7923-6335-4?otherVersion=978-90-481-5468-5]
- **Volume 12 (2003): Photosynthetic Nitrogen Assimilation and Associated Carbon and Respiratory Metabolism,** edited by Christine H. Foyer and Graham Noctor, from UK and France. Sixteen chapters, 304 pp, Hardcover, ISBN: 978-07923-6336-1 [http://www.springer.com/life+sciences/plant+sciences/book/978-0-7923-6336-1]
- **Volume 11 (2001): Regulation of Photosynthesis,** edited by Eva-Mari Aro and Bertil Andersson, from Finland and Sweden. Thirty-two chapters, 640 pp, Hardcover, ISBN: 978-0-7923-6332-3 [http://www.springer.com/life+sciences/plant+sciences/book/978-0-7923-6332-3]
- **Volume 10 (2001): Photosynthesis: Photobiochemistry and Photobiophysics**, authored by Bacon Ke, from USA. Thirty-six chapters, 792 pp, Softcover, ISBN: 978-0-7923-6791-8; Hardcover: ISBN: 978-0-7923-6334-7 [http://www.springer.com/life+sciences/plant+sciences/book/978-0-7923-6334-7]
- **Volume 9 (2000): Photosynthesis: Physiology and Metabolism,** edited by Richard C. Leegood, Thomas D. Sharkey and Susanne von Caemmerer, from UK, USA and Australia. Twenty-four chapters, 644 pp, Hardcover, ISBN: 978-07923-6143-5 [http://www.springer.com/life+sciences/plant+sciences/book/978-0-7923-6143-5]
- **Volume 8 (1999): The Photochemistry of Carotenoids,** edited by Harry A. Frank, Andrew J. Young, George Britton and Richard J. Cogdell, from USA and UK. Twenty chapters, 420 pp, Hardcover, ISBN: 978-0-7923-5942-5 [http://www.springer.com/life+sciences/plant+sciences/book/978-0-7923-5942-5]
- **Volume 7 (1998): The Molecular Biology of Chloroplasts and Mitochondria in**

Chlamydomonas, edited by Jean David Rochaix, Michel Goldschmidt-Clermont and Sabeeha Merchant, from Switzerland and USA. Thirty-six chapters, 760 pp, Hardcover, ISBN: 978-0-7923-5174-0 [http://www.springer.com/life+sciences/plant+sciences/book/978-0-7923-5174-0]
- **Volume 6 (1998): Lipids in Photosynthesis: Structure, Function and Genetics**, edited by Paul-André Siegenthaler and Norio Murata, from Switzerland and Japan. Fifteen chapters, 332 pp, Hardcover, ISBN: 978-0-7923-5173-3 [http://www.springer.com/life+sciences/plant+sciences/book/978-0-7923-5173-3]
- **Volume 5 (1997): Photosynthesis and the Environment,** edited by Neil R. Baker, from UK. Twenty chapters, 508 pp, Hardcover, ISBN: 978-07923-4316-5 [http://www.springer.com/life+sciences/plant+sciences/book/978-0-7923-4316-5]
- **Volume 4 (1996): Oxygenic Photosynthesis: The Light Reactions**, edited by Donald R. Ort and Charles F. Yocum, from USA. Thirty-four chapters, 696 pp, Softcover: ISBN: 978-0-7923-3684–6; Hardcover, ISBN: 978-0-7923-3683-9 [http://www.springer.com/life+sciences/plant+sciences/book/978-0-7923-3683-9]
- **Volume 3 (1996): Biophysical Techniques in Photosynthesis**, edited by Jan Amesz and Arnold J. Hoff, from The Netherlands. Twenty-four chapters, 426 pp, Hardcover, ISBN: 978-0-7923-3642-6 [http://www.springer.com/life+sciences/plant+sciences/book/978-0-7923-3642-6]
- **Volume 2 (1995): Anoxygenic Photosynthetic Bacteria**, edited by Robert E. Blankenship, Michael T. Madigan and Carl E. Bauer, from USA. Sixty-two chapters, 1331 pp, Hardcover, ISBN: 978-0-7923-3682-8 [http://www.springer.com/life+sciences/plant+sciences/book/978-0-7923-3681-5]
- **Volume 1 (1994): The Molecular Biology of Cyanobacteria**, edited by Donald R. Bryant, from USA. Twenty-eight chapters, 916 pp, Hardcover, ISBN: 978-1-4020-2400-9 [http://www.springerlink.com/content/978-1-4020-2400-9]

Further information on these books and ordering instructions can be found at http://www.springer.com/series/5599. Contents of volumes 1–31 can also be found at http://www.life.uiuc.edu/govindjee/photosynSeries/ttocs.html. (For volumes 33–35, pdf files of the entire Front Matter are available.)

Special 25 % discounts are available to members of the International Society of Photosynthesis Research, ISPR http://www.photosynthesisresearch.org/. See http://www.springer.com/ispr.

Future Advances in Photosynthesis and Respiration and Other Related Books

The readers of the current series are encouraged to watch for the publication of the forthcoming books (not necessarily arranged in the order of future appearance):

- *Canopy Photosynthesis: From Basics to Applications* (Editors: Kouki Hikosaka, Ülo Niinemets and Niels P.R. Anten)
- *Non-Photochemical Quenching (NPQ) and Thermal Energy Dissipation In Plants, Algae and Cyanobacteria* (Editors: Barbara Demmig-Adams, Győző Garab and Govindjee)
- *Cytochromes* (Editors: William A. Cramer and Tovio Kallas)

In addition to the above contracted books, the following topics are under consideration:

- Algae, Cyanobacteria: Biofuel and Bioenergy
- Artificial Photosynthesis
- ATP Synthase and Proton Translocation
- Bacterial Respiration II
- Biohydrogen Production
- Carotenoids II
- Cyanobacteria II
- Ecophysiology
- Evolution of Photosynthesis
- Global Aspects of Photosynthesis
- Green Bacteria and Heliobacteria
- Interactions between Photosynthesis and other Metabolic Processes
- Limits of Photosynthesis: Where do we go from here

- Photosynthesis, Biomass and Bioenergy
- Photosynthesis under Abiotic and Biotic Stress
- Plant Respiration II

If you have any interest in editing/co-editing any of the above listed books, or being an author, please send an E-mail to Tom Sharkey (tsharkey@msu.edu) and/or to Govindjee at gov@illinois.edu. Suggestions for additional topics are also welcome.

In view of the interdisciplinary character of research in photosynthesis and respiration, it is our earnest hope that this series of books will be used in educating students and researchers not only in plant sciences, molecular and cell biology, integrative biology, biotechnology, agricultural sciences, microbiology, biochemistry, chemical biology, biological physics, and biophysics, but also in bioengineering, chemistry, and physics.

Acknowledgments

We take this opportunity to thank and congratulate Martin F. Hohmann-Marriott for his outstanding editorial work; he has done a fantastic job, not only in editing, but also in organizing this book for all of us, and for his highly professional dealing with the reviewing process. We thank all the 49 authors of this book (see the list above): without their authoritative chapters, there would be no such volume. We give special thanks to Mr. Prakash Marudhu, SPi Global, India for directing the typesetting of this book; his expertise has been crucial in bringing this book to completion. We owe Jacco Flipsen, Andre Tournois, and Ineke Ravesloot (of Springer) thanks for their friendly working relation with us that led to the production of this book.

January 1, 2014 **Govindjee**

Department of Plant Biology,
Department of Biochemistry and Center
of Biophysics & Quantitative Biology
University of Illinois at Urbana-Champaign,
Urbana, IL 61801, USA
gov@illinois.edu

Thomas D. Sharkey
Department of Biochemistry
and Molecular Biology
Michigan State University,
East Lansing, MI, 48824, USA
tsharkey@msu.edu

Series Editors

A 2013 photograph of Govindjee (on the right) with his wife Rajni. Photo by Zsuzsanna Deaĭk.

Govindjee, who uses one name only, was born on October 24, 1932, in Allahabad, India. Since 1999, he has been Professor Emeritus of Biochemistry, Biophysics and Plant Biology at the University of Illinois at Urbana-Champaign (UIUC), Urbana, IL, USA. He obtained his B.Sc. (Chemistry and Biology) and M.Sc. (Botany; Plant Physiology) in 1952 and 1954, from the University of Allahabad. He studied 'Photosynthesis' at the UIUC, under two pioneers of photosynthesis Robert Emerson, and Eugene Rabinowitch, obtaining his Ph.D. in 1960, in Biophysics. He is best known for his research on excitation energy transfer, light emission (prompt and delayed fluorescence, and thermoluminescence), primary photochemistry and electron transfer in "Photosystem II" (PS II, water-plastoquinone oxido-reductase). His research, with many collaborators, has included the discovery of a short-wavelength form of chlorophyll (Chl) *a* functioning in what is now called PS II; of the two-light effect in Chl *a* fluorescence; and, with his wife Rajni Govindjee, of the two-light effect (Emerson Enhancement) in NADP reduction in chloroplasts. His major achievements, together with several other researchers, include an understanding of the basic relationship between Chl *a* fluorescence and photosynthetic reactions; an unique role of bicarbonate/carbonate on the electron acceptor side of PS II, particularly in the protonation events involving the Q_B binding region; the theory of thermoluminescence in plants; the first picosecond measurements on the primary photochemistry of PS II; and the use of Fluorescence Lifetime Imaging Microscopy (FLIM) of Chl *a* fluorescence in understanding *photoprotection*, by plants, against excess light. His current focus is on the 'History of Photosynthesis Research', and in 'Photosynthesis Education'. He has served on the faculty of the UIUC for ~40 years. Govindjee's honors include: Fellow of the American Association of Advancement of Science (AAAS); Distinguished Lecturer of the School of Life Sciences, UIUC; Fellow and Lifetime member of the National Academy of Sciences (India); President of the American Society for Photobiology (1980–1981); Fulbright Scholar (1956), Fulbright Senior Lecturer (1997), and Fulbright Specialist (2012); Honorary President of the 2004 International Photosynthesis Congress (Montréal, Canada); the first recipient of the Lifetime Achievement Award of the Rebeiz Foundation for Basic Biology, 2006; Recipient of the Communication Award of the International Society of Photosynthesis Research, 2007; and the Liberal Arts & Sciences Lifetime Achievement Award of the UIUC, 2008. Further, Govindjee was honored **(1)** in 2007, through two special volumes of Photosynthesis Research, celebrating his 75th birthday and for his 50-year dedicated research in 'Photosynthesis' (Guest Editor: Julian Eaton-Rye); **(2)** in 2008, through a special International Symposium on 'Photosynthesis in a Global Perspective', held in November, 2008, at the University of Indore, India; **(3)** Volume 34 of this Series "*Photosynthesis: Plastid Biology, Energy Conversion and Carbon Assimilation*", edited by Julian Eaton-Rye, Baishnab C. Tripathy, and one of us (TDS), was dedicated to Govindjee, celebrating his academic career; and **(4)** in 2013, through two special volumes of Photosynthesis Research, celebrating his 80th birthday (Guest Editors: Suleyman Allakhverdiev, J.-R. Shen, and Gerald T. Edwards). Of special note is the article by Julian Eaton-Rye (2013). Govindjee at 80: more than 50 years of free energy for photosynthesis. Photosynthesis Research 116:111–144.

Govindjee is coauthor of *Photosynthesis* (John Wiley, 1969); and editor of many books, published by several publishers including Academic Press and Kluwer Academic Publishers (now Springer). Each year a Govindjee and Rajni Govindjee Award (http://www.life.illinois.edu/plantbio/PlBiogiving.html; http://sib.illinois.edu/grants_Govindjee.htm) is being given to graduate students, by the Department of Plant Biology (odd years) or by the Department of Biochemistry (even years), at the UIUC, to recognize Excellence in Biological Sciences. For further information on Govindjee, see his web site at http://www.life.illinois.edu/govindjee.

Thomas D. (Tom) Sharkey obtained his Bachelor's degree in Biology in 1974 from Lyman Briggs College, a residential science college at Michigan State University, East Lansing, Michigan, USA. After 2 years as a research technician in the federally funded Plant Research Laboratory at Michigan State University under the mentorship of Professor Klaus Raschke, Tom entered the PhD program in the same lab, and graduated in 1980. Postdoctoral research was carried out with Professor Graham Farquhar at the Australian National University, in Canberra, where he co-authored a landmark review on photosynthesis and stomatal conductance that continues to receive much attention 30 years after its publication. For 5 years, Tom worked at the Desert Research Institute together with Professor Barry Osmond, followed by 20 years as a professor of botany at the University of Wisconsin in Madison. In 2008, Tom became Professor and Chair of the Department of Biochemistry and Molecular Biology at Michigan State University. Tom's research interests center on the biochemistry and biophysics of gas exchange between plants and the atmosphere. Photosynthetic gas exchange, especially carbon dioxide uptake and use, and isoprene emission from plants, are the two major research topics in his laboratory. Among his contributions are measurements of the carbon dioxide concentration inside leaves, studies of the resistance to diffusion of carbon dioxide within the mesophyll of leaves of C_3 plants, and an exhaustive study of short-term feedback effects on carbon metabolism. As part of the study of short-term feedback effects, Tom's research group demonstrated that maltose is the major form of carbon export from chloroplasts at night, and made significant contributions to the elucidation of the pathway by which leaf starch is converted to sucrose at night. In the isoprene research field, Tom is recognized as the leading advocate for thermotolerance of photosynthesis as the explanation for why plants emit isoprene. In addition, his laboratory has cloned many of the genes that underlie isoprene synthesis, and he has published many papers on the biochemical regulation of isoprene synthesis. Tom has coedited three books: T.D. Sharkey, E.A. Holland and

H.A. Mooney (Eds.) *Trace Gas Emissions from Plants*, Academic, San Diego, CA, 1991; R.C. Leegood, T.D. Sharkey, and S, von Caemmerer (Eds.) *Physiology and Metabolism, Advances in Photosynthesis (and Respiration)*, Volume 9 of this Series, Kluwer (now Springer), Dordrecht, 2000; and Volume 34 of this series *Photosynthesis: Plastid Biology, Energy Conversion and Carbon Assimilation*, Advances in Photosynthesis and Respiration *Including Bioenergy and Related Processes,* Julian J. Eaton-Rye, Baishnab C. Tripathy and Thomas D. Sharkey (Eds) Springer. Tom joined the series founder Govindjee as Series Co-editor from volume 31 of this series. Tom is currently the Chairperson of the Department of Biochemistry and Molecular Biology, Michigan State University, East Lansing, Michigan. For further information see his web page at: http://www.bmb.msu.edu/faculty/sharkey/Sharkey/index.html.

Contents

Preface

This book, *The Structural Basis of Biological Energy Generation*, provides a detailed overview of the structural foundation for bioenergetics in bacteria, algae and plants from the molecular to the organism level. The authors of the chapters review our current understanding of how organisms channel energy gradients into generating the living world. Thus the book illustrates the mechanisms employed in these organisms to efficiently capture light energy, transport electrons and protons, and fix carbon.

In addition to chlorophyll-based photoconverters, a fundamentally different type of photoconverters, rhodopsins, convert light energy into a trans-membrane proton gradient using conformational changes of the carotenoid retinal. The molecular mechanisms that underlie the light conversion by rhodopsins is the topic of Chap. 1. The capture and utilization of light is the energetic basis of all organisms that inhabit ecosystems on the surface of Earth. A careful description of the molecular mechanisms that facilitate light harvesting is provided in chapters reviewing light-harvesting complexes of heterokont algae (Chap. 2), peridinin-chlorophyll proteins (Chap. 3), phycobilisomes (Chap. 4), and chlorosomes (Chap. 5). Current models for the biogenesis of chlorosomes and phycobilisomes are also reviewed in detail in Chaps. 4 and 5. An in-depth analysis of how light penetrates the complex organ for light capture by plants is provided in Chap. 19. The influence of morphological and environmental factors on photosynthesis in lichens is discussed in Chap. 20.

Several chapters provide molecular insights into electron and proton transport within protein complexes. Chapter 6 explores structure-function relationships of ATPases found in different groups of organisms. Coupling electron transfer to proton-translocation is the topic of a chapter on Cytochrome b_6f and *bc* complexes (Chap. 8). Electron transport to oxygen in quinone oxidases (Chap. 9) and cytochrome oxidases (Chap. 10) are explored using structural and computational approaches. Methionine sulfoxide reductase recovers proteins damaged by electron transport to oxygen (Chap. 11). In addition to their destructive action in mitochondria, reactive oxygen species are also an important signaling molecule in mitochondria (Chap. 24).

Biological systems employ compartmentalization to efficiently execute cellular functions. Protein complexes can provide sophisticated compartments for light-harvesting (Chaps. 4 and 5), carbon fixation (Chap. 7), and coordinated proton and electron transport (Chap. 12). In addition, membrane systems also compartmentalize the respiratory (Chaps. 15 and 21) and photosynthetic (Chap. 16) energy generation machinery. Structural and physiological insights into energy generation of bacteria and algae are provided in chapters with a focus on heliobacteria (Chap. 13), cyanobacteria (Chap. 14), mycobacteria (Chap. 15) and green algae (Chap. 17).

The ability to convert inorganic carbon into organic molecules is the defining characteristic of autotrophic organisms. Carboxysomes are complex protein assemblies with the function to locally increase the amount of inorganic carbon. The molecular structure of carboxysomes is reviewed in Chap. 7, and insights obtained from imaging carboxysomes in cyanobacteria are provided in Chap. 14. The mode of concentrating inorganic carbon is also the focus of a chapter on diatoms (Chap. 18). In addition, the potential utilization of fixed carbon for the production of biofuels and carotenoids in different green algae is reviewed in Chap. 17.

Mitochondria and chloroplasts are the result of endosymbiotic events. Therefore many functional elements can only be understood in an evolutionary context. This is especially true for the machinery employed in the biogenesis of chloroplasts (Chap. 23) and mitochondria (Chap. 24). Physiological and structural characteristics of the red algal chloroplast (Chap. 16), mitochondrion (Chap. 21) and the mitochondrion-derived hydrogenosome (Chap. 22) exemplify the transformation of bacteria into organelles specialized for light energy conversion and energy generation in the presence or absence of oxygen.

November 14, 2013

Martin F. Hohmann-Marriott
Department of Biotechnology,
Norwegian University of Science
and Technology, NTNU,
N-7491 Trondheim, Norway
martin.hohmann-marriott@ntnu.no

The Editor

Martin F. Hohmann-Marriott was born on July 21, 1971 in Mannheim, Germany. He graduated with a Diplom in Biology from the Julius-Maximilians-Universität Würzburg, in 1998, having worked with Laurens Mets (University of Chicago, USA) and Ulrich Schreiber (Julius-Maximilians-Universität Würzburg) on his Diplom thesis.

Martin completed his PhD in Plant Biology in 2005, working with Robert E. Blankenship, Robert R. Roberson and Wim F. Vermaas at Arizona State University, Tempe, Arizona, USA. He had obtained Postdoctoral Fellowships from the National Research Council (USA), the Japan Society for Promotion of Science (Japan) and the Foundation for Research, Science and Technology (New Zealand). These fellowships enabled Martin to work with Richard D. Leapman (National Institute of Biomedical Imaging and Bioengineering, Bethesda MD, USA), Jun Minagawa (Hokkaido University, Sapporo, Japan) and Julian J. Eaton-Rye (University of Otago, New Zealand). Martin was appointed Associate Professor in the Department of Biotechnology at the Norwegian University of Science and Technology (Trondheim, Norway) in 2011. Here he lectures in courses on biochemistry, molecular genetics and bioinformatics.

Martin's research interest is in bioenergetics, especially the evolution of the photosynthetic machinery in different organisms. Martin has worked on the photosynthetic machinery and physiology of green sulfur bacteria, cyanobacteria, green algae and lately heterokont algae. His current research focus is renewable energy and the application of structural approaches to address open questions in bioenergetics.

At the Norwegian University of Science and Technology, Martin established the PhotoSynLab (http://photosynlab.org). PhotoSynLab is an 'open-science' laboratory that shares all protocols and publishes all experimental results online. PhotoSynLab is intended to apply synthetic biology approaches and laboratory automation to explore photosynthetic organisms.

Contributors

Noam Adir
Schulich Faculty of Chemistry,
Technion, Israel Institute of Technology,
Technion City, Haifa 32000, Israel
nadir@tx.technion.ac.il

Marie Asao
Department of Microbiology,
The Ohio State University,
Columbus, OH 43210-1292, USA
asao.1@osu.edu

Miloš Barták
Faculty of Science, Department of
Experimental Biology, Masaryk University,
62500 Brno, Czech Republic
mbartak@sci.muni.cz

Frauke Baymann
Bioénergétique et Ingénierie des protéines,
Unité mixte de récherche 7281,
Fédération de récherche 3479,
Centre National de la Récherche
Scientifique/Aix-Marseille Université,
13009 Marseille, France
baymann@imm.cnrs.fr

Marlene Benchimol
Laboratório de Ultraestrutura Celular,
Universidade Santa Úrsula,
22231-010 Rio de Janeiro, Brazil
marlenebenchimol@gmail.com

Michael Berney
Department of Microbiology and Immunology,
Albert Einstein College of Medicine,
Bronx, NY 10461, USA
michael.berney@einstein.yu.edu

Egbert J. Boekema
Department of Biophysical Chemistry,
GBB, University of Groningen,
9747AG Groningen, The Netherlands
e.j.boekema@rug.nl

Bettina Böttcher
School of Biological Sciences,
University of Edinburgh,
King's Buildings, Edinburgh EH9 3JR, UK
bettina.boettcher@ed.ac.uk

Zenilda L. Bouzon
Plant Cell Biology Laboratory,
Department of Cell Biology,
Embryology and Genetics,
Federal University of Santa Catarina,
88040-900 Florianópolis, SC, Brazil

Central Laboratory of Electron Microscopy,
Federal University of Santa Catarina,
88040-900 Florianópolis, SC, Brazil
zenilda.bouzon@ufsc.br

Hans-Peter Braun
Institute of Plant Genetics, University of
Hannover, D-30419 Hannover, Germany
braun@genetik.uni-hannover.de

Leonid S. Brown
Department of Physics,
University of Guelph,
Guelph, ON, N1G 2W1, Canada
lebrown@uoguelph.ca

Claudia Büchel
Institute of Molecular Biosciences,
Goethe Universität Frankfurt,
D-60438 Frankfurt, Germany
c.buechel@bio.uni-frankfurt.de

Sarah J. Butcher
Institute of Biotechnology, University
of Helsinki, 00014 Helsinki, Finland
Sarah.Butcher@helsinki.fi

Aaron M. Collins
Center for Integrated Nanotechnologies,
Los Alamos National Laboratory,
Los Alamos, NM 87185-1413, USA
aaronmcollins@gmail.com

Gregory M. Cook
Department of Microbiology
and Immunology, University of Otago,
9054 Dunedin, New Zealand
gregory.cook@otago.ac.nz

Vangelis Daskalakis
Department of Environmental
Science and Technology,
Cyprus University of Technology,
3603 Lemesos, Cyprus
evangelos.daskalakis@cut.ac.cy

Liron David
Schulich Faculty of Chemistry,
Technion, Israel Institute of Technology,
Technion City, Haifa 32000, Israel
liron.david@childrens.harvard.edu

Alexander Dikiy
Department of Biotechnology,
Norwegian University of Science
and Technology,
N-7491 Trondheim, Norway
alex.dikiy@biotech.ntnu.no

Olena Dobrovolska
Department of Biotechnology,
Norwegian University of Science
and Technology, N-7491 Trondheim, Norway
olena.dobrovolska@biotech.ntnu.no

Natalya V. Dudkina
Department of Biological Sciences,
Birkbeck, University of London,
London WC1E 7HX, UK
n.dudkina@mail.cryst.bbk.ac.uk

Daniela Gargano
Faculty of Science and Technology,
Centre for Organelle Research,
University of Stavanger,
N-4036 Stavanger, Norway
daniela.gargano@uis.no

Maria Luisa Genova
Dipartimento di Scienze
Biomediche e Neuromotorie,
Università di Bologna,
40126 Bologna, Italy
marialuisa.genova@unibo.it

Holly L. Gorton
Department of Biology,
St. Mary's College of Maryland,
St. Mary's City, MD 20686, USA
hlgorton@smcm.edu

Peter Gräber
Institut für Physikalische Chemie,
Albert-Ludwigs-Universität Freiburg,
D-79104 Freiburg, Germany
peter.graeber@physchem.uni-freiburg.de

Kathi Gundermann
Institute of Molecular Biosciences,
Goethe Universität Frankfurt,
D-60438 Frankfurt, Germany
gunderma@stud.uni-frankfurt.de

Eckhard Hofmann
Faculty of Biology and Biotechnology,
Department of Biophysics,
Ruhr-University Bochum,
D-44780 Bochum, Germany
eckhard.hofmann@bph.ruhr-uni-bochum.de

Matthew S. Kimber
Department of Molecular
and Cellular Biology,
University of Guelph,
Guelph, ON, N1G 2W1, Canada
mkimber@uoguelph.ca

Peter G. Kroth
Fachbereich Biologie,
Universität Konstanz,
D-78457 Konstanz,
Germany
Peter.Kroth@uni-konstanz.de

Jodi Maple
Faculty of Science and Technology,
Centre for Organelle Research,
University of Stavanger,
N-4036 Stavanger, Norway
jodi.maple@uis.no

Ailie Marx
Schulich Faculty of Chemistry,
Technion, Israel Institute of Technology,
Technion City, Haifa 32000, Israel
ailie@techunix.technion.ac.il

Yusuke Matsuda
Department of Bioscience,
Research Center for Environmental Bioscience, Kwansei Gakuin University,
Sanda, Hyogo 669-1337, Japan
yusuke@kwansei.ac.jp

Allison E. McDonald
Department of Biology, Wilfrid Laurier University, Science Building, Waterloo,
ON, N2L 3C5, Canada
amcdonald@wlu.ca

Jason A. Mears
Department of Pharmacology
and Center for Mitochondrial Disease,
Case Western Reserve University
School of Medicine,
Cleveland, OH 44106, USA
jam348@case.edu

Simon Geir Møller
Department of Biological Sciences,
St. John's University,
New York, NY 11439, USA
mollers@stjohns.edu

Tomáš Polívka
Faculty of Sciences,
Department of Physics and Biophysics,
University of South Bohemia,
370-05 České Budějovice,
Czech Republic
tpolivka@jcu.cz

Jakub Pšenčík
Faculty of Mathematics and Physics,
Charles University in Prague,
121 16 Prague 2, Czech Republic
psencik@karlov.mff.cuni.cz

W. Matthew Sattley
Division of Natural Sciences,
Indiana Wesleyan University,
Marion, IN 46953-4974, USA
matthew.sattley@indwes.edu

Eder C. Schmidt
Postdoctoral Research of Postgraduate
Program in Cell Biology and Development,
Department of Cell Biology,
Embryology and Genetics,
Federal University of Santa Catarina,
88040-900 Florianópolis, SC, Brazil
edcash@ccb.ufsc.br

Maria Schmidt
Department of Plant Physiology,
Institute of Biology, University of Leipzig,
D-04103 Leipzig, Germany
maria.schmidt@uni-leipzig.de

Elena Shumilina
Department of Biotechnology,
Norwegian University of Science
and Technology, N-7491 Trondheim, Norway
elena.shumilina@ntnu.no

Carmen Simioni
Post-Graduate Program
in Cell Biology and Development,
Department of Cell Biology,
Embryology and Genetics,
Federal University
of Santa Catarina,
88040-900 Florianópolis,
SC, Brazil
carmensimioni@hotmail.com

Joseph Kuo-Hsiang Tang
Carlson School of Chemistry
and Biochemistry, Department of Biology,
Clark University, Worcester, MA 01610, USA
jtang@clarku.edu

Felix ten Brink
Bioénergétique et Ingénierie des protéines,
Unité mixte de récherche 7281,
Fédération de récherche 3479,
Centre National de la Récherche
Scientifique/Aix-Marseille Université,
13009 Marseille, France
tenbrink@imm.cnrs.fr

Claire S. Ting
Department of Biology,
Williams College, Williamstown,
MA 01267, USA
cting@williams.edu

Roman Tuma
The Astbury Centre for Structural
Molecular Biology, University of Leeds,
Leeds, LS2 9JT, UK
r.tuma@leeds.ac.uk

Greg C. Vanlerberghe
Department of Biological
Sciences and Department of Cell
and Systems Biology,
University of Toronto Scarborough,
Toronto, ON, M1C1A4, Canada
gregv@utsc.utoronto.ca

Constantinos Varotsis
Department of Environmental
Science and Technology,
Cyprus University of Technology,
3603 Lemesos, Cyprus
c.varotsis@cut.ac.cy

Thomas C. Vogelmann
Department of Plant Biology,
University of Vermont,
Burlington, VT 05405, USA
thomas.vogelmann@uvm.edu

Christian Wilhelm
Department of Plant Physiology,
Institute of Biology,
University of Leipzig,
D-04103 Leipzig, Germany
cwilhelm@rz.uni-leipzig.de

Author Index

Chapter 1

Proton-Pumping Microbial Rhodopsins – Ubiquitous Structurally Simple Helpers of Respiration and Photosynthesis

Leonid S. Brown*
Department of Physics, University of Guelph, Guelph, ON, N1G 2W1, Canada

Summary

For almost four decades, bacteriorhodopsin has served as a classic example of the simplest standalone proton gradient generator. Bacteriorhodopsin-based bioenergetics was viewed as the most basic type of photosynthesis, becoming useful under limiting oxygen conditions only in a small group of extremophilic haloarchaea. With the advent of genomic and metagenomic high-throughput sequencing, the taxonomic and ecological diversity of bacteriorhodopsin-related proteins (microbial rhodopsins) appeared to be large. In this chapter, we survey structural and taxonomic diversity of proton-pumping microbial rhodopsins, describing haloarchaeal, fungal, algal, and eubacterial representatives, including those in photosynthetic organisms. Comparison of both primary and 3-D structures is made, and common structural trends are pointed out. Finally, we outline the main structural blocks involved in light-driven proton-transport mechanism, and discuss its conserved and variable parts.

I. Introduction

In the last three decades of the past century, membrane bioenergetics based on bacteriorhodopsin (BR) and related microbial rhodopsins serving as light-driven retinal-binding proton pumps was considered to be a "poor relative" of chlorophyll photosynthesis and respiration. Even though BR has entered almost every biochemistry textbook as an elegant and simple stand-alone photochemical energy converter, which nicely confirmed the chemiosmotic theory (Oesterhelt and Stoeckenius 1973; Racker and Stoeckenius

*Author for correspondence, e-mail: lebrown@uoguelph.ca

M.F. Hohmann-Marriott (ed.), The Structural Basis of Biological Energy Generation, Advances in Photosynthesis and Respiration 39, pp. 1–20, DOI 10.1007/978-94-017-8742-0_1,

1974), it was viewed as a rarity, both taxonomically and ecologically. Indeed, these retinal-binding proteins were found only in a few species of archaeal extremophiles (halobacteria) restricted to salterns and lakes with similarly high salinity, and as such could not be assigned any global bioenergetic significance (Oren 1999; Schafer et al. 1999). The situation has been changing during the last 12–13 years with the growing realization that proton-pumping microbial rhodopsins are omnipresent geographically and widespread taxonomically (Spudich et al. 2000; Brown and Jung 2006; Sharma et al. 2006; Klare et al. 2008; DeLong and Beja 2010). Due to the unprecedented advances in genomic and environmental sequencing, we are now aware of thousands of new variants of homologs of BR, not just in haloarchaea and in numerous species of *Eubacteria*, but also in eukaryotic microbes, including fungi and algae (Brown 2004; Tsunoda et al. 2006; Slamovits et al. 2011; Wada et al. 2011; Zhang et al. 2011). Furthermore, proton-pumping rhodopsins can be useful not only in heterotrophic microbes, under the conditions when carbon or oxygen are scarce, but, interestingly, are even found in some algae and bacteria, which possess chlorophyll-based photosynthesis (Tsunoda et al. 2006; Miranda et al. 2009; Raven 2009; Wada et al. 2011). Numerous microbial rhodopsins, both proton-pumping and not, are found in terrestrial ecosystems, in freshwater habitats, and in the oceans (Atamna-Ismaeel et al. 2008, 2011; Sharma et al. 2008; Martinez-Garcia et al. 2012). In marine ecosystems, they span the range from polar to equatorial waters, and were even suggested to account for a substantial fraction of the total solar energy conversion in the biosphere (Beja et al. 2001; Venter et al. 2004; Sabehi et al. 2005; Jung et al. 2008; Zubkov 2009; Koh et al. 2010).

Abbreviations: ACR-2 – *Acetabularia* rhodopsin-2; AR – Archaerhodopsin; BR – Bacteriorhodopsin; GPR – Green-absorbing proteorhodopsin; GR – *Gloeobacter* rhodopsin; PR – Proteorhodopsin; RD – Rhodopsin; TM – Transmembrane; XR – Xanthorhodopsin

It is not surprising that along with the expanding taxonomic and ecological diversity of microbial rhodopsins we are becoming aware of their new functions, or modifications of the known functions (proton pumping, chloride pumping, and photosensory transduction by haloarchaeal rhodopsins). The arsenal of functions of microbial rhodopsins now includes light-gated cation channels (channelrhodopsins), light-activated enzymes (enzymerhodopsins), and several new types of photosensors (Jung et al. 2003; Spudich 2006; Hegemann 2008). While we are not going to discuss these functional subclasses of microbial rhodopsins here, concentrating on proton pumps instead, it should be mentioned that the distinction between ion pumps and photosensors is sometimes vague. It has been shown that under certain conditions both sensory rhodopsins and channelrhodopsins can work as proton pumps (Sasaki and Spudich 2000; Feldbauer et al. 2009). Thus, if proton pumping is demonstrated only *in vitro*, such results should be treated with caution until their physiological relevance is confirmed. Therefore, some of the novel proton pumping rhodopsins we will discuss here may be reclassified as photosensors in the future, when more physiological data becomes available.

In this review, we will first survey the emerging taxonomic and structural diversity of novel proton-pumping microbial rhodopsins. Next, we will analyze the common structural principles underlying their light-driven transmembrane proton-pumping mechanism and its variations in different subgroups of rhodopsins.

II. Taxonomic and Structural Diversity of Proton-Pumping Microbial Rhodopsins

While the general protein architecture of proton-pumping microbial rhodopsins is well-conserved, their primary structure shows large variability. Typically, they are

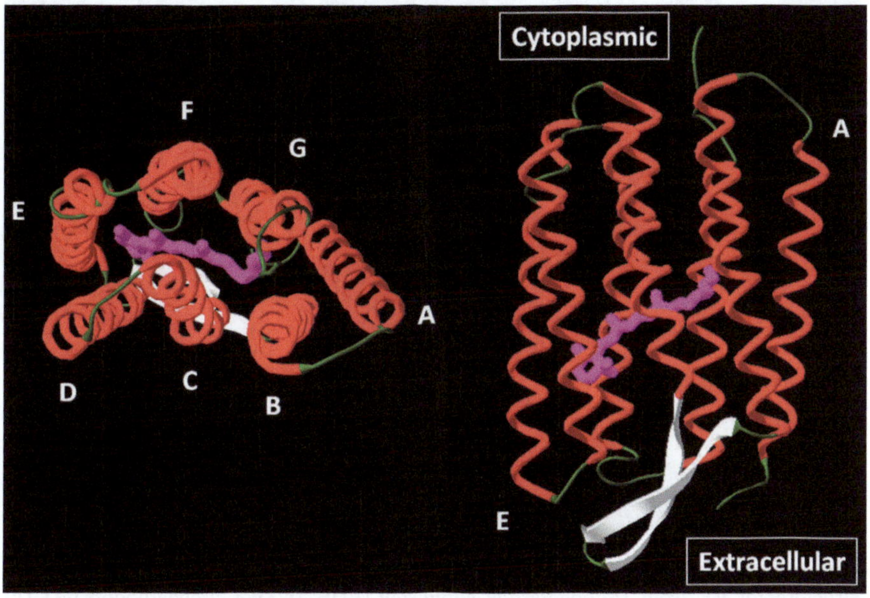

Fig. 1.1. Typical 7-transmembrane architecture of proton-pumping microbial rhodopsins exemplified by BR from *Halobacterium salinarum* (PDB 1QHJ, rendered by Swiss-PDB viewer, Guex and Peitsch 1997). The α-helices are shown in *red* and labeled with *white* capital letters, β-strands in *white*, unstructured segments in *green*, all-trans-retinal chromophore with the connected Lys side-chain in *purple*. *Top view* from the cytoplasmic side is shown on the *left, side view* with helices *B*, *C*, and *D* facing the viewer is shown on the *right*.

seven-transmembrane (7-TM) helical bundles with relatively small extramembrane loops and tails (the TM helices 1–7 are usually referred to as helices A-G). The all-*trans*-retinal chromophore is covalently bound via the Schiff base linkage to a lysine side-chain in the middle of the seventh helix (Fig. 1.1). The protonated retinal Schiff base connects two proton-conducting semi-channels, the extracellular semi-channel with its primary proton acceptor and the cytoplasmic semi-channel with its primary proton donor, and serves as the central active element in the mechanism of proton transport (see Sect. III for more details). The intramembrane proton-conducting pathways are mainly composed by polar residues from the helices B, C, F, and G (aided by several water molecules) (Subramaniam et al. 2002; Lanyi 2004; Luecke et al. 2008; Kouyama and Murakami 2010; Wada et al. 2011), while helices D and E interact with the distal part of retinal and participate in tuning of its spectral and isomeric properties (Ihara et al. 1994; Shimono et al. 2003). For proton-pumping rhodopsins, the core of their transmembrane regions, which constitutes the retinal-binding pocket and proton-conducting pathways, is strongly conserved (Ihara et al. 1999; Ruiz-Gonzalez and Marin 2004; Adamian et al. 2006; Brown and Jung 2006), while peripheral parts of TM helices, extramembrane loops and tails can be highly divergent (Fig. 1.2).

The variety of primary structures of proton-pumping microbial rhodopsins mirrors that of microbial rhodopsins in general. Roughly, the tree of microbial rhodopsins can be subdivided into the three major clusters (Fig. 1.3): haloarchaeal/fungal, proteorhodopsin (PR)-like, and xanthorhodopsin (XR)/actinorhodopsin-like (McCarren and DeLong 2007; Sharma et al. 2009). The taxonomic composition of these clusters is often non-uniform, probably due to frequent lateral gene transfer (Ruiz-Gonzalez and Marin 2004; Frigaard et al. 2006; Sharma et al. 2006; Slamovits et al. 2011; Martinez-Garcia

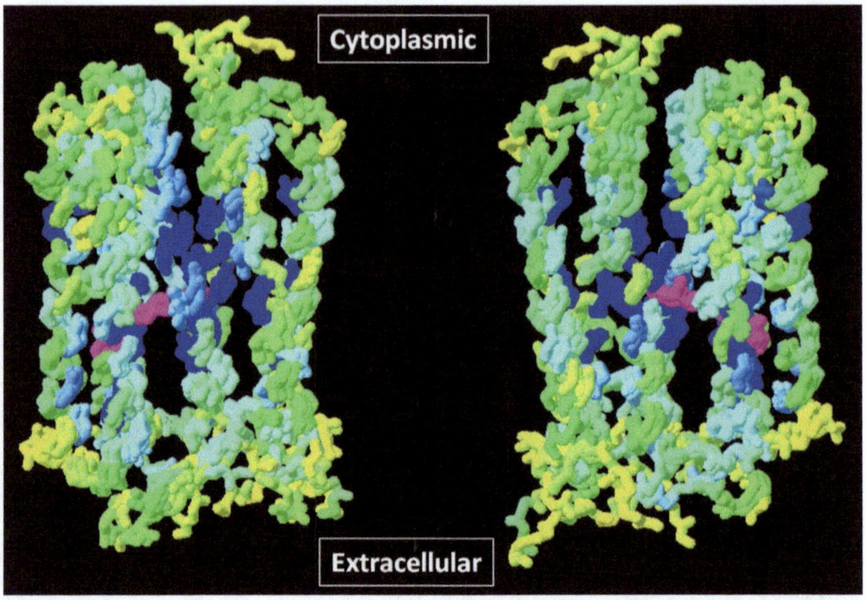

Fig. 1.2. Conservation of the membrane core of proton-pumping rhodopsins with known X-ray structures. Backbone structures of BR from *Halobacterium salinarum* (PDB 1C3W), archaerhodopsin-1 (PDB 1UAZ), archaerhodopsin-2 (PDB 2EI4), xanthorhodopsin (*XR*) from *Salinibacter ruber* (PDB 3DDL), and *Acetabularia* rhodopsin-2 (PDB 3AM6) were superimposed using the iterative magic fit of Swiss-PDB viewer (Guex and Peitsch 1997) and colored according to the alignment diversity *(dark blue* corresponds to the most conserved residues, *yellow* to the least conserved, with *cyan and green* showing intermediate degrees of conservation). Retinal chromophore is shown in *purple*. *Side view* with helices *B*, *C*, and *D* facing the viewer is shown on the *left*, and that with helices *E*, *F*, and *G* on the *right*.

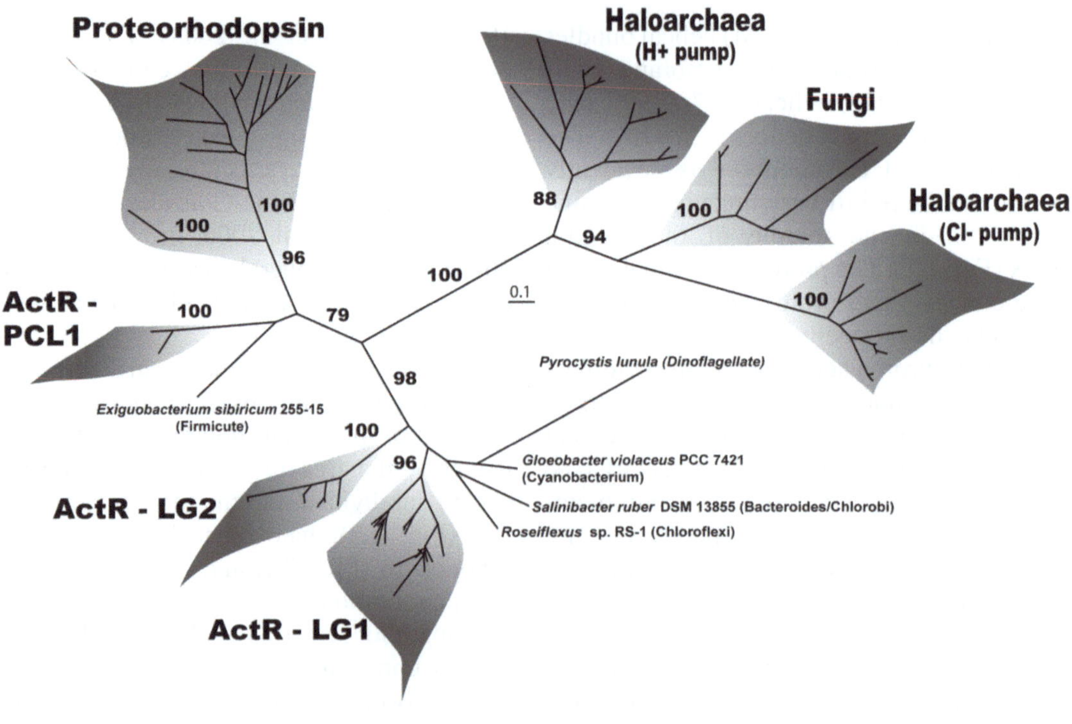

Fig. 1.3. Three major clusters of microbial rhodopsins shown on a best maximum likelihood tree constructed from 81 selected sequences (From Sharma et al. 2009, with permission of the Nature Publishing Group). The haloarchaeal/fungal cluster is located on the *right*, the PR-like cluster on the *left*, and the XR-like (actinorhodopsin-like) cluster at the *bottom* of the tree, with *Salinibacter* XR shown as *Salinibacter ruber* DSM 13855.

et al. 2012). The haloarchaeal/fungal cluster is mainly represented by close homologs of BR, halorhodopsin-like chloride pumps, haloarchaeal sensory rhodopsins and various fungal rhodopsins, but representatives from eubacteria and microscopic algae can be found in this cluster as well (Sharma et al. 2006; Slamovits et al. 2011; Zhang et al. 2011). The PR-like cluster mainly encompasses proton-pumping rhodopsins from marine proteobacteria and *Bacteroidetes*, but examples of PR-like proteins can be found in *Archaea* and dinoflagellate algae also (Frigaard et al. 2006; Slamovits et al. 2011). The XR-like (actinorhodopsin-like) cluster is probably the most taxonomically diverse, as it contains rhodopsins from many freshwater actinobacterial species, as well as from cyanobacteria, proteobacteria, *Bacteroidetes*, *Chloroflexi*, *Deinococci*, *Planctomycetes*, dinoflagellates and a few other microscopic algae (Sharma et al. 2009; Martinez-Garcia et al. 2012). XR from *Salinibacter* is known to bind a second auxiliary chromophore, carotenoid salinixanthin, but it is not obvious if all members of the cluster do the same, even though it is very likely for some of them (Balashov et al. 2005, 2010). A few species of rhodopsins intermediate between the XR-like and PR-like clusters have been observed, some of which could be ascribed to recombination with laterally transferred genes (Martinez-Garcia et al. 2012). The lateral gene transfer sometimes results in the co-existence of rhodopsins from different clusters in the same host. For example, bacterium *Salinibacter ruber* possesses rhodopsins from BR-like and XR-like clusters, while dinoflagellate *Oxyrrhis marina* has rhodopsins of XR-like and fungal-like types (Mongodin et al. 2005; Slamovits et al. 2011).

Focusing on the microbial rhodopsins, for which proton-pumping function has been proven or strongly suggested, one can observe a very similar clustering pattern. It is easy to distinguish haloarchaeal (BR-like), fungal, PR-like, and XR-like structural clusters (some of which have mixed taxonomic composition) with a very few intermediate species (Fig. 1.4). Below, we will briefly survey taxonomic and general structural features of these groups.

The haloarchaeal (BR-like) group of proton-pumping rhodopsins includes BR and its close homologs (sometimes referred to as archaerhodopsins, cruxrhodopsins, and deltarhodopsins), all belonging to the members of *Halobacteriaceae* (Ihara et al. 1999). It was in these facultative aerobes, where the bioenergetic role of BR-like pigments under limiting oxygen conditions was proven for the first time (Danon and Stoeckenius 1974). Numerous X-ray structures for BR in various states and photointermediates, including several mutants (Neutze et al. 2002; Lanyi and Schobert 2004; Hirai et al. 2009), as well as for archaerhodopsin-1 and archaerhodopsin-2 (Enami et al. 2006) have been solved. Overall structures of BR and archaerhodopsins are very similar, consistent with their 55–60 % sequence identity, despite the fact that archaerhodopsin-2 binds bacterioruberin, which participates in trimer formation (Fig. 1.5) (Yoshimura and Kouyama 2008). As BR tends to form native 2D crystals (so-called purple membrane), good resolution EM structures of BR in the native lipid environment are available as well (Hirai and Subramaniam 2009). As this is the most studied group of retinal-binding light-driven proton pumps, many reviews are available on all aspects of BR functioning as well as on comparison of BR with its close homologs, and the reader is referred to these papers for more information (Haupts et al. 1999; Ihara et al. 1999; Lanyi 2004; Sharma et al. 2007; Klare et al. 2008). Here we would like to mention that while most of halobacterial species have only one gene encoding a BR-like proton pump, recently it was found that both *Haloarcula marismortui* and *Haloquadratum walsbyi* possess two different BR-like proteins (Baliga et al. 2004; Bolhuis et al. 2006). Unexpectedly, it appears that only in the former species both proteins function as proton pumps, while in the latter only one of the forms is active in ion transport (Fu et al. 2010; Sudo et al. 2011).

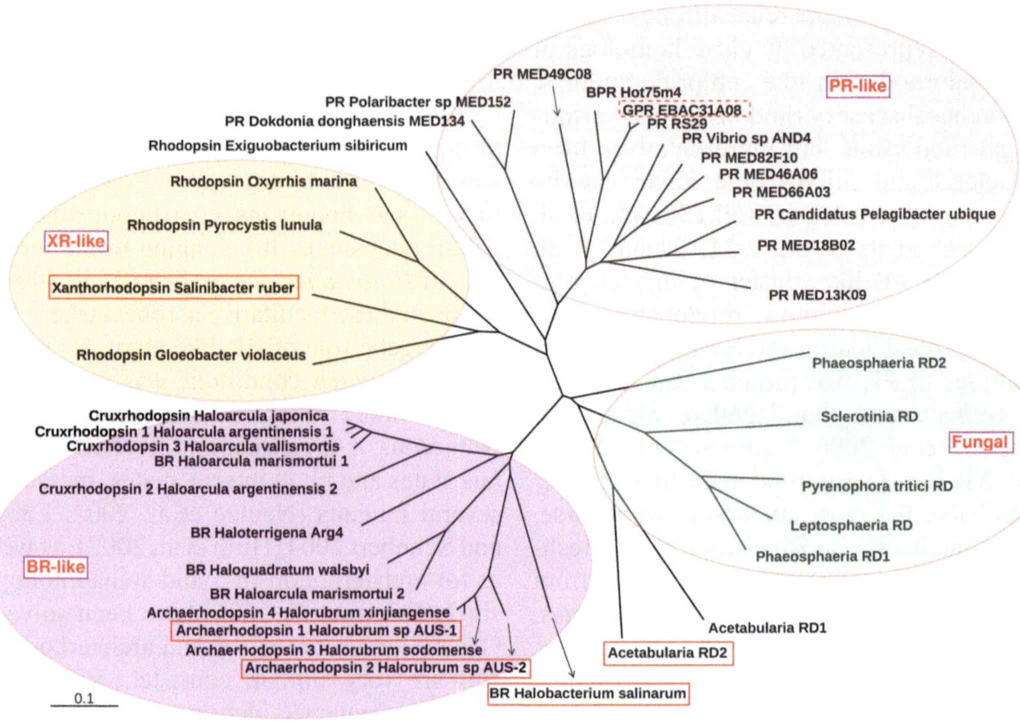

Fig. 1.4. Unrooted CLUSTALW (Thompson et al. 2002) guide tree (Rendered by TreeView, Page 2002) of selected microbial rhodopsins with proven or strongly suggested proton-pumping function (see text for details). Four major groups are marked by ovals, rhodopsins with known X-ray structures are boxed with a *red solid line*, and rhodopsin with NMR structure is boxed with *red dotted line*.

The fungal group of proton-pumping rhodopsins, closest to the haloarchaeal cluster, may represent only a small fraction of all rhodopsins and opsin-related proteins (in fungi proteins similar to opsins which lost their retinal-binding lysine) (Brown 2004; Idnurm et al. 2010; Fan et al. 2011). The TM regions of these ion-transporting fungal rhodopsins are very similar to that of BR, so it is not surprising that their photochemistry and proton-pumping ability are reminiscent of those of halobacterial rhodopsins as well. The first proton-pumping fungal rhodopsin characterized belongs to *Leptosphaeria*. It was expressed in the methylotrophic yeast *Pichia pastoris*, and pumps protons, when reconstituted in liposomes (Waschuk et al. 2005). Its proton pumping activity, along with that of a number of related fungal rhodopsins, was confirmed electrophysiologically upon heterologous expression in neurons (Chow et al. 2010; Boyden et al. 2011). Even then, the physiological function of these proton-pumping fungal rhodopsins is not clear, unlike that of their haloarchaeal homologs, and they may not have a significant bioenergetic role. Interestingly, some of fungal rhodopsins, which could be expected to show proton-pumping activity based on the sequence similarity to BR (such as those from *Neurospora* and *Podospora*), turned out to be inactive in ion transport and possess slow photochemical cycles suggestive of their photosensory role (Bieszke et al. 1999; Brown et al. 2001; Chow et al. 2010). Recently, we characterized a representative of a new subgroup of fungal rhodopsins (so-called auxiliary subgroup named after the fact that these rhodopsins are present in addition to other rhodopsin forms in the same host). Interestingly, this rhodopsin from *Phaeosphaeria* (*Phaeosphaeria*

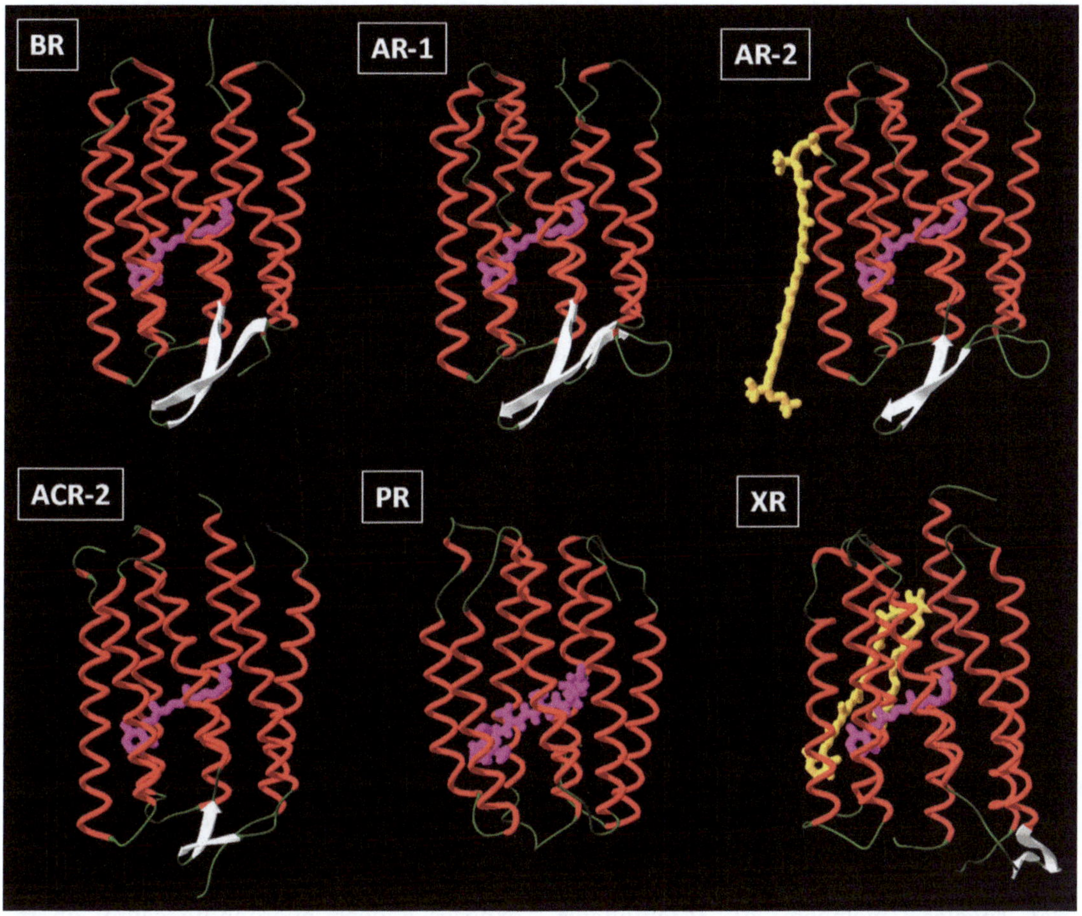

Fig. 1.5. Comparison of the overall structures (all from X-ray crystallographic data except for the solution NMR structure for PR) of proton-pumping microbial rhodopsins as rendered by Swiss-PDB viewer (Guex and Peitsch 1997). Backbones and chromophores of BR, AR-1, AR-2, *Acetabularia* rhodopsin-2 (ACR-2), PR, and XR are shown (PDB accession numbers 1QHJ, 1UAZ, 2EI4, 3AM6, 2L6X, 3DDL). The cytoplasmic sides are on *top*, α-helices are shown in *red*, β-strands in *white*, unstructured segments in *green*, all-*trans*-retinal chromophore with the connected Lys side-chain in *purple*, bound carotenoids in *orange* (bacterioruberin of AR-2 and salinixanthin of XR).

rhodopsin 2) demonstrated photochemistry highly suggestive of its proton-pumping ability, showing that fungi, similar to some halobacteria mentioned above, may have multiple forms of proton-pumping rhodopsins (Fan et al. 2011).

Structurally related to the fungal cluster (Fig. 1.4) are two proton-pumping rhodopsins from the green alga *Acetabularia acetabulum*, for which ion transport was demonstrated in oocytes (Tsunoda et al. 2006; Wada et al. 2011). On the other hand, a similar rhodopsin from a cryptomonad *Guillardia theta* did not show the light-driven proton transport in *E. coli* cells (Sineshchekov et al. 2005). The physiological role of proton-pumping by rhodopsins from *Acetabularia* (and possibly other algal homologs (Zhang et al. 2011)) is intriguing, as these rhodopsins are present in addition to powerful chlorophyll-based electron-coupled proton transporters, the photosystems. Recent X-ray structure for one of the *Acetabularia* rhodopsins (Wada et al. 2011) shows significant resemblance to that of BR (Fig. 1.5), as could be expected from the sequence similarity of the TM regions.

The cluster of PR-like proton pumps may be the most abundant in terms of the number of species that house such a rhodopsin.

Since the discovery of PR published in 2000 (Beja et al. 2000), close to a thousand of variants of PR-like proteins were found, mainly in uncultured marine picoplankton composed by proteo- and flavo- bacteria (Venter et al. 2004; Stingl et al. 2007; DeLong and Beja 2010). While the original GPR (shown as GPR EBAC31A08 in Fig. 1.4) is well-studied *in vitro* and its proton-pumping abilities were demonstrated in several systems (liposomes, oocytes, *E. coli* cells) (Beja et al. 2000; Friedrich et al. 2002; Dioumaev et al. 2003), only a few other members of the cluster were characterized photochemically or electrophysiologically (Man-Aharonovich et al. 2004; Sineshchekov and Spudich 2004; Sabehi et al. 2005; Desiderio et al. 2007; Stingl et al. 2007; Kralj et al. 2008). There is an ongoing dispute on whether the vectoriality of proton-pumping by GPR is reversed at low pH (Dioumaev et al. 2003; Sineshchekov and Spudich 2004; Lorinczi et al. 2009). However, the behavior at low pH may not be physiologically relevant, as *in situ* conditions are mildly alkaline. The physiological importance of proton pumping by PR-like proteins *in vivo* was demonstrated for several bacterial species (members of *Pelagibacter*, *Dokdonia*, *Vibrio*, and *Polaribacter* genera) and uncultured variants, especially under carbon starvation (Gonzalez et al. 2008; Gomez-Consarnau et al. 2010; Kimura et al. 2011; Steindler et al. 2011). It was also modeled in *E. coli* and *Shewanella* cells, where significant levels of photophosphorylation and proton-motive force were achieved (Martinez et al. 2007; Walter et al. 2007; Johnson et al. 2010). On the other hand, there is an evidence that other PR-like proteins, some of which were shown to possess slow photocycles, may have photosensory rather than ion-pumping functions (Spudich 2006; Fuhrman et al. 2008). As PR did not produce well-diffracting crystals (Gourdon et al. 2008), no high-resolution structure is available for members of this cluster, but its architecture was explored by solid-state and solution NMR (Pfleger et al. 2009; Shi et al. 2009a; Reckel et al. 2011). The latter yielded a structural model of the low pH form of GPR shown in Fig. 1.5, which shows some differences from the haloarchaeal template, such as the prominent lack of β-hairpin in the BC loop on the extracellular side.

The XR-like (actinorhodopsin-like) proteins have been often referred to as PR-like, but several recent analyses show that they form a very distinct separate cluster (Sharma et al. 2009; Martinez-Garcia et al. 2012). The first characterized member of this group was XR from *Salinibacter ruber*, which has an auxiliary carotenoid chromophore, salinixanthin (Fig. 1.5), serving as a unique antenna (Balashov et al. 2005). The light-driven proton-pumping by XR was demonstrated directly in *Salinibacter* cell-membrane vesicles as well as by photoinhibition of respiration in cell suspensions, and the action spectra of both processes confirm the efficient migration of light from the carotenoid to the retinal. The X-ray structure of XR (Fig. 1.5) shows a number of interesting differences from those of BR and its close homologs. The lengths, tilts, and rotations of several helices are different, and the outward orientation of the β-hairpin in the BC loop exposes a deep hydrophilic cavity on the extracellular side of XR (Luecke et al. 2008). There are also several functionally important differences related to the composition and location of proton donors and acceptors, which will be discussed in the next section. The second XR-like protein with a proven proton-pumping ability comes from the photosynthetic thylakoidless cyanobacterium *Gloeobacter violaceous*, where it could possibly aid the less efficient chlorophyll photosynthesis (Raven 2009; Mimuro et al. 2011). *Gloeobacter* rhodopsin (GR) shows the light-driven proton transport and a fast photocycle in *E. coli* cell membranes (Kawanabe et al. 2009; Miranda et al. 2009), where it exists in a carotenoid-free form, but it is able to bind both salinixanthin and *Gloeobacter*-specific carotenoid echinenone and to use them as light-harvesting antennae (Imasheva et al. 2009; Balashov et al. 2010). Interestingly, similar proteins are found both in photosynthetic (*Pyrocystis*) and non-photosynthetic (*Oxyrrhis*) dinoflagellates (Okamoto and Hastings 2003; Slamovits et al. 2011). For the latter, the proton-pumping role was suggested based on protein localization, sequence, and expression levels, while for the former it was directly shown in *E. coli* cells (A.R. Choi and K.-H. Jung, personal communication 2009). Finally, even though numerous members of this cluster (so-called actinorhodopsins) were found in freshwater actinobacteria and suggested to have proton-pumping role (Sharma et al. 2008, 2009), they were not included in Fig. 1.4 for the lack of their photochemical or physiological characterization.

In the following section, we will discuss common features and modifications of the structural basis of proton pumping by various microbial rhodopsins surveyed above.

III. The Common Structural Basis of Proton-Pumping by Microbial Rhodopsins

Taking advantage of many detailed studies of proton transport by BR (Rothschild 1992; Krebs and Khorana 1993; Haupts et al. 1999; Balashov 2000; Herzfeld and Lansing 2002; Neutze et al. 2002; Lanyi and Schobert 2004; Maeda et al. 2005; Hirai et al. 2009), it is possible to break down the mechanism of haloarchaeal transmembrane light-driven proton translocation into several major steps and corresponding structural blocks (Fig. 1.6). The central part is the protonated retinal Schiff base with its complex counterion, which includes the carboxylic proton acceptor and functionally important water molecules (Luecke et al. 1999b; Kandori

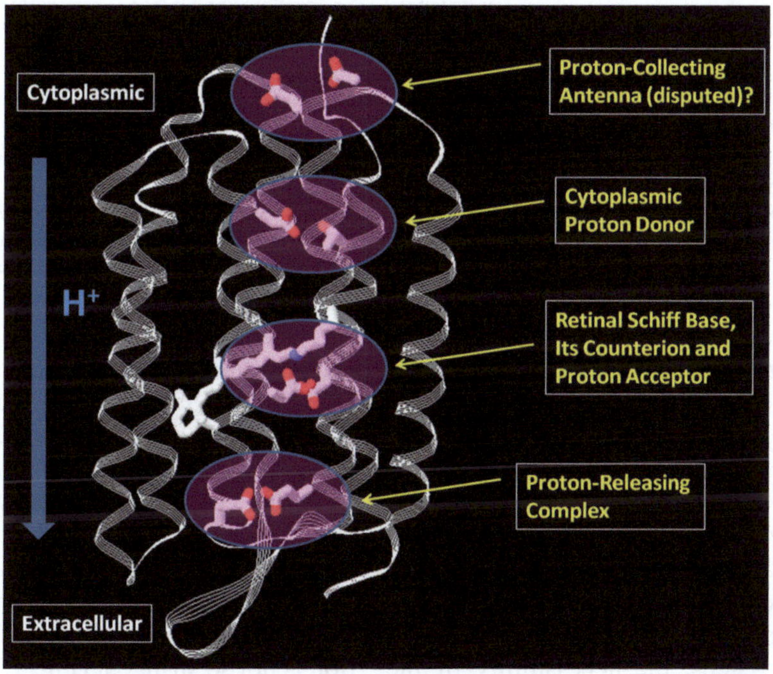

Fig. 1.6. The main (proton acceptor and donor) and auxiliary (proton-collecting antenna and proton-releasing complex) structural blocks believed to participate in the light-driven proton transport by microbial rhodopsins, shown on the structure of BR (rendered by Swiss-PDB viewer (Guex and Peitsch 1997), PDB accession number 1QHJ). The backbone are shown as a *white ribbon*, *blue arrow* indicates the direction of light-driven proton transport, the selected side-chains are shown in CPK colors (D36 and D38 for the putative proton-collecting antenna; D96 and T46 for the proton donor complex; retinal Schiff base with K216, D85, and D212 for the proton acceptor complex; E194 and E204 for the proton release complex).

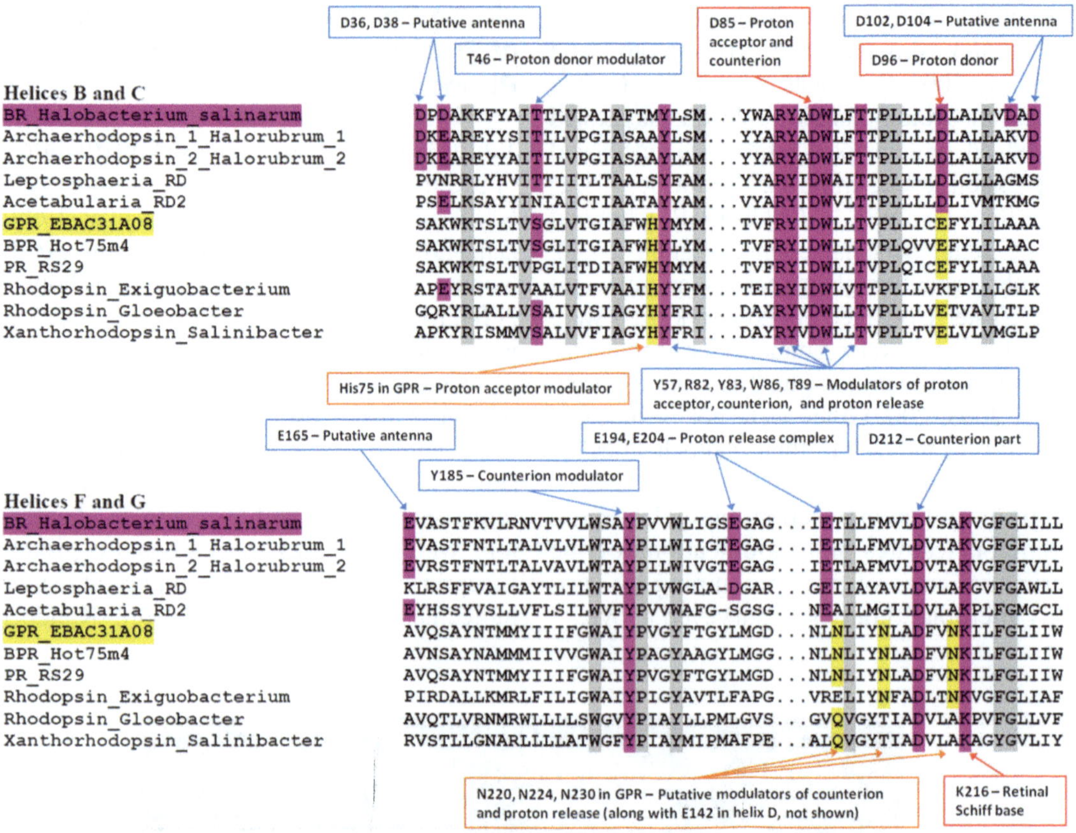

Fig. 1.7. Partial CLUSTALW (Thompson et al. 2002) alignment of selected sequences of proton-pumping microbial rhodopsins showing the extent of conservation and location of the residues believed to be important for proton transport. Residues proven or suggested to be important for proton transport in BR are *highlighted purple*, for PR – *yellow*, universally conserved residues not directly involved in proton translocation are shown in *grey*. Unless indicated otherwise, the residues are numbered according to the *H. salinarum* BR sequence.

2000). The retinal photoisomerization induces proton transfer inside this complex (from the Schiff base to the primary proton acceptor, normally represented by Asp), which is believed to be the most important event in the photocyle. The subsequent reprotonation of the Schiff base occurs from the carboxylic proton donor (usually, Asp or Glu) located on the cytoplasmic side. To ensure vectoriality of these steps, the accessibility of the Schiff base switches from the extracellular to the cytoplasmic side. Several hypotheses have been put forward on the nature of this switch, ranging from pure changes in the retinal geometry to modulation of the pK_a values of proton donors and acceptors (so-called affinity switch) (Haupts et al. 1997; Brown et al. 1998; Herzfeld and Tounge 2000; Subramaniam and Henderson 2000; Kandori 2004; Lorenz-Fonfria and Kandori 2009). From the comparison of BR with its homologs, it appears that while these primary acceptor and donor blocks are obligatory and strongly conserved, the other blocks may be optional or modifiable and their conservation is not so stringent (Fig. 1.7) (Brown and Jung 2006). These optional blocks include the proton-releasing complex on the extracellular side and possibly a proton-collecting antenna on the cytoplasmic side, even though the mere existence of the latter is disputed (Brown et al. 1999; Checover et al. 2001).

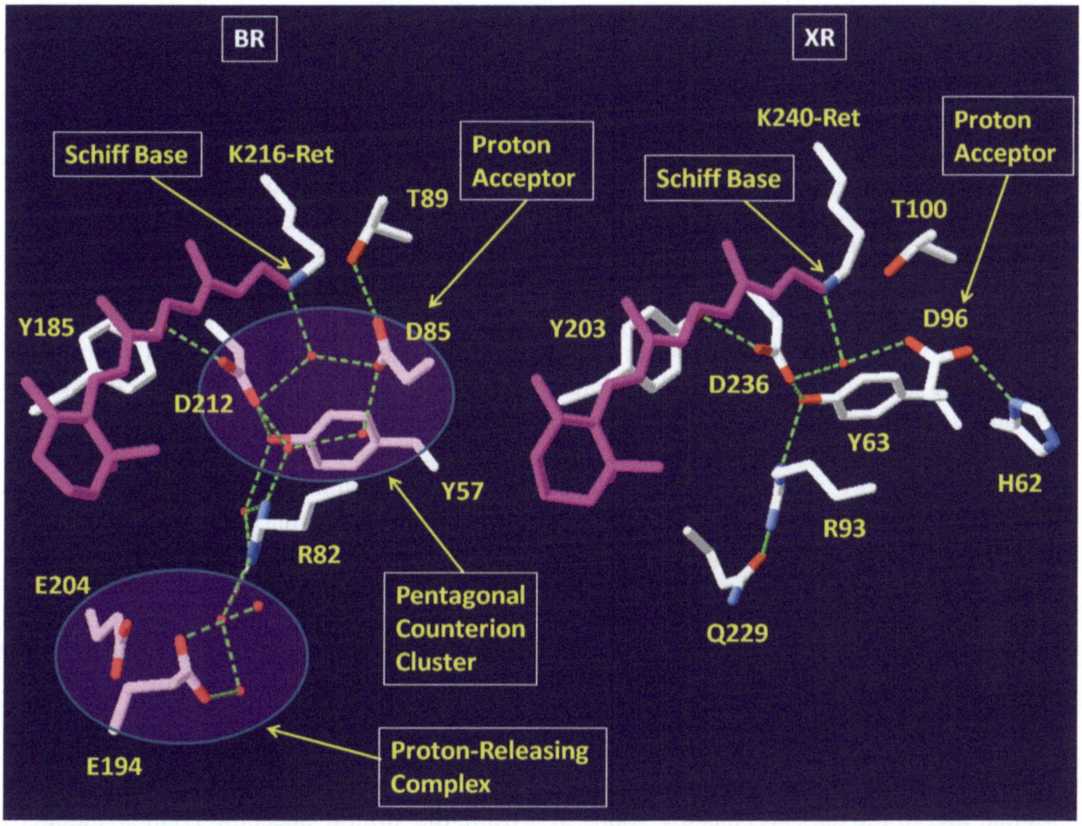

Fig. 1.8. Structures of the Schiff base counterion/proton acceptor complexes of BR and XR (rendered by Swiss-PDB viewer (Guex and Peitsch 1997), PDB accession numbers 1C3W and 3DDL) with the adjacent extracellular hydrogen-bonded networks (including the proton release complex in the case of BR). Water molecules are shown as *red spheres*, hydrogen bonds are shown in *green*, retinal in *purple*, only side-chains are shown for simplicity. The cytoplasmic sides are on *top*.

The established BR-based model of the Schiff base counterion/primary proton acceptor complex is that of a "pentagonal hydrogen-bonded cluster" (Fig. 1.8) (Luecke et al. 1999b; Shibata et al. 2003). This cluster includes the Schiff base nitrogen, two negatively charged aspartates with extremely low pK_a values (Asp85 and Asp212 in BR), and three water molecules, one of which is very strongly hydrogen-bonded and highly polarized (Kandori 2004). It has been suggested that the presence of this strongly hydrogen-bonded water in the immediate vicinity of the Schiff base is a very reliable signature of proton-pumping rhodopsins, and, indeed, its characteristic low-frequency infrared band nicely correlates with the proton transport function (Kandori 2011). The Schiff base counterion complex of microbial rhodopsins is modulated by interactions with several conserved polar side-chains, such as Arg, Tyr, Trp, and Thr (Figs. 1.7 and 1.8). In the case of BR, it is extended by several additional water molecules, which form a hydrogen-bonded network connecting it with the proton-releasing complex on the extracellular surface (Fig. 1.8). Interestingly, XR- and PR-like proteins adopted somewhat different structures for this complex, adding a unique histidine side-chain hydrogen-bonded to the proton-accepting aspartate (Fig. 1.8) (Luecke et al. 2008; Bergo et al. 2009; Hempelmann et al. 2011), which resulted in the higher pK_a values of the counterion than those found for BR-like proteins (Dioumaev et al. 2002; Imasheva et al. 2006; Miranda et al. 2009). While most of the water molecules present in the pentagonal cluster and the extended hydrogen-bonded network have not been observed in XR, the key water molecule coordinated by the Schiff base and the two aspartates is conserved (Lanyi and Balashov 2011). In agreement with this, two related eubacterial proton pumps, GPR and GR, show the infrared signature of strongly hydrogen-bonded water (Furutani et al. 2006a; Hashimoto et al. 2010). It is worth mentioning that PR-like (but not XR-like) rhodopsins introduced another polar residue into the counterion complex (Asn230 in GPR, Fig. 1.7), but its exact role is not known (Bergo et al. 2004).

The primary proton transfer from the protonated Schiff base to its carboxylic acceptor (Asp85 in BR) requires a dramatic light-induced decrease of the pK_a of the donor and a moderate-to-large increase in the proton affinity of the acceptor (depending on its original pK_a in the dark) (Balashov 2000). It has been established that in BR this primary proton transfer is coupled to the proton release from the extracellular side. A super-conserved arginine (Arg82 in BR) plays the key role in this electrostatic coupling by moving its positive charge from the proton acceptor towards the proton-releasing complex (Balashov et al. 1995; Richter et al. 1996; Luecke et al. 1999a). While in BR and its close homologs the proton releasing complex is formed mainly by the dyad of two glutamates (Glu194 and Glu204) helped by Arg82 and the intervening water molecules (Brown et al. 1995; Balashov et al. 1997; Garczarek and Gerwert 2006) (Fig. 1.8), these acidic residues are not conserved in most of eubacterial and some of eukaryotic proton pumps (Fig. 1.7). Consistent with the different nature of proton-releasing complexes in these proteins, the coupling of the counterion and the neighbouring arginine side-chain may not exist in other microbial rhodopsins. The absence of this coupling is indicated by the outward orientation of side-chains of the homologs of Arg82 in XR and ACR-2 (Luecke et al. 2008; Wada et al. 2011) and confirmed by mutational studies on GPR (Partha et al. 2005).

Even in the most well-studied case of BR, there is still no firm consensus concerning the source of the released proton. Under physiological conditions, the so-called "early" sub-millisecond proton release in BR was suggested to occur from one of the glutamic acid residues (Glu194 or Glu204), or from Arg82, or from water molecules coordinated by them (Brown et al. 1995; Balashov et al. 1997; Xiao et al. 2004; Garczarek and Gerwert 2006; Lorenz-Fonfria and Kandori 2009). It is not clear whether this early release is essential, as it is not observed until the much later stages in the photocycles of many proton-pumping microbial rhodopsins (including some BR-like ones) (Lukashev et al. 1994; Dioumaev et al. 2002; Waschuk et al. 2005; Luecke et al. 2008; Miranda et al. 2009; Kikukawa et al. 2011). It was suggested that the "late" proton release can occur directly from the proton acceptor (Asp85 aided by Asp212 in BR or the Asp-His complex in XR and PR) or from other surface residues (Balashov et al. 1999; Dioumaev et al. 1999; Bergo et al. 2009). XR- and PR-like rhodopsins contain a number of unique polar residues in their TM regions (Fig. 1.7), some of which could participate in the proton release, notably a glutamic acid at the extracellular end of helix D (Glu142 in GPR, shown to be protonated) (Kralj et al. 2008; Shi et al. 2009b).

After the completion of the deprotonation, the Schiff base must be reprotonated from the cytoplasmic side, provided that the accessibility switch is functional. In some non-pumping microbial rhodopsins and in proton-pumping microbial rhodopsins under certain conditions (e.g., at low hydration or low temperature, as well as in some mutants, often combined with secondary photochemistry), the proton can return to the Schiff base from the extracellular side, resulting in futile shuttling (Ormos 1991; Ganea et al. 1997; Hessling et al. 1997; Sasaki and Spudich 2000; Brown et al. 2001). Compared

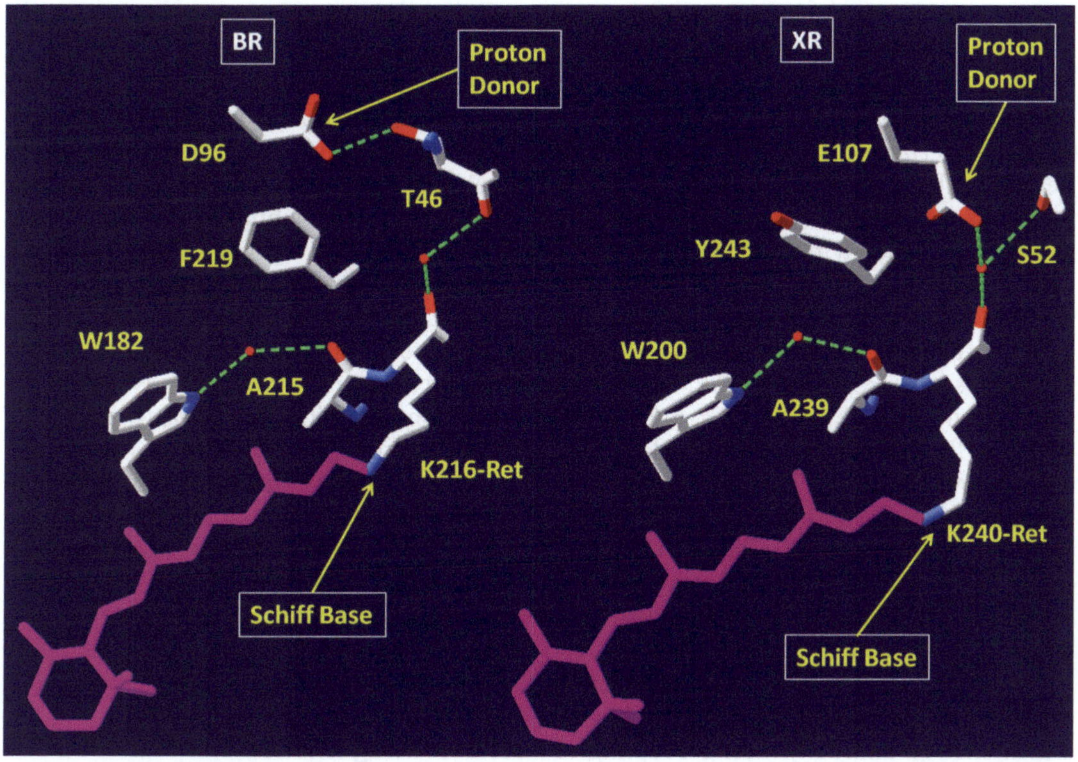

Fig. 1.9. Structures of the Schiff base proton donor complexes of BR and XR (rendered by Swiss-PDB viewer (Guex and Peitsch 1997), PDB accession numbers 1C3W and 3DDL). Water molecules are shown as *red spheres*, hydrogen bonds are shown in *green*, retinal in *purple*, only selected side-chains are shown for simplicity. The cytoplasmic sides are on *top*.

to the extracellular halves of proton-pumping microbial rhodopsins, their cytoplasmic halves are much less polar and contain very few water molecules (Fig. 1.9), at least in the dark state (Lanyi 2004; Kouyama and Murakami 2010; Freier et al. 2011). This makes perfect physiological sense, as the cytoplasmic side must be proton-impermeable between the Schiff base and the cytoplasmic proton donor to prevent the proton backflow. Additionally, the environment of the carboxylic cytoplasmic proton donor should be hydrophobic to keep its pK_a high, as it must be protonated even at mildly alkaline pH (Balashov 2000; Luecke et al. 2000). In BR and its close homologs this is achieved by surrounding the donor (Asp-96) by Leu and Phe residues, one of which insulates it from the Schiff base (Phe-219, Fig. 1.9) (Schobert et al. 2003). In haloarchaeal and fungal proton-pumping rhodopsins, the aspartic proton donor is usually modulated by hydrogen bonding with Thr (Thr46 in BR) (Rothschild et al. 1992; Brown et al. 1994; Lanyi and Schobert 2006; Fan et al. 2007). The conservation of the geometry of this complex is probably quite important, as it was found that in a fungal rhodopsin (from *Leptosphaeria*) the replacement of Asp by Glu results in the non-functional proton donor, a phenotype similar to that of non-pumping fungal rhodopsin from *Neurospora* (Furutani et al. 2006b). Interestingly, in the related algal ACR-2, the proton donor complex is quite different, as the threonine is replaced by asparagine and an additional cysteine interacts with the aspartic donor (Wada et al. 2011). It should be noted that when the proton donor is disabled by mutation (e.g., Asp to Asn), the proton transport still occurs, albeit in a very inefficient manner, due to the slow photocycle turnover (Butt et al. 1989). Related to this, there is a recent intriguing finding that proton-pumping rhodopsin from *Exiguobacterium sibiricum* (intermediate between XR and PR-like groups, Fig. 1.4) lacks the carboxylic cytoplasmic proton donor altogether, having replaced it with Lys (Fig. 1.7) (Petrovskaya et al. 2010).

The common proton-donating Asp/Thr pair on the cytoplasmic side of BR-like proteins is often replaced by Glu/Ser pair in XR-like and PR-like rhodopsins, but conservation of the serine is not absolute (Fig. 1.7). The geometry of this Glu-Ser pair is quite different from that in BR, as can be judged from the structure of XR (Fig. 1.9) (Luecke et al. 2008). It appears that the proton of the donor is located closer to the Schiff base and that proton-conducting pathways are partially pre-built in the dark. This tendency is probably even more pronounced in GR, which showed unusually strong coupling of the retinal and the proton donor, as demonstrated by vibrational spectroscopy on mutants (Miranda et al. 2009; Hashimoto et al. 2010). Normally, for the communication of the proton donor and the Schiff base to occur, a conformational change allowing water molecules to enter the cytoplasmic side and to build proton pathways is required (Luecke et al. 2000; Schobert et al. 2003; Hirai and Subramaniam 2009; Freier et al. 2011). Numerous studies showed that an important part of this conformational change is an outward tilt of the cytoplasmic half of the helix F, at least in BR and PR (Subramaniam et al. 1999; Kataoka and Kamikubo 2000; Klare et al. 2004; Lanyi and Schobert 2004; Andersson et al. 2009). Additional conformational changes are required to disconnect the Schiff base and the proton donor again and to restore the high pK_a of the donor to ensure its reprotonation from the cytoplasm. It was suggested that in BR several surface Asp and Glu residues may help in this process forming a proton-collecting antenna (Figs. 1.6 and 1.7) (Riesle et al. 1996; Checover et al. 2001), but mutagenesis and bioinformatic analysis do not give strong support to this idea (Brown et al. 1999).

The mechanism of proton transport by BR and related rhodopsins is probably the most studied among membrane proton-translocating machines and has inspired researchers from many other subfields of bioenergetics (Wikstrom 1998; Decoursey 2003; Mills and Ferguson-Miller 2003; Mulkidjanian et al. 2005). Despite its seeming simplicity, we see many interesting variations of the proton transport mechanism among newly discovered rhodopsins exhibiting an amazing taxonomic and structural diversity. With this comes the realization that proton-pumping microbial rhodopsins may serve as common bioenergetically important helpers of respiration and photosynthesis that are especially useful when these processes are inefficient.

Acknowledgements

This work is supported by Natural Sciences and Engineering Research Council of Canada and the University of Guelph.

References

Adamian L, Ouyang Z, Tseng YY, Liang J (2006) Evolutionary patterns of retinal-binding pockets of type I rhodopsins and their functions. Photochem Photobiol 82:1426–1435

Andersson M, Malmerberg E, Westenhoff S, Katona G, Cammarata M, Wohri AB, Johansson LC, Ewald F, Eklund M, Wulff M, Davidsson J, Neutze R (2009) Structural dynamics of light-driven proton pumps. Structure 17:1265–1275

Atamna-Ismaeel N, Sabehi G, Sharon I, Witzel KP, Labrenz M, Jurgens K, Barkay T, Stomp M, Huisman J, Beja O (2008) Widespread distribution of proteorhodopsins in freshwater and brackish ecosystems. ISME J 2:656–662

Atamna-Ismaeel N, Finkel OM, Glaser F, Sharon I, Schneider R, Post AF, Spudich JL, von Mering C, Vorholt JA, Iluz D, Beja O and Belkin S (2011) Microbial rhodopsins on leaf surfaces of terrestrial plants. Environ Microbiol 14:40–46

Balashov SP (2000) Protonation reactions and their coupling in bacteriorhodopsin. Biochim Biophys Acta 1460:75–94

Balashov SP, Govindjee R, Imasheva ES, Misra S, Ebrey TG, Feng Y, Crouch RK, Menick DR (1995) The two pKa's of aspartate-85 and control of thermal isomerization and proton release in the arginine-82 to lysine mutant of bacteriorhodopsin. Biochemistry 34:8820–8834

Balashov SP, Imasheva ES, Ebrey TG, Chen N, Menick DR, Crouch RK (1997) Glutamate-194 to cysteine mutation inhibits fast light-induced proton release in bacteriorhodopsin. Biochemistry 36:8671–8676

Balashov SP, Lu M, Imasheva ES, Govindjee R, Ebrey TG, Othersen B 3rd, Chen Y, Crouch RK, Menick DR (1999) The proton release group of bacteriorhodopsin controls the rate of the final step of its photocycle at low pH. Biochemistry 38:2026–2039

Balashov SP, Imasheva ES, Boichenko VA, Anton J, Wang JM, Lanyi JK (2005) Xanthorhodopsin: a proton pump with a light-harvesting carotenoid antenna. Science 309:2061–2064

Balashov SP, Imasheva ES, Choi AR, Jung KH, Liaaen-Jensen S, Lanyi JK (2010) Reconstitution of *Gloeobacter* rhodopsin with echinenone: role of the 4-keto group. Biochemistry 49:9792–9799

Baliga NS, Bonneau R, Facciotti MT, Pan M, Glusman G, Deutsch EW, Shannon P, Chiu Y, Weng RS, Gan RR, Hung P, Date SV, Marcotte E, Hood L, Ng WV (2004) Genome sequence of *Haloarcula marismortui*: a halophilic archaeon from the Dead Sea. Genome Res 14:2221–2234

Beja O, Aravind L, Koonin EV, Suzuki MT, Hadd A, Nguyen LP, Jovanovich S, Gates CM, Feldman RA, Spudich JL, Spudich EN, DeLong EF (2000) Bacterial rhodopsin: evidence for a new type of phototrophy in the sea. Science 289:1902–1906

Beja O, Spudich EN, Spudich JL, Leclerc M, DeLong EF (2001) Proteorhodopsin phototrophy in the ocean. Nature 411:786–789

Bergo V, Amsden JJ, Spudich EN, Spudich JL, Rothschild KJ (2004) Structural changes in the photoactive site of proteorhodopsin during the primary photoreaction. Biochemistry 43:9075–9083

Bergo VB, Sineshchekov OA, Kralj JM, Partha R, Spudich EN, Rothschild KJ, Spudich JL (2009) His-75 in proteorhodopsin, a novel component in light-driven proton translocation by primary pumps. J Biol Chem 284:2836–2843

Bieszke JA, Spudich EN, Scott KL, Borkovich KA, Spudich JL (1999) A eukaryotic protein, NOP-1, binds retinal to form an archaeal rhodopsin-like photochemically reactive pigment. Biochemistry 38:14138–14145

Bolhuis H, Palm P, Wende A, Falb M, Rampp M, Rodriguez-Valera F, Pfeiffer F, Oesterhelt D (2006) The genome of the square archaeon *Haloquadratum walsbyi*: life at the limits of water activity. BMC Genom 7:169

Boyden ES, Chow BY, Han X, Qian X, Klapoetke NC, Kwon AH (2011) Light-activated proton pumps, particular microbial rhodopsins, for use in adjusting cell pH or voltage, or in proton release. USA Patent US 2,011,016,568,1

Brown LS (2004) Fungal rhodopsins and opsin-related proteins: eukaryotic homologues of bacteriorhodopsin with unknown functions. Photochem Photobiol Sci 3:555–565

Brown LS, Jung KH (2006) Bacteriorhodopsin-like proteins of eubacteria and fungi: the extent of conservation of the haloarchaeal proton-pumping mechanism. Photochem Photobiol Sci 5:538–546

Brown LS, Yamazaki Y, Maeda A, Sun L, Needleman R, Lanyi JK (1994) The proton transfers in the cytoplasmic domain of bacteriorhodopsin are facilitated by a cluster of interacting residues. J Mol Biol 239:401–414

Brown LS, Sasaki J, Kandori H, Maeda A, Needleman R, Lanyi JK (1995) Glutamic-acid-204 is the terminal proton release group at the surface of bacteriorhodopsin. J Biol Chem 270:27122–27126

Brown LS, Dioumaev AK, Needleman R, Lanyi JK (1998) Local-access model for proton transfer in bacteriorhodopsin. Biochemistry 37:3982–3993

Brown LS, Needleman R, Lanyi JK (1999) Functional roles of aspartic acid residues at the cytoplasmic surface of bacteriorhodopsin. Biochemistry 38:6855–6861

Brown LS, Dioumaev AK, Lanyi JK, Spudich EN, Spudich JL (2001) Photochemical reaction cycle and proton transfers in *Neurospora* rhodopsin. J Biol Chem 276:32495–32505

Butt HJ, Fendler K, Bamberg E, Tittor J, Oesterhelt D (1989) Aspartic acids 96 and 85 play a central role in the function of bacteriorhodopsin as a proton pump. EMBO J 8:1657–1663

Checover S, Marantz Y, Nachliel E, Gutman M, Pfeiffer M, Tittor J, Oesterhelt D, Dencher NA (2001) Dynamics of the proton transfer reaction on the cytoplasmic surface of bacteriorhodopsin. Biochemistry 40:4281–4292

Chow BY, Han X, Dobry AS, Qian X, Chuong AS, Li M, Henninger MA, Belfort GM, Lin Y, Monahan PE, Boyden ES (2010) High-performance genetically targetable optical neural silencing by light-driven proton pumps. Nature 463:98–102

Danon A, Stoeckenius W (1974) Photophosphorylation in *Halobacterium halobium*. Proc Natl Acad Sci U S A 71:1234–1238

Decoursey TE (2003) Voltage-gated proton channels and other proton transfer pathways. Physiol Rev 83:475–579

DeLong EF, Beja O (2010) The light-driven proton pump proteorhodopsin enhances bacterial survival during tough times. PLoS Biol 8:e1000359

Desiderio RA, Laney SR, Letelier RM, Giovannoni SJ (2007) Using lasers to probe the transient light absorption by proteorhodopsin in marine bacterioplankton. Appl Opt 46:7329–7336

Dioumaev AK, Brown LS, Needleman R, Lanyi JK (1999) Fourier transform infrared spectra of a late intermediate of the bacteriorhodopsin photocycle suggest transient protonation of Asp-212. Biochemistry 38:10070–10078

Dioumaev AK, Brown LS, Shih J, Spudich EN, Spudich JL, Lanyi JK (2002) Proton transfers in the photochemical reaction cycle of proteorhodopsin. Biochemistry 41:5348–5358

Dioumaev AK, Wang JM, Balint Z, Varo G, Lanyi JK (2003) Proton transport by proteorhodopsin requires that the retinal Schiff base counterion Asp-97 be anionic. Biochemistry 42:6582–6587

Enami N, Yoshimura K, Murakami M, Okumura H, Ihara K, Kouyama T (2006) Crystal structures of archaerhodopsin-1 and -2: common structural motif in archaeal light-driven proton pumps. J Mol Biol 358:675–685

Fan Y, Shi L, Brown LS (2007) Structural basis of diversification of fungal retinal proteins probed by site-directed mutagenesis of *Leptosphaeria* rhodopsin. FEBS Lett 581:2557–2561

Fan Y, Solomon P, Oliver RP, Brown LS (2011) Photochemical characterization of a novel fungal rhodopsin from *Phaeosphaeria nodorum*. Biochim Biophys Acta 1807:1457–1466

Feldbauer K, Zimmermann D, Pintschovius V, Spitz J, Bamann C, Bamberg E (2009) Channelrhodopsin-2 is a leaky proton pump. Proc Natl Acad Sci U S A 106:12317–12322

Freier E, Wolf S, Gerwert K (2011) Proton transfer via a transient linear water-molecule chain in a membrane protein. Proc Natl Acad Sci U S A 108: 11435–11439

Friedrich T, Geibel S, Kalmbach R, Chizhov I, Ataka K, Heberle J, Engelhard M, Bamberg E (2002) Proteorhodopsin is a light-driven proton pump with variable vectoriality. J Mol Biol 321:821–838

Frigaard NU, Martinez A, Mincer TJ, DeLong EF (2006) Proteorhodopsin lateral gene transfer between marine planktonic Bacteria and Archaea. Nature 439:847–850

Fu HY, Lin YC, Chang YN, Tseng H, Huang CC, Liu KC, Huang CS, Su CW, Weng RR, Lee YY, Ng WV, Yang CS (2010) A novel six-rhodopsin system in a single archaeon. J Bacteriol 192: 5866–5873

Fuhrman JA, Schwalbach MS, Stingl U (2008) Proteorhodopsins: an array of physiological roles? Nat Rev Microbiol 6:488–494

Furutani Y, Ikeda D, Shibata M, Kandori H (2006a) Strongly hydrogen-bonded water molecule is observed only in the alkaline form of proteorhodopsin. Chem Phys 324:705–708

Furutani Y, Sumii M, Fan Y, Shi LC, Waschuk SA, Brown LS, Kandori H (2006b) Conformational coupling between the cytoplasmic carboxylic acid and the retinal in a fungal light-driven proton pump. Biochemistry 45:15349–15358

Ganea C, Gergely C, Ludmann K, Varo G (1997) The role of water in the extracellular half channel of bacteriorhodopsin. Biophys J 73:2718–2725

Garczarek F, Gerwert K (2006) Functional waters in intraprotein proton transfer monitored by FTIR difference spectroscopy. Nature 439:109–112

Gomez-Consarnau L, Akram N, Lindell K, Pedersen A, Neutze R, Milton DL, Gonzalez JM, Pinhassi J (2010) Proteorhodopsin phototrophy promotes survival of marine bacteria during starvation. PLoS Biol 8:e1000358

Gonzalez JM, Fernandez-Gomez B, Fernandez-Guerra A, Gomez-Consarnau L, Sanchez O, Coll-Llado M, Del Campo J, Escudero L, Rodriguez-Martinez R, Alonso-Saez L, Latasa M, Paulsen I, Nedashkovskaya O, Lekunberri I, Pinhassi J, Pedros-Alio C (2008) Genome analysis of the proteorhodopsin-containing marine bacterium *Polaribacter* sp. MED152 (Flavobacteria). Proc Natl Acad Sci U S A 105:8724–8729

Gourdon P, Alfredsson A, Pedersen A, Malmerberg E, Nyblom M, Widell M, Berntsson R, Pinhassi J, Braiman M, Hansson O, Bonander N, KarlssonG, Neutze R (2008) Optimized *in vitro* and *in vivo* expression of proteorhodopsin: a seven-transmembrane proton pump. Protein Expr Purif 58:103–113

Guex N, Peitsch MC (1997) SWISS-MODEL and the Swiss-PDB viewer: an environment for comparative protein modeling. Electrophoresis 18:2714–2723

Hashimoto K, Choi AR, Furutani Y, Jung KH, Kandori H (2010) Low-temperature FTIR study of *Gloeobacter* rhodopsin: presence of strongly hydrogen-bonded water and long-range structural protein perturbation upon retinal photoisomerization. Biochemistry 49:3343–3350

Haupts U, Tittor J, Bamberg E, Oesterhelt D (1997) General concept for ion translocation by halobacterial retinal proteins: the isomerization/switch/transfer (IST) model. Biochemistry 36:2–7

Haupts U, Tittor J, Oesterhelt D (1999) Closing in on bacteriorhodopsin: progress in understanding the

molecule. Annu Rev Biophys Biomol Struct 28:367–399
Hegemann P (2008) Algal sensory photoreceptors. Annu Rev Plant Biol 59:167–189
Hempelmann F, Holper S, Verhoefen MK, Woerner AC, Kohler T, Fiedler SA, Pfleger N, Wachtveitl J, Glaubitz C (2011) His75-Asp97 cluster in green proteorhodopsin. J Am Chem Soc 133:4645–4654
Herzfeld J, Lansing JC (2002) Magnetic resonance studies of the bacteriorhodopsin pump cycle. Annu Rev Biophys Biomol Struct 31:73–95
Herzfeld J, Tounge B (2000) NMR probes of vectoriality in the proton-motive photocycle of bacteriorhodopsin: evidence for an 'electrostatic steering' mechanism. Biochim Biophys Acta 1460:95–105
Hessling B, Herbst J, Rammelsberg R, Gerwert K (1997) Fourier transform infrared double-flash experiments resolve bacteriorhodopsin's M1 to M2 transition. Biophys J 73:2071–2080
Hirai T, Subramaniam S (2009) Protein conformational changes in the bacteriorhodopsin photocycle: comparison of findings from electron and X-ray crystallographic analyses. PLoS One 4:e5769
Hirai T, Subramaniam S, Lanyi JK (2009) Structural snapshots of conformational changes in a seven-helix membrane protein: lessons from bacteriorhodopsin. Curr Opin Struct Biol 19:433–439
Idnurm A, Verma S, Corrochano LM (2010) A glimpse into the basis of vision in the kingdom Mycota. Fungal Genet Biol 47:881–892
Ihara K, Amemiya T, Miyashita Y, Mukohata Y (1994) Met-145 is a key residue in the dark adaptation of bacteriorhodopsin homologs. Biophys J 67:1187–1191
Ihara K, Umemura T, Katagiri I, Kitajima-Ihara T, Sugiyama Y, Kimura Y, Mukohata Y (1999) Evolution of the archaeal rhodopsins: evolution rate changes by gene duplication and functional differentiation. J Mol Biol 285:163–174
Imasheva ES, Balashov SP, Wang JM, Lanyi JK (2006) pH-Dependent transitions in xanthorhodopsin. Photochem Photobiol 82:1406–1413
Imasheva ES, Balashov SP, Choi AR, Jung KH, Lanyi JK (2009) Reconstitution of *Gloeobacter violaceus* rhodopsin with a light-harvesting carotenoid antenna. Biochemistry 48:10948–10955
Johnson ET, Baron DB, Naranjo B, Bond DR, Schmidt-Dannert C, Gralnick JA (2010) Enhancement of survival and electricity production in an engineered bacterium by light-driven proton pumping. Appl Environ Microbiol 76:4123–4129
Jung KH, Trivedi VD, Spudich JL (2003) Demonstration of a sensory rhodopsin in eubacteria. Mol Microbiol 47:1513–1522
Jung JY, Choi AR, Lee YK, Lee HK, Jung KH (2008) Spectroscopic and photochemical analysis of proteorhodopsin variants from the surface of the Arctic Ocean. FEBS Lett 582:1679–1684
Kandori H (2000) Role of internal water molecules in bacteriorhodopsin. Biochim Biophys Acta 1460: 177–191
Kandori H (2004) Hydration switch model for the proton transfer in the Schiff base region of bacteriorhodopsin. Biochim Biophys Acta 1658:72–79
Kandori H (2011) Protein–controlled ultrafast photoisomerization in rhodopsin and bacteriorhodopsin. In: Ramamurthy V, Inoue Y (eds) Supramolecular photochemistry: controlling photochemical processes. Wiley, Hoboken, pp 571–595
Kataoka M, Kamikubo H (2000) Structures of photointermediates and their implications for the proton pump mechanism. Biochim Biophys Acta 1460: 166–176
Kawanabe A, Furutani Y, Jung KH, Kandori H (2009) Engineering an inward proton transport from a bacterial sensor rhodopsin. J Am Chem Soc 131:16439–16444
Kikukawa T, Shimono K, Tamogami J, Miyauchi S, Kim SY, Kimura-Someya T, Shirouzu M, Jung KH, Yokoyama S, Kamo N (2011) Photochemistry of *Acetabularia* rhodopsin II from a marine plant, *Acetabularia acetabulum*. Biochemistry 50: 8888–8898
Kimura H, Young CR, Martinez A, Delong EF (2011) Light-induced transcriptional responses associated with proteorhodopsin-enhanced growth in a marine flavobacterium. ISME J 5:1641–1651
Klare JP, Bordignon E, Engelhard M, Steinhoff HJ (2004) Sensory rhodopsin II and bacteriorhodopsin: light activated helix F movement. Photochem Photobiol Sci 3:543–547
Klare JP, Chizhov I, Engelhard M (2008) Microbial rhodopsins: scaffolds for ion pumps, channels, and sensors. Results Probl Cell Differ 45:73–122
Koh EY, Atamna-Ismaeel N, Martin A, Cowie RO, Beja O, Davy SK, Maas EW, Ryan KG (2010) Proteorhodopsin-bearing bacteria in Antarctic sea ice. Appl Environ Microbiol 76:5918–5925
Kouyama T, Murakami M (2010) Structural divergence and functional versatility of the rhodopsin superfamily. Photochem Photobiol Sci 9:1458–1465
Kralj JM, Bergo VB, Amsden JJ, Spudich EN, Spudich JL, Rothschild KJ (2008) Protonation state of Glu142 differs in the green- and blue-absorbing variants of proteorhodopsin. Biochemistry 47: 3447–3453
Krebs MP, Khorana HG (1993) Mechanism of light-dependent proton translocation by bacteriorhodopsin. J Bacteriol 175:1555–1560
Lanyi JK (2004) Bacteriorhodopsin. Annu Rev Physiol 66:665–688

Lanyi JK, Balashov SP (2011) Xanthorhodopsin. In: Ventosa A (ed) Halophiles and hypersaline environments. Springer, Dordrecht, pp 319–340

Lanyi JK, Schobert B (2004) Local-global conformational coupling in a heptahelical membrane protein: transport mechanism from crystal structures of the nine states in the bacteriorhodopsin photocycle. Biochemistry 43:3–8

Lanyi JK, Schobert B (2006) Propagating structural perturbation inside bacteriorhodopsin: crystal structures of the M state and the D96A and T46V mutants. Biochemistry 45:12003–12010

Lorenz-Fonfria VA, Kandori H (2009) Spectroscopic and kinetic evidence on how bacteriorhodopsin accomplishes vectorial proton transport under functional conditions. J Am Chem Soc 131:5891–5901

Lorinczi E, Verhoefen MK, Wachtveitl J, Woerner AC, Glaubitz C, Engelhard M, Bamberg E, Friedrich T (2009) Voltage- and pH-dependent changes in vectoriality of photocurrents mediated by wild-type and mutant proteorhodopsins upon expression in *Xenopus* oocytes. J Mol Biol 393: 320–341

Luecke H, Schobert B, Richter HT, Cartailler JP, Lanyi JK (1999a) Structural changes in bacteriorhodopsin during ion transport at 2 Angstrom resolution. Science 286:255–260

Luecke H, Schobert B, Richter HT, Cartailler JP, Lanyi JK (1999b) Structure of bacteriorhodopsin at 1.55 angstrom resolution. J Mol Biol 291:899–911

Luecke H, Schobert B, Cartailler JP, Richter HT, Rosengarth A, Needleman R, Lanyi JK (2000) Coupling photoisomerization of retinal to directional transport in bacteriorhodopsin. J Mol Biol 300:1237–1255

Luecke H, Schobert B, Stagno J, Imasheva ES, Wang JM, Balashov SP, Lanyi JK (2008) Crystallographic structure of xanthorhodopsin, the light-driven proton pump with a dual chromophore. Proc Natl Acad Sci U S A 105:16561–16565

Lukashev EP, Govindjee R, Kono M, Ebrey TG, Sugiyama Y, Mukohata Y (1994) pH dependence of the absorption spectra and photochemical transformations of the archaerhodopsins. Photochem Photobiol 60:69–75

Maeda A, Gennis RB, Balashov SP, Ebrey TG (2005) Relocation of water molecules between the Schiff base and the Thr46-Asp96 region during light-driven unidirectional proton transport by bacteriorhodopsin: an FTIR study of the N intermediate. Biochemistry 44:5960–5968

Man-Aharonovich D, Sabehi G, Sineshchekov OA, Spudich EN, Spudich JL, Beja O (2004) Characterization of RS29, a blue-green proteorhodopsin variant from the Red Sea. Photochem Photobiol Sci 3:459–462

Martinez A, Bradley AS, Waldbauer JR, Summons RE, DeLong EF (2007) Proteorhodopsin photosystem gene expression enables photophosphorylation in a heterologous host. Proc Natl Acad Sci U S A 104:5590–5595

Martinez-Garcia M, Swan BK, Poulton NJ, Gomez ML, Masland D, Sieracki ME, Stepanauskas R (2012) High-throughput single-cell sequencing identifies photoheterotrophs and chemoautotrophs in freshwater bacterioplankton. ISME J 6:113–123

McCarren J, DeLong EF (2007) Proteorhodopsin photosystem gene clusters exhibit co-evolutionary trends and shared ancestry among diverse marine microbial phyla. Environ Microbiol 9:846–858

Mills DA, Ferguson-Miller S (2003) Understanding the mechanism of proton movement linked to oxygen reduction in cytochrome *c* oxidase: lessons from other proteins. FEBS Lett 545:47–51

Mimuro M, Tsuchiya T, Koyama K, Peschek GA (2011) Bioenergetics in a primordial cyanobacterium *Gloeobacter violaceus* PCC 7421. In: Peschek GA (ed) Bioenergetic processes of cyanobacteria. Springer, The Netherlands, pp 211–238

Miranda MR, Choi AR, Shi L, Bezerra AG Jr, Jung KH, Brown LS (2009) The photocycle and proton translocation pathway in a cyanobacterial ion-pumping rhodopsin. Biophys J 96:1471–1481

Mongodin EF, Nelson KE, Daugherty S, Deboy RT, Wister J, Khouri H, Weidman J, Walsh DA, Papke RT, Sanchez Perez G, Sharma AK, Nesbo CL, MacLeod D, Bapteste E, Doolittle WF, Charlebois RL, Legault B, Rodriguez-Valera F (2005) The genome of *Salinibacter ruber*: convergence and gene exchange among hyperhalophilic bacteria and archaea. Proc Natl Acad Sci U S A 102:18147–18152

Mulkidjanian AY, Cherepanov DA, Heberle J, Junge W (2005) Proton transfer dynamics at membrane/water interface and mechanism of biological energy conversion. Biochemistry (Mosc) 70:251–256

Neutze R, Pebay-Peyroula E, Edman K, Royant A, Navarro J, Landau EM (2002) Bacteriorhodopsin: a high-resolution structural view of vectorial proton transport. Biochim Biophys Acta 1565:144–167

Oesterhelt D, Stoeckenius W (1973) Functions of a new photoreceptor membrane. Proc Natl Acad Sci U S A 70:2853–2857

Okamoto OK, Hastings JW (2003) Novel dinoflagellate clock-related genes identified through microarray analysis. J Phycol 39:519–526

Oren A (1999) Bioenergetic aspects of halophilism. Microbiol Mol Biol Rev 63:334–348

Ormos P (1991) Infrared spectroscopic demonstration of a conformational change in bacteriorhodopsin involved in proton pumping. Proc Natl Acad Sci U S A 88:473–477

Page RD (2002) Visualizing phylogenetic trees using TreeView. Current Protocols in Bioinformatics, Chapter 6: Unit 6.2

Partha R, Krebs R, Caterino TL, Braiman MS (2005) Weakened coupling of conserved arginine to the proteorhodopsin chromophore and its counterion implies structural differences from bacteriorhodopsin. Biochim Biophys Acta 1708:6–12

Petrovskaya LE, Lukashev EP, Chupin VV, Sychev SV, Lyukmanova EN, Kryukova EA, Ziganshin RH, Spirina EV, Rivkina EM, Khatypov RA, Erokhina LG, Gilichinsky DA, Shuvalov VA, Kirpichnikov MP (2010) Predicted bacteriorhodopsin from *Exiguobacterium sibiricum* is a functional proton pump. FEBS Lett 584:4193–4196

Pfleger N, Worner AC, Yang J, Shastri S, Hellmich UA, Aslimovska L, Maier MS, Glaubitz C (2009) Solid-state NMR and functional studies on proteorhodopsin. Biochim Biophys Acta 1787:697–705

Racker E, Stoeckenius W (1974) Reconstitution of purple membrane vesicles catalyzing light-driven proton uptake and adenosine triphosphate formation. J Biol Chem 249:662–663

Raven JA (2009) Functional evolution of photochemical energy transformations in oxygen-producing organisms. Funct Plant Biol 36:505–515

Reckel S, Gottstein D, Stehle J, Lohr F, Verhoefen MK, Takeda M, Silvers R, Kainosho M, Glaubitz C, Wachtveitl J, Bernhard F, Schwalbe H, Guntert P, Dotsch V (2011) Solution NMR structure of proteorhodopsin. Angew Chem Int Ed Engl 50: 11942–11946

Richter HT, Brown LS, Needleman R, Lanyi JK (1996) A linkage of the pK(a)'s of asp-85 and glu-204 forms part of the reprotonation switch of bacteriorhodopsin. Biochemistry 35:4054–4062

Riesle J, Oesterhelt D, Dencher NA, Heberle J (1996) D38 is an essential part of the proton translocation pathway in bacteriorhodopsin. Biochemistry 35:6635–6643

Rothschild KJ (1992) FTIR difference spectroscopy of bacteriorhodopsin: toward a molecular model. J Bioenerg Biomembr 24:147–167

Rothschild KJ, He YW, Sonar S, Marti T, Khorana HG (1992) Vibrational spectroscopy of bacteriorhodopsin mutants. Evidence that Thr-46 and Thr-89 form part of a transient network of hydrogen bonds. J Biol Chem 267:1615–1622

Ruiz-Gonzalez MX, Marin I (2004) New insights into the evolutionary history of type 1 rhodopsins. J Mol Evol 58:348–358

Sabehi G, Loy A, Jung KH, Partha R, Spudich JL, Isaacson T, Hirschberg J, Wagner M, Beja O (2005) New insights into metabolic properties of marine bacteria encoding proteorhodopsins. PLoS Biol 3:1409–1417

Sasaki J, Spudich JL (2000) Proton transport by sensory rhodopsins and its modulation by transducer-binding. Biochim Biophys Acta 1460:230–239

Schafer G, Engelhard M, Muller V (1999) Bioenergetics of the archaea. Microbiol Mol Biol Rev 63:570–620

Schobert B, Brown LS, Lanyi JK (2003) Crystallographic intermediates of structures of the M and N bacteriorhodopsin: assembly of a hydrogen-bonded chain of water molecules between Asp-96 and the retinal Schiff base. J Mol Biol 330:553–570

Sharma AK, Spudich JL, Doolittle WF (2006) Microbial rhodopsins: functional versatility and genetic mobility. Trends Microbiol 14:463–469

Sharma AK, Walsh DA, Bapteste E, Rodriguez-Valera F, Ford Doolittle W, Papke RT (2007) Evolution of rhodopsin ion pumps in haloarchaea. BMC Evol Biol 7:79

Sharma AK, Zhaxybayeva O, Papke RT, Doolittle WF (2008) Actinorhodopsins: proteorhodopsin-like gene sequences found predominantly in non-marine environments. Environ Microbiol 10:1039–1056

Sharma AK, Sommerfeld K, Bullerjahn GS, Matteson AR, Wilhelm SW, Jezbera J, Brandt U, Doolittle WF, Hahn MW (2009) Actinorhodopsin genes discovered in diverse freshwater habitats and among cultivated freshwater Actinobacteria. ISME J 3:726–737

Shi L, Lake EM, Ahmed MA, Brown LS, Ladizhansky V (2009a) Solid-state NMR study of proteorhodopsin in the lipid environment: secondary structure and dynamics. Biochim Biophys Acta 1788:2563–2574

Shi L, Ahmed MA, Zhang W, Whited G, Brown LS, Ladizhansky V (2009b) Three-dimensional solid-state NMR study of a seven-helical integral membrane proton pump–structural insights. J Mol Biol 386:1078–1093

Shibata M, Tanimoto T, Kandori H (2003) Water molecules in the Schiff base region of bacteriorhodopsin. J Am Chem Soc 125:13312–13313

Shimono K, Hayashi T, Ikeura Y, Sudo Y, Iwamoto M, Kamo N (2003) Importance of the broad regional interaction for spectral tuning in Natronobacterium pharaonis phoborhodopsin (sensory rhodopsin II). J Biol Chem 278:23882–23889

Sineshchekov OA, Spudich JL (2004) Light-induced intramolecular charge movements in microbial rhodopsins in intact E. coli cells. Photochem Photobiol Sci 3:548–554

Sineshchekov OA, Govorunova EG, Jung KH, Zauner S, Maier UG, Spudich JL (2005) Rhodopsin-mediated photoreception in cryptophyte flagellates. Biophys J 89:4310–4319

Slamovits CH, Okamoto N, Burri L, James ER, Keeling PJ (2011) A bacterial proteorhodopsin proton pump in marine eukaryotes. Nat Commun 2:183

Spudich JL (2006) The multitalented microbial sensory rhodopsins. Trends Microbiol 14:480–487

Spudich JL, Yang CS, Jung KH, Spudich EN (2000) Retinylidene proteins: structures and functions from archaea to humans. Annu Rev Cell Dev Biol 16:365–392

Steindler L, Schwalbach MS, Smith DP, Chan F, Giovannoni SJ (2011) Energy starved candidatus pelagibacter ubique substitutes light-mediated ATP production for endogenous carbon respiration. PLoS One 6:e19725

Stingl U, Desiderio RA, Cho JC, Vergin KL, Giovannoni SJ (2007) The SAR92 clade: an abundant coastal clade of culturable marine bacteria possessing proteorhodopsin. Appl Environ Microbiol 73:2290–2296

Subramaniam S, Henderson R (2000) Molecular mechanism of vectorial proton translocation by bacteriorhodopsin. Nature 406:653–657

Subramaniam S, Lindahl I, Bullough P, Faruqi AR, Tittor J, Oesterhelt D, Brown L, Lanyi J, Henderson R (1999) Protein conformational changes in the bacteriorhodopsin photocycle. J Mol Biol 287:145–161

Subramaniam S, Hirai T, Henderson R (2002) From structure to mechanism: electron crystallographic studies of bacteriorhodopsin. Philos Trans A Math Phys Eng Sci 360:859–874

Sudo Y, Ihara K, Kobayashi S, Suzuki D, Irieda H, Kikukawa T, Kandori H, Homma M (2011) A microbial rhodopsin with a unique retinal composition shows both sensory rhodopsin II and bacteriorhodopsin-like properties. J Biol Chem 286:5967–5976

Thompson JD, Gibson TJ, Higgins DG (2002) Multiple sequence alignment using ClustalW and ClustalX. Current Protocols in Bioinformatics, Chapter 2: Unit 2 3

Tsunoda SP, Ewers D, Gazzarrini S, Moroni A, Gradmann D, Hegemann P (2006) H+-pumping rhodopsin from the marine alga *Acetabularia*. Biophys J 91:1471–1479

Venter JC, Remington K, Heidelberg JF, Halpern AL, Rusch D, Eisen JA, Wu D, Paulsen I, Nelson KE, Nelson W, Fouts DE, Levy S, Knap AH, Lomas MW, Nealson K, White O, Peterson J, Hoffman J, Parsons R, Baden-Tillson H, Pfannkoch C, Rogers YH, Smith HO (2004) Environmental genome shotgun sequencing of the Sargasso Sea. Science 304:66–74

Wada T, Shimono K, Kikukawa T, Hato M, Shinya N, Kim SY, Kimura-Someya T, Shirouzu M, Tamogami J, Miyauchi S, Jung KH, Kamo N, Yokoyama S (2011) Crystal structure of the eukaryotic light-driven proton-pumping rhodopsin, *Acetabularia rhodopsin* II, from marine alga. J Mol Biol 411:986–998

Walter JM, Greenfield D, Bustamante C, Liphardt J (2007) Light-powering *Escherichia coli* with proteorhodopsin. Proc Natl Acad Sci U S A 104:2408–2412

Waschuk SA, Bezerra AG, Shi L, Brown LS (2005) *Leptosphaeria* rhodopsin: bacteriorhodopsin-like proton pump from a eukaryote. Proc Natl Acad Sci U S A 102:6879–6883

Wikstrom M (1998) Proton translocation by bacteriorhodopsin and heme-copper oxidases. Curr Opin Struct Biol 8:480–488

Xiao YW, Hutson MS, Belenky M, Herzfeld J, Braiman MS (2004) Role of arginine-82 in fast proton release during the bacteriorhodopsin photocycle: a time-resolved FT-IR study of purple membranes containing N-15-labeled arginine. Biochemistry 43:12809–12818

Yoshimura K, Kouyama T (2008) Structural role of bacterioruberin in the trimeric structure of archaerhodopsin-2. J Mol Biol 375:1267–1281

Zhang F, Vierock J, Yizhar O, Fenno LE, Tsunoda S, Kianianmomeni A, Prigge M, Berndt A, Cushman J, Polle J, Magnuson J, Hegemann P, Deisseroth K (2011) The microbial opsin family of optogenetic tools. Cell 147:1446–1457

Zubkov MV (2009) Photoheterotrophy in marine prokaryotes. J Plankton Res 31:933–938

Chapter 2

Structure and Functional Heterogeneity of Fucoxanthin-Chlorophyll Proteins in Diatoms

Kathi Gundermann and Claudia Büchel*
Institute of Molecular Biosciences, Goethe Universität Frankfurt, D-60438 Frankfurt, Germany

Summary

Fucoxanthin-chlorophyll proteins (FCPs) of diatoms are divided into three groups, the main light harvesting antennas Lhcf, the photosystem I-specific Lhcr, and Lhcx involved in photoprotection. All are closely related to higher plant light harvesting complexes (LHCs) when comparing sequences, albeit smaller and more hydrophobic. However, pigmentation differs from higher plant LHCs with around eight chlorophyll *a*, two chlorophyll *c* and six fucoxanthin per monomer. Fucoxanthin, with a carbonyl moiety conjugated to the polyene backbone, undergoes extreme bathochromic shifts upon protein binding, dividing the different fucoxanthins into more red, green and blue absorbing ones. Excitation energy transfer is extremely efficient, either directly from chlorophyll *c* to chlorophyll *a* or from fucoxanthin to chlorophyll *a* involving the S_1/ICT state of fucoxanthin. Most Lhcf assemble into trimers, whereby only in centric diatoms Lhcx was found in trimers as well, and specific oligomeric FCP complexes are present. Whereas the arrangement of FCPs around the photosystems is largely unknown, spectroscopic measurements together with homology considerations allow for a first rough model of the pigment arrangement in trimeric and oligomeric FCP complexes. Blue fucoxanthin is bound analogously to lutein in LHCII, surrounded by the same four chlorophyll *a*, since binding sites are conserved. Additionally, chlorophyll *a* can be found in

*Author for correspondence, e-mail: c.buechel@bio.uni-frankfurt.de

M.F. Hohmann-Marriott (ed.), The Structural Basis of Biological Energy Generation, Advances in Photosynthesis and Respiration 39, pp. 21–37, DOI 10.1007/978-94-017-8742-0_2,

a604, a614, b605 and a611, although binding of the latter has to be different due to the lack of long wavelength absorption in FCPs. Chlorophyll *c* is most probably bound in b609 and a613. The red fucoxanthin cluster around helix 2, which has less sequence homology to LHCII. The green fucoxanthins are most probably located around the violaxanthin and b601 binding sites of LHCII, whereby the former is probably a mixed site for fucoxanthins and diadinoxanthin/diatoxanthin.

I. Introduction

Membrane-intrinsic light-harvesting proteins belonging to the same family as higher plant LHC are wide-spread amongst other eukaryotic photoautotrophic organism like e.g. stramenopiles, including brown algae and diatoms. This review will focus on diatom light- harvesting proteins, which, due to their main carotenoid, are also called fucoxanthin-chlorophyll proteins (FCPs).

Diatoms are unicellular photosynthetic organisms characterized by an ornamental cell wall made of silica (Raven and Waite 2004) (Fig. 2.1a, b). They fall into two main groups, the so-called pennate diatoms and centric diatoms (Medlin et al. 1996). Whereas the former are longish in shape and usually contain one or two plastids per cell, the centrics have a rotational symmetry and are characterized by more than two plastids (Medlin et al. 1996) (Fig. 2.1c–f). In the public data bases, two analyzed genomes are available at the moment, one of the pennate diatom *Phaeodactylum tricornutum* and one from the centric *Thalassiosira pseudonana* (Armbrust et al. 2004; Bowler et al. 2008). See also Chap. 18 for information on carbon fixation in Diatoms.

Stramenopiles plastids are derived from a secondary endosymbiosis (for recent review see Green (2011)). Ancient red algae are postulated to be the endosymbionts, i.e. the ancestors of the plastids (Archibald and Keeling 2002; Wolfe et al. 1994), but genes typical of green algae and plants can be found as well (Moustafa et al. 2009). Due to the secondary endosymbiosis, chloroplasts are enveloped by four membranes instead of two membranes indicating a primary endosymbiotic event. The outer chloroplast membrane is in connection with the nuclear endoplasmatic reticulum that possesses ribosomes (Gibbs 1970).

The thylakoid structure in brown algae and diatoms is different as well: no grana – stroma differentiation (Gibbs 1970) can be found and thus no lateral heterogeneity concerning the distribution of photosystem (PS) I and PS II was detected so far (Pyszniak and Gibbs 1992). Thylakoids are instead organized in long bands of three thylakoids (i.e. six membranes) each, running along the whole length of the plastids (Pyszniak and Gibbs 1992) (Fig. 2.1g, h). The thylakoid membranes contain the multi-protein

Abbreviations: Chl – Chlorophyll; Dd – Diadinoxanthin; Dt – Diatoxanthin; FCP – Fucoxanthin-chlorophyll protein; Fx – Fucoxanthin; LHC – Light-harvesting complex; NPQ – Non-photochemical quenching; PS – Photosystem

Fig. 2.1. Diatoms are unicellular organisms characterized by an ornamental silica cell wall as seen by scanning electron microscopy for the centric diatom *C. meneghiniana* (**a**, **b**). These diatoms contain more than two chloroplasts per cell as shown in the light micrograph (**c**) and the transmission electron microscopy picture (**d**) of *C. meneghiniana.* The morphology of pennate diatoms, here *P. tricornutum*, is demonstrated in (**e**, light microscopy) and (**f**, transmission electron microscopy). The arrangement of thylakoid membranes in bands of three thylakoids each are shown in thin sectioning in (**g**) and in a schematic overview of the plastid in (**h**). *Bars* represent 1 μm in (**a**), (**b**), (**d**) and (**f**), 2 μm in (**c**) and (**e**), and 0.2 μm in (**g**).

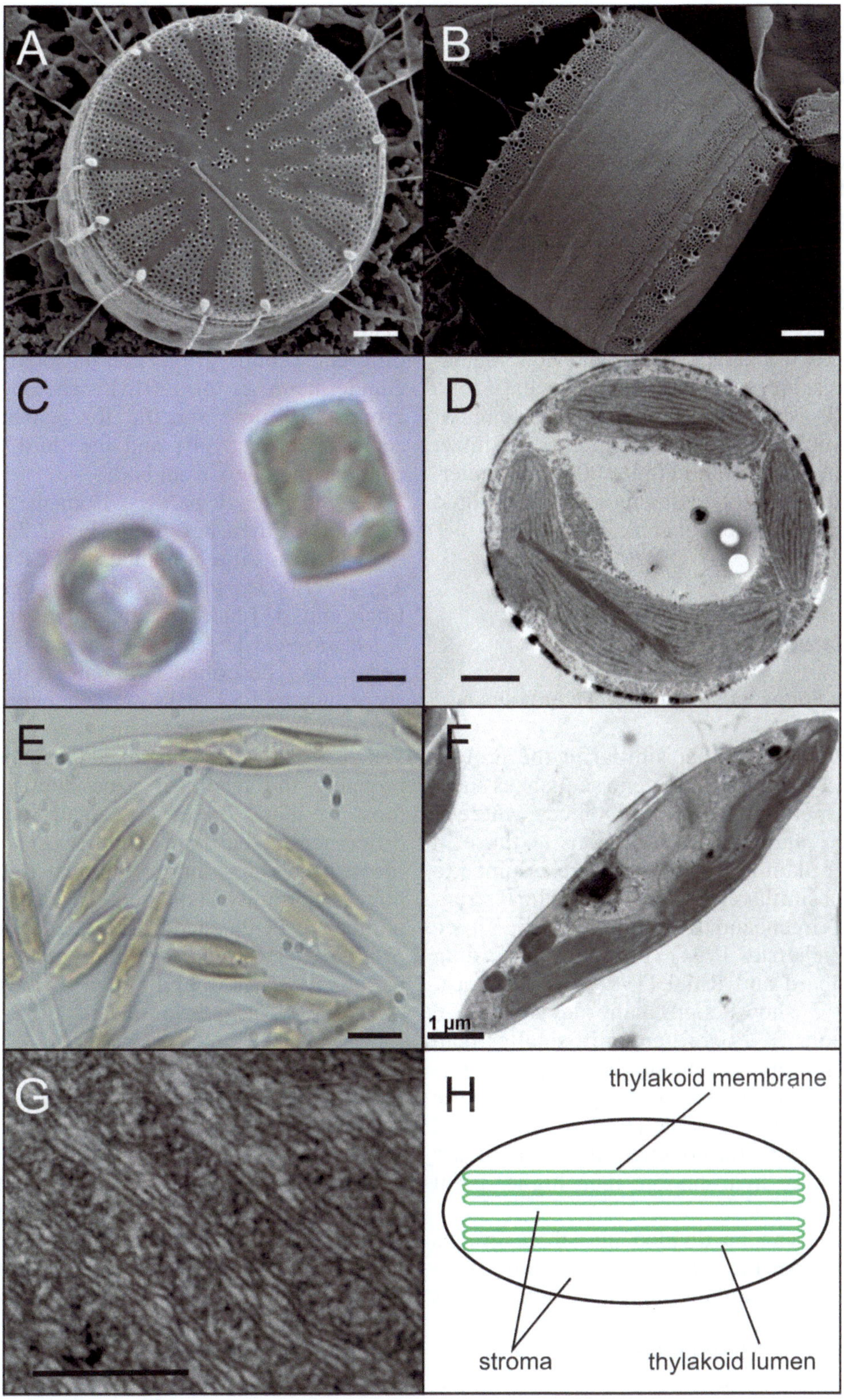
A
B
C
D
E
F
1 µm
G
H
thylakoid membrane
stroma
thylakoid lumen

complexes usually found in eukaryotic thylakoids, i.e. photosystem I and II, cytochrome b_6/f complexes (see Chap. 8), ATP synthases (see Chap. 6) and antenna systems. Photosystems are homologous to those of higher plants except for some minor subunits (Ikeda et al. 2008; Nagao et al. 2007, 2010; Veith et al. 2009; Veith and Büchel 2007), and in general light capture and electron transport thus resemble those of spermatophytes. However, the lack of grana poses the questions of how energy is distributed between the photosystems and how regulation is achieved. Some indications exist for a domain structure, e.g. results of circular dichroism (CD) measurements of intact diatom cells (Szabó et al. 2008) were interpreted to indicate domains of different lipid composition (Lepetit et al. 2010).

II. The Light Harvesting Proteins of Diatoms

A. The Fucoxanthin-Chlorophyll Proteins

All FCPs are rather similar at the protein level. Prediction from gene sequences show three membrane spanning helices, whereby helix 1 and 3 are homologous to those in higher plant LHC and therefore assumed to form a similar cross-like superhelical structure (Green and Kühlbrandt 1995; Green and Pichersky 1994) (Fig. 2.2b, according to Eppard and Rhiel (1998)). In contrast, helix 2 shows significant deviations. In addition, FCPs are generally smaller than LHC proteins due to smaller loops and termini. Thus, molecular weights are in the range of 18–21 kDa (Eppard and Rhiel 1998). As a consequence, the proteins are even more hydrophobic than higher plant LHCs.

FCPs can be divided into three groups (Fig. 2.2a): the main light harvesting proteins, called Lhcf nowadays, are homologous between diatoms and their relatives, e.g. brown algae. These proteins are very abundant and were the first to be identified at the gene-level (Bhaya and Grossman 1993; Eppard et al. 2000; Eppard and Rhiel 1998, 2000). A second group is called Lhcr, due to their similarity to red algal LHCI genes (Durnford et al. 1996). The third group is related to LhcSR (former LI818) proteins of *Chlamydomonas reinhardtii*, which are involved in light protection (Eppard and Rhiel 1998; Peers et al. 2009; Richard et al. 2000; Zhu and Green 2010), and are called Lhcx in diatoms. Whereas the Lhcr and Lhcx proteins are very similar in centric and pennate diatoms, the Lhcf proteins fall into three different groups (Fig. 2.2, adapted from Gundermann et al. 2013), whereby two groups are more specific for pennates or centrics, respectively, and the third group contains members from both.

As can already be seen from the list of members of the Lhcf group (Fig. 2.2a), diatoms possess quite a number of FCP genes, e.g. for *P. tricornutum* 17 Lhcf genes, 14 Lhcr and 4 Lhcx are annotated and for *T. pseudonana* 11 Lhcf, 14 Lhcr and 7 Lhcx genes are described. All are expressed according to EST data or their expression was proven otherwise (Nymark et al. 2009). Whereas a few of those are identical and thus probably the result of gene duplications, most of them show very small sequence differences. The high number of almost identical proteins might be related to the huge adaptability of diatoms. The even larger number of FCPs encoded by the genome of *Fragilariopsis cylindrus* (http://genome.jgi-psf.org/), a psychrophilic alga living in the Antarctic ice, argues for this.

B. Supramolecular Organization of Fucoxanthin-Chlorophyll Proteins

Biochemical work on FCPs dates back to the 80s, when mainly sucrose density centrifugation was used to separate photosystems from FCPs as a whole (Alberte et al. 1981; Brown 1988; Caron and Brown 1987; Fawley et al. 1986; Friedman and Alberte 1984, 1986; Gugliemelli 1984; Owens 1986, 1988; Owens and Wold 1986). Most FCPs can be isolated as 'free' FCPs, i.e. separate from the photosystems, by many biochemical methods

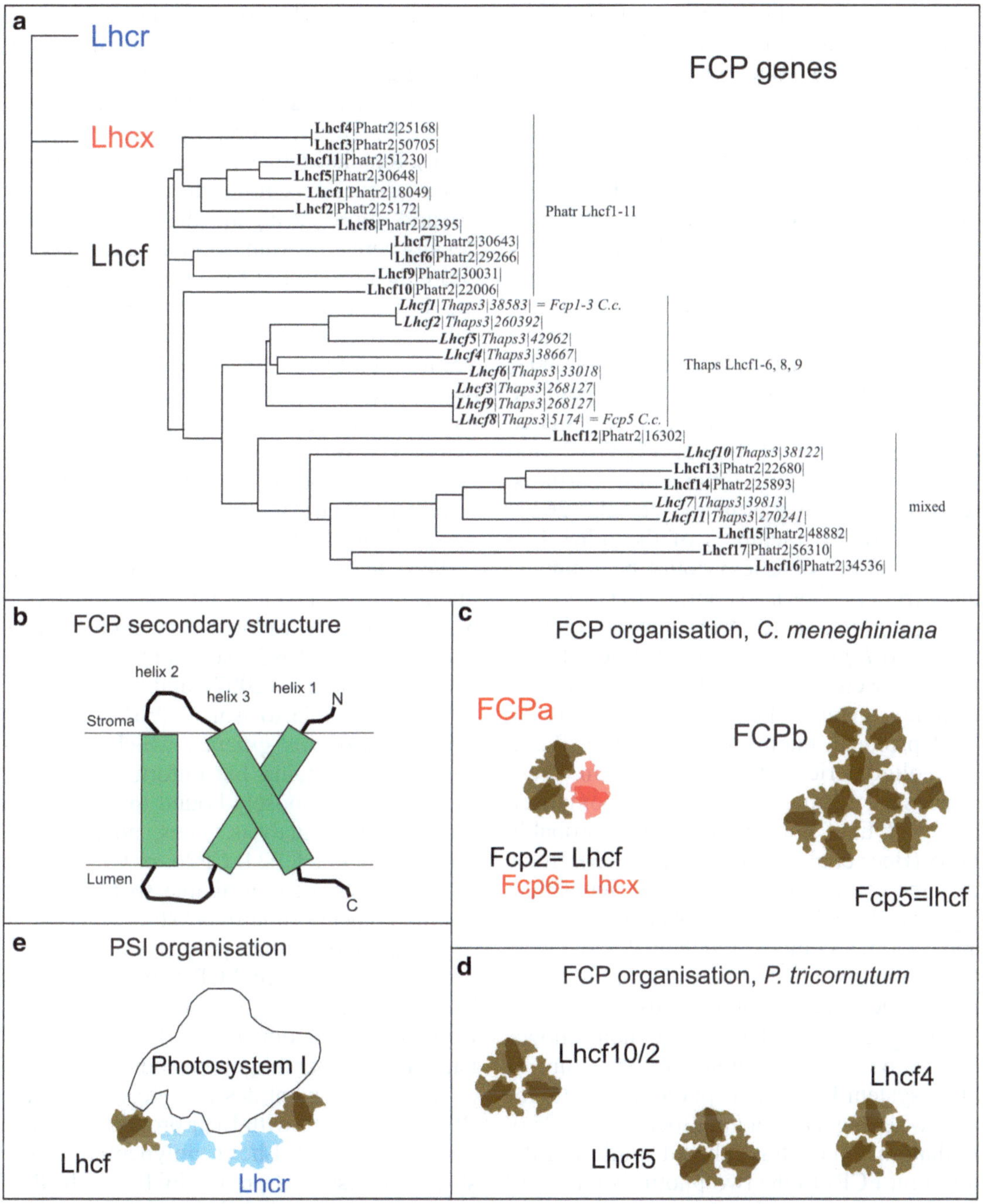

Fig. 2.2. Diatom light-harvesting polypeptides can be differentiated into three groups, Lhcr, Lhcf and Lhcx proteins. Concerning pennate and centric diatoms, most of the Lhcf genes are different between the two groups as demonstrated by the phylogenetic tree (Adapted from Gundermann et al. 2013) (**a**, Phatr = *P. tricornutum*, Thaps = *T. pseudonana*). All of them share the same topology with three membrane-spanning helices predicted to show the same arrangement as in LHCII, as depicted in the cartoon as side view into the membrane (**b**). However, the oligomeric state and polypeptide composition of the FCP complexes is different between centrics and pennates: whereas in the centric diatom *C. meneghiniana* (**c**) trimers (FCPa) and higher oligomers (FCPb) make up the main antenna complexes, only trimers can be found in the pennate *P. tricornutum* (**d**), whereby the trimer population is quite inhomogeneous. Labels refer to the main polypeptides only. In the trimers found in *C. meneghiniana*, also Lhcx polypeptides are present, which have so far not been reported for *P. tricornutum*. On the other hand the composition of the photosystem I antenna (**e**) seems to be similar with mainly Lhcr and Lhcf proteins bound.

(sucrose density centrifugation, ion exchange chromatography, blue native polyacrylamide electrophoresis) (Beer et al. 2006, 2011; Berkaloff et al. 1990; Brakemann et al. 2006; Büchel 2003; Grouneva et al. 2011; Guglielmi et al. 2005; Gundermann and Büchel 2008; Lavaud et al. 2003). This 'pool' of FCPs consists of trimeric FCP complexes (Büchel 2003; Lepetit et al. 2007), whereby in centrics specific complexes of higher oligomeric state were found as well (Beer et al. 2006, 2011; Büchel 2003; Grouneva et al. 2011; Gundermann and Büchel 2008) (Fig. 2.2c). In *P. tricornutum*, a pennate diatom, three major trimers were isolated lately, composed of Lhcf5, Lhcf10/2 and Lhcf4 with different interaction partners, respectively (Gundermann et al. 2013) (Fig. 2.2d). No members of the other Lhc families (Lhcr or Lhcx) could be found in these trimers. This is in contrast to *Cyclotella meneghiniana*, a centric diatom closely related to *T. pseudonana*, where Lhcx (Fcp6) proteins were found in the major trimeric complex, named FCPa, accompanied by Lhcf proteins, mainly by Fcp2 (Fig. 2.2c). The oligomeric complex found in these organisms, named FCPb, was composed solely of Lhcf polypeptides, most probably Fcp5 (Beer et al. 2006; Büchel 2003). Since diatoms contain many more Lhc genes than higher plants a larger hetereogeneity in trimer and oligomer composition than elucidated so far may be revealed using more sophisticated biochemical separation methods.

In higher plants there is excellent knowledge about the supramolecular structure of photosystem I as well as photosystem II (for review see e.g. Dekker and Boekema (2005)). In diatoms, in contrast, the attribution of the different FCPs to the two photosystems and/or their supramolecular structure remains ill-defined. Photosystem I supercomplexes were isolated early (Berkaloff et al. 1990) and from several organisms (Brakemann et al. 2006; Grouneva et al. 2011; Ikeda et al. 2008; Veith et al. 2009; Veith and Büchel 2007). Since gene sequences became available, Lhcr proteins were always supposed to serve as PSI antennas, which could indeed be shown at the protein level as well (Grouneva et al. 2011; Veith et al. 2009) (Fig. 2.2e). There is still some controversy, as to whether this PSI antenna is exclusively composed of Lhcr proteins (Lepetit et al. 2010). Results by blue native polyacrylamide electrophoresis (Grouneva et al. 2011), a method more stringent than e.g. sucrose density centrifugation, supported earlier data about the presence of Lhcf proteins as part of the PSI antenna (Brakemann et al. 2006; Juhas and Büchel 2012; Veith et al. 2009; Veith and Büchel 2007). In addition, Lhcx polypeptides were found, but solely in pennates (Grouneva et al. 2011). Like in higher plants, PSI of diatoms is a monomer (Veith and Büchel 2007), but no data about the arrangement of the antenna proteins are currently available.

Very little is known about PSII-specific interactions of FCPs so far, since supercomplexes as can be isolated from higher plants (Boekema et al. 1995) have not been obtained yet. Nagao et al. (2007) and Nagao et al. (2010) were able to isolate PSII complexes from *Chaetoceros gracilis*, which still contained FCP proteins, but unfortunately these FCPs were too loosely bound to allow for a more detailed analysis. Thus, no proof for minor Lhcs like CP24, CP26, or CP29 found in plant systems, is available so far, despite immunological similarities (Rhiel et al. 1997).

In summary, Lhcf polypeptides are the main constituents of FCP trimers, whereas Lhcx proteins were found only in trimers from centric diatoms so far. In those organisms also higher oligomers built from specific Lhcf polypeptides are present as well. PSI is associated with Lhcr proteins, but also with Lhcf (and Lhcx) polypeptides. The structural association of FCPs with PSII remains unresolved. Only some data obtained in *C. meneghiniana*, points to FCPb complexes being more closely associated with PSI, and FCPa serving PSII (Szabó et al. 2010; Veith and Büchel 2007). This severe lack of precise knowledge concerning protein associations also extends to the overall arrangement of the complexes in the thylakoid membranes.

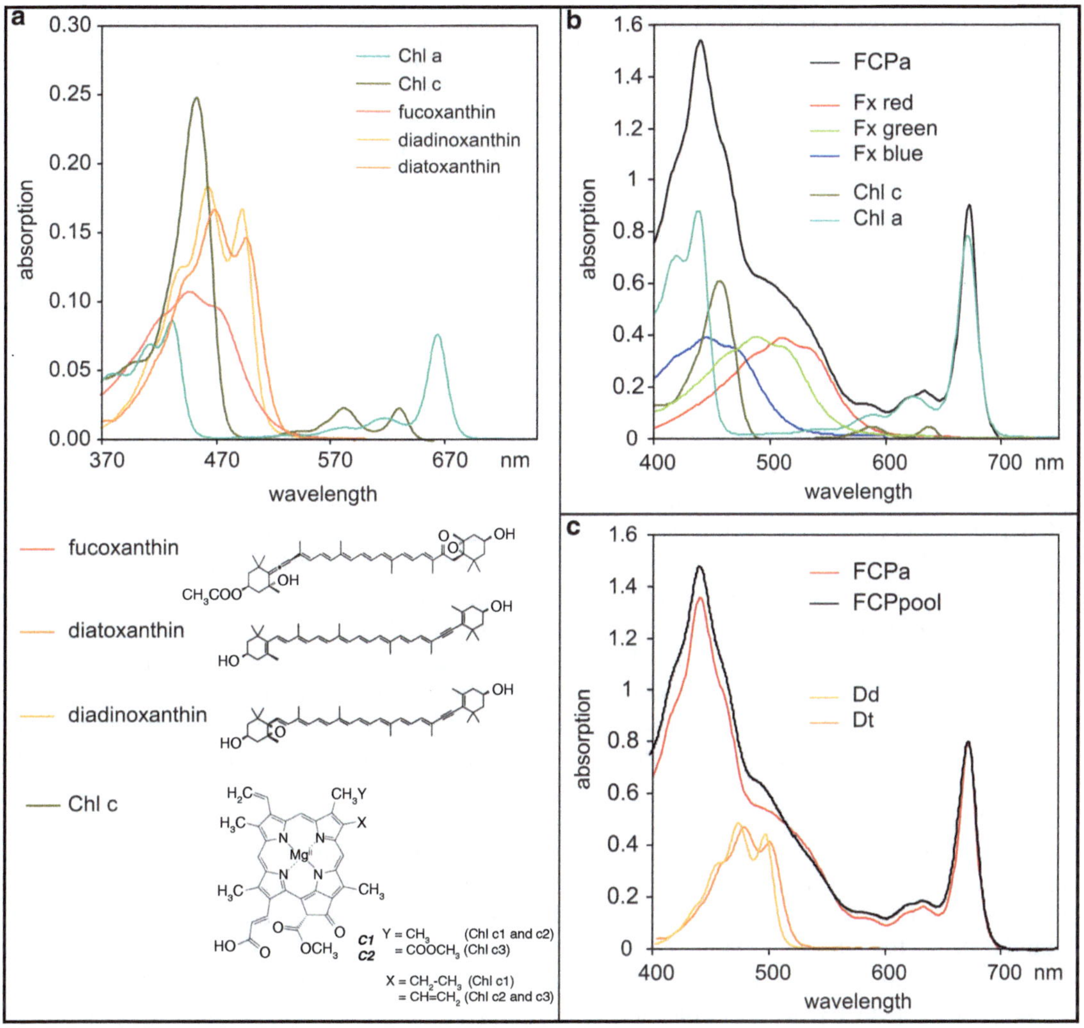

Fig. 2.3. Chl *a*, Chl *c* and Fx are bound to FCP complexes besides the xanthophyll cycle pigments Dd and Dt. In (**a**) the absorbance spectra of all pigments (1 mM in 80 % acetone) are shown. Upon binding to the protein, Fx undergoes extreme bathochromic shifts, whereby several populations can be distinguished, i.e. '*red*', '*green*' and '*blue*' Fx (**b**) (Adopted from Premvardhan et al. 2008). Depending on the isolation method (sucrose gradient centrifugation, FCP pool, or ion exchange chromatography, FCPa) different amounts of Dd and Dt can be found in the preparations as demonstrated by the absorbance spectra in (**c**).

C. Pigmentation of Fucoxanthin-Chlorophyll Proteins

When comparing plant LHC with FCPs the most obvious difference is the pigmentation. Diatoms exhibit a brownish color, which is due to the carotenoid fucoxanthin (Fx) bound to FCPs. This carotenoid is found in much higher amounts in FCPs than carotenoids in LHCII: the molar Chl/carotenoid ratio is almost 1:1 in FCPs compared to the 14:4 in LHCII (Beer et al. 2006; Papagiannakis et al. 2005). Fucoxanthin is a rather peculiar carotenoid with a carbonyl moiety in conjugation with the polyene backbone that is also found in peridinin (Damjanović et al. 2000; Frank et al. 2000; Katoh et al. 1991; Zigmantas et al. 2004) (Fig. 2.3a). Fx displays an extreme bathochromic shift upon protein binding, extending the absorption from

390 nm up to 580 nm (Premvardhan et al. 2009). This light-absorbing capability in the blue-green range is used for successful photosynthesis in the aquatic environments. Since the bathochromic shift depends strongly on the polarity of the protein environment (Koyama et al. 1996), different shifts are exhibited by the different Fx found in a FCP monomer. Using stark and resonance Raman spectroscopy, more 'blue', 'green' and 'red' absorbing Fx molecules could be detected (Fig. 2.3b) in FCPa as well as in FCPb (Premvardhan et al. 2008, 2009, 2010). The existence of differently absorbing Fx was also demonstrated in whole cells using electrochromic shift measurements (Szabó et al. 2010). In carotenoids in general, due to their symmetry, absorption from the ground state to the lowest-energy singlet S_1 state (2^1A_g) is symmetry-forbidden. Instead, excitation of the ground state (1^1A_g) results in the S_2 state (1^1B_u). In carbonyl-containing carotenoids it was shown that after excitation into S_2 the so-called ICT (intramolecular charge transfer) state is populated (Zigmantas et al. 2004). This state is strongly coupled to the S_1 state and more generally referred to as S_1/ICT. Indeed, the excitation energy transfer from Fx to Chl *a* was shown to proceed mainly via the S_1/ICT state in FCPs (Gildenhoff et al. 2010a; Papagiannakis et al. 2005). Stark spectroscopy reveals that upon photon absorption by Fx in solvent, a huge change in the static dipole moment of 17 D takes place, indicating photo-induced charge transfer of the Fx molecules. When Fx in FCPs is examined, two population with changes of 17 D and up to 40 D can be distinguished, again underlining the different properties of the different Fx molecules bound (Premvardhan et al. 2008).

As accessory chlorophyll Chl *c* is found instead of Chl *b*, but only 1 Chl *c* per 4 Chl *a* is bound. In contrast to Chl *b*, Chl *c* is characterized by a huge Soret-band absorption and little Q_Y or Q_X absorption (Fig. 2.3a, b). The absorption in the Soret is shifted to the red in comparison to Chl *a*. Three different Chl *c* exist, distinguished by their residues at the porphyrin ring as shown in Fig. 2.3a. Since this residue contains a double bond, which is part of the conjugated system in Chl c_2 and c_3, these Chl *c*'s absorb at slightly longer wavelengths as compared to Chl c_1. In prymnesiophytes Chl c_3 and in dinoflagellates Chl c_2 was described as the major Chl *c*, whereas diatoms usually contain Chl c_1 and some Chl c_2 as well (Fawley 1989; Jeffrey and Humphrey 1975; Kraay et al. 1992). In all Chl *c*'s, the lack of a phytol chain makes the molecule much more polar and, from a structural point of view, much smaller than Chl *a* or Chl *b*.

In addition to these pigments diatoms contain diadinoxanthin (Dd) and diatoxanthin (Dt). Dd is de-epoxidised to Dt under conditions of increased light intensities in the so-called xanthophyll cycle (Lavaud et al. 2002; Lohr and Wilhelm 1999). The amount of Dd or Dt found in the various FCP preparations differs tremendously depending on the isolation method (Fig. 2.3c). Usually the more xanthophyll cycle pigments are found, the more lipids are still contained in the samples (Beer et al. 2006; Büchel 2003; Lepetit et al. 2010). However, this does not necessarily argue for a localization of xanthophyll pigments in the lipid phase, i.e. not bound to the protein. Using more rigid methods, which get rid of almost all lipid, Dd or Dt is still found in e.g. FCPa preparations, and the amount depends on the presence of Fcp6, a Lhcx protein (Beer et al. 2006, 2011). However, most probably part of the Dd is found in the lipid shell around the FCP complexes (Gundermann and Büchel 2012; Lepetit et al. 2010), serving as a pool of xanthophylls pigment. Dt is thought to play an important role in the protection against an surplus of light, and the Dt (and Dd) that is newly synthesized in response to high-light was recently demonstrated to be protein-bound (Alexandre et al. 2014).

A long ongoing debate is the actual pigment to protein stoichiometry inside FCPs. Most isolated complexes exhibit rather similar absorption spectra, whereby only the amount of xanthophyll cycle pigments differs, visible as a shoulder around 485 nm (Fig. 2.3c). In LHCII, 8 Chl *a*, 6 Chl *b* and

4 carotenoids are bound per monomer (Liu et al. 2004; Standfuss et al. 2005). The pigment ratio of FCPs is around 3–4 Chl *a* : 1 Chl *c* : 3–4 Fx, depending on isolation procedure and species, whereby the FCPs of *P. tricornutum* are characterized by a higher amount of Fx (Beer et al. 2006; Gundermann et al. 2013; Joshi-Deo et al. 2010; Lepetit et al. 2007; Papagiannakis et al. 2005). Because the Q_Y absorption of FCPs is at a relatively short wavelength (671 nm, see Fig. 2.3a, b) it was argued that especially Chl *a* molecules have to be further apart than in LHCII in order to avoid excitonic interactions, which result in longer wavelength absorbing Chls. This argument was strengthened by CD spectra, where no excitonic interactions are visible in the Q_Y band (Büchel 2003; Joshi-Deo et al. 2010; Szabó et al. 2008). Using resonance Raman spectroscopy, two differently bound Chl *c* molecules were identified in FCPa, as well as in FCPb, by their signature 'ring-breathing modes' at ~1,360 cm^{-1} (Premvardhan et al. 2010). Thus, most likely around 8 Chl *a*, 2 Chl *c* and up to 8 Fx are bound per FCP monomer.

D. Excitation Energy Transfer Between Pigments in Fucoxanthin-Chlorophyll Proteins

Light-harvesting systems are first of all characterized by their efficiency to absorb and transfer light energy. Early on the S_1/ICT state was identified as one major state upon absorption of solar energy by fucoxanthin, being responsible for up to 60 % of the energy transfer to Chl *a* (Papagiannakis et al. 2005). No transfer into Chl *c* could be observed within the limit of the instrumentation (<100 fs), giving rise to the assumption that Fx transfers its absorbed energy directly (or via other Fx) to Chl *a* molecules. On the other hand this means that Chl *c* absorption will lead to direct transfer to Chl *a* (Gildenhoff et al. 2010a; Papagiannakis et al. 2005; Premvardhan et al. 2009).

In intact systems the transfer to Chl *a* from fucoxanthin is extremely fast. Gildenhoff et al. (2010a) determined lifetimes for FCPa of <150 fs for the fucoxanthin S_2 state (transferring directly into the Q_X state of Chl *a*), and 0.6/0.9 ps for the unrelaxed and 2.6/4.2 ps for the relaxed S_1/ICT state (transferring into Q_Y), respectively, whereby the higher values represent the lifetimes of the 'red' Fx molecules and the lower ones those of the 'blue' Fx. Thus, when exciting the 'blue/green' and the 'red' Fx molecules to different extents, the observed dynamics change. For FCPa an additional time constant at around 25 ps was found after excitation at 500 nm and assigned to the intrinsic lifetime of the blue absorbing fucoxanthins engaged in Fx – Fx excitation energy transfer (Gildenhoff et al. 2010a). Using anisotropy measurements it was concluded that one Fx'red' and two of the Fx'blue/green' transfer their energy directly to Chl *a*, whereas a further Fx'blue/green' is depending on another Fx molecule for excitation energy transfer to Chl *a* (Gildenhoff et al. 2010b).

When comparing the different oligomeric states of FCPa and FCPb, the trimeric FCPa has the more efficient energy transfer, which is also reflected in its higher Chl a fluorescence quantum yield. On the other hand the oligomeric FCPb is intrinsically less fluorescent (Gundermann and Büchel 2008), and thus some of the lifetimes mentioned above are even shorter (Gildenhoff et al. 2010a).

E. Lhcx and Photoprotection

Diatoms, like higher plants, protect themselves against fast changing levels of light intensities. Energy fluxes that exceed the conversion capacity of the photosynthetic machinery can cause damage. To avoid impairment the surplus of energy is very efficiently dissipated as heat, a mechanism which is called non-photochemical quenching (NPQ) due to its simultaneous reduction in fluorescence emission. In diatoms NPQ is more pronounced than in higher plants (Ruban et al. 2004), correlates with the de-epoxidation of Dd to Dt (Lavaud et al. 2002, 2003), is pH-dependent as in higher plants but no PsbS protein or homologues exists in diatoms. Here we will focus mainly on the contribution of Lhcx proteins to NPQ.

Lhcx proteins, whether from pennates or centrics, are up-regulated during prolonged high light (Bailleul et al. 2010; Becker and Rhiel 2006; Beer et al. 2006; Janssen et al. 2001; Lepetit et al. 2010; Nymark et al. 2009; Oeltjen et al. 2002, 2004; Zhu and Green 2010). A direct relationship between the amount of Lhcx proteins expressed and the capability for NPQ was also demonstrated (Bailleul et al. 2010; Zhu and Green 2010). However, when comparing Lhcx proteins in pennate and centric diatoms, a crucial difference becomes obvious. In centrics, a Lhcx protein (Fcp6) was found as a constituent of a trimeric complex (FCPa). The amount of Fcp6 depended on the light intensity during growth, as did the diatoxanthin content (Beer et al. 2006). In contrast, in the pennate *P. tricornutum*, no Lhcx proteins could be found in trimeric complexes so far (Grouneva et al. 2011; Gundermann et al. 2013). Lepetit et al. (2012) were the first to propose that it is unlikely that Lhcx1 of *P. tricornutum* binds Dd or Dt, based on their interpretation of the work of Bailleul et al. (2010). Lepetit and coworkers hypothesized that Lhcx1 might play a role in NPQ through the induction of conformational changes, which in turn influence the energy transfer to the photosystems, in analogy to what is hypothesized about psbS in higher plants (for review see Szabò et al. 2005; Ruban et al. 2012). The same role was attributed to Lhcx1 from a centric diatom, *T. pseudonana*, because its amount did not change dramatically upon high-light stress, in contrast to Lhcx6 (Zhu and Green 2010). No protein comparable to Lhcx6 is described so far in *C. meneghiniana*, since Fcp6 closely resembles Lhcx1. Gundermann and Büchel (2008) were able to show that the Fcp6 (Lhcx1)-containing FCPa of *C. meneghiniana* changes its fluorescence yield in dependence on Dt and Fcp6 content, implying a function of Lhcx1 in the regulation of fluorescence emission in centrics.

Since antenna aggregation was proposed as a mechanism to reduce fluorescence yield *in vivo* (Miloslavina et al. 2009), influence of protein distance, pH and content of xanthophyll cycle pigments on the Lhcx containing FCPa in proteoliposomes was analysed (Gundermann and Büchel 2012) as well. Recently it had already been demonstrated that FCPb is changing fluorescence yield depending on aggregation, but not according to Dt content or pH (Gundermann and Büchel 2008). Indeed, FCPa protein aggregation led to reduced fluorescence yields, which was in addition strongly influenced by the pH. On the other hand Dt reduced fluorescence emission in addition, but independently of pH and protein aggregation. Thus it seems that in centric diatoms a constituent of the trimeric complexes is an active player in NPQ, whereas in pennates at least Lhcx1 is working in a more independent manner. This is rather surprising, since, when comparing Lhcx proteins, Lhcx1 of *P. tricornutum* is most closely related to Fcp6 of *C. meneghiniana*. Since differences mainly concern helix 2 of those proteins (data not shown) they might directly relate to the pigment binding capacities of Lhcx1 of *P. tricornutum* (see below).

F. Structural Model Based on Spectroscopic Data

No structural data are available on other membrane-intrinsic light-harvesting proteins besides LHCII (Liu et al. 2004; Standfuss et al. 2005) and the quite similar CP29 from higher plants (Pan et al. 2011). Since FCPa and FCPb complexes from the centric diatom *C. meneghiniana* are by far the most studied diatom antenna systems using several spectrometric methods, our model here (Fig. 2.4) is based on results obtained from this organism.

Eppard and Rhiel (1998) were the first to model sequences of FCP using the LHCII structure as template. They identified five conserved Chl *a* binding sites (a602, a603, a610, a612 and a613, nomenclature according to Liu et al. (2004)). With more sequences available, Premvardhan et al. (2010) were able to identify two further conserved binding sites, a614 and b609. Premvardhan and coworkers built a model, which had to fit some major requirements: (i) the pigment content of FCPa and FCPb is based on 2 Chl *c* per monomer, i.e. the FCP contain 6–8 Chl *a* : 2 Chl *c* : 5–6 Fx in total; (ii) the Chl *a* molecules are not allowed to be arranged in a way to favor excitonic interactions; (iii) Fx is not transferring energy to Chl *c* and Fx and Chl *c* should thus be in an appropriate distance and/or orientation; (iv) Chl *c* to Chl *a* transfer is extremely fast, i.e. Chl *c* has to be in close proximity to Chl *a* molecule(s).

The two 'blue'Fx were proposed to be located homologous to the central luteins of LHCII (Premvardhan et al. 2009) clustering around helix 1 and 3. In close distance four symmetrically related Chl-binding sites can be found, which are conserved in FCPs (a610, a612, a602, a603). The frequently observed fast energy transfer from Fx to Chl *a* supports the occupation of these chlorophyll-binding sites by Chl *a*.

Further away two additional Chl *a*'s are found in LHCII, namely a613 (binding site conserved) and a604, the latter being ligated by a water molecule in LHCII. Thus, no prove for binding or non-binding is available from sequence comparison only. However, the quite similar structure prediction around those central helices makes it very likely that this symmetry-related site is occupied by Chl in FCP, as well. This also holds for the binding sites a614 and b609, where the amino acids responsible for Chl binding are conserved.

When comparing FCPa and FCPb, spectral differences are quite low. However, the precise stoichiometry of pigments is slightly different. 6 Fx : 6–7 Chl *a*: 2 Chl *c* can be found in the trimeric FCPa, whereas 5 Fx : 8 Chl *a* : 2 Chl *c* are bound to the oligomeric FCPb. Thus, FCPa has a slightly reduced Chl *a* content as compared to FCPb, which in turn is reduced in Fx, but is characterized by 'red' Fx absorbing at even longer wavelength than those in FCPa (Premvardhan et al. 2009). This difference in Chl *a* content can either be due to missing binding sites, which is unlikely due to the high similarity of the protein sequences, or to loss of pigments during the isolation procedures. If the difference in Chl *a* content is due to pigment loss, this may indicate that one to two of the Chl *a* molecules are more loosely, i.e. peripherally, bound. The most peripheral Chls in the LHCII structure are b605 and a611, the latter being slightly more central but only bound by a lipid. Those Chls might thus be present in FCPb, but missing in FCPa preparations. The Chl *a* molecule of LHCII bound at a611 participates in forming a dimer with Chl *a*612, which can be detected by its red shifted absorption (Georgakopoulou et al. 2007), which is missing in all FCP complexes. Thus, if the a611 binding site is present and occupied, the Chl *a* will have to adopt a different orientation in order to avoid excitonic dimer formation with Chl *a*612.

Thus, the ten Chl sites in FCPs could be identified with some probability, whereby only six (a602, a603, a610, a611, a612 and b605) are likely to be occupied by Chl *a*. In LHCII, the Chl *b* in b609 is strongly coupled to Chl *a* (a603) and both binding sites are predicted to exist in FCPs as well. In contrast, no direct interaction partner to a604 is conserved in FCPs. This renders b609 a probable candidate for Chl *c* binding in FCPs, whereas a604 is then probably occupied by Chl *a*. The other two sites left, a614 and a613, are rather close together, but further away from other Chls. One of these Chls is most probably a Chl *c* in FCPs, since in LHCII the two Chl *a* molecules interact exitonically (Georgakopoulou et al. 2007), but nothing like that can be seen in FCPs. Thus, to break this excitonic couple, one site is most probably occupied by Chl *c*. The Chl *c* molecules in FCPs can also be distinguished as more 'red' and 'blue'-absorbing. It has been argued that if one Chl *c* occupies b609, the lowest energy site in LHCII, the remaining Chl *c* should absorb at higher energies, which would make a613, the lower energy site in LHCII, less probable (Premvardhan et al. 2010). However, this argument implies almost identical protein environments in LHCII and FCP complexes, which is rather unlikely. On the other hand the porphyrin in a614 is in closer contact to the violaxanthin binding site, a scenario to be avoided, because of the lack of Fx – Chl *c* excitation energy transfer. Thus, in Fig. 2.4 Chl *c* is presented in a613.

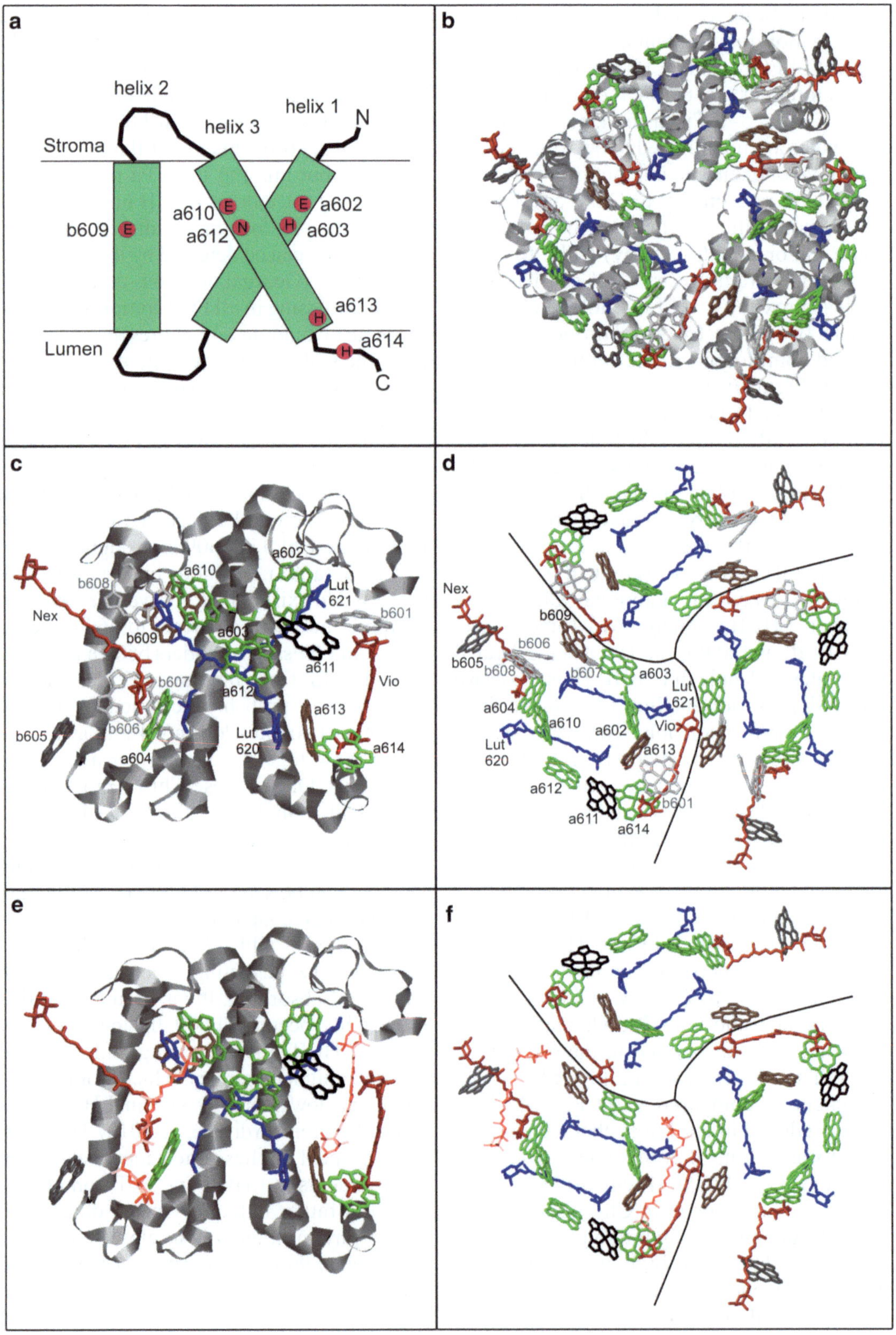
a
helix 2
helix 3
helix 1
N
Stroma
b609
a610
a612
a602
a603
a613
a614
Lumen
C
b
c
Nex
b608
a610
a602
Lut 621
b601
b609
a603
a611
b607
a612
Vio
a613
b605
b606
a604
Lut 620
a614
d
Nex
b609
b606
b605
b608
b607
a603
Lut 621
a604
a610
a602
Vio
Lut 620
a613
a612
b601
a611
a614
e
f

Since only ten Chls are bound to the FCPs, the space occupied by the remaining 4 Chls and the two additional phytol chains in LHCII can be filled by carotenoids in FCPs. The minimal number of Fx is five (in case of FCPb) or six (in case of FCPa), whereby the absorption properties differentiate them into blue Fx (2 molecules), green Fx (1–2 molecules) and red Fx (2 molecules). The blue Fx should be located in the lutein binding sites as outlined above. Using Stark and resonance Raman spectroscopy (Premvardhan et al. 2008, 2009), a model for the location of the different Fx molecules was developed. In this model, red Fx molecules are located near helix 2, whereby the structural differences between neoxanthin and fucoxanthin make it unlikely that one Fx is exactly bound like neoxanthin. One of the green Fx most probably occupies the violaxanthin binding site, and the other green Fx is located in the b601 site (Premvardhan et al. 2009). These data were supported by Gildenhoff et al. (2010b) demonstrating fast energy transfer from two of the blue/green Fx, as well as from one Fx'red' to Chl *a*, whereas a further Fx'blue/green' was depending on another Fx molecule. The other 2–3 Fx found in FCPs isolated using more gentle preparations, or in FCPs from *P. tricornutum*, have then to be more loosely bound and are not shown in Fig. 2.4 The model holds for FCPb, as well as FCPa, whereby in the latter the Fcp6 (Lhcx1) polypeptide carries Dd as well. This Dd was assumed to be located in the violaxanthin pocket as a mixed binding site with the green Fx (Premvardhan et al. 2010).

One implication of the relatively short wavelength absorption (671 nm) and the lack of excitonic Chl *a*/Chl *a* interactions is striking: a610/a611/a612 cannot work together as final emitters as shown for higher plants (Novoderezhkin et al. 2004). Thus the energy landscape of pigments in FCPs must also differ from that of LHCII, not only because of the replacement of Chl *b* by Chl *c*, but also concerning the Chl *a* molecules, despite some of them being bound in comparable sites.

Although the models based on spectroscopic data are quite elaborate by now, many questions remain, which only a proper structure determination would solve. One of the unresolved questions it the arrangement of the Fx molecules, which - in contrast to the symmetric lutein - is highly asymmetric. This implies that always two orientations exist, that are not accounted for by the current model, but are important in excitation energy transfer. Another important open question is the precise localization and function of Dd and Dt.

III. Conclusions

Despite all homologies, FCP complexes of diatoms are quite different from higher plant LHCs. Obvious deviations concern pigmentation, and thus the overall energy landscape. Not so obvious differences slowly emerge with our increasing knowledge about the structure of FCP complexes. The three different groups of FCP polypeptides, Lhcf, Lhcr and Lhcx, have no direct counterparts and plant Lhc proteins, except for the main light harvester Lhcf. Even Lhcr polypeptides, which might be compared to LHCI, are not the sole PSI antenna components. Lhcx proteins have a counterpart only in green algae and functional similarities to psbS, but are 3-helix proteins and, at least in centric

Fig. 2.4. Homology model based on the LHCII structure (pdb 1RWT) and spectroscopic analyses (Modified after Premvardhan et al. 2010). Graphics were made using RasMol and pigments are labeled according to Liu et al. (2004). In (**a**) potential Chl *a*-binding sites are shown, identified by sequence alignment. (**b**) Gives the *top view* of a trimer with pigments and protein backbone, the latter is removed in (**d**) and (**f**) for clarity. In (**c**) and (**e**) *side views* of one monomer are shown. Pigments are colored according to their attribution in FCPs: *green* = Chl *a*, *brown* = Chl *c*, *white* = Chls not present in FCPs, *gray* and *black* = Chl *a* missing in FCPa but present in FCPb, *blue* = 'blue' Fx and *red* = all other Fx molecules. In the *lower panel* two carotenoids per monomer are added as compared to LHCII in the volumes left by the missing Chl molecules. Note that the precise arrangement of each of these pigments is not known, especially in the case of the carotenoids and pigments around helix 2.

diatoms, contain the full complement of pigments. Thus, obtaining a molecular structure of FCPs to further elucidate pigment-binding and excitation energy transfer, as well as data about supercomplex formation are needed. This additional information will be extremely useful in the future to understand two major features of diatoms: their extreme ecological success, being responsible for up to 25 % of the biomass production on earth (Falkowski et al. 1998), and the related ability to regulate light harvesting versus photoprotection very efficiently.

Acknowledgments

CB would like to express her extreme thankfulness to her present and former group members, colleagues and collaborators, without whom the vast increase in our knowledge about FCP structure and function during the last years would not have been possible. CB and KG gratefully acknowledge continuous support by the Deutsche Forschungsgemeinschaft (Bu812: grants 4–8) and the European Union (MRTN-CT-2003-505069 "Intro2"; MC-ITN-2009-238017 "Harvest").

References

Alberte RS, Friedman AL, Gustafson DL, Rudnick MS, Lyman H (1981) Light-harvesting systems of brown algae and diatoms. Isolation and characterization of chlorophyll *a*/*c* and chlorophyll *a*/fucoxanthin pigment-protein complexes. Biochim Biophys Acta 635:304–316

Alexandre MTA, Gundermann K, Pascal AA, Grondelle R, Büchel C, Robert B (2014) Probing the carotenoid content of intact *Cyclotella* cells by resonance Raman spectroscopy. Photosynth Res 119(3):273–281

Archibald JM, Keeling PJ (2002) Recycled plastids: a 'green movement' in eukaryotic evolution. Trends Genet 18:577–584

Armbrust EV, Berges JA, Bowler C, Green BR, Martinez D, Putnam NH et al (2004) The genome of the diatom *Thalassiosira pseudonana*: ecology, evolution, and metabolism. Science 306:79–86

Bailleul B, Rogato A, De Martino A, Coesel S, Cardol P, Bowler C et al (2010) An atypical member of the light-harvesting complex stress-related protein family modulates diatom responses to light. Proc Natl Acad Sci U S A 107:18214–18219

Becker F, Rhiel E (2006) Immuno-electron microscopic quantification of the fucoxanthin chlorophyll a/c binding polypeptides Fcp2, Fcp4, and Fcp6 of *Cyclotella cryptica* grown under low- and high-light intensities. Int Microbiol 9:29–36

Beer A, Gundermann K, Beckmann J, Büchel C (2006) Subunit composition and pigmentation of fucoxanthin-chlorophyll proteins in diatoms: evidence for a subunit involved in diadinoxanthin and diatoxanthin binding. Biochemistry 45: 13046–13053

Beer A, Juhas M, Büchel C (2011) Influence of different light intensities and different iron nutrition on the photosynthetic apparatus in the diatom *Cyclotella meneghiniana* (*Bacillariophyceae*). J Phycol 47:1266–1273

Berkaloff C, Caron L, Rousseau B (1990) Subunit organization of PS I particles from brown algae and diatoms: polypeptide and pigment analysis. Photosynth Res 23:181–193

Bhaya D, Grossman AR (1993) Characterization of gene clusters encoding the fucoxanthin chlorophyll proteins of the diatom *Phaeodactylum tricornutum*. Nucleic Acids Res 21:4458–4466

Boekema EJ, Hankamer B, Bald D, Kruip J, Boonstra AF, Barber J, Rögner M (1995) Supramolecular structure of photosystem II complex from green plants and cyanobacteria. Proc Natl Acad Sci U S A 92:175–179

Bowler C, Allen AE, Badger JH, Grimwood J, Jabbari K, Kuo A et al (2008) The Phaeodactylum genome reveals the evolutionary history of diatom genomes. Nature 456:239–244

Brakemann T, Schlörmann W, Marquardt J, Nolte M, Rhiel E (2006) Association of fucoxanthin chlorophyll a/c-binding polypeptides with photosystems and phosphorylation in the centric diatom *Cyclotella cryptica*. Protist 157:463–475

Brown JS (1988) Photosynthetic pigment organization in diatoms (*Bacillariophyceae*). J Phycol 24:96–102

Büchel C (2003) Fucoxanthin-chlorophyll proteins in diatoms: 18 and 19 kDa subunits assemble into different oligomeric states. Biochemistry 42: 13027–13034

Caron L, Brown JS (1987) Chlorophyll-carotenoid protein complexes from the diatom, *Phaeodactylum tricornutum*: spectrophotometric, pigment and polypeptide analyses. Plant Cell Physiol 28: 775–785

Damjanović A, Ritz T, Schulten K (2000) Excitation transfer in the peridinin-chlorophyll-protein of *Amphidinium carterae*. Biophys J 79:1695–1705

Dekker JP, Boekema EJ (2005) Supramolecular organization of thylakoid membrane proteins in green plants. Biochim Biophys Acta 1706:12–39

Durnford DG, Aebersold R, Green BR (1996) The fucoxanthin-chlorophyll proteins from a chromophyte alga are part of a large multigene family: structural and evolutionary relationships to other light-harvesting antennae. Mol Gen Genet 253: 377–386

Eppard M, Rhiel E (1998) The genes encoding light-harvesting subunits of *Cyclotella cryptica* (*Bacillariophyceae*) constitute a complex and heterogeneous family. Mol Gen Genet 260:335–345

Eppard M, Rhiel E (2000) Investigation on gene copy number, introns and chromosomal arrangements of genes encoding the fucoxanthin chlorophyll a/c-binding proteins of the centric diatom *Cyclotella cryptica*. Protist 151:27–39

Eppard M, Krumbein WE, von Haesler A, Rhiel E (2000) Characterization of fcp4 and fcp12, two additional genes encoding light harvesting proteins of *Cyclotella cryptica* (*Bacillariophyceae*) and phylogenetic analysis of this complex gene family. Plant Biol 2:283–289

Falkowski PG, Barber RT, Smetacek V (1998) Biogeochemical controls and feedbacks on ocean primary production. Science 281:200–206

Fawley MW (1989) A new form of chlorophyll c involved in light-harvesting. Plant Physiol 91: 727–732

Fawley MW, Grossman AR (1986) Polypeptides of a light-harvesting complex of the diatom *Phaeodactylum tricornutum* are synthesized in the cytoplasm of the cell as precursors. Plant Physiol 81:149–155

Frank HA, Bautista JA, Josue J, Pendon Z, Hiller RG, Sharples FP et al (2000) Effect of the solvent environment on the spectroscopic properties and dynamics of the lowest excited states of carotenoids. J Phys Chem B 104:4569–4577

Friedman AL, Alberte RS (1984) A diatom light-harvesting pigment-protein complex. Plant Physiol 76:483–489

Friedman AL, Alberte RS (1986) Biogenesis and light regulation of the major light harvesting chlorophyll-protein of diatoms. Plant Physiol 80:43–51

Georgakopoulou S, van der Zwan G, Bassi R, van Grondelle R, van Amerongen H, Croce R (2007) Understanding the changes in the circular dichroism of light harvesting complex II upon varying its pigment composition and organization. Biochemistry 46:4745–4754

Gibbs SP (1970) The comparative ultrastructure of the algal chloroplast. Ann NY Acad Sci 175:454–473

Gildenhoff N, Amarie S, Gundermann K, Beer A, Büchel C, Wachtveitl J (2010a) Oligomerization and pigmentation dependent excitation energy transfer in fucoxanthin-chlorophyll proteins. Biochim Biophys Acta 1797:543–549

Gildenhoff N, Herz J, Gundermann K, Büchel C, Wachtveitl J (2010b) The excitation energy transfer in the trimeric fucoxanthin-chlorophyll protein from *Cyclotella meneghiniana* analyzed by polarized transient absorption spectroscopy. Chem Phys 373:104–109

Green BR (2011) After the primary endosymbiosis: an update on the chromalveolate hypothesis and the origins of algae with Chl *c*. Photosynth Res 107:103–115

Green BR, Kühlbrandt W (1995) Sequence conservation of light-harvesting and stress-response proteins in relation to the three-dimensional molecular structure of LHCII. Photosynth Res 44:139–148

Green BR, Pichersky E (1994) Hypothesis for the evolution of three-helix Chl *a*/*b* and Chl *a*/*c* light-harvesting antenna proteins from two-helix and four-helix ancestors. Photosynth Res 39:149–162

Grouneva I, Rokka A, Aro E (2011) The thylakoid membrane proteome of two marine diatoms outlines both diatom-specific and species-specific features of the photosynthetic machinery. J Proteome Res: 111109140228006

Guglielmi G, Lavaud J, Rousseau B, Etienne A, Houmard J, Ruban AV (2005) The light-harvesting antenna of the diatom *Phaeodactylum tricornutum*. Evidence for a diadinoxanthin-binding subcomplex. FEBS J 272:4339–4348

Gugliemelli A (1984) Isolation and characterization of pigment-protein particles from the light-harvesting complex of *Phaeodactylum tricornutum*. Biochim Biophys Acta 766:45–50

Gundermann K, Büchel C (2008) The fluorescence yield of the trimeric fucoxanthin-chlorophyll-protein FCPa in the diatom *Cyclotella meneghiniana* is dependent on the amount of bound diatoxanthin. Photosynth Res 95:229–235

Gundermann K, Büchel C (2012) Factors determining the fluorescence yield of fucoxanthin-chlorophyll complexes (FCP) involved in non-photochemical quenching in diatoms. Biochim Biophys Acta 1817:1044–1052

Gundermann K, Haufe A, Schmidt M, Weisheit W, Mittag M, Büchel C (2013) Identification of several sub-populations in the pool of light harvesting proteins in the pennate diatom *Phaeodactylum tricornutum*. Biochim Biophys Acta 1827:303–310

Ikeda Y, Komura M, Wanatabe M, Minami C, Koike H, Itoh S et al (2008) Photosystem I complexes associated with fucoxanthin-chlorophyll-binding proteins from a marine centric diatom, *Chaetoceros gracilis*. Biochim Biophys Acta 1777:351–361

Janssen M, Bathke L, Marquardt J, Krumbein WE, Rhiel E (2001) Changes in the photosynthetic apparatus of diatoms in response to low and high light intensities. Int Microbiol 4:27–33

Jeffrey SW, Humphrey GF (1975) New spectrometric equations for determining chlorophyll *a*, *b*, c_1 and c_2 in higher plants, algae and natural phytoplankton. Biochem Physiol Pflanz 167:191–194

Joshi-Deo J, Schmidt M, Gruber A, Weisheit W, Mittag M, Kroth PG, Büchel C (2010) Characterization of a trimeric light-harvesting complex in the diatom *Phaeodactylum tricornutum* built of FcpA and FcpE proteins. J Exp Bot 61:3079–3087

Juhas M, Büchel C (2012) Properties of photosystem I antenna protein complexes of the diatom *Cyclotella meneghiniana*. J Exp Bot 63:3673–3681

Katoh T, Nagashima U, Mimuro M (1991) Fluorescence properties of the allenic carotenoid fucoxanthin: implication for energy transfer in photosynthetic pigment systems. Photosynth Res 27:221–226

Koyama Y, Kuki M, Andersson PO, Gillbro T (1996) Singlet excited states and the light-harvesting function of carotenoids in bacterial photosynthesis. Photochem Photobiol 63:243–256

Kraay GW, Zapata M, Veldhuis MJW (1992) Separation of chlorophylls c_1 c_2, and c_3 of marine phytoplankton by reversed-phase-C18-high-performance liquid chromatography. J Phycol 28:708–712

Lavaud J, Rousseau B, van Gorkom HJ, Etienne A (2002) Influence of the diadinoxanthin pool size on photoprotection in the marine planktonic diatom *Phaeodactylum tricornutum*. Plant Physiol 129:1398–1406

Lavaud J, Rousseau B, Etienne A (2003) Enrichment of the light-harvesting complex in diadinoxanthin and implications for the nonphotochemical fluorescence quenching in diatoms. Biochemistry 42:5802–5808

Lepetit B, Volke D, Szabó M, Hoffmann R, Garab G, Wilhelm C, Goss R (2007) Spectroscopic and molecular characterization of the oligomeric antenna of the diatom *Phaeodactylum tricornutum*. Biochemistry 46:9813–9822

Lepetit B, Volke D, Gilbert M, Wilhelm C, Goss R (2010) Evidence for the existence of one antenna-associated lipid-dissolved and two protein-bound pools of diadinoxanthin cycle pigments in diatoms. Plant Physiol 154:1905–1920

Lepetit B, Goss R, Jakob T, Wilhelm C (2012) Molecular dynamics of the diatom thylakoid membrane under different light conditions. Photosynth Res 111:245–257

Liu Z, Yan H, Wang K, Kuang T, Zhang J, Gui L et al (2004) Crystal structure of spinach major light-harvesting complex at 2.72 Å resolution. Nature 428:287–292

Lohr M, Wilhelm C (1999) Algae displaying the diadinoxanthin cycle also possess the violaxanthin cycle. Proc Natl Acad Sci U S A 96:8784–8789

Medlin LK, Kooistra WH, Gersonde R, Wellbrock U (1996) Evolution of the diatoms (*Bacillariophyta*). II. Nuclear-encoded small-subunit rRNA sequence comparisons confirm a paraphyletic origin for the centric diatoms. Mol Biol Evol 13:67–75

Miloslavina Y, Grouneva I, Lambrev PH, Lepetit B, Goss R, Wilhelm C, Holzwarth AR (2009) Ultrafast fluorescence study on the location and mechanism of non-photochemical quenching in diatoms. Biochim Biophys Acta 1787:1189–1197

Moustafa A, Beszteri B, Maier UG, Bowler C, Valentin K, Bhattacharya D (2009) Genomic footprints of a cryptic plastid endosymbiosis in diatoms. Science 324:1724–1726

Nagao R, Ishii A, Tada O, Suzuki T, Dohmae N, Okumura A et al (2007) Isolation and characterization of oxygen-evolving thylakoid membranes and photosystem II particles from a marine diatom *Chaetoceros gracilis*. Biochim Biophys Acta 1767:1353–1362

Nagao R, Tomo T, Noguchi E, Nakajima S, Suzuki T, Okumura A et al (2010) Purification and characterization of a stable oxygen-evolving photosystem II complex from a marine centric diatom, *Chaetoceros gracilis*. Biochim Biophys Acta 1797:160–166

Novoderezhkin VI, Palacios MA, van Amerongen H, van Grondelle R (2004) Energy-transfer dynamics in the LHCII complex of higher plants: modified Redfield approach. J Phys Chem B 108:10363–10375

Nymark M, Valle KC, Brembu T, Hancke K, Winge PW, Andresen K et al (2009) An integrated analysis of molecular acclimation to high light in the marine diatom *Phaeodactylum tricornutum*. PLoS ONE 4:e7743

Oeltjen A, Krumbein WE, Rhiel E (2002) Investigations on transcript sizes, steady state mRNA concentrations and diurnal expression of genes encoding fucoxanthin chlorophyll *a/c* light harvesting polypeptides in the centric diatom *Cyclotella cryptica*. Plant Biol 4:250–257

Oeltjen A, Marquardt J, Rhiel E (2004) Differential circadian expression of genes fcp2 and fcp6 in *Cyclotella cryptica*. Int Microbiol 7:127–131

Owens TG (1986) Light-harvesting function in the diatom *Phaeodactylum tricornutum* – II. Distribution of excitation energy between the photosystems. Plant Physiol 80:739–746

Owens TG (1988) Light-harvesting antenna systems in the chlorophyll *a*/*c*-containing algae. In: Stevens SE, Bryant DA (eds) Light-energy transduction in photosynthesis: higher plants and bacterial models. American Society of Plant Physiologists, Rockville, MD, pp 122–136

Owens TG, Wold ER (1986) Light-harvesting function in the diatom *Phaeodactylum tricornutum* – I. Isolation and characterization of pigment-protein complexes. Plant Physiol 80:732–738

Pan X, Li M, Wan T, Wang L, Jia C, Hou Z et al (2011) Structural insights into energy regulation of light-harvesting complex CP29 from spinach. Nat Struct Mol Biol 18:309–315

Papagiannakis E, van Stokkum IHM, Fey H, Büchel C (2005) Spectroscopic characterization of the excitation energy transfer in the fucoxanthin-chlorophyll protein of diatoms. Photosynth Res 86:241–250

Peers G, Truong TB, Ostendorf E, Busch A, Elrad D, Grossman AR et al (2009) An ancient light-harvesting protein is critical for the regulation of algal photosynthesis. Nature 462:518–521

Premvardhan L, Sandberg D, Fey H, Birge R, Büchel C, van Grondelle R (2008) The charge-transfer properties of the S_2 state of fucoxanthin in solution and in fucoxanthin chlorophyll-a/c_2 protein (FCP) based on Stark spectroscopy and molecular orbital theory. J Phys Chem B 112:11838–11853

Premvardhan L, Bordes L, Beer A, Büchel C, Robert B (2009) Carotenoid structures and environments in trimeric and oligomeric fucoxanthin chlorophyll *a*/c_2 proteins from resonance Raman spectroscopy. J Phys Chem B 113:12565–12574

Premvardhan L, Robert B, Beer A, Büchel C (2010) Pigment organization in fucoxanthin chlorophyll *a*/c_2 proteins (FCP) based on resonance Raman spectroscopy and sequence analysis. Biochim Biophys Acta 1797:1647–1656

Pyszniak AM, Gibbs SP (1992) Immunocytochemical localization of photosystem I and the fucoxanthin-chlorophyll *a*/*c* light-harvesting complex in the diatom *Phaeodactylum tricornutum*. Protoplasma 166:208–217

Raven JA, Waite AM (2004) The evolution of silicification in diatoms: inescapable sinking and sinking as escape? New Phytol 162:45–61

Rhiel E, Marquardt J, Eppard M, Mörschel E, Krumbein WE (1997) The light-harvesting system of the diatom *Cyclotella cryptica*. Isolation and characterisation of the main light harvesting complex and evidence for the existence of minor pigment proteins. Bot Acta 110:109–117

Richard C, Ouellet H, Guertin M (2000) Characterization of the LI818 polypeptide from the green unicellular alga *Chlamydomonas reinhardtii*. Plant Mol Biol 42:303–316

Ruban AV, Lavaud J, Rousseau B, Guglielmi G, Horton P, Etienne A (2004) The super-excess energy dissipation in diatom algae: comparative analysis with higher plants. Photosynth Res 82:165–175

Ruban AV, Johnson MP, Duffy CDP (2012) The photoprotective molecular switch in the photosystem II antenna. Biochim Biophys Acta 1817:167–181

Standfuss J, van Scheltinga ACT, Lamborghini M, Kühlbrandt W (2005) Mechanisms of photoprotection and nonphotochemical quenching in pea light-harvesting complex at 2.5 Å resolution. EMBO J 24:919–928

Szabò I, Bergantino E, Giacometti GM (2005) Light and oxygenic photosynthesis: energy dissipation as a protection mechanism against photo-oxidation. EMBO Rep 6:629–634

Szabó M, Lepetit B, Goss R, Wilhelm C, Mustárdy L, Garab G (2008) Structurally flexible macro-organization of the pigment-protein complexes of the diatom *Phaeodactylum tricornutum*. Photosynth Res 95:237–245

Szabó M, Premvardhan L, Lepetit B, Goss R, Wilhelm C, Garab G (2010) Functional heterogeneity of the fucoxanthins and fucoxanthin-chlorophyll proteins in diatom cells revealed by their electrochromic response and fluorescence and linear dichroism spectra. Chem Phys 373:110–114

Veith T, Büchel C (2007) The monomeric photosystem I-complex of the diatom *Phaeodactylum tricornutum* binds specific fucoxanthin chlorophyll proteins (FCPs) as light-harvesting complexes. Biochim Biophys Acta 1767:1428–1435

Veith T, Brauns J, Weisheit W, Mittag M, Büchel C (2009) Identification of a specific fucoxanthin-chlorophyll protein in the light harvesting complex of photosystem I in the diatom *Cyclotella meneghiniana*. Biochim Biophys Acta 1787:905–912

Wolfe GR, Cunningham FX, Durnford D, Green BR, Gantt E (1994) Evidence for a common origin of chloroplasts with light-harvesting complexes of different pigmentation. Nature 367:566–568

Zhu S, Green BR (2010) Photoprotection in the diatom *Thalassiosira pseudonana*: role of LI818-like proteins in response to high light stress. Biochim Biophys Acta 1797:1449–1457

Zigmantas D, Hiller RG, Sharples FP, Frank HA, Sundström V, Polívka T (2004) Effect of a conjugated carbonyl group on the photophysical properties of carotenoids. Phys Chem Chem Phys 6:3009–3016

Chapter 3

Structure-Function Relationship in Peridinin-Chlorophyll Proteins

Tomáš Polívka*
Faculty of Science, Department of Physics and Biophysics, University of South Bohemia, 370-05 České Budějovice, Czech Republic

and

Eckhard Hofmann
Faculty of Biology and Biotechnology, Department of Biophysics, Ruhr-University Bochum, D-44780 Bochum, Germany

Summary

An important component of the photosynthetic apparatus is a light-harvesting system that captures light energy and transfers it efficiently to the reaction center. Depending on environmental conditions, photosynthetic antennas have adopted various strategies for this function. The water soluble antenna complex of dinoflagellates, peridinin–chlorophyll *a* protein (PCP), represents a unique light-harvesting strategy because, unlike other antenna systems which have a preponderance of chlorophyll, the carotenoid peridinin serves in PCP as the major light-harvesting pigment. The key structural feature of peridinin is a conjugated

*Author for correspondence, e-mail: tpolivka@jcu.cz

M.F. Hohmann-Marriott (ed.), The Structural Basis of Biological Energy Generation, Advances in Photosynthesis and Respiration 39, pp. 39–58, DOI 10.1007/978-94-017-8742-0_3,

carbonyl group which makes the spectroscopic properties of peridinin very sensitive to its local environment. This property is a crucial factor for maintaining the high efficiency of energy transfer between peridinin and Chl *a* in PCP. PCP is also amenable to site-directed mutagenesis and reconstitution with different pigments, allowing to study effects of both pigment and amino acid exchange on energy transfer pathways within the complex. Since high resolution structures of native, reconstituted and mutated PCP complexes are now available, this knowledge provides an ideal platform to relate structural motifs to energy transfer pathways and efficiencies in PCP. This Chapter summarizes results of structural and spectroscopic investigations of PCP and related proteins, emphasizing the specific light-harvesting strategy developed by dinoflagellates.

I. Introduction

In order to efficiently capture sunlight and utilize it in photochemical reactions occurring in reaction centers, all photosynthetic organisms rely on their light-harvesting proteins whose central function is to absorb light and funnel its energy to the reaction centers. Unlike reaction centers, light-harvesting systems exhibit large variability among photosynthetic organisms. The large amount of different antenna types and light-harvesting strategies reflects how different organisms have adapted for light conditions specific for their natural habitat.

Absorption bands of photosynthetic light-harvesting systems cover essentially the whole sunlight spectrum extending from 350 to 1,000 nm. Such broad coverage is achieved by only three types of antenna pigments, (bacterio) chlorophylls, carotenoids and phycobilins. Although absorption bands of these pigments in solution do not cover the whole sunlight spectrum, further tuning of their spectral properties is achieved via pigment-protein interaction within the light-harvesting proteins (Green and Parson 2003; Polívka and Frank 2010). Precise positions of the antenna pigments in proteins provide optimal distances and orientations among the pigments ensuring efficient energy transfer both within light-harvesting complexes and between antennas and reaction centers.

Some photosynthetic organisms have developed strategies for enhancing their light-harvesting capacity, especially in the blue-green spectral region which is not accessible by chlorophyll, yet is the region of highest solar irradiance in water. Cyanobacteria, red algae and cryptophytes use phycobiliproteins in their light-harvesting antenna to capture blue-green light (Glazer 1985), whereas many groups of marine eukaryotic algae employ carotenoids, such as peridinin, fucoxanthin, siphonaxanthin, and prasinoxanthin (Macpherson and Hiller 2003). These carotenoids contain a conjugated carbonyl group that is known to extend carotenoid absorption far beyond 500 nm, thus embedding these carotenoids into light-harvesting systems helps to capture the green light which is of vital importance for underwater photosynthesis.

Thus, in contrast to higher plants, where the major light-harvesting function is performed by chlorophylls, carotenoids often become important light-harvesting pigments in many photosynthetic microorganisms.

Abbreviations: BChl – Bacteriochlorophyll; CD – Circular dichroism; Chl – Chlorophyll; ENDOR – Electron nuclear double resonance; EPR – Electron paramagnetic resonance; FCP – Fucoxanthin chlorophyll protein; HSPCP – High salt PCP; ICT – Intramolecular charge transfer; LD – Linear dichroism; LHC – Light harvesting complex; MFPCP – Main form PCP used when needed to distinguish from other forms; MNDO-PSCDI – Modified neglect of differential overlap with partial single and double configuration interaction; NIR – Near infra red; PCP – Peridinin chlorophyll protein; Per – Peridinin; RFPCP – Refolded PCP construct equivalent to the N-terminal domain of MFPCP; VIS – Visible

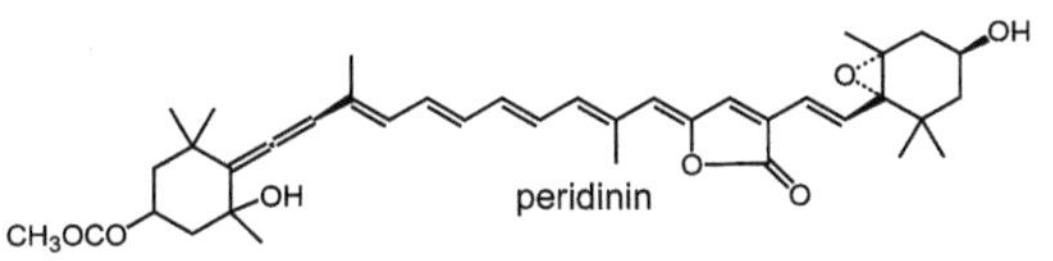

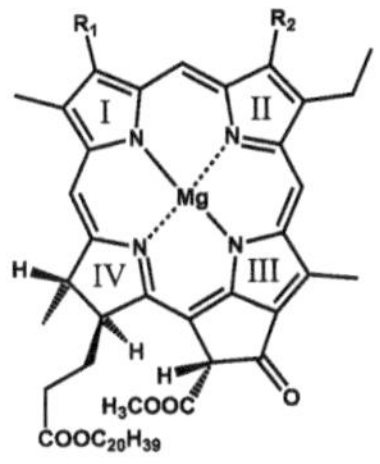

Chlorophyll	R$_1$	R$_2$
Chl-*a*	-CH=CH$_2$	-CH$_3$
Chl-*b*	-CH=CH$_2$	-CHO
acetyl Chl-*a*	-COCH$_3$	-CH$_3$
Chl-*d*	-CHO	-CH$_3$
BChl-*a**	-COCH$_3$	-CH$_3$

*The C=C bond in ring II is saturated in BChl-*a*

Fig. 3.1. Structure of peridinin and chlorophyll. Table shows the functional groups in positions R_1 and R_2 for various chlorophylls.

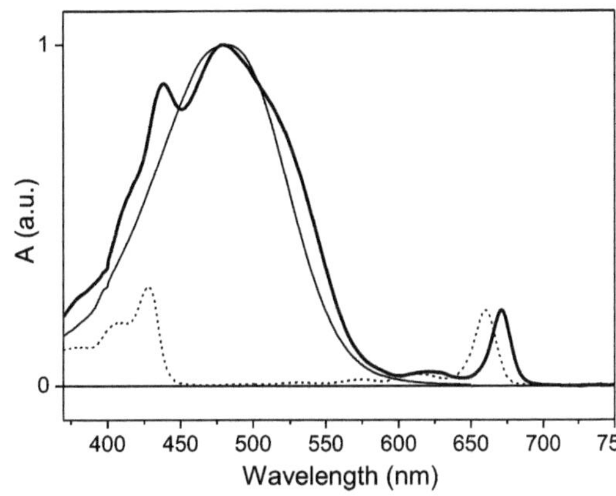

Fig. 3.2. Absorption spectrum of native PCP (*thick solid line*) and its pigments, peridinin in methanol (*thin solid line*) and Chl *a* in acetone (*dotted line*). Spectra of PCP and peridinin are normalized to maximum, spectrum of Chl *a* is normalized to match the maximum of the Q_y band in PCP.

This is typical for example in purple bacteria which utilize BChl *a* and BChl *b* that have essentially no absorption in the 450–550 nm spectral region. Therefore, purple bacteria employ a variety of carotenoids that cover the blue-green spectral region and effectively transfer absorbed energy to BChls (Polívka and Frank 2010). In extreme cases, however, the carotenoids even become the major light-harvesting pigments as they outnumber chlorophylls in antenna proteins.

Such an example of a carotenoid-based light-harvesting antenna is the peridinin-chlorophyll *a* protein (PCP) of dinoflagellates described a long time ago (Prezelin and Haxo 1976; Haxo et al. 1976; Song et al. 1976). While higher plants and eukaryotic algae typically utilize membrane-bound light-harvesting proteins (Green and Parson 2003), PCP is a water soluble antenna protein, located within the thylakoid lumen, containing only two pigments: Chl *a* and the carotenoid peridinin (Koka and Song 1977; Hofmann et al. 1996) (Figs. 3.1 and 3.2). It was shown nearly 40 years ago that PCP exhibits an unusual stoichiometric ratio of pigments, accommodating two Chl *a* molecules and eight peridinins, which efficiently transfer energy to chlorophylls (Song et al. 1976). Interestingly, several dinoflagellate species host a PCP gene encoding only for half of the protein. It was postulated, that these small forms do form homodimers and are the ancestors of the larger forms, which are the result of a gene duplication event (Hofmann et al. 1996). Twenty years later, the structure of the main form of PCP (MFPCP) from the dinoflagellate *Amphidinium carterae* was determined to 2.0 Å (Hofmann et al. 1996). Recently, structures of other PCP constructs were refined to 1.5 Å (Schulte et al. 2009a), making this light-harvesting complex ideal for studies of energy transfer. Another attractive advantage of working with the PCP complex is that it is amenable to mutation and reconstitution with different pigments (Miller et al. 2005; Schulte et al. 2010), which provides an opportunity to examine systematically the factors controlling the light-harvesting function of PCP.

II. Peridinin-Chlorophyll Protein Complex of *Amphidinium carterae*

A. Structure

One fundamental question on the PCP structure arose, as soon as the high pigment content (about 25 % w/w) was established. While in

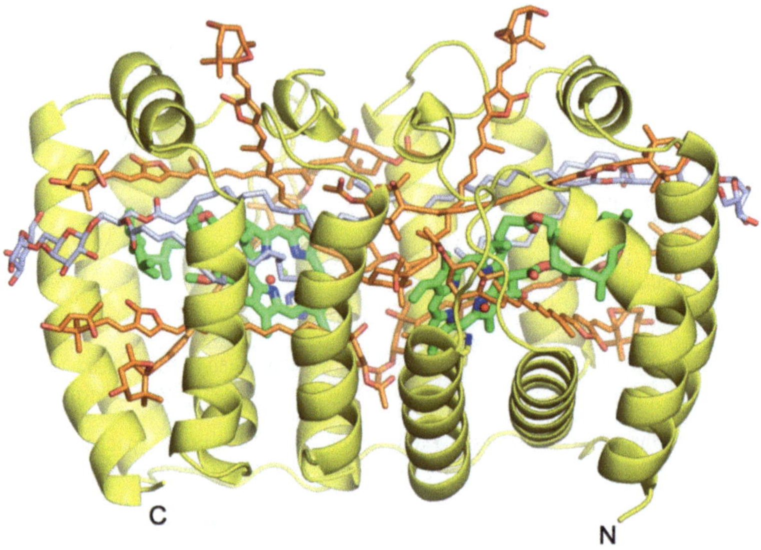

Fig. 3.3. Structure of MFPCP (PDB 1PPR, chain M). The protein is shown as *yellow* cartoon. The cofactors are shown as stick in *green* (Chl *a*), *orange* (peridinin) and *light-blue* (digalactosylglycerol). N- and C-termini are labelled.

membrane-bound complexes hydrophobic pigments like chlorophylls and carotenoids can at least partially be embedded in the hydrophobic interface to the lipid bilayer, in water soluble complexes the protein has to completely encapsule the pigments to protect them from the solvent. In the FMO protein, this task is managed by formation of an amphipathic β-barrel-cage, enclosing seven bacteriochlorophyll molecules (Matthews et al. 1979; Tronrud et al. 2009). For PCPs, sequence analysis suggested a mainly α-helical structure, requiring a completely different protein organization. Indeed the structure of the MFPCP can be classified as a new all-α fold (Fig. 3.3) (Hofmann et al. 1996). Making an analogy to a ship, the helices form the frames or ribs of the hull, with the pigments enclosed in the hold. Interestingly, in the structure of MFPCP, a pseudo-two-fold axis is present, relating the N-terminal with the C-terminal part of the protein, with a short unstructured linker connecting both (in the position of the keel of the boat). The two-fold axis is also found in the arrangement of the pigment molecules: Four peridinin molecules are clustered in van

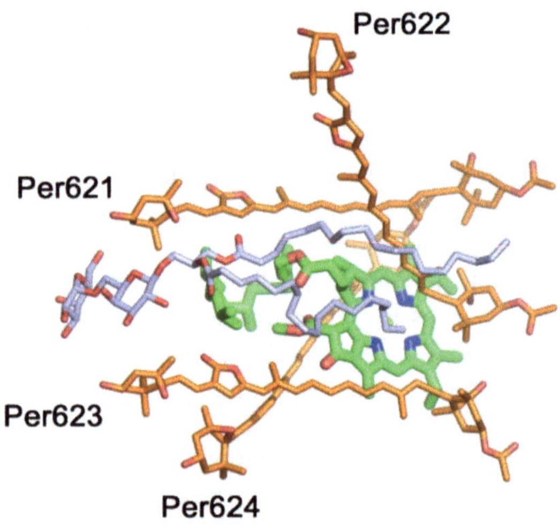

Fig. 3.4. The C-terminal pigment cluster (PDB 1PPR, chain M). The cofactors are shown as stick in *green* (Chl *a*), *orange* (peridinin) and *light-blue* (digalactosylglycerol). Peridinin molecules are labelled with their respective residue number.

der Waals distance around a central chlorophyll molecule in each domain (Fig. 3.4). In addition, in each cluster one lipid molecule (digalactosyldiacylglycerol) is an integral part of the cofactor packing. While the same lipids are found in all PCP structures solved to date, they are not further considered in

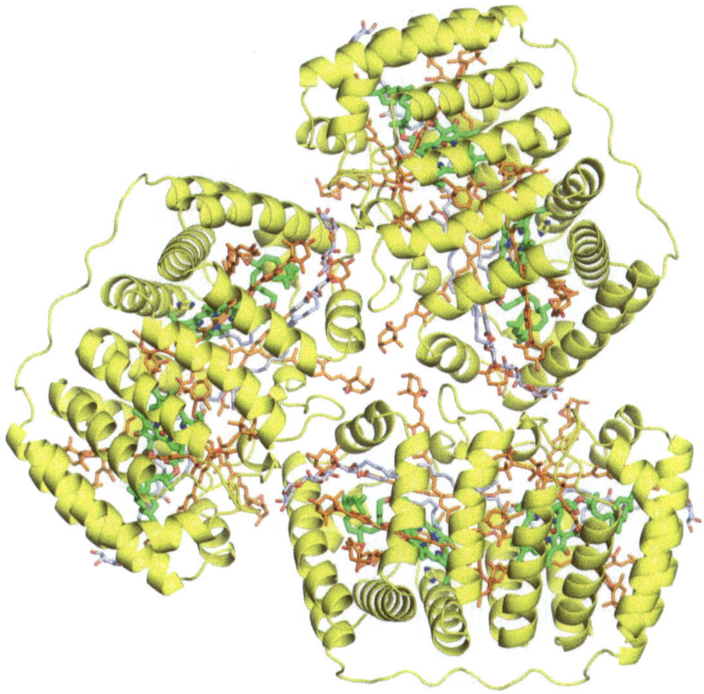

Fig. 3.5. The MFPCP trimer (PDB 1PPR). The protein is shown as *yellow* cartoon. The cofactors are shown as stick in *green* (Chl *a*), *orange* (peridinin) and *light-blue* (digalactosylglycerol).

this review, as they are spectroscopically silent and no data concerning their functional role are available yet. The interface between both domains is formed to a large extend by the head groups of intercalating pigments, making the interface partly polar and suggesting an independent folding of both halves. This concept would be in agreement with the existence of small forms of PCPs in different algae, which consist of only half of the apoprotein and form homodimers (Schulte, unpublished). Taken together, these observations suggest a historic gene duplication event, leading to the large PCP forms. Interestingly, in several different crystal packings analyzed to date, MFPCP always is present as a non-crystallographic trimer (Fig. 3.5). It has been suggested on the basis of analytical ultracentrifugation data, that this form is also relevant for the *in vivo* situation in the thylakoid lumen, where very high concentrations of PCP complexes are expected.

B. Light-Harvesting Function of Peridinin-Chlorophyll Proteins

Precise positions and mutual orientations of the pigments obtained from the MFPCP structure provided an ideal platform for studies of energy transfer pathways within the complex. Fluorescence excitation experiments demonstrated a long time ago that peridinins in PCP transfer energy to Chl *a* with efficiency approaching 100 % (Song et al. 1976). Yet, resolving subtle details concerning the energy transfer pathways and the influence of the protein environment had to wait until the development of time-resolved spectroscopy. In the late 90s of the past century this technique achieved sub-100 fs time resolution in the spectral region extending from UV to NIR region.

The first experiment addressing the excited-state lifetimes and energy transfer pathways in MFPCP was carried out by Bautista et al. (1999a). These authors showed that the lifetime of the lowest peridinin excited state, S_1, has a lifetime of 3.5 ps in MFPCP, while a markedly longer S_1 lifetime of 13.5 ps was obtained for peridinin in solution. This was a clear indication of an efficient energy transfer pathway via the S_1 state of peridinin. By comparing the efficiency of the S_1 route with total peridinin-to-Chl energy transfer efficiency (88 %), it was concluded that the peridinin S_1 state is the major energy donor in MFPCP. However, this pioneering experiment also showed that properties of the peridinin S_1 state in solution deviate from those known for other carotenoids, because the peridinin S_1 lifetime strongly depends on solvent polarity (Bautista et al. 1999b).

Since knowledge of the intrinsic (without energy transfer) S_1 lifetime of peridinin in MFPCP is crucial for determining the energy transfer efficiency, a number of subsequent studies addressed the polarity dependence of the peridinin lifetime in solution (Bautista et al. 1999b; Frank et al. 2000; Zigmantas et al. 2001, 2003; Papagiannakis et al. 2004). It was established that due to the conjugated carbonyl group of peridinin (Fig. 3.1), peridinin has an intramolecular charge transfer (ICT) state whose coupling to the S_1 state causes the observed polarity-dependent S_1 lifetime. Yet, the precise nature of the S_1-ICT coupling remains unresolved. The coupling increases with polarity of the environment and can be quantified by magnitude of the characteristic ICT-like bands in transient absorption spectra (Frank et al. 2000; Zigmantas et al. 2004). As these bands are also identified in MFPCP (Zigmantas et al. 2002), it is obvious that the ICT state plays a role in peridinin-to-Chl energy transfer in MFPCP (Fig. 3.6).

The peridinin ICT state was first included in the analysis of energy transfer pathways by Krueger et al. (2001) who showed by global analysis that although the S_1 and ICT states of peridinin could be separated spectrally, they always exhibit identical dynamics. This lead to a notion of a strongly-coupled S_1/ICT state (Zigmantas et al. 2001), playing the role of energy donor in MFPCP. While

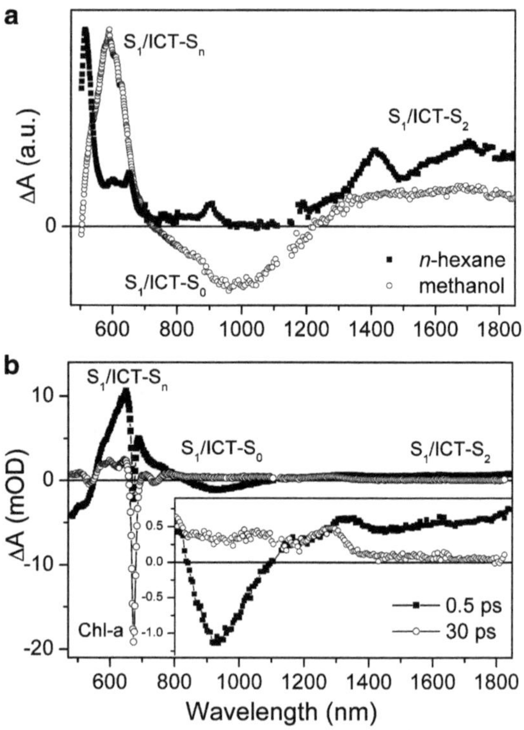

Fig. 3.6. (**a**) Transient absorption spectra of peridinin in polar (methanol) and non-polar (*n*-hexane) solvent measured at 1 ps after excitation at 490 nm. (**b**) Transient absorption spectra of PCP after excitation of peridinin at 530 nm. Spectra are recorded at time delays corresponding to peridinin signal (0.5 ps) and Chl *a* signal (30 ps).

Krueger et al. (2001) focused exclusively on the visible region of transient absorption spectra, a follow-up study explored the near-IR region that provides information about peridinin ICT stimulated emission (Zigmantas et al. 2002). This was the decisive experiment proving that the ICT state has an important role in peridinin-to-Chl energy transfer, because the decay of ICT-stimulated emission matches perfectly the rise of Chl *a* bleaching signal (Fig. 3.7).

These studies provided basic information about energy transfer pathways and rates in MFPCP. The lifetime of the S_1/ICT state in MFPCP is 2.7 ± 0.3 ps (Zigmantas et al. 2002; Krueger et al. 2001; Linden et al. 2004; Ilagan et al. 2006a; van Stokkum et al. 2009), which is significantly shorter than in any solvent (Bautista et al. 1999b). Since the intrinsic peridinin lifetime in

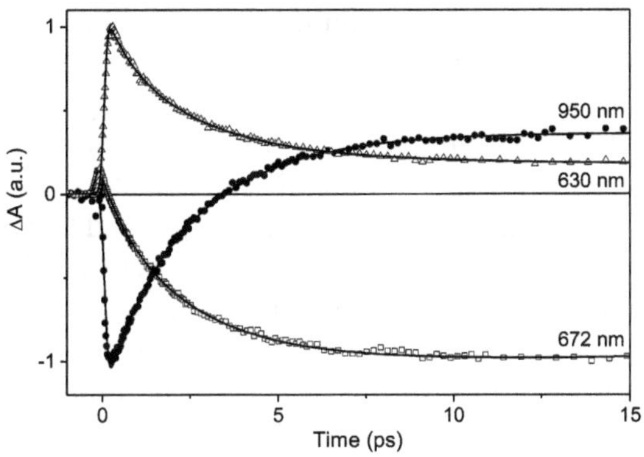

Fig. 3.7. Kinetics of PCP at 630, 672 and 950 nm. Rise of Chl *a* signal at 672 nm matches the decay of the ICT state at 950 nm.

MFPCP is not known, it had to be calculated using the overall energy transfer efficiency of ~88 % obtained from fluorescence excitation experiments (Bautista et al. 1999a). Based on experimental evidence that the S_2 route accounts for 25–50 % of total energy transfer (Krueger et al. 2001; Zigmantas et al. 2002; Linden et al. 2004), the S_1/ICT lifetime of peridinin yields 15–20 ps, categorizing the protein environment of peridinins in MFPCP to be of moderate polarity.

Further insight into energy transfer processes in MFPCP was provided by time-resolved studies of peridinin and MFPCP in the mid-IR spectral region that follows dynamics of specific vibrational modes (van Tassle et al. 2007; Bonetti et al. 2010). Contrary to data from the VIS/NIR regions, focusing on dynamics of C–C, C=C and C=O vibrations between 1,000 and 1,800 cm^{-1} suggests that it is indeed possible to separate S_1 and ICT dynamics of peridinin both in solution (van Tassle et al. 2007; Bonetti et al. 2010) and in MFPCP (Bonetti et al. 2010). Global analysis of mid-IR data favors the ICT state as main energy donor while the role of the S_1 state in energy transfer is only marginal (Bonetti et al. 2010). In addition, selectivity of the mid-IR data allowed separating contributions from individual peridinins for the first time. Contrary to the VIS/NIR data, which treated all four peridinins equally, analysis of mid-IR transient spectra identified Per611/621 and/or Per613/623 as those carrying the ICT state and are consequently the main energy donors, while Per612/622 andPer614/624 are the locations of the S_1 and the triplet state, respectively.

It should be also noted that the efficiency of the S_2 channel remains an open question. Krueger et al. (2001) estimated the efficiency of the S_2 pathway to be 25–50 % based on global analysis of transient absorption data, while a lower limit of 25 % was obtained from analysis of the Chl *a* kinetics at 670 nm (Zigmantas et al. 2002). Later, fluorescence up-conversion experiments on peridinin in methanol and in MFPCP yielded S_2 lifetimes of 130 and 66 fs, respectively, resulting in ~50 % efficiency of the S_2 pathway (Linden et al. 2004). This fluctuation of values demonstrates the complications in determining the efficiency of the S_2 channel, which – based on calculations – was not predicted to exist (Damjanovic et al. 2000). Moreover, later experiments showed that even the S_2 state has a charge-transfer character (Premvardhan et al. 2005), putting the same constraints on the S_2 state as on the S_1 state.

In addition to the picosecond kinetics of singlet-singlet energy transfer, the photoprotective function of peridinin is achieved via efficient triplet-triplet energy transfer from Chl *a* to peridinin. Peridinin triplet states were characterized in detail by EPR/ENDOR spectroscopies that assigned Per 614/624 as the one involved in quenching Chl *a* triplet (di Valentin et al. 2008a, b; Niklas et al. 2007), in agreement with the assignment based on mid-IR transient absorption spectra (Bonetti et al. 2010). Dynamics of triplet-triplet transfer is, however, likely more complicated than singlet energy transfer. Two distinct peridinin triplet populations were identified decaying with a lifetime of 13 μs and 42 μs, respectively (Alexandre et al. 2007; Bonetti et al. 2009). Moreover, simultaneous presence of peridinin and Chl *a* signals during triplet decay suggests excitonic interaction leading to a delocalization of triplet state over Per614/624 and Chl *a*. This interaction was hypothesized

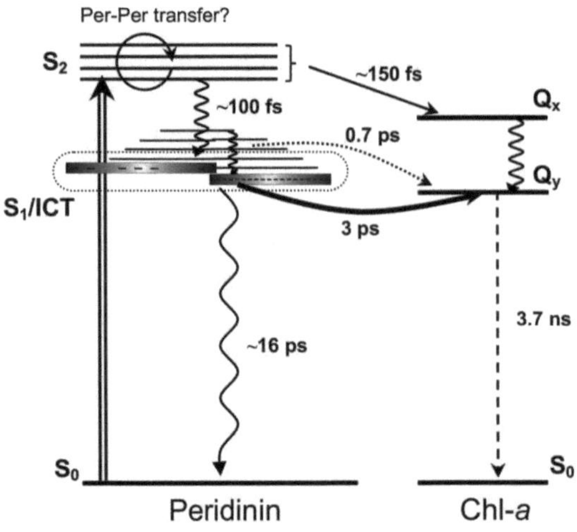

Fig. 3.8. Scheme of energy levels and energy transfer pathways between peridinin and Chl *a* in the MFPCP complex. Intramolecular relaxation processes are denoted by *wavy arrows*, while the *dashed arrow* represents the long-lived Chl *a* fluorescence. *Solid arrows* represent main energy transfer channels. The *dotted arrows* represent possible minor energy transfer channel involving higher vibrational levels of the S_1/ICT state. Excitation is shown as a *double arrow*. All processes are labeled by the corresponding time constant.

to lower the energy of the peridinin triplet, thereby facilitating efficient quenching of the Chl *a* triplet (Alexandre et al. 2007).

The large amount of data collected during the past decade led to the energy transfer scheme depicted in Fig. 3.8. It is beyond doubt that the main channel is the S_1/ICT state, but the exact role of the ICT state remains to be elucidated. Bonetti et al. (2010) suggested that the ICT state acts as the sole energy donor. However, a problem with such a mechanism is that the ICT emission indicates an energy level too low to transfer energy to the Q_y band of Chl *a*. An alternative hypothesis by Zigmantas et al. (2002) assumes the ICT-S_1 coupling to be the crucial factor; via this coupling the ICT state increases the transition dipole moment of the S_1/ICT state. Thus, the S_1 state is still the energy donor, while the ICT coupling enhances energy transfer rate of the Förster-type energy transfer mechanism. It must be noted that besides increasing energy transfer efficiency there is another important function of the ICT state. Comparing a number of carotenoids with a conjugated carbonyl group, which is important for activating the ICT state, with their non-carbonyl counterparts, showed that presence of the conjugated C=O group decreases the energy gap between S_2 and S_1/ICT states. This allows for effective capture of green light by moving the absorption spectrum of the S_2 state of peridinin to lower energies while keeping the S_1/ICT state high enough to enable efficient energy transfer to Chl *a* (Zigmantas et al. 2004; Polívka et al. 2007a, b). MFPCP therefore represents an example of a system where energy transfer pathways and their efficiencies are finely tuned, not only by the protein scaffold ensuring a proper orientation of donor and acceptor molecules, but also by adjusting the polarity of the local environment.

III. Peridinin-Chlorophyll Protein Reconstituted with Different Chlorophylls

While the atomic structure of native MFPCP provided a basis for the understanding of its light-harvesting function, the complicated nature of peridinin excited-state dynamics prevented determining the precise mechanisms of energy transfer. Theoretical studies based on the MFPCP structure suggested that the Coulomb coupling (Förster-type mechanism) facilitates the energy transfer via the S_1/ICT state (Damjanovic et al. 2000). The same study, however, did not predict the transfer rates measured for the S_2 route. Even the Förster excitation transfer mechanism suggested for the S_1/ICT transfer was later challenged due to the low energy of the ICT state, hypothesized to be the main energy donor (Papagiannakis et al. 2006).

One successful approach to resolve issues concerning the energy transfer mechanism is pigment exchange (Herek et al. 2000), which alters certain parameters in equations for calculation of energy transfer rates, providing

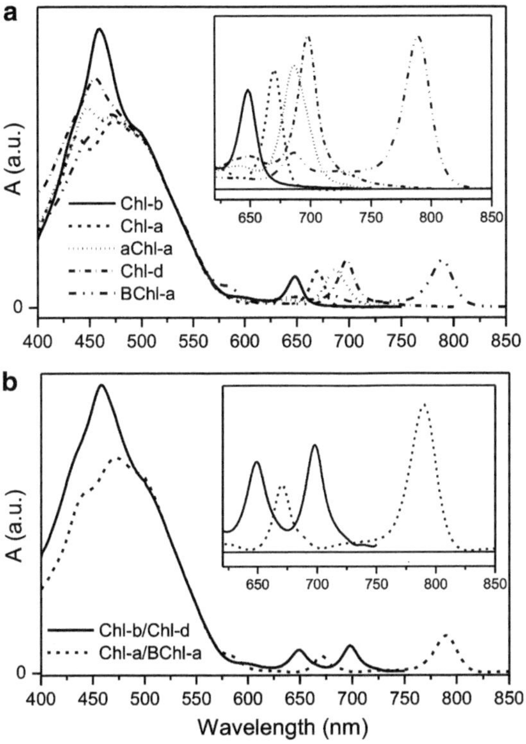

Fig. 3.9. Absorption spectra of (**a**) PCP complexes reconstituted with different Chl species, and (**b**) PCP complexes with mixed Chl sites.

the structural changes induced by the pigment exchange are known. The PCP complex provides an ideal opportunity to systematically examine the factors controlling energy transfer, because it is amenable to reconstitution with different chlorophylls. In 2005, Miller et al. demonstrated successful reconstitution of PCP with five different chlorophylls (Fig. 3.1) while maintaining the functionality of the reconstituted PCP complexes. Absorption spectra of these complexes shown in Fig. 3.9 demonstrate how reconstitution tunes the energy of the acceptor state in PCP.

A. Structure of Reconstituted Peridinin-Chlorophyll Proteins

To demonstrate the full integrity of these reconstituted complexes, consisting of a construct of the N-terminal domain of MFPCP (denoted refolded PCP, RFPCP), their atomic structure was determined by Schulte and coworkers (Schulte et al. 2009a). For the first time, an *in vitro* refolded caroteno-chlorophyll-protein complex was resolved at atomic resolution, thereby ultimately proving the successful reconstruction of the protein scaffold. The structure of RFPCP superposes perfectly with the N-terminal part of the MFPCP structure, thereby underlining the existence of a minimal building block of about 150 aminoacids, one chlorophyll and four carotenoids (Fig. 3.10). The protein forms crystallographic homodimers, which again superpose nicely with the pseudo-dimer in MFPCP. This truncated MFPCP version therefore mimics the small PCP forms observed in other dinoflagellates mentioned above. As the interface in the RFPCP homodimer is identical to the domain interface in the MFPCP structure, truly monomeric RFPCP is not expected to exist in solution. Therefore the term PCP monomer is frequently used for both MFPCP (with two domains covalently linked) or the small-form PCP homodimers (not covalently linked), thereby referring to a holoprotein with two domains harbouring two separate pigment clusters. Trimerization of the RFPCP homodimers is not observed either in crystals or in solution, suggesting an evolutionary adaptation of the MFPCP in the 'deck' -region (of the previously used PCP ship analogy) to optimize the trimer interfaces.

In addition, structures of chlorophyll-substituted complexes were solved at resolutions beyond 2 Å (Schulte et al. 2010). While Chl *b* and Chl *d* replace Chl *a* in an almost identical position, BChl *a* is found in a slightly shifted orientation with an overall RMSD of 0.5 Å for the macrocycle.

B. Tuning Energy Transfer Rate by Reconstitution

Time-resolved experiments on reconstituted PCP complexes revealed how reconstitution affects the dynamics of energy transfer from

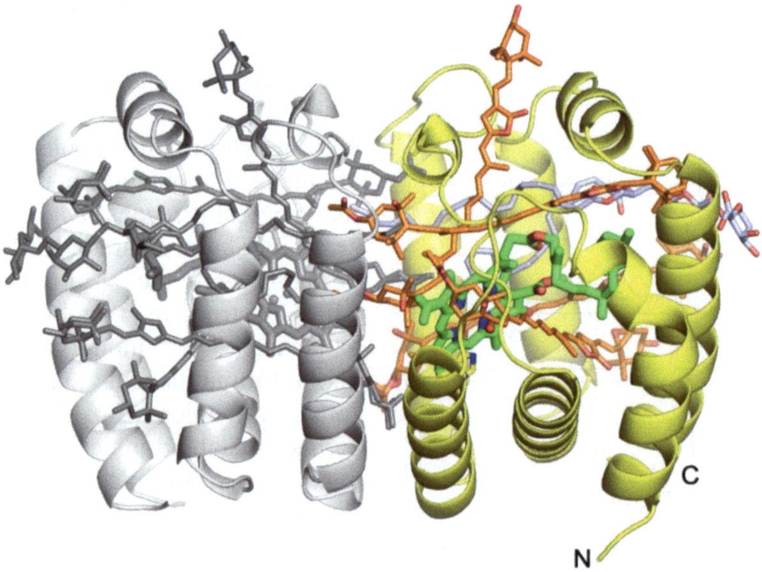

Fig. 3.10. Structure of RFPCP (PDB 3IIS). The protein is shown as *yellow* cartoon. The cofactors are shown as stick in *green* (chlorophyll), *orange* (peridinin) and *light-blue* (digalactosylglycerol). N- and C-terminus are labeled. Shown in *gray* is the crystallographic neighboring molecule forming the biological homodimeric unit.

peridinin to Chl (Polívka et al. 2005). All reconstituted complexes were excited at 535 nm where only peridinin absorbs. Energy transfer dynamics monitored either as rise of Chl bleaching or as decay of the ICT emission in near-IR, are depicted in Fig. 3.11. It is evident that with increasing energy of the acceptor state (Chl Q_y band) the S_1/ICT lifetime increases, indicating decrease of energy transfer efficiency. The S_1/ICT lifetimes for different chlorophylls are 5.9 ps (Chl *b*), 2.9 ps (Chl *a*), 2.2 ps (acetyl Chl *a*), 1.7 ps (Chl *d*) and 0.5 ps (BChl *a*). Providing that the intrinsic peridinin lifetime is the same in all RFPCP complexes, these lifetimes yield energy transfer times of 9.4 ps (Chl *b*), 3.5 ps (Chl *a*), 2.5 ps (acetyl Chl *a*), 1.9 ps (Chl *d*) and 0.5 ps (Bchl-*a*).

It is interesting to compare these values with the native MFPCP complex. First of all, it is obvious that RFPCP reconstituted with its native pigment, Chl *a*, fully maintains its light-harvesting function as the energy transfer rate changed only marginally from 3.2 ps in native MFPCP to 3.5 ps in RFPCP reconstituted with Chl *a*. It may be surprising, however, that RFPCP reconstituted with Chls having their Q_y band at lower energy than Chl *a*, perform even better than native PCP. This feature was further confirmed by measurements of total energy transfer efficiency using fluorescence excitation spectra (Ilagan et al. 2006b). Overall efficiency increases from ~80 % in RFPCP with Chl *b* to nearly 100 % when RFPCP is reconstituted with the 'reddest' (longest wavelengths) pigment, BChl *a*. A comparison of the S_1/ICT transfer efficiency with the total energy transfer efficiency clearly indicates that the S_2 route must be also dependent on the acceptor pigment. Indeed, the analysis of Chl kinetics (Fig. 3.11) indicates that the efficiency of the S_2 pathway exhibits a reverse dependence to the S_1/ICT route: the S_2 efficiency is highest for Chl *b* (45 %) and it decreases to less than 30 % for Chl *d* (Polívka et al. 2005).

The dependence of energy transfer rates via the S_1/ICT state on the type of acceptor Chl obtained from experiment were compared with the expected dependence of a Förster-type energy transfer mechanism operating for this route. Calculations were carried out under the assumption that distance and mutual

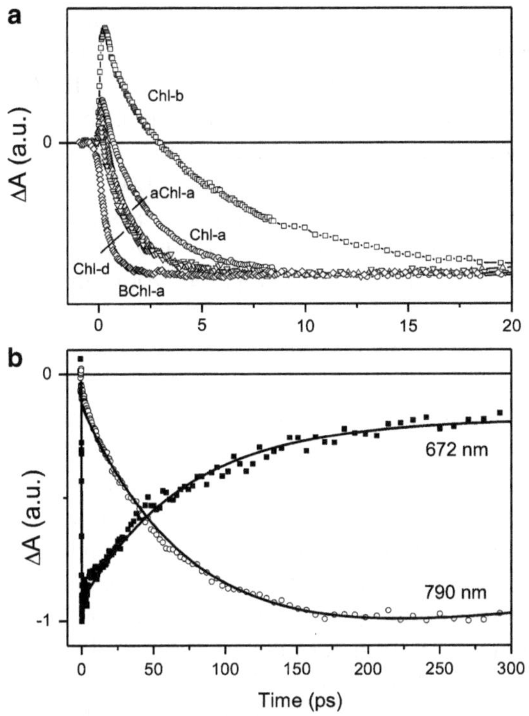

Fig. 3.11. (**a**) Rise of Chl signal in PCP complexes reconstituted with different Chl species after excitation of peridinin at 530 nm. (**b**) Kinetics of PCP complex with Chl sites occupied by Chl *a* and BChl *a* after excitation of Chl *a* at 670 nm. Kinetic at 672 monitors decay of Chl *a* while the kinetic at 790 nm reflects arrival of excitation energy at Bchl *a*.

orientation of donor and acceptor are the same in all reconstituted complexes, which is justified by the structures described in the previous section. In this case, the governing parameter is the spectral overlap between peridinin S_1/ICT emission, known from steady-state fluorescence in solution (Bautista et al. 1999b), and Q_y absorption band of chlorophylls. If different transition dipole moments of the acceptor chlorophylls are taken into account, the trend observed in the experiments is well reproduced in calculations, supporting the notion that energy transfer between peridinin and Chl in PCPs proceeds via the Förster mechanism (Polívka et al. 2005). It should be noted, however, that in its original form the Förster-mechanism energy transfer only considers dipole-dipole interaction (Förster 1948), while it is now established that full Coulombic coupling is necessary to reproduce energy transfer rates involving the S_1 state of carotenoids (Damjanovic et al. 2000; Scholes 2003).

Thus, the calculations confirm the observed trend, indicating that for chlorophylls with a Q_y band lower in energy than Chl *a*, the energy transfer is even more efficient than for the native MFPCP complex. Though this may seem surprising, the broad and featureless S_1/ICT emission of peridinin peaking in the 700–740 nm region (Bautista et al. 1999b) must favour chlorophylls having their Q_y band above 700 nm, which precisely matches the experimental data (Polívka et al. 2005). For the S_2-mediated energy transfer, the opposite trend is likely caused by the variation of overlap between S_2 emission and Q_x absorption that increases with shifting the Q_x band to higher energy. Thus, the S_1/ICT-mediated energy transfer contribution is smallest to Chl *b* compared to other chlorophylls with lower energy Q_x and Q_y, but this is at least partially compensated by an increased efficiency of the S_2 pathway in energy transfer to Chl *b*.

Reconstitution of RFPCP with different chlorophylls also enabled to study energy transfer between the two Chl *a* molecules located in separate domains of the RFPCP homodimer. In native PCP these Chl *a* molecules have slightly different energies (Krikunova et al. 2006; Wormke et al. 2007). Fluorescence anisotropy decay measurements that monitor Chl-Chl energy transfer within the MFPCP monomer revealed a time constant of 6 ps (Kleima et al. 2000a). Essentially the same time constant was obtained from transient absorption data recorded at low temperature (Ilagan et al. 2004; Ilagan et al. 2006a). Interestingly, a large difference between energies of two Chl *a* molecules in PCP was revealed in a new PCP form, the so-called high-salt PCP (HSPCP, see below) that exhibits a clear splitting of the Q_y bands of Chls *a* at low temperature (Ilagan et al. 2004, 2006a). Accordingly, Chl-Chl energy transfer was slower in HSPCP, yielding 23 ps (Ilagan et al. 2006a).

Successful reconstitution of RFPCP with chlorophyll mixtures produces a significant fraction of RFPCP dimers with mixed chlorophyll sites (Miller et al. 2005). This feature leads to more detailed exploration of inter-domain Chl-Chl energy transfer. The existence of RFPCP dimers with mixed Chl sites was also confirmed by single molecule spectroscopy (Brotosudarmo et al. 2006; Wormke et al. 2007, 2008). Absorption spectra of two such complexes, containing either Chl *b*/Chl *d* pairs or Chl *a*/Bchl-*a* pairs within a single RFPCP homodimer, are shown in Fig. 3.9b. Since the Q_y bands of chlorophylls are well separated, selective excitation of the higher-lying chlorophyll enables monitoring the Chl-Chl energy transfer rate (Polívka et al. 2008). The resulting energy transfer times are 40 ± 5 and 59 ± 3 ps for Chl *b*/Chl *d* pair and Chl *a*/BChl *a* pair, respectively. Modeling the energy transfer rates in reconstituted and native PCPs yields values in very good agreement with experimental data (Kleima et al. 2000a; Ilagan et al. 2006a; Polívka et al. 2008), confirming that the inter-domain Chl-Chl energy transfer proceeds via the Förster mechanism.

IV. Single-Point Mutation of Peridinin-Chlorophyll Protein

While the reconstituted PCP complexes provide information about the effect of different pigments on the function of PCP, they do not reveal anything concerning the role of the protein in tuning the spectroscopic properties of PCP. It has been known for a long time that absorption bands of individual peridinins in PCP can be separated in low temperature absorption and CD spectra. Analysis of the second derivative of a MFPCP absorption spectrum together with CD, linear dichroism (LD) and triplet-minus-singlet spectra at 4 K (Kleima et al. 2000b) provided evidence for at least three spectrally distinct peridinins with lowest absorption bands at 520, 537 and 555 nm. Additionally, a blue peridinin with a peak wavelength at 485 nm was found in CD and absorption spectra at 20 K (Carbonera et al. 1999). Following the pigment notation used in the MFPCP structure, it is now established that the peridinin with blue-shifted absorption spectrum is Per612/622 (Ilagan et al. 2004, 2006a) while the reddest peridinin is assigned to Per614/624 (Carbonera et al. 1999; Alexandre et al. 2007; van Stokkum et al. 2009; Schulte et al. 2009a).

Such clear distinction of absorption spectra of the four peridinins in a single PCP domain contrasts with the inability to separate energy transfer channels of individual peridinins. With the exception of mid-IR transient absorption experiment, which suggests distinct energy transfer channels associated with Per621/611 and Per623/613 Bonetti et al. (2010), no evidence for different energy transfer rates of individual peridinins has been found. Moreover, experiments using different excitation wavelengths spanning the region from 475 to 535 nm (Krueger et al. 2001; Zigmantas et al. 2002; Linden et al. 2004; Ilagan et al. 2006a; van Stokkum et al. 2009; Schulte et al. 2009a) did not find any significant difference in energy transfer dynamics.

A. Structure of the N89L Mutant

The large collection of results described above suggests that PCP represents a robust system, in which energy transfer efficiency is not sensitive to the spectral properties of individual peridinins. To test this hypothesis, Schulte et al. (2009a) performed site-directed mutagenesis on the RFPCP apo-protein, changing asparagine-89, which is located close to Per-614, to leucine. This modified protein (N89L) was refolded in the presence of peridinin and Chl *a*, and its structure was solved at 1.45 Å resolution (Schulte et al. 2009a). Not surprisingly, the protein scaffold remained unchanged, and also the pigments in the vicinity of the mutation were not changed in position or conformation within the experimental errors.

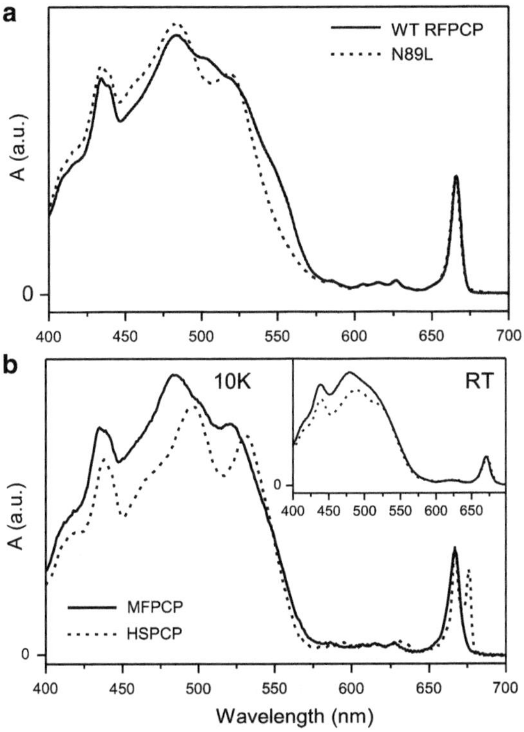

Fig. 3.12. (**a**) Absorption spectra of wild type RFPCP and N89L mutant measured at 10 K. (**b**) Comparison of absorption spectra of MFPCP and HSPCP at 10 K and at room temperature (*inset*).

B. Spectroscopic Properties of the N89L Mutant

In contrast to the observed structural conservation of the N89L mutant, the absorption spectrum measured at 10 K, which is compared to that of wild type RFPCP (in Fig. 3.12a), exhibits dramatic changes. Whereas spectral features assigned to Chl *a*, Q_y band at 672 nm and Soret band at 435 nm remain unchanged, the 'reddest' (longest wavelength) part of the peridinin absorption is missing in the N89L mutant. Fitting the N89L absorption spectrum as a sum of absorption spectra of individual peridinins revealed that the change can be reproduced by a ~24 nm blue shift of the reddest peridinin, Per614 (Schulte et al. 2009a).

Transient absorption spectroscopy of the N89L mutant confirmed the robustness of PCP as a light-harvesting antenna. It exhibits essentially the same energy transfer dynamics as wild type RFPCP (Schulte et al. 2009a). Following the conclusions of Bonetti et al. (2010), this observation could be explained by postulating that Per614 is only marginally involved in singlet energy transfer and its major role is to quench Chl *a* triplet state. If this were true, however, it would be hard to reconcile how Per614, which is expected to have the strongest interaction with Chl *a* (Carbonera et al. 1999; Damjanovic et al. 2000; Niklas et al. 2007), could be prevented from being an efficient energy donor. In addition, fluorescence excitation spectra show that the 'reddest' peridinin in native PCP transfers energy with high efficiency, directly confirming that Per614 must be an efficient energy donor. Thus, the high energy transfer efficiency of Per614 is rather the result of the broad and featureless peridinin emission (Bautista et al. 1999b), which implies that changes in spectral overlap will have minor consequences on energy transfer efficiency. Moreover, the major energy transfer route is via the S_1/ICT state, whose energy is much less sensitive to local environment than the energy of S_2. Thus, even though the S_2 state of Per614 is shifted by 24 nm, the S_1/ICT energy may remain the same, providing the basis for the identical energy transfer rates observed in N89L and native RFPCP.

Transient absorption spectra of the N89L mutant also resolved the long discussed issue of peridinin response to Chl *a* excitation. It is known that even after Chl *a* excitation, there is a clear bleaching signal in the spectral region of peridinin absorption (Krueger et al. 2001; Polívka et al. 2005; van Stokkum et al. 2009). The origin of this signal was suggested to be either due to excitonic interaction or electrochromic response of peridinin (van Stokkum et al. 2009). Low temperature transient absorption spectra of the N89L mutant showed that the bleaching signal was blue-shifted as compared to native RFPCP, unequivocally identify Per614 as the source of the signal (Schulte et al. 2009a). Moreover, analysis of the second derivative

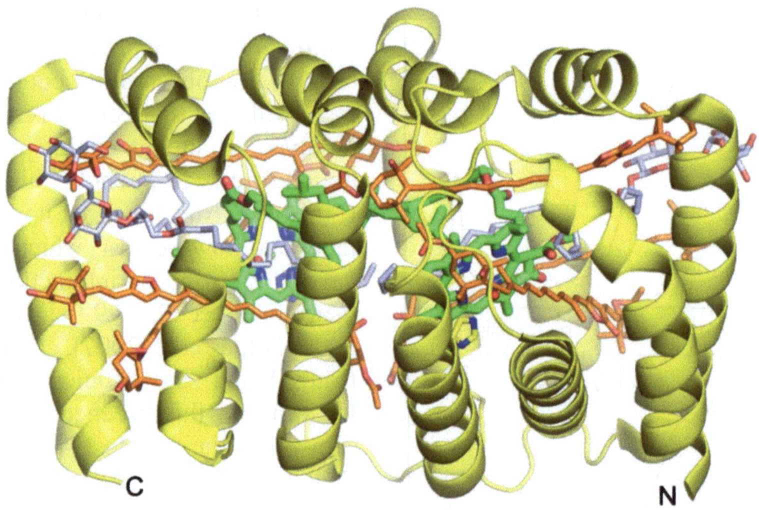

Fig. 3.13. Structure of HSPCP (PDB 2C9E). The protein is shown as *yellow* cartoon. The cofactors are shown as stick in *green* (chlorophyll), *orange* (peridinin) and *light-blue* (digalactosylglycerol). N- and C-terminus are labeled.

spectra demonstrated that the bleaching signal is consistent with an electrochromic shift of Per614 caused by the nearby Chl *a*.

V. High-Salt Peridinin-Chlorophyll Protein

Besides the MFPCP, a naturally-occurring variant was described in 1996 (Sharples et al. 1996). This form eluted at higher salt concentrations than MFPCP during ion-exchange chromatography and represented only about 2 % of the PCP fraction. Accordingly, it was denoted high-salt PCP (HSPCP).

The structure of HSPCP was solved at a resolution of 2.1 Å (Schulte et al. 2009b). This structure revealed a scaffold very similar to MFPCP with an RMSD of 1.89 Å for 286 Cα-atoms (Fig. 3.13). Major differences are only observed in the 'deck'-region (of the previously used PCP ship analogy) involved in trimer contacts in the MFPCP. Indeed HSPCP is found as a monomeric protein in the crystals. In contrast to the rather conserved protein moiety, the pigment arrangement shows some remarkable changes. Most notably, Per612 and Per622 from MFPCP are missing in HSPCP. As these two peridinins are traversing the 'deck'-region in MFPCP, their removal also contributes to the structural changes observed in the HSPCP variant. The remaining peridinins are found in almost identical positions as in MFPCP. However, a dramatic rearrangement is observed for the two chlorophyll molecules. While the tetrapyrrole macrocycle remains in a similar position as in MFPCP, the phytol chains are folded back into the interdomain interface.

Absorption spectra of HSPCP do not significantly differ from those of MFPCP at room temperature, except for a decrease of the peridinin absorption band due to missing Per612/622 (Fig. 3.12b). When cooled down to 10 K, however, it is obvious that the missing peridinin contributes mainly to the blue part of overall peridinin absorption. Another interesting difference between MFPCP and HSPCP complexes is a splitting of the Chl *a* Q_y band at low temperature in the HSPCP complex (Ilagan et al. 2004). Whereas the MFPCP complex exhibits only one unresolved Q_y band signaling similar spectral properties of the two Chl *a* molecules, in HSPCP the Q_y band splits into two upon cooling to 10 K (Fig. 3.12b). This splitting is

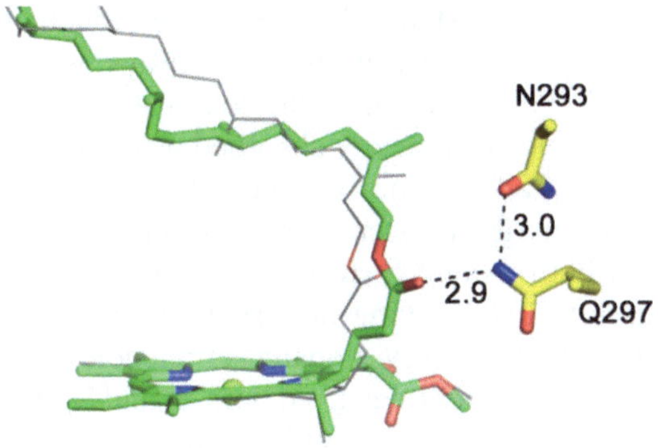

Fig. 3.14. Chlorophyll binding in HSPCP (PDB 2C9E). Superposed are the C- and N-terminally located chlorophyll molecules (*green* and *gray* sticks, respectively). Also shown and labelled are two residues of the C-terminal binding pocket in *yellow* sticks. Two hydrogen bonds are show as *dashed lines* together with distances in Å.

a result of heterogeneity of chlorophyll binding within the two HSPCP domains (Fig. 3.14). The phytol carbonyl group is hydrogen bonded by Q297 only in the C-terminal binding niche, resulting in out of plane deformation of ring D of the tetrapyrrole macrocycle. In comparison, the chlorophyll in the N-terminal binding pocket retains a more planar conformation. Preliminary MNDO-PSCDI calculations confirmed that this deformation can explain the observed band split for HSPCP (R. Birge, personal communications 2008). Another factor contributing to the observed split might be different electrostatic interactions within the two binding pockets, possibly induced by two surface exposed positively charged amino acid residues (R113 in the N-terminal domain, K284 in the C-terminal domain) which have been suggested by MNDO-PSCDI calculations (Ilagan et al. 2006a). While this result has been controversially discussed (Schulte et al. 2009b), a full understanding of the protein effects clearly requires new calculations, including the complete chromophores, more of the residues lining the binding pocket, and the inclusion of counterions at the protein surface.

Transient absorption spectra recorded after excitation of peridinin in HSPCP demonstrate the robustness of PCP. Despite some changes in spectral bands, dynamics of peridinin-to-Chl *a* energy transfer in HSPCP is very similar to that in the main form PCP (Ilagan et al. 2006a). The S_1/ICT lifetime is ~3 ps in HSPCP, which is slightly slower than ~2.7 ps obtained from various experiments on PCP, but this difference has essentially no effect on overall energy transfer efficiency (Ilagan et al. 2006a). Low temperature time-resolved experiments on HSPCP revealed another interesting feature of PCP. Contrary to all other light-harvesting complexes studied so far, energy transfer via the S_1/ICT route is faster at low temperature. At 10 K, the S_1/ICT lifetime of peridinin in HSPCP is 1.9 ps, thus markedly shorter than 3 ps at room temperature (Ilagan et al. 2006a). The same trend was also observed in native RFPCP and the N89L mutant (Schulte et al. 2009a), indicating that it is a common feature of PCP complexes. The origin of this anomalous energy transfer rate – temperature dependence has not been credibly explained so far, but it is likely that it is related to the temperature dependence of S_1-ICT coupling in peridinin. Comparison of transient absorption data taken for peridinin in 2-MTHF solution at 273 K and 77 K showed that ICT-like features are significantly enhanced at 77 K (Ilagan et al. 2006a). This suggests that the effect of the ICT state increases with decreasing temperature, which may be the reason for the increased energy transfer rate via the S_1/ICT state at low temperature.

VI. Connection to Reaction Center: Intrinsic LHC of *Amphidinium carterae*

While a vast amount of spectroscopic, structural and theoretical data collected for PCP complexes during the past decade provided deep insight into light-harvesting function of PCP, much less is known about energy transfer from PCP to the reaction center. It has been postulated, that energy transfer might proceed via the major membrane bound light-harvesting complex LHC (Mimuro et al. 1990). This complex is a member of the CAB (chlorophyll *a*/*b* binding) family of light-harvesting proteins also present in higher plants (Hiller et al. 1993) (see also Chap. 2 for related light harvesting complexes in diatoms). Based on sequence comparison and pigment analysis, a similar structure to the LHCII from pea or spinach is to be expected (Hiller et al. 1995), though LHC binds Chl c_2 instead of Chl *b*. This replacement is probably accommodated by a rearrangement of the structural scaffold in the membrane due to the missing phytol moiety. Indeed, while LHC from *Amphidinium* could be crystallized in our laboratory (EH), structure solution using the higher plant LHCII models as templates failed so far, possibly indicating a larger structural difference to *Amphidinium* LHC.

Theoretical calculations based on a homology model of LHC trimers together with the atomic structure of MFPCP trimers indicate an efficient energy transfer by Förster transfer with a slight orientation dependency (Hofmann 1999). Based on this finding a biophysical interaction between both proteins has been postulated, but so far no biochemical evidence could be presented to substantiate this. The rather restricted space in the thylakoid lumen is likely to position PCP complexes in rather close vicinity of the membrane, thereby alleviating the need for specific interaction. Another puzzling factor, possibly connected to the interaction with the membrane, is the heterogeneity of PCP complexes within a single algal species. Usually different isoforms are observed (Haxo et al. 1976), but the coexistence of different forms (i.e. HSPCP and MFPCP in *Amphidinium*), and even of small and large PCP forms has also been reported (Govind et al. 1990). As the relative ratio seems to be variable, the forms might represent a means to control energy flow into the photosynthetic system.

The only study (but see Note Added in Proof) of energy transfer pathways within LHC (Polívka et al. 2006) demonstrated that nearly 90 % of energy absorbed by peridinins in the 520–550 nm region is transferred to Chl. Energy transfer from peridinin proceeds via both S_2 and S_1/ICT states. Surprisingly, despite LHC having no structural similarity with PCP, energy transfer efficiencies and energy transfer rates from both S_2 and S_1/ICT states are strikingly similar in these two systems (Polívka et al. 2006). The decrease in energy transfer efficiency below 525 nm is due to light absorption by another carotenoid, diadinoxanthin. This carotenoid contributes only marginally to energy transfer and is likely to have a role in photoprotection. On the other hand, nearly 100 % efficient energy transfer occurs between Chl c_2 and Chl *a* within the LHC. The time constant characterizing the Chl c_2 to Chl *a* energy transfer is 1.4 ps. Carotenoids in the *Amphidinium* LHC also quench Chl *a* triplet states via triplet-triplet energy transfer and at least two different carotenoid triplets were found after excitation of Chl *a* (Di Valentin et al. 2010). It should be noted that comparable energy transfer patterns were also observed in a protein closely related to the *Amphidinium* LHC, the fucoxanthin-chlorophyll protein (FCP) (see Chap. 4) of *Cyclotella meneghiniana*, which binds the carotenoid fucoxanthin instead of peridinin (Papagiannakis et al. 2005; Gildenhoff et al. 2010).

VII. Mimicking Peridinin-Chlorophyll Protein Function

The specific spectroscopic properties of peridinin in PCP have also become the inspiration for syntheses of supramolecular complexes mimicking PCP antenna function. From the point of view of potential application in artificial photosynthesis, the most attractive properties of carbonyl carotenoids are: (1) shrinking the S_2–S_1 energy gap, leading to extension of the absorption spectrum beyond 550 nm while keeping the S_1 state high enough in energy to transfer energy efficiently to the Q_y band of porphyrin-like acceptors, and (2) their behavior offers a possibility for tuning excited state properties by changing the solvent polarity. The successful synthesis of carotenoid-pyropheophorbide dyads utilizing carbonyl carotenoids peridinin and fucoxanthin was achieved more than a decade ago (Shinoda et al. 1995; Osuka and Kume 1998). In both systems, energy transfer from carotenoid to pyropheophorbide was demonstrated by measurements of fluorescence excitation spectra (Osuka and Kume 1998; Osuka et al. 1999). In addition, it was shown that both fucoxanthin and peridinin are able to quench triplet states of the attached pyropheophorbide (Osuka et al. 1999) mimicking the photoprotective function of carotenoids in natural systems.

Time-resolved spectroscopy on peridinin- and fucoxanthin-pyropheophorbide dyads in solvents with different polarity confirmed energy transfer between carotenoid and pyropheophorbide, but also showed that energy transfer efficiency can be tuned by solvent polarity (Polívka et al. 2007b). For the dyad containing peridinin, energy transfer efficiency decreased from 80 % in benzene to 69 % in tetrahydrofuran and further to 22 % in acetonitrile. Lower efficiencies were observed when fucoxanthin was used as the energy donor; the highest efficiency in benzene was 27 %, dropping to 19 % in tetrahydrofuran and 13 % in acetonitrile (Polívka et al. 2007b). The polarity-dependent energy transfer efficiency was caused by competition of energy transfer between the S_1/ICT and the intrinsic S_1/ICT lifetime which decreases with increasing solvent polarity (Frank et al. 2000; Zigmantas et al. 2004). Yet, the energy transfer was faster in a solvent with higher polarity, as energy transfer time decreased from 44 ps in benzene to 31 ps in acetonitrile. These energy transfer times are still an order of magnitude slower than peridinin-chlorophyll energy transfer in PCP, demonstrating the optimal arrangements of pigments in PCP complex. In the synthetic dyad containing fucoxanthin the energy transfer times are even slower, yielding values of ~200 ps (Polívka et al. 2007b). However, as shown for a series of fucoxanthin-pyropheophorbide dyads the actual energy transfer efficiency depends on where fucoxanthin is attached to pyropheophorbide, indicating that proper orientation of the molecules in the dyad may increase energy transfer efficiency substantially (Debreczeny et al. 1997).

Note Added in Proof

Two new studied of energy transfer appeared after submission of this Chapter. Šlouf et al. (2013) carried out transient absorption spectroscopy on LHC from *Amphidinium carterae* at cryogenic temperatures and Niedzwiedzki et al. (2013) performed experiments with LHC from *Symbiodinium*.

Acknowledgment

The authors thank Roger Hiller and Harry Frank for many years of collaboration on research described in this chapter. Harry Frank is greatly acknowledged for providing data for Fig. 3.12. TP thanks the Czech Science Foundation (P205/11/1164) for financial support. EH acknowledges support by the collaborative research center SFB480 (project C6) of the Deutsche Forschungsgemeinschaft.

References

Alexandre MTA, Luhrs DC, van Stokkum IHM, Hiller R, Groot ML, Kennis JTM, van Grondelle R (2007) Triplet state dynamics in peridinin-chlorophyll a-protein: a new pathway of photoprotection in LHCs? Biophys J 93:2118–2128

Bautista JA, Hiller RG, Sharples FP, Gosztola D, Wasielewski MR, Frank HA (1999a) Singlet and triplet energy transfer in the peridinin-chlorophyll *a* protein from *Amphidinium carterae*. J Phys Chem A 103:2267–2273

Bautista JA, Connors RE, Raju BB, Hiller RG, Sharples FP, Gosztola D, Wasielewski MR, Frank HA (1999b) Excited state properties of peridinin: observation of a solvent dependence of the lowest excited singlet state lifetime and spectral behavior unique among carotenoids. J Phys Chem B 103:8751–8758

Bonetti C, Alexandre MTA, Hiller RG, Kennis JTM, Grondelle R (2009) Chl a triplet quenching by peridinin in H-PCP and organic solvent revealed by step-scan FTIR time-resolved spectroscopy. Chem Phys 357:63–69

Bonetti C, Alexandre MTA, van Stokkum IHM, Hiller RG, Groot ML, van Grondelle R, Kennis JTM (2010) Identification of excited-state energy transfer and relaxation pathways in the peridinin-chlorophyll complex: an ultrafast mid-infrared study. Phys Chem Chem Phys 12:9256–9266

Brotosudarmo THP, Hofmann E, Hiller RG, Wormke S, Mackowski S, Zumbusch A, Brauchle C, Scheer H (2006) Peridinin-chlorophyll-protein reconstituted with chlorophyll mixtures: preparation, bulk and single molecule spectroscopy. FEBS Lett 580: 5257–5262

Carbonera D, Giacometti G, Segre U, Hofmann E, Hiller RG (1999) Structure-based calculations of the optical spectra of the light-harvesting peridinin-chlorophyll-protein complexes from *Amphidinium carterae* and *Heterocapsa pygmaea*. J Phys Chem B 103:6349–6356

Damjanovic A, Ritz T, Schulten K (2000) Excitation transfer in the peridinin-chlorophyll-protein of *Amphidinium carterae*. Biophys J 79:1695–1705

Debreczeny MP, Wasielewski MR, Shinoda S, Osuka A (1997) Singlet-singlet energy transfer mechanisms in covalently-linked fucoxanthin- and zeaxanthin-pyropheophorbide molecules. J Am Chem Soc 119: 6407–6414

Di Valentin M, Ceola S, Salvadori E, Agostini G, Carbonera D (2008a) Identification by time-resolved EPR of the peridinins directly involved in chlorophyll triplet quenching in the peridinin-chlorophyll *a*-protein from *Amphidinium carterae*. Biochim Biophys Acta 1777:186–195

Di Valentin M, Ceola S, Agostini G, Giacometti GM, Angerhofer A, Crescenzi O, Barone V, Carbonera D (2008b) Pulse ENDOR and density functional theory on the peridinin triplet state involved in the photo-protective mechanism in the peridinin-chlorophyll *a*-protein from *Amphidinium carterae*. Biochim Biophys Acta 1777:295–307

Di Valentin M, Salvadori E, Agostini G, Biasibetti F, Ceola S, Hiller R, Giacometti GM, Carbonera D (2010) Triplet-triplet energy transfer in the major intrinsic light-harvesting complex of *Amphidinium carterae* as revealed by ODMR and EPR spectroscopies. Biochim Biophys Acta 1797:1759–1767

Förster T (1948) Zwischenmolekulare Energiewanderung und Fluoreszenz. Ann Phys 437:55–75

Frank HA, Bautista JA, Josue J, Pendon Z, Hiller RG, Sharples FP, Gosztola D, Wasielewski MR (2000) Effect of the solvent environment on the spectroscopic properties and dynamics of the lowest excited states of carotenoids. J Phys Chem B 104:4569–4577

Gildenhoff N, Herz J, Gundermann K, Büchel C, Wachtveitl J (2010) The excitation energy transfer in the trimeric fucoxanthin-chlorophyll protein from *Cyclotella meneghiniana* analyzed by polarized transient absorption spectroscopy. Chem Phys 373:104–109

Glazer AN (1985) Light harvesting by phycobilisomes. Annu Rev Biophys Chem 14:47–77

Govind N, Roman S, Iglesias-Prieto R, Trench R, Triplett E, Prezelin BB (1990) An analysis of the light-harvesting peridinin-chlorophyll *a*-proteins from dinoflagellates by immunoblotting techniques. Proc R Soc Lond B Biol Sci 240:187–195

Green RB, Parson WW (eds) (2003) Light-harvesting antennas in photosynthesis. Kluwer Academic Publishers, Dordrecht

Haxo FT, Kycia JH, Somers GF, Bennett A, Siegelman HW (1976) Peridinin-chlorophyll *a* proteins of the dinoflagellate *Amphidinium carterae* (Plymouth 450). Plant Physiol 57:297–303

Herek JL, Fraser NJ, Pullerits T, Martinsson P, Polívka T, Scheer H, Cogdell RG, Sundström V (2000) B800-B850 energy transfer mechanism in bacterial LH2 complexes investigated by B800 pigment exchange. Biophys J 78:2590–2596

Hiller RG, Wrench PM, Gooley AP, Shoebridge G, Breton J (1993) The major intrinsic light-harvesting protein of *Amphidinium*: characterization and relation to other light-harvesting proteins. Photochem Photobiol 57:125–131

Hiller RG, Wrench PM, Sharples FP (1995) The light-harvesting chlorophyll *a-c*-binding protein of dinoflagellates: a putative polyprotein. FEBS Lett 363:175–178

Hofmann E (1999) Strukturanalyse der Lichtsammler Peridinin-Chlorophyll a-Proteine (PCPs) von Amphidinium carterae und Heterocapsa pygmaea. Doctoral dissertation, University Konstanz

Hofmann E, Wrench P, Sharples F, Hiller R, Welte W, Diederichs K (1996) Structural basis of light-harvesting by carotenoids: peridinin-chlorophyll-protein from *Amphidinium carterae*. Science 272:1788–1791

Ilagan RP, Shima S, Melkozernov A, Lin S, Blankenship RE, Sharples FP, Hiller RG, Birge RR, Frank HA (2004) Spectroscopic properties of the main-form and high-salt peridinin-chlorophyll a proteins from *Amphidinium carterae*. Biochemistry 43:1478–1487

Ilagan RP, Koscielecki JF, Hiller RG, Sharples FP, Gibson GN, Birge RR, Frank HA (2006a) Femtosecond time-resolved absorption spectroscopy of main-form and high-salt peridinin-chlorophyll a-proteins at low temperatures. Biochemistry 45:14052–14063

Ilagan RP, Chapp TW, Hiller RG, Sharples FP, Polívka T, Frank HA (2006b) Optical spectroscopic studies of light-harvesting by pigment-reconstituted peridinin-chlorophyll-proteins at cryogenic temperatures. Photosynth Res 90:5–15

Kleima FJ, Hofmann E, Gobets B, van Stockum IHM, van Grondelle R, Diederichs K, van Amerongen H (2000a) Förster excitation energy transfer in in peridinin-chlorophyll-*a*-protein. Biophys J 78:344–353

Kleima FJ, Wendling M, Hofmann E, Peterman EJG, van Grondelle R, van Amerongen H (2000b) Peridinin chlorophyll *a* protein: relating structure and steady-state spectroscopy. Biochemistry 39:5184–5195

Koka P, Song PS (1977) Chromophore topography and binding environment of peridinin-chlorophyll *a*-protein complexes from marine dinoflagellate algae. Biochim Biophys Acta 495:220–231

Krikunova M, Lokstein H, Leupold D, Hiller RG, Voigt B (2006) Pigment-pigment interactions in PCP of *Amphidinium carterae* investigated by non-linear polarization spectroscopy in the frequency domain. Biophys J 90:261–271

Krueger BP, Lampoura SS, van Stokkum IHM, Papagiannakis E, Salverda JM, Gradinaru CC, Rutkauskas D, Hiller RG, van Grondelle R (2001) Energy transfer in the peridinin chlorophyll a protein of *Amphidinium carterae* studied by polarised transient absorption and target analysis. Biophys J 80:2843–2855

Linden PA, Zimmermann J, Brixner T, Holt NE, Vaswani H, Hiller RG, Fleming GR (2004) Transient absorption study of peridinin and peridinin-chlorophyll-a-protein after two photon excitation. J Phys Chem B 108:10340–10345

Mackowski S, Wormke S, Brotosudarmo THP, Scheer H, Brauchle C (2008) Fluorescence spectroscopy of reconstituted peridinin-chlorophyll-protein complexes. Photosynth Res 95:253–260

Macpherson A, Hiller RG (2003) Light harvesting systems in chlorophyll-c containing algae. In: Green RB, Parson WW (eds) Light-harvesting antennas in photosynthesis. Kluwer Academic Publishers, Dordrecht, pp 323–352

Matthews BW, Fenna RE, Bolognesi MC, Schmid MF, Olson JM (1979) Structure of a bacteriochlorophyll *a*-protein from the green photosynthetic bacterium *Prosthecochloris aestuarii*. J Mol Biol 131:259–285

Miller DJ, Catmull J, Puskeiler R, Tweedale H, Sharples FP, Hiller RG (2005) Reconstitution of the peridinin-chlorophyll *a* protein (PCP): evidence for functional flexibility in chlorophyll binding. Photosynth Res 86:229–240

Mimuro M, Tamai N, Ishimaru T, Yamazaki I (1990) Characteristic fluorescence components in photosynthetic pigment system of a marine dinoflagellate, *Protogonyaulax tamarensis*, and excitation energy flow among them. Studies by means of steady-state and time-resolved fluorescence spectroscopy. Biochim Biophys Acta 1016:280–287

Niedzwiedzki DM, Jiang J, Lo CS, Blankenship RE (2013) Spectroscopic properties of the chlorophyll a–chlorophyll c_2–peridinin–protein-complex (acpPC) from the coral symbiotic dinoflagellate *Symbiodinium*. Photosynth Res. doi:10.1007/s11120-013-9794-5

Niklas J, Schulte T, Prakash S, van Gastel M, Hofmann E, Lubitz W (2007) Spin-density distribution of the carotenoid triplet state in the peridinin-chlorophyll-protein antenna. A Q-band pulse electron-nuclear double resonance and density functional theory study. J Am Chem Soc 129:15442–15443

Osuka A, Kume T (1998) Fucoxanthin- and peridinin-pheophorbide-alpha molecules as a new light-harvesting model. Tetrahedron Lett 39:655–658

Osuka A, Kume T, Haggquist GW, Jávorfi T, Lima JC, Melo E, Naqvi KR (1999) Photophysical characteristics of two model antenna systems: a fucoxanthin-pyropheoporbide dyad and its peridinin analogue. Chem Phys Lett 313:499–504

Papagiannakis E, Larsen DS, van Stokkum IHM, Vengris M, Hiller RG, van Grondelle R (2004) Resolving the excited state equilibrium of peridinin in solution. Biochemistry 43:15303–15309

Papagiannakis E, van Stokkum IHM, Fey H, Büchel C, van Grondelle R (2005) Spectroscopic characterization of the excitation energy transfer in the fucoxanthin-chlorophyll protein of diatoms. Photosynth Res 86:241–250

Papagiannakis E, Vengris M, Larsen DS, van Stokkum IHM, Hiller RG, van Grondelle R (2006) Use of ultrafast dispersed pump-dump-probe and pump-repump-probe spectroscopies to explore the light-induced dynamics of peridinin in solution. J Phys Chem B 110:512–521

Polívka T, Frank HA (2010) Molecular factors controlling photosynthetic light harvesting by carotenoids. Acc Chem Res 43:1125–1134

Polívka T, Pascher T, Sundström V, Hiller RG (2005) Tuning energy transfer in the peridinin-chlorophyll complex by reconstitution with different chlorophylls. Photosynth Res 86:217–227

Polívka T, van Stokkum IHM, Zigmantas D, van Grondelle R, Sundström V, Hiller RG (2006) Energy transfer in the major intrinsic light-harvesting complex from *Amphidinium carterae*. Biochemistry 45:8516–8526

Polívka T, Hiller RG, Frank HA (2007a) Spectroscopy of the peridinin – chlorophyll-*a* protein: insight into light-harvesting strategy of marine algae. Arch Biochem Biophys 458:111–120

Polívka T, Pellnor M, Melo E, Pascher T, Sundström V, Osuka A, Naqvi KR (2007b) Polarity-tuned energy transfer efficiency in artificial light-harvesting antennae containing carbonyl carotenoids peridinin and fucoxanthin. J Phys Chem C 111:467–476

Polívka T, Pascher T, Hiller RG (2008) Energy transfer in the peridinin-chlorophyll protein complex reconstituted with mixed chlorophyll sites. Biophys J 94:3198–3207

Premvardhan L, Papagiannakis E, Hiller RG, van Grondelle R (2005) The charge-transfer character of the S_0-S_2 transition in the carotenoid peridinin is revealed by stark spectroscopy. J Phys Chem B 109:15589–15597

Prézelin BB, Haxo FT (1976) Purification and characterization of peridinin-chlorophyll a-proteins from the marine dinoflagellates *Glenodinium* sp. and *Gonyaulax polyedra*. Planta 128:133–141

Scholes GD (2003) Long-range resonance energy transfer in molecular systems. Annu Rev Phys Chem 54:57–87

Schulte T, Niedzwiedzki DM, Birge RR, Hiller RG, Polívka T, Hofmann E, Frank HA (2009a) Identification of a single peridinin sensing Chl *a* excitation in reconstituted PCP by crystallography and spectroscopy. Proc Natl Acad Sci U S A 106:20764–20769

Schulte T, Sharples FP, Hiller RG, Hofmann E (2009b) X-ray structure of the high-salt form of the peridinin-chlorophyll-a protein from the dinoflagellate *Amphidinium carterae*: modulation of the spectral properties of pigments by the protein environment. Biochemistry 48:4466–4475

Schulte T, Hiller RG, Hofmann E (2010) X-ray structures of the peridinin-chlorophyll-protein reconstituted with different chlorophylls. FEBS Lett 584:973–978

Sharples FP, Wrench PM, Ou KL, Hiller RG (1996) Two distinct forms of the peridinin-chlorophyll alpha-protein from *Amphidinium carterae*. Biochim Biophys Acta 1276:117–123

Shinoda S, Osuka A, Nishimura Y, Yamazaki I (1995) Synthesis of natural carotenoid-modified pyropheophorbide dyads for investigation of carotenoid-chlorophyll excited state interactions. Chem Lett 12:1139–1140

Šlouf V, Fuciman M, Johanning S, Hofmann E, Frank HA, Polívka T (2013) Low temperature time-resolved spectroscopic study of the major light-harvesting complex of *Amphidinium carterae*. Photosynth Res 117:257–265

Song PS, Koka P, Berzelin BB, Haxo FT (1976) Molecular topology of photosynthetic light-harvesting pigment complex, peridinin-chlorophyll-a-protein, from marine dinoflagellates. Biochemistry 15:4422–4427

Tronrud DE, Wen J, Gay L, Blankenship RE (2009) The structural basis for the difference in absorbance spectra for the FMO antenna protein from various green sulfur bacteria. Photosyn Res 100:79–87

van Stokkum IHM, Papagiannakis E, Vengris M, Salverda JM, Polívka T, Zigmantas D, Larsen DS, Lampoura SS, Hiller RG, van Grondelle R (2009) Inter-pigment interactions in the peridinin chlorophyll protein studied by global and target analysis of time resolved absorption spectra. Chem Phys 357:70–78

van Tassle AJ, Prantil MA, Hiller RG, Fleming GR (2007) Excited state structural dynamics of the charge transfer state of peridinin. Isr J Chem 47:17–24

Wormke S, Mackowski S, Brotosudarmo THP, Jung C, Zurnbusch A, Ehrl M, Scheer H, Hofmann E, Hiller RG, Brauchle C (2007) Monitoring fluorescence of individual chromophores in peridinin chlorophyll-protein complex using single molecule spectroscopy. Biochim Biophys Acta 1767:956–964

Zigmantas D, Polívka T, Hiller RG, Yartsev A, Sundström V (2001) Spectroscopic and dynamic properties of the peridinin lowest singlet excited states. J Phys Chem A 105:10296–10306

Zigmantas D, Hiller RG, Polívka T, Sundström V (2002) Carotenoid to chlorophyll energy transfer in the peridinin-chlorophyll-a-protein complex involves an intramolecular charge transfer state. Proc Natl Acad Sci U S A 99:16760–16765

Zigmantas D, Hiller RG, Yartsev A, Sundström V, Polívka T (2003) Dynamics of excited states of the carotenoid peridinin in polar solvents: dependence on excitation wavelength, viscosity, and temperature. J Phys Chem B 107:5339–5348

Zigmantas D, Hiller RG, Sharples FP, Frank HA, Sundström V, Polívka T (2004) Effect of a conjugated carbonyl group on the photophysical properties of carotenoids. Phys Chem Chem Phys 6:3009–3016

Chapter 4

Piecing Together the Phycobilisome

Ailie Marx, Liron David, and Noam Adir*
Schulich Faculty of Chemistry, Technion, Israel Institute of Technology, Technion City, Haifa 32000, Israel

Summary

Photosynthesis is driven by the absorption of light by arrays of pigments bound within protein complexes called antennas, followed by the efficient transfer of energy to the photochemical reaction centers. Cyanobacteria, red algae and cyanelles contain phycobilisomes (PBS) as their major antenna complex. The PBS is an extremely large (3–7 MDa), multi-layered complex bound to the stromal side of the photosynthetic membrane. In this review, we will describe the important structural and functional characteristics of the phycobilisome complex experimentally obtained over the past 40 years, especially in relation to the phycobilisomes unique absorption characteristics and its ability to self-assemble and disassemble.

I. The Phycobilisome Antenna – An Enormous Pigment-Protein Complex

Photosynthesis is the essential life supporting process which converts light energy into chemical energy and, in oxygenic species, the concomitant splitting of water producing molecular oxygen. Organisms that have evolved in different environmental surroundings have different molecular strategies on how to perform the first step in this process, light harvesting. The efficient capture of the ambient light followed by transfer of the absorbed energy into the photosynthetic reaction centers (RCs) are performed by pigment-protein complexes called antennas or light-harvesting complexes. The variety of architectures in the antenna systems is more varied than that of the RCs, establishing both the need to preserve the internal functional details required for photochemistry in the RCs and the evolutionary liberty to find

*Author for correspondence, e-mail: nadir@tx.technion.ac.il

M.F. Hohmann-Marriott (ed.), The Structural Basis of Biological Energy Generation, Advances in Photosynthesis and Respiration 39, pp. 59–76, DOI 10.1007/978-94-017-8742-0_4,

different solutions for actual light absorption (Neilson and Durnford 2010; Cogdell et al. 2004; Blankenship et al. 1995; Adir 2005, 2008; Adir et al. 2006; Grossman et al. 1995). Among the different antenna systems which have evolved to carry out this task, the phycobilisome (PBS) is unique in that it is a truly enormous pigment-protein complex found outside the membrane. The PBS anchored to the stromal side of the thylakoid membrane in cyanobacteria (see Chap. 14) and red algae (see Chap. 16) can reach molecular weights of 3–7 MDa and dimensions of over 50–75 nm in size – an order of magnitude larger than the photosystem II (PSII) complex to which it primarily funnels light energy.

Early structural characterization of isolated PBSs was carried out on the level of the entire complex and the earliest electron microscopy (EM) pictures revealed an aesthetic structure made up of a core surrounded by rods (Gantt and Conti 1966a, b; Gantt and Lipschultz 1972; Teale and Dale 1970; Edwards and Gantt 1971; Bryant et al. 1976; Tandeau de Marsac and Cohen-Bazire 1977; Bryant and Cohen-Bazire 1981; Glazer 1989). A precise characterization of the macrostructure of the PBS has however proved not so trivial since original models of radiating rods would appear at odds with what is now known about the dimensions of the phycobiliprotein (PBP) disks which compose the core and rods. Not visualized in the EM pictures were the unpigmented protein components of the PBS, assumed to be buried within the PBS complex occupying the hollow cavities within the rods and core (Liu et al. 2005). Their presumed position, and the overall architecture of the PBS led to the proposal that these proteins play a role in complex assembly and stability and were thus termed linker proteins (LPs).

Since these early glimpses were obtained, more advanced techniques have provided vastly improved images of the subunits, but little more specific detail on the architecture of the entire PBS. Various sub-assemblies of the basic PBS units have been studied extensively and numerous high-resolution structures of individual PBP components have been determined by X-ray crystallography (Schirmer et al. 1986; Adir 2005). These structures have simultaneously revealed both the structural basis for functional details and also raised questions as to the manner by which the complex undergoes assembly. The wealth of information obtained from the structures has been combined with spectroscopic and biochemical techniques to reveal facets of energy absorption and transfer (Sauer and Scheer 1988; Beck and Sauer 1992), complex stability, self-assembly properties and possible mechanisms of disassembly (MacColl 1983; MacColl et al. 2003; McGregor et al. 2008). Significant questions which remain are the arrangement of the rods around the core, what form of rod-rod interactions exist and what the mode of attachment of the rods to the core is. The fashion of attachment has been attributed to the presence of LPs, however, one of the well-known properties of the PBS, is that the complex disintegrates into trimeric rings (and lose their association with the LPs) when isolated in the absence of high concentrations of buffered phosphate (Gantt and Lipschultz 1972). It appears that the LPs are only weakly bound to the PBPs, and thus their role in stabilization of the entire PBS structure cannot be easily rationalized. This observation is puzzling, but may be due to the physiological phenomenon of PBS disassembly under nutrient stress (Collier and Grossman 1992; Grossman et al. 1993). The loss of the linkers in isolated PBS sub-assemblies has led to only limited structural information, however as we will describe below, partial LPs structures have recently become available. In this

Abbreviations: AFM – Atomic force microscopy; APC – Allophycocyanin; EM – Electron microscopy; L_C –Core linkers; L_{CM} – Core-membrane linker; LP – Linker proteins; L_R – Rod linker; L_{RC} – Rod-core membrane linker; PBP – Phycobiliprotein; PBS – Phycobilisome; PC – Phycocyanin; PCB – Phycocyanobilin cofactor; PE – Phycoerythrin; PEB – Phycoerythrobilin cofactor; PEC–Phycoerythrocyanin; PUB – Phycourobilin cofactor; PVB – Phycobiliviolin cofactor; RC – Reaction center; TEM – Transmission electron microscopy

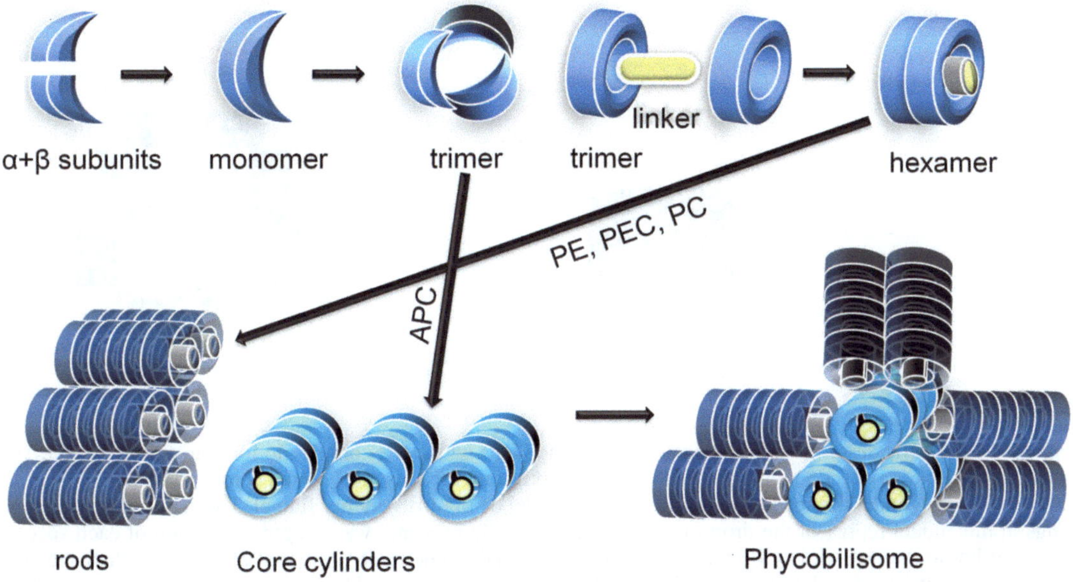

Fig. 4.1. Schematic representation of the ordered association of PBPs and LPs to form the PBS complex. The *arrows* indicate the direction of assembly. All PBP components form trimers, with the PE, PEC or PC components forming hexamers that then further assemble into rods. Four APC trimers (containing both major and minor species) form cylinders, and 2–5 such cylinders form the PBS core. The six rod sub-structures bind to the core with the terminal ring flush with the outer cylinder circumference. In this depiction the rods are slightly displaced to reveal their inner cavity, proposed to house the rod-core linker proteins.

review we will describe the basic facets of PBS subunit structure and function based on various experimental evidence and attempt to build a picture of the entire complex consistent with these experiments.

II. Building Blocks – Crystal Structures of Individual Components

The PBS is made up of phycobiliproteins (PBPs), the pigmented component of the complex, and five types of mostly unpigmented proteins commonly referred to collectively as linker proteins. Two PBPs are always present in PBS complexes – allophycocyanin (APC), which forms the basic building block for the core, and phycocyanin (PC), which is the major component of rods that is always proximal to the core. Some species contain additional PBP rod components, phycoerythrin (PE) or phycoerythrocyanin (PEC), which occupy positions more distal than PC in relation to the core (MacColl 1998). Each of the different PBPs have highly similar three dimensional structures and in each case the higher order aggregates are assembled analogously from a basic building block, the monomer (Fig. 4.1). The monomer in actuality is made up of two similar but not identical polypeptide chains (always denoted as the α and β subunits). The association of three monomers forms a trimeric ring, and two such rings can further associate through contacts between the α subunits, into hexamers (in PC, PE and PEC). Four APC trimers associate into a cylinder and 2–5 cylinders, depending on the species, bundle together to form the core. Both the cylinders and hexamers form structures which resemble a tube with ample empty space to accommodate the LPs. The rods are composed of hexamers which extend by stacking face to face one onto the other through mostly β subunit interactions. In those species which contain them, the hexamers of PE or PEC seamlessly extend out from the PC elongating the rods further. The specific placement of PBPs within the overall PBS

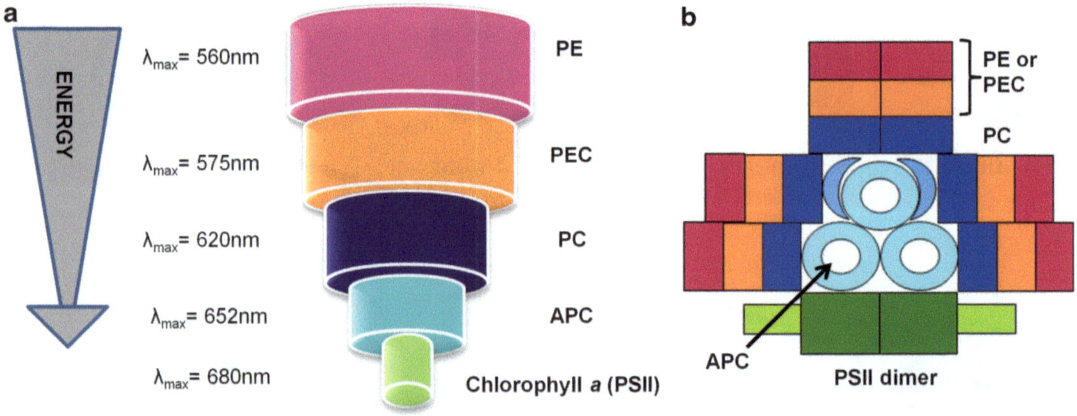

Fig. 4.2. Schematic description of the positioning of the PBPs within the complex to form an energy funnel. The rods of all species contain either PE or PEC (but not both) and (or only) PC. Cores contain only APC, however the number of cylinders is species dependent. (**a**) A model of the PBS energy funnel. The *size* and *color* of the rings in this model represent the drop in energy content as the maximal wavelength absorption of each species shifts further to the *red*, while all rings are physically of equal diameter. (**b**) A model of the entire PBS showing the relative positions of each component. The length of the rods, determined by the number of hexamers, and their PBP content is species dependent.

structure is functionally crucial as it forms the basis for the creation of an energy funnel (Fig. 4.2) whereby the higher energy absorbing components (PE (λ_{max}=560 nm), PEC (λ_{max}=575 nm) and PC (λ_{max}=620 nm)) are more distal to the core which is comprised of the lowest energy absorbing PBP species, APC (λ_{max}=650 nm). The core also contains variants of APC subunits (α^B, β^{16} and core-membrane linker (L_{CM})) which have red-shifted absorbing pigments (MacColl 2004; Lundell et al. 1981; Glazer and Bryant 1975; Ducret et al. 1998) that closely overlap the absorption spectrum of chlorophyll *a*, the primary pigment found in cyanobacterial photosystems to which the energy is ultimately transferred.

All PBS pigments are linear tetrapyrrole bilins (Glazer 1989; Adir 2008), connected covalently to the PBPs through thioether bonds to conserved cysteine residues (Fig. 4.3). APC and PC contain a common pigment, phycocyanobilin (PCB), bound to analogous positions on the α and β subunits. PC binds an additional third PCB on a short loop, inserted towards the end of the C-terminal of the β chain, positioned on the outer disk surface. Other bilin types can be found bound to PE and PEC and these differ from PCB in the number and arrangement of conjugated double bonds. That APC and PC bind chemically identical bilins and yet possess distinctive absorption properties highlights the role of the protein scaffold in influencing the structure and chemical environment, which tunes the spectroscopic characteristics of these pigments (McGregor et al. 2008).

X-ray crystal structures of each of the PBPs have revealed a highly conserved fold highlighting the subtlety with which they are able to tune absorption properties (Fig. 4.3). This allows the PBPs to act as highly similar "building blocks" which are capable of assembling analogously and yet structural subtleties (further detailed in following sections) allow for specificity in binding partners with no intermixing between the various kinds of PBPs and specific tuning of the cofactor environment to create different absorption and fluorescence properties necessary to funnel energy down the complex. Indeed numerous crystal structures for each of the individual PBPs are now available in the Protein Data Bank (www.pdb.org, reviewed in Adir 2005). The structures of the different PBP types are highly similar, which is a direct result of the relatively high homology

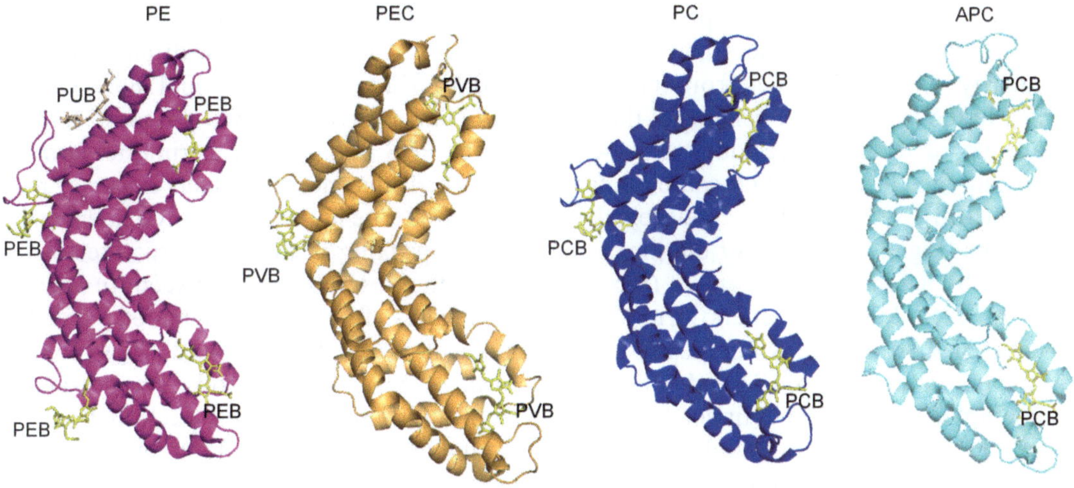

Fig. 4.3. Structures of the different PBPs in monomeric form. The position of attachment of the different bilins are shown and the *color* of the structure approximately represents the color of the isolated protein. *APC* allophycocyanin, *PC* phycocyanin, *PCB* phycocyanobilin cofactor, *PEB* phycoerythrobilin cofactor, *PEC* phycoerythrocyanin, *PE* phycoerythrin, *PUB* phycourobilin cofactor, *PVB* phycobiliviolin cofactor.

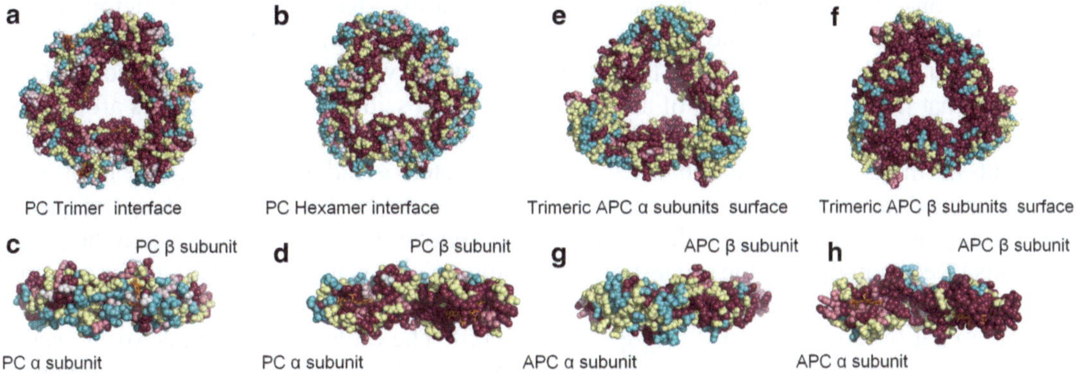

Fig. 4.4. Evolutionary conservation mapped onto trimeric PC and APC structures. Conservation of residues on (i) the disk face composed of primarily α-subunits of PC (**a**) and APC (**e**); (ii) residues on the disk face composed of primarily β-subunits of PC (**b**) and APC (**f**); (iii) the outer circumference of the PC (**c**) and APC (**g**) disk and (iv) the inner surface of the PC (**d**) and APC (**h**) disk. *Red* signifies a high level of conservation, *blue* no conservation and *yellow* not enough data to determine the level of conservation. The analysis indicates that the interaction interfaces (between trimers, between hexamers or between linkers and trimers/hexamers) are more highly conserved than the outer circumference of the discs and that APC is more highly conserved than PC.

within the primary sequences. In each case the structure forms an eight α-helical, globin-like structure. Consideration of the most conserved areas of any structure is useful in elucidating functionally significant areas. Given that the homology among the PBPs of the same type from different species and also between the different PBP types is quite high, it is interesting to observe that the areas of particular structural conservation in the different PBPs is subtlety variant. The ConSurf Server (Ashkenazy et al. 2010) (http://consurf.tau.ac.il/) was used to map evolutionary conservation as determined from sequence alignment onto the three dimensional structure of PC and APC from *T. vulcanus* (Fig. 4.4).

In the process of forming trimers, the PBPs present two different faces, one made up primarily of residues of the β subunits of each of the three monomers and the other primarily made up residues from the α subunits. In the formation of hexamers, the face made up of α subunit residues (Fig. 4.4a) associate face to face (which can be notated as: [βα][αβ]) while contacts between the β PC faces (Fig. 4.4b) allow for the formation of rods by assembly of hexamers ([βα][αβ][βα][αβ]). In APC the trimers form similar faces (Fig. 4.4c and Fig. 4.4d, respectively); however, on the basis of EM and X-ray crystallographic studies, the cylinder formed by four APC trimers is not thought to be composed of two hexamers, due to looser interactions between trimers, and perhaps also due the influence of the L_{CM} linker that serves to provide contacts for core formation and to bind the entire PBS to PSII (Capuano et al. 1991). This may be the reason for differences in sequence conservation between PC and APC. In PC, the α subunits face of the PC trimer are more conserved than the β subunits face. This suggests that the hexameric level of assembly ([βα][αβ]) cannot tolerate extensive changes, without harming other functionalities. On the other hand, in APC the β subunits face of the trimers are more conserved than that of the α subunits face, strengthening the suggestion that APC cylinders form by association of four trimers in a different order: [αβ][βα][αβ][βα]. This order enables the positioning of the L_C linker (Reuter et al. 1999; Ducret et al. 1996, 1998) to terminate the core cylinder on both sides. For both PC and APC the inner cavity is far more conserved that the outer circumference and this may allude to the importance of specific interaction (or perhaps the need to avoid strong interactions) between PBPs and LPs.

One of the interesting aspects of the different crystal structures is the lattice packing (Fig. 4.5). In 16 of the 17 PC structures available in 2011, hexamers are formed *in crystal*. The single structure which does not form hexamers is one of several structures of PC isolated from *Thermosynechococcus vulcanus* (Adir and Lerner 2003). This particular structure (PDB code 1ON7) originated from a unique PC fraction which was shown to be a blue-shifted (PC_{612}) form with an unmethylated Asnβ72 residue (Klotz et al. 1986; Swanson and Glazer 1990) and was proposed to be a minor PBS component, a trimeric form of PC which may form contacts between rods and cores. The importance of considering the crystal packing in the PBP structures stems from the resemblance between *in crystal* aggregates formed by symmetry related molecules and the assumed physiological form of hexamers and rods. The hexamers from different structures pack within the crystal in a number of different ways, among them with the formation of infinitely long rods.

In those structures where rods are formed *in crystal*, a tantalizing question arises as to what extent these crystal rods resemble biological rods, particularly in terms of the stacking orientation and the lateral contacts formed by adjacent rods both factors which could significantly affect energy transfer pathways within the complex and perhaps hint at possible rod arrangements around the core in complex. Whilst initial PBS models suggested a radiating arrangement of rods (Glauser et al. 1992; Yamanaka et al. 1982) subsequent models have suggested the possibility of close lateral contact with for example the formation of rod pairs (Glazer 1989; Adir 2005). The biological relevance of *in crystal* rod formation is not certain, since numerous alternative forms of crystal packing have been observed in the structures. This may be a result of the crystallization process or indicates that several hexamer interactions are possible, or is the result of the lack of LPs. A rather unique result has been recently reported, whereby several essentially identical structures from the same PC source have been obtained from crystallization and cryoprotection under vastly different conditions (David et al. 2011). This appears to suggest that the contacts formed may be indicative of the forces governing the self-assembly of the PBS complex and not a mere artifact of crystallization. Indeed, a recent structure of a functional PC rod and PC trimer from the

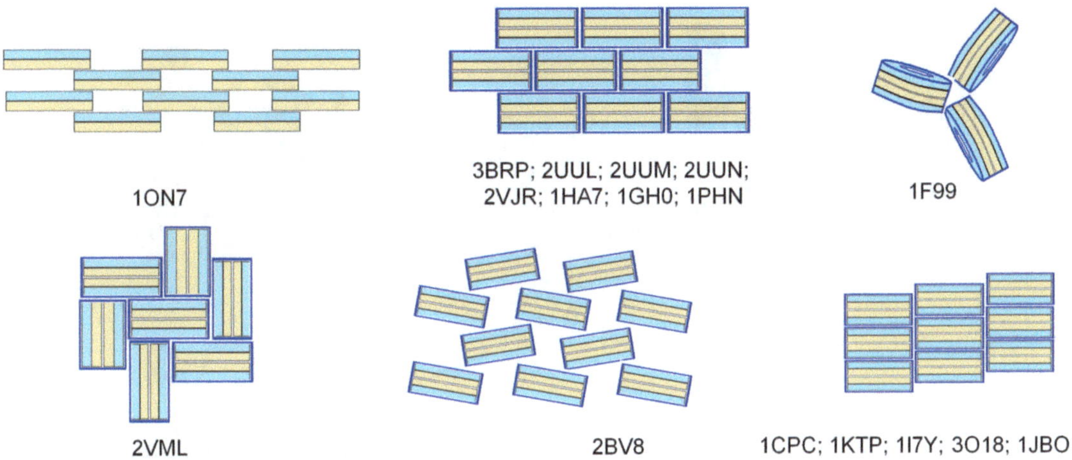

Fig. 4.5. Schematic representation of the different types of crystal packing for each of the PC structures available in the PDB to date. In all structures the minimal physical unit is a trimer with the α-subunits shown in *yellow* and the β-subunits in *blue*. In all structures except 1ON7, the trimers associate into hexamers, and in some cases into infinite rods.

same species (David et al. 2011) has further provided evidence to support the biological relevance of interactions between symmetry related molecules in the crystal structures of isolated components to subcomplexes of the PBS (see below).

PBP structures have of course not only provided insight into PBS assembly but have also provided high-resolution detail on structural fine points. The geometries and orientations of the different bilins within the protein scaffold have been well described by high-resolution X-ray crystallography (Fig. 4.6), and the role of structurally essential water molecules has also been addressed. Increasingly higher resolution structures (better that 1.0 Å) will hopefully facilitate the description of protonation states allowing for an accurate description of energy states.

In addition to the PBPs, around 15 % of the PBS is made up of LPs, dubbed as such after one of their two proposed functions, to guide the association of the PBPs into larger subcomplexes (Liu et al. 2005). Their second function has been clearly shown as performing modification or tuning of the spectroscopic properties of the PBPs within the complex. The LPs are generally subgrouped according to their assumed position within the complex; the core-membrane linker (L_{CM}), core linkers (L_C), rod-core linkers (L_{RC}) and rod linkers (L_R). The LPs are much less conserved across species than the PBPs and indeed the number and type of LPs varies from species to species. A recent report on the interaction between linkers and a homologue of a HSP90 chaperone (Sato et al. 2010) would indicate that linkers cannot associate with PBPs spontaneously. Whether the higher order oligomers of PBPs assemble around LPs is not known, however reconstitution studies (Ducret et al. 1998) do not indicate this to be the case.

In stark contrast to the PBPs, structures of the LPs have remained elusive. The highly hydrophobic nature of these proteins makes the isolation of the homogeneous, soluble and concentrated solutions of LPs that are required for crystallization or NMR difficult. Indeed, of the LP structures available to date, only one was actually visualized in actual association with a PBP (the APC structure, 1B33). Very recently the Northeast Structural Genomics Consortium in the USA have solved several structures of domains of the $L_{CM,}$ using NMR techniques (PDB codes 2L06 and 2KY4) and X-ray crystallography (PDB codes 3OSJ and 3OHW) as well as a

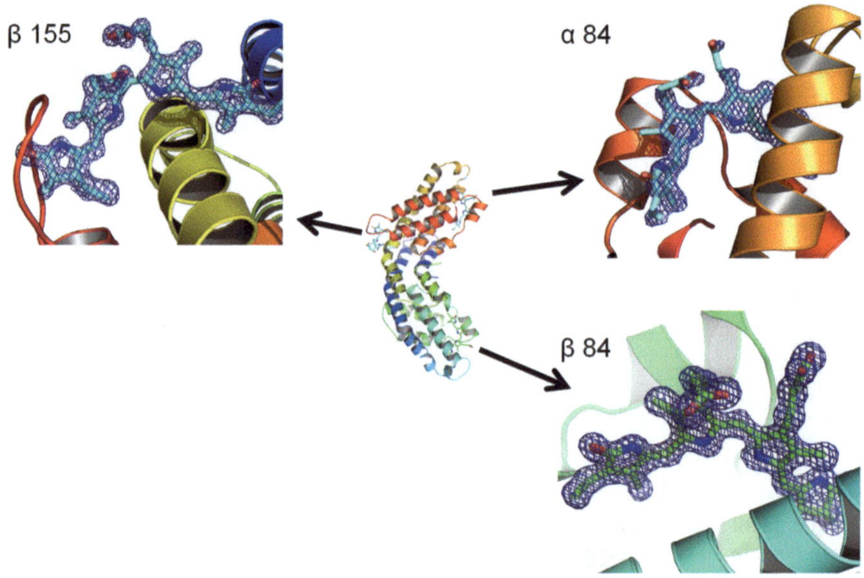

Fig. 4.6. 2Fo – Fc omit electron density maps contoured at 1.5 σ around each of the three phycyanobilin cofactors from the 3O18 PC structure determined to 1.35 Å.

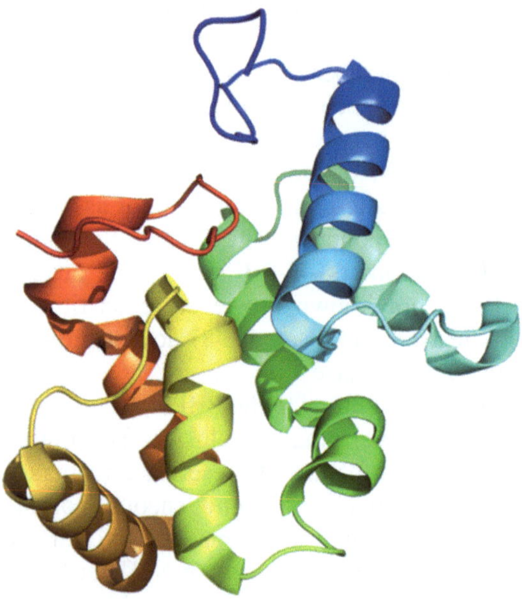

Fig. 4.7. X-ray crystal structure of the N-terminal domain of the rod linker isolated from *Synechocystis sp.* PCC 6803 (PDB code 3PRU). The protein is depicted in *colors* of the spectrum running from the N-terminal (*blue*) to the carboxyl-terminal (*red*). The portion of the protein crystallized includes residues 14–158 out of the total 291 residues in the full length protein.

fragment of the L_R also using NMR (PDB code 2L3W) and X-ray crystallography (PDB code 3PRU, Fig. 4.7). The structures of the L_R represent the N-terminal half of the protein from two different species and the four structures of L_{CM} domains are homologous to this portion of the L_R meaning that these six new structures have a certain degree of structural redundancy (root mean square deviation of the aligned α-carbon positions are between 1 and 1.5 Å). Elucidation of the complete structure of these LPs, along with their exact positioning within the complex, seems again to be a problem that can only be solved by total structural characterization of the PBS complex or subcomplexes. Even a low to medium resolution structure could provide a wealth of information particularly considering that it may be merged with the high-resolution detail now available for the PBPs.

III. A Problem of Symmetry – Crystallizing the Complex and Subcomplexes

The PBS complex was first visualized by EM experiments as a central core surrounded by rods and led to the first PBS model which suggested an arrangement whereby the rods are arranged in radial fashion around the core. A second model, already mentioned above, suggests a parallel rod pair arrangement in which the rods are arranged as two rod doublets that are parallel to the thylakoid membrane and one rod doublet which is perpendicular to the membrane. Cryo-EM micrographs (Yi et al. 2005) of PBSs have supported the parallel pair model which is also consistent with the crystal structures of PC (Adir 2005). Atomic force microscopy (AFM) studies performed on isolated PBSs and native thylakoid membranes from red algae (Arteni et al. 2008) visualized the isolated PBS as a dimer in opposition to previous TEM (transmission electron microscopy) micrographs that have only ever shown the isolated PBS as a monomeric unit. The native architecture of PBS attached to the thylakoid membrane and their crowding distribution under different light intensities was also shown by AFM and TEM (Liu et al. 2008). Under low light intensity the PBSs are organized in ordered rows whereas under medium light the PBS shows a different pattern of organization, namely a random arrangement with the tendency to form clusters (Arteni et al. 2008, 2009). These studies have demonstrated that the PBS can form a kind of crystalline lattice and suggest perhaps that there is a natural tendency of the PBS organize in an ordered fashion, a feature conducive to crystallizing the entire complex despite its enormity.

The use of X-ray crystallography to characterize the structure of subcomplexes of the PBS to higher resolutions has been hindered by the same symmetry and self assembly properties that facilitates the crystallization of PBPs. Most of the PBPs characterized to date have crystallized in space groups with three-fold symmetry. Since the LPs traverse the internal cavity of the PBP complexes along the symmetry axis, the result is a super-position of at least six different LP orientations, there by cancelling out the contribution of the LPs to the calculated electron density. The anomalous nature of the crystal packing in the 1B33 structure (Reuter et al. 1999), in which two full trimers occupy the asymmetric unit at almost right angle, overcomes this problem and so this APC-L_C complex is the only complete PBP-LP complex structure available to date. The LP in this structure is composed of both β-sheets and α-helices and is asymmetrically positioned within the cavity, interacting with two out of the three monomers within the trimer. The LPs presence induced a flattening effect on the overall trimer structure of APC, within the crystal lattice.

Recently the X-ray crystal structure of an intact and functional PBS rod was solved (David et al. 2011). The presence of all rod components (α and β PC subunits as well as three rod LPs, the rod capping $L_R^{8.7}$, L_R and one L_{RC} (the *cpcG4* gene product) was confirmed by SDS-PAGE, single crystal confocal microscopy and mass spectrometry and the functional integrity of the rod was demonstrated using spectroscopy on both solutions and crystals, which confirmed the characteristic red shift in absorption and fluorescence emission as compared to trimeric PC (Fig. 4.8). The crystal structure of the rod was identical to that of trimeric PC – a monomer in the asymmetric unit cell which extends in the crystal lattice to form hexamers which extend into infinite rods. While the biochemical evidence from the crystal analysis left little doubt as to the presence of LPs in the crystal the absence of clear electron density prevented the building of an ordered structure. Nonetheless, the PC rod structure, within which the linker is enclosed, showed subtle hints of the linkers' presence. Comparison of the isolated PC and rod PC structures led to the identification of several amino acids located on the wall of the inner cavity, αGlu7, αArg15, βLys7, and βGln113, with improved electron density in the rod structure. This improved electron density indicates a stabilizing force, presumably interaction with the LPs, and indeed homology modeling to the 1B33 APC-L_C structure shows that these residues are in analogous positions to those APC residues which interact with the L_C

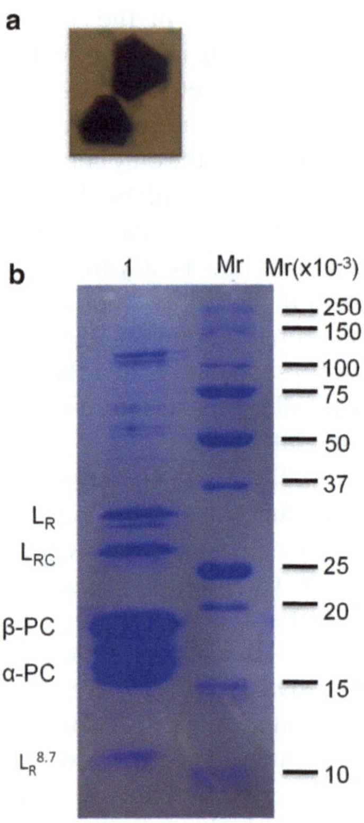

c

Subunit	Gene	Function	%Pep
α-PC	*cpcA*	PBP	89
β-PC	*cpcB*	PBP	78
L_R^{32}	*cpcC*	Rod linker	38
$L_R^{8.7}$	*cpcD*	Rod cap	50
L_{RC}^{30}	*cpcG4*	Rod-core linker	29

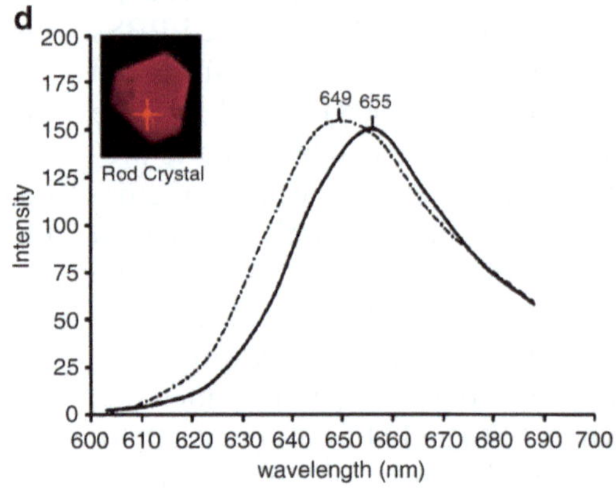

Fig. 4.8. Analysis of rod crystals (David et al. 2011). (**a**) Rod crystals have a similar morphology to that of trimeric PC and X-ray crystallography shows that PC in rod form is nearly identical to that of trimeric PC (not shown). (**b**) SDS-PAGE of the isolated rod shows the presence of three linker proteins (L_R, L_{RC} and $L_R^{8.7}$) in addition to the α and β subunits of PC, however core components are still present. (**c**) Solubilized crystals were analysed by mass spectrometry showing that only a single rod type, containing CpcG4 crystallized. %Pep, percentage of possible peptides obtained by mass spectrometry. (**d**) Confocal fluorescence microscopy (at 1 μm^3 resolution) analysis of rod crystals (*full line*) and trimeric PC crystals (*dashed line*) demonstrates the expected *red* shift in emission from PC in rod assembly. The same results were obtained from all depths of the crystal showing that the linkers were present throughout the crystal.

C-terminal. Further comparison of crystallographic B factors between the isolated PC and PC rod structures further hinted at the stabilizing influence of the linkers' presence. Whilst on average the B factors were similar, comparison between individual atoms showed that there were fewer cases of significantly higher B factors in the rod structure. Several side chains distributed throughout the PC monomer were identified as having reduced B factors in the rod structure and among them αGlu109 was noted as significant due to its positioning on the inner side of the rod and its possible interaction with a LP based again on homology modeling to the 1B33 structure.

This PC rod structure is not the first PBP-linker complex to be crystallized. Aside from the APC-Lc structure mentioned above, two structures from crystals of PE containing the γ linker subunits exist, and again in these cases, the electron density from the linker was also averaged out during structure refinement and so its particular structure remained largely unassigned (Ficner et al. 1992; Ritter et al. 1999).

Elucidating the complete LP structures and locating their position within the rods and core, challenges plagued by solubility and symmetry problems, may require the use of more indirect methods or a combination of techniques. Using the available structures for L_C (which is homologous to both the C terminal portion of L_R and L_{RC} as well as the rod capping linker, $L_R^{8.7}$) and the N terminal portion of L_R (which is also homologous to the N terminal section of L_{RC}) models may be created to aid in understanding the nature of the LP structures and their general interactions with the PBPs. Such models were created for the LPs present, but not visualized, in the above mention PC rod structure. Calculations for the hexamer cavity volume and the modeled LP volume demonstrated that there is ample space available within the central channel of the rod to accommodate the linkers without protruding from either end of the rod, or indeed without actually linking adjacent hexamers. It was suggested that since the major role of the linkers may be in fine-tuning the spectral properties of the PBPs, they might be renamed "tuning proteins" (David et al. 2011).

IV. Essential Functionalities Revealed by Structural Subtleties

One of the elegant aspects of the PBS structure is the way in which the various PBPs are similar enough to use the same principles of association and yet dissimilar enough to maintain specificity. The α and β subunits of each PBP type associate to form structurally similar monomers followed by the formation of similarly shaped trimers. There have been no reports of isolation of mixed PBP monomers or trimers *in vivo* (or in mutants lacking one of the subunits) indicating a tightly controlled assembly mechanism (Anderson and Toole 1998). Heterologus expression of single PBPs in *E. coli* enables the formation of some homodimeric species (Tooley et al. 2001; Ge et al. 2009), but when both subunits are expressed, then the natural monomer assembles (Arciero et al. 1988; Biswas et al. 2010). In addition, APC and PC bind the same cofactor and yet have significantly different absorption and fluorescence properties. These are both functionally critical properties since the correct association of PBPs and the differing absorption properties create the energy funnel which facilitates the efficient collection and transfer of light into the photosystems.

In order to understand the mechanisms of energy capture and transfer in the complex it would be sage to question the origin of the different absorption properties of the PBPs. PC and APC bind chemically identical chromophores, (the PCB), and form highly similar three dimensional structures. The monomeric forms of these two PCBs have nearly identical absorption and fluorescence properties (although PC has an additional PCB). However, during trimer assembly the α84 PCB of one monomer comes into proximity with the β84 PCB of an adjacent monomer radically changing the electronic environment of both PBPs (Fig. 4.9). While the geometries and orientations of these proximal PCBs are nearly identical, the absorption maxima of the trimeric form of PC is 620 nm while trimeric APC absorbs at 652 nm (with a shoulder at 620 nm) (MacColl 2004; MacColl et al. 2003).

The significant red-shift upon the trimerization in APC had long been rationalized in general terms as a result of conformational changes in the PCB, a change in the chemical environment of the PCB or a delocalization of excited states resulting from the nearing of two cofactors on adjacent monomers. Recent experiments however showed that only a slight loosening of the trimer (as opposed to complete monomerization) is enough to revert the APC absorption spectra to one resembling that of PC or monomeric APC (McGregor et al. 2008). This means that the simple proximity between two cofactors on adjacent monomers is not sufficient to afford the change alone. A careful structural analysis and comparison between PC and APC and the chemical environment surrounding the β84 PCB, which is positioned at the trimer interface has provided a more complete and detailed explanation for the absorption shift (Fig. 4.10). In the monomeric form of the PBP the β84 PCB is half

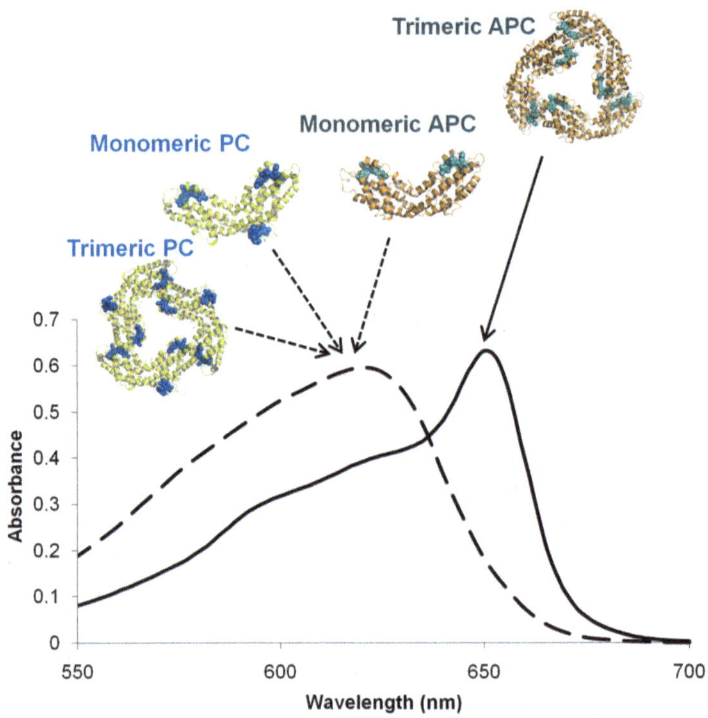

Fig. 4.9. Absorption characteristics of monomers and trimers of the phycocyanobilin containing PC and APC subunits. The monomeric form of both PC and APC as well as the trimeric form of PC have an absorption spectrum with $\lambda_{max} = 620$ nm (*broken line*) whilst the trimeric form of APC has a *blue* shifted absorption spectrum with $\lambda_{max} = 650$ nm and a shoulder at 620 nm (*line*).

embedded in protein and half solvent exposed. Upon trimerization the solvent exposed portion becomes enclosed by a protein pocket provided by the adjacent monomer and the chemical nature this pocket in PC and APC is significantly different in both shape and chemical character. The PC pocket is more shallow and polar than the deeper APC pocket which is created by aromatic and hydrophobic residues.

Surrounding the cofactor with a pocket of hydrophobic character would ordinarily induce a small blue-shift in its absorption, and not the observed strong (30 nm) red-shift. It was proposed that the source of the red shift could only be the result of a strong coupling between the two PCBs, although 20 Å separate their centers. The strong coupling was suggested to be a result of a unique structural environment (with respect to PBPs) that surrounds the hydrophobic pocket with a second shell of charged amino acids and it is this combination that induces coupling. This would explain why even slight loosening of the trimer interaction domain (without actual disassembly), would allow a certain amount of solvent exposure, resulting in the loss of a hydrophobic environment, and loss of coupling.

Since the integrity of the higher order complexes is important to the functioning of the PBS it is interesting to consider the mechanisms by which different species maintain complex assembly in challenging environmental niches such as extreme cold or heat, high salinity and deep within the water column where light exposure is limited. Each of these challenges requires special survival mechanisms and often this is found at the level of the protein (Inoue et al. 2000). Among the numerous PBP structures are several isolated from thermophiles and it has

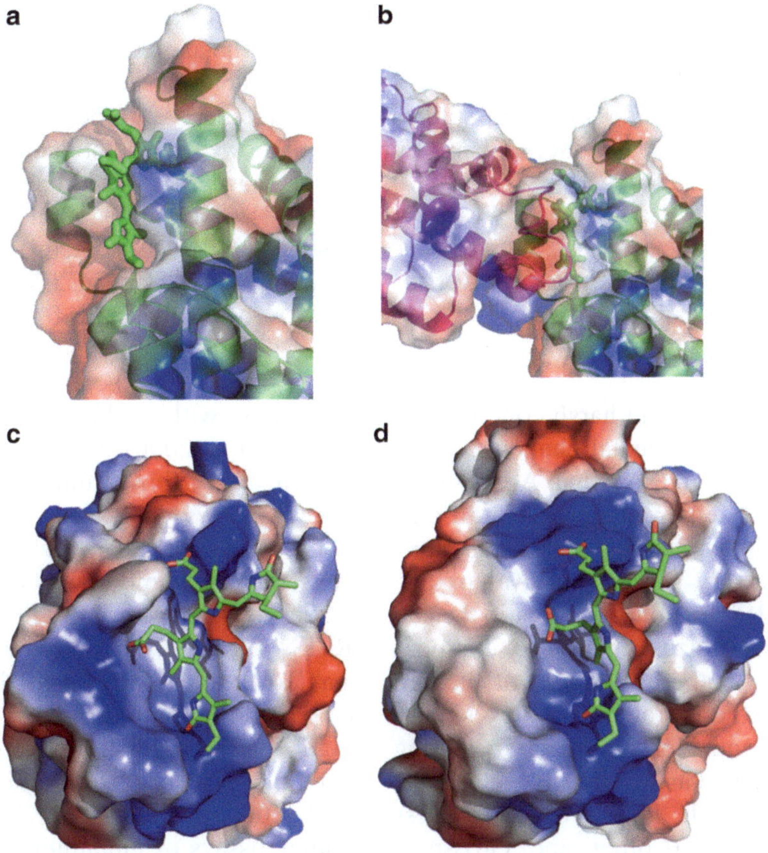

Fig. 4.10. Monomeric APC showing the solvent exposed position of the $\alpha 84$ cofactor (*panel* **a**), which becomes enclosed during trimerization (*panel* **b**). The nature of the pocket around $\alpha 84$ created by the adjacent monomer during trimerization is shown for APC (*panel* **c**) and PC (*panel* **d**). All proteins were depicted with their molecular surfaces *colored* by the calculated electrostatic potential of the surface. *Red* and *blue* patches indicate negative and positive potentials, respectively.

been shown that indeed the isolated protein itself is resistant to heat denaturation. However a comparison of the primary sequences and structures originating from mesophiles and thermophiles shows very little variation and it is only structural nuances which can be attributed to affording thermostability. *T. vulcanus* is a cyanobacterium which grows optimally at 55–60 °C and both APC and PC structures from protein isolated from this species have been determined. Comparison of these structures together with those of other PC and APC from thermophiles and mesophiles deposited in the PDB has enabled the identification of possible sources of the thermostability. In the case of PC, studies have indicated that general characteristics such as cavities, hydrogen bonding, ion pairs, secondary structure surface polarity and amino acid composition are not consistent with conferring thermostability. A single conserved amino acid change, αPhe28 to aspartic acid, was suggested as providing an additional polar contact at both the monomer and hexamer interaction interfaces and that this is perhaps enough to provide the extra stability required to prevent denaturation at higher temperatures (Adir et al. 2001, 2002). In the case of APC, it was suggested that the presence of an extra charged residue in the second shell surrounding the PCB binding pocket at the trimer interface (described above) may provide extra stability (McGregor et al. 2008). Under certain circumstances however the ordered disassembly of the PBS is necessary and here too structural studies have helped elucidate the mechanisms by which this occurs.

V. Disassembly of the Phycobilisome – A David and Goliath Battle at the Molecular Level

The ability to survive in harsh conditions is not the only environmental challenge faced by cyanobacteria and red alga. Adaptability to changes in environmental conditions, particularly the availability of essential nutrients, is necessary for the survival of any organism. PBSs are both abundant in the cell and enormous in size. In fact, they compose around 50 % of the dry weight of the cell meaning that these protein complexes comprise a substantial energy reserve. During times of nutrient starvation cells are capable of disassembling the PBS in order to access this energy store (Collier and Grossman 1992, 1994) and probably also as a mechanism important for protecting against overexcitation of the photosystems. Cyanobacteria, formerly known as blue-green alga, owe their distinctive color to the presence of both the PBSs and chlorophyll-containing complexes and the disassembly of the PBS results in a bleached phenotype whereby the cells take on a yellow-green color (Grossman et al. 1993). An ensemble of genes, including *nblA*, *nblB*, *nblC*, *nblR*, *nblS* and *ald* have been shown to be involved in initiating and carrying out this disassembly. However, a single small protein (only 6–7 kDa) appears to be the active player in the actual disassembly mechanism (Collier and Grossman 1994) (Fig. 4.11). Expression of the gene encoding for this protein, the NblA, is increased 50-fold under conditions of nitrogen and in some cases sulfur starvation (Collier and Grossman 1994). Despite this, NblA itself has not been shown to possess any proteolytic activity and presumably further proteins are required to carry out the actual degradation.

The exact means by which the disassembly occurs has yet to be conclusively established; however, mutational and structural studies have offered different mechanistic proposals. Two studies in recent years have provided the NblA protein structures from three different organisms. The first structure of NblA to be solved was from *Anabaena* sp. PCC 7120 and revealed an open four helical bundle formed by the dimerization of two monomers with a helix-loop-helix motif (Bienert et al. 2006). The subsequent crystal structures of NblA from *T. vulcanus* and *Synechococcus elongatus* sp. PCC 7942 (Dines et al. 2008) demonstrated that whilst the sequence homology between different species is low the structural motif is similar (RMS of 1.0–1.7 Å). These structures also enabled the development of hypothesis as to possible mechanisms by which NblA might interact with the phycobilisome and perform disassembly (Dines et al. 2008; Bienert et al. 2006; Karradt et al. 2008). *In vitro* binding assays of glutathione S-transferase-tagged NblA from *Anabaena sp.* PCC 7120 demonstrated that NblA binds PC and PE. Mutation experiments on NblA originally proposed that this association in *Anabaena sp.* PCC 7120 occurred through interaction with the C- terminus of the NblA protein; however, a more generally applicable model was later proposed. This alternative model takes into account the fact that NblA proteins have low homology (<30 %) in contrast to their PBP binding partners (80–90 %). Indeed an NblA protein from one species can compliment function in a deletion strain from another species despite this relatively low sequence homology. Given the similarity between the three NblA structures it appears that the structural motif is important for function and yet an analysis of electrostatics showed variance between both NblA and PBP proteins that could not support a consistent model. The nature of the NblA helix-loop-helix motif is reminiscent of the PBPs themselves

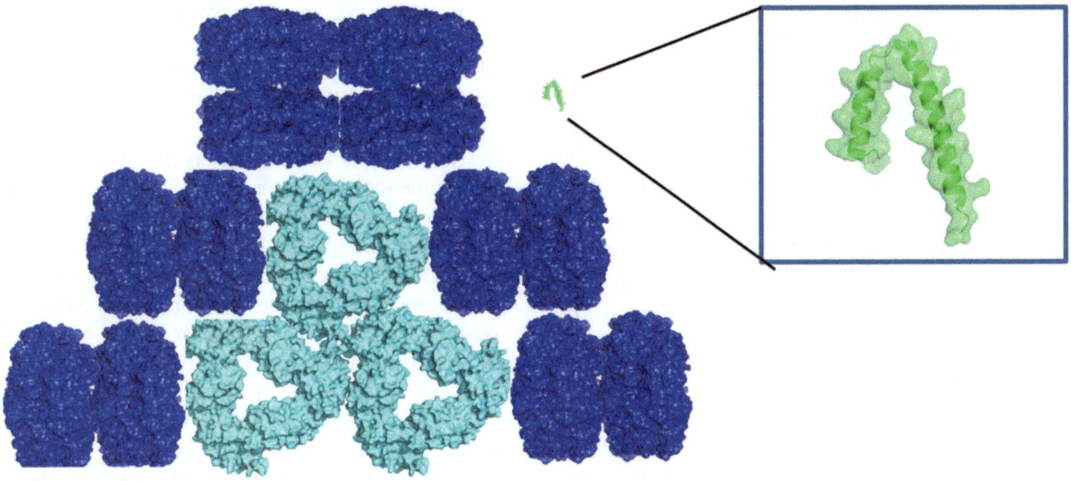

Fig. 4.11. The PBS complex is disassembled by interaction with the NblA protein. The relative sizes of the NblA protein (*green*, and shown enlarged in the *inset*) and a model of the PBS complex (*blue*) are shown.

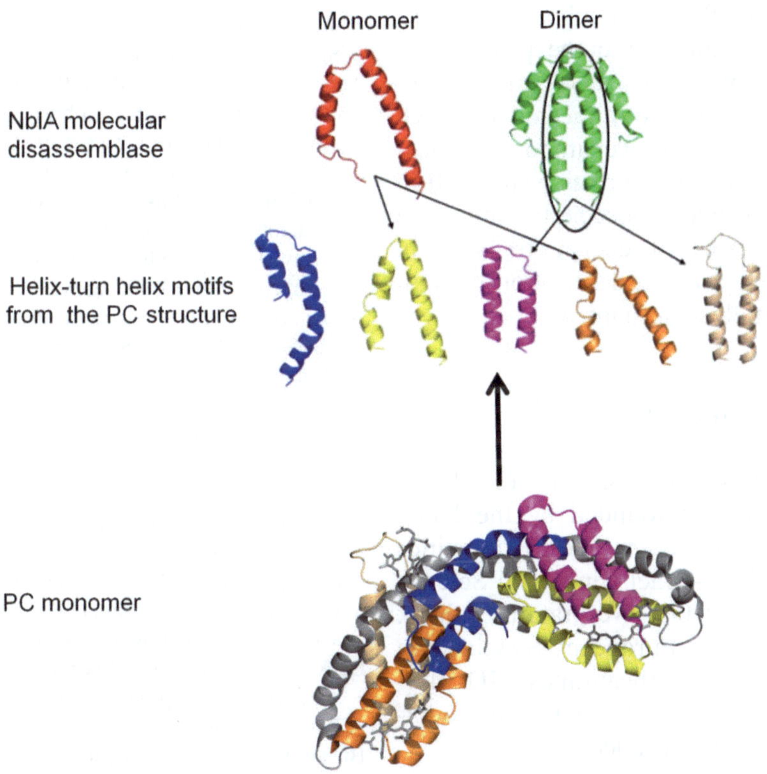

Fig. 4.12. Source of NblA protein mimicry of phycobiliproteins lies in the presence of similar helix-turn helix motifs. Each motif of the PC subunit is shown in a different color on both the monomeric structure (*bottom*) and also shown carved out of the structure with corresponding colors (*middle*). The NblA is shown at the *top* in both monomeric (*red*) or dimer (*green*) form. *Thin arrows* show elements of the PC subunit with angles similar to those found in the two NblA assemblies.

which are in essence bundles of helix-loop-helix motifs (Fig. 4.12). Comparison of these motifs with either the NblA monomer or the non-covalently bonded helix-loop-helix formed by dimerization, clearly demonstrates pairs (NblA and PBP) with similarity in terms of helices' length and separation angle. Based on this observation it has been suggested that a clever trick of structural mimicry may be the key to the NblA disassembly action. The first actions of disassembly may be actively coupled to proteolysis by specific proteases as has been suggested (Karradt et al. 2008), or by the battery of less-specific proteases in the cell.

The phycobilisome, enchanting researchers with its brilliant colors, symmetry and molecular enormity is slowly being pieced together by studies on both the individual components and the larger subcomplexes, as well as investigations into the proteins by which assembly and disassembly are afforded. A continued multifaceted approach will simultaneously answer important questions on the macrostructure as well as fill in fine details, especially those pertaining to the exact mechanisms of energy transfer, in order to give a detailed description of light harvesting's molecular giant.

Acknowledgements

This work was supported by the Israel Science Foundation founded by the Israel Academy of Sciences and Humanities (1045/06) and US-Israel Binational Science Foundation (2009406). We gratefully thank the staff of the European Synchrotron Radiation Facility (beamlines ID-14-1, ID23-1) for provision of synchrotron radiation facilities and assistance.

References

Adir N (2005) Elucidation of the molecular structures of components of the phycobilisome: reconstructing a giant. Photosynth Res 85:15–32

Adir N (2008) Structure of the phycobilisome antennae in cyanobacteria and red algae. In: Fromme P (ed) Photosynthetic protein complexes: a structural approach. WILEY-VCH Verlag GmbH & Co. KGaA, Weinheim, pp 243–274

Adir N, Lerner N (2003) The crystal structure of a novel unmethylated form of C-phycocyanin, a possible connector between cores and rods in phycobilisomes. J Biol Chem 278:25926–25932

Adir N, Dobrovetsky Y, Lerner N (2001) Structure of C-phycocyanin from the thermophilic cyanobacterium *Synechococcus vulcanus* at 2.5 A: structural implications for thermal stability in phycobilisome assembly. J Mol Biol 313:71–81

Adir N, Vainer R, Lerner N (2002) Refined structure of C-phycocyanin from the cyanobacterium Synechococcus vulcanus at 1.6 A: insights into the role of solvent molecules in thermal stability and co-factor structure. Biochim Biophys Acta 1556:168–174

Adir N, Dines M, Klartag M, McGregor A, Melamed-Frank M (2006) Assembly and disassembly of phycobilisomes. In: Shively JM (ed) Microbiology monographs: inclusions in prokaryotes, vol 2. Springer, Berlin, pp 47–77

Anderson LK, Toole CM (1998) A model for early events in the assembly pathway of cyanobacterial phycobilisomes. Mol Microbiol 30:467–474

Arciero DM, Bryant DA, Glazer AN (1988) In vitro attachment of bilins to apophycocyanin. I. Specific covalent adduct formation at cysteinyl residues involved in phycocyanobilin binding in C-phycocyanin. J Biol Chem 263:18343–18349

Arteni AA, Liu LN, Aartsma TJ, Zhang YZ, Zhou BC, Boekema EJ (2008) Structure and organization of phycobilisomes on membranes of the red alga *Porphyridium cruentum*. Photosynth Res 95:169–174

Arteni AA, Ajlani G, Boekema EJ (2009) Structural organisation of phycobilisomes from *Synechocystis* sp. strain PCC 6803 and their interaction with the membrane. Biochim Biophys Acta 1787:272–279

Ashkenazy H, Erez E, Martz E, Pupko T, Ben-Tal N (2010) ConSurf 2010: calculating evolutionary conservation in sequence and structure of proteins and nucleic acids. Nucleic Acids Res 38(Suppl):W529–W533

Beck WF, Sauer K (1992) Energy-transfer and exciton-state relaxation processes in allophycocyanin. J Phys Chem 96:4658–4666

Bienert R, Baier K, Volkmer R, Lockau W, Heinemann U (2006) Crystal structure of NblA from *Anabaena* sp. PCC 7120, a small protein playing a key role in phycobilisome degradation. J Biol Chem 281:5216–5223

Biswas A, Vasquez YM, Dragomani TM, Kronfel ML, Williams SR, Alvey RM, Bryant DA, Schluchter WM (2010) Biosynthesis of cyanobacterial phycobiliproteins in *Escherichia coli*: chromophorylation

efficiency and specificity of all bilin lyases from *Synechococcus* sp. strain PCC 7002. Appl Environ Microbiol 76:2729–2739
Blankenship RE, Olson JM, Miller M (1995) Antenna complexes from green photosynthetic bacteria. In: Blankenship RE, Madigan MT, Bauer CE (eds) Anoxygenic photosynthetic bacteria. Kluwer Academic Publishers, Dordrecht, pp 399–435
Bryant DA, Cohen-Bazire G (1981) Effects of chromatic illumination on cyanobacterial phycobilisomes. Evidence for the specific induction of a second pair of phycocyanin subunits in *Pseudanabaena* 7409 grown in red light. Eur J Biochem 119:415–424
Bryant DA, Glazer AN, Eiserling FA (1976) Characterization and structural properties of the major biliproteins of *Anabaena* sp. Arch Microbiol 110:61–75
Capuano V, Braux AS, Tandeau de Marsac N, Houmard J (1991) The "anchor polypeptide" of cyanobacterial phycobilisomes. Molecular characterization of the *Synechococcus* sp. PCC 6301 apce gene. J Biol Chem 266:7239–7247
Cogdell RJ, Gardiner AT, Roszak AW, Law CJ, Southall J, Isaacs NW (2004) Rings, ellipses and horseshoes: how purple bacteria harvest solar energy. Photosynth Res 81:207–214
Collier JL, Grossman AR (1992) Chlorosis induced by nutrient deprivation in *Synechococcus* sp. strain PCC 7942: not all bleaching is the same. J Bacteriol 174:4718–4726
Collier JL, Grossman AR (1994) A small polypeptide triggers complete degradation of light-harvesting phycobiliproteins in nutrient-deprived cyanobacteria. EMBO J 13:1039–1047
David L, Marx A, Adir N (2011) High-resolution crystal structures of trimeric and rod phycocyanin. J Mol Biol 405:201–213
Dines M, Sendersky E, David L, Schwarz R, Adir N (2008) Structural, functional, and mutational analysis of the NblA protein provides insight into possible modes of interaction with the phycobilisome. J Biol Chem 283:30330–30340
Ducret A, Sidler W, Wehrli E, Frank G, Zuber H (1996) Isolation, characterization and electron microscopy analysis of a hemidiscoidal phycobilisome type from the cyanobacterium *Anabaena* sp. PCC 7120. Eur J Biochem 236:1010–1024
Ducret A, Muller SA, Goldie KN, Hefti A, Sidler WA, Zuber H, Engel A (1998) Reconstitution, characterization and mass analysis of the pentacylindrical allophycocyanin core complex from the cyanobacterium *Anabaena* sp. PCC 7120. J Mol Biol 278:369–388
Edwards MR, Gantt E (1971) Phycobilisomes of the thermophilic blue-green alga *Synechococcus lividus*. J Cell Biol 50:896–900
Ficner R, Lobeck K, Schmidt G, Huber R (1992) Isolation, crystallization, crystal structure analysis and refinement of B-phycoerythrin from the red alga *Porphyridium sordidum* at 2.2 A resolution. J Mol Biol 228:935–950
Gantt E, Conti SF (1966a) Granules associated with the chloroplast lamellae of *Porphyridium cruentum*. J Cell Biol 29:423–434
Gantt E, Conti SF (1966b) Phycobiliprotein localization in algae. Brookhaven Symp Biol 19:393–405
Gantt E, Lipschultz CA (1972) Phycobilisomes of *Porphyridium cruentum*. I. Isolation. J Cell Biol 54:313–324
Ge B, Sun H, Feng Y, Yang J, Qin S (2009) Functional biosynthesis of an allophycocyan beta subunit in *Escherichia coli*. J Biosci Bioeng 107:246–249
Glauser M, Bryant DA, Frank G, Wehrli E, Rusconi SS, Sidler W, Zuber H (1992) Phycobilisome structure in the cyanobacteria *Mastigocladus laminosus* and *Anabaena* sp. PCC 7120. Eur J Biochem 205:907–915
Glazer AN (1989) Light guides. Directional energy transfer in a photosynthetic antenna. J Biol Chem 264:1–4
Glazer AN, Bryant DA (1975) Allophycocyanin B (lambdamax 671, 618 nm): a new cyanobacterial phycobiliprotein. Arch Microbiol 104:15–22
Grossman AR, Schaefer MR, Chiang GG, Collier JL (1993) The phycobilisome, a light-harvesting complex responsive to environmental conditions. Microbiol Rev 57:725–749
Grossman AR, Bhaya D, Apt KE, Kehoe DM (1995) Light-harvesting complexes in oxygenic photosynthesis: diversity, control, and evolution. Annu Rev Genet 29:231–288
Inoue N, Emi T, Yamane Y, Kashino Y, Koike H, Satoh K (2000) Effects of high-temperature treatments on a thermophilic cyanobacterium *Synechococcus vulcanus*. Plant Cell Physiol 41:515–522
Karradt A, Sobanski J, Mattow J, Lockau W, Baier K (2008) NblA, a key protein of phycobilisome degradation, interacts with ClpC, a HSP100 chaperone partner of a cyanobacterial Clp protease. J Biol Chem 283:32394–32403
Klotz AV, Leary JA, Glazer AN (1986) Post-translational methylation of asparaginyl residues. Identification of beta-71 gamma-N-methylasparagine in allophycocyanin. J Biol Chem 261:15891–15894
Liu LN, Chen XL, Zhang YZ, Zhou BC (2005) Characterization, structure and function of linker polypeptides in phycobilisomes of cyanobacteria and red algae: an overview. Biochim Biophys Acta 1708:133–142
Liu LN, Aartsma TJ, Thomas JC, Lamers GE, Zhou BC, Zhang YZ (2008) Watching the native supramo-

lecular architecture of photosynthetic membrane in red algae: topography of phycobilisomes and their crowding, diverse distribution patterns. J Biol Chem 283:34946–34953

Lundell DJ, Yamanaka G, Glazer AN (1981) A terminal energy acceptor of the phycobilisome: the 75,000-dalton polypeptide of Synechococcus 6301 phycobilisomes–a new biliprotein. J Cell Biol 91:315–319

MacColl R (1983) Stability of allophycocyanin's quaternary structure. Arch Biochem Biophys 223:24–32

MacColl R (1998) Cyanobacterial phycobilisomes. J Struct Biol 124:311–334

MacColl R (2004) Allophycocyanin and energy transfer. Biochim Biophys Acta 1657:73–81

MacColl R, Eisele LE, Menikh A (2003) Allophycocyanin: trimers, monomers, subunits, and homodimers. Biopolymers 72:352–365

McGregor A, Klartag M, David L, Adir N (2008) Allophycocyanin trimer stability and functionality are primarily due to polar enhanced hydrophobicity of the phycocyanobilin binding pocket. J Mol Biol 384:406–421

Neilson JA, Durnford DG (2010) Evolutionary distribution of light-harvesting complex-like proteins in photosynthetic eukaryotes. Genome 53:68–78

Reuter W, Wiegand G, Huber R, Than ME (1999) Structural analysis at 2.2 A of orthorhombic crystals presents the asymmetry of the allophycocyanin-linker complex, AP.LC7.8, from phycobilisomes of *Mastigocladus laminosus*. Proc Natl Acad Sci U S A 96:1363–1368

Ritter S, Hiller RG, Wrench PM, Welte W, Diederichs K (1999) Crystal structure of a phycourobilin-containing phycoerythrin at 1.90-A resolution. J Struct Biol 126:86–97

Sato T, Minagawa S, Kojima E, Okamoto N, Nakamoto H (2010) HtpG, the prokaryotic homologue of Hsp90, stabilizes a phycobilisome protein in the cyanobacterium *Synechococcus elongatus* PCC 7942. Mol Microbiol 76:576–589

Sauer K, Scheer H (1988) Exitation transfer in C-phycocyanin. Förster transfer rate and exciton calculations based on new crystal structure data for C-phycocyanins from *Agmenellum quadruplaticum* and *Mastigocladus laminosus*. Biochim Biophys Acta 936:157–170

Schirmer T, Huber R, Schneider M, Bode W, Miller M, Hackert ML (1986) Crystal structure analysis and refinement at 2.5 A of hexameric C-phycocyanin from the cyanobacterium *Agmenellum quadruplicatum*. The molecular model and its implications for light-harvesting. J Mol Biol 188:651–676

Swanson RV, Glazer AN (1990) Phycobiliprotein methylation. Effect of the gamma-N-methylasparagine residue on energy transfer in phycocyanin and the phycobilisome. J Mol Biol 214:787–796

Tandeau de Marsac N, Cohen-Bazire G (1977) Molecular composition of cyanobacterial phycobilisomes. Proc Natl Acad Sci U S A 74:1635–1639

Teale FW, Dale RE (1970) Isolation and spectral characterization of phycobiliproteins. Biochem J 116:161–169

Tooley AJ, Cai YA, Glazer AN (2001) Biosynthesis of a fluorescent cyanobacterial C-phycocyanin holo-alpha subunit in a heterologous host. Proc Natl Acad Sci U S A 98:10560–10565

Yamanaka G, Lundell DJ, Glazer AN (1982) Molecular architecture of a light-harvesting antenna. Isolation and characterization of phycobilisome subassembly particles. J Biol Chem 257:4077–4086

Yi ZW, Huang H, Kuang TY, Sui SF (2005) Three-dimensional architecture of phycobilisomes from *Nostoc flagelliforme* revealed by single particle electron microscopy. FEBS Lett 579:3569–3573

Chapter 5

Chlorosomes: Structure, Function and Assembly

Jakub Pšenčík*
Faculty of Mathematics and Physics, Charles University in Prague, 121 16 Prague 2, Czech Republic

Sarah J. Butcher
Institute of Biotechnology, University of Helsinki, 00014 Helsinki, Finland

and

Roman Tuma
The Astbury Centre for Structural Molecular Biology, University of Leeds, Leeds, LS2 9JT, UK

*Author for correspondence, e-mail: psencik@karlov.mff.cuni.cz

M.F. Hohmann-Marriott (ed.), The Structural Basis of Biological Energy Generation, Advances in Photosynthesis and Respiration 39, pp. 77–109, DOI 10.1007/978-94-017-8742-0_5,

Summary

Chlorosomes are light-harvesting complexes found in photosynthetic bacteria belonging to three diverse phyla: *Chlorobi*, *Chloroflexi* and *Acidobacteria*. They are composed of bacteriochlorophylls with minor contributions from proteins, lipids, carotenoids and quinones. Proteins are confined to the surface of the chlorosome while most bacteriochlorophyll molecules are found within the interior where they assemble into aggregates. These aggregates consist of lamellar structures, in which bacteriochlorophylls form curved layers while hydrophobic esterifying alcohols of bacteriochlorophylls from adjacent layers interdigitate and hold the system together. Such an arrangement supports strong excitonic coupling between the pigments within a layer and enables efficient excitation energy transfer. This chapter surveys general features of the chlorosome, including structure, energy transfer, photoprotective mechanisms and assembly.

I. Introduction

Chlorosomes were discovered in 1964 and described as oblong bodies attached to the inner side of the cytoplasmic membrane in thin sections of cells from *Chlorobi* species (Fig. 5.1) (Cohen-Bazire et al. 1964). Subsequently, chlorosomes were found in three bacterial phyla: most of the known species of *Chlorobi* (all known members of the family *Chlorobiaceae*, green sulfur bacteria), certain *Chloroflexi* species (the majority of the known members of the class *Chloroflexi* and "*Candidatus* Chlorothrix halophila" (Klappenbach and Pierson, 2004), filamentous anoxygenic phototrophs, formerly known as green non-sulfur bacteria), and one bacterium belonging to *Acidobacteria* (Bryant et al. 2007). While chlorosomes from all these bacteria are rather similar, the rest of their photosynthetic apparatus exhibits substantial differences, which suggests that a horizontal genetic transfer may have been responsible for spreading chlorosomes among these diverse phyla (Frigaard and Bryant 2004).

A unique property of chlorosomes is that their main pigments, bacteriochlorophylls (BChl) *c*, *d* or *e*, are organized in the form of an aggregate, in contrast to other photosynthetic light-harvesting complexes where proteins maintain the distances and mutual orientations between pigments. The aggregates inside a chlorosome are composed of many thousands of tightly associated BChl molecules making the chlorosome the largest light-harvesting complex known to date. The aggregation also leads to strong excitonic coupling between the BChls which provides the basis for high light-harvesting efficiency. Consequently, some of the chlorosome-possessing bacteria can grow under extremely low light conditions (Overmann et al. 1992; Beatty et al. 2005) and chlorosomes are considered the most efficient antenna found in nature.

The first full genomic sequence of a chlorosome-containing bacterium was obtained from the *Chlorobi* species *Chlorobaculum* (*Cba.*) *tepidum* (Eisen et al. 2002), and it quickly became a model organism and is used as such throughout this review. This organism was originally known as *Chlorobium* (*Chl.*) *tepidum*, but it has been renamed on the basis of new phylogenetic classification that employed 16S rRNA and Fenna–Matthews–Olson (FMO) complex gene sequences (Imhoff 2003; Imhoff and Thiel 2010). *Chloroflexus* (*Cfl.*) *aurantiacus* is the most intensively studied member of the *Chloroflexi*.

This chapter reviews the structure, function and assembly of chlorosomes with emphasis on the results obtained since the last review

Abbreviations: BChl – Bacteriochlorophyll; Cba – Chlorobaculum; CD – Circular dichroism; Cfl – Chloroflexus; Chl. – Chlorobium

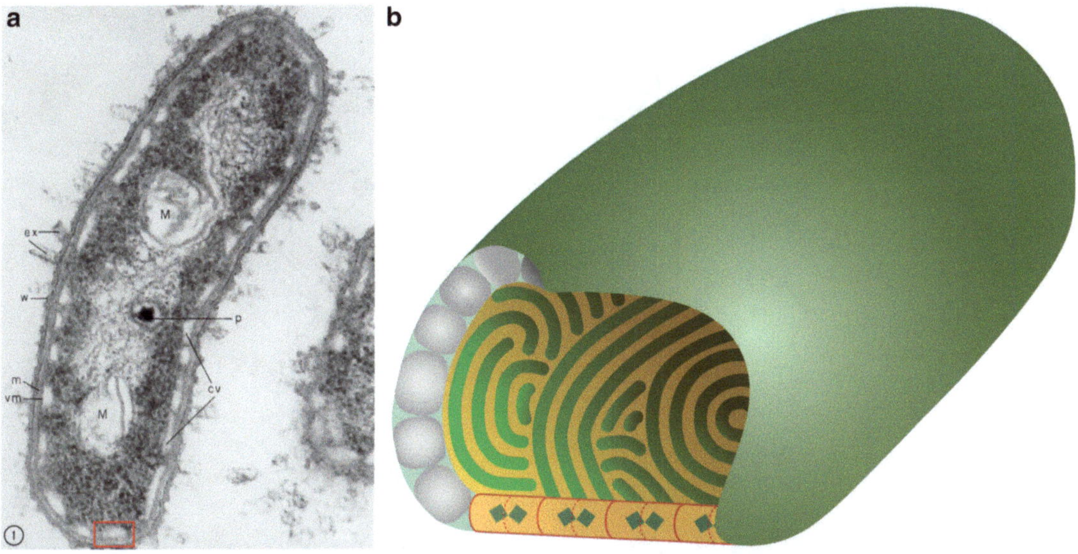

Fig. 5.1. (**a**) Longitudinal section of the cell *Chlorobium thiosulfatophilum* T, which should correspond to *Chlorobaculum thiosulfatiphilum* according to the current nomenclature (Imhoff and Thiel 2010), showing chlorosomes (white bodies labeled as 'cv', chlorobium vesicles; one of the chlorosomes is highlighted by the *red rectangle*) attached to the cytoplasmic membrane. The figure is from the original paper in which chlorosomes were first described (Cohen-Bazire et al. 1964, reproduced with permission from Rockefeller University Press). (**b**) Schematic diagram of the shape and interior of a typical chlorosome. The lamellar layers (*green*) consist of BChl aggregates and are held together by hydrophobic interactions between interdigitated esterifying alcohols of BChls (see also Fig. 5.8) forming the hydrophobic space (*orange*) between the chlorin layers which is also occupied by carotenoids and quinones. The baseplate at the *bottom* of the chlorosome is shown as made of CsmA protein dimers, each containing one BChl *a* molecule (*blue-green rectangle*), the orange background of the baseplate alludes to the presence of carotenoids in the baseplate. The envelope is formed mainly by proteins (the *gray particles*), while lipids fill the space between the proteins (*cyan*).

in this series was published (Blankenship and Matsuura 2003). During the intervening period several additional reviews concerning chlorosomes have been published (Frigaard and Bryant 2004, 2006; Oostergetel et al. 2010). Related topics such as self-assembling aggregates of chlorosomal BChls and their synthetic analogues have been reviewed elsewhere (Balaban et al. 2005; Miyatake and Tamiaki 2005, 2010).

II. Composition

A chlorosome contains pigments, proteins and lipids. The dry weight of a typical chlorosome encompasses 50 % BChl *c*, *d* or *e*, 30 % protein and about 10 % lipids, the rest being carotenoids, quinones and BChl *a* (Blankenship and Matsuura 2003). In terms of molar ratio with respect to BChl *c* in *Cba. tepidum* chlorosomes this corresponds to approximately 0.1 carotenoids, 0.1 quinones, 0.1 lipids, 0.03 protein molecules and 0.01 BChl *a* (Frigaard and Bryant 2006).

A. Bacteriochlorophylls

The main constituent of the chlorosome is BChl *c*, *d* or *e* (Fig. 5.2), which is the principal light-harvesting pigment (for a review see Scheer 2006). Using direct methods, it has been estimated that a single chlorosome of *Cba. tepidum* contains between 100,000 and 250,000 BChl *c* molecules (Montano et al. 2003a; Saga et al. 2007). Another estimate of BChl *c* number per chlorosome can be obtained from the volume of a typical *Cba. tepidum* chlorosome (an ellipsoid $200 \times 50 \times 25$ nm) and density (approx.

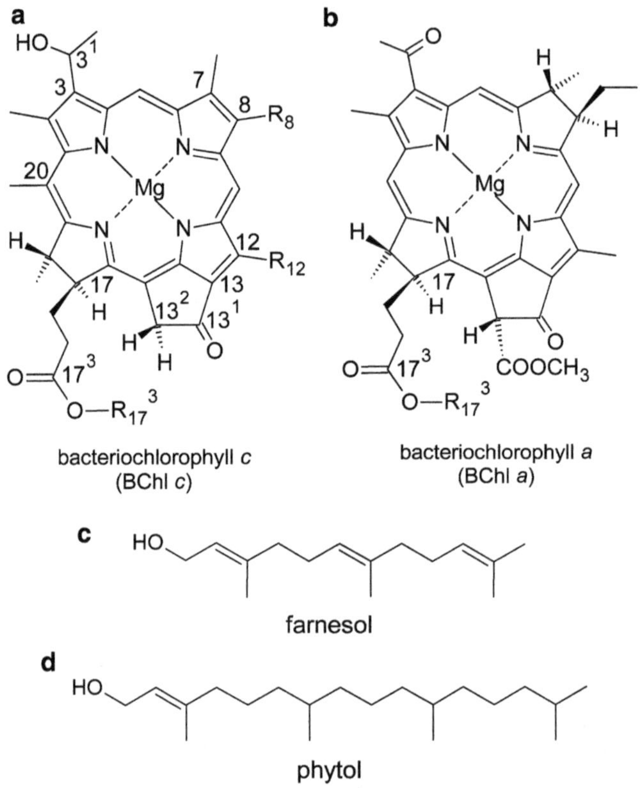

Fig. 5.2. (**a**) Structure of bacteriochlorophyll (BChl) *c*. BChl *d* differs from BChl *c* by lacking the methyl group at C_{20}, BChl *e* has an aldehyde group at C_7 instead of methyl. R_8 substituents are mainly ethyl, propyl, or isobutyl while R_{12} can be methyl or ethyl. $R_{17}{}^3$ stands for the alkyl group of the esterifying alcohol. BChl *c* in chlorosomes from *Cba. tepidum* is esterified mainly with farnesol. (**b**) Structure of BChl *a*. $R_{17}{}^3$ is phytyl. (**c**, **d**) Structure of farnesol and phytol.

1.16 g/cm^3 (Psencik et al. 2004)). Such a chlorosome would contain about 100,000 BChl *c* (Mr ~800) molecules, if it was filled by BChl *c* alone. Considering that about 30 % of the chlorosome dry weight is comprised of proteins (Blankenship and Matsuura 2003) (protein density ~1.3 g/ cm^3) a number between 60 and 80,000 BChl *c* molecules is obtained, depending on the amount of the remaining chlorosome components (e.g. carotenoids, quinones, lipids). The differences in these estimates most likely reflect batch to batch variation in chlorosome dimensions and various assumptions in the methods used to determine the volume and BChl fraction within the chlorosome. Nevertheless, all these estimates far exceed the amounts of pigments found in other photosynthetic light-harvesting complexes.

Unlike 'true' bacteriochlorophylls, such as BChl *a*, that are derived from bacteriochlorin, BChl *c*, *d* and *e* are based on chlorin, which is the skeleton of chlorophylls and differs from the bacteriochlorin by the presence of a double bond between C7 and C8 (Fig. 5.2). Historically, they were isolated from bacteria and got their trivial name prior to the determination of their chemical structure. BChl *c*, *d* and *e* differ from each other at the C20 position where BChl *c* and *e* have methyl, and at the C7 position where BChl *e* has an aldehyde group. The fourth possible combination, BChl *f* (aldehyde group at C7, hydrogen at C20) has not yet been found to occur naturally, but was recently synthesized and spectroscopically characterized (Tamiaki et al. 2011). Another interesting feature of BChl *c*, *d* and *e* in *Chlorobi* is that they are found as a mixture of several homologs differing in both the alkyl substituent at C8 and C12, and in the alcohol esterified at C17^3. The degree of alkylation at C8 and C12 increases at low-light conditions (Borrego and Garcia-Gil 1995; Borrego et al. 1999a). BChl *c* homologs differing in the substituent at C8 and C12, and in the alcohol esterified at C17^3 were also identified in *Acidobacteria* member "*Candidatus* Chloracidobacterium (Cab.) thermophilum" (Costas et al. 2012). In contrast, *Chloroflexi* usually possess a single BChl *c* homolog containing ethyl at C8 and methyl at C12, although different types of esterifying alcohols are attached at C17^3.

BChl *c*, *d* and *e* have several unique properties. They are found exclusively in chlorosomes and, most significantly, they have the propensity to self-assemble into large aggregates. Aggregation is mediated by their unique structural features, namely the presence of a hydroxy group at asymmetric C3^1 and the absence of a bulky methoxycarbonyl group at C13^2 (Fig. 5.2). The prevailing stereochemistry at C3^1 changes from R to S with increasing size of the C8 substituent (Scheer 2006).

The key interaction for aggregate formation is the intermolecular coordination of the central Mg^{2+} ion of one BChl molecule to the hydroxy group at C3^1 of a second BChl. In addition, a hydrogen bond is formed between the hydroxy group at C3^1 of the second BChl and the keto group at C13^1 of a third BChl molecule (Hildebrandt et al. 1994). The importance of the latter interaction has been recently questioned and it was instead proposed that the C13^1 keto group of the second BChl is weakly coordinated to the central Mg ion of yet another BChl within the aggregate (Jochum et al. 2008). Nevertheless, the prevailing opinion is that the hydrogen bond is formed in the aggregate and it seems that its main role is to stabilize parallel orientation of the Q_y transition dipole moments, which is required for strong exciton coupling (Sect. IV.A), while an important role in stabilization of the aggregate is played by stacking (π–π) interactions (Alster et al. 2012). Altogether, these interactions are in principle compatible with many possible molecular arrangements, and some of them are discussed in the structural context in Sect. III.C.3.

In addition to BChl *c*, *d* and *e*, chlorosomes from all species also contain BChl *a* (Fig. 5.2), which does not form aggregates. BChl *a* is found in a pigment-protein complex within the baseplate (Sect. III.A). BChl *a* represents typically 1 % of the total BChl content in *Chlorobi*, and about 5 % in *Chloroflexi*.

B. Carotenoids

Chlorosomes also contain a significant, albeit variable amount of carotenoids. They partition into two different pools: (1) carotenoids in the chlorosome interior (Sect. III.C) and (2) carotenoids associated with BChl *a* in the baseplate (Sect. III.A). Carotenoids in chlorosomes, as in other photosynthetic complexes, have several roles: photoprotective, light-harvesting and structural (for a review see Polivka and Sundstrom 2004). The amount of carotenoids depends on the bacterial species, growth phase, light and temperature conditions (Oelze and Golecki 1995). For example in *Chlorobi*, the stoichiometry between BChl and carotenoids can vary between approximately 2:1 and 20:1 (Borrego and Garcia-Gil 1995; Borrego et al. 1999a). Although some of the analyses were done on whole cells, they are nevertheless relevant to chlorosomes since the majority of the carotenoids in *Chlorobi* are present in chlorosomes (Frigaard et al. 1997; Frigaard and Bryant 2006). Such a generalization does not apply to *Cfl. aurantiacus* which exhibits an increase of the carotenoid fraction under high-light conditions (Schmidt et al. 1980; Larsen et al. 1994) but most of the additional carotenoids are localized within structures known as wax oleosomes (Blankenship et al. 1995).

Most of the carotenoids in chlorosomes of *Chlorobi* and *Chloroflexi* are carotenes (i.e. carotenoids that do not contain oxygen). Recently, chlorosomes from acidobacterium Cab. thermophilum were shown to contain significant amounts of xanthophylls (carotenoids containing oxygen), namely the ketocarotenoids echinenone and canthaxanthin (Fig. 5.3). These ketocarotenoids were suggested to play a protective role in the aerobic conditions, which are natural to this bacterium (Costas et al. 2012). Chlorosomes from *Chlorobi* mainly possess carotenoids with one or two aryl rings, namely conjugated ϕ-end-group (1,2,5-trimethylphenyl) (Fig. 5.3), which are rare in other photosynthetic organisms. Green-colored species of *Chlorobi* (i.e. containing BChl *c* or *d*, e.g. *Cba. tepidum*), contain mainly the carotenoids chlorobactene and 1′,2′-dihydrochlorobactene (Takaichi and Oh-Oka 1999), both with one aryl ring. On the other hand, brown-colored species (containing BChl *e*) contain isorenieratene and β-isorenieratene, with two or one aryl ring, respectively (Imhoff 1995).

Chlorosomes of *Chloroflexi*, e.g. *Cfl. aurantiacus* contain mainly β- and γ-carotene (Fig. 5.3) and their derivatives, which instead of an ϕ-end-group, possess one or two β-rings (Halfen et al. 1972; Maresca et al. 2008). It is interesting to note that β-carotene is a precursor for isorenieratene, while γ-carotene leads to chlorobactene (Takaichi 1999) (Fig. 5.3). This poses a question as to why *Chlorobi* species take the carotenoid biosynthesis further and introduce the ϕ-ring? The introduction of the ϕ-end-group has no effect on the visible absorption spectrum of the carotenoid, because the conjugated system of the ϕ-ring is effectively decoupled from the polyene backbone and does not contribute to the overall conjugation length (Fuciman et al. 2010). Thus, the ϕ-ring does not extend the light-harvesting spectral coverage. This is exemplified by a *crtU* null mutant of *Cba. tepidum* that does not convert γ-carotene into chlorobactene but forms chlorosomes with very similar absorption spectra. In addition the shape and structure of the mutant chlorosomes are indistinguishable from those of the wild type (Ikonen et al. 2007), suggesting that the ϕ-ring is likewise not essential for chlorosome morphogenesis. However, chlorosomes from this mutant photo-degrade at a rate almost twice that of wild type (Kim et al. 2007) suggesting that the presence of the ϕ-ring enables more efficient photoprotection.

Both photoprotection of BChls by triplet-triplet quenching and energy transfer from the short lived S_2 state of carotenoids to BChls (see Sects. IV.C.2 and IV.B.1) require close proximity and a suitable orientation of the respective pigments. In other light-harvesting complexes this is achieved via association with a protein scaffold, which is absent from the chlorosome interior. The possession of the ϕ-ring by carotenoids from *Chlorobi* species may provide the necessary structural constraints to support electronic coupling between carotenoids and BChls through the π–π interaction between the planar conjugated system of the carotenoid ϕ-ring and that of the BChls' chlorin ring (Fuciman et al. 2010). The CH-π interactions of the β-ring of carotenoids with the conjugated system of BChl molecules (Wang et al. 2004) are likely to play a similar role in chlorosomes containing β-ring carotenoids such as those from the *Chloroflexi* species. However, CH-π interactions are expected to be weaker than π–π stacking. Thus, perhaps stronger and more efficient coupling via the ϕ-ring is required in *Chlorobi* for photoprotection and fast energy transfer. The fact that chlorobactene is more efficient in inducing self-assembly of BChl *c in vitro* than β-carotene indeed suggests that the ϕ-ring interacts more strongly with BChl *c* than the β-ring (Klinger et al. 2004; Alster et al. 2010). It remains to be seen how ketocarotenoids in *Acidobacteria* are interacting with BChls.

C. Quinones

Another group of compounds found in chlorosomes are isoprenoid quinones (for a recent review see Nowicka and Kruk 2010). These amphiphilic molecules consist of a polar head group and a hydrophobic isoprenoid side chain (Fig. 5.4). The polar head group is a conjugated (poly)cyclic dione

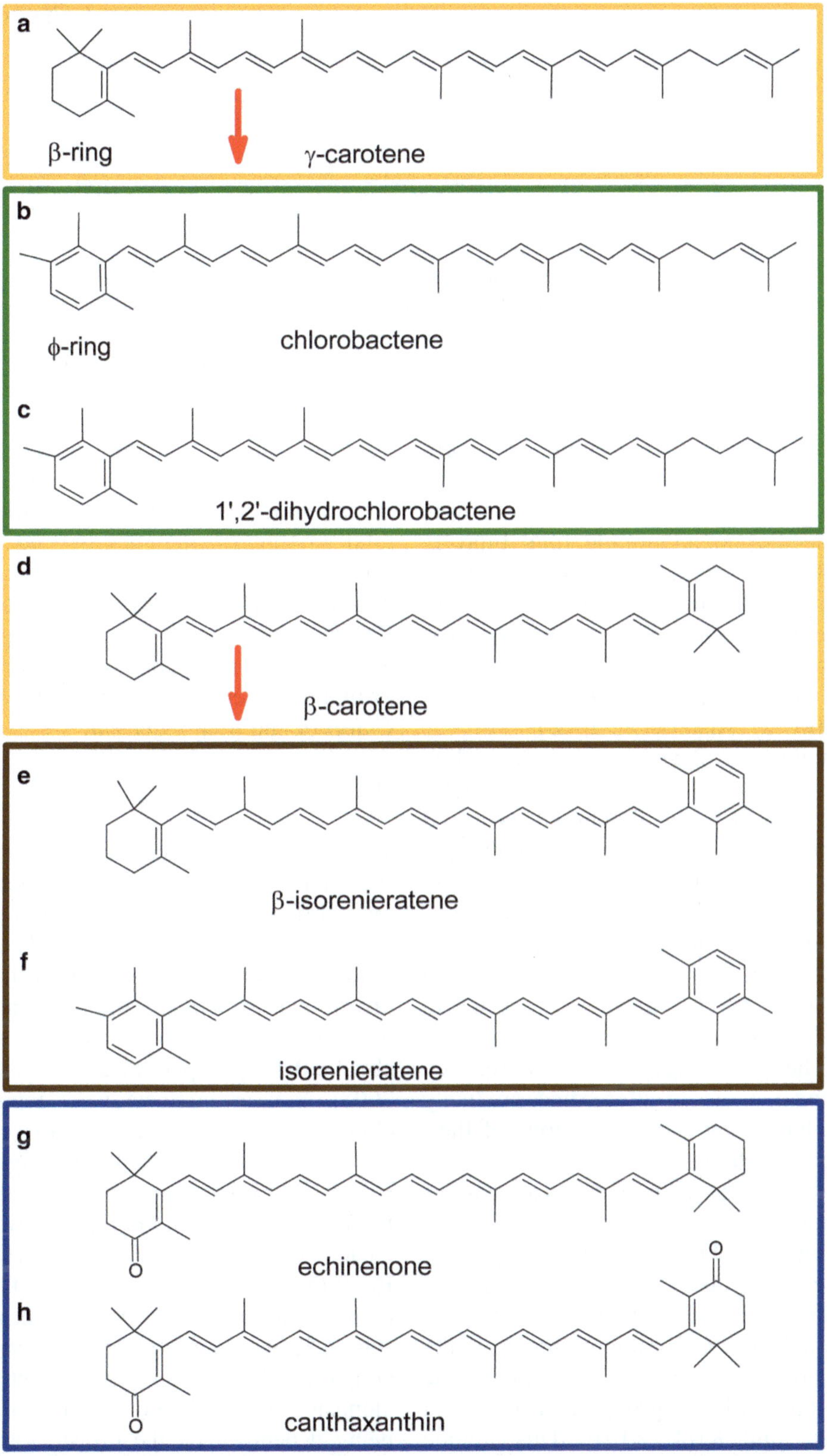

Fig. 5.3. Structures of the most abundant chlorosomal carotenoids. (**a**) γ-carotene, (**b**) chlorobactene, (**c**) 1′,2′-dihydrochlorobactene, (**d**) β-carotene, (**e**) β-isorenieratene, (**f**) isorenieratene, (**g**) echinenone, (**h**) canthaxanthin. The *arrows* indicate the order of synthesis in the biosynthetic pathway. The *boxes* group carotenoids found in chlorosomes from different bacteria: *Chloroflexi* (*orange*), green-colored *Chlorobi* (*green*), brown-colored *Chlorobi* (*brown*) and *Acidobacteria* (*blue*). A β-ring and ϕ-ring are delineated in *panel* (**a**) and (**b**), respectively.

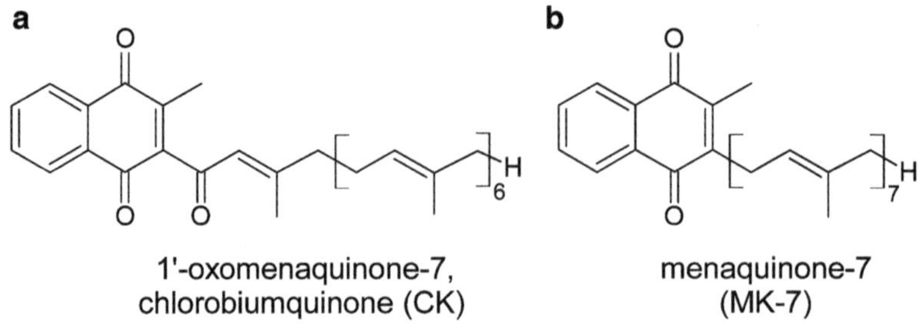

Fig. 5.4. Structures of the most abundant quinones in *Cba. tepidum* chlorosomes. (**a**) 1′-oxomenaquinone-7 (chlorobiumquinone) (**b**) menaquinone-7.

structure. The apolar isoprenoid side chain of varying length gives the molecules a lipid-soluble character and usually anchors the molecules in membrane lipid bilayers (Nowicka and Kruk 2010). In chlorosomes, it seems that the role of the side chain is to increase the hydrophobicity of the molecule and allow its incorporation into the hydrophobic space within the lamellar assembly (Sect. III.C.2) (Alster et al. 2008).

Quinones are known to play a role as electron transporters (Nowicka and Kruk 2010). The polar head is the functional part, and can undergo a two-step, reversible reduction leading to quinol. Isoprenoid quinones play a principal role in the redox-dependent protective excitation quenching in chlorosomes of *Chlorobi* species (Sect. IV.C.1) (Frigaard et al. 1997), which was recently observed also for an *Acidobacteria* member (Costas et al. 2011). The close-range association between BChls and quinones in chlorosomes are likely to be achieved by π–π interactions of their conjugated systems.

In *Chlorobi*, the main quinones are chlorobiumquinone (1′-oxo-menaquinone-7) and menaquinone-7 (Fig. 5.4), which are present at a fraction of about 0.1 mol per 1 mol of BChl (Frigaard et al. 1997; Blankenship and Matsuura 2003). Chlorobiumquinone is the only known isoprenoid naphthoquinone containing a carbonyl group in its side chain (Nowicka and Kruk 2010). This unique structure may be important for effective excitation quenching. Dihydrogenated menaquinone-8 was detected in chlorosomes from acidobacterium Cab. thermophilum and proposed to act as an excitation quencher (Costas et al. 2012). *Chloroflexi* contain mostly menaquinone-10 in an amount similar to the total amount of quinones in *Cba. tepidum* (Frigaard et al. 1997). The function of quinones in *Chloroflexi* is currently unknown. Based on the arguments summarized in Sect. IV.C.1, they are not involved in the excitation quenching mechanism found in *Chlorobi* and *Acidobacteria*.

D. Lipids

The chlorosome lipid composition differs from that of other photosynthetic bacteria. The main difference is the prevalence of glycolipids at the expense of phospholipids and other lipid types. For chlorosomes from the model organism *Cba. tepidum* the fraction of polar lipids contains 60 % glycolipids, and only about 30 % phospholipids (Sorensen et al. 2008). Similar numbers were reported for chlorosomes from other species (for reviews see Blankenship et al. 1995; Frigaard and Bryant 2006). The main lipid originally found in *Chloroflexi* and *Chlorobi* species was monogalactosyl diacylglyceride (MGDG) (Fig. 5.5). Recently a rhamnose derivative of MGDG, rhamnosylgalactosyl diacylglyceride (RGDG) (Fig. 5.5), was identified as the prevalent glycolipid in *Cba. tepidum* (Sorensen et al. 2008) alongside with MGDG, the ratio of the two being dependent on the growth temperature (Mizoguchi et al. 2011). MGDG is also the main lipid found in chloroplast thylakoid membranes. It is a neutral lipid with a small polar headgroup, which does not form bilayers,

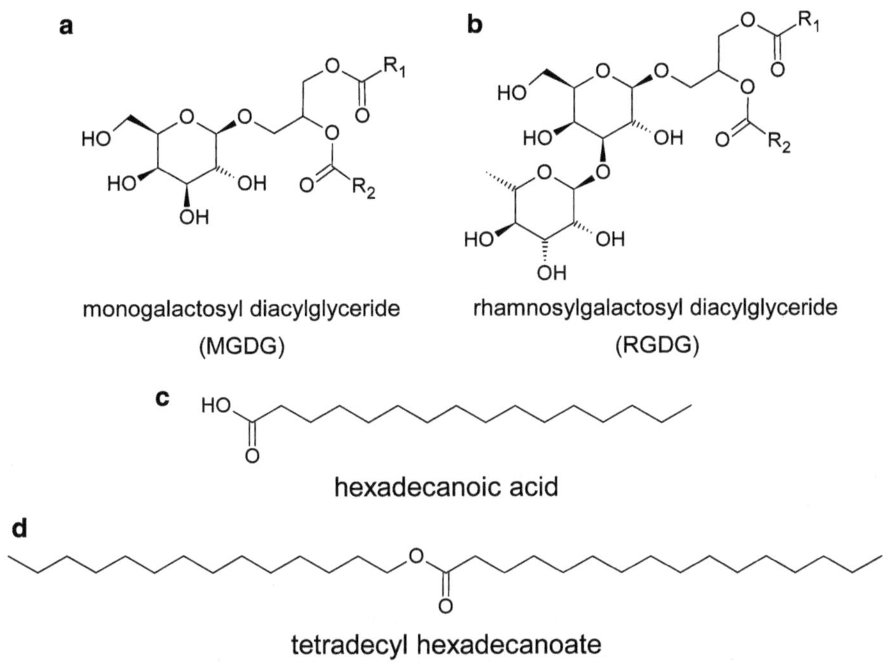

Fig. 5.5. Structure of the most abundant lipids in *Cba. tepidum* chlorosomes. (**a**) monogalactosyl diacylglyceride (MGDG) and (**b**) rhamnosylgalactosyl diacylglyceride (RGDG) as the representatives of polar lipids. Substituents R_1 and R_2 represent esterifying fatty acids. The major fatty acid found in *Cba. tepidum* chlorosomes is (**c**) hexadecanoic acid (Sorensen et al. 2008). (**d**) Tetradecyl hexadecanoate is the most abundant non-polar lipid (wax ester).

but an inverted hexagonal phase in water. It has been suggested that one of its functions in chloroplasts could be to facilitate optimal packing of large integral membrane proteins (Gounaris and Barber 1983). Perhaps it plays a similar role in the protein-rich chlorosome envelope (Sect. III.B). In addition to MGDG, *Cfl. aurantiacus* also contains digalactosyl diacylglyceride (DGDG) (Knudsen et al. 1982). The chlorosomes from acidobacterium Cab. thermophilum contain diacylglyceryl hydroxymethyltrimethyl-β-alanine (DGTA) as the main lipid (Costas et al. 2012).

Chlorosomes from thermophilic species (both from *Chloroflexi* and *Chlorobi*) also contain non-polar lipids (hydrophobic wax-esters), which are most probably located in the chlorosome interior and their amount exceeds that of the polar lipids (Blankenship et al. 1995; Sorensen et al. 2008). The most abundant non-polar lipid in *Cba. tepidum* is tetradecyl hexadecanoate (Fig. 5.5). The non-polar lipids were suggested to be involved in stabilization of pigment aggregates at elevated temperatures (Sorensen et al. 2008).

E. Proteins

Ten proteins, with molar masses ranging from 6.2 to 26 kDa (Persson et al. 2000), have been found in chlorosomes from *Cba. tepidum* (CsmA, CsmB, CsmC, CsmD, CsmE, CsmF, CsmH, CsmI, CsmJ, and CsmX) and can be clustered into four structural groups (Vassilieva et al. 2002; for a review see Frigaard and Bryant 2006). Four similar groups are found in *Cfl. aurantiacus* (Frigaard and Bryant 2006). In the first group are CsmA and CsmE. CsmA, and possibly also CsmE, binds BChl *a* via a conserved histidine residue (Frigaard and Bryant 2006; Zobova et al. 2011). CsmA, the most abundant chlorosomal protein (up to ~50 % of all protein mass in chlorosomes (Frigaard et al. 2004)), forms the baseplate (Sect. III.A). It is the only protein, which is essential for the formation of chlorosomes (Frigaard et al. 2004). However the genetic

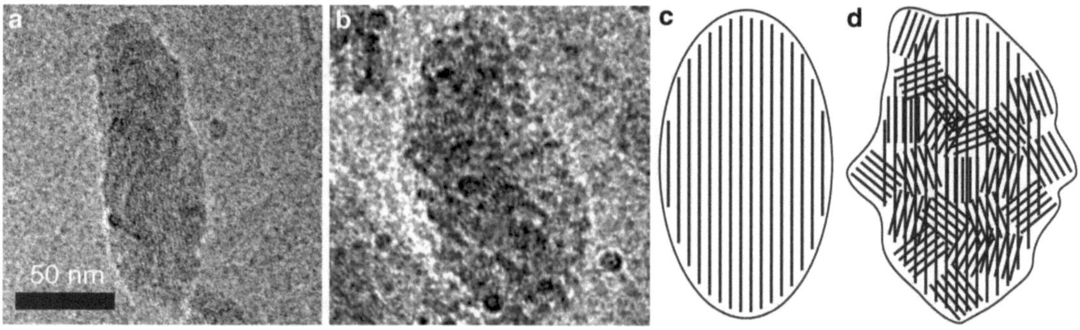

Fig. 5.6. Electron cryomicroscopy (cryo-EM) images of chlorosomes. (**a**) A smooth chlorosome from *Cba. tepidum* (Psencik et al. 2004) and (**b**) rough chlorosome from *Chl. phaeovibrioides*. It should be noted that chlorosomes from *Chl. phaeovibrioides* are usually significantly larger compared to that from *Cba. tepidum*. However, chlorosomes with similar sizes were selected to aid visual comparison of their internal organization. (**c**) Schematic representation of the lamellar system in smooth chlorosomes, where the prevailing orientation of the lamellae is parallel to the long axis of the chlorosome and (**d**) rough chlorosomes, where a part of the lamellar system forms domains with random orientation (Psencik et al. 2006).

deletion of all proteins from one structural group severely affects the function and morphology of the chlorosomes (Frigaard and Bryant 2006; Li and Bryant 2009). Crosslinking experiments have shown that CsmA forms dimers and larger oligomers (Bryant et al. 2002; Li et al. 2006). CsmA is the only chlorosomal protein for which structural information is available and is discussed in Sect. III.A. The other proteins are located in the chlorosome envelope (Sect. III.B).

CsmC, CsmD, CsmH, CsmI, CsmJ, and CsmX proteins were found to form homomultimers and interactions between different proteins have also been reported (Li et al. 2006). The second structural group encompasses CsmI, CsmJ and CsmX, which are iron-sulfur proteins with sequence similarity in their amino-terminal domains to [2Fe-2S] ferredoxins (Li et al. 2009). They form mixed multimers with midpoint electrochemical potentials varying between −350 and +90 mV, depending on their subunit composition (Frigaard and Bryant 2006). These proteins are probably involved in restoring energy transfer upon return to anaerobic conditions (Sect. IV.C.1) (Frigaard and Bryant 2006).

The remaining two groups (CsmB/CsmF and CsmC/CsmD) encompass proteins that affect the size and shape of the chlorosome and also the absorption maxima of the BChl aggregates (Li and Bryant 2009). The sequence of CsmH consists of two domains exhibiting similarities to CsmB/CsmF and CsmC/CsmD, respectively. In addition, two proteins, CsmS and CsmT, which exhibit no sequence similarity to other known chlorosome proteins, were found to be abundant in the envelope of *Acidobacteria* member Cab. thermophilum (Costas et al. 2011).

III. Structure

Chlorosomes form ellipsoid bodies with typical dimensions for *Cba. tepidum* of 150–200 × 50 × 25 nm (Fig. 5.1). However, shape and dimensions may vary considerably between species and with growing conditions. The chlorosomes from green-colored *Chlorobi* or *Chloroflexi* species (containing BChl *c* or *d*, e.g. *Cba. tepidum*) appear to have a smooth surface and an ellipsoidal shape while chlorosomes from the brown-colored *Chlorobi* (BChl *e* containing, e.g. *Chl. phaeovibrioides*) are usually larger, irregular and exhibit a rough surface (Fig. 5.6) (Martinez-Planells et al. 2002). As explained in Sect. III.C.1, the rough surface reflects the

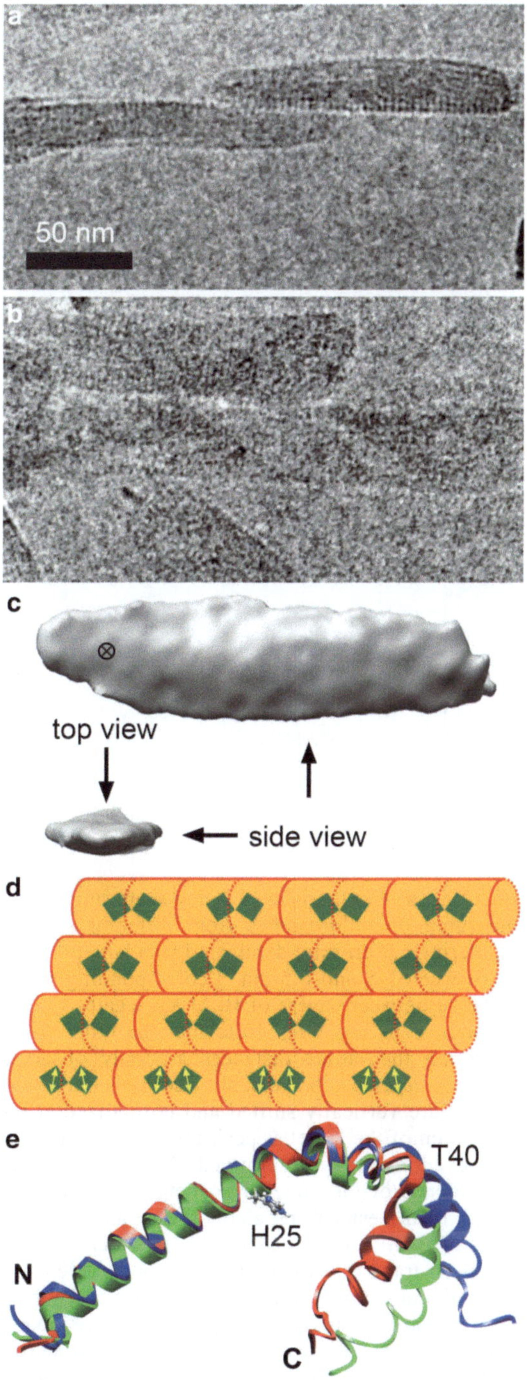

Fig. 5.7. (**a**) Baseplate as observed in cryo-EM images of chlorosomes from *Cfl. aurantiacus* in a side-view (the string of *dots* on one side of the chlorosome) and (**b**) *top-view* (perpendicular striation). (**c**) Tomography of a single chlorosome defining the *side-view* and *top-view* (Psencik et al. 2009, EMD-1642). (continued)

internal organization of pigment aggregates into domains. The overall architecture consists of the baseplate, envelope and interior (Fig. 5.1). Proteins are located only in the envelope and the baseplate while the interior is composed of pigment assemblies. The chlorosome attaches to the cytoplasmic membrane either directly (*Chloroflexi*) or via the FMO complex (*Chlorobi, Acidobacteria*).

A. Baseplate

The baseplate is located on one side of the chlorosome (for a recent review see Pedersen et al. 2010) and mediates excitation energy transfer towards the reaction centers located in the cytoplasmic membrane. The baseplate is built from multiple copies of CsmA protein. Each CsmA protein specifically coordinates at least one BChl *a* molecule and further associates with carotenoids (Fig. 5.1) (Montano et al. 2003b; Frigaard and Bryant 2006). In addition, it has been proposed that the ordered baseplate imposes long-range order within lamellar BChl aggregates (Ikonen et al. 2007; Arellano et al. 2008), presumably via nucleating or directing pigment assembly on its surface. Carotenoids have been shown to be important for baseplate

Fig. 5.7. (continued) (**d**) Schematic of the baseplate explaining cryo-EM images and X-ray diffraction. The baseplate is shown consisting of dimers in a row. The rows are observable in cryo-EM and the corresponding spacing in X-ray data. No spacing in the direction along the row is observed in cryo-EM or X-ray data presumably due to tight packing of helices in this direction. BChl *a* molecule is depicted by the *blue-green rectangle* and the presence of carotenoids by the *orange* background. Tentative orientation of the Q_y transition dipole moments of BChl *a* is shown by *yellow arrows*. (**e**) Structure of *Cba. tepidum* CsmA protein determined by NMR in organic solvent (Pedersen et al. 2008, PDB 2k37). Three representative conformations from the solution ensemble (*red, green, blue*) were aligned using the hydrophobic N-terminal helix (part embedded in the chlorosome) to illustrate the flexibility of the hydrophilic C-terminus (FMO binding), which pivots around Thr 40. His 25 coordinates BChl *a*.

morphogenesis (Foidl et al. 1997; Arellano et al. 2001; Ikonen et al. 2007).

The mature CsmA protein consists of 59 amino acids. It is produced by post-translational cleavage of the last 20 residues from a 79 amino acid long precursor during chlorosome morphogenesis (Pedersen et al. 2010). The atomic model of CsmA, determined by NMR in an organic solvent, revealed a kinked hydrophobic N-terminal helix (residues 2–40) connected to a shorter hydrophilic C-terminal helix (residues 40–49) (Fig. 5.7). In the chlorosome, the N-terminus points into the chlorosome interior and the C-terminus to the exterior, where it can interact with the membrane or FMO complex (Pedersen et al. 2008).

CsmA subunits are arranged into a two dimensional crystalline lattice, first seen in freeze-fracture electron microscopy (EM) images (Staehelin et al. 1978) and more recently directly imaged by cryo-EM (Fig. 5.7) (Psencik et al. 2009; Oostergetel et al. 2010). The lattice spacing of 3.2–3.3 nm (q=1.90–1.95 nm^{-1}) was deduced from X-ray diffraction (Fig. 5.8) (Ikonen et al. 2007) and analysis of electron micrographs (Psencik et al. 2009; Oostergetel et al. 2010). The baseplate lattice in *Cfl. aurantiacus* is approximately perpendicular to the long axis of the chlorosome (Fig. 5.7) (Staehelin et al. 1978; Psencik et al. 2009), while that in *Cba. tepidum* intersects the long axis at an ~40° angle (Oostergetel et al. 2010).

A model of the baseplate from *Cba. tepidum* was built using CsmA dimers as a basic unit (Pedersen et al. 2010). The model describes coordination of two BChl *a* at the dimer interface by histidine side chains within a conserved patch of residues (GHW). Carotenoids that are present in the baseplate are expected to be in close proximity to the BChl *a* pair in order to provide a protective function (quenching of the BChl *a* triplet states), which requires an overlap of molecular orbitals (Arellano et al. 2000a).

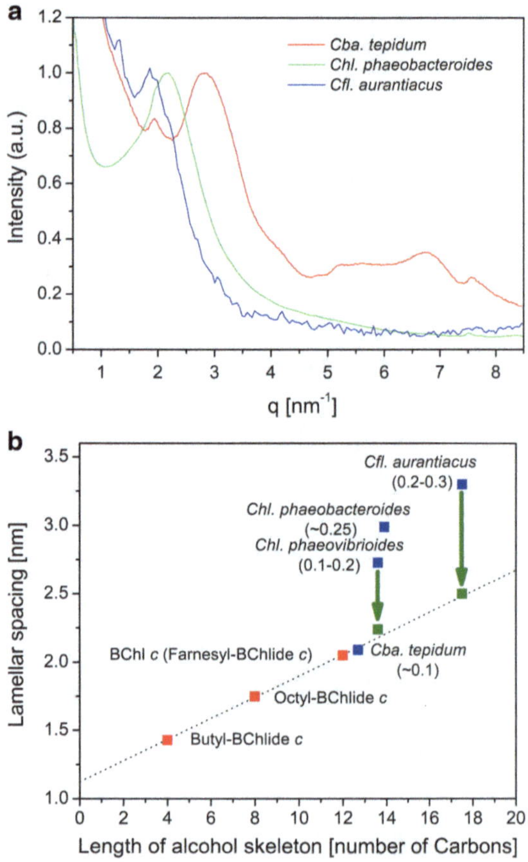

Fig. 5.8. (**a**) Comparison of X-ray scattering curves measured on chlorosomes from different species. Scattering patterns are dominated by a lamellar peak (q between 1.5 and 3.5 nm^{-1}) corresponding to the distance between layers of BChl aggregates (lattice constant c in Fig. 5.9). Note that the other peaks reflecting the lattice distances are only observed for *Cba. tepidum*, presumably due to better order. Curves are vertically shifted and normalized at the maximum of the lamellar peak for visualization. (**b**) The experimentally determined (X-ray diffraction) dependence of the lamellar spacing on the average length of the esterifying alcohol. Spacing in aggregates of BChl *c* and its analogues, in which farnesyl has been trans-esterified with two shorter alcohols, butanol and octanol, and assembled *in vitro* by hexane (*red*) are compared with spacing in chlorosomes from different species (*blue*) and after carotenoid removal by hexane (*green*). The numbers in parenthesis are carotenoid to BChl molar ratios determined for native chlorosomes (Psencik et al. 2006; Psencik et al. 2010; Psencik et al. 2013).

B. Envelope

The envelope is defined here as the surface layer of the chlorosome with the exception of the baseplate (Fig. 5.1). The envelope was first observed as a ~3 nm thick surface layer in freeze fracture EM (Staehelin et al. 1980) and recent cryo-EM images are consistent with this observation (Oostergetel et al. 2010). These high resolution EM images, however, have not revealed any regular structure within the envelope, ruling out the presence of a lipid bilayer and crystalline protein arrays. The envelope contains the polar lipids found in the chlorosome, with their polar heads oriented towards the cytoplasm (Chung and Bryant 1996). It has been assumed for a long time that these lipids form a monolayer around the chlorosome. However, recent estimates suggest that the amount of lipids is sufficient to cover only about 5 % of the surface, and therefore more than 90 % of the surface is most likely covered by proteins with lipids filling the remaining space (Fig. 5.1) (Sorensen et al. 2008).

C. Interior

The interior of the chlorosome is filled with self-assembled pigment aggregates mostly composed of BChl *c*, *d* or *e* with minor, but essential, contributions from carotenoids, quinones (Figs. 5.1 and 5.8), and non-polar lipids in the case of thermophilic species. The first structural models of the chlorosome interior were based on freeze fracture micrographs. It was proposed that the BChl aggregates were organized in tightly packed rods (resembling hexagonal phase) with a diameter of ~5 nm in *Cfl. aurantiacus* and ~10 nm in *Chlorobi* (Staehelin et al. 1978, 1980). However, recent higher resolution cryo-EM images and X-ray diffraction results revealed that the pigments form a lamellar phase (Psencik et al. 2004; for a recent review see Oostergetel et al. 2010), while the rod-like appearance in freeze-fracture was due to the intrinsic curvature of the lamellae (Oostergetel et al. 2007).

1. Long-Range Organization of Bacteriochlorophyll Aggregates

Cryo-EM images revealed a striation pattern often parallel to the long axis of the chlorosome with spacing ranging from ~2 nm for *Cba. tepidum* (Fig. 5.6) (Psencik et al. 2004; Oostergetel et al. 2007) to 2.6–4.0 nm for *Cfl. aurantiacus* (Fig. 5.7) (Psencik et al. 2009). Corresponding solution X-ray scattering (Fig. 5.8), which reports average structure of all the chlorosomes in the sample, exhibited features with comparable spacings (from 2.1 nm for *Cba. tepidum* to 3.3 nm for *Cfl. aurantiacus*) and a scattering pattern typical of the lamellar phase seen for other amphipathic self-assembling molecules (lipids, block copolymers etc.) (Hamley 1998). The lamellar arrangement has been so far observed in green- and brown-colored *Chlorobi* species and *Cfl. aurantiacus* and recently also for acidobacterium Cab. thermophilum (Costas et al. 2011) suggesting that this is a universal feature of chlorosome structure.

From cryo-EM observation of the same chlorosome from different projection angles (tilt series) it became evident that the lamellar system cannot be planar but must be curved (Psencik et al. 2004). Important insight into the arrangement of aggregates came from visualization of end-on views of chlorosomes from *Cba. tepidum*, which revealed a variety of curved lamellae ranging from disordered structures prevailing in the wild type chlorosomes (Fig. 5.9), to ordered multilayered cylinders observed mainly in chlorosomes from the *Cba. tepidum* triple mutant bchQRU which harbors a single BChl *d* homolog (Oostergetel et al. 2007; Oostergetel et al. 2010). Based on these images an idealized internal structure model consisting of lamellar cylinders has been proposed (Ganapathy et al. 2009). However, given the results of hexane-wash experiments (Sect. III.C.2), the reported anisotropy of X-ray scattering (Psencik et al. 2010), the extent of the observed intrinsic disorder and existence of domains (see below), it is unlikely that the completely closed cylin-

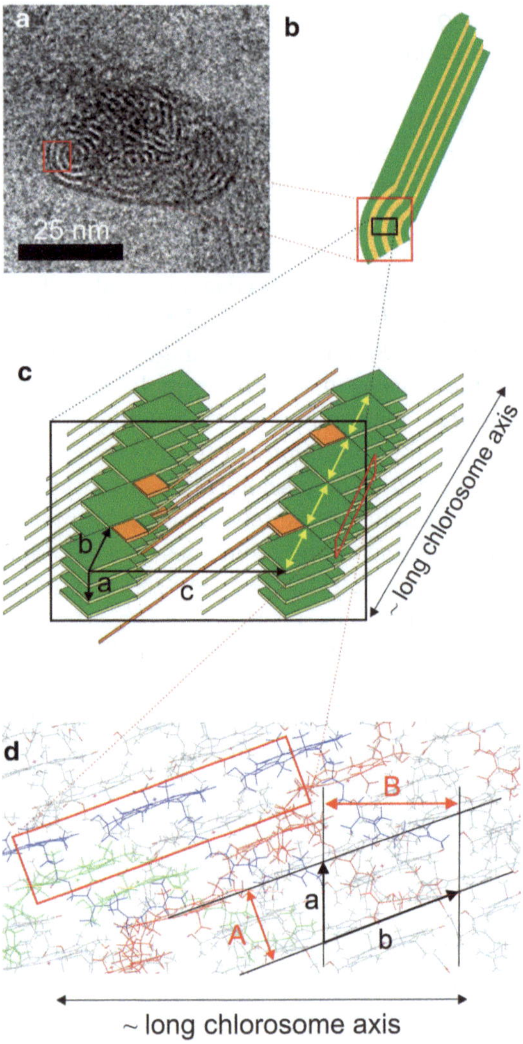

Fig. 5.9. (**a**) Curved lamellar system as seen in end-on-views of chlorosomes from wild-type *Cba. tepidum* (Oostergetel et al. 2010) (Figure courtesy of Dr. Gert T. Oostergetel and Prof. Egbert J. Boekema). (**b**) Lamellar quarter-cylinder as a basic structural motif of both open curved lamellar structures prevailing in wild-type chlorosomes and multilayered cylinders observed mostly in a bchQRU mutant of *Cba. tepidum*. (**c**) Schematics of BChl (*green*) stacking in aggregates (compatible with both antiparallel and syn-anti dimer) together with carotenoids and quinones (*orange*) localized in the hydrophobic space between the BChl layers. An example of a possible molecular arrangement within the layer is shown in (**d**) for the antiparallel dimer (*green*) as a building block (Model courtesy of Dr. Jan Alster). A plausible three-dimensional lattice with constants **a, b, c** (**c**) and the corresponding two-dimensional lattice (**d**) is also shown together with (continued)

drical structures are the prevailing feature of wild-type chlorosomes.

Despite the variability in the observed higher order arrangements, the organization of BChl aggregates always consists of repeating layers, i.e. it indeed has a lamellar character (Feng and Ruckenstein 2006), and differs only in the extent of the long range order. The lamellar structure reflects the alternation of relatively polar layers of stacked chlorin rings and interdigitated hydrophobic esterifying alcohol chains (Fig. 5.9). A key feature of this arrangement is that the esterifying alcohols protrude from the layer containing stacked chlorin rings in both directions, thus enabling interaction between esterifying alcohols from neighboring layers. Such an arrangement provides a plausible explanation for the formation of lamellae by hydrophobic interaction and is most likely achieved by alternating orientations of the chlorin rings within the stacks (Sect. III.C.3). This is reminiscent of block copolymers, which also exhibit polar and apolar regions and commonly adopt lamellar structures under a variety of conditions (Spontak and Patel 2004).

The lamellar curvature effectively transforms the linear chlorin ring stacks into helices. The helicity explains the strong circular dichroism (CD) that has been observed for chlorosomes (Sect. IV.A). However, the CD signal would vanish in lamellar layers, exhibiting continuous regions with alternating curvature (effectively changing the handedness of the helical stacks, as in Fig. 4A in Psencik et al. 2004). In chlorosomes such regions with opposite curvature are separated by fragmentation as observed in the end-on views (Oostergetel et al. 2007).

Fig. 5.9. (continued) the observed Bragg spacing (*red* **A**,**B**). Another plausible arrangement based on the syn-anti dimer as the building block is shown in Fig. 5 from (Ganapathy et al. 2009). The alignment of the Q_y transition dipole moments (*yellow arrows* in *panel* **c**) leads to strong excitonic coupling between the BChl molecules. The intrinsic curvature of the layers (*panels* **a**, **b**) causes helical arrangement of BChl molecules within each layer (molecules forming two selected helices are shown in *red* and *blue* in *panel* (**d**)).

Considerable fragmentation of the lamellar system in the direction perpendicular to the long-axis of the chlorosome was revealed by the above-mentioned tilt series and end-on views even in the well-ordered *Cba. tepidum* chlorosomes. In addition, the cryo-EM projections revealed segregation of the lamellar system into distinct domains with different orientations with respect to the long-axis of the chlorosome. This segregation into domains was occasionally observed in *Cba. tepidum* chlorosomes, and was particularly pronounced in BChl *e*-containing *Chl. phaeovibrioides* chlorosomes (Fig. 5.6) (Psencik et al. 2006). The existence of such domains may have interesting implications for light-harvesting. Since BChl transition dipole moments are oriented along the lamellar layers (Fig. 5.9), the orientational disorder resulting from domains effectively increases capture of photons with different polarizations within the same chlorosome. Physiologically, this may be important under low light conditions (Psencik et al. 2006). The extensive presence of domains causes the rough surface appearance of the chlorosome, as observed by atomic force microscopy and EM (Fig. 5.6) (Martinez-Planells et al. 2002; Psencik et al. 2006).

2. Lamellar Spacing and Location of Carotenoids and Quinones

The lamellar model presented in Fig. 5.9 predicts that the spacing should increase with the length of the esterifying alcohol chain. This has been confirmed experimentally by observing a linear dependence of the lamellar spacing on the chain length for *in vitro*-assembled BChl *c* analogues (Fig. 5.8) (Psencik et al. 2010). The extrapolation to zero length yields spacing of about 1.1 nm which is the approximate diameter of the chlorin ring. The experimental spacing for *Cba. tepidum* (2.1 nm) matches the predicted value (Fig. 5.8). The slope (0.08 nm/carbon) of this dependence corresponds to the projection of the hydrocarbon chain into the direction of spacing. In comparison, the projection of C–C bond length (average 0.154 nm) into the direction of the chain yields an increment of 0.12 nm per carbon in an all-trans chain, i.e. 50 % higher. Hence, the esterifying alcohol chains are either extensively twisted (harboring gauche isomers) and/or their average direction is at an angle to the spacing.

Compared to the green-colored members of *Chlorobi*, e.g. *Cba. tepidum*, chlorosomes from the brown-colored species (e.g. *Chl. phaeovibrioides*) generally contain higher amounts of BChl homologs with esterifying alcohols longer than farnesyl. *Cfl. aurantiacus* chlorosomes harbors BChls with octadecanol, which is the longest esterifying alcohol found so far. Interestingly, the experimentally observed spacings for chlorosomes from these species fall well above the predicted line (Fig. 5.8). The explanation of this observation lies in the fact that these chlorosomes also contain larger amounts of carotenoids than e.g. those from *Cba. tepidum* and the lamellar spacing further increases with the content of carotenoids. It has been shown that carotenoids can be effectively removed from intact chlorosomes by washing with hexane (Brune et al. 1987). Comparison of native and hexane-washed chlorosomes by X-ray diffraction and cryo-EM revealed that the spacing significantly decreases upon carotenoid removal (Fig. 5.8) while the chlorosomes remained intact (Psencik et al. 2006, 2013). This observation is consistent with the proposed localization of carotenoids (and most likely other lipophilic molecules such as quinones) within the hydrophobic inter-lamellar space where they interdigitate with esterifying alcohols (Fig. 5.9). Notably, after carotenoid removal, the resulting lamellar spacing closely follows the predicted linear dependence (Fig. 5.8). It is likely that the closed concentric cylinders observed in the bchQRU mutant of *Cba. tepidum* cannot undergo a transition connected with a shrinking of the lamellar spacing.

Figure 5.8 also illustrates that the chlorosomes which accumulate large amounts of carotenoids tend to employ longer esterifying alcohols, presumably to expand the interlamellar space. Longer alcohols and larger carotenoid contents are found in BChl *e*-containing bacteria, which often live at low light conditions in deep water layers, where the available light falls mainly into the spectral region of the carotenoid absorption. Therefore, the higher carotenoid content is important for light-harvesting in these bacteria. Generally, the larger the spacing the more disordered the lamellar system becomes (Psencik et al. 2006, 2009). The increasing disorder is manifested by a broader distribution of the observed spacings as well as by the presence of domains (Sect. III.C.1).

3. Short-Range Organization of Bacteriochlorophylls

Chemical groups that are involved in self-assembly of BChl aggregates (Sect. II.A) were identified mainly by vibrational spectroscopy (for reviews see Blankenship et al. 1995; Blankenship and Matsuura 2003). *In vitro* assembly studies with BChl *c* and their derivatives were instrumental in pointing out the critical role of the hydrophobic effect and esterifying chains in assembly (Klinger et al. 2004; Zupcanova et al. 2008).

To reveal the actual molecular arrangements, mainly methods based on NMR were used, and many structural models have been proposed (early studies are reported in Blankenship and Matsuura 2003; for newer models see Egawa et al. 2007; Jochum et al. 2008; Kakitani et al. 2009; Ganapathy et al. 2009). In addition, X-ray diffraction studies from partially oriented chlorosomes and BChl *c* aggregates have yielded information about the size of the asymmetric unit and the approximate orientation of the lattice with respect to the long axis of the chlorosome (Psencik et al. 2010). Here we attempt to compare the NMR-based models with the X-ray data.

The previously proposed models generally fall into three classes: (a) parallel monomer (Holzwarth and Schaffner 1994; van Rossum et al. 2001; Jochum et al. 2008) (b) antiparallel piggy-back dimer (Nozawa et al. 1994; Egawa et al. 2007; Kakitani et al. 2009) (c) syn-anti parallel dimer (Ganapathy et al. 2009). The parallel monomer model is unlikely considering that the hydrophobic esterifying alcohol chains stabilize the lamellar system and thus need to point in opposite directions (Klinger et al. 2004). This requirement is fulfilled by the latter two arrangements (Fig. 5.10), which are also compatible with the results of X-ray experiments. In addition to the most intense lamellar peak (spacing about 2.1 nm), the X-ray scattering patterns obtained for *Cba. tepidum* chlorosomes exhibit two peaks (Bragg spacing ~0.94 and ~1.17 nm) that have been assigned to the chlorin ring lattice within individual lamellar layers (Fig. 5.9) (Psencik et al. 2004). The three values define a monoclinic unit cell, which is too large to contain just a single BChl molecule, but it can accommodate two BChl molecules. Thus, the basic building unit of the aggregate seems to be a dimer. While the interpretation of recent NMR data favors the syn-anti dimer model (Ganapathy et al. 2009), the antiparallel model is also compatible with Stark spectroscopy results which indicate the absence of a permanent dipole moment difference between the ground and the excited state, a condition readily explained by the symmetry of the antiparallel dimer (Frese et al. 1997).

Anisotropic X-ray scattering was obtained for *Cba. tepidum* chlorosomes that partially oriented with their long axis parallel to each other (Psencik et al. 2010). Given that in *Cba. tepidum* lamellar spacing (lattice vector **c**) is perpendicular to the long axis of the chlorosome (Figs. 5.6 and 5.8) correlations between the intensity of the lamellar and other two peaks, respectively, yield an approximate orientation of the lattice (Fig. 5.9). These correlations suggest that the lattice vector **a** is nearly normal to the main axis of the chlorosome, while axis **b** is close to parallel to the main axis. However, the orientation of the lattice is approximate and other arrangements

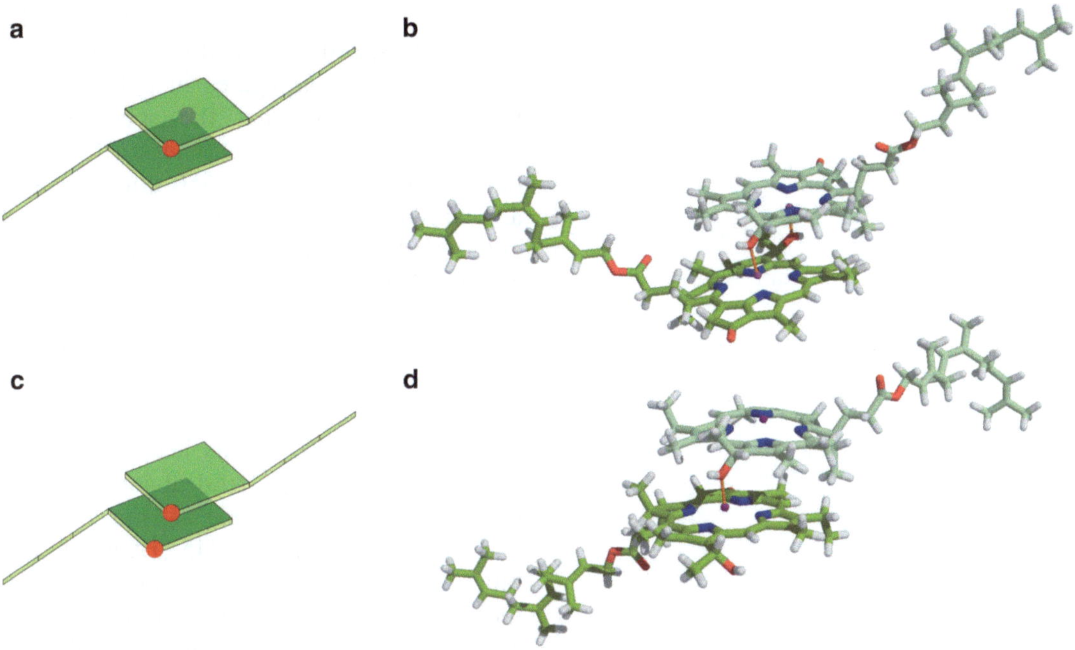

Fig. 5.10. (**a**) Schematic representation of the antiparallel piggyback dimer (Nozawa et al. 1994; Egawa et al. 2007) and (**c**) *syn-anti* dimer (Ganapathy et al. 2009). The *red dots* denote the positions of the hydroxy groups. Corresponding molecular models are shown in the *right panels* (**b** and **d**) (Models courtesy of Dr. Jan Alster).

with similar spacings are possible as proposed in Fig. 5 of Ganapathy et al. (2009). In addition, accounting for the lattice curvature proved essential for reaching good agreement between the model and the experimental diffraction data (Alster et al. 2012).

IV. Function

Being a light-harvesting complex, the main function of the chlorosome is to absorb photons and transfer their energy towards reaction centers. Within the chlorosome, energy transfer comprises the following steps indicated by arrows:

Carotenoids in the chlorosome interior				Carotenoids in the baseplate		
↓				↓		
BChl agg.	→	BChl agg.	→	BChl *a*	→	towards RC

It follows the well-known funnel concept: pigments at the periphery of the photosynthetic system absorb light of higher energies and transfer the resulting excitation energy via several intermediate complexes containing fewer pigments (baseplate, and FMO complex in *Chlorobi* or B806-866 complex in *Chloroflexi*, Sect. IV.B.3) with their excited state energies gradually decreasing. In addition, photoprotective mechanisms, which deal with excess energy and unwanted side reactions, also operate in the chlorosome.

A. Spectral Properties

A prerequisite for understanding energy transfer within the chlorosome is knowledge of the spectral properties of the aggregated BChls. The dense packing of the BChl molecules in the aggregate and the alignment of their Q_y dipole moments by hydrogen bonding network (Fig. 5.9) leads to strong excitonic coupling between pigments, which

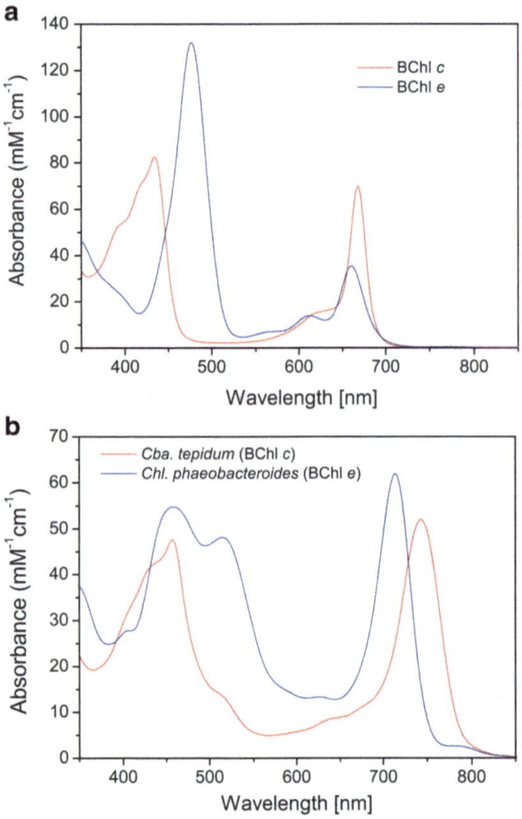

Fig. 5.11. (**a**) Absorption spectra of BChl *c* and *e* monomers in methanol. Spectra are scaled to their extinction coefficients at Q_y maxima: 70 mM^{-1} cm^{-1} for BChl *c* (Stanier and Smith 1960) and 35.5 mM^{-1} cm^{-1} for BChl *e* (Borrego et al. 1999b). (**b**) Absorption spectra of chlorosomes from *Cba. tepidum* containing aggregates of BChl *c* and from *Chl. phaeobacteroides* containing aggregates of BChl *e*, both in Tris-HCl buffer, pH 8.0. Spectra are scaled to their extinction coefficients at Q_y maxima: 52 mM^{-1} cm^{-1} for BChl *c* (Chew et al. 2007) and 62 mM^{-1} cm^{-1} for BChl *e* (Arellano et al. 2000b). Note that the absorption spectra of chlorosomes also contain contributions from carotenoids (mainly between 425 and 525 nm) and BChl *a* (Q_y maximum at ~795 nm).

in turn causes excitation delocalization over a number of BChl molecules. Interaction between BChls also leads to splitting of the monomeric energy levels into a number of so-called exciton states and redistribution of the oscillator strength among them. The resulting optical spectra are thus remarkably different compared to those of monomers (spectral properties of aggregates are reviewed in van Amerongen et al. 2000). This is illustrated in Fig. 5.11, which compares the absorption spectra of BChl *c* and BChl *e* as monomers and aggregates in chlorosomes, respectively. Note that the spectroscopic properties of the two monomeric BChls (Q_y band maxima between ~650 and 670 nm) are similar to those of chlorophyll *a* and chlorophyll *b*, respectively, and reflect their structural similarity.

1. Absorption and Fluorescence Spectroscopy

The aggregation of BChls leads to a large red-shift of their Q_y band absorption (up to ~80 nm, ~1,600 cm^{-1}) to ~715 nm (BChl *e*), ~730 nm (BChl *d*), or ~745 (BChl *c*). The blue-green region of spectra is dominated by the so called Soret band (consisting of B_x and B_y bands) with a variable contribution from carotenoids (mainly between 425 and 525 nm). The Soret band is also affected by aggregation, however to a lesser degree. The changes in the Soret band are more pronounced in the case of BChl *e*, where excitonic coupling between rather strong B transition dipole moments (mainly between B_x dipoles, and between B_y dipoles, respectively) leads to a pronounced splitting of the Soret band and a shift of the main part of the B_y band to ~515 nm (Shibata et al. 2010). Consequently, absorption coverage is improved in the spectral region between 500 and 600 nm (Fig. 5.11), which corresponds to the main portion of the light reaching the habitat of these bacteria (Vila and Abella 1994).

As mentioned above, the red shift of the Q_y band is a consequence of the strong excitonic coupling between BChl monomers (estimated coupling energy from approximately −750 to −550 cm^{-1} (Oostergetel et al. 2010)). The resulting spectra resemble those of J-aggregates (linear stacks of pigments with their transition dipole moments arranged head-to-tail) with the exception of exchange line-narrowing (Oostergetel et al. 2010). The spectra are relatively broad even if aggregates are prepared from pure single

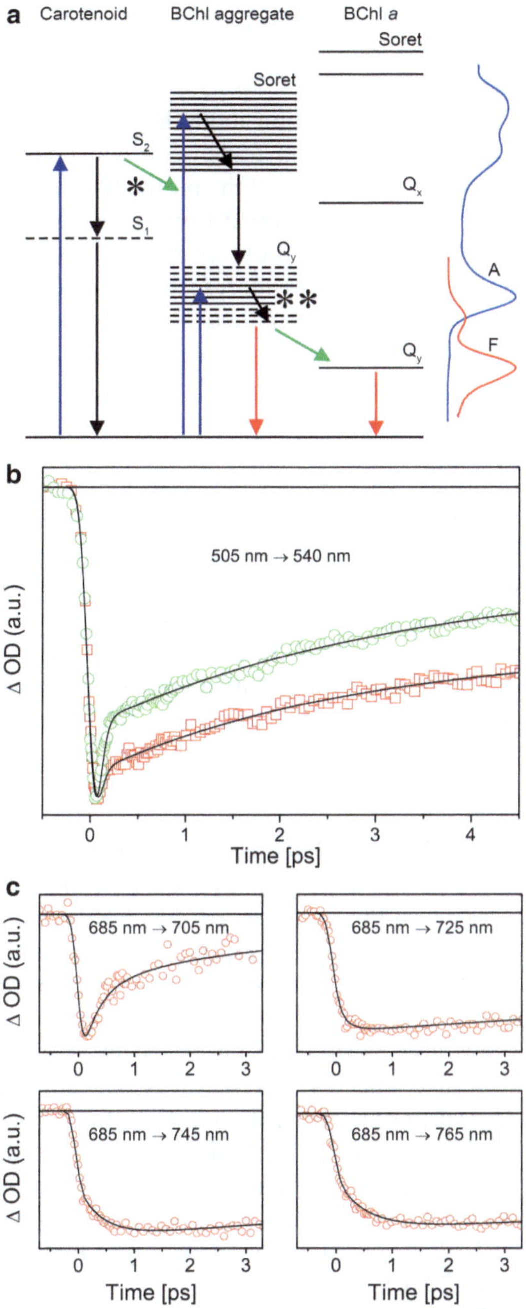

Fig. 5.12. (**a**) Simplified representation of the main transitions and energy transfer steps in chlorosomes. *Horizontal lines* represent states with a high (*solid line*) and low (*dashed*) oscillator strength (including forbidden states). *Vertical arrows* represent absorption (*blue*), fluorescence (*red*) and internal conversion (*black*). *Oblique arrows* represent exciton relaxation (*black*) and energy transfer (*green*). Absorption (*blue*) and fluorescence (*red*) spectra of BChl *e* containing bacterium *Chl. phaeobacteroides* are shown on the *right* (continued)

Fig. 5.12. (continued) for comparison. Note that the fluorescence spectrum is dominated by the emission from BChl *a*, (maximum at ~810 nm) despite the fact that these chlorosomes contain only about ~1 % fraction of BChl *a*. This is due to the efficient energy transfer from BChl *e* aggregates to the baseplate. (**b**) Transient absorption kinetics measured on chlorosomes with normal (*red*) and low (*green*) carotenoid content, excited and probed in the carotenoid absorption region. The initial rapid decay corresponds to very fast (<100 fs) energy transfer from carotenoids to BChl aggregates (*arrow* labeled with * in *panel* **a**) (Psencik et al. 2002). (**c**) Transient absorption kinetics measured on chlorosomes after excitation of the blue edge of the Q_y band of BChl aggregates. The fast initial rise component becomes slower and more prominent with an increasing probe wavelength, which is a manifestation of the exciton relaxation within the Q_y manifold of BChl aggregates (*arrow* labeled with ** in *panel* **a**) (Psencik et al. 2003). The excitation and probe wavelengths are indicated in each panel.

isomers in non-polar solvents *in vitro* (Steensgaard et al. 2000b). The presence of different BChl homologs and epimers in chlorosomes further increases inhomogeneous broadening and thus also the spectral width. The broadening due to the content of homologs with different alkyl substituents at C8 and C12 can be observed when the Q_y bands are compared: For chlorosomes from *Cfl. aurantiacus* containing a single type of BChl *c* homolog, the full width at half maximum (FWHM) is ~30 nm (~500 cm^{-1}) while the FWHM is ~50 nm (~900 cm^{-1}) for *Cba. tepidum* containing several homologues.

Another difference between the BChl aggregates and J-aggregates is the large red shift of the fluorescence peak with respect to the absorption maximum which is observed even at room temperature (~30 nm, ~500 cm^{-1}, Fig. 5.12). A plausible explanation is that the maximal oscillator strength in the BChl aggregate absorption does not, in contrast to J-aggregates, correspond to the lowest exciton level. This is corroborated by hole-burning and low temperature fluorescence measurements, which revealed the distribution of the lowest exciton levels (at temperatures close to absolute zero, fluorescence occurs from the lowest exciton levels after exciton relaxation, but at higher temperatures also from thermally populated upper exciton levels). These states are found only in the red edge of the broad absorption band and within the whole low temperature fluorescence band (Fetisova and Mauring 1992; Fetisova et al. 1994; Psencik et al. 1994, 2003), i.e. they are significantly red shifted from the main absorbing states and possess small oscillator strength. Such a redistribution of the oscillator strength upon aggregation seems to be an effective light-harvesting strategy under low light conditions. The probability of a photon absorption is high due to a large number of pigments in the chlorosome (the chlorosome-containing species exhibit the largest antenna pigment to reaction center ratio known among photosynthetic organisms (Frigaard et al. 2003)), while the subsequent fast exciton relaxation (Sect. IV.B.2) to the red-shifted lowest exciton levels with a small oscillator strength decreases the probability of excitation loss by fluorescence. Similar reasoning was used for B850 aggregates of BChl *a* in purple bacteria (Hu et al. 1998). Aggregates with a high excitation to fluorescence conversion efficiency, such as pure J-aggregates, would not be suitable for light-harvesting.

The broad absorption bands of BChl aggregates, which improve the spectral coverage of the chlorosome, cannot be explained simply by convolution of the inhomogeneous broadening of the lowest exciton level distribution (estimated to be 100–250 cm^{-1} from the above mentioned hole-burning and low temperature fluorescence spectra) with the homogeneous line-width of a single upper exciton level ($\leq$350 cm^{-1} (Dostal et al. 2012)). A plausible explanation proposes the existence of aggregate units inside the chlorosome (caused by fragmentation and disorder within the lamellar system, Sect. III.C.1), each of which exhibits several allowed transitions, with significant spectral variability between them.

2. Polarized-Light and Single Molecule Spectroscopy

Important information about the polarization of the main exciton transitions and consequently organization of the pigments in chlorosomes can be obtained from experiments, which use linearly polarized light, such as linear dichroism or fluorescence anisotropy. As the hydrogen bonding pattern tends to keep the Q_y transition dipole moments of BChl monomers parallel (Fig. 5.9), and the lamellar system is often parallel to the long axis of the chlorosomes, the strongest exciton transitions of the Q_y band should also be polarized close to parallel with the long axis of the chlorosome. For instance, for perfect cylinders with helically arranged monomers the strongest transition is polarized precisely parallel to the axis of the cylinder (Didraga et al. 2002) although this does not necessarily hold true for the transition moments of the individual monomers. Linear dichroism was measured for partially oriented *Cfl. aurantiacus* chlorosomes and values ranging from 15 to 30° have been obtained for the angle between the main exciton Q_y transition dipole and the long axis of the chlorosomes (for a recent review see Oostergetel et al. 2010). Similar results were obtained from the polarization measurements of the strongest transitions within the Q_y band for various *Chlorobi* species (Arellano et al. 2000b; Tamiaki et al. 2010). In addition, the polarization of the short- and long-wavelength parts of the Soret band determined by linear dichroism were used to assign them to B_x and B_y transitions, respectively (Tamiaki et al. 2010).

The ensemble measurements discussed above provide only average values. Information about disorder and heterogeneity in the orientation of the transition dipole moments within individual chlorosomes can be obtained by single molecule spectroscopy (for a recent review see Saga et al. 2010). Relatively broad emission spectra with their widths comparable to that of ensemble spectra were obtained from single chlorosomes indicating significant disorder already within individual chlorosomes. Furthermore, emission polarization provided evidence that multiple transition dipoles with different orientations are responsible for the observed fluorescence from a single chlorosome at low temperature (Shibata et al. 2007, 2009). Using three-dimensional linear dichroism it was shown that the Q_y exciton transition dipoles exhibit larger angles to the long axis of the chlorosome, when monitored from the top than from the side of the wild type chlorosomes of *Cba. tepidum* (Furumaki et al. 2011). In other words, while the dominant component of the transition dipole is assumed to be parallel to the long axis of the chlorosome, the component in the direction of the width of the chlorosome is larger than that in the direction of its height, as concluded also by Shibata et al. (2009). This demonstrates a lack of rotational symmetry in the pigment arrangement and reflects the asymmetric orientation of the curved lamellar system. However, a more symmetric distribution was observed by Furumaki et al. (2011) for the triple mutant bchQRU, in agreement with the proposed structure of that mutant containing lamellar cylinders. While the broad distribution of the transition dipole orientations reflects structural disorder, it is not correlated with observed large static excitonic disorder (distribution of peak wavelengths and widths of the Q_y bands) (Furumaki et al. 2011).

Linear dichroism and fluorescence anisotropy, both in ensemble and in single chlorosome measurements, showed the orientation of the BChl *a* transition moment to be nearly perpendicular to the long axis of the chlorosome and to the plane of the baseplate in both *Chloroflexi* and *Chlorobi* species (Fig. 5.7) (Matsuura et al. 1993; Mimuro et al. 1994; Arellano et al. 2000b; Shibata et al. 2007, 2009; Tamiaki et al. 2010). Such an orientation would be unfavorable for energy transfer from the main absorbing states of BChl aggregates, which are, as discussed above, polarized close to parallel with the long axis of the chlorosome. However, the analysis of fluorescence polarization from a single chlorosome indicates presence of a sufficient Q_y transition dipole component in the direction of that of BChl *a* (Shibata et al. 2007, 2009; Furumaki et al. 2011). Previous measurements also showed that the low energy states of BChl aggregates have a larger component of their dipole moment parallel to that of BChl *a* (Matsuura et al. 1993; Mimuro et al. 1994). These observations are important for understanding efficient energy transfer from BChl aggregates to the baseplate.

The excitonic interaction also leads to the appearance of strong CD, which is a consequence of lamellar curvature that induces local helicity in the organization of the transition dipoles (Fig. 5.9). The curvature is essential, since close to planar structures yield only weak CD (Linnanto and Korppi-Tommola 2008), although completely closed tubular structures are not necessary for appearance of the large CD signal. It has been shown for the closed, helical cylindrical aggregates (Didraga et al. 2002) and certain open curved structures (Linnanto and Korppi-Tommola 2008) that their CD depends on the length of the aggregate. This dependence may explain why CD spectra are very sensitive to growth conditions and sometimes different preparations of chlorosomes from the same species exhibit spectra with opposite signs.

B. Excitation Energy Transfer

Excitation energy transfer in chlorosomes has been extensively studied using various spectroscopy techniques during the last ~25 years (for reviews see Blankenship et al. 1995; Blankenship and Matsuura 2003; Oostergetel et al. 2010). The evaluation has been complicated by the fact that the excitation dynamics depend on species, growth conditions, redox state and excitation intensity. However, a consistent picture of the excitation relaxation has emerged. The principal steps on the energy pathway are discussed below while the relevant energy levels are schematically shown in a simplified form in Fig. 5.12.

1. From Carotenoids to Bacteriochlorophylls

Two distinct sets of carotenoids contribute to light-harvesting. Excitation energy absorbed by the carotenoids within the chlorosome interior is transferred to BChl aggregates with efficiencies estimated from fluorescence excitation spectra to be most often between 50 % and 80 % depending on the type of the chlorosome and measuring conditions (van Dorssen et al. 1986; Melo et al. 2000; Psencik et al. 2002). Energy from carotenoids in the baseplate is transferred directly to BChl *a* within the baseplate with an efficiency of about 30 % for *Cfl. aurantiacus* (Montano et al. 2003b), however, these carotenoids are assumed to have mainly photoprotective function (Sect. IV.C.2).

Absorption into the S_1 state is dipole-forbidden for carotenoids, hence the main absorbing state is S_2 (Fig. 5.12, for a review see Polivka and Sundstrom 2004). For isorenieratene- and β-isorenieratene-containing chlorosomes from *Chl. phaeobacteroides*, the carotenoid S_1 state lifetime within chlorosomes was found to be about 10 ps (Psencik et al. 2002), i.e. similar to that in solvent (Fuciman et al. 2010). Therefore this state cannot be significantly involved in the excitation energy transfer. On the other hand, excitation is transferred to BChls from the very short-lived S_2 state (lifetime of ~150 fs for carotenoids with a similar conjugated system, e.g. β-carotene, in solution), as judged from the shortening of its lifetime to ~50 fs in chlorosomes (Fig. 5.12). When considering the measured energy transfer efficiency from carotenoids to BChls of 60–70 %, a transfer time of 65–100 fs can be determined (Psencik et al. 2002). The energy transfer from carotenoids to BChls thus constitutes the fastest energy transfer process in chlorosomes observed so far.

2. Within Bacteriochlorophyll Aggregates

The Q_y band of the BChl aggregates can be populated either by direct excitation, or by internal relaxation from the Soret band, or by energy transfer from carotenoids (Fig. 5.12). The internal relaxation from the Soret to Q_y band is fast (relaxation time <100 fs, (Psencik et al. 2002)). As mentioned in Sect. IV.A, the Q_y band of the BChl aggregate consists of a number of exciton levels. To understand excitation energy transfer within the chlorosome interior it is useful to begin with description of exciton delocalization and exciton relaxation.

From femtosecond transient spectra of the chlorosomes from *Cfl. aurantiacus* it was deduced that the transient changes in the BChl *c* spectral region are 7–8 times larger than that of BChl *a* (Savikhin et al. 1998). A very similar result was obtained for BChl *e* and BChl *a* in *Chl. phaeobacteroides* (Psencik et al. 2003). Both results are consistent with a larger oscillator strength of the main transition of the aggregate compared to that of the monomer, and substantial exciton delocalization in BChl aggregates. More specifically, delocalization over 10–12 pigments in the *Cfl. aurantiacus* and over 2–3 pigments in *Cba. tepidum* chlorosomes has been suggested from low temperature photon echo studies (Prokhorenko et al. 2000), although the latter might be underestimated (Prokhorenko et al. 2002).

Many different decay components have been found on the 100 fs–100 ps time scale (for reviews see Blankenship et al. 1995; Blankenship and Matsuura 2003; Oostergetel et al. 2010). The kinetics may easily become affected by singlet-singlet excitation annihilation, which is a direct consequence of the high pigment density in chlorosomes. In order to obtain kinetics which reveal the intrinsic relaxation times it is therefore necessary to apply sufficiently low excitation intensities, and also to observe a corresponding rise component in the spectral region of the excitation acceptor. Such an approach enabled resolution of an energy transfer from the blue portion of the Q_y absorption band of aggregated BChls to their red shifted states. This process corresponds to exciton relaxation from higher to lower exciton levels. It was first observed in chlorosomes from *Cfl. aurantiacus* as a 300 fs rise at 758 nm after excitation at 731 nm at 19 K (Savikhin et al. 1996). Later it was also observed in two studies at room temperature, with the transfer time increasing with the energetic separation between the excitation and probe wavelengths from approximately 200 to 1000 fs for BChl *e* containing *Chl. phaeobacteroides* (Fig. 5.12) (Psencik et al. 2003) and from 150 to 250 fs in BChl *c* containing *Cfl. aurantiacus* (Martiskainen et al. 2009). The smaller span of values for the latter most likely reflects a narrower exciton manifold in *Cfl. aurantiacus* compared to *Chlorobi* species.

As mentioned above, the excitation is delocalized over several pigments, not the whole aggregate, which is a consequence of the existing disorder. Each of the exciton states is partly localized in a different part of the aggregate, and therefore the exciton relaxation is connected with the spatial redistribution of the energy, leading to spatio-temporal energy transfer. The exciton levels in the middle of the exciton energy level manifold usually have a more delocalized character than those on the edges (Pullerits 2000). Thus the exciton relaxation is connected also with the spatial localization of an exciton. Förster-type excitation hopping between the localized sites governs further energy transfer steps.

The presence of different spectral forms of aggregated BChls has been observed in many studies, and it has been proposed that they are spatially separated to ensure the 'energy-funneling' effect even within the chlorosome interior (i.e. aggregates absorbing more to the red would be closer to the baseplate and those with absorption shifted to the blue on the opposite side). For a review on this topic see Blankenship and Matsuura (2003). However, such a spatial separation has never been demonstrated and, since the organization of the chlorosome interior is most probably determined by self-assembly, it seems unlikely. It appears most probable that after exciton relaxation, only those excitations that are found in close contact with the baseplate can be transferred directly to BChl *a*, but the bulk of excitation energy travels randomly within the whole chlorosome until it is close to the baseplate from where it can be transferred to BChl *a* (Psencik et al. 2003). As the excitons are found in low-energy states with a small oscillator strength during this random walk (Sect. IV.A), the probability of excitation loss by fluorescence is low. The energy differences between lowest exciton levels (100–250 cm^{-1}, Sect. IV.A.1) can be overcome with the help of thermal energy (~200 cm^{-1} at ambient temperatures).

3. From Aggregates to Bacteriochlorophyll a

Energy transfer from BChl aggregates to BChl *a* in the baseplate strongly depends on the BChl type. As stated in the previous section, it is likely that the transfer occurs from spatially localized states of BChl aggregates and can be adequately described by the Förster mechanism, i.e. it depends on the overlap between the absorption spectrum of BChl *a* and the emission spectrum of the aggregated BChl. This overlap decreases from BChl *c*, through BChl *d*, to BChl *e* and therefore also the rate constants of energy transfer between the aggregates and BChl *a* decrease in the same order (Causgrove et al. 1992). Consequently the fastest energy transfer has been observed for *Cfl. aurantiacus* chlorosomes (~10 ps), which contain BChl *c* (for a review see Blankenship et al. 1995; Martiskainen et al. 2009). Somewhat slower rates (30–40 ps) were observed for BChl *c* containing *Chlorobi* members that exhibit thicker chlorosomes than the ones found in *Cfl. aurantiacus*, and thus indicating that the transfer within the chlorosome also contributes to the rates observed. The even slower transfer times that have been observed for BChl *d* containing chlorosomes (65–75 ps), and the wide range of values observed for BChl *e* containing species including significant components in the range of 90–190 ps are apparently due to the further decreasing overlap between the fluorescence of the BChl *d* and *e* aggregates and absorption of BChl *a*, and the larger size of BChl *e*-containing chlorosomes (Causgrove et al. 1992; van Walree et al. 1999; Steensgaard et al. 2000a; Psencik et al. 2003, and references therein).

During subsequent steps the excitation energy is transferred from BChl *a* in the baseplate (absorption around 795 nm, emission ~810 nm) to the outside of the chlorosome. Briefly, in the case of *Chlorobi* and *Acidobacteria* the energy is transferred via the out-of-membrane FMO complex to reaction center complexes in the cytoplasmic membrane (the electron donor is called P840 for *Chlorobi*), whereas in the case of *Chloroflexi* it is transferred via an in membrane complex B808-866 to reaction centers with an electron donor denoted as P865 or P870. The reader is referred to the reviews of Blankenship et al. (1995) and Blankenship and Matsuura (2003) for further detail.

C. Photoprotective Mechanisms

Excitation quenching occurring under aerobic conditions (redox-dependent quenching) in *Chlorobi* species has been suggested to contribute to the protective mechanisms preventing photooxidative damage. In addition, it has been demonstrated that chlorosomes possess photoprotective mechanisms preventing generation of harmful singlet oxygen under aerobic conditions, in which several processes have been proposed to be involved.

1. Redox-Dependent Excitation Quenching

Chlorobi species are obligate anaerobes, although they may get exposed to oxygen (e.g. aquatic species during changes in the position of the chemocline). However, the efficient excitation energy transfer towards the reaction centers only occurs in the absence of oxygen (for reviews see Blankenship and Matsuura, 2003; Frigaard and Bryant 2006). Under aerobic conditions, the excited states of the antenna pigments are rapidly quenched, both in whole cells and in isolated chlorosomes, by a mechanism that prevents energy transfer towards the reaction centers. The type I reaction centers of *Chlorobi* contain low potential electron acceptors (iron sulfur centers) that directly reduce ferredoxin. The reduced ferredoxin may react with oxygen to produce superoxide, which in turn results in a variety of damaging photooxidative products (Blankenship and Matsuura 2003).

Chlorosomal quinones were shown to play a key role in the excitation quenching (Frigaard et al. 1997). In addition, oxidized BChl radicals (van Noort et al. 1997) were suggested to be involved in quenching. In fact, it is not the presence of oxygen itself, which triggers the excitation-quenching, but the change of the redox potential. Redox titrations revealed a pH-dependent midpoint potential of about −150 mV for the induction of the quenching in isolated chlorosomes (Blankenship et al. 1993). Interestingly, redox-dependent excitation quenching was observed also for the *Acidobacteria* member Cab. thermophilum (Costas et al. 2011), which also harbors type I reaction centers, but lives at aerobic conditions (Bryant et al. 2007).

Chlorosomes of *Cfl. aurantiacus* exhibit only a weak fluorescence quenching under aerobic conditions (about ten times weaker compared to that in chlorosomes from *Cba. tepidum* (Frigaard et al. 1998) although they contain a similar amount of quinones as *Cba. tepidum* (Frigaard et al. 1997). However, chlorosomes of *Cfl. aurantiacus* contain mainly menaquinone-10 and no chlorobiumquinone. It seems that the midpoint potential of the quinone is the crucial parameter for understanding the extent of fluorescence quenching in chlorosomes from various bacteria (Frigaard and Bryant 2006). More importantly, in contrast to *Cba. tepidum* the excitation quenching is insignificant in whole cells of *Cfl. aurantiacus* (Frigaard and Matsuura 1999), presumably due to their ability to maintain the redox potential.

2. Triplet State Quenching

BChl triplet states are involved in light-induced damage through generating harmful singlet oxygen under aerobic conditions. Carotenoids are known, apart from their light-harvesting function, to protect photosynthetic complexes against damage caused by singlet oxygen (for a review see Polivka and Sundstrom 2004). This is achieved either by quenching the triplet states of chlorophylls or by direct scavenging of singlet oxygen. An efficient quenching of the triplet states of both BChl aggregates and BChl *a* by triplet-triplet energy transfer to carotenoids was observed in chlorosomes from *Chlorobi* and *Chloroflexi* at room temperature (Melo et al. 2000; Arellano et al. 2000a). In contrast, optically detected magnetic resonance revealed triplet states of BChl aggregates and BChl *a* at low temperature (Psencik et al. 1994; Carbonera et al. 2001) and only evidence for quenching of BChl *a* triplets by carotenoids was obtained (Carbonera et al. 2001). The triplet quenching is more important for BChl *a* in the baseplate, which is the bottleneck in energy transfer from the chlorosome, and where the probability of triplet state formation is increased by a longer excited state lifetime compared to BChl aggregates. Thus, the key protective role of carotenoids is in the baseplate while damage of BChl aggregates is mainly prevented by the self-protective properties of BChl aggregates which are described in the next paragraph. However, BChl *c* photodegradation is three times faster in carotenoid-free-chlorosomes than in wild type chlorosomes (Kim et al. 2007). Therefore, carotenoids indeed protect the pigments within the chlorosome interior as well. Accommodation of carotenoids in the inter-lamellar space provides an effective way to position these protective molecules in the vicinity of BChl rings without disrupting excitonic coupling between them.

The aggregation-mediated protection mechanisms are based on the excited state properties of BChls modified by excitonic interaction between the monomers in the aggregate. The significance of the self-protection afforded by aggregation has been demonstrated by comparing photodegradation of monomeric BChl *c* in methanol with BChl in carotenoid-deficient chlorosomes from *Cba. tepidum* mutants. The former is 3 orders of magnitude faster than the latter (Kim et al. 2007). For BChl *d* dimers it was shown that the quantum yield of the singlet oxygen photogeneration is six times smaller than that for the monomer, while no singlet oxygen was detected in aggregates (Krasnovsky et al. 1994). This behavior was explained by the shortening of the excited states lifetimes, both singlet and triplet, upon aggregation (Arellano et al. 2002). While the shorter S_1 lifetimes decrease the probability of intersystem crossing and thus reduce the triplet state population, the probability of singlet oxygen formation is further decreased by the shorter triplet state lifetime.

The existence of another triplet-like state in BChl aggregates has been proposed on the basis of experiments with carotenoid-deficient mutants of *Cba. tepidum*. This long-lived state is supposed to be a triplet exciton state formed by coupling between the monomeric triplet states in the densely packed aggregate, and its energy was estimated to be below that necessary for singlet oxygen generation (Kim et al. 2007). It still remains to be seen whether such a state can be observed spectroscopically. Its energy was predicted to be close to 1,400 nm (Kim et al. 2007), i.e. at longer wavelengths than the phosphorescence from the triplet states of BChl *c* monomers (960 nm) and aggregates (~1,050 nm) *in vitro* (Krasnovsky and Litvin 1975), energies of which are both above that of singlet oxygen (~1,270 nm).

V. Assembly

Only a few lines of evidence about chlorosome biogenesis have been obtained experimentally. In *Cfl. aurantiacus*, the development of chlorosomes was followed under different conditions, e.g. after changing from aerobic to semi anaerobic, or from chemotrophic to phototrophic conditions, which induce chlorosome formation (Sprague et al. 1981; Foidl et al. 1998 and references therein). However the observations have not yielded a complete description of the morphogenesis. Hence, the two existing models were built using rather indirect lines of evidence.

Since chlorosomes are found in bacteria belonging to unrelated phyla it is plausible to conclude that they appeared as a result of horizontal gene transfer. However, that would require the transfer of the whole chlorosomal synthetic and assembly machinery (Frigaard and Bryant 2006; Hohmann-Marriott and Blankenship 2007). Since only one of the characterized chlorosomal proteins, CsmA, appears necessary for chlorosome morphogenesis the existence of a specialized assembly machinery is unlikely. Thus morphogenesis most likely relies on physical properties and self-assembly of the components. Based on this and other arguments a plausible and simple model of biogenesis was proposed by Hohmann-Marriott and Blankenship (2007). This hypothesis regards the chlorosome as a special lipid body, and puts forward a mechanism similar to that proposed for lipid body biogenesis, which are commonly found in various types of cells. The model suggests that chlorosomal pigments and quinones accumulate between the two leaflets of the cytoplasmic membrane, and glycolipids accumulate in their vicinity due to their affinity for the pigments. In addition, chlorosomal proteins associate with the membrane presumably due to their amphipathic character. The growing body then separates from the membrane and the baseplate is formed last (Fig. 5.13).

However, since the baseplate protein CsmA is essential for chlorosome morphogenesis, it seems more likely that the baseplate forms first as supposed in a second model (Pedersen et al. 2010) and supported by EM (Sprague et al. 1981). The chlorosomal lipids then accumulate on the top of the baseplate forming an empty chlorosome. The growth of the chlorosome proceeds by accumulation of the BChl *c* inside and chlorosome proteins other than CsmA in the envelope (Fig. 5.13). The formation of the baseplate prior to aggregate assembly is also consistent with the fact that mutants defective in BChl *c* synthesis still develop the baseplate (Frigaard et al. 2005), and with the observed correlation between baseplate defects and increased disorder in the lamellar structure observed in chlorosomes that are deficient in carotenoid synthesis, or after hexanol treatment (Psencik et al. 2006; Ikonen et al. 2007; Arellano et al. 2008). Most likely the 2D lattice of the baseplate together with the associated BChl *a* acts as a nucleation site for BChl *c* assembly, perhaps even without the envelope formed (see below). That would be consistent with observations that the chlorosomal carotenoids, quinones and lipids promote BChl *c* assembly in aqueous solutions in which BChl *c* alone remains partially soluble (Hirota et al. 1992; Klinger et al. 2004; Alster et al. 2008).

The second model leaves one to puzzle how the chlorophylls would be transported to the chlorosome via the baseplate, if their final biosynthesis stage occurs in the cytoplasmic membrane. It would require a pigment transporter as suggested by Frigaard et al. (2003) which has not yet been identified. Without such a transporter it is easier to understand the filling up of the chlorosome with pigments using the model of Hohmann-Marriott and Blankenship (2007).

An interesting question arises as to whether or not the aggregates are formed in a hydrophobic pocket of a chlorosome, or before the closed envelope is formed. In the latter case the aggregation would be driven

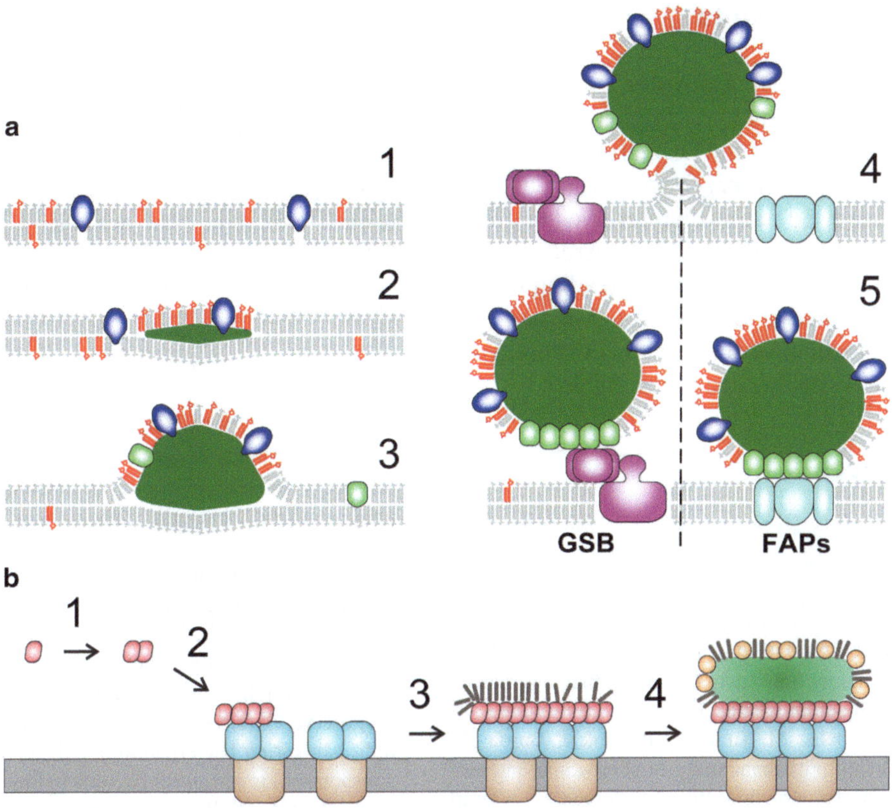

Fig. 5.13. Two proposed schemes for chlorosome morphogenesis. (**a**) A model by Hohmann-Marriott and Blankenship (2007): The process starts at the cytoplasmic membrane (*1*) by accumulation of pigments and quinones (*dark green*) (*2*) and is followed by recruitment of envelope lipids (*red*) with affinity for pigments and proteins (*blue*) with affinity for envelope lipids forming the emerging chlorosome (*3*). Further accumulation of the chlorosome components leads to the completion of the body (*4*), which finally separates in a chlorosome connected with a cytoplasmic membrane either via FMO complex (in *Chlorobi* and *Acidobacteria*, *left*) or directly (in *Chloroflexi*, *right*) (Figure courtesy of Dr. Martin Hohmann-Marriott). (**b**) A model by Pedersen et al. (2010) for *Chlorobi* species: CsmA-BChl *a* is synthesized (*1*) and assembles into a baseplate (*2*) on the FMO protein layer (*3*), which is followed by attachment of envelope lipids (*4*) and accumulation of BChl and envelope proteins (Figure courtesy of Dr. Niels-Ulrik Frigaard).

by the hydrophobic effect due to esterifying alcohols, and the same effect would ensure incorporation of carotenoids and quinones into the hydrophobic inter-lamellar space. If the aggregates were formed inside the hydrophobic pocket, the driving force would be π–π stacking interaction, Mg coordination and hydrogen bonding between the polar chlorin heads. It has been shown that the esterifying alcohols (in addition to the presence of either lipids, carotenoids or quinones) are necessary for the formation of the aggregates in an aqueous environment, but not in a non-polar solvent, such as hexane (Zupcanova et al. 2008). Therefore it would be more difficult to envision the need for the esterifying alcohols that are formed during the last step of chlorosomal BChls biosynthesis, (transesterification, Frigaard and Bryant 2004), if the assembly was driven solely by chlorin-chlorin interactions. It is likely that the main thermodynamic driving force in pigment organization is the hydrophobic effect, with finely tuned interactions between the chlorin rings being important for the short-range arrangement.

Acknowledgements

We are indebted Dr. Jan Alster, Prof. Egbert J. Boekema, Dr. Niels-Ulrik Frigaard, Dr. Martin Hohmann-Marriott and Dr. Gert T. Oostergetel (in alphabetical order) for providing us with the figures used in this chapter. We are grateful to Dr. Jan Alster, Dr. Juan Arellano and Dr. Tomáš Mančal for critical reading of this text and valuable discussions. We are also indebted to all our collaborators over the years that have made this chapter possible.

References

Alster J, Zupcanova A, Vacha F, Psencik J (2008) Effect of quinones on formation and properties of bacteriochlorophyll c aggregates. Photosynth Res 95:183–189

Alster J, Polivka T, Arellano JB, Chabera P, Vacha F, Psencik J (2010) Beta-carotene to bacteriochlorophyll *c* energy transfer in self-assembled aggregates mimicking chlorosomes. Chem Phys 373:90–97

Alster J, Kabelac M, Tuma R, Psencik J, Burda JV (2012) Computational study of short-range interactions in bacteriochlorophyll aggregates. Comp Theor Chem 998:87–97

Arellano JB, Melo TB, Borrego CM, Garcia-Gil J, Naqvi KR (2000a) Nanosecond laser photolysis studies of chlorosomes and artificial aggregates containing bacteriochlorophyll *e*: evidence for the proximity of carotenoids and bacteriochlorophyll *a* in chlorosomes from *Chlorobium phaeobacteroides* strain CL1401. Photochem Photobiol 72:669–675

Arellano JB, Psencik J, Borrego CM, Ma YZ, Guyoneaud R, Garcia-Gil J, Gillbro T (2000b) Effect of carotenoid biosynthesis inhibition on the chlorosome organization in *Chlorobium phaeobacteroides* strain CL1401. Photochem Photobiol 71:715–723

Arellano JB, Borrego CM, Martinez-Planells A, Garcia-Gil LJ (2001) Effect of carotenoid deficiency on cells and chlorosomes of *Chlorobium phaeobacteroides*. Arch Microbiol 175:226–233

Arellano JB, Melo TB, Borrego CM, Naqvi KR (2002) Bacteriochlorophyll *e* monomers, but not aggregates, sensitize singlet oxygen: implications for a self-protection mechanism in chlorosomes. Photochem Photobiol 76:373–380

Arellano JB, Torkkeli M, Tuma R, Laurinmäki P, Melo TB, Ikonen TP, Butcher SJ, Serimaa RE, Psencik J (2008) Hexanol-induced order-disorder transitions in lamellar self-assembling aggregates of bacteriochlorophyll *c* in *Chlorobium tepidum* chlorosomes. Langmuir 24:2035–2041

Balaban TS, Tamiaki H, Holzwarth AR (2005) Chlorins programmed for self-assembly. Top Curr Chem 258:1–38

Beatty JT, Overmann J, Lince MT, Manske AK, Lang AS, Blankenship RE, Van Dover CL, Martinson TA, Plumley FG (2005) An obligately photosynthetic bacterial anaerobe from a deep-sea hydrothermal vent. Proc Natl Acad Sci U S A 102: 9306–9310

Blankenship RE, Matsuura K (2003) Antenna complexes from green photosynthetic bacteria. In: Green BR, Parson WW (eds) Light-harvesting antennas in photosynthesis. Kluwer Academic Publishers, Dordrecht, pp 195–217

Blankenship RE, Cheng P, Causgrove TP, Brune DC, Wang SH, Chon J, Wang J (1993) Redox regulation of energy transfer efficiency in antennas of green photosynthetic bacteria. Photochem Photobiol 57:103–107

Blankenship RE, Olson JM, Miller M (1995) Antenna complexes from green photosynthetic bacteria. In: Blankenship RE, Madigan MT, Bauer CE (eds) Anoxygenic photosynthetic bacteria. Kluwer Academic Publisher, Dordrecht, pp 399–435

Borrego CM, Garcia-Gil LJ (1995) Rearrangement of light harvesting bacteriochlorophyll homologues as a response of green sulfur bacteria to low light intensities. Photosynth Res 45:21–30

Borrego CM, Gerola PD, Miller M, Cox RP (1999a) Light intensity effects on pigment composition and organisation in the green sulfur bacterium *Chlorobium tepidum*. Photosynth Res 59:159–166

Borrego CM, Arellano JB, Abella CA, Gillbro T, Garcia-Gil J (1999b) The molar extinction coefficient of bacteriochlorophyll e and the pigment stoichiometry in *Chlorobium phaeobacteroides*. Photosynth Res 60:257–264

Brune DC, Nozawa T, Blankenship RE (1987) Antenna organization in green photosynthetic bacteria. 1. Oligomeric bacteriochlorophyll c as a model for the 740 nm absorbing bacteriochlorophyll *c* in *Chloroflexus aurantiacus* chlorosomes. Biochemistry 26:8644–8652

Bryant DA, Vassilieva EV, Frigaard NU, Li H (2002) Selective protein extraction from *Chlorobium tepidum* chlorosomes using detergents. Evidence that CsmA forms multimers and binds bacteriochlorophyll *a*. Biochemistry 41:14403–14411

Bryant DA, Costas AMG, Maresca JA, Chew AGM, Klatt CG, Bateson MM, Tallon LJ, Hostetler J, Nelson WC, Heidelberg JF, Ward DM (2007) Candidatus Chloracidobacterium thermophilum: an aerobic phototrophic acidobacterium. Science 317: 523–526

Carbonera D, Bordignon E, Giacometti G, Agostini G, Vianelli A, Vannini C (2001) Fluorescence and absorption detected magnetic resonance of chlorosomes from green bacteria *Chlorobium tepidum* and *Chloroflexus aurantiacus*. A comparative study. J Phys Chem B 105:246–255

Causgrove TP, Brune DC, Blankenship RE (1992) Förster energy transfer in chlorosomes of green photosynthetic bacteria. J Photochem Photobiol B 15:171–179

Chew AGM, Frigaard NU, Bryant DA (2007) Bacteriochlorophyllide *c* C-8(2) and C-12(1) methyltransferases are essential for adaptation to low light in *Chlorobaculum tepidum*. J Bacteriol 189:6176–6184

Chung S, Bryant DA (1996) Characterization of the *csmD* and *csmE* genes from *Chlorobium tepidum*. The CsmA, CsmC, CsmD, and CsmE proteins are components of the chlorosome envelope. Photosynth Res 50:41–59

Cohen-Bazire G, Pfennig N, Kunisawa R (1964) The fine structure of green bacteria. J Cell Biol 22:207–225

Costas AMG, Tsukatani Y, Romberger SP, Oostergetel GT, Boekema EJ, Golbeck JH, Bryant DA (2011) Ultrastructural analysis and identification of envelope proteins of "Candidatus Chloracidobacterium thermophilum" chlorosomes. J Bacteriol 193:6701–6711

Costas AMG, Tsukatani Y, Rijpstra WI, Schouten S, Welander PV, Summons RE, Bryant DA (2012) Identification of the bacteriochlorophylls, carotenoids, quinones, lipids, and hopanoids of "Candidatus Chloracidobacterium thermophilum". J Bacteriol 194:1158–1168

Didraga C, Klugkist JA, Knoester J (2002) Optical properties of helical cylindrical molecular aggregates: the homogeneous limit. J Phys Chem B 106:11474–11486

Dostal J, Mancal T, Augulis R, Vacha F, Psencik J, Zigmantas D (2012) Two-dimensional electronic spectroscopy reveals ultrafast energy diffusion in chlorosomes. J Am Chem Soc 134:11611–11617

Egawa A, Fujiwara T, Mizoguchi T, Kakitani Y, Koyama Y, Akutsu H (2007) Structure of the light-harvesting bacteriochlorophyll *c* assembly in chlorosomes from *Chlorobium limicola* determined by solid-state NMR. Proc Natl Acad Sci U S A 104:790–795

Eisen JA, Nelson KE, Paulsen IT, Heidelberg JF, Wu M, Dodson RJ, Deboy R, Gwinn ML, Nelson WC, Haft DH, Hickey EK, Peterson JD, Durkin AS, Kolonay JL, Yang F, Holt I, Umayam LA, Mason T, Brenner M, Shea TP, Parksey D, Nierman WC, Feldblyum TV, Hansen CL, Craven MB, Radune D, Vamathevan J, Khouri H, White O, Gruber TM, Ketchum KA, Venter JC, Tettelin H, Bryant DA, Fraser CM (2002) The complete genome sequence of *Chlorobium tepidum* TLS, a photosynthetic, anaerobic, green-sulfur bacterium. Proc Natl Acad Sci U S A 99:9509–9514

Feng J, Ruckenstein E (2006) Morphology transitions of AB diblock copolymer melts confined in nanocylindrical tubes. J Chem Phys 125:1164911

Fetisova ZG, Mauring K (1992) Experimental evidence of oligomeric organization of antenna bacteriochlorophyll *c* in green bacterium *Chloroflexus aurantiacus* by spectral hole burning. FEBS Lett 307:371–374

Fetisova ZG, Mauring K, Taisova AS (1994) Strongly exciton-coupled BChl *e* chromophore system in the chlorosomal antenna of intact cells of the green bacterium *Chlorobium phaeovibrioides*: a spectral hole burning study. Photosynth Res 41:205–210

Foidl M, Golecki JR, Oelze J (1997) Phototrophic growth and chlorosome formation in *Chloroflexus aurantiacus* under conditions of carotenoid deficiency. Photosynth Res 54:219–226

Foidl M, Golecki JR, Oelze J (1998) Chlorosome development in *Chloroflexus aurantiacus*. Photosynth Res 55:109–114

Frese R, Oberheide U, van Stokkum IHM, van Grondelle R, Foidl M, Oelze J, van Amerongen H (1997) The organization of bacteriochlorophyll *c* in chlorosomes from *Chloroflexus aurantiacus* and the structural role of carotenoids and protein – an absorption, linear dichroism, circular dichroism and Stark spectroscopy study. Photosynth Res 54:115–126

Frigaard NU, Bryant D (2004) Seeing green bacteria in a new light: genomics-enabled studies of the photosynthetic apparatus in green sulfur bacteria and filamentous anoxygenic phototrophic bacteria. Arch Microbiol 182:265–276

Frigaard NU, Bryant DA (2006) Chlorosomes: antenna organelles in photosynthetic green bacteria. In: Shively JM (ed) Complex intracellular structures in prokaryotes, vol 2, Microbiology monographs. Springer, Berlin, pp 79–114

Frigaard NU, Matsuura K (1999) Oxygen uncouples light absorption by the chlorosome antenna and photosynthetic electron transfer in the green sulfur bacterium *Chlorobium tepidum*. Biochim Biophys Acta 1412:108–117

Frigaard NU, Takaichi S, Hirota M, Shimada K, Matsuura K (1997) Quinones in chlorosomes of green sulfur bacteria and their role in the redox-dependent fluorescence studied in chlorosome-like bacteriochlorophyll *c* aggregates. Arch Microbiol 167:343–349

Frigaard NU, Matsuura K, Hirota M, Miller M, Cox RP (1998) Studies of the location and function of isoprenoid quinones in chlorosomes from green sulfur bacteria. Photosynth Res 58:81–90

Frigaard NU, Chew AGM, Li H, Maresca JA, Bryant DA (2003) *Chlorobium tepidum*: insights into the structure, physiology, and metabolism of a green sulfur bacterium derived from a complete genome sequence. Photosynth Res 78:93–117

Frigaard NU, Li H, Milks KJ, Bryant DA (2004) Nine mutants of *Chlorobium tepidum* each unable to synthesize a different chlorosome protein still assemble functional chlorosomes. J Bacteriol 186:646–653
Frigaard NU, Li H, Martinsson P, Das SK, Frank HA, Aartsma TJ, Bryant DA (2005) Isolation and characterization of carotenosomes from a bacteriochlorophyll *c*-less mutant of *Chlorobium tepidum*. Photosynth Res 86:101–111
Fuciman M, Chabera P, Zupcanova A, Hribek P, Arellano JB, Vacha F, Psencik J, Polivka T (2010) Excited state properties of aryl carotenoids. Phys Chem Chem Phys 12:3112–3120
Furumaki S, Vacha F, Habuchi S, Tsukatani Y, Bryant DA, Vacha M (2011) Absorption linear dichroism measured directly on a single light-harvesting system: the role of disorder in chlorosomes of green photosynthetic bacteria. J Am Chem Soc 133:6703–6710
Ganapathy S, Oostergetel GT, Wawrzyniak PK, Reus M, Chew AGM, Buda F, Boekema EJ, Bryant DA, Holzwarth AR, de Groot HJM (2009) Alternating syn-anti bacteriochlorophylls form concentric helical nanotubes in chlorosomes. Proc Natl Acad Sci U S A 106:8525–8530
Gounaris K, Barber J (1983) Monogalactosyldiacyglycerol: the most abundant polar lipid in nature. Trends Biochem Sci 8:378–381
Halfen LN, Pierson BK, Francis GW (1972) Carotenoids of a gliding organism containing bacteriochlorophylls. Arch Mikrobiol 82:240–246
Hamley IW (1998) The Physics of Block Copolymers. Oxford University Press, Oxford
Hildebrandt P, Tamiaki H, Holzwarth AR, Schaffner K (1994) Resonance Raman-spectroscopic study of metallochlorin aggregates. Implications for the supramolecular structure in chlorosomal BChl c antennae of green bacteria. J Phys Chem 98:2192–2197
Hirota M, Moriyama T, Shimada K, Miller M, Olson JM, Matsuura K (1992) High degree of organization of bacteriochlorophyll *c* in chlorosome-like aggregates spontaneously assembled in aqueous solution. Biochim Biophys Acta 1099:271–274
Hohmann-Marriott MF, Blankenship RE (2007) Hypothesis on chlorosome biogenesis in green photosynthetic bacteria. FEBS Lett 581:800–803
Holzwarth AR, Schaffner K (1994) On the structure of bacteriochlorophyll molecular aggregates in the chlorosomes of green bacteria. A molecular modelling study. Photosynth Res 41:225–233
Hu XC, Damjanovic A, Ritz T, Schulten K (1998) Architecture and mechanism of the light-harvesting apparatus of purple bacteria. Proc Natl Acad Sci U S A 95:5935–5941
Ikonen TP, Li H, Psencik J, Laurinmäki PA, Butcher SJ, Frigaard NU, Serimaa RE, Bryant DA, Tuma R (2007) X-ray scattering and electron cryomicroscopy study on the effect of carotenoid biosynthesis to the structure of *Chlorobium tepidum* chlorosomes. Biophys J 93:620–628
Imhoff JF (1995) Taxonomy and physiology of phototrophic purple bacteria and green sulfur bacteria. In: Blankenship RE, Madigan MT, Bauer CE (eds) Anoxygenic photosynthetic bacteria. Kluwer Academic Publishers, Dordrecht, pp 1–15
Imhoff JF (2003) Phylogenetic taxonomy of the family *Chlorobiaceae* on the basis of 16S rRNA and fmo (Fenna Matthews-Olson protein) gene sequences. Int J Sys Evol Microbiol 53:941–951
Imhoff JF, Thiel V (2010) Phylogeny and taxonomy of *Chlorobiaceae*. Photosynth Res 104:123–136
Jochum T, Reddy CM, Eichhofer A, Buth G, Szmytkowski J, Kalt H, Moss D, Balaban TS (2008) The supramolecular organization of self-assembling chlorosomal bacteriochlorophyll *c*, *d*, or *e* mimics. Proc Natl Acad Sci U S A 105:12736–12741
Kakitani Y, Koyama Y, Shimoikeda Y, Nakai T, Utsumi H, Shimizu T, Nagae H (2009) Stacking of bacteriochlorophyll *c* macrocycles in chlorosome from *Chlorobium limicola* as revealed by intermolecular C-13 magnetic-dipole correlation, X-ray diffraction, and quadrupole coupling in Mg-25 NMR. Biochemistry 48:74–86
Kim H, Li H, Maresca JA, Bryant DA, Savikhin S (2007) Triplet exciton formation as a novel photoprotection mechanism in chlorosomes of *Chlorobium tepidum*. Biophys J 93:192–201
Klappenbach JA, Pierson BK (2004) Phylogenetic and physiological characterization of a filamentous anoxygenic photoautotrophic bacterium 'Candidatus Chlorothrix halophila' gen. nov., sp nov., recovered from hypersaline microbial mats. Arch Microbiol 181:17–25
Klinger P, Arellano JB, Vacha FE, Hala J, Psencik J (2004) Effect of carotenoids and monogalactosyl diglyceride on bacteriochlorophyll *c* aggregates in aqueous buffer: implications for the self-assembly of chlorosomes. Photochem Photobiol 80:572–578
Knudsen E, Jantzen E, Bryn K, Ormerod JG, Sirevag R (1982) Quantitative and structural characteristics of lipids in *Chlorobium* and *Chloroflexus*. Arch Microbiol 132:149–154
Krasnovsky AA Jr, Litvin FF (1975) Mechanisms of delayed luminescence of photosynthetic pigments. Bull USSR Acad Sci Phys 39:1968–1971
Krasnovsky AA, Lopez J, Cheng P, Liddell PA, Blankenship RE, Moore TA, Gust D (1994)

Generation and quenching of singlet molecular-oxygen by aggregated bacteriochlorophyll *d* in model systems and chlorosomes. Photosynth Res 40:191–198

Larsen KL, Cox RP, Miller M (1994) Effects of illumination intensity on bacteriochlorophyll *c* homolog distribution in *Chloroflexus aurantiacus* grown under controlled conditions. Photosynth Res 41:151–156

Li H, Bryant DA (2009) Envelope proteins of the CsmB/CsmF and CsmC/CsmD motif families influence the size, shape, and composition of chlorosomes in *Chlorobaculum tepidum*. J Bacteriol 191:7109–7120

Li H, Frigaard NU, Bryant DA (2006) Molecular contacts for chlorosome envelope proteins revealed by cross-linking studies with chlorosomes from *Chlorobium tepidum*. Biochemistry 45:9095–9103

Li H, Jubelirer S, Costas AMG, Frigaard NU, Bryant DA (2009) Multiple antioxidant proteins protect *Chlorobaculum tepidum* against oxygen and reactive oxygen species. Arch Microbiol 191:853–867

Linnanto JM, Korppi-Tommola JEI (2008) Investigation on chlorosomal antenna geometries: tube, lamella and spiral-type self-aggregates. Photosynth Res 96:227–245

Maresca JA, Graham JE, Bryant DA (2008) The biochemical basis for structural diversity in the carotenoids of chlorophototrophic bacteria. Photosynth Res 97:121–140

Martinez-Planells A, Arellano JB, Borrego CM, Lopez-Iglesias C, Gich F, Garcia-Gil JS (2002) Determination of the topography and biometry of chlorosomes by atomic force microscopy. Photosynth Res 71:83–90

Martiskainen J, Linnanto J, Kananavicius R, Lehtovuori V, Korppi-Tommola J (2009) Excitation energy transfer in isolated chlorosomes from *Chloroflexus aurantiacus*. Chem Phys Lett 477:216–220

Matsuura K, Hirota M, Shimada M, Mimuro M (1993) Spectral forms and orientation of bacteriochlorophylls c and a in chlorosomes of green photosynthetic bacterium *Chloroflexus aurantiacus*. Photochem Photobiol 57:92–97

Melo TB, Frigaard NU, Matsuura K, Naqvi KR (2000) Electronic energy transfer involving carotenoid pigments in chlorosomes of two green bacteria: *Chlorobium tepidum* and *Chloroflexus aurantiacus*. Spectrochim Acta A Mol Biol Spectrosc 56: 2001–2010

Mimuro M, Hirota M, Nishimura Y, Moriyama T, Yamazaki I, Shimada K, Matsuura K (1994) Molecular organization of bacteriochlorophyl in chlorosomes of the green photosynthetic bacterium *Chloroflexus aurantiacus*: studies of fluorescence depolarization accompanied by energy transfer processes. Photosynth Res 41:181–191

Miyatake T, Tamiaki H (2005) Self-aggregates of bacteriochlorophylls *c*, *d* and *e* in a light-harvesting antenna system of green photosynthetic bacteria: effect of stereochemistry at the chiral 3-(1-hydroxyethyl) group on the supramolecular arrangement of chlorophyllous pigments. J Photochem Photobiol C 6:89–107

Miyatake T, Tamiaki H (2010) Self-aggregates of natural chlorophylls and their synthetic analogues in aqueous media for making light-harvesting systems. Coord Chem Rev 254:2593–2602

Mizoguchi T, Yoshitomi T, Harada J, Tamiaki H (2011) Temperature- and time-dependent changes in the structure and composition of glycolipids during the growth of the green sulfur photosynthetic bacterium *Chlorobaculum tepidum*. Biochemistry 50:4504–4512

Montano GA, Bowen BP, LaBelle JT, Woodbury NW, Pizziconi VB, Blankenship RE (2003a) Characterization of *Chlorobium tepidum* chloromes: a calculation of bacteriochlorophyll *c* per chlorosome and oligomer modeling. Biophys J 85:2560–2565

Montano GA, Wu HM, Lin S, Brune DC, Blankenship RE (2003b) Isolation and characterization of the B798 light-harvesting baseplate from the chlorosomes of *Chloroflexus aurantiacus*. Biochemistry 42:10246–10251

Nowicka B, Kruk J (2010) Occurrence, biosynthesis and function of isoprenoid quinones. Biochim Biophys Acta 1797:1587–1605

Nozawa T, Ohtomo K, Suzuki M, Nakagawa H, Shikama Y, Konami H, Wang ZY (1994) Structures of chlorosomes and aggregated BChl *c* in *Chlorobium tepidum* from solid state high resolution CP/MAS ^{13}C NMR. Photosynth Res 41:211–223

Oelze J, Golecki JR (1995) Membranes and chlorosomes of green bacteria: structure, composition, and development. In: Blankenship RE, Madigan MT, Bauer CE (eds) Anoxygenic photosynthetic bacteria. Kluwer Academic Publishers, Dordrecht, pp 259–278

Oostergetel GT, Reus M, Gomez Maqueo Chew A, Bryant DA, Boekema EJ, Holzwarth AR (2007) Long-range organization of bacteriochlorophyll in chlorosomes of *Chlorobium tepidum* investigated by cryo-electron microscopy. FEBS Lett 581: 5435–5439

Oostergetel GT, van Amerongen H, Boekema EJ (2010) The chlorosome: a prototype for efficient light harvesting in photosynthesis. Photosynth Res 104:245–255

Overmann J, Cypionka H, Pfennig N (1992) An extremely low-light-adapted phototrophic sulfur

bacterium from the Black Sea. Limnol Oceanogr 37:150–155
Pedersen MO, Underhaug J, Dittmer J, Miller M, Nielsen NC (2008) The three-dimensional structure of CsmA: a small antenna protein from the green sulfur bacterium *Chlorobium tepidum*. FEBS Lett 582:2869–2874
Pedersen MO, Linnanto J, Frigaard NU, Nielsen NC, Miller M (2010) A model of the protein-pigment baseplate complex in chlorosomes of photosynthetic green bacteria. Photosynth Res 104:233–243
Persson S, Sonksen CP, Frigaard NU, Cox RP, Roepstorff P, Miller M (2000) Pigments and proteins in green bacterial chlorosomes studied by matrix-assisted laser desorption ionization mass spectrometry. Eur J Biochem 267:450–456
Polivka T, Sundstrom V (2004) Ultrafast dynamics of carotenoid excited states-from solution to natural and artificial systems. Chem Rev 104:2021–2071
Prokhorenko VI, Steensgaard DB, Holzwarth AR (2000) Exciton dynamics in the chlorosomal antennae of the green bacteria *Chloroflexus aurantiacus* and *Chlorobium tepidum*. Biophys J 79:2105–2120
Prokhorenko VI, Holzwarth AR, Muller MG, Schaffner K, Miyatake T, Tamiaki H (2002) Energy transfer in supramolecular artificial antennae units of synthetic zinc chlorins and co-aggregated energy traps. A time- resolved fluorescence study. J Phys Chem B 106:5761–5768
Psencik J, Vacha M, Adamec F, Ambroz M, Dian J, Bocek J, Hala J (1994) Hole-burning study of excited-state structure and energy- transfer dynamics of bacteriochlorophyll-*c* in chlorosomes of green sulfur photosynthetic bacteria. Photosynth Res 42:1–8
Psencik J, Ma YZ, Arellano JB, Garcia-Gil J, Holzwarth AR, Gillbro T (2002) Excitation energy transfer in chlorosomes of *Chlorobium phaeobacteroides* strain CL1401: the role of carotenoids. Photosynth Res 71:5–18
Psencik J, Ma YZ, Arellano JB, Hala J, Gillbro T (2003) Excitation energy transfer dynamics and excited-state structure in chlorosomes of *Chlorobium phaeobacteroides*. Biophys J 84:1161–1179
Psencik J, Ikonen TP, Laurinmäki P, Merckel MC, Butcher SJ, Serimaa RE, Tuma R (2004) Lamellar organization of pigments in chlorosomes, the light harvesting complexes of green photosynthetic bacteria. Biophys J 87:1165–1172
Psencik J, Arellano JB, Ikonen TP, Borrego CM, Laurinmäki PA, Butcher SJ, Serimaa RE, Tuma R (2006) Internal structure of chlorosomes from brown-colored *Chlorobium* species and the role of carotenoids in their assembly. Biophys J 91:1433–1440
Psencik J, Collins AM, Liljeroos L, Torkkeli M, Laurinmäki P, Ansink HM, Ikonen TP, Serimaa RE, Blankenship RE, Tuma R, Butcher SJ (2009) Structure of chlorosomes from the green filamentous bacterium *Chloroflexus aurantiacus*. J Bacteriol 191:6701–6708
Psencik J, Torkkeli M, Zupcanova A, Vacha F, Serimaa RE, Tuma R (2010) The lamellar spacing in self-assembling bacteriochlorophyll aggregates is proportional to the length of the esterifying alcohol. Photosynth Res 104:211–219
Psencik J, Arellano JB, Collins AM, Laurinmäki P, Torkkeli M, Löflund B, Serimaa RE, Blankenship RE, Tuma R, Butcher SJ (2013) The structural and functional role of carotenoids in chlorosomes. J Bacteriol 195:1727–1734
Pullerits T (2000) Exciton states and relaxation in molecular aggregates: numerical study of photosynthetic light harvesting. J Chin Chem Soc 47:773–784
Saga Y, Shibata Y, Ltoh S, Tamiaki H (2007) Direct counting of submicrometer-sized photosynthetic apparatus dispersed in medium at cryogenic temperature by confocal laser fluorescence microscopy: estimation of the number of bacteriochlorophyll c in single light-harvesting antenna complexes chlorosomes of green photosynthetic bacteria. J Phys Chem B 111:12605–12609
Saga Y, Shibata Y, Tamiaki H (2010) Spectral properties of single light-harvesting complexes in bacterial photosynthesis. J Photochem Photobiol C 11:15–24
Savikhin S, Zhu Y, Blankenship RE, Struve WS (1996) Intraband energy transfers in the BChl *c* antenna of chlorosomes from the green photosynthetic bacterium Chloroflexus aurantiacus. J Phys Chem 100:17978–17980
Savikhin S, Buck DR, Struve WS, Blankenship RE, Taisova AS, Novoderezhkin VI, Fetisova ZG (1998) Excitation delocalization in the bacteriochlorophyll *c* antenna of the green bacterium *Chloroflexus aurantiacus* as revealed by ultrafast pump-probe spectroscopy. FEBS Lett 430:323–326
Scheer H (2006) An overview of chlorophylls and bacteriochlorophylls: biochemistry, biophysics, functions and applications. In: Grimm B, Porra RJ, Rudiger W, Scheer H (eds) Chlorophylls and bacteriochlorophylls. Springer, Dordrecht, pp 1–26
Schmidt K, Maarzahl M, Mayer F (1980) Development and pigmentation of chlorosomes in *Chloroflexus aurantiacus* strain Ok-70-fl. Arch Microbiol 127:87–97
Shibata Y, Saga Y, Tamiaki H, Itoh S (2007) Polarized fluorescence of aggregated bacteriochlorophyll c and baseplate bacteriochlorophyll a in single chlorosomes isolated from *Chloroflexus aurantiacus*. Biochemistry 46:7062–7068

Shibata Y, Saga Y, Tamiaki H, Itoh S (2009) Anisotropic distribution of emitting transition dipoles in chlorosome from *Chlorobium tepidum*: fluorescence polarization anisotropy study of single chlorosomes. Photosynth Res 100:67–78

Shibata Y, Tateishi S, Nakabayashi S, Itoh S, Tamiaki H (2010) Intensity borrowing via excitonic couplings among soret and Q(y) transitions of bacteriochlorophylls in the pigment aggregates of chlorosomes, the light-harvesting antennae of green sulfur bacteria. Biochemistry 49:7504–7515

Sorensen PG, Cox RP, Miller M (2008) Chlorosome lipids from *Chlorobium tepidum*: characterization and quantification of polar lipids and wax esters. Photosynth Res 95:191–196

Spontak RJ, Patel NP (2004) Phase behaviour of copolymer blends. In: Hamley IW (ed) Developments in block copolymers science and technology. Wiley, Chichester, pp 159–212

Sprague SG, Staehelin LA, Dibartolomeis MJ, Fuller RC (1981) Isolation and development of chlorosomes in the green bacterium *Chloroflexus aurantiacus*. J Bacteriol 147:1021–1031

Staehelin LA, Golecki JR, Fuller RC, Drews G (1978) Visualization of the supramolecular architecture of chlorosome (Chlorobium type vesicles) in freeze-fractured cells of *Chloroflexus aurantiacus*. Arch Microbiol 119:269–277

Staehelin LA, Golecki JR, Drews G (1980) Supramolecular organization of chlorosome (Chlorobium vesicles) and of their membrane attachment site in *Chlorobium limicola*. Biochim Biophys Acta 589:30–45

Stanier RY, Smith JHC (1960) The chlorophylls of green bacteria. Biochim Biophys Acta 41:478–484

Steensgaard DB, van Walree CA, Permentier H, Baneras L, Borrego CM, Garcia-Gil J, Aartsma TJ, Amesz J, Holzwarth AR (2000a) Fast energy transfer between BChl *d* and BChl *c* in chlorosomes of the green sulfur bacterium *Chlorobium limicola*. Biochim Biophys Acta 1457:71–80

Steensgaard DB, Wackerbarth H, Hildebrandt P, Holzwarth AR (2000b) Diastereoselective control of bacteriochlorophyll *e* aggregation. 3(1)-S-BChl *e* is essential for the formation of chlorosome-like aggregates. J Phys Chem B 104:10379–10386

Takaichi S (1999) Carotenoids and carotenogenesis in anoxygenic photosynthetic bacteria. In: Frank HA, Young AJ, Britton G, Cogdell RJ (eds) The photochemistry of carotenoids. Kluwer, Dordrecht, pp 39–69

Takaichi S, Oh-Oka H (1999) Pigment composition in the reaction center complex from the thermophilic green sulfur bacterium, *Chlorobium tepidum*: carotenoid glucoside esters, menaquinone and chlorophylls. Plant Cell Physiol 40:691–694

Tamiaki H, Tateishi S, Nakabayashi S, Shibata Y, Itoh S (2010) Linearly polarized light absorption spectra of chlorosomes, light-harvesting antennas of photosynthetic green sulfur bacteria. Chem Phys Lett 484:333–337

Tamiaki H, Komada J, Kunieda M, Fukai K, Yoshitomi T, Harada J, Mizoguchi T (2011) In vitro synthesis and characterization of bacteriochlorophyll-*f* and its absence in bacteriochlorophyll-*e* producing organisms. Photosynth Res 107:133–138

van Amerongen H, Valkunas L, van Grondelle R (2000) Photosynthetic excitons. World Scientific, Singapore

van Dorssen RJ, Vasmel H, Amesz J (1986) Pigment organization and energy transfer in the green photosynthetic bacterium *Chloroflexus aurantiacus*. II. The chlorosome. Photosynth Res 9:33–45

van Noort PI, Zhu Y, LoBrutto R, Blankenship RE (1997) Redox effects on the excited-state lifetime in chlorosomes and bacteriochlorophyll *c* oligomers. Biophys J 72:316–325

van Rossum BJ, Steensgaard DB, Mulder FM, Boender GJ, Schaffner K, Holzwarth AR, de Groot HJM (2001) A refined model of the chlorosomal antennae of the green bacterium *Chlorobium tepidum* from proton chemical shift constraints obtained with high-field 2-D and 3-D MAS NMR dipolar correlation spectroscopy. Biochemistry 40:1587–1595

van Walree CA, Sakuragi Y, Steensgaard DB, Frigaard NU, Cox RP, Holzwarth AR, Miller M (1999) Effect of alkaline treatment on bacteriochlorophyll a, quinones and energy transfer in chlorosomes from *Chlorobium tepidum* and *Chlorobium phaeobacteroides*. Photochem Photobiol 69:322–328

Vassilieva EV, Stirewalt VL, Jakobs CU, Frigaard NU, Inoue-Sakamoto K, Baker MA, Sotak A, Bryant DA (2002) Subcellular localization of chlorosome proteins in *Chlorobium tepidum* and characterization of three new chlorosome proteins: CsmF, CsrnH, and CsmX. Biochemistry 41:4358–4370

Vila X, Abella CA (1994) Effects of Light quality on the physiology and the ecology of planktonic green sulfur bacteria in lakes. Photosynth Res 41:53–65

Wang YL, Mao LS, Hu XC (2004) Insight into the structural role of carotenoids in the photosystem I: a quantum chemical analysis. Biophys J 86: 3097–3111

Zobova A, Taisova A, Lukashev E, Fedorova N, Baratova L, Fetisova Z (2011) CsmA protein is associated with BChl *a* in the baseplate subantenna of chlorosomes of the photosynthetic green filamentous bacterium *Oscillochloris trichoides* belonging to the family *Oscillochloridaceae*. J Biophys 2011:1860382

Zupcanova A, Arellano JB, Bina D, Kopecky J, Psencik J, Vacha F (2008) The length of esterifying alcohol affects the aggregation properties of chlorosomal bacteriochlorophylls. Photochem Photobiol 84: 1187–1194

Chapter 6

The Structure of ATPsynthases in Photosynthesis and Respiration

Bettina Böttcher*
School of Biological Sciences, University of Edinburgh, King's Buildings, Edinburgh EH9 3JR, UK

and

Peter Gräber
Institut für Physikalische Chemie, Albert-Ludwigs-Universität Freiburg, D-79104 Freiburg, Germany

Summary

A large number of biochemical reactions in cells are coupled with the hydrolysis of ATP, e.g. biosynthesis, ion translocation, muscle contraction. The largest amount of ATP is generated by F-type H^+-ATPases, which are membrane integrated enzymes occurring in the cytoplasma membranes of bacteria, inner mitochondrial membranes and thylakoid membranes. Together with A-type and V-type ATPases they form a specific protein family, which couples proton (Na^+) translocation across the membrane with the synthesis or hydrolysis of ATP. ATPases of this family are found in all taxonomic kingdoms and comprise F-ATPases, A-ATPases in Archaea and some bacteria and V-ATPases in the inner membranes of eukaryotes.

*Author for correspondence, e-mail: bettina.boettcher@ed.ac.uk

M.F. Hohmann-Marriott (ed.), The Structural Basis of Biological Energy Generation, Advances in Photosynthesis and Respiration 39, pp. 111–132, DOI 10.1007/978-94-017-8742-0_6,

A- and F-ATPases can function in ATP hydrolysis and ATP synthesis, although their main physiological role is ATP synthesis in most organisms. In contrast, V-ATPases are dedicated proton pumps, which work only in the direction of ATP hydrolysis. The predecessor of these ATPases was already present in the last universal common ancestor and probably might have evolved from RNA transporting translocases, which were located in the primordial membranes. Due to this evolutionary relationship, A-, V- and F-ATPases have a common architecture, which comprises a membrane-embedded part (F_0, A_0 or V_0) and an extrinsic part (F_1, A_1 or V_1) separated by a connecting region. This architecture supports a rotary mechanism, in which the rotation of the membrane- embedded part is concomitant with the transport of H^+ (or Na^+) across the membrane. The rotation of the membrane-embedded rotor is conveyed via a central stalk in the connecting region to the extrinsic part, where it is coupled to conformational changes in the catalytic nucleotide binding sites that support ATP synthesis/hydrolysis. An additional peripheral connection between the extrinsic part and the membrane bound part prevents the co-rotation of the extrinsic part and thus supports efficient coupling. Although, atomic structures of the holoenzymes are still missing, atomic models of sub-complexes and subunits give detailed insights into how the rotary mechanism couples ion-translocation with ATP synthesis/hydrolysis in this protein family.

I. Introduction

F-type H^+-ATPsynthases are the central enzymes of the energy metabolism in modern cells. They are membrane integrated enzymes, which occur in the energy-transducing cytoplasmic membranes of bacteria and Archaea, in the thylakoid membrane of cyanobacteria and eukaryotic oxygenic phototrophs, and in the inner mitochondrial membranes. Respiratory as well as the photosynthetic electron transport chain, located in these membranes, generate a transmembrane electrochemical potential by translocating protons across the membranes. The H^+-ATPsynthase can use the Gibbs Free Energy derived from the backflow of protons to synthesize ATP from ADP and inorganic phosphate ("chemiosmotic theory", Mitchell 1961). The two functionalities of the enzymes (proton translocation and ATP synthesis) are reflected by their structure. ATPsynthases consist of: (1) the hydrophilic, extrinsic F_1-part, which carries the nucleotide and phosphate binding sites and supports ATP synthesis and ATP hydrolysis and the hydrophilic; (2) membrane integrated F_0-part, which carries the proton binding sites and performs the proton translocation across the membrane. Both parts are coupled structurally and functionally by a central and a peripheral stalk (Fig. 6.1).

The bacterial H^+-ATPsynthase from *E. coli* has the simplest known subunit composition and will be described here as the prototype: The F_1-part contains five subunits with the stoichiometry $\alpha_3\beta_3\gamma\delta\varepsilon$; the F_0-part contains three subunits with the stoichiometry ab_2c_{10}. The catalytic nucleotide binding sites are located on the β-subunits. The "binding change" theory describes the cooperativity between these sites (Boyer 1993). Accordingly, a catalytic site can exist in three different conformations "open" (containing no bound nucleotides), "closed" (bound ADP and P_i) and "tight" (tightly bound ATP). This model was later corroborated by the first crystal structure of the mitochondrial F_1 (Abrahams et al. 1994).

During catalysis the sequence open → closed → tight → open leads to the binding of substrates, formation of a tightly bound ATP, and finally to the release of the product.

Abbreviations: aa – Amino acid; A-ATPase – Archaeal ATPase/ATPsynthase; ADP – Adenosine diphosphate; ATP – Adenosine triphosphate; EM – Electron microscopy; F-ATPase – F-type ATPase/ATPsynthase; FRET – Fluorescence resonance energy transfer; LUCA – Last universal common ancestor; P_i – Inorganic phosphate; V-ATPase – Vacuolar-type H+-ATPase

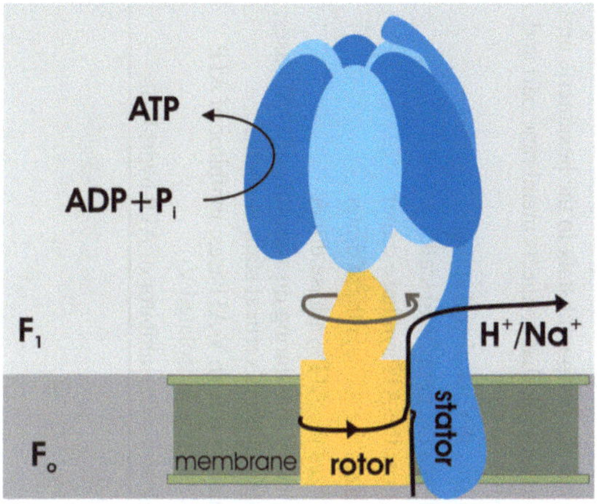

Fig. 6.1. Relationship between function and the bi-partite organization of F-ATPases. F-ATPases consist of a membrane-embedded F_0-part and an extrinsic F_1-part. During ATP synthesis, H^+ (or Na^+) pass between the membrane-embedded rotor (*orange*) and the stator (*dark blue*) and drive rotation of the rotor (*orange*). This induces conformational changes in the three catalytic nucleotide binding sites of $(\alpha\beta)_3$ (*light blue, dark blue*), which enable ATP synthesis form ADP and P_i. During a complete rotation of the rotor, each of the three sites produces one ATP.

Each of the three catalytic sites is in one of these conformations and during catalysis all three sites change their conformation simultaneously. This cooperativity is achieved by the different interactions of the γ-subunit with the three β-subunits, which gives rise to the different conformations of the catalytic sites. A rotational movement of the centrally located γ-subunit was proposed (Boyer 1993) and later experimentally confirmed for ATP hydrolysis catalyzed by immobilized F_1-parts (Noji 1998) and for ATP synthesis catalyzed by F_0F_1 reconstituted into liposome (Diez et al. 2004).

The γ-subunit together with the ε-subunit form the central stalk in F_0F_1 and, both subunits interact with the c-subunits which form a ring like structure (c_{10}) in the membrane. Each c-subunit carries a proton binding site (glutamic acid or aspartic acid) located in the centre of the membrane. Protons from the high proton concentration side can reach this binding site via a proton conducting entrance half channel. Protons can leave this site by an exit half channel, which is connected with the low proton concentration side. This proton translocation leads to a rotation of the c-ring together with the connected γ/ε-complex (Junge et al. 1997).

Thus, F-type ATPsynthases are two structurally and functionally coupled nanomotors. The F_1-part catalyses ATP hydrolysis and this leads to a rotational movement of the γ-subunit. The F_0-part catalyses proton transport which leads to a rotation of the c-ring. The coupling of both motors in F_0F_1 allows the catalysis of the coupled reaction in both directions, proton transport-coupled ATP hydrolysis, and proton transport-coupled ATP synthesis.

In addition to the well characterized F-type ATPases, A-type (Archaea) ATPases and V-type (vacuolar) ATPases have been characterized. These enzymes also couple ATP synthesis/hydrolysis with the translocation of H^+ (in some bacteria also Na^+-ions) across a membrane and are evolutionary related to F-type ATPases. The two functionalities (ATP synthesis/hydrolysis and ion-translocation) of these enzymes are reflected by a similar architecture, which shows two distinct parts (1) the extrinsic part (V_1 or A_1), which carries the nucleotide binding sites and supports ATP synthesis/hydrolysis; (2) the membrane-embedded part (V_0 or A_0), which enables the ion translocation.

II. Relation Between V-, A- and F-ATPases

Although V-, A- and F-ATPases have a similar architecture their subunit composition is different (see Table 6.1). F-ATPases function primarily in the direction of ATP synthesis but can also operate in the opposite direction using ATP hydrolysis for actively transporting H^+ across the membrane, thereby generating a transmembrane electrochemical potential. With only eight different subunits, bacteria have the simplest F-ATPases. Most of the extrinsic part, F_1, is formed by the three α- and the three β-subunits, which are arranged alternating in a hexamer and carry the nucleotide

Table 6.1. Composition of A-, V- and F-ATPases.

Building block	F_1F_o bacterial	F_1F_o chloroplast	F_1F_o mitochondrial	A_1A_0 thermus	V_1V_0 yeast	Function
Hexameric unit						
	α(3)	α (3)	α (3)	B(3)	B(3)	Non-catalytic nucleotide-binding subunits
	β(3)	β (3)	β (3)	A(3)	A(3)	Catalytic nucleotide-binding subunits
Rotating central stalk						
	γ	γ	γ	D	D	Axle
	ε	ε	δ	F	F	Foot, regulatory role
			E			Stabilization of central stalk
				C	d	Adaptor to the rotor
Rotor						
	c(11–13)	III(14–15)	C(8–10)	L (10?)	c(8?),c',c"	Rotating part of ion-channel
Peripheral stalk						
	a	IV	a	C-term I	C-term a	Static part of ion channel
				N-term I	N-term a	Interconnects vertical peripheral stalk elements
	δ	δ	OSCP	C-term E(2)	C-term E(3)	Attaches the peripheral stalk to the hexameric unit
	b(2)	I,II	b	G(2) N-term E(2)	G(3) N-term E(3)	Peripheral stalk that connects membrane part and extrinsic part
			d			Stabilizes the peripheral stalk
			F6			Stabilizes the peripheral stalk
			e			Stabilizes the ATPsynthase dimer
			f			Stabilizes the ATPsynthase dimer
			g			Stabilizes the ATPsynthase dimer
					C	Regulatory role during disassembly of V-ATPases; interconnects vertical stalk elements
					H	Regulatory role in V-ATPases; inhibitor of ATP hydrolysis in isolated V_1
			IF6			Regulatory role, inhibitor of ATP hydrolysis

Stoichiometries are given in *brackets*. Membrane-embedded subunits are *shaded grey*; related subunits are shown in the same row binding sites. The α- and β-subunits are evolutionary related and have probably arisen by gene-duplication. While the β-subunits hold the catalytic nucleotide binding sites at the interface to the α-subunits, the α-subunits carry the regulatory nucleotide binding sites at the interface to the β-subunits. Inside the hexamer, the γ- and ε-subunits from the central stalk that rotates during catalysis. At the outside of the hexamer subunit δ provides an anchor for the peripheral stator, which is formed by the two amphiphilic b-subunits of the membrane integrated F_0. The b-subunits span the whole length of the bacterial F-ATPase. In F_0, the b-subunits attach to the a-subunit, which provides the static part for the H^+/Na^+-translocation path. The other part of the translocation path is formed by the c-ring rotor. The number of c-subunits varies between species. The smallest copy number is reported for the bovine mitochondrial F-ATPase (eight copies, Watt et al. 2010) and the largest for a cyanobacterial F-ATPase (15 copies in *Spirulina platensis*, Pogoryelov et al. 2005). If we assume that one ATP molecule is synthesized per β-subunit during a 360° rotation of the γ-subunit in F_1 and that each c-subunit translocates one proton the H^+/ATP ratio is identical with the subunit stoichiometry c/β. Since all F-type ATPsynthases analyzed so far contain three β-subunits, the variable number of c-subunits determines the H^+/ATP conversion efficiency of an organism. The general tendency is that more protons per synthesized ATP are necessary in enzymes from photosynthetic membranes than from respiratory membranes.

F-ATPases from chloroplasts and mitochondria are more complex than the bacterial F-ATPases. Most alterations and additional subunits are located in the peripheral stator. In the chloroplast F-ATPase the two copies of subunit b are replaced by the related subunits I and II. Mitochondrial F-ATPases have the most complex stator that consists of subunits b, d, e, f, g, A6L, F6 and OSCP (Collinson et al. 1994; Meyer et al. 2007; Rees et al. 2009; Förster et al. 2010). The central stalk of mitochondrial F-ATPases is formed by three subunits (γ, δ and ε) compared to only two in bacteria and chloroplasts (subunits γ and ε; with ε being homologous to the mitochondrial δ-subunit, see Table 6.1).

A- and V-ATPases are more distantly related to F-ATPases than they are to each other. Due to their close relationship, some authors refer to A-ATPases also as bacterial V-ATPases. However, A-ATPases and eukaryotic V-ATPases play a different physiological role and have a different subunit composition. While A-ATPases in Archaea fulfil a similar physiological role in ATP synthesis as F-ATPases in bacteria, eukaryotic V-ATPases are dedicated proton-pumps that do not synthesize ATP under physiological conditions. Parallel with the different physiological role, sequence comparisons of the catalytic subunits and of the rotor subunits consistently cluster A-ATPases separately form eukaryotic V-ATPases (Gogarten et al. 1992; Kibak et al. 1992). This justifies considering A-ATPases and V-ATPases as separate classes.

A-ATPases are found in Archaea and also in some bacteria, which have acquired them via horizontal gene-transfer. Presumably, in these bacteria the A-ATPases have replaced the respective F-ATPase (Olendzenski et al. 2000; Lapierre et al. 2006). Similarly, as in F-ATPases, some A-ATPases of certain bacteria transport Na^+ instead of H^+. However, these Na^+-translocating ATPases have also the capability to transport H^+, whereas vice versa, primarily H^+-transporting ATPases have somewhat smaller ion binding sites and therefore are dedicated H^+-translocators.

In A-ATPases, the extrinsic A_1 is formed by the catalytic nucleotide binding subunits A, which are homologous to the β-subunits in F-ATPases and by the non-catalytic subunits B, which are related to the α-subunits in F-ATPases. The central stalk in A-ATPases consists of three subunits, C, D and F, which are functionally, but probably not evolutionary, related to the subunits of the central stalk in F-ATPases. The peripheral stalk elements are formed by subunits E and G. These are related to the bacterial b and δ-subunits of F_1, respectively. Subunit E may constitute a fusion product of subunits b and δ. In contrast to the b-subunits in bacterial F-ATPases, subunits E and G have no membrane anchor of their own. They attach to subunit I in A_0. Subunit I forms the static part of the proton translocation path. This subunit is much larger than subunit a in F-ATPases and contains a large N-terminal extrinsic part, which provides the membrane anchor to the soluble stator subunits E and G. Similar to F-ATPases the other part of the proton pathway is contributed by a rotor formed of multiple copies of subunit L (see Table 6.1).

Eukaryotic V-ATPases have two additional subunits, C and H (see Table 6.1 for composition), which are involved in the regulation of the dissociation of V_1 from V_0. This regulatory mechanism is specific to V-ATPases and occurs upon nutrient shortage, which causes V_0 and V_1 to detach (Kane 1995; Sumner et al. 1995).

III. Evolution

The fact that F-, V- and A-ATPases are found in all taxonomic kingdoms indicates their early appearance in evolution. Sequence analysis suggests that their predecessor was already present in the last universal common ancestor (LUCA, Tuller et al. 2010). The ATPase in LUCA had already two different types of nucleotide binding subunits, of which only one was catalytically active. This early ATPase was probably not yet involved in generating ATP. It is likely that the ATPase had evolved from RNA translocases, which arose by the fusion of an RNA-helicase with a pore for translocation (Mulkidjanian et al. 2007). These early RNA-translocases were probably placed in the primordial membranes to enable exchange of RNA between different cells. The early membranes were most likely not yet tight for ions but sequestered RNA and protein from the outside world. Exchange of RNA frequently occurred between different cells. This required an efficient translocation system. Indications of the involvement of the ancestors of the nucleotide binding subunits of the ATPases in RNA or DNA transport are still given by their sequence similarity to RNA helicase and DNA helicases. These helicases are members of the same AAA-ATPase clan that encompasses the nucleotide binding subunits of modern A-, V- and F-ATPases. This sequence similarity to the helicases is concomitant with strong structural similarities. This is evident, for example, from the similarity of the hexameric assembly of TrwB, which is a DNA-dependent ATPase involved in DNA transport during bacterial conjugation (Gomis-Ruth et al. 2001) with the $(\alpha\beta)_3$-assembly of F-ATPases (Fig. 6.2a). This early RNA/DNA-transport system might have evolved into a protein-translocation machinery probably similar to the membrane bound type III secretion system, where the structure of the nucleotide binding subunits are also highly reminiscent of the nucleotide binding subunits in F-ATPases (Fig. 6.2b), (Zarivach et al. 2007).

Protein and DNA/RNA translocases propel a substrate through a central hole of the hexameric assembly of the nucleotide binding subunits. While the transported molecule probably rotates during transport the translocation channel does not. So in the early translocase system the translocation channel was most likely statically connected to the ATPase moiety in contrast to its later transition into a rotor in V-, A- and F-ATPases. The early translocases might have acquired additional links between the translocation channel and the hexameric ATPase moiety to stabilize the complex and to prevent the co-rotation of the pore while the transported molecule was propelled through the pore. Some of these static elements might have been retained in F-, V- and A-ATPases as peripheral stator elements. This is supported by the homology between essential subunits of the peripheral stator with proteins of the type III secretion system (Pallen et al. 2006).

The central shaft is essential for the rotational coupling of ATP synthesis/hydrolysis to H^+/Na^+ translocation. In contrast to the peripheral stator, the central shaft is not conserved between F- and V-/A-ATPases. This suggests that the different types of ATPases evolved independently from a

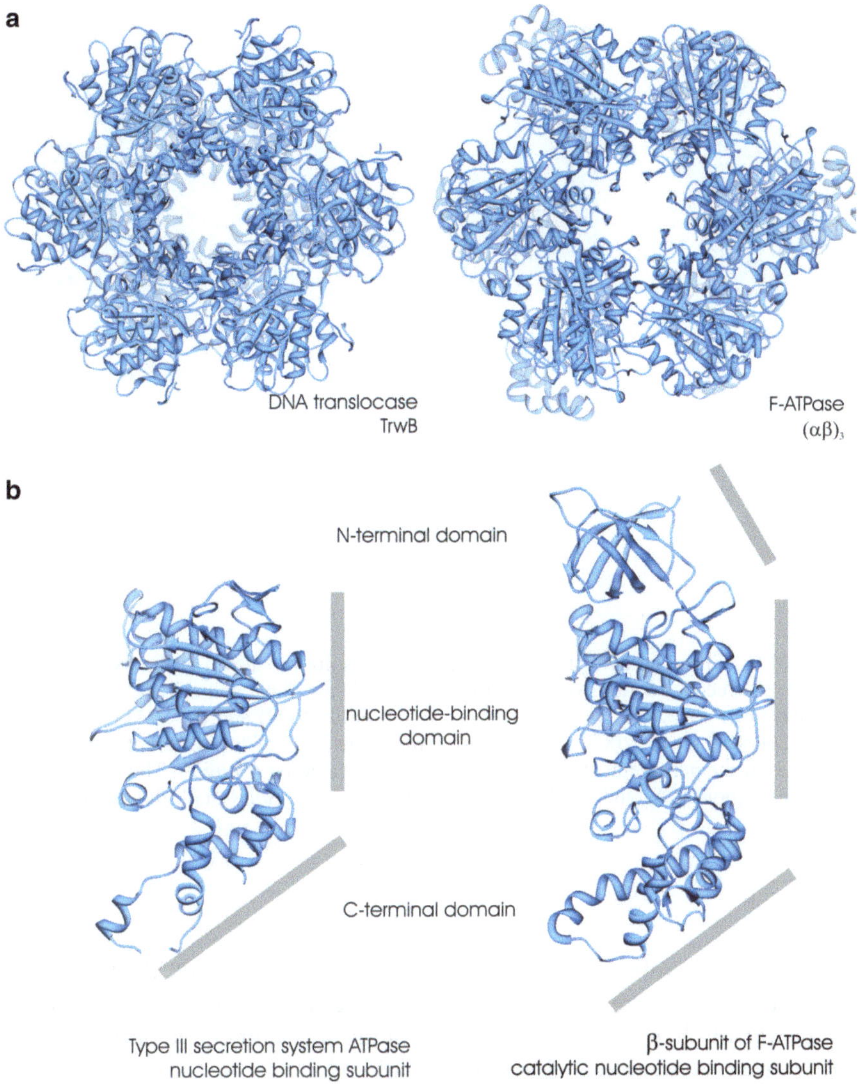

Fig. 6.2. Structural relationship between nucleotide binding subunits in F-ATPase and evolutionary related ATPases involved in DNA-transport. (**a**) The soluble part of the membrane bound DNA translocase TrwB, which transports DNA across membranes in conjugation (TrwB, 1GKI, Gomis-Ruth et al. 2001) and the $(\alpha\beta)_3$-assembly of F-ATPases (1SKY, Shirakihara et al. 1997) have a similar hexagonal arrangement. (**b**) The nucleotide binding subunit of Type III secretion ATPase (2OBM, Zarivach et al. 2007) and the catalytic nucleotide binding subunit of F-ATPase (1SKY, Shirakihara et al. 1997) have a similar architecture.

protein translocase, before the rotational coupling of the two functions was established (Mulkidjanian et al. 2007). Later the central stalk was acquired independently, probably originating from a translocated protein. This scenario could explain the lack of homology in an element that is central for the rotational coupling.

IV. Structure

A. Overall Structure

Structural studies on F-, A- and V-ATPases that reveal atomic resolution of the holoenzymes are still missing. Therefore, our structural knowledge of the architecture of

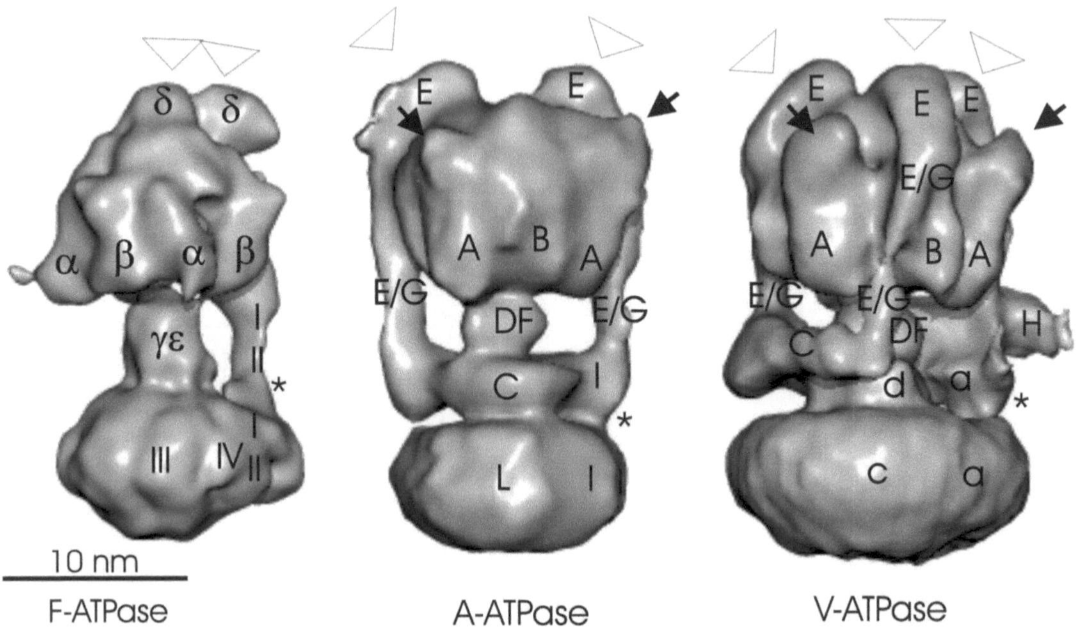

Fig. 6.3. F-, A- and V-ATPase holoenzymes at intermediate resolution shown at the same scale. Surface representations of the EM-reconstructions of the chloroplast F-ATPase form spinach (*left*, Mellwig and Böttcher 2003), the A-ATPase of *Pyrococcus furiosus* (*center*, Vonck et al. 2009) and the V-ATPase of *Saccharomyces cerevisiae* (*right*, Diepholz et al. 2008b). The approximate positions of the different subunits are indicated. The attachment sites of the peripheral stalks to $F_1/A_1/V_1$ are indicated by *triangles*. The approximate position of the non-homologous region in A- and V-ATPases is shown by *arrows*. *Stars* mark the membrane anchor of the peripheral stalk.

the holoenzymes still relies on electron microscopic studies at intermediate resolution (Mellwig and Böttcher 2003; Rubinstein et al. 2003; Venzke et al. 2005; Diepholz et al. 2008b; Lau et al. 2008; Zhang et al. 2008; Muench et al. 2009; Vonck et al. 2009; Lau and Rubinstein 2010).

The EM-reconstructions show that A- and V- ATPases are much more similar to each other than they are to F-ATPases, in accordance with their evolutionary relationship (Fig. 6.3). Main differences between A- and V-ATPases on one hand and F-ATPases on the other hand are: (1) the overall length, which is shorter in F-ATPases; (2) the shape of the membrane distal side of the extrinsic part, which is pointed in F-ATPases and more flat and broader in A- and V-ATPases; (3) the length of the central stalk, which is shorter in F-ATPases and longer and more symmetrical in A- and V-ATPases; (4) the static connections between extrinsic part and membrane-embedded part, which are simpler in F-ATPases.

1. Extrinsic Part

At intermediate resolution the α- and β-subunits in F-ATPases are almost indistinguishable in shape, which gives the F_1-part a regular appearance (Böttcher et al. 2000; Mellwig and Böttcher 2003; Rubinstein et al. 2003, 2005). The single, peripheral static element is anchored in the centre of the αβ-assembly, at the membrane distal side of the F_1-part and gives rise to the pointed appearance of F_1.

In A- and V-ATPases the A-subunits have an insert, which is located at the membrane distal side of V_1/A_1 (Wilkens et al. 2005).

This insert is absent in the B-subunit and also in the α-and β-subunits of F-ATPases and thus was termed the 'non-homologous region'. The non-homologous region increases the coupling efficiency and the stability of the enzyme (Shao et al. 2003), which suggests a role in stator binding. However, the EM-studies of the holoenzymes (Bernal and Stock 2004; Diepholz et al. 2008b; Zhang et al. 2008; Muench et al. 2009) do not show any direct interactions between the non-homologous region and the stator elements. The vertical stator elements are exclusively attached to the smaller B-subunits at the membrane distal side of V_1/A_1. In V-ATPases, a stator element is attached to each of the B-subunits (Diepholz et al. 2008b; Zhang et al. 2008; Muench et al. 2009), whereas in A-ATPases the stator elements are only attached to two of the three B-subunits (Bernal and Stock 2004; Vonck et al. 2009; Lau and Rubinstein 2010). The non-homologous region and the static connectors contribute a large mass to the membrane distal side of V_1/A_1, which explains the larger diameter of this region compared to F-ATPases and its flatter and broader appearance.

The intermediate resolution structures also shed light onto the conformational states of the nucleotide binding subunits. Although individual nucleotides cannot be resolved, the relative packing of the nucleotide binding subunits and the rotational position of the stator are resolved. The expectation is that in an enzyme, which is working with a rotational mechanism, the catalytic nucleotide binding subunits are equivalent. In this case the enzyme should stop in an arbitrary conformation. This would give rise to at least three different conformations which occur with equal frequency. However, none of the intermediate resolution structures of the A-, V- and F-ATPases shows indications for such a polymorphism in conformation. For F- and V-ATPases it was argued that they stop in a unique resting position (Mellwig and Böttcher 2003; Diepholz et al. 2008a). This is supported by single molecule FRET measurements that report a preferred conformation for inactive chloroplast F-ATPases (Bienert et al. 2009, 2011). According to the FRET measurements the resting conformation is different to the predominant conformations during catalysis. This suggests that ATPases have different conformational modes for catalysis and idling.

2. Stator

Despite the sequence similarities of subunits in the peripheral stator, the peripheral stator is astonishingly diverse in F-, V-, and A-ATPases (Fig. 6.3). While F-ATPases have only one extrinsic peripheral stator element (Böttcher et al. 1998; Wilkens and Capaldi 1998; Karrasch and Walker 1999), A-ATPases have two (Bernal and Stock 2004; Esteban et al. 2008; Vonck et al. 2009; Lau and Rubinstein 2010) and eukaryotic V-ATPases have three (Diepholz et al. 2008b; Kitagawa et al. 2008; Zhang et al. 2008; Muench et al. 2009). Irrespective of the number of extrinsic stator elements, the stator assembly is anchored in the membrane at a single site at the periphery of the rotor.

In F-ATPases the stator forms a single long arm, which connects the membrane distal side of F_1 to the periphery of the membrane-embedded F_0 (Böttcher et al. 1998, 2000; Mellwig and Böttcher 2003; Rubinstein et al. 2003, 2005). The stator is tightly attached to the $(\alpha\beta)_3$-assembly and thus does not form a prominent separate unit. In general the stator of mitochondrial F-ATPases is thicker (Rubinstein et al. 2003, 2005) than the stator of F-ATPases in chloroplasts (Mellwig and Böttcher 2003) and bacteria (Böttcher et al. 2000). These differences in mass reflect the more complex subunit composition of the stator in mitochondria.

In A- and V-ATPases two or three identical, vertical stator elements attach to the non-catalytic B-subunits. For an effective anchorage of the vertical stator elements to the membrane, they are interconnected by one or two horizontal arms, which are absent in F-ATPases. In A-ATPases the interconnection is achieved by a single horizontal arm (subunit I in *Thermus thermophilus*), which is directly inserted into the membrane and provides sufficient connectivity for the whole stator assembly.

V-ATPases also have this membrane-anchored arm (formed by subunit *a*) and two more structural elements that are involved in the horizontal interconnection of the vertical stator elements. One of the additional arms establishes the connection between the second and third vertical stator element. This arm is formed by subunit C (Wilkens et al. 2005), which plays a pivotal role in the regulatory disassembly. During disassembly subunit C is lost form the holoenzyme, which is concomitant with a detachment of V_1 from V_0. This loss of the C-subunit removes the anchorage of the third vertical stator element and thus weakens the connection between V_1 and V_0.

The other additional horizontal arm binds to the first vertical stator element and to the membrane anchored horizontal arm but does not interconnect the vertical stator elements. This element is formed by subunit H (Wilkens et al. 2004), which after regulatory disassembly of V-ATPases inhibits ATP hydrolysis of the detached V_1 (Parra et al. 2000). In the holoenzyme, subunit H is located at a strategic position, where after detachment from the horizontal arm, it could change its conformation and connect to the central stalk (Jefferies and Forgac 2008). This would generate a connection between the first vertical stator element and the central rotating stalk and thus would disable rotational catalysis, which is required for ATP hydrolysis.

B. Atomic Models of F_1, A_1 and V_1

Much progress has been made in the high-resolution structure determination of F_1. First fundamental insights were gained in the early 1990 form structures of large portions of the mitochondrial F_1 (Abrahams et al. 1994). These structures show the arrangement of the α- and β-subunits around the central shaft. The α- and β-subunits have a similar fold, which consists of three domains: The N-terminal domain forms a β-barrel that interacts with neighbouring β-barrels in a crown-like structure in the $(\alpha\beta)_3$-assembly. The crown stabilizes the interaction between the α- and β-subunits in the assembly. The central domain of α and β contains the nucleotide binding sites and is referred to as nucleotide binding domain. Finally, the C-terminal domain is closest to the membrane and changes its conformation in response to the orientation of the central stalk.

The arrangement of αβ-pairs around the central stalk was surprisingly asymmetric, with different occupancies in the nucleotide binding sites (Abrahams et al. 1994). One αβ-pair ($\alpha_E\beta_E$) is wide open with no nucleotide bound, one αβ-pair ($\alpha_{DP}\beta_{DP}$) is partly closed with an ADP bound and one ($\alpha_{TP}\beta_{TP}$) is tightly closed with a non-hydrolysable ATP analogue (AMPPNP) bound (Fig. 6.4). The differences in the conformation of the nucleotide binding sites are caused by the C-terminal domain of the catalytic β-subunit, which moves in respect to the rest of the subunit (change in tertiary structure) and thus opens or closes the nucleotide binding sites. In contrast to the β-subunits, the C-terminal domains of the non-catalytic α-subunits remain much more rigid. In the centre of the αβ-assembly the γ-subunit gives rise to different interactions with the β-subunits and, thereby, forms a large portion of the central stalk. The orientation of the γ-subunit determines the conformations of the three nucleotide binding sites, which are even maintained in the absence of bound nucleotides (Kabaleeswaran et al. 2009). Recent atomic force microscopy investigations with immobilized single $(\alpha\beta)_3$-complexes show, that the different conformations are observed even in the absence of the γ-subunit, provided that nucleotides are present (Uchihashi et al. 2011).

The central stalk plays an essential role in translating the rotational movement of the rotor in F_0 into conformational changes in the three, catalytic nucleotide binding sites on the β-subunits. In F-ATPases the central shaft is an elongated assembly, which is formed by subunits γ and ε in chloroplasts and bacteria, and by subunits γ, δ and ε in mitochondria. The structure of the whole stalk together with the αβ-assembly of the

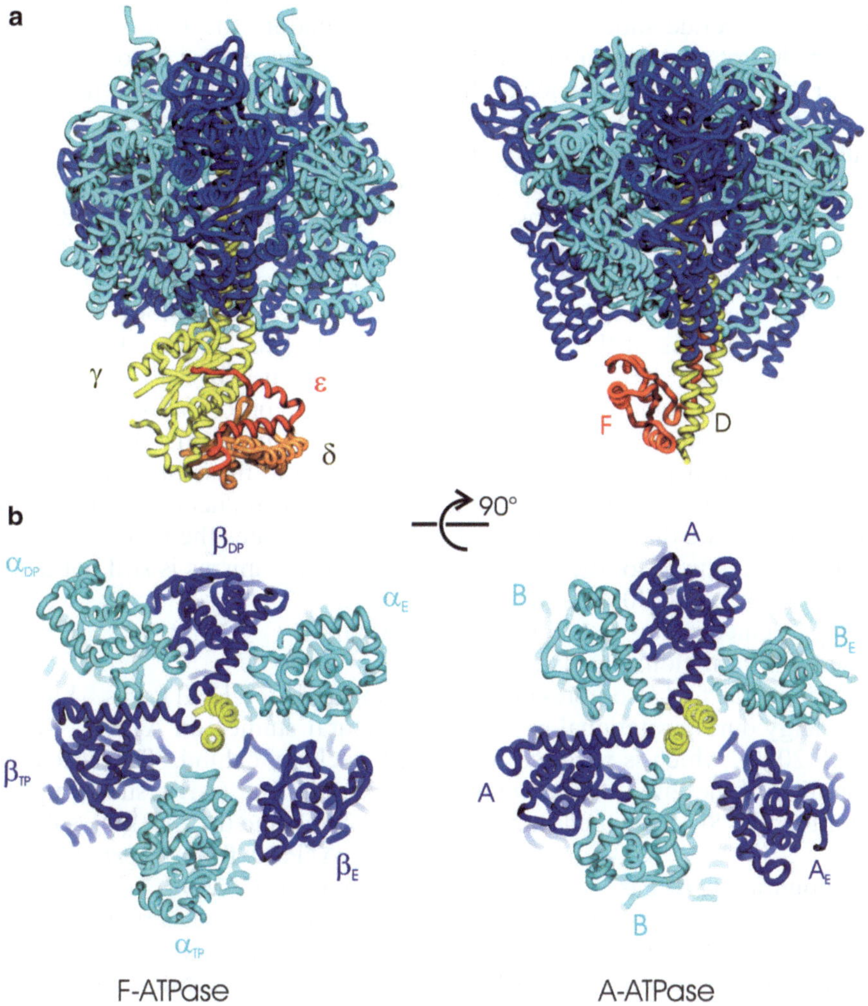

Fig. 6.4. Comparison of mitochondrial F_1-ATPase (*left*; 1e79; Gibbons et al. 2000) and A_1-ATPase of *Thermus thermophilus* (*right*; 3A5D, Numoto et al. 2009) without stalk subunits. (**a**) Side views and (**b**) views form the membrane proximal side. Subunits, which form the empty nucleotide binding sites are indexed with "E".

mitochondrial F_1 (Gibbons et al. 2000) gives insights into the structural elements that are important for the rotational coupling-mechanism. The stalk has an elongated shape with a maximal length of 114 Å. It is bulky at one end and thin at the other end. The bulky part (foot) bridges the gap between the membrane proximal side of the αβ-assembly and the rotor in F_0. The foot has an oval cross section with a maximal extension of 54 Å × 41 Å towards the rotor. The thin part of the stalk, which is formed by two long helices, extends 47 Å into the centre of the αβ-assembly. The longer of the two helices is the C-terminal helix of the γ-subunit (residues 199–272), and spans almost the entire length of the stalk. The shorter helix is the N-terminal helix of subunit γ (residues 2–49). Both helices are considerably bent, providing an eccentric axle, which differently interacts with the C-terminal domains of the three αβ-pairs in F_1. These differences in interaction provide a suitable mechanism for inducing conformational changes in the three catalytic nucleotide binding subunits depending on the rotational position of the central stalk. The coupling of the rotation of the γ-subunit to the changes in the conformation of the β-subunits is probably achieved by the conserved Arg 75 in the foot of the γ-subunit, which interacts in turn with the conserved DELSEED region of the three β-subunits (Gibbons et al. 2000).

The rotational mechanism is surprisingly robust. Of the two helices of the γ-subunit, the shorter N-terminal helix is already sufficient to support rotational catalysis (Mnatsakanyan et al. 2009). Rotational catalysis is even maintained if both helices are truncated and do not penetrate into the αβ-assembly. This shows that no rigid axle or rigid pivot point is required for the rotational mechanism (Furuike et al. 2008) and shows that the key elements for the rotational coupling and conformational cross-talk are located in the foot. The foot also stores most of the elastic energy during rotational catalysis (Sielaff et al. 2008), which is required for a smooth coupling of the three stepped motor in the F_1-part to the 8–15 stepped motor in the F_0-part.

In mitochondrial F-ATPase, the foot is formed by contributions of all three subunits of the central stalk. These subunits extensively interact and stabilize the base of the N- and C-terminal helix of the γ-subunit. The foot also contains elements that control the activity of the F-ATPase. Major players are the ε-subunit in the bacterial and chloroplast F-ATPase (60 % sequence identity to the mitochondrial δ-subunit) and a sulfur-sulfur-bridge in the γ-subunit of the chloroplast F-ATPase. Although the ε-subunit is essential for coupling of proton translocation to ATP synthesis, it also has an inhibitory effect on ATP hydrolysis (Nelson et al. 1972; Smith and Sternweis 1977) in bacteria and chloroplasts. Crystal structures of the isolated central shaft of *E. coli* F-ATPase show that in addition the related δ-subunit of the mitochondrial F-ATPase, the C-terminal helices of subunit ε in the bacterial F-ATPase can adopt an extended conformation (Rodgers and Wilce 2000, Fig. 6.5). In this extended conformation, the C-terminal domain of ε probably penetrates into the αβ-assembly (Hausrath et al. 2001) and interacts with the conserved DELSEED-region of the β-subunits. In this conformation ATP hydrolysis is inhibited, while ATP synthesis is still maintained. Therefore, it has been proposed, that subunit ε is a conformational switch (Tsunoda et al. 2001), which enables either only ATP synthesis (extended conformation) or ATP synthesis and hydrolysis (contracted conformation). The structure of F_0F_1 from *E. coli* with different ε conformations has been reported recently (Cingolani and Duncan 2011). In contrast to these insights, the mechanism by which the reduction of the disulfide in the subunit activates the F-ATPase in the photosynthetic organisms is still not understood.

1. A_1 and V_1

Compared to F-ATPases, high-resolution structural studies on A- and V-ATPases are still lacking behind. The structures of the nucleotide-free $(AB)_3$ sub-complex (Maher et al. 2009) and the $(AB)_3DF$ sub-complex of the A-ATPase from *Thermus thermophilus* (Numoto et al. 2009) (Fig. 6.4) have been determined. These structures provide the first insights into the relative arrangement of the catalytic nucleotide-binding subunits and how they interact. In the nucleotide-free $(AB)_3$-subcomplex the AB-pairs are arranged by three-fold symmetry and are stabilized by β-barrel domains in the A- and B-subunits, which form a crown similar as in F-ATPases (Abrahams et al. 1994).

Further insights into ATP hydrolysis-coupled rotation come from the $(AB)_3DF$ sub-complex (Numoto et al. 2009), which shows the central stalk composed of subunits D and F (Figs. 6.4 and 6.5). Compared to F-ATPases the central shaft is less bent and the foot is considerably smaller. Consequently, the central stalk interferes less with the C-terminal domains of the AB-subunits than the central stalk in F-ATPases. Therefore, in contrast to the function of the stalk in F-ATPases, the stalk

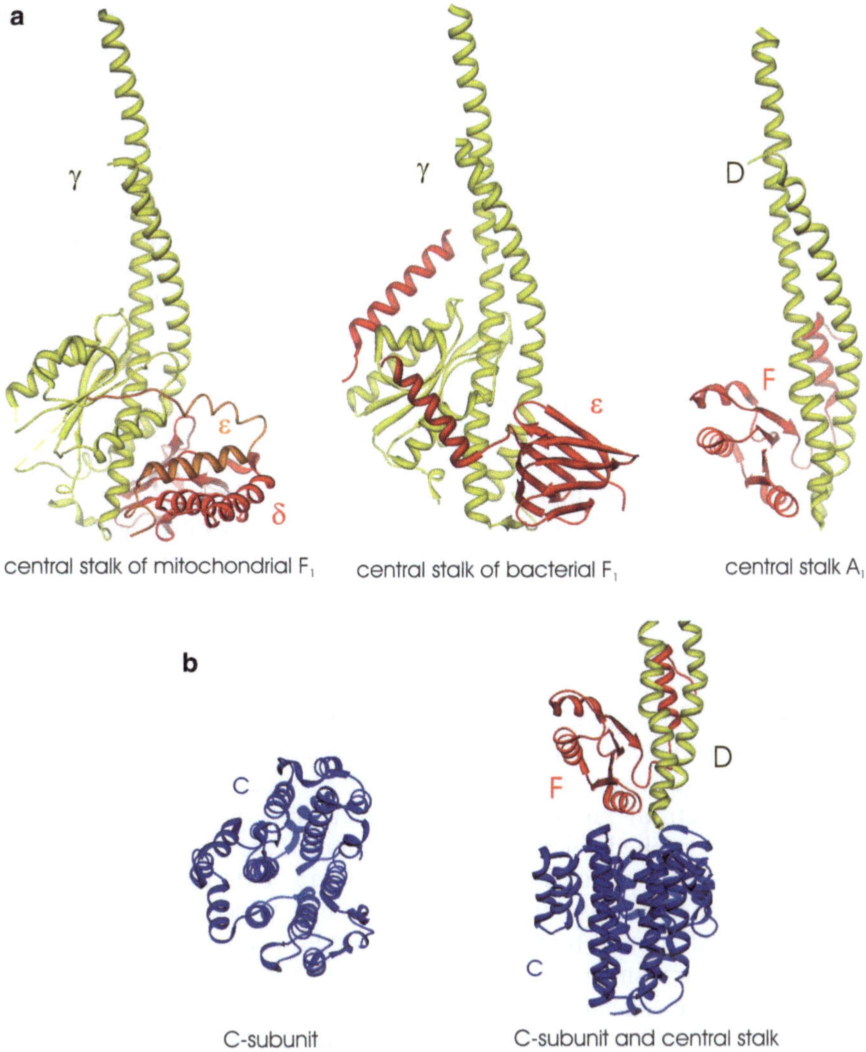

Fig. 6.5. Central stalks of A-, V- and F-ATPases. (**a**) Ribbon representations of the central stalks of the mitochondrial F-ATPase (*left*, 1e79, Gibbons et al. 2000), of the *E. coli* F-ATPase (*centre*, 1jnv, Hausrath et al. 2001) and of the *Thermus thermophilus* A-ATPase (*right*, 3a5d, Numoto et al. 2009). (**b**) A-ATPases have an additional subunit C (*blue*, 1r5z, Iwata et al. 2004), which is absent in F-ATPases. This subunit forms an adaptor between rotor and central stalk. A possible arrangement of central stalk and adaptor subunit is shown on the *right*.

is unlikely to drive major conformational changes in the tertiary structure of the A/B-subunits in A- and V-ATPases.

Similar to the stalk-containing F_1-complexes (Abrahams et al. 1994; Gibbons et al. 2000) the three AB-pairs in $(AB)_3DF$ are in different conformational states (Fig. 6.4). One of the AB-pairs forms a wide-open nucleotide binding site that is empty and the other two AB-pairs have more closed nucleotide binding sites with bound nucleotides. The catalytic A-subunits show almost no differences in their tertiary structure in the three AB-pairs. The non-catalytic B-subunits are somewhat more variable in the tertiary structure of their C-terminal domains, but much less variable than in the β-subunits of F-ATPases. Thus the conformational differences between the different AB-pairs in $(AB)_3DF$ are mainly driven by changes in the packing of A and B (changes in the quaternary structure), rather than by changes in the tertiary structure of the subunits. This suggests that the mechanic details of the rotational coupling might be different between A-ATPases and F-ATPases.

The stalk, which is resolved in the $(AB)_3DF$-complex contacts the rotor via an additional subunit (d in V-ATPases and C in A-ATPases), which acts as an adaptor between the rotor and the central shaft. The adaptor subunit is mainly α-helical and has a triangular cross-section (Fig. 6.5), which is large enough to seal the large pore of the rotor and to provide an extended, flat interface for binding the central stalk (Iwata et al. 2004).

C. The Peripheral Stalk

1. F-ATPase

The rotational mechanism requires a peripheral static connection between the F_1-part and the F_0-part, which prevents co-rotation. For a long time, structural knowledge of the stator of F-ATPases relied solely on the structural characterization of stator fragments or individual subunits. In bacteria, the N-terminal fragment of subunit b (aa 1–34) forms a single, straight transmembrane helix followed by a short hinge and another helical stretch (Dmitriev et al. 1999, Fig. 6.6). The soluble C-terminal part of subunit b (aa 62–122) also forms a long α-helix (Del Rizzo et al. 2002, Fig. 6.6), which probably dimerizes as a coiled-coil, in which the helices are staggered in respect to each other (Claggett et al. 2009; Steigmiller et al. 2005). The dimers have a boomerang-like shape with a similar curvature as the surface of the αβ-assembly (Priya et al. 2008). Thus they could be tightly attached to the αβ-assembly, without introducing conformational strain into the holoenzyme.

The structure of the soluble part of the bovine mitochondrial peripheral stalk together with F_1 has been determined (Rees et al. 2009) and shows most of the extrinsic part of the F-ATPase including the central shaft and the peripheral stalk subunit OSCP (homologous to the δ subunit in bacteria and chloroplasts), and parts of the subunits b, F6 and d. This information is complemented by a partly overlapping structure of the soluble part of subunit b together with subunits d and F6 (Dickson et al. 2006). Both structures together allow generating a mosaic structure of the whole extrinsic part (F_1 and peripheral stalk) of the mitochondrial F-ATPase (Fig. 6.6).

The peripheral stalk in mitochondrial F-ATPase is much more complex than anticipated form the studies of the fragments of the bacterial b-subunit. It consists of several long helices, which are oriented parallel. The helices are interrupted by loops. These loops are paralleled by helices of other subunits giving the whole stator a rigid nature. The rigid parts are interrupted by two more flexible regions in OSCP and b. In OSPC the flexible link is located between the C- and N-terminal domains. The N-terminal domain of OSCP binds via electrostatic interactions to the crown formed by the α- and β-subunits and does not directly interact with other subunits. In contrast, the C-terminal domain of OSPC forms an extensive interface with the b-subunit, which is further stabilized by subunit F6 wrapping around it. This results in a stable, rigid link to the rest of the peripheral stalk. The flexible linkage within OSPC probably enables following the changing shape of F_1 during catalysis.

The other flexible region is located at amino acid 146 in the b-subunit, approximately halfway between the top and the bottom of the αβ-assembly. This area supports a hinge movement and enables subtle changes in the curvatures of the stator by approximately 9° (Rees et al. 2009). Taken together most of the stator is rather stiff and can probably only store little or no elastic energy.

The peripheral stalk adds asymmetry to the crystallized F_1 and makes the three αβ-pairs distinguishable. In the crystallized conformation the peripheral stalk is attached to the non-catalytic interface next to an α-subunit in the DP state and to a β-subunit in the TP state (Rees et al. 2009). These findings

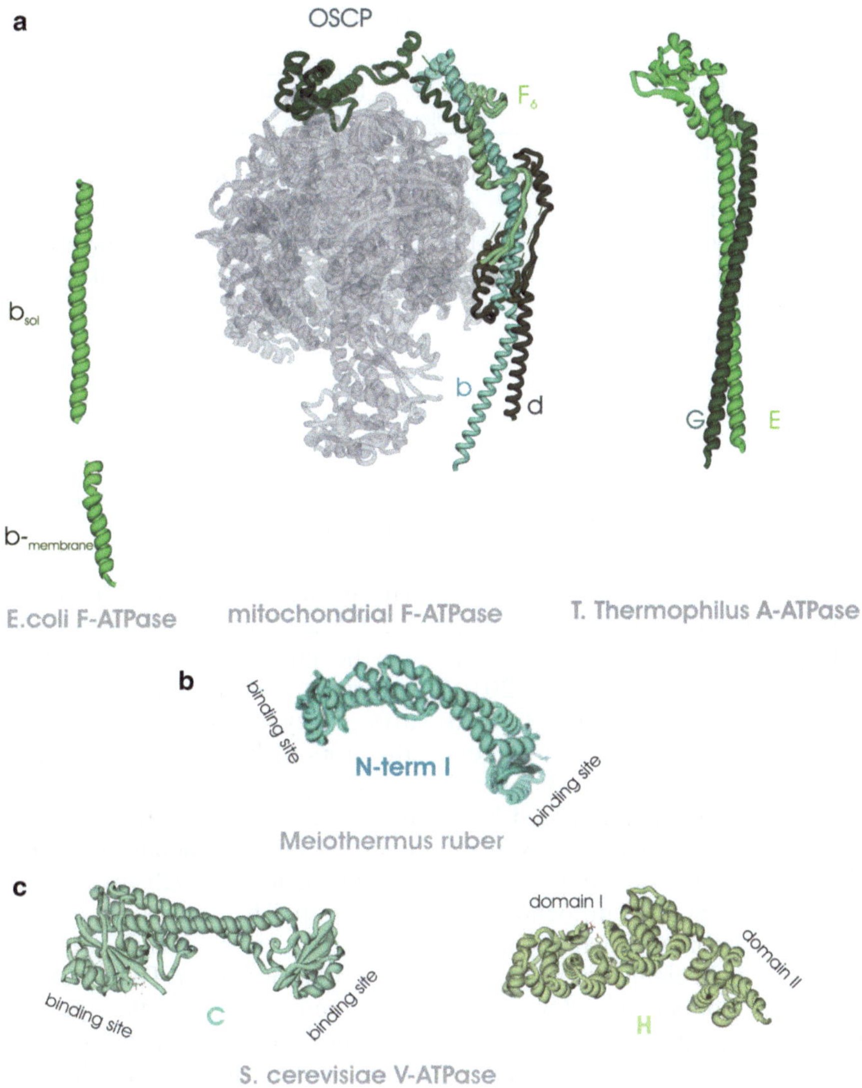

Fig. 6.6. Peripheral stalk elements of F-, V- and A- ATPases. (**a**) Peripheral stalk elements. *Left*: Soluble part (1L2P, Del Rizzo et al. 2002) and membrane-integrated part (1b9u, Dmitriev et al. 1999) of b-subunit of *E. coli* F-ATPase. *Centre*: Soluble part of the peripheral stalk of mitochondrial F-ATPase attached to $(\alpha\beta)_3\gamma\delta\varepsilon$ (*grey*). The peripheral stalk structure is a mosaic (2wss, Rees et al. 2009; and 2cly, Dickson et al. 2006). *Right*: Vertical stator element of subunits E and G of the A-ATPase from *Thermus thermophilus* (3k5b, Lee et al. 2010). (**b**) The N-terminal domain of subunit I, which is the horizontal peripheral stalk elements that has a counterpart in V-ATPases and interconnects two vertical peripheral stalk elements (3rrk, Srinivasan et al. 2011) (**c**) The V-ATPase specific horizontal peripheral stalk elements of the V-ATPase are formed by subunits H (1ho8, Sagermann et al. 2001) and C (1u7l, Drory et al. 2004).

further support that the F-ATPase assumes a defined resting state and that the three catalytic nucleotide binding sites have different thermodynamic properties in the inactive enzyme.

2. A- and V-ATPase

A-ATPases and V-ATPases have two and three vertical peripheral stalk elements, respectively. These vertical elements do not

Table 6.2. High-resolution structures of the membrane-embedded rotors or rotors together with the central stalk and $(\alpha\beta)_3$.

Species	Type	Subunits in rotor	Transported ion	pdb	References
Spinach	Chloroplast F-ATPase	14	H+	2w5j	Vollmar et al. (2009)
Spirulina platensis	Chloroplast F-ATPase	15	H+	2wie	Pogoryelov et al. (2009)
Ilyobacter tartaricus	Bacterial F-ATPase	11	Na+	2wgm	Meier et al. (2005, 2009)
Bacillus pseudofirmus	Bacterial F-ATPase	13	H+	2x2v	Preiss et al. (2010)
Saccharomyces cerevisiae	Mitochondrial F-ATPase	10	H+	2x0k, 1q01	Stock et al. (1999) Dautant et al. (2010)
Bos taurus	Mitochondrial F-ATPase	8	H+	2xnd	Watt et al. (2010)
Enterococcus hirae	Bacterial A-ATPase	10	Na+	2bl2	Murata et al. (2005)

have a membrane anchor and attach to the non-catalytic B-subunits. The vertical peripheral stalk elements consist of subunits E and G. The N-terminal domains of E and G form a long right-handed coiled-coil domain of 140 Å length, which is connected to a more globular head domain (Lee et al. 2010, Fig. 6.6). The head domain is mainly formed by the C-terminal domain of subunit E and a short helical stretch of subunit G. Head domain and the coiled-coil are joined by a kink, which probably provides flexibility between the two domains, similar to the flexible linkage within OSCP in mitochondrial F-ATPase. The membrane anchor for the vertical peripheral stalk elements is provided by the N-terminal domain of subunit a in V-ATPases and of subunit I in A-ATPases. This N-terminal domain is formed by a helical bundle, which is capped on both ends by two similar domains that bind to the vertical peripheral stalk elements (Srinivasan et al. 2011, Fig. 6.6).

The two V-ATPase-specific subunits C and H form horizontal peripheral stalk elements. Subunit C interconnects two vertical stator elements, which is reflected by similar protein binding motifs at opposite ends of the subunit (Fig. 6.6). The structure of subunit C is related to the N-terminal domain of subunit I, although less bent. Subunit H is important for inhibiting ATP hydrolysis in V_1 but not in the holoenzyme. This is achieved by a conformational rearrangement of its two domains (Fig. 6.6).

D. Structure of the Membrane-Embedded Rotor

The membrane-embedded rotors of A-, V- and F-ATPases form stable assemblies, which can be purified as a whole. This stability greatly facilitates the crystallization of different types of rotors (Table 6.2). So far, all rotors of known structure have an hourglass shape, which has the smallest diameter in the centre of the membrane (Fig. 6.7). In F-ATPases, the subunits in the rotor consist of two helices, which are connected by a short loop. The helices are arranged in two concentric rings, which are staggered in respect to each other. The N-terminal helices form the inner ring and are generally longer than the C-terminal helices in the outer ring. The ion-binding sites are located between two C-terminal helices and an N-terminal helix close to the centre of the membrane, where the diameter of the rotor is smallest. In all rotors a conserved Glu contributes to the coordination of the ion. In Na^+-transporting F-ATPases, selectivity for Na^+ is achieved by a somewhat larger ion-binding pocket and by a water molecule that contributes to Na^+-co-ordination. The ion-binding pocket opens only towards the hydrophobic environment of the membrane, which makes it unfavourable for the ion to leave the rotor. In the holoenzyme, the rotor interfaces with a static half-channel, which is formed by the membrane-integrated subunit. This interface between rotor and static half channel could

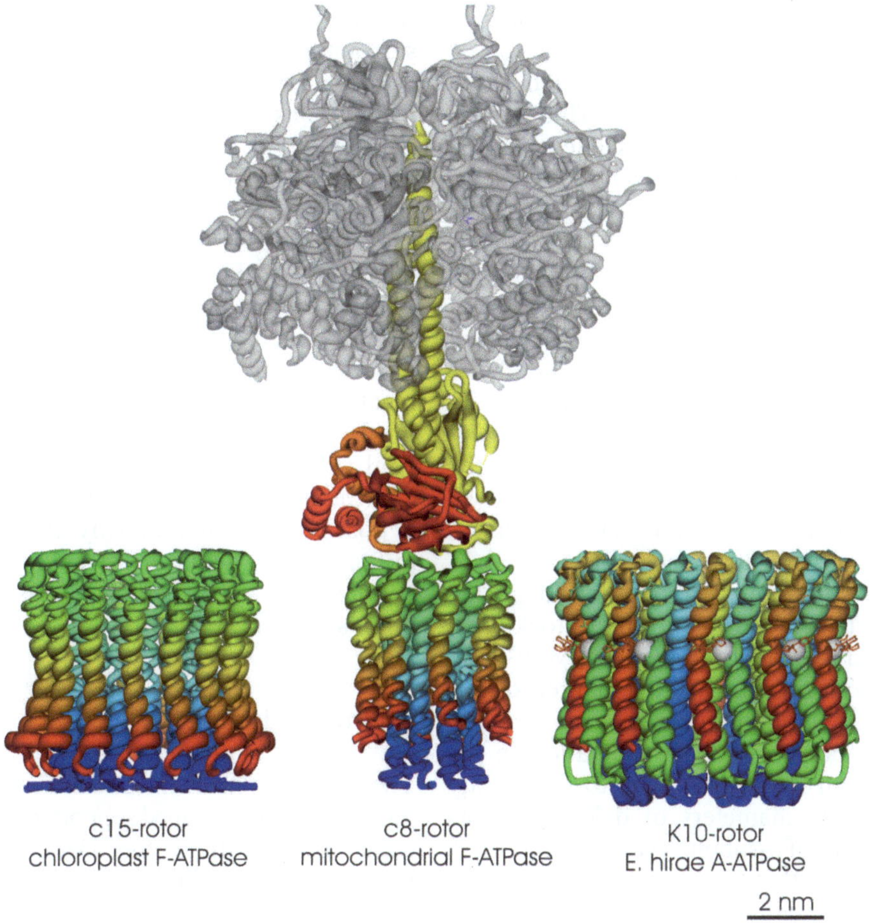

Fig. 6.7. Membrane-embedded rotor. The subunits in the rotor are colored according to their residue number (form N-terminus [*blue*] to C-terminus [*red*]). *Left*: The rotor of the chloroplast F-ATPase form *Spirulina platensis* is formed by 15 copies of subunit c (2wie, Pogoryelov et al. 2009) and is the largest F-ATPase rotor that has been reported so far. *Centre*: The bovine mitochondrial F-ATPase has the rotor with the smallest copy number. It is shown together with the central stalk (γ [*yellow*], δ [*red*] and ε [*orange*]) and $(\alpha\beta)_3$ [*grey*] (2xnd, Watt et al. 2010). The F-ATPase rotor subunits in chloroplasts and in mitochondria consist of two transmembrane helices each. *Right*: The rotor of the A-ATPase form *E. hirae* (2db4, Murata et al. 2003) consists of ten subunits, with four transmembrane helices each. This A-ATPase pumps Na^+ across the membrane. The bound Na^+-ions are shown as white spheres.

provide a suitable environment for the transport of the ions across the membrane.

In V- and in some A-ATPases the rotor subunits have arisen by a gene duplication and fusion event leading to four transmembrane helices per subunit. For a long time it was assumed that this gene-duplication and fusion had led to a reduction in the copy number of rotor subunits keeping the total number of helices in the rotor similar to F-ATPases. However, the only known structure of an Na^+-pumping A-ATPase rotor of *E. hirae* shows ten subunits (Murata et al. 2005) with 40 transmembrane helices (Fig. 6.7), suggesting that no reduction in the copy number of the rotor subunits took place. In contrast to F-ATPases, the ion binding site in *E. hirae* is located within a single subunit. This increases the distance between the ion-binding sites in the rotor. As in F-ATPases the Na^+-binding site contains a conserved Glu, but no water molecule that is involved in the coordination of the Na^+ is present (Meier et al. 2009).

All rotors have a large hole, which forms a pore through the membrane. This pore is probably an evolutionary consequence of the translocase-ancestors. To maintain the membrane potential the pore has to be efficiently sealed. From the lumenal side this happens with a lipidic plug (Meier et al. 2001; Seelert et al. 2003). From the cytoplasmic side the pore is closed by the central shaft. Structures of the mitochondrial F-ATPase sub-complexes of the rotor together with the central stalk and the αβ-assembly (Stock et al. 1999; Watt et al. 2010) show that the stalk binds to the loop regions of the rotor subunits mainly via the δ-subunit, but also with smaller contributions from the ε and γ subunits (Fig. 6.7). The interaction interface between loops and foot is between 600 Å^2 in yeast (Stock et al. 1999) and 790 Å^2 in *Bos taurus* (Watt et al. 2010). Both structures show marked differences in the binding mode between foot and rotor. So far, no obvious adaptation in the structures of the central stalk to the greatly variable pore diameters of different rotors (8–15 copies of c-subunit) has been reported. This suggests that the lipidic plug is the key element for sealing the pores of the rotor against leakage rather than the foot of the central stalk.

E. Higher Order Structures in Native Membranes

In mitochondria most of the complexes of the respiratory chain are organized in supercomplexes (Schägger and Pfeiffer 2000; Dudkina et al. 2005a, 2011; Schäfer et al. 2007; Bultema et al. 2009, see Chaps. 12 and 21), which bring functional elements of respiration into close spatial proximity. These supercomplexes do not contain the F-ATPase, which aggregates into separate higher order structures. F-ATPases form dimers, which are stabilized by three dimer-specific subunits (Arnold et al. 1998). The dimers are joined in the F_0 at the side of the stator (Dudkina et al. 2005b; Minauro-Sanmiguel et al. 2005) and form an arc, which gives rise to a strong curvature of the surrounding membrane.

In the inner membranes of mitochondria the F-ATPase dimers are stacked in long rows (Dudkina et al. 2010; Strauss et al. 2008) and induce the formation of the highly-curved cristae (Giraud et al. 2002). Simulations show that the curvature increases the electric field close to F_0 (Strauss et al. 2008). The ATPase dimers only occur in mitochondria, whereas in chloroplasts the F-ATPases form monomers or random aggregates that are located in the minimally curved membranes of the grana and stroma lamellae (Daum et al. 2010). These differences in the micro-environment (curved vs. flat) could provide a possible explanation why F-ATPases of mitochondria require fewer protons per synthesized ATP than F-ATPases in the photosynthetic apparatus.

Acknowledgements

The authors thank the Deutsche Forschungsgemeinschaft for continued support. BB acknowledges support from the Darwin Trust of Edinburgh and the EU NoE 3D-Repertoire.

References

Abrahams JP, Leslie AG, Lutter R, Walker JE (1994) Structure at 2.8 A resolution of F1-ATPase from bovine heart mitochondria. Nature 370: 621–628

Arnold I, Pfeiffer K, Neupert W, Stuart RA, Schägger H (1998) Yeast mitochondrial F1F0-ATP synthase exists as a dimer: identification of three dimer-specific subunits. EMBO J 17:7170–7178

Bernal RA, Stock D (2004) Three-dimensional structure of the intact *Thermus thermophilus* H+-ATPase/synthase by electron microscopy. Structure (Camb) 12:1789–1798

Bienert R, Rombach-Riegraf V, Diez M, Gräber P (2009) Subunit movements in single membrane-bound H+-ATP synthases from chloroplasts during ATP synthesis. J Biol Chem 284:36240–36247

Bienert R, Zimmermann B, Rombach-Riegraf V, Gräber P (2011) Time-dependent FRET with single enzymes: domain motions and catalysis in H(+) -ATP Synthases. Chemphyschem 12:507–510

Böttcher B, Schwarz L, Gräber P (1998) Direct indication for the existence of a double stalk in CF0F1. J Mol Biol 281:757–762

Böttcher B, Bertsche I, Reuter R, Gräber P (2000) Direct visualisation of conformational changes in EF0F1 by electron microscopy. J Mol Biol 296:449–457

Boyer PD (1993) The binding change mechanism for ATP synthase–some probabilities and possibilities. Biochim Biophys Acta 1140:215–250

Bultema JB, Braun HP, Boekema EJ, Kouril R (2009) Megacomplex organization of the oxidative phosphorylation system by structural analysis of respiratory supercomplexes from potato. Biochim Biophys Acta 1787:60–67

Cingolani G, Duncan TM (2011) Structure of the ATP synthase catalytic complex (F(1)) from *Escherichia coli* in an autoinhibited conformation. Nat Struct Mol Biol 18:701–707

Claggett SB, O'Neil Plancher M, Dunn SD, Cain BD (2009) The b subunits in the peripheral stalk of F1F0 ATP synthase preferentially adopt an offset relationship. J Biol Chem 284:16531–16540

Collinson IR, Runswick MJ, Buchanan SK, Fearnley IM, Skehel JM, van Raaij MJ, Griffiths DE, Walker JE (1994) Fo membrane domain of ATP synthase from bovine heart mitochondria: purification, subunit composition, and reconstitution with F1-ATPase. Biochemistry 33:7971–7978

Daum B, Nicastro D, Austin J 2nd, McIntosh JR, Kühlbrandt W (2010) Arrangement of photosystem II and ATP synthase in chloroplast membranes of spinach and pea. Plant Cell 22:1299–1312

Dautant A, Velours J, Giraud MF (2010) Crystal structure of the Mg.ADP-inhibited state of the yeast F1c10-ATP synthase. J Biol Chem 285:29502–29510

Del Rizzo PA, Bi Y, Dunn SD, Shilton BH (2002) The "second stalk" of *Escherichia coli* ATP synthase: structure of the isolated dimerization domain. Biochemistry 41:6875–6884

Dickson VK, Silvester JA, Fearnley IM, Leslie AG, Walker JE (2006) On the structure of the stator of the mitochondrial ATP synthase. EMBO J 25:2911–2918

Diepholz M, Börsch M, Böttcher B (2008a) Structural organization of the V-ATPase and its implications for regulatory assembly and disassembly. Biochem Soc Trans 36:1027–1031

Diepholz M, Venzke D, Prinz S, Batisse C, Flörchinger B, Rössle M, Svergun DI, Böttcher B, Fethiere J (2008b) A different conformation for EGC stator subcomplex in solution and in the assembled yeast V-ATPase: possible implications for regulatory disassembly. Structure 16:1789–1798

Diez M, Zimmermann B, Börsch M, König M, Schweinberger E, Steigmiller S, Reuter R, Felekyan S, Kudryavtsev V, Seidel CA, Gräber P (2004) Proton-powered subunit rotation in single membrane-bound F0F1-ATP synthase. Nat Struct Mol Biol 11:135–141

Dmitriev O, Jones PC, Jiang W, Fillingame RH (1999) Structure of the membrane domain of subunit b of the *Escherichia coli* F0F1 ATP synthase. J Biol Chem 274:15598–15604

Drory O, Frolow F, Nelson N (2004) Crystal structure of yeast V-ATPase subunit C reveals its stator function. EMBO Rep 5:1148–1152

Dudkina NV, Eubel H, Keegstra W, Boekema EJ, Braun HP (2005a) Structure of a mitochondrial supercomplex formed by respiratory-chain complexes I and III. Proc Natl Acad Sci U S A 102:3225–3229

Dudkina NV, Heinemeyer J, Keegstra W, Boekema EJ, Braun HP (2005b) Structure of dimeric ATP synthase from mitochondria: an angular association of monomers induces the strong curvature of the inner membrane. FEBS Lett 579:5769–5772

Dudkina NV, Kudryashev M, Stahlberg H, Boekema EJ (2011) Interaction of complexes I, III, and IV within the bovine respirasome by single particle cryoelectron tomography. Proc Natl Acad Sci U S A 108:15196–15200

Esteban O, Bernal RA, Donohoe M, Videler H, Sharon M, Robinson CV, Stock D (2008) Stoichiometry and localization of the stator subunits E and G in *Thermus thermophilus* H+-ATPase/synthase. J Biol Chem 283:2595–2603

Förster K, Turina P, Drepper F, Haehnel W, Fischer S, Gräber P, Petersen J (2010) Proton transport coupled ATP synthesis by the purified yeast H+-ATP synthase in proteoliposomes. Biochim Biophys Acta 1797:1828–1837

Furuike S, Hossain MD, Maki Y, Adachi K, Suzuki T, Kohori A, Itoh H, Yoshida M, Kinosita K Jr (2008) Axle-less F1-ATPase rotates in the correct direction. Science 319:955–958

Gibbons C, Montgomery MG, Leslie AG, Walker JE (2000) The structure of the central stalk in bovine F1-ATPase at 2.4 Å resolution. Nat Struct Biol 7:1055–1061

Giraud MF, Paumard P, Soubannier V, Vaillier J, Arselin G, Salin B, Schaeffer J, Brethes D, di Rago JP, Velours J (2002) Is there a relationship between the supramolecular organization of the mitochondrial ATP synthase and the formation of cristae? Biochim Biophys Acta 1555:174–180

Gogarten JP, Starke T, Kibak H, Fishman J, Taiz L (1992) Evolution and isoforms of V-ATPase subunits. J Exp Biol 172:137–147

Gomis-Ruth FX, Moncalian G, Perez-Luque R, Gonzalez A, Cabezon E, de la Cruz F, Coll M (2001) The bacterial conjugation protein TrwB resembles ring helicases and F1- ATPase. Nature 409:637–641

Hausrath AC, Capaldi RA, Matthews BW (2001) The conformation of the epsilon- and gamma-subunits within the *Escherichia coli* F1 ATPase. J Biol Chem 276:47227–47232

Iwata M, Imamura H, Stambouli E, Ikeda C, Tamakoshi M, Nagata K, Makyio H, Hankamer B, Barber J, Yoshida M, Yokoyama K, Iwata S (2004) Crystal structure of a central stalk subunit C and reversible association/dissociation of vacuole-type ATPase. Proc Natl Acad Sci U S A 101:59–64

Jefferies KC, Forgac M (2008) Subunit H of the vacuolar (H+) ATPase inhibits ATP hydrolysis by the free V1 domain by interaction with the rotary subunit F. J Biol Chem 283:4512–4519

Junge W, Lill H, Engelbrecht S (1997) ATP synthase: an electrochemical transducer with rotatory mechanics. Trends Biochem Sci 22:420–423

Kabaleeswaran V, Shen H, Symersky J, Walker JE, Leslie AG, Mueller DM (2009) Asymmetric structure of the yeast F1 ATPase in the absence of bound nucleotides. J Biol Chem 284:10546–10551

Kane PM (1995) Disassembly and reassembly of the yeast vacuolar H+-ATPase in vivo. J Biol Chem 270:17025–17032

Karrasch S, Walker JE (1999) Novel features in the structure of bovine ATP synthase. J Mol Biol 290:379–384

Kibak H, Taiz L, Starke T, Bernasconi P, Gogarten JP (1992) Evolution of structure and function of V-ATPases. J Bioenerg Biomembr 24:415–424

Kitagawa N, Mazon H, Heck AJ, Wilkens S (2008) Stoichiometry of the peripheral stalk subunits E and G of yeast V1-ATPase determined by mass spectrometry. J Biol Chem 283:3329–3337

Lapierre P, Shial R, Gogarten JP (2006) Distribution of F- and A/V-type ATPases in *Thermus scotoductus* and other closely related species. Syst Appl Microbiol 29:15–23

Lau WC, Rubinstein JL (2010) Structure of intact *Thermus thermophilus* V-ATPase by cryo-EM reveals organization of the membrane-bound VO motor. Proc Natl Acad Sci U S A 107:1367–1372

Lau WC, Baker LA, Rubinstein JL (2008) Cryo-EM structure of the yeast ATP synthase. J Mol Biol 382:1256–1264

Lee LK, Stewart AG, Donohoe M, Bernal RA, Stock D (2010) The structure of the peripheral stalk of *Thermus thermophilus* H+-ATPase/synthase. Nat Struct Mol Biol 17:373–378

Maher MJ, Akimoto S, Iwata M, Nagata K, Hori Y, Yoshida M, Yokoyama S, Iwata S, Yokoyama K (2009) Crystal structure of A3B3 complex of V-ATPase from Thermus thermophilus. EMBO J 28:3771–3779

Meier T, Matthey U, Henzen F, Dimroth P, Müller DJ (2001) The central plug in the reconstituted undecameric c cylinder of a bacterial ATP synthase consists of phospholipids. FEBS Lett 505:353–356

Meier T, Polzer P, Diederichs K, Welte W, Dimroth P (2005) Structure of the rotor ring of F-Type Na+-ATPase from *Ilyobacter tartaricus*. Science 308:659–662

Meier T, Krah A, Bond PJ, Pogoryelov D, Diederichs K, Faraldo-Gomez JD (2009) Complete ion-coordination structure in the rotor ring of Na+-dependent F-ATP synthases. J Mol Biol 391:498–507

Mellwig C, Böttcher B (2003) A unique resting position of the ATP-synthase from chloroplasts. J Biol Chem 278:18544–18549

Meyer B, Wittig I, Trifilieff E, Karas M, Schägger H (2007) Identification of two proteins associated with mammalian ATP synthase. Mol Cell Proteomics 6:1690–1699

Minauro-Sanmiguel F, Wilkens S, Garcia JJ (2005) Structure of dimeric mitochondrial ATP synthase: novel F0 bridging features and the structural basis of mitochondrial cristae biogenesis. Proc Natl Acad Sci U S A 102:12356–12358

Mitchell (1961) Coupling of phosphorylation to electron and hydrogen transfer by a chemi-osmotic type of mechanism. Nature 191:144–148

Mnatsakanyan N, Hook JA, Quisenberry L, Weber J (2009) ATP synthase with its gamma subunit reduced to the N-terminal helix can still catalyze ATP synthesis. J Biol Chem 284:26519–26525

Muench SP, Huss M, Song CF, Phillips C, Wieczorek H, Trinick J, Harrison MA (2009) Cryo-electron microscopy of the vacuolar ATPase motor reveals its mechanical and regulatory complexity. J Mol Biol 386:989–999

Mulkidjanian AY, Makarova KS, Galperin MY, Koonin EV (2007) Inventing the dynamo machine: the evolution of the F-type and V-type ATPases. Nat Rev Microbiol 5:892–899

Murata T, Arechaga I, Fearnley IM, Kakinuma Y, Yamato I, Walker JE (2003) The membrane domain of the Na+-motive V-ATPase from *Enterococcus hirae* contains a heptameric rotor. J Biol Chem 278:21162–21167

Murata T, Yamato I, Kakinuma Y, Leslie AG, Walker JE (2005) Structure of the rotor of the V-Type Na+-ATPase from *Enterococcus hirae*. Science 308:654–659

Nelson N, Nelson H, Racker E (1972) Partial resolution of the enzymes catalyzing photophosphorylation. XII. Purification and properties of an inhibitor isolated from chloroplast coupling factor 1. J Biol Chem 247:7657–7662

Noji H (1998) The rotary enzyme of the cell: the rotation of F1-ATPase. Science 282:1844–1845

Numoto N, Hasegawa Y, Takeda K, Miki K (2009) Inter-subunit interaction and quaternary rearrangement defined by the central stalk of prokaryotic V1-ATPase. EMBO Rep 10:1228–1234

Olendzenski L, Liu L, Zhaxybayeva O, Murphey R, Shin DG, Gogarten JP (2000) Horizontal transfer of archaeal genes into the deinococcaceae: detection by molecular and computer-based approaches. J Mol Evol 51:587–599

Pallen MJ, Bailey CM, Beatson SA (2006) Evolutionary links between FliH/YscL-like proteins from bacterial type III secretion systems and second-stalk components of the FoF1 and vacuolar ATPases. Protein Sci 15:935–941

Parra KJ, Keenan KL, Kane PM (2000) The H subunit (Vma13p) of the yeast V-ATPase inhibits the ATPase activity of cytosolic V1 complexes. J Biol Chem 275:21761–21767

Pogoryelov D, Yu J, Meier T, Vonck J, Dimroth P, Müller DJ (2005) The c15 ring of the *Spirulina platensis* F-ATP synthase: F1/F0 symmetry mismatch is not obligatory. EMBO Rep 6:1040–1044

Pogoryelov D, Yildiz O, Faraldo-Gomez JD, Meier T (2009) High-resolution structure of the rotor ring of a proton-dependent ATP synthase. Nat Struct Mol Biol 16:1068–1073

Preiss L, Yildiz O, Hicks DB, Krulwich TA, Meier T (2010) A new type of proton coordination in an F(1) F(o)-ATP synthase rotor ring. PLoS Biol 8:e1000443

Priya R, Tadwal VS, Roessle MW, Gayen S, Hunke C, Peng WC, Torres J, Gruber G (2008) Low resolution structure of subunit b (b (22-156)) of Escherichia coli F(1)F(O) ATP synthase in solution and the b-delta assembly. J Bioenerg Biomembr 40:245–255

Rees DM, Leslie AG, Walker JE (2009) The structure of the membrane extrinsic region of bovine ATP synthase. Proc Natl Acad Sci U S A 21:21597

Rodgers AJW, Wilce MC (2000) Structure of the gamma-epsilon complex of ATP synthase. Nat Struct Biol 7:1051–1054

Rubinstein JL, Walker JE, Henderson R (2003) Structure of the mitochondrial ATP synthase by electron cryomicroscopy. EMBO J 22:6182–6192

Rubinstein JL, Dickson VK, Runswick MJ, Walker JE (2005) ATP synthase from Saccharomyces cerevisiae: location of subunit h in the peripheral stalk region. J Mol Biol 345:513–520

Sagermann M, Stevens TH, Matthews BW (2001) Crystal structure of the regulatory subunit H of the V-type ATPase of Saccharomyces cerevisiae. Proc Natl Acad Sci U S A 98:7134–7139

Schäfer E, Dencher NA, Vonck J, Parcej DN (2007) Three-dimensional structure of the respiratory chain supercomplex I1III2IV1 from bovine heart mitochondria. Biochemistry 46:12579–12585

Schägger H, Pfeiffer K (2000) Supercomplexes in the respiratory chains of yeast and mammalian mitochondria. EMBO J 19:1777–1783

Seelert H, Dencher NA, Müller DJ (2003) Fourteen protomers compose the oligomer III of the proton-rotor in spinach chloroplast ATP synthase. J Mol Biol 333:337–344

Shao E, Nishi T, Kawasaki-Nishi S, Forgac M (2003) Mutational analysis of the non-homologous region of subunit A of the yeast V-ATPase. J Biol Chem 278:12985–12991

Shirakihara Y, Leslie AG, Abrahams JP, Walker JE, Ueda T, Sekimoto Y, Kambara M, Saika K, Kagawa Y, Yoshida M (1997) The crystal structure of the nucleotide-free alpha 3 beta 3 subcomplex of F1-ATPase from the thermophilic Bacillus PS3 is a symmetric trimer. Structure 5:825–836

Sielaff H, Rennekamp H, Wächter A, Xie H, Hilbers F, Feldbauer K, Dunn SD, Engelbrecht S, Junge W (2008) Domain compliance and elastic power transmission in rotary F(O)F(1)-ATPase. Proc Natl Acad Sci U S A 105:17760–17765

Smith JB, Sternweis PC (1977) Purification of membrane attachment and inhibitory subunits of the proton translocating adenosine triphosphatase from *Escherichia coli*. Biochemistry 16:306–311

Srinivasan S, Vyas NK, Baker ML, Quiocho FA (2011) Crystal structure of the cytoplasmic N-terminal domain of subunit I, a homolog of subunit a, of V-ATPase. J Mol Biol 412:14–21

Steigmiller S, Börsch M, Gräber P, Huber M (2005) Distances between the b-subunits in the tether domain of F(0)F(1)-ATP synthase from *E. coli*. Biochim Biophys Acta 1708:143–153

Stock D, Leslie AG, Walker JE (1999) Molecular architecture of the rotary motor in ATP synthase. Science 286:1700–1705

Strauss M, Hofhaus G, Schröder RR, Kühlbrandt W (2008) Dimer ribbons of ATP synthase shape the inner mitochondrial membrane. EMBO J 27:1154–1160

Sumner JP, Dow JA, Earley FG, Klein U, Jäger D, Wieczorek H (1995) Regulation of plasma membrane V-ATPase activity by dissociation of peripheral subunits. J Biol Chem 270:5649–5653

Tsunoda SP, Rodgers AJ, Aggeler R, Wilce MC, Yoshida M, Capaldi RA (2001) Large conformational changes of the epsilon subunit in the bacterial F1F0 ATP synthase provide a ratchet action to regulate this rotary motor enzyme. Proc Natl Acad Sci U S A 98:6560–6564

Tuller T, Birin H, Gophna U, Kupiec M, Ruppin E (2010) Reconstructing ancestral gene content by coevolution. Genome Res 20:122–132

Uchihashi T, Iino R, Ando T, Noji H (2011) High-speed atomic force microscopy reveals rotary catalysis of rotorless F-ATPase. Science 333:755–758

Venzke D, Domgall I, Köcher T, Fethiere J, Fischer S, Böttcher B (2005) Elucidation of the stator organization in the V-ATPase of *Neurospora crassa*. J Mol Biol 349:659–669

Vollmar M, Schlieper D, Winn M, Büchner C, Groth G (2009) Structure of the c14 rotor ring of the proton translocating chloroplast ATP synthase. J Biol Chem 284:18228–18235

Vonck J, Pisa KY, Morgner N, Brutschy B, Müller V (2009) Three-dimensional structure of A1A0 ATP synthase from the hyperthermophilic archaeon *Pyrococcus furiosus* by electron microscopy. J Biol Chem 284:10110–10119

Watt IN, Montgomery MG, Runswick MJ, Leslie AG, Walker JE (2010) Bioenergetic cost of making an adenosine triphosphate molecule in animal mitochondria. Proc Natl Acad Sci U S A 107:16823–16827

Wilkens S, Capaldi RA (1998) Electron microscopic evidence of two stalks linking the F1 and F0 parts of the *Escherichia coli* ATP synthase. Biochim Biophys Acta 1365:93–97

Wilkens S, Inoue T, Forgac M (2004) Three-dimensional structure of the vacuolar ATPase. Localization of subunit H by difference imaging and chemical cross-linking. J Biol Chem 279:41942–41949

Wilkens S, Zhang Z, Zheng Y (2005) A structural model of the vacuolar ATPase from transmission electron microscopy. Micron 36:109–126

Zarivach R, Vuckovic M, Deng W, Finlay BB, Strynadka NC (2007) Structural analysis of a prototypical ATPase from the type III secretion system. Nat Struct Mol Biol 14:131–137

Zhang Z, Zheng Y, Mazon H, Milgrom E, Kitagawa N, Kish-Trier E, Heck AJ, Kane PM, Wilkens S (2008) Structure of the yeast vacuolar ATPase. J Biol Chem 283:35983–35995

Chapter 7

Carboxysomes – Sequestering RubisCO for Efficient Carbon Fixation

Matthew S. Kimber*
Department of Molecular and Cellular Biology, University of Guelph, Guelph, ON, N1G 2W1, Canada

Summary

Carbon fixation in cyanobacteria (and some chemoautotrophs) occurs in large cytoplasmic bodies known as carboxysomes. Carboxysomes are organized as an enzymatic core comprised of RubisCO and carbonic anhydrase, encapsulated within a thin protein shell. The carbonic anhydrase converts concentrated bicarbonate into carbon dioxide, which can then be fixed by RubisCO. The shell is required as it presents a semi-permeable barrier; in its absence, CO_2 would escape too rapidly to allow efficient fixation. Despite the complexity of this arrangement, and large size of the resulting particle (90–400 nm), carboxysomes are self-organizing and require as few as eight different polypeptides (including the enzymes) for proper self-assembly. This chapter explores discusses the structure and functional roles of the proteins that build up the carboxysome, and analyses the unique architecture of these particles.

*Author for correspondence, e-mail: mkimber@uoguelph.ca

M.F. Hohmann-Marriott (ed.), The Structural Basis of Biological Energy Generation, Advances in Photosynthesis and Respiration 39, pp. 133–148, DOI 10.1007/978-94-017-8742-0_7,

I. Introduction

While the dark reactions of photosynthesis in cyanobacteria (see Chap. 14) utilize much the same enzymatic machinery as higher plants, cyanobacteria's high surface-area-to-volume ratio has allowed the evolution of a unique strategy for optimizing carbon fixation efficiency. Active pumping of inorganic carbon into the cytosol (using a variety of ATP or NADH dependent pumps (Price 2011)) allows cyanobacteria to largely compensate for RubisCO's inherently poor CO_2 affinity. However, this approach is only efficient if CO_2 is fixed faster than it escapes the cell. Cellular membranes are of little help in this regard, as CO_2 is very membrane-permeable, and cyanobacteria, being generally unicellular and micrometer sized, cannot exploit large-scale anatomical adaptations such as those found, for example, in C4 plants. Instead, cyanobacteria construct a protein barrier to prevent CO_2 escape, exploiting the relative impermeability of this material. Inorganic bicarbonate is accumulated in the cytosol, while RubisCO is sequestered with carbonic anhydrase behind a semi-permeable protein shell layer (Fig. 7.1). Once HCO_3^- is within this barrier, it can be converted to CO_2, and is then trapped in an environment that provides maximal opportunities for fixation (Dou et al. 2008). This is necessarily a complex arrangement, and one might intuitively expect that assembling such a complex would require a dedicated molecular infrastructure, such as that required for organizing and maintaining eukaryotic organelles. However, the proteins that comprise the shell turn out to be inherently self-organizing (Bonacci et al. 2012), so they – along with the encapsulated enzymes and a very few ancillary proteins that appear to play a primarily organizational role – spontaneously assemble into large polyhedral bodies; these bodies are known as carboxysomes.

While carboxysomes may be highly unusual objects by eukaryotic standards, they are by no means unique among bacteria. Rather, large scale sequencing efforts have revealed that up to 20 % of bacterial species genomes encode related shell proteins that encapsulate enzymes from quite metabolic different processes (Kerfeld et al. 2010); this larger set of particles are known as "bacterial microcompartments" (Yeates et al. 2010). Very few of these have been studied to date; the best characterized examples are the ethanolamine (etu) and propanediol utilization (pdu) microcompartments from *Salmonella*, where the shell is thought to help sequester volatile and potentially toxic intermediates from the bulk cytosol (Penrod and Roth 2006). Interestingly, sequence comparisons have also led to the realization that "carboxysomes" are paraphyletic, with related, but quite different particles being found in the two major cyanobacterial groups. The β-carboxysomes are characterized by the presence of type Ib RubisCO, and their genes are designated by "ccm" (carbon concentrating mechanism). These carboxysomes are found in β-cyanobacteria, a diverse group found in essentially every conceivable habitat with light. Alpha-carboxysomes are characterized by the presence of type Ia RubisCO, and the genes are designated "cso". Alpha-carboxysomes are found in the α-cyanobacterial groups *Prochlorococcus* and *Synechoccoccus*, which are largely found in nutrient depleted open ocean waters, which they ecologically dominate. In addition to cyanobacteria, α-carboxysomes are also found in the proteobacterial chemoautotrophs; because these organisms are experimentally more tractable (lacking, for example, the extensive thylakoid membranes that complicate the purification of carboxysomes from β-cyanobacteria), much of what we know about carboxysomes is based on work performed in these organisms. In this chapter we discuss the structural biology of

---

Abbreviations: BMC – Bacterial microcompartment; CA – Carbonic anhydrase; DPB – Doubled permuted bacterial microcompartment; etu – Ethanolamine utilization microcompartment; pdu – Propanediol utilization microcompartment; 3PGA – 3-Phosphoglycerate; RuBP – Ribulose 1,5-bisphosphate; RubisCO – Ribulose 1,5-bisphosphate carboxylase/oxygenase

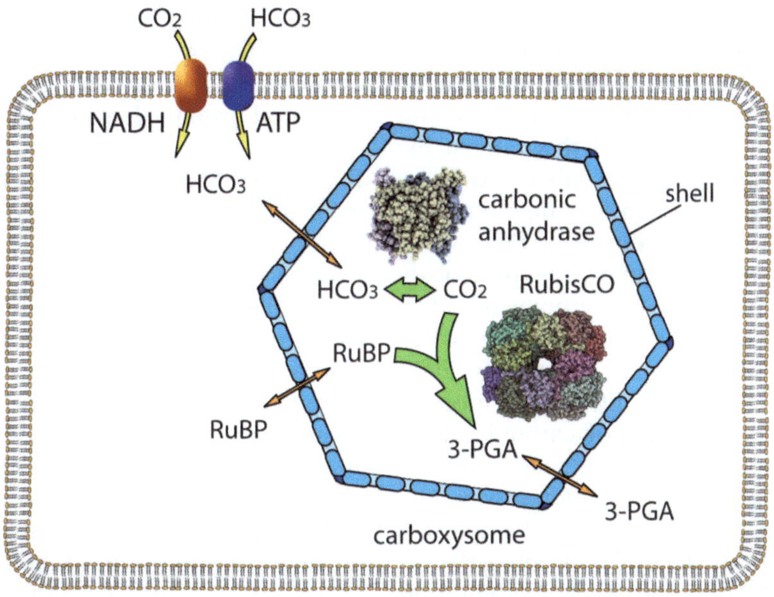

Fig. 7.1. Overview of the role of the carboxysome in cyanobacterial carbon fixation. Cyanobacteria actively concentrate HCO_3^- in their cytosol using a variety of ATP and NADH dependent pumps. Essentially all RubisCO and carbonic anhydrase in the cell are sequestered in carboxysomes, where they are surrounded by a protein shell. Bicarbonate can enter the carboxysome through small pores in the shell, and is then converted to CO_2 by the encapsulated carbonic anhydrase. The presence of the shell impedes the escape of CO_2, allowing it to accumulate and promoting RubisCO catalysed carboxylation of ribulose-1,5-bisphosphate (RuBP). The resulting two molecules of 3-phosphoglycerate can escape back into the cytosol, where RuBP can be regenerated by the Calvin cycle.

α- and β-carboxysomes, both at the level of the structure of the individual proteins (where known) and the overall organization of the supermolecular complex.

II. Carboxysomal Carbonic Anhydrases

A. CcaA – The Classical β-Carboxysomal β-Carbonic Anhydrase

As stated above, carboxysomes have an absolute requirement for an encapsulated carbonic anhydrase to equilibrate their internal inorganic carbon pool. In β-carboxysomes, CcaA was identified in early studies of *Synechoccoccus* sp. PCC 7942 as a gene homologous to plant β-carbonic anhydrase that is required for carbon fixation under atmospheric CO_2 conditions (Fukuzawa et al. 1992; Yu et al. 1992). CcaA, is highly active in recombinant form (So and Espie 1998) and is most similar to the plant chloroplast β-CAs such as the well-characterized pea enzyme (Kimber and Pai 2000); however, it has a carboxysome-specific C-terminal extension of around 60 amino acids (Fig. 7.2a). Most of this extension is hydrophilic and highly variable in sequence, but a 16 amino acid motif at the very end is more conserved (So and Espie 1998). This motif can be deleted without loss of activity *in vitro*, suggesting it may play a role in protein-protein interactions (So et al. 2002). In analogy with similar microcompartmental motifs, it was recently suggested that this motif may help recruit CcaA to the carboxysome shell (Kinney et al. 2012).

B. CcmM – A β-Carboxysomal γ-Carbonic Anhydrase

Shortly after the early work establishing that CcaA is a carboxysomal carbonic anhydrase in β-cyanobacteria, a surprising second candidate carboxysomal carbonic anhydrase was noted in these organisms. In announcing the discovery of a new "γ-class" of carbonic anhydrase from *Methanosarcina thermophila* in 1994, the authors noted that almost the only significant homolog (Alber and Ferry 1994) was a 35 % identical domain from CcmM, from *Synechococcus* sp. PCC 7942. This open reading frame was previously found to be important for carboxysome formation (Price et al. 1993). However, when tested, both the *Synechococcus* sp. PCC 7942 and *Synechocystis* sp. PCC6803 proteins proved to lack any trace of carbonic anhydrase activity when expressed recombinantly (So and Espie 2005). This finding was puzzling in light of the fact that all important catalytic residues (Ferry 2010) appear to be conserved in almost all CcmM homologs, although it did reflect a larger failure to find activity in any γ-CA homolog except those from *Methanosarcina* itself. The uncertainty was further heightened when sequencing of multiple β-cyanobacterial genomes revealed that about a third of species lack a CcaA homolog, and in some species, including *Thermosynechococcus elongatus* (Nakamura et al. 2002), CcmM is the only homolog of a known CA present. This paradox was recently resolved by the finding that the CcmM from *T. elongatus* is highly active in recombinant form (Peña et al. 2010). Interestingly, the structure of this protein (Fig. 7.2b) revealed that in addition to the long β-helix and catalytic site covering α-helix seen in the *T. thermophila* enzyme (Kisker et al. 1996), CcmM has an additional C-terminal α-helix that is stabilized by a disulfide bond. This motif proved essential to activity in the enzyme; constructs lacking the cysteine residues are inactive, as are longer constructs under reducing conditions. Since the reducing potential of the cytosol is sufficient to inactivate the enzyme, this implies that the interior of the carboxysome is necessarily oxidizing (Peña et al. 2010). This switching motif is not conserved in many strains, including *Synechocytis* sp. PCC 6803 and *Synechococcus* sp. PCC7942, that contain a CcaA ortholog in their genomes. In these species, CcmM appears to have lost its enzymatic function and acts solely as a structural protein (Peña et al. 2010).

C. CsoCA – The Highly Divergent α-Carboxysomal β-Carbonic Anhydrase

Alpha-cyanobacteria have neither CcaA nor CcmM-like proteins, and generally lack any clear homologs of the established α-, β- and γ-carbonic anhydrase families. However, systematic testing showed that *csoS3* in the carboxysome operon is an active carbonic anhydrase when expressed recombinantly; this gene was subsequently renamed *csoSCA* (So et al. 2004). Inactivating this gene results in carboxysomes that do not support growth at ambient CO_2 concentrations, the expected phenotype for a protein essential to carboxysome functioning (Dou et al. 2008). CsoSCA has no detectable sequence similarity to known carbonic anhydrases, and it was therefore initially described as the founding member of a new class – the ε-CAs (So et al. 2004). However, when the structure was eventually determined, this protein proved to be a β-CA, albeit a highly divergent example (Fig. 7.2c). The protein is actually built as a head-to-tail fusion of two β-CA domains, along with an additional α-helical N-terminal domain. The second β-CA domain is highly degenerate and does not contain a catalytic site, contributing only a few motifs to the catalytic site formed predominantly by the other domain. The divergence from canonical β-CAs is underscored by the fact that only five core catalytic residues – the three zinc binding residues, and a critical aspartate and arginine pair – are conserved with the rest of the protein family (Sawaya et al. 2006).

III. Structure and Organization of the Shell Proteins

Carboxysome and other bacterial microcompartment shells are built from small proteins from three protein families, two of which are

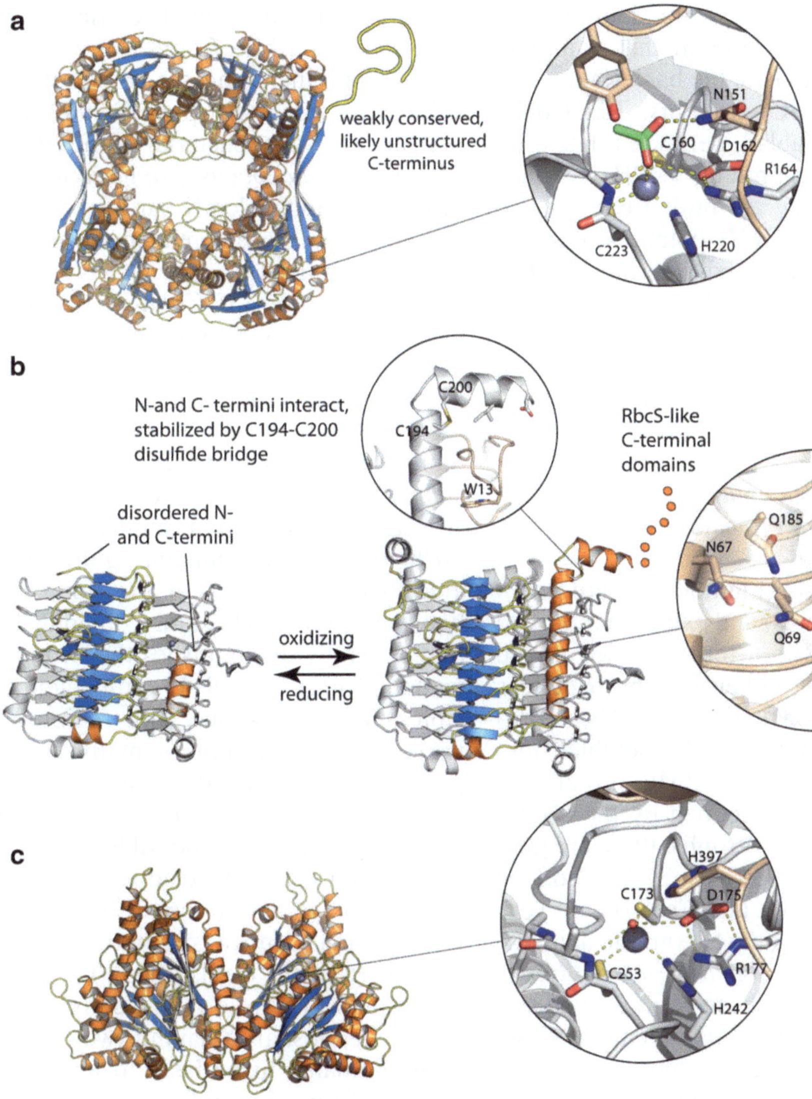

Fig. 7.2. Carboxysomal carbonic anhydrases. (**a**) CcaA is a β-carbonic anhydrase found in many, but by no means all β-carboxysomes. It is about 30 % identical to the plant β-CAs (shown; 1EKJ); plant β-CAs, however, lack a 60 amino acid C-terminal extension unique to carboxysomal β-CAs. It should also be noted that the oligomeric state of β-CAs are quite variable, though all variants are built as larger associations of dimers. This is important as CcaA interacts with CcmM, and, for example, a hexameric CcaA might better match the trimeric symmetry of CcmM. Note that the catalytic site (*inset*) has only five residues conserved when compared with CsoSCA. (**b**) CcmM is an essential structural protein in β-carboxysomes, although it is not an active carbonic anhydrase in all strains. The structure (3KWC) differs from the canonical γ-CA, Cam, in having an additional C-terminal helix which is stabilized by a disulfide bond between Cys194 and Cys200 (*inset*). This disulfide bond is stable only under oxidizing conditions. Under reducing conditions (*left*), mimicked by the truncation at residue 193, the protein rearranges itself into an inactive configuration, with the long active site-covering helix largely disordered (3KWD). This structural transition depends upon motifs at the N- and C-terminus which are not conserved in many β-cyanobacterial species that have an active CcaA; this truncated structure is likely a good approximation of the structure of CcmM in such species. (**c**) Structure of CsoSCA (2FGY), the only carbonic anhydrase known to be associated with α-carboxysomes. The protein has a highly modified β-CA fold, built from two copies of the β-CA domain fused head to tail. These second of these domains has become highly degraded, so that only one catalytic site is present. The active site (*inset*) shows residues contributed from the second, degraded β-CA domain in beige.

distantly related. The most abundant of these (and presumably the protein that comprise the bulk of the shell) are the bacterial microcompartment (BMC) proteins. Related to these are the doubled permuted bacterial microcompartment proteins (DPB); as their name suggests, they are built from head-to tail-fusions of two BMC domains, each of which has been circularly permuted. The third family are the CcmL/CsoS4 proteins, which are generally pentameric. All known microcompartments appear to require at least one protein from each of these families. We consider each group in turn.

A. The Canonical Bacterial Microcompartment Proteins

The BMC proteins (pfam00936) are small (90–110 amino acids) and exhibit a modified ferridoxin fold (with a four stranded mixed β-sheet flanked on either side by α-helices (Kerfeld et al. 2005)) (Fig. 7.3). These protomers are invariably found assembled into six-membered rings. The BMC proteins in α-carboxysomes are encoded by the CsoS1 paralogs. In chemoautotrophs, there are generally three paralogs – CsoS1A, CsoS1B, and CsoS1C. CsoS1A and CsoS1C are very closely related, with only two amino acid differences in *Halothiobacillus neapolitanus* and the structures of these two proteins are, not surprisingly, essentially identical (Tsai et al. 2007, 2009). The structure for CsoS1B is not known, but its sequence very closely resembles CsoS1A and CsoS1C, except for a short C-terminal extension. In *Prochlorococcus* MED4 CsoS1A is the only single domain BMC protein found, and this would appear to be the minimal functional complement (Kerfeld et al. 2005).

Beta-carboxysomes show an even higher BMC paralog diversity than α-carboxysomes, with the most commonly observed complement of BMC proteins being CcmK1, CcmK2, CcmK3, CcmK4 and CcmO. However, the composition of BMC paralogs in β-carboxysomes can be quite variable, and only CcmK2 and CcmO are universally conserved: CcmK3 and CcmK4 are absent in the most basal cyanobacteria, such as *Gloeobacter*, while CcmK1 is absent in a few strains, including the important model strain *Synechococcus* sp. PCC 7942. In addition, a few cyanobacterial strains (such as *Thermosynechococcus elongatus*) have one or two additional CcmK paralogs that do not appear to be closely related to any of the standard paralogs.

The structures of CcmK4, CcmK1 and CcmK2 very closely resemble CsoS1A and CsoS1C, but have an additional helix αD at the C-terminus that packs on the convex face of the hexamer (Kerfeld et al. 2005; Samborska and Kimber 2012). CcmK1 and CcmK2 are closely related and very conserved, with the two proteins having over 90 % identity when comparing different paralogs within a strain, or the same paralog between strains. The most obvious feature that generally distinguishes the two proteins is that CcmK1 has a 9–11 amino acid extension at the C-terminus. This region is only weakly conserved, but normally contains a pair of arginine residues and is generally basic. Interestingly, CcmK4 has a similar extension, hinting at a possible commonality in function. These residues were either deleted, or found to be disordered in all CcmK1 and CcmK4 structures determined to date (Kerfeld et al. 2005; Tanaka et al. 2009), and their role is presently unknown. CcmO is unusual in having two CcmK-like BMC domains fused head to tail (and presumably forms a pseudo-hexameric trimeric ring) in addition to a long C-terminal extension comprised of a handful of hydrophilic amino acids that is predicted to be intrinsically disordered. CcmK3 is almost wholly unstudied, but its close association with CcmK4 (the two genes form an operon, and no strains are known where only one occurs) suggests that these two proteins may be functionally related.

The crystals of BMC proteins from carboxysomes, as well as other microcompartments, show a clear preference for packing as single layers of hexamers, forming continuous flat, largely gapless sheets (Fig. 7.3). A similar packing behavior has also been

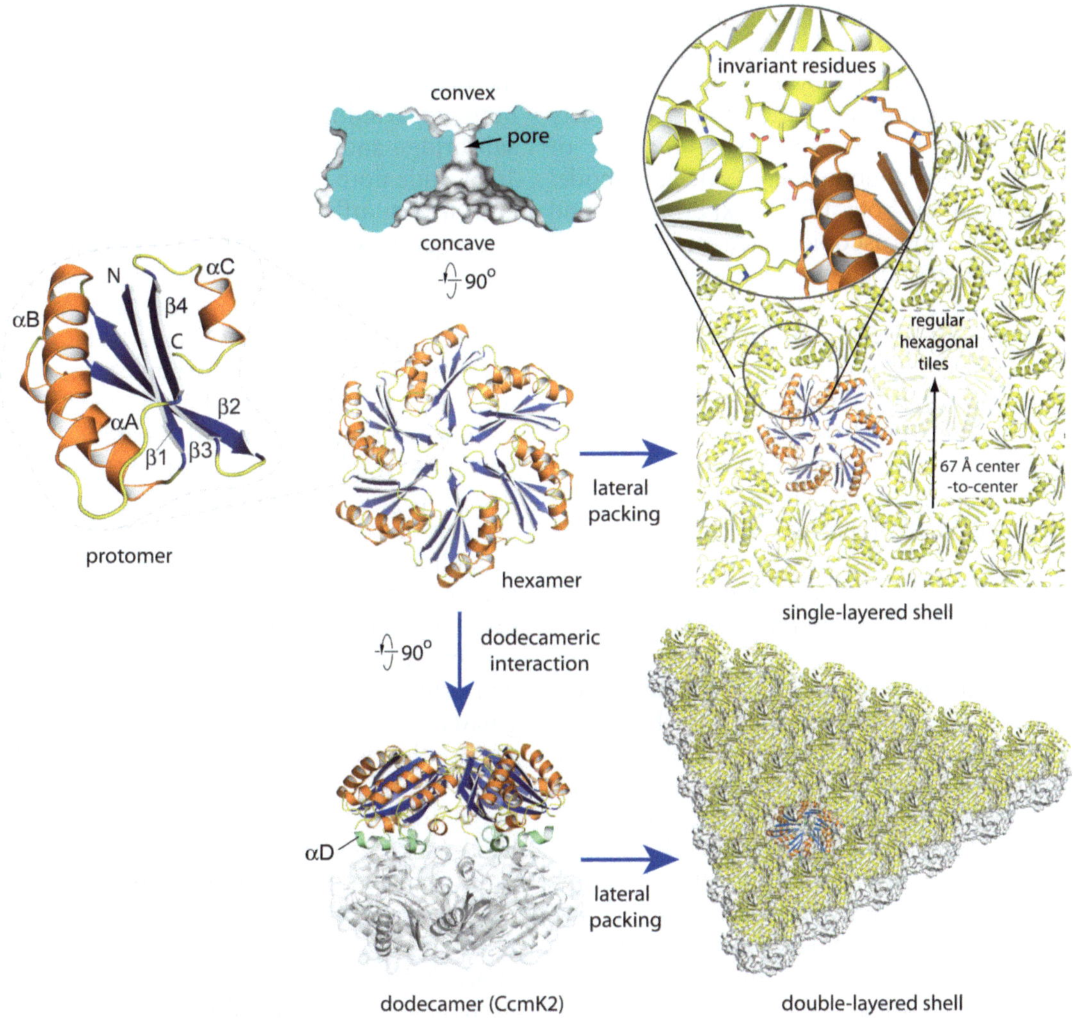

Fig. 7.3. The structure and interaction modes of the BMC proteins, and the inferred organization of the facets of the shell. The shell is built from multiple copies of the BMC proteins – CsoS1A-C in α-carboxysomes, and CcmK1-4 and CcmO in β-carboxysomes. The BMC proteins adopt a four stranded ferredoxin-like fold. Six protomers associate to form a hexameric ring, characterized by having an outline that conforms closely in shape to that of a regular *hexagon*, and perforated by a *small central pore*. These hexamers interact with one another via conserved residues along their edges to form a continuous sheet, with P6 planar symmetry. CcmK2 was recently shown to have an additional interaction mode, forming a dodecamer by interactions mediated through αD on the concave side of the hexamer. CcmK2 appears to form a double-layered shell-like arrangement *in vitro*. Depending on the crystal form used as a model, this sheet can have no, or only small pores between hexamers; the central pore is suspected, but not proven, to be the main channel conducting metabolites into and out of the particle.

shown in two-dimensional CcmK crystals, grown on a membrane (Dryden et al. 2009). This packing mode is thought to reflect the functional interaction mode of these proteins, as such extended sheets of BMC proteins forming the facets of the microcompartment shell would parallel the organization of high T number viruses (Kerfeld et al. 2005). Consistent with a role in forming the bulk of the shell, BMC proteins are highly abundant in carboxysomes, with only RubisCO being present in greater quantities (Long et al. 2007; Menon et al. 2008). However, recent studies of *Thermosynechococcus* CcmK2 show that this protein has an additional interaction mode, forming a convex-to-convex dodecamer mediated by the C-terminal helix. This dodecameric interaction is exactly positioned to fuse two single-layers of CcmK2 into double-layered sheet, a model supported by both FRET experiments and electron microscopic measurements of β-carboxysomal shell thickness (Samborska and Kimber 2012). However, given that current models do not explain features such as the requirement for diversity of BMC protein paralogs, or the enhanced stability of the mature shell, much work remains to be done before we can be confident that we have a good working understanding of microcompartment shell organization.

B. CsoS1D – A Circularly Permuted Bacterial Microcompartment Protein

The doubled permuted BMC proteins were discovered as very distant BMC homologs in α-cyanobacteria, and subsequently found to be present in all microcompartment-containing organisms (Klein et al. 2009). CsoS1D is built from a variant BMC domain that is circularly permuted relative to the canonical BMC domain topology (Fig. 7.4a). This permuted topology has also been seen in other single domain microcompartment proteins including EutL and PduU (Crowley et al. 2008; Sagermann et al. 2009). Here, the residues corresponding to the N- and C-termini of the canonical BMC domain are connected, with new N- and C- termini introduced between αB and β4 (Klein et al. 2009). In CsoS1D and related proteins, two such domains are fused, N-to-C- terminus, resulting in a protein with a pseudo-six-fold symmetry; the protein also has a 50 amino acid N-terminal extension in α-cyanobacteria (though not in chemoautotrophs). This motif proved disordered in both available structures. Two trimeric rings interact through their concave sides, forming a pseudo-dodecameric hexamer. However, this arrangement is asymmetric, with the two rings in distinct conformations. In one ring, three copies of Arg120 meet in the center, forming a salt bridge with Glu121 (Fig. 7.4b, c), and resulting in a pore that is wholly closed. In the opposite ring, the protomers are rotated outward, resulting in the disruption of the salt bridge and the consequent opening of a 14 Å wide pore (Fig. 7.3d, e). This pattern is seen in two different crystal forms, suggesting negative cooperativity in the conformation of the opposing rings, as seen, for example, in GroEL (Danziger 2003). The closed form of the ring appears to be compatible with the packing arrangement seen for BMC proteins, including CsoS1A, but the opened ring is expanded, meaning that it would likely require an adjustment in the lattice. Intriguingly, comparison of different crystals of CcmK2, for example, suggest that accommodations on this order may be accomplish-able. The function of CsoS1D is presently unclear. It was originally suggested that by cycling between the open and closed conformations, the protein could act as a gated pore that would allow larger molecules to pass through the shell (Klein et al. 2009). However, it is unclear how specificity would be maintained in this model, or how the shell-packing would accommodate the structural changes required to transition between the conformers. It was recently reported that CosS1D is detectable in the mature carboxysome particle (Roberts et al. 2012), and moreover, that recombinantly expressed α-carboxysomes are irregular in form if CsoS1D is absent (Bonacci et al. 2012). The implication is that this protein plays an important architectural role in organizing the shell, though the details are presently obscure.

An equivalent DPB protein in β-cyanobacteria is provisionally named CcmP. Sequence conservation would argue that this protein likely has a structure and a functional role that parallels CsoS1D, but no experimental data has been published on this protein to date.

C. CcmL/CsoS4 – The Vertex Protein

The third family of proteins invariably associated with microcompartment operons

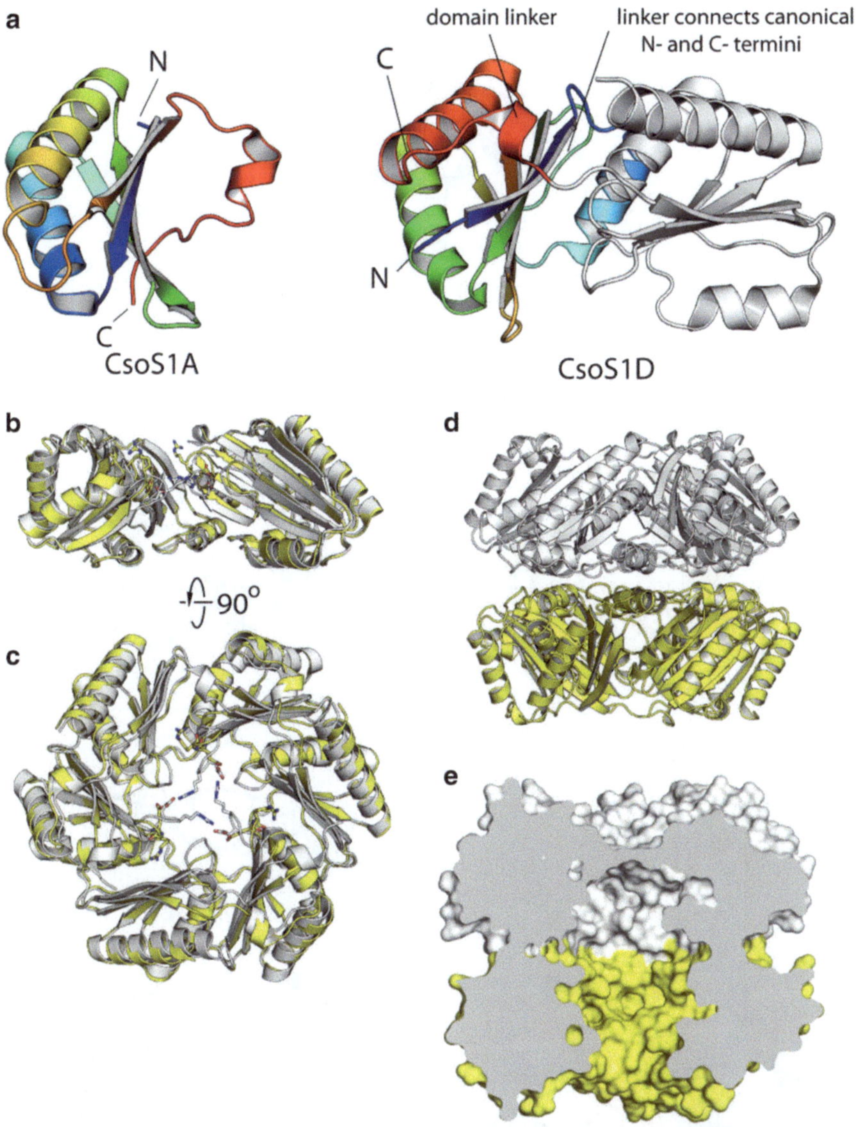

Fig. 7.4. The structure of the doubled permuted BMC protein, CsoS1D. (**a**) Organization of the protomer. The N-terminal BMC domain is shown in progression from *red* to *blue*, N- to C-terminus. The C-terminal domain is shown in *white*. The CsoS1A protomer is shown on the *left* for comparison. CsoS1D BMC domains are organized as a circular permutation of the canonical BMC domain. The N-terminal most β-strand in CsoS1D corresponds to β4 in CsoS4A. (**b**) Superposition of the two independent trimeric CsoS1D (3 F56) rings, viewed in cross-sectional view with the front most protomers omitted for clarity. The open conformation is shown in *yellow*, and the closed conformation in *white*. In the open conformation, the BMC domains rotate up and outwards, with the concave face remaining almost fixed, and maximal movement of the convex face. Maximal difference in the backbone is around 2 Å, except in the pore lining loops, which undergo a large structural rearrangement to create a large (14 Å) pore. (**c**) The same superposition as in 7.4b), but viewed from the convex face. (**d**) The CsoS1D hexamer, viewed orthogonal to the three-fold axis. (**e**) Molecular surface, cut through the center. Note the open conformation presents a wide pore, while the *ring* in the closed conformation is sealed.

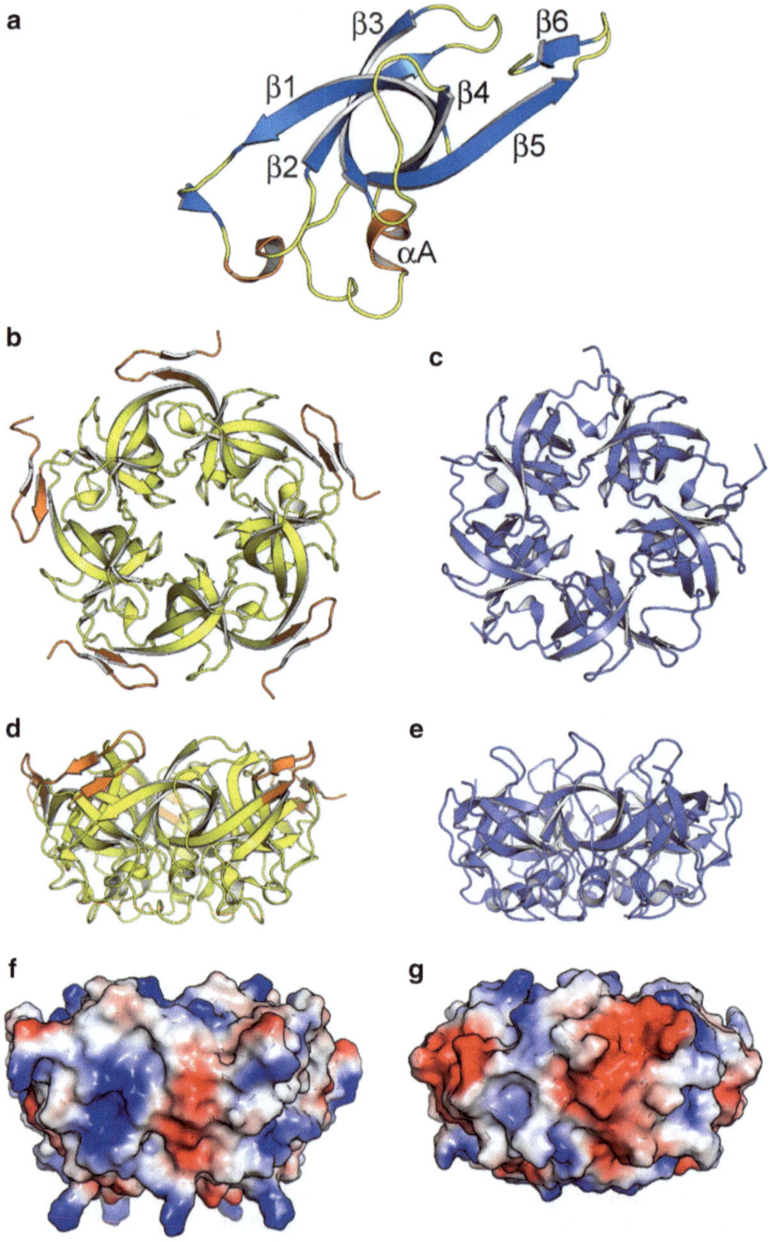

Fig. 7.5. The structure and role of CsoS4 and CcmL. (**a**) Organization of the CcmL protomer (2QW7), with the five-fold symmetry axis oriented vertical in the plane of the page. The protein is built as a 5-stranded up-down β-barrel. (**b**, **d** and **f**) The CcmL pentamer shown along and oblique to the five-fold axis in ribbon representation, and oblique to the five-fold as an electrostatic surface. (**c**, **e**, and **g**) The CsoS4A pentamer (2RCF), shown from equivalent orientations. Note that CcmL is slightly longer than CsoS4, with an extension to β5 and an additional strand, β6 (*orange*).

is the pfam03319 family, known CsoS4 in α-carboxysomes, and as CcmL in β-carboxysomes. These proteins are small – 80–100 amino acids, with a protomer structure built around a five-stranded β-barrel motif (Fig. 7.5) (Tanaka et al. 2008). This motif assembles into pentamers, with an overall regular pentagonal prism shape.

Icosahedra have five-fold symmetry axis where facets meet at the vertices; in microcompartment shells, these special positions would be expected to require five-fold symmetric "plugs" to seal the shell. Modeling shows that CcmL/CsoS4 may be of a suitable size and shape to occupy the vertices of the microcompartment, interacting with five neighboring BMC proteins (Tanaka et al. 2008). Consistent with this role, these proteins are considerably less abundant in the α-carboxysome than the CsoS1 paralogs (Cai et al. 2009). Interestingly, though, the role of these two proteins seems subtly different in the two carboxysomes. In β-carboxysomes, CcmL is always present as a single copy, and is about 20 amino acids longer than CsoS4 (with these additional residues forming a β-hairpin motif) and deletion of the gene results in rod-shaped elongated carboxysomes (Price and Badger 1991). In α-carboxysomes, there are always two copies of the protein – CsoS4A and CsoS4B – and deleting both results in carboxysomes that look similar to wild type carboxysomes (with a minority slightly elongated), but which leak CO_2 and fail to efficiently fix carbon (Cai et al. 2009). It should also be noted that EutN (the CcmL homolog in the *eut* microcompartment) is hexameric (Tanaka et al. 2008), implying that this protein family may be capable of playing more than one role in shell organization.

IV. Organization of the α-Carboxysome Interior

Cryo-electron tomographic reconstructions have been reported for the α-carboxysomes from *Synechococcus* WH8102 (Iancu et al. 2007) and *Halothiobacillus neapolitanus* (Schmid et al. 2006). These studies show α-carboxysomes to be organized as regular icosahedra, with an approximately 4 nm thick shell. Interestingly, the diameters of particles isolated from a given strain vary widely, from 88 to 108 nm, suggesting that the process by which α-carboxysomes assemble allows for considerable variability in diameter, while preserving the overall icosahedral geometry.

CsoSCA is tightly associated with the α-carboxysome shell (Baker et al. 2000; So et al. 2004) (See Fig. 7.6b). The protein is not required structurally in shell formation, as CsoSCA knockout strains still form carboxysomes that closely resemble their wild type counterparts. An interesting outstanding question is how exactly this complex is mediated, as the CsoSCA most likely forms a dimer, with a clear symmetry mismatch to the six-fold symmetric CsoS1 shell proteins (Dou et al. 2008).

RubisCO distribution is semi-organized in α-carboxysomes, being packed in roughly spherical shells; this organization corresponds to the default geometry for hard spheres packing randomly within confined icosahedral spaces, and may therefore require no specific interaction to drive it (Iancu et al. 2007). RubisCO in the α-carboxysome appears to be only loosely associated with the shell, as it readily leaks out from particles ruptured by freeze-thaw treatment (So et al. 2004; Menon et al. 2008). Consistent with this, assembly of the α-carboxysome shell is independent of the expression of RubisCO (Menon et al. 2008). The determinants targeting RubisCO to the carboxysome reside on the large subunit, but have not been specifically identified (Menon et al. 2008). The mechanism by which RubisCO is recruited is unclear, but may be related to CsoS2, which immuno-gold staining maps to the interior of the particle (Baker et al. 1999). This protein, is expressed in two distinct isoforms, migrating on SDS PAGE gels at 85 and 130 kDa (Baker et al. 1999). The protein is highly hydrophilic, and characterized by a [ILVTMF]-[ST]-[G] sequence repeat that is separated by hydrophilic spacers (Cannon et al. 2002). The high fraction of hydrophilic residues suggests that this protein is most likely intrinsically disordered. The most obvious candidate function for CsoS2 would be protein-protein interactions, and indeed, CsoS2 may interact with CbbL, CbbS, CsoS4 and itself based on results from bacterial two-hybrid assays. However, most

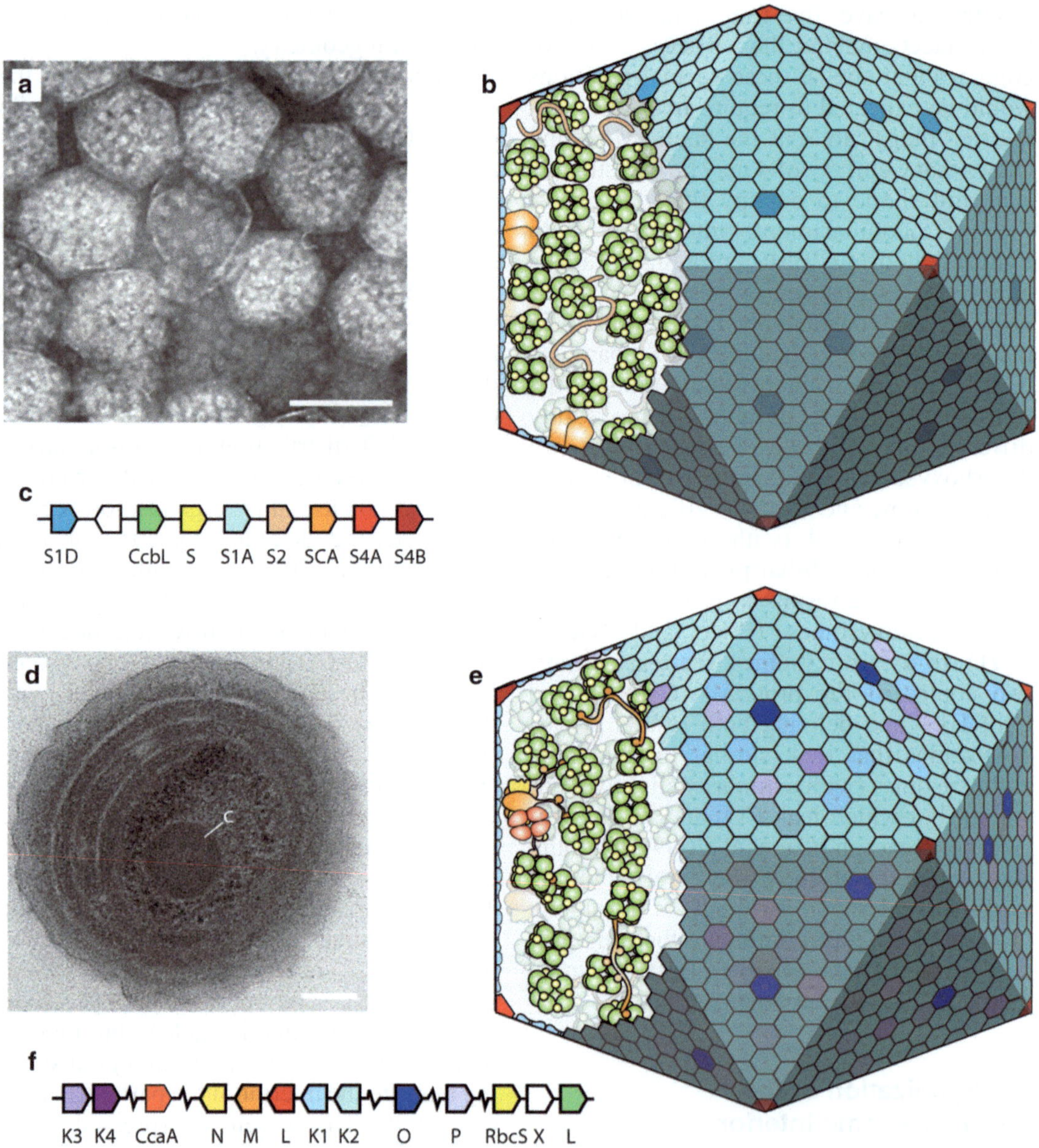

Fig. 7.6. Organization of carboxysomes. (**a**) Electron micrographs of purified α-carboxysomes from *Prochlorococcus marinus* MED4. Individual RubisCO molecules can be readily visualized as small, light colored particles both within the carboxysomes, as well as more diffusely scattered through the media. Scale bar is 100 nm (Image courtesy of Evan Roberts (Cannon/Heinhorst laboratory, University of Southern Mississippi)) (*Panels a* and *d* previously appeared in Espie and Kimber (2011), and are reproduced with permission). (**b**) Overall organization of the α-carboxysome (the minimalist carboxysome from *Prochlorococcus* MED4 is used as an example). The carboxysomes is shown in schematic, with part of the shell omitted to show interior organization. Note that many details shown in this figure are not addressed by available data, and are therefore largely speculative. E.g. the organization of the facet edges, and the distribution of different BMC paralogs within the facet. (**c**) Genomic organization of genes known to encode carboxysomal proteins, colored to correspond to the proteins shown in the *upper panel*. (**d**) A β-carboxysome (C) in the cytoplasm of *Synechococcus* sp. PCC 7942. Scale bar is 100 nm (Image courtesy of Prof. Tammy Sage and Kathy Saults (Ecology and Evolutionary Biology, University of Toronto)). (**e**) Overall organization of a β-carboxysome with the most commonly observed protein complement (e.g. *Synechocystis* sp. PCC 6803). Note that, while the carboxysomes are shown as the same size to facilitate comparison, β-carboxysomes are typically double the dimensions of α-carboxysomes. (**f**) Carboxysome gene organization in *Synechocystis* sp. PCC 6803.

of these results could not be replicated when bait and target were switched (Gonzales et al. 2005), leaving this as an intriguing possibility that requires further investigation.

V. Organization of the β-Carboxysome Interior

To date, there are no electron tomographic studies of β-carboxysomes, mostly because protocols to obtain pure preparations of β-carboxyomes have not yet been sufficiently developed. Most imaging has been done in the context of intact cells, where observations are consistent with a polyhedral, and possibly icosahedral organization (Fig. 7.6a). Like α-carboxysomes, β-carboxysomes appear to vary considerably in size, with those from *Synechococcus* sp. PCC7942 having a standard deviation of 26 nm around the 172 nm mean diameter (Long et al. 2007).

The most important protein for organizing the β-carboxysome interior seems to be CcmM (see Fig. 7.5b). As mentioned above, this protein has a left handed β-helical N-terminal domain, which in some species, is an active carbonic anhydrase. This protein also has a series of three to five repeats of a small C-terminal domain, spaced out by hydrophilic linkers on the order of 40 amino acids in length. These domains, whose very similar sequence clearly indicates a gene duplication event, are clear homologs of the RubisCO small subunit, with about 24 % sequence identity (Price et al. 1993). Interestingly, this protein is produced in two variants: a 58 kDa full length variant, and a smaller 35 kDa variant that encompasses only the RubisCO-like domains and linkers (Long et al. 2005). The 35 kDa variant is produced at about four times the level of the full length variant (Long et al. 2007). This variant is produced by translation from an internal ribosome entry site, located at Val226, and both isoforms are required for functional carboxysomes (Long et al. 2010). Interestingly, additional candidate translation sites are present in equivalent positions in the other small subunit repeats (Long et al. 2010), and are possibly used in *Synechocystis* sp. PCC 6803 to produce additional small isoforms (Cot et al. 2008).

CcmM interacts strongly with the large subunit of RubisCO; this interaction requires the small C-terminal subunits, but not the large subunits (Cot et al. 2008). In CcmM IMAC pull down experiments, RubisCO associates strongly with both isoforms; in these complexes and RbcS is depleted compared to its usual 1:1 ratio with RbcL (Long et al. 2007). Together, this evidence suggests that CcmM may interact with RbcL by displacing RbcS from one or more of its sites in the complex.

CcmM appears to be the central protein for organizing the β-carboxysome as it is able to interact with all encapsulated proteins, as well as possibly the shell. In yeast two hybrid studies, the N-terminal domain of CcmM interacts with the β-carbonic anhydrase, CcaA; this interaction is independent of CcaA's C-terminal extension (Cot et al. 2008). Interestingly, CcmM from organisms that do not contain CcaA retains the ability to bind this protein, suggesting that CcaA binds a universally conserved site on CcmM (Peña et al. 2010). The same yeast two-hybrid experiments indicate that CcmM also binds CcmN (Cot et al. 2008). CcmN is an approximately 23 kDa protein, with an N-terminal domain with homology to left-handed β-helical proteins including CcmM (though lacking the zinc binding residues, and all other known functional determinants). Yeast two-hybrid experiments show that CcmN interacts with CcmM (Cot et al. 2008), an interaction, subsequently confirmed by co-purification, to be mediated by the N-terminal β-helical domains of both proteins (Kinney et al. 2012). CcmN's C-terminus is characterized by an extended intrinsically unstructured region, followed by a short, conserved sequence. Kinney et al. show that this motif can interact at least weakly with CcmK2, suggesting that this motif may be important for recruiting the CcmM/CcmN complex to the shell (Kinney et al. 2012). An essential role for this motif is consistent with a point mutation, Gly145Asp, being sufficient to completely abolish carboxysome function (Friedberg et al. 1989).

It should be noted that although CcmM provides an obvious mechanism for recruiting RubisCO to the β-carboxysome, only a fraction of RubisCO interacts with this protein, with the majority of RubisCO being recruited without participating in such tight complexes (Long et al. 2007). In this context, it is interesting to note that RubisCO in β-carboxysomes appears much more structured than in α-carboxysomes, taking on a distinctly crystalline organization (Kaneko et al. 2006). This possibly suggests that weak RubisCO-RubisCO interactions may be sufficient to drive self-assembly, with CcmM perhaps providing a mechanism to associate the overall complex with the shell.

VI. Conclusions

Cyanobacteria have their own unique adaptations for carbon fixation, and their organization of RubisCO into carboxysomes is one of the most unusual aspects of their carbon concentrating mechanism. With as few as eight distinct proteins (nine for β-carboxysomes) comprising these bodies, they nevertheless manage to organize themselves as a highly complicated three dimensional structure that occupies a substantial fraction of the cytoplasmic volume of the cells that house them. Despite much progress in our study of carboxysomes and related microcompartments, especially over the past decade, our understanding remains fairly fragmentary, and potentially still subject to major revisions.

Acknowledgements

This work was funded by a Discovery Grant from the National Science and Engineering Research Council of Canada to MSK (# 327280).

References

Alber BE, Ferry JG (1994) A carbonic anhydrase from the archaeon *Methanosarcina thermophila*. Proc Natl Acad Sci U S A 91:6909–6913

Baker SH, Lorbach SC, Rodriguez-Buey M, Williams DS, Aldrich HC, Shively JM (1999) The correlation of the gene csoS2 of the carboxysome operon with two polypeptides of the carboxysome in *Thiobacillus neapolitanus*. Arch Microbiol 172:233–239

Baker SH, Williams DS, Aldrich HC, Gambrell AC, Shively JM (2000) Identification and localization of the carboxysome peptide Csos3 and its corresponding gene in *Thiobacillus neapolitanus*. Arch Microbiol 173:278–283

Bonacci W, Teng PK, Afonso B, Niederholtmeyer H, Grob P, Silver PA, Savage DF (2012) Modularity of a carbon-fixing protein organelle. Proc Natl Acad Sci U S A 109:478–483

Cai F, Menon BB, Cannon GC, Curry KJ, Shively JM, Heinhorst S (2009) The pentameric vertex proteins are necessary for the icosahedral carboxysome shell to function as a CO_2 leakage barrier. PLoS ONE 4:e7521

Cannon GC, Heinhorst S, Bradburne CE, Shively JM (2002) Carboxysome genomics: a status report. Funct Plant Biol 29:175–182

Cot SS-W, So AKC, Espie GS (2008) A multiprotein bicarbonate dehydration complex essential to carboxysome function in cyanobacteria. J Bacteriol 190:936–945

Crowley CS, Sawaya MR, Bobik TA, Yeates TO (2008) Structure of the PduU shell protein from the Pdu microcompartment of *Salmonella*. Structure 16:1324–1332

Danziger O (2003) Conversion of the allosteric transition of GroEL from concerted to sequential by the single mutation Asp-155 -> Ala. Proc Natl Acad Sci U S A 100:13797–13802

Dou Z, Heinhorst S, Williams EB, Murin CD, Shively JM, Cannon GC (2008) CO_2 fixation kinetics of Halothiobacillus neapolitanus mutant carboxysomes lacking carbonic anhydrase suggest the shell acts as a diffusional barrier for CO_2. J Biol Chem 283:10377–10384

Dryden KA, Crowley CS, Tanaka S, Yeates TO, Yeager M (2009) Two-dimensional crystals of carboxysome shell proteins recapitulate the hexagonal packing of three-dimensional crystals. Protein Sci 18:2629–2635

Espie GS, Kimber MS (2011) Carboxysomes: cyanobacterial RubisCO comes in small packages. Photosynth Res 109:7–20

Ferry JG (2010) The gamma class of carbonic anhydrases. Biochim Biophys Acta 1804:374–381

Friedberg D, Kaplan A, Ariel R, Kessel M, Seijffers J (1989) The 5′-flanking region of the gene encoding the large subunit of ribulose-1,5-bisphosphate carboxylase/oxygenase is crucial for growth of the cyanobacterium *Synechococcus* sp. strain PCC 7942 at the level of CO_2 in air. J Bacteriol 171:6069–6076

Fukuzawa H, Suzuki E, Komukai Y, Miyachi S (1992) A gene homologous to chloroplast carbonic anhydrase (icfA) is essential to photosynthetic carbon dioxide fixation by *Synechococcus* PCC7942. Proc Natl Acad Sci U S A 89:4437–4441

Gonzales AD, Light YK, Zhang Z, Iqbal T, Lane TW, Martino A (2005) Proteomic analysis of the CO_2-concentrating mechanism in the open-ocean cyanobacterium *Synechococcus* WH8102. Can J Bot 83:735–745

Iancu CV, Ding HJ, Morris DM, Dias DP, Gonzales AD, Martino A, Jensen GJ (2007) The structure of isolated *Synechococcus* strain WH8102 carboxysomes as revealed by electron cryotomography. J Mol Biol 372:764–773

Kaneko Y, Danev R, Nagayama K, Nakamoto H (2006) Intact carboxysomes in a cyanobacterial cell visualized by Hilbert differential contrast transmission electron microscopy. J Bacteriol 188:805–808

Kerfeld CA, Sawaya MR, Tanaka S, Nguyen CV, Phillips M, Beeby M, Yeates TO (2005) Protein structures forming the shell of primitive bacterial organelles. Science 309:936–938

Kerfeld CA, Heinhorst S, Cannon GC (2010) Bacterial microcompartments. Annu Rev Microbiol 64:391–408

Kimber MS, Pai EF (2000) The active site architecture of *Pisum sativum* beta-carbonic anhydrase is a mirror image of that of alpha-carbonic anhydrases. EMBO J 19:1407–1418

Kinney JN, Salmeen A, Cai F, Kerfeld CA (2012) Elucidating the essential role of the conserved carboxysomal protein CcmN reveals a common feature of bacterial microcompartment assembly. J Biol Chem 287:17729–17736

Kisker C, Schindelin H, Alber BE, Ferry JG, Rees DC (1996) A left-hand beta-helix revealed by the crystal structure of a carbonic anhydrase from the archaeon *Methanosarcina thermophila*. EMBO J 15:2323–2330

Klein MG, Zwart P, Bagby SC, Cai F, Chisholm SW, Heinhorst S, Cannon GC, Kerfeld CA (2009) Identification and structural analysis of a novel carboxysome shell protein with implications for metabolite transport. J Mol Biol 392:319–333

Long BM, Price GD, Badger MR (2005) Proteomic assessment of an established technique for carboxysome enrichment from *Synechococcus* PCC7942. Can J Bot 83:746–757

Long BM, Badger MR, Whitney SM, Price GD (2007) Analysis of carboxysomes from *Synechococcus PCC7942* reveals multiple Rubisco complexes with carboxysomal proteins CcmM and CcaA. J Biol Chem 282:29323–29335

Long BM, Tucker L, Badger MR, Price GD (2010) Functional cyanobacterial beta-carboxysomes have an absolute requirement for both long and short forms of the CcmM protein. Plant Physiol 153:285–293

Menon BB, Dou Z, Heinhorst S, Shively JM, Cannon GC (2008) *Halothiobacillus neapolitanus* carboxysomes sequester heterologous and chimeric RubisCO species. PLoS ONE 3:e3570

Nakamura Y, Kaneko T, Sato S, Ikeuchi M, Katoh H, Sasamoto S, Watanabe A, Iriguchi M, Kawashima K, Kimura T, Kishida Y, Kiyokawa C, Kohara M, Matsumoto M, Matsuno A, Nakazaki N, Shimpo S, Sugimoto M, Takeuchi C, Yamada M, Tabata S (2002) Complete genome structure of the thermophilic cyanobacterium *Thermosynechococcus elongatus* BP-1 (supplement). DNA Res 9:135–148

Peña KL, Castel SE, De Araujo C, Espie GS, Kimber MS (2010) Structural basis of the oxidative activation of the carboxysomal -carbonic anhydrase, CcmM. Proc Natl Acad Sci U S A 107:2455–2460

Penrod JT, Roth JR (2006) Conserving a volatile metabolite: a role for carboxysome-like organelles in *Salmonella enterica*. J Bacteriol 188:2865–2874

Price GD (2011) Inorganic carbon transporters of the cyanobacterial CO_2 concentrating mechanism. Photosynth Res 109:47–57

Price GD, Badger MR (1991) Evidence for the role of carboxysomes in the cyanobacterial CO_2-concentrating mechanism. Can J Bot 69:963–973

Price GD, Howitt SM, Harrison K, Badger MR (1993) Analysis of a genomic DNA region from the cyanobacterium *Synechococcus* sp. strain PCC7942 involved in carboxysome assembly and function. J Bacteriol 175:2871–2879

Roberts EW, Cai F, Kerfeld CA, Cannon GC, Heinhorst S (2012) Isolation and characterization of the Prochlorococcus carboxysome reveal the presence of the novel shell protein CsoS1D. J Bacteriol 194:787–795

Sagermann M, Ohtaki A, Nikolakakis K (2009) Crystal structure of the EutL shell protein of the ethanolamine ammonia lyase microcompartment. Proc Natl Acad Sci U S A 106:8883–8887

Samborska B, Kimber MS (2012) A dodecameric CcmK2 structure suggests β-carboxysomal shell facets have a double-layered organization. Structure 20:1353–1362

Sawaya MR, Cannon GC, Heinhorst S, Tanaka S, Williams EB, Yeates TO, Kerfeld CA (2006) The structure of beta-carbonic anhydrase from the carboxysomal shell reveals a distinct subclass with one active site for the price of two. J Biol Chem 281:7546–7555

Schmid MF, Paredes AM, Khant HA, Soyer F, Aldrich HC, Chiu W, Shively JM (2006) Structure of *Halothiobacillus neapolitanus* carboxysomes by cryo-electron tomography. J Mol Biol 364:526–535

So AK, Espie GS (1998) Cloning, characterization and expression of carbonic anhydrase from the cyanobacterium *Synechocystis* PCC6803. Plant Mol Biol 37:205–215

So AK, Cot SS-W, Espie GS (2002) Characterization of the C-terminal extension of carboxysomal carbonic anhydrase from *Synechocystis* sp. PCC6803. Funct Plant Biol 29:183–194

So AK, Espie GS (2005) Cyanobacterial carbonic anhydrases. Can J Bot 83:721–734

So AK, Espie GS, Williams EB, Shively JM, Heinhorst S, Cannon GC (2004) A Novel evolutionary lineage of carbonic anhydrase (ε class) is a component of the carboxysome shell. J Bacteriol 186:623

Tanaka S, Kerfeld CA, Sawaya MR, Cai F, Heinhorst S, Cannon GC, Yeates TO (2008) Atomic-level models of the bacterial carboxysome shell. Science 319:1083–1086

Tanaka S, Sawaya MR, Phillips M, Yeates TO (2009) Insights from multiple structures of the shell proteins from the beta-carboxysome. Protein Sci 18:108–120

Tsai Y, Sawaya MR, Cannon GC, Cai F, Williams EB, Heinhorst S, Kerfeld CA, Yeates TO (2007) Structural analysis of CsoS1A and the protein shell of the *Halothiobacillus neapolitanus* carboxysome. PLoS Biol 5:e144

Tsai Y, Sawaya MR, Yeates TO (2009) Analysis of lattice-translocation disorder in the layered hexagonal structure of carboxysome shell protein CsoS1C. Acta Crystallogr D Biol Crystallogr 65:980–988

Yeates TO, Crowley CS, Tanaka S (2010) Bacterial microcompartment organelles: protein shell structure and evolution. Annu Rev Biophys 39:185–205

Yu JW, Price GD, Song L, Badger MR (1992) Isolation of a putative carboxysomal carbonic anhydrase gene from the cyanobacterium *Synechococcus* PCC7942. Plant Physiol 100:794–800

Chapter 8

Rieske/Cytochrome *b* Complexes: The Turbo Chargers of Chemiosmosis

Felix ten Brink and Frauke Baymann*
Bioénergétique et Ingénierie des protéines, Unité mixte de récherche 7281, Fédération de récherche 3479, Centre National de la Récherche Scientifique/Aix-Marseille Université, 31, chemin Joseph Aiguier, 13009 Marseille, France

Summary

Rieske/cytochrome *b* complexes are transmembrane enzymes of many bioenergetic reaction chains. They operate according to a Q-cycle mechanism that allows the transfer of an extra proton over the membrane per electron transferred along the chain. Thus they optimize the ATP yield from the terminal electron donor/acceptor couples of the respective reaction chains. Two members of this family of enzymes, cytochrome bc_1 complex from proteobacteria and mitochondria and cytochrome b_6f complex from cyanobacteria and chloroplast are well studied and crystal structures are available. While revealing unexpected features of the complexes, as the domain movement of the Rieske iron/sulfur protein and the presence of an unusual heme cofactor in a quinone binding pocket, they opened up new questions

*Author for correspondence, e-mail: baymann@imm.cnrs.fr

M.F. Hohmann-Marriott (ed.), The Structural Basis of Biological Energy Generation, Advances in Photosynthesis and Respiration 39, pp. 149–165, DOI 10.1007/978-94-017-8742-0_8,

concerning the function of the enzyme. Few structural and functional studies are available for Rieske/cyt*b* complexes from further species, which is the vast majority of bacterial and archeael phyla, and most information comes from genome analysis. Here we discuss structure and function of the complex from an evolutionary perspective.

I. Introduction

Rieske/cytochrome *b* complexes are transmembrane proton pumps that oxidize quinol and reduce soluble electron carriers such as cytochrome *c*, plastocyanin or high potential iron-sulfur proteins on the outer/stromal/periplasmic site of the membrane. They are involved in bioenergetic reaction chains of most organisms in oxygen respiration, photosynthesis, nitrate and nitrite respiration, sulfide oxidation and hydrogen oxidation (Fig. 8.1). Whenever an energy difference of about 250 mV between donor quinols and acceptor proteins is available Rieske/cytochrome *b* complexes can make use of it to pump one extra proton across the membrane per electron transferred along the bioenergetic electron transfer chain thus optimizing the outcome of the redox chain.

The b_6f complex of chloroplasts was discovered in 1954 (Hill), mitochondrial bc_1 complex first isolated in 1962 (Hatefi et al.). In 1964 Rieske and coworkers described the typical EPR signal of the iron-sulfur subunit of the mitochondrial complex that later was named 'Rieske protein'. In 1976 Peter Mitchell in his ground-breaking work developed the Q-cycle mechanism to describe the function of the bc_1 complex (Fig. 8.2, see below).

The discovery of homology between cytochrome bc_1 and cytochrome b_6f complex appeared slowly. Nelson and Neumann (1972) pointed out the similarity in co-factors, Clark and Hind (1983) described both complexes as functional dimers, Widger and coworkers (1984) discovered that the transmembranous cytochrome *b* (cyt*b*) subunits are homologous. Biochemical and biophysical studies and recent genome sequencing projects showed the wide distribution of this type of enzyme among species (Fig. 8.3). In 2000 Schütz and coworkers (2000) described the core of the enzyme (the Rieske subunit and cytochrome *b*, see below) as homologous and proposed the name Rieske/cyt*b* complexes for the entire enzyme family.

Although most representatives from the large variety of Rieske/cyt*b* complexes are poorly studied, some complexes from widely distributed organisms are well characterized. Among them are the bc_1 complex from proteobacteria and mitochondria (see Chap. 21) and the b_6f complex from cyanobacteria (see Chap. 14) and chloroplasts. Some data are available on the complexes from Actinobacteria (Sone et al. 2001, 2003, see also Chap. 15), Heliobacteria (reviewed in Baymann and Nitschke 2010, see Chap. 13), Chlorobiaceae (Brugna et al. 1998b), Aquificales (Schütz et al. 2003), Bacilli (Yu and Le Brun 1988; Sone and Fujiwara 1991; Tanaka et al. 1996) and *Sulfolobus* (Iwasaki et al. 1995; Komorowski et al. 2002).

II. Structural Properties of Rieske/Cytochrome *b* Complexes

X-ray crystallographic structures of cytochrome *f*, bc_1 complex from mitochondria and proteobacteria (Figs. 8.4 and 8.5), and b_6f complex from cyanobacteria and chloroplasts (Fig. 8.6) confirmed the identical architecture of the core of the complex (Rieske subunit and cytochrome *b*) and many of the structural predictions made before, based on biochemical

Abbreviations: cyt*b* – Cytochrome *b*; DBMIB – 2,5-dibromo-3-methyl-6-isopropyl-p-benzoquinone; EPR – Electron Paramagnetic Resonance; FNR – Ferredoxin NAD(P)H Reductase; LUCA – Last Universal Common Ancestor; NQNO – 2-*n*-nonyl-4-hydroxylquinoline *N*-oxide; PS – Photosystem; RC – Photosynthetic reaction center; UHDBT – 5-n-undecyl-6-hydroxy-4,7-dioxobenzothiazole

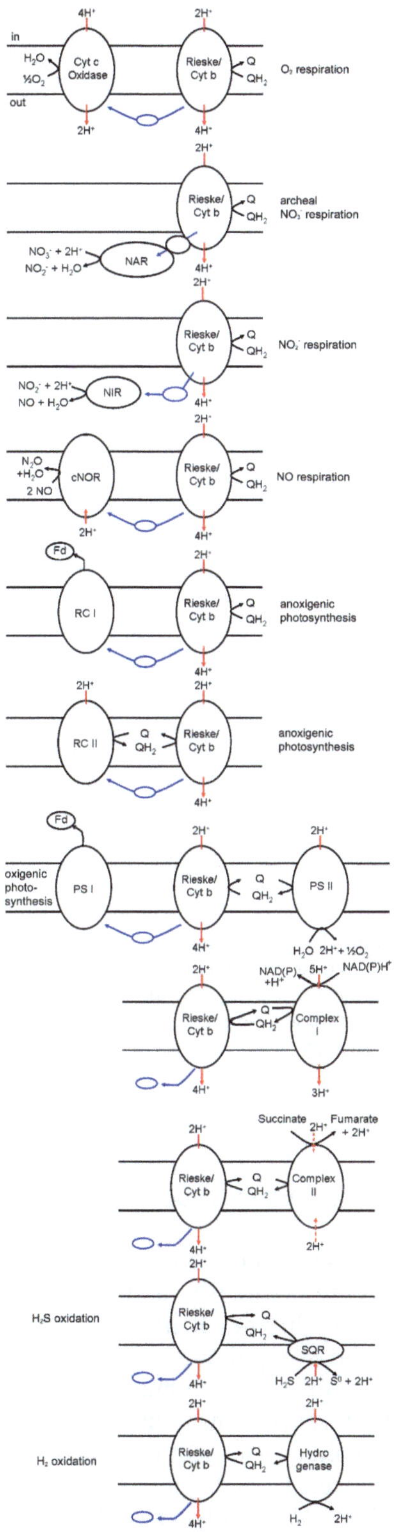

Fig. 8.1. Schematic representation of enzymes involved in bioenergetic reaction chains with a Rieske/cyt*b* complex. *Blue lines* represent electron transfer, *red lines* proton transfer, *dotted red lines* proton uptake by complex II from the outer side of the membrane for menaquinone reduction and from the inner side for ubiquinone reduction. *NIR* symbolizes the structurally unrelated copper- and cytochrome cd_1-nitrite reductases, *NAR* nitrate reductase, *SQR* sulfide quinone reductase, *cNOR* cytochrome NO-reductase, *RCI/II* photosynthetic reaction center I and II, *PSI/II* photosystem I and II.

and biophysical experiments. Each structure also revealed a completely unexpected feature of the complex. The cytochrome *f* structure showed that cytochromes c_1 and *f* (Martinez et al. 1994) have unrelated structures. The structure of the bc_1 complex showed that the head domain of the Rieske subunit is mobile (Zhang et al. 1998) (Fig. 8.7). The structure of the b_6f complex unveiled three co-factors that are absent from bc_1 complexes: a chlorophyll, a carotenoid and a heme (Stroebel et al. 2003; Kurisu et al. 2003).

A. Rieske Subunit

A feature shared by all Rieske/cyt*b* complexes is the presence of a Rieske iron/sulfur subunit with a single membrane-spanning α-helix and a head domain on the outer side of the membrane. This head domain binds a 2Fe2S-cluster with two cysteine ligands and two histidine ligands and features mainly β-sheets connected by loops that have variable length among different species. The histidine ligands induce a characteristic EPR signal of the reduced cluster at $g_y = 1.89$ that distinguishes the Rieske protein from other iron-sulfur proteins. The histidine ligands also contribute to confer the iron-sulfur cluster a positive redox midpoint potential. A pivoting movement of the head domain of the Rieske protein has been detected by crystallographic analysis of the bc_1 complex, displacing the iron-sulfur cluster by 16 Å between a position close to cytochrome c_1 to a position close to the quinol oxidation site.

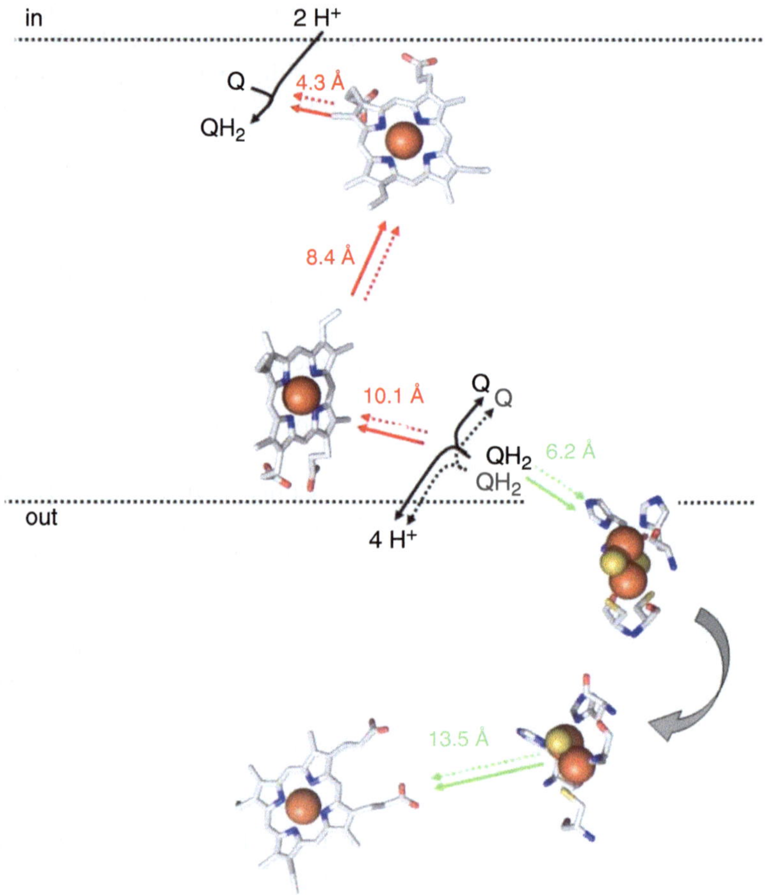

Fig. 8.2. Schematic representation of the Q-cycle in the bc_1 complex. Electrons from the $QH_2/Q^{\circ-}$ transition in the Q_o site are transferred to the Rieske protein and to cytochrome c_1, electrons from the $Q^{\circ-}/Q$ transition proceed across the membrane via the two *b*-hemes to the Q_i site. Electron transfer pathways of the first electron (*green*) and the second electron (*red*) are indicated in *solid lines* (first quinol oxidized in Q_0) and *dotted lines* (second quinol oxidized in Q_0). *Numbers*: edge-to-edge distances between co-factors. *Bended arrow*: pivoting movement of the Rieske head domain.

The former position was observed in native crystals, whereas the latter has been obtained in the presence of the inhibitor stigmatellin. It is called 'proximal position' in crystallography papers.

This movement was subsequently observed by EPR studies on partially ordered membrane multilayers of spinach chloroplasts (Schoepp et al. 1999a), *Chlorobium* (Brugna et al. 1998b), proteobacteria (Brugna et al. 2000) and *Sulfolobus* (Brugna et al. 1998a) and may therefore be a general property of Rieske/cyt*b* complexes.

B. Transmembrane Cytochrome b Subunit

The second constitutive element of all Rieske/cyt*b* complexes is a transmembrane *b*-type cytochrome. It harbours two *b*-type hemes and two quinone binding sites, one on either site of the membrane and is composed of at least seven transmembrane helices that can be split in two subunits with four and three helices, named cytochrome b_6 and subunit IV, respectively. The four histidine residues that are ligands to the hemes are on helix II and IV and are spaced by 14 amino

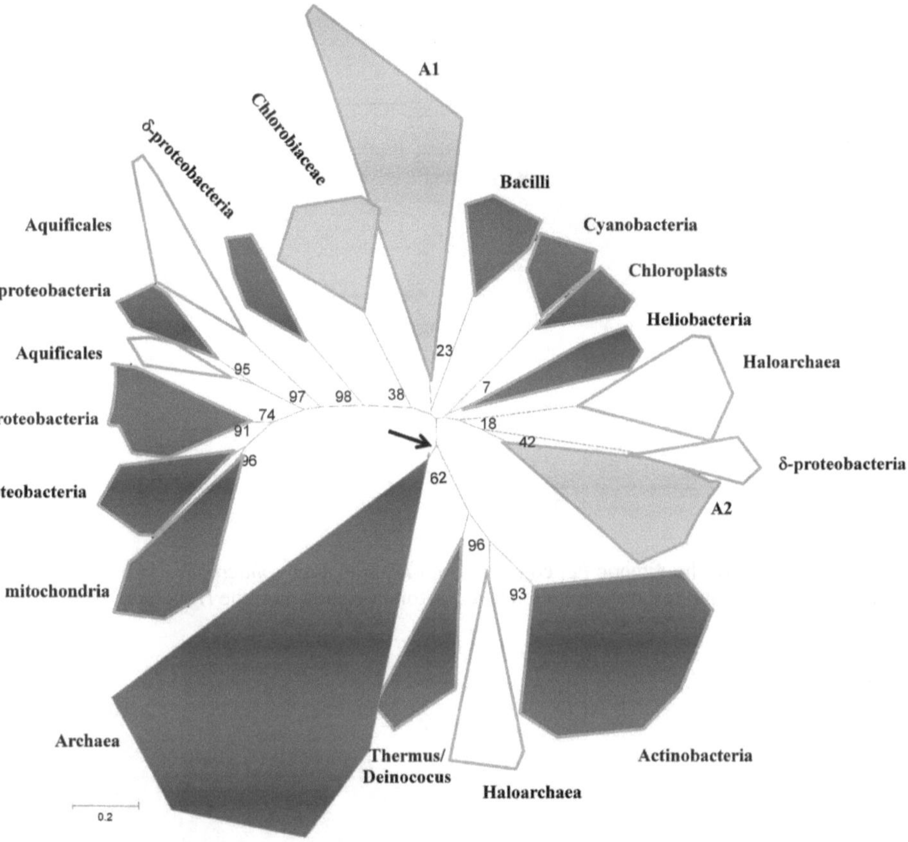

Fig. 8.3. Phylogenetic tree of the cytochrome *b* subunit. *Dark areas*: phyla that inherited the Rieske/cyt*b* complex vertically; *white areas*: phyla that acquired it by lateral gene transfer; *light gray areas*: phyla under discussion. The *arrow* indicates the root of the tree, i.e. the position of LUCA. The numbers are bootstrap values (out of 100). Rieske/cyt*b* complexes were found in genomes of most phyla of Crenarchaeota (with the exception of Acidolobales and Ferridicoccales), Halobacteria and Thermoplasmata among the Euryarchaeota (but not in methanogens, Archaeoglobi and Thermococci), in a representative of Thaumarchaeota and Korarchaeota, but not in Nanoarchaeota. Among Bacteria they have not been found in the genomes of Fibrobacterias, Elusiomicrobia, Dictyoglomi, Fusobacteria, Synergistetes, Tenericutes, Thermotogae, Enterobacteria. Among the other phyla at least a few representatives have the complex in their genomes, even if most Chloroflexaceae and δ-proteobacteria (with the exception of the *Geobacter* species) are devoid of a Rieske/cyt*b* complex. Group A1 and A2 comprise members of Planctomycetes, Acidobacteria, δ-proteobacteria, Gematomonadetes, Nitrospiraceae, Chloroflexaceae and Chlamydiae.

acid residues in most species. In mitochondria, α-proteobacteria and some Archaea the His residues of helix IV are 13 amino acid residues apart. Functional consequences of this difference are not known. If the cytochrome *b* subunit is split into two proteins the split always occurs after helix IV.

The quinone oxidation site, Q_o site, also named Q_z site or Q_p site to indicate that it is localized close to the positive site of the membrane, is lined by residues from cytochrome *b* or b_6/subunit IV (Fig. 8.8). One of the histidine ligands to the Rieske cluster and a cysteine residue from the Rieske protein also surfaces in the binding pocket. In b_6f complexes the tail of the chlorophyll molecule is localized in the entrance channel to the quinone binding site (Fig. 8.8 panel b). The porphyrin ring of the chlorophyll molecule occupies the space where in bc_1

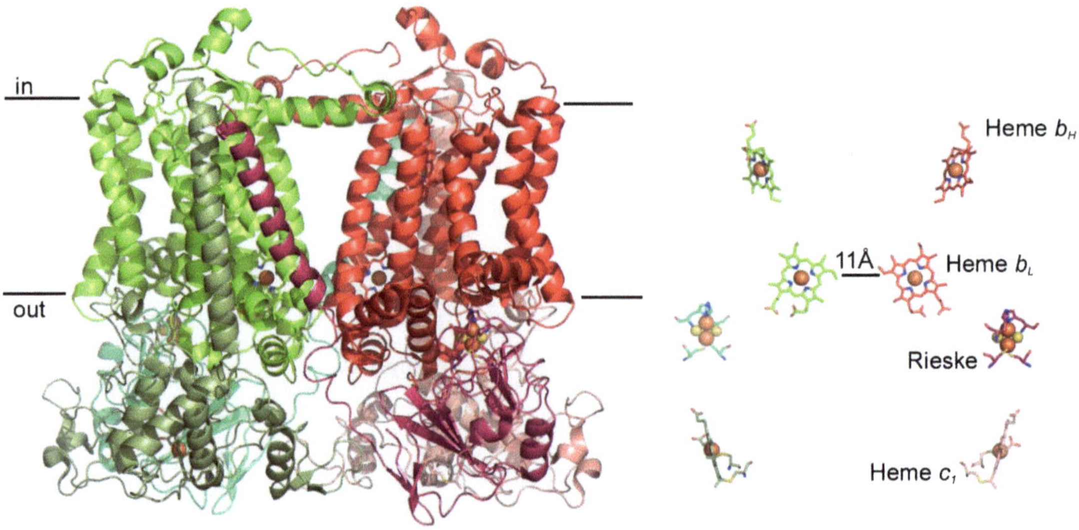

Fig. 8.4. X-ray structure of the dimeric bc_1 complex from *Rhodobacter sphaeroides* (pdb: 2QJP). The two monomers are colored in green and red respectively. The cofactors are shown on the *right hand* site.

Fig. 8.5. X-ray structure of the dimeric bc_1 complex from chicken (pdb: 3H1J). Cytochrome *b*, c_1 and the Rieske protein are shown in *grey*, the additional subunits are colored (*red* and *magenta* for the 'core' subunits).

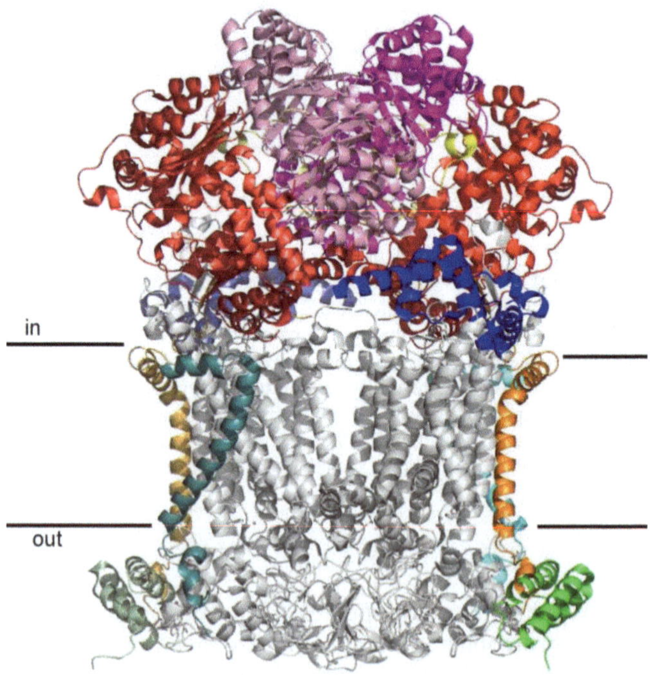

complexes the eighth transmembrane helix of cytochrome *b* is localized (Fig. 8.4).

The Q_i site, also named Q_c site or Q_n site to indicate that it is localized close to the negative site of the membrane, is lined by residues from cytochrome *b* or b_6/subunit IV. In b_6f complexes a heme, named heme c_i or c_n, is exposed to the Q_i binding site (Fig. 8.9 panel b). The heme c_i iron is situated 3.5 Å from the propionate of heme b_H (Figs. 8.6

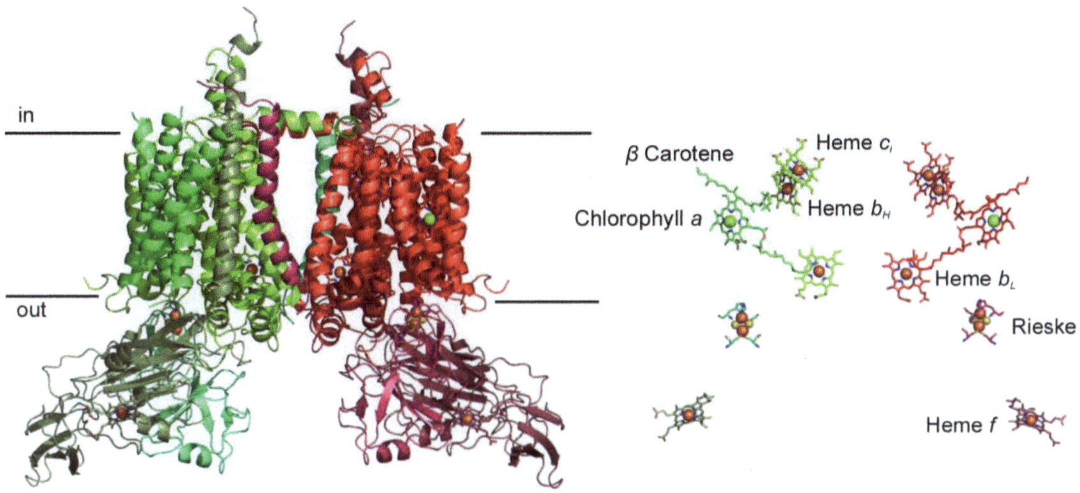

Fig. 8.6. X-ray structure of the dimeric b_6f complex from *Chlamydomonas reinhardtii* (pdb:1Q90). The two monomers are colored in *green* and *red* respectively. On the *right hand* site, the cofactors are shown with the heme iron in *red* and the chlorophyll manganese in *green*.

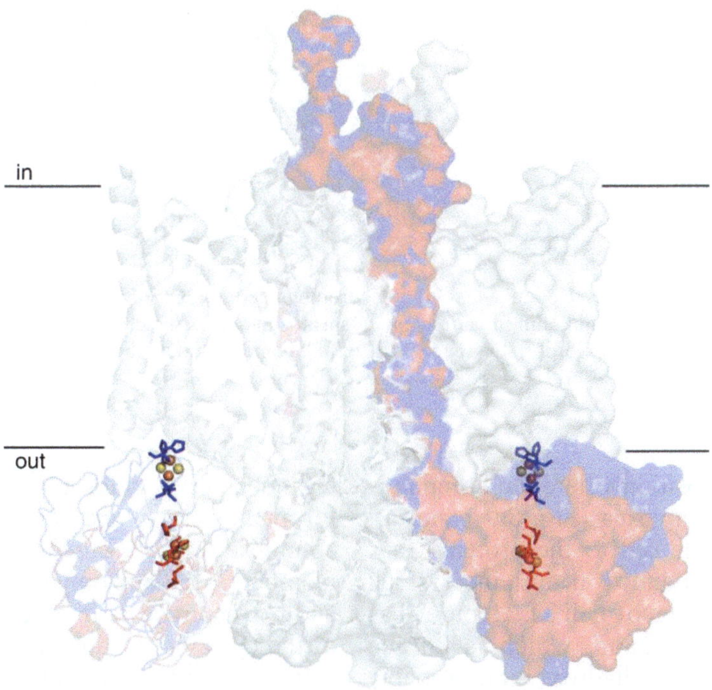

Fig. 8.7. Superposition of X-ray structures of the bc_1 complex with the Rieske head domain close to cytochrome c_1 (pdb 3H1J in *red*) or close to the Q_o site (pdb 3H1H in *blue*). The Rieske iron-sulfur clusters and their ligands are represented in the ball and stick model, the Rieske subunits are colored (*blue* and *red* for the two positions of the globular domain) and represented in the space-fill mode for one monomer and in the ribbon mode for the other monomer. The remaining subunits of the complex are shown in *grey* (space-fill for one monomer and ribbon for the other monomer).

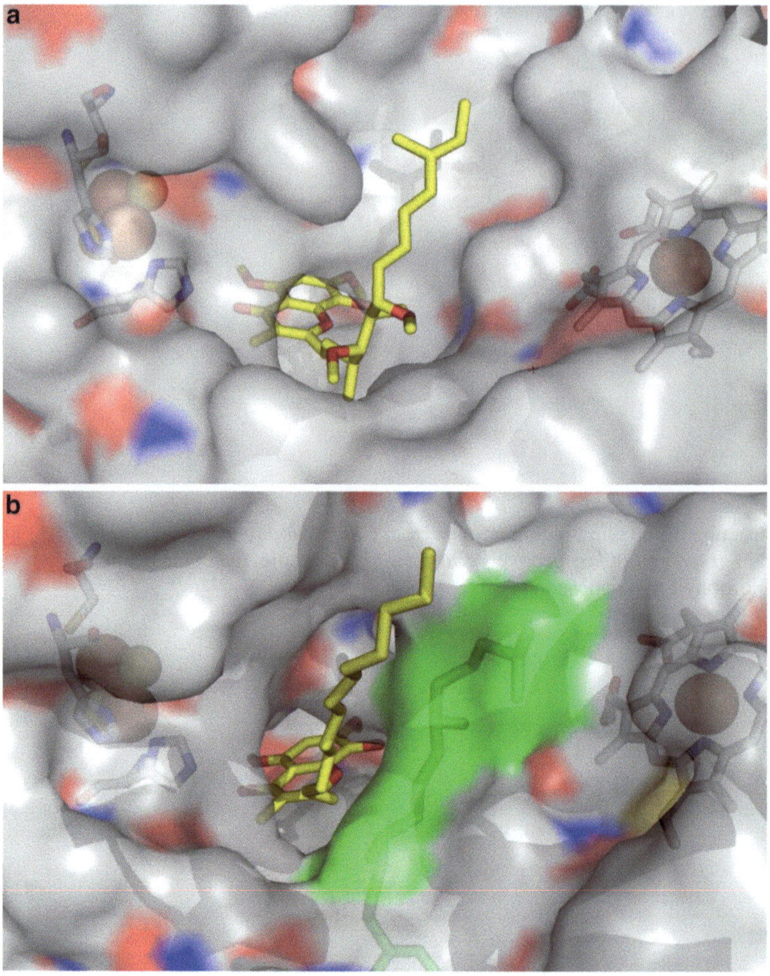

Fig. 8.8. (**a**) Q_o site of Rieske/cyt*b* complexes. (**a**) cytochrome bc_1 complex (pdb 3H1J); (**b**) cytochrome b_6f complex (pdb: 1Q90). Tridecylstigmatellin in *yellow*, *green* surface indicates the chlorophyll (its electron density is not well defined, suggesting flexibility). On the *right hand site* heme b_L and on the *left hand site* the Rieske 2Fe2S cluster are indicated.

and 8.9b) in a position where the quinone is found in bc_1 complexes (Fig. 8.9 panel a). The heme c_i is linked to the protein via a single thioether bond to a cysteine residue near the N-terminus of helix I. Heme c_i has no axial protein ligand, which is unique for heme proteins. A hydroxyl ion probably forms a ligand to the iron on one side of the heme whereas a phenylalanine residue shields access to the iron from the quinone binding site. Genome analysis showed that the cysteine ligand is conserved in the amino acid sequence of cytochrome b_6 from Cyanobacteria, chloroplasts, Heliobacteria, Firmicutes and in one of the two Rieske/cyt*b* complexes of *Geobacter* species and in a few members of Planctomycetes, Acidobacteria and Nitrospiraceae. In all cases the presence of the cysteine residue is accompanied by a split cytochrome *b* with a total of seven transmembrane helices. So far EPR spectra of heme c_i from Cyanobacteria (Zhang et al. 2004), chloroplasts (Baymann et al. 2007) and Heliobacteria (Ducluzeau et al. 2008) have been obtained. Peroxidase activity on SDS-PAGE indicates its presence in *Bacillus*

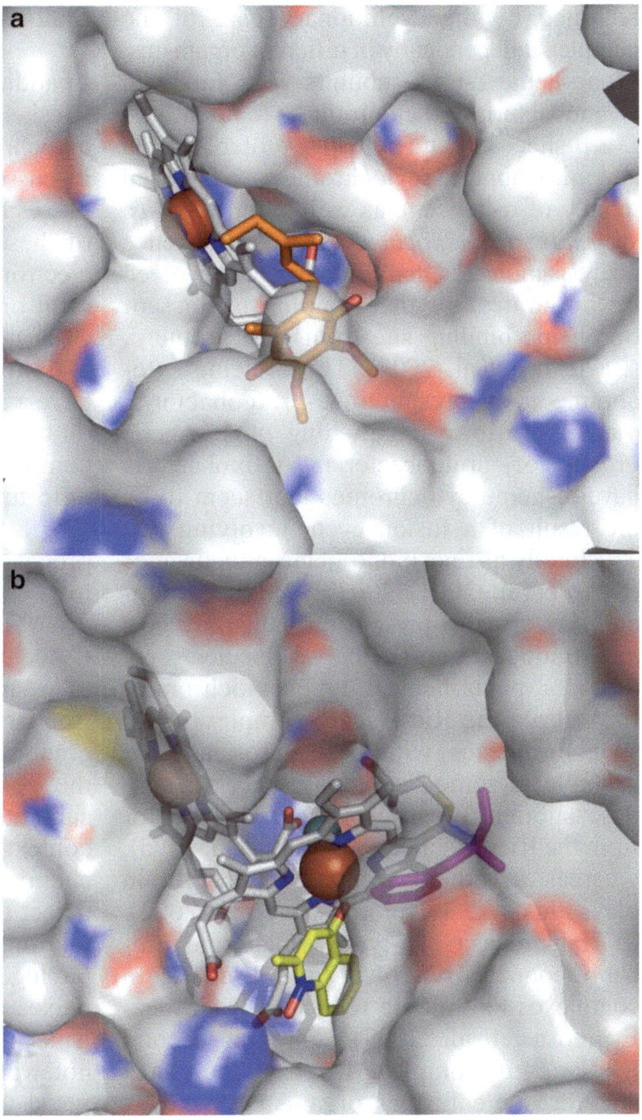

Fig. 8.9. Q_i site of Rieske/cyt*b* complexes (**a**) cytochrome bc_1 complex with quinone in *orange* and heme b_H in *grey* (pdb 3H1J); (**b**) b_6f complex with NQNO in *yellow*, heme b_H (*left, grey*), heme c_i (*middle, grey*) covalently attached to a cysteine (in *yellow*), water molecule (*dark-cyan*) and the phenylalanine (*magenta*) that covers part of the surface of heme c_i in the Q_i site (D. Stroebel and D. Picot, unpublished).

(Yu and Le Brun 1988). The crystal structures of the b_6f complex of cyanobacteria and green alga (Stroebel et al. 2003; Kurisu et al. 2003) showed that the cytochrome *b* subunits also harbors a carotenoid of unknown function.

C. Electron Acceptor Subunit of the Rieske Protein

In addition to the Rieske subunit and cytochrome *b*, most Rieske/cyt*b* complexes have a third cofactor-containing subunit – a cytochrome that accepts electrons from the Rieske cluster and transfers them to periplasmic electron carriers such as cytochromes, plastocyanin or high-potential iron-sulfur proteins. In Bacteria, *c*-type cytochromes with one to seven hemes have been observed in genomic context with the genes coding for the Rieske protein and cytochrome *b*, in Archaea *b*-type cytochromes are present. The fact that these heme-containing proteins represent a wide variety of structures and number of hemes and are evolutionary unrelated, indicates that they have been recruited by the evolutionary core of the complex independently in different organisms. Mono-heme *c*-type subunits were found in Cyanobacteria and chloroplasts (cytochrome *f*, a β-sheet structure), α-β-γ-proteobacteria and mitochondria (cytochrome c_1, composed mainly of α-helices), as well as in *Bacillus*, *Thermus* and *Deinococcus* species. Sequence analysis indicated that cytochrome c_1 evolved in the context of the complex by loss of one heme binding site from cytochrome c_4 (Baymann et al. 2004). The di-heme cytochrome c_4 is present in ε-proteobacteria, one of the lowest branching phylogenetic groups of proteobacteria. Actinobacteria have di-heme cytochromes that seem to be the result of a fusion event involving two structurally distinct mono-heme cytochromes (Sone et al. 2001, 2003). Di-heme cytochromes are also present in Planctomycetes and Acidobacteria. *Geobacter* species show tetra- or penta-heme *c*-type cytochromes. In Archaea the Rieske/cyt*b* complex forms supercomplexes with other proteins. For example in some *Sulfolobus* species a oxidase subunit II acts as electron acceptor from the Rieske subunit (Iwasaki et al. 1995), whereas in others a blue-copper protein plays this role (Komorowski et al. 2002). In Haloarchaea a *b*-type cytochrome probably connects one of the two Rieske/cyt*b* complexes to a nitrate reductase (Martinez-Espinosa et al. 2007). For the second Rieske/cyt*b* complex from Haloarchaea and for some further organisms, such as *Chlorobiaceae* and no electron acceptor subunit is known.

Cytochrome c_1 and *f* are embedded in the membrane part of the complex by a single transmembrane helix. Crystal structures showed that this helix is localized in identical positions in bc_1 and b_6f complexes. Additional structures of Rieske/cyt*b* complexes from other organisms are required to decide, whether this helix is part of the evolutionary core of the enzyme. A special arrangement of the Rieske/cyt*b* complexes exists in *Bacillus*, where a cytochrome *c* is fused to the N-terminus of the transmembrane subunit IV.

D. Additional Subunits

In addition to the cytochrome *b* subunit, the Rieske subunit and the electron acceptor cytochrome subunit of the Rieske protein discussed above, many Rieske/cyt*b* complexes have additional subunits. So far none of theses subunits appear to be involved in electron transport, but they may have structural or regulatory roles. The most spectacular addition are the two biggest subunits of mitochondrial bc_1 complex, which each have a mass of 40–50 kDa, and are localized on the cytoplasmic/luminal/inner-site of the complex. As the biggest subunits they were unfortunately named core subunit I and II (Fig. 8.5). They are homologous to the matrix processing peptidase and have been shown to possess that function in plant mitochondria (Glaser and Dessi 1999 for a review). Cytochrome bc_1 and b_6f complexes show up to 6 small subunits per monomer that are frequently composed of a single transmembrane helix and localized on the periphery of the complex core. Since these additional subunits are not always encoded in genomic context with the core of the complex, biochemical investigations are necessary to detect them and we have therefore no information about their presence among the phylogenetic diversity of Rieske/cyt*b* complexes represented in poorly studied organisms. Alternative configurations are also noteworthy, as 2-3 transmembrane α-helices fused to the transmembrane α-helix of the Rieske subunit in Actinobacteria and additional α-helices at the C-terminus of cytochrome *b* in Archaea.

E. A Structural Dimer: Crossing over of the Helix of the Rieske Protein

Both, bc_1 and b_6f complexes are homo-dimers. The Rieske head domain of each monomer interacts with the quinone oxidation site and cytochrome c_1/f of the same monomer whereas its transmembrane α-helixes cross over to the other monomer and thus joins the two monomers (Figs. 8.4 and 8.6). Consequently, loss of the Rieske protein during a purification procedure generally induces monomerization of the remaining complex. Since the way the dimer is formed is conserved between bc_1 and b_6f complexes, and since the movement of the Rieske protein head domain was found in all complexes studied so far, the dimeric structure is most likely a common characteristic of all Rieske/cyt*b* complexes. Also conserved between bc_1 and b_6f complexes are two cavities at the monomer-monomer interface in the membrane-embedded part of the complex. Each cavity provides access to the Q_i site of one monomer and the Q_0 site of the other monomer (Xia et al. 1997; Stroebel et al. 2003; Kurisu et al. 2003).

III. Function of Rieske/Cytochrome *b* Complexes: The Q-cycle

Figure 8.2 illustrates the Q-cycle proposed by Mitchell (1976) and later modified by Crofts and coworkers (1983). During turnover of the Rieske/cyt*b* complex a quinol is reduced in the Q_o site on the periplasmic/outer side of the membrane. One of the electrons of this quinol reduces the Rieske iron-sulfur protein, the second electron is transferred to the *b*-hemes. When a second quinol is oxidized in the Q_0 site, again, one electron is transferred to the Rieske protein and the second one to the *b*-hemes. The two electrons that entered the *b*-hemes reduce a quinone bound to the Q_i-binding site on the cytoplasmic/inner side of the membrane. Reduction of a quinone in the Q_i site results in uptake of two protons from the inner side of the membrane whereas four protons have been liberated on the outer side of the membrane from the two quinols oxidized in the Q_o site. These quinone reactions contribute to the transmembrane proton gradient.

A particularity of the reaction mechanism of Rieske/cyt*b* complexes resides in the fact that the substrate to be oxidized by the complex, quinol, and the molecule to be reduced in one of the electron transfer pathways of the enzyme, quinone, are chemically identical. The redox reactions of these quinones are catalyzed by one-electron co-factors and therefore imply a potentially highly reactive semiquinone intermediate. For the reaction to proceed, the properties of the two quinones should therefore be modulated by the protein environment in a way to favor forward electron transfer and to avoid labile and therefore highly reactive semiquinone intermediates. The nature of the quinol depends on the pool quinone present in the parent organism. Its redox midpoint potential is at about +100 mV for ubiquinone (Moret et al. 1961) (mitochondria and α-β- and some γ-proteobacteria), plastoquinone (Okayama 1976) (cyanobacteria and chloroplasts) and caldariellaquinone (Anemüller and Schäfer 1990) (Sulfolobales) and – 70 mV for menaquinone (Wagner et al. 1974) (all the other species).

A. Bi-furcated Quinol Oxidation Reaction in the Q_o Site

During quinol oxidation in the Q_o site the quinol is deprotonated to the outer side of the membrane. One electron is transferred to the Rieske iron-sulfur cluster and further on to its electron acceptor cytochrome, implying a pivoting movement of the Rieske head domain. The second electron is transferred from the resulting semiquinone to the closest *b*-heme and further to the second *b*-heme close to the inner side of the membrane (Fig. 8.2). The two half reactions of this quinol/quinone couple have different redox midpoint potentials: the first electron leaves the quinol at a potential substantially more positive than the second electron. Correspondingly the life-time of the reactive semiquinone intermediate is short and the redox midpoint potentials of the acceptor co-factors are positive for the Rieske cluster and negative for the *b*-hemes. The redox midpoint potential of the Rieske cluster is modulated by two specific amino acid residues in close contact to the iron sulfur center in a range of +150 to +380 mV (Schoepp et al. 1999b) to adjust it to the midpoint potential of pool quinone species of the parent organism. The redox midpoint potential of the close-by *b*-heme (named b_L for low-potential or b_p for positive side of the membrane) is at −50 mV in bc_1 complexes and at −150 mV in b_6f complexes and below –300 mV in Rieske/cytb complexes from Heliobacteria (to be published). The second *b*-heme on the inner side of the membrane (named b_H for high potential or b_n for negative side of the membrane) has a 100 mV more positive redox midpoint potential than heme b_L in all studied complexes, facilitating electron transfer across the membrane against the proton motive force.

The first oxidation of the quinol is the rate limiting step of the overall redox reaction of the complex as shown at least in the case of the bc_1 complex (Crofts et al. 2003). The evacuation of the second electron across the membrane is fast if the *b*-hemes are oxidized so that there has been a long-lasting discussion whether the two electron transfer reactions are concerted, concomitant or consecutive. In any case the life time of the semiquinone intermediate is short, minimizing harmful side reactions. If the quinol reduction site is blocked by an inhibitor and electrons pile up in the *b*-hemes, oxidation of the semiquinone in the quinol oxidation site slows down and superoxide production can occur.

Under normal turnover conditions, the bi-furcated electron transfer of the Q_o site seems to be an efficient and astonishingly reliable process. Many tricks have been played to assess bypass reactions that circumvent the strict bifurcation at the Q_o site. The experiments on bc_1 complexes are summarized in Kramer and coworkers (2009). Recently a mutant of the *Chlamydomonas reinhardtii* b_6f complex has been reported that had only heme b_L in the cytochrome *b* subunit (hemes c_i and b_H were absent). This mutant still grows photosynthetically, albeit faintly, which implies electron transfer from quinol to plastocyanin and thus turnover of the b_6f complex. In this mutant, oxidation of the quinol in the Q_o site in a bifurcated manner was proposed to be followed by re-reduction of the product quinone in the quinone oxidation side Q_o by heme b_L and electron transfer from the so-formed semiquinone to the Rieske protein (Malnoë et al. 2011).

Extensive inhibitor studies were carried out for bc_1 complexes and several structures with inhibitors in the quinone oxidation site are available (reviewed in Esser et al. 2004). Q_o site inhibitors bind in two different positions in the binding site. Stigmatellin and UHDBT bind close to the Rieske cluster forming a hydrogen bond to the ε-N of the His ligand to the cluster and fix the Rieske protein in its position close to the binding site. These inhibitors, when present in the Q_0 site, may raise the redox midpoint potential of the iron sulfur cluster by up to 250 mV. Another group of inhibitors, including myxothiazol bind closer to heme b_L and induce enhanced mobility of the Rieske cluster. The different binding properties of inhibitors in the Q_o site raised the question whether there are two quinones present in the site or whether the semiquinone is moving from one niche to the other to accommodate the bifurcated electron transfer. So far this remains an open question and no crystal structure with a quinone in the Q_o site has been obtained.

Investigated b_6f complexes showed a somewhat different sensitivity to Q_o site inhibitors than bc_1 complex, but several molecules, such as stigmatellin, act on both complexes. Inhibitor studies on Rieske/cyt*b* complexes other than bc_1 and b_6f complexes are scarce. DBMIB was shown to affect the orientation of the Rieske protein and its EPR signal in *Chlorobium* (Brugna et al. 1998b); DBMIB, UHDBT and stigmatellin induced a shift of the Rieske EPR signal in Heliobacteria, as did stigmatellin in *Bacillus*, DBMIB and stigmatellin inhibited turnover of the complex *in vivo* in Heliobacteria (reviewed in Baymann and Nitschke 2010).

B. Quinone Reduction Reactions in the Q_i Site

Quinone reduction reactions are best understood for bc_1 complexes. The electron is transferred from heme b_H to the quinone in the Q_i site and the resulting semiquinone is stabilized in the binding pocket until a second electron arrives from oxidation of another quinol in the Q_o site (Fig. 8.2). Stabilization of the semiquinone in the binding site translates to shifted redox midpoint potentials of the two half waves that were determined to +50 ($Q/Q^{\circ-}$) and −20 mV ($Q^{\circ-}/QH_2$) at pH 7.6 for mitochondrial bc_1 complex (Robertson et al. 1984). Both midpoint potentials are above the midpoint potential of heme b_L and its electron donor semiquinone, thereby favoring forward electron transfer. The structure of the quinone in the Q_i binding site of mitochondrial bc_1 complexes has been resolved, confirming strong interactions of the quinone with its binding site. The presence of quinone in the Q_i binding site induces in some cases an up-shift of 100 mV in the redox midpoint potential of heme b_H, then called heme b_{150}, an effect that remains to be interpreted (Gray et al. 1994).

Antimycin A is a potent inhibitor of the Q_i site of bc_1 complexes that have been crystallized with the inhibitor bound. In b_6f complexes Antimycin A is inefficient and no stable semiquinone species has been detected. Here, heme c_i is localized between heme b_H and the quinone binding site and the redox midpoint potential of heme b_H is by 100 mV lower than in bc_1 complexes, suggesting a different reaction mechanisms. In the b_6f complex, the surface of heme c_i towards the Q_i site is partially covered by a Phe residue that hinders access to the heme iron (Fig. 8.9b). The inhibitor NQNO can become a ligand to the oxidized heme c_i, as reflected by altered EPR characteristics (Baymann et al. 2007), a drastically lowered redox midpoint potential of the heme (Alric et al. 2005), as well as the position of the inhibitor in the crystal structure (Fig. 8.9b and Yamashita et al. 2007) that partially overlaps with the Phe side chain. On the other side of the heme the hydroxyl- or water-ligand to the heme iron forms a hydrogen bridge to one of the propionates of heme b_H. This hydrogen bond connection between the hemes suggest redox interactions between hemes b_H and c_i.

The observations in the absence and presence of NQNO indicate that binding/unbinding of quinone as a function of the redox state of heme c_i and/or the quinone may play a role in the reaction mechanism of the Q_i site of b_6f and related complexes. The presence of the Phe residue and modulation of the redox properties of heme c_i may be at work to avoid reactions of oxygen with the heme iron. The last aspect seems crucial, since b_6f complexes are part of the O_2-producing photosynthetic electron transport chain and reduced iron catalyzes reduction reactions of oxygen, that can result in radical oxygen species and heme damage.

No information on the mechanism of quinone reduction in other Rieske/cyt*b* complexes is available. Sequence analysis showed that the Phe residue, conserved among chloroplasts and cyanobacterial b_6f complexes, is replaced by a glutamic or aspartic acid residue in Rieske/cyt*b* complexes of Heliobacteria and Firmicutes. In these species the carboxylate group of the amino acid may provide a (temporary) ligand to the heme iron.

C. Oxidant–Induced Reduction

The so-called oxidant–induced reduction is a consequence of the Q-cycle mechanism. If the Rieske cluster and its electron acceptor subunit (when present) are reduced and if quinols are present, addition of an oxidant that oxidizes the Rieske center, will result in the oxidation of a quinol that in turn reduces the *b*-hemes with its second electron. Addition of an oxidant thus induces the reduction of a *b*-heme – a unique assay to detect the presence of a Rieske/cyt*b* complex in membranes in addition to the characteristic EPR signal of the Rieske cluster.

D. Reversed and Cyclic Electron Transfer

The entire redox reactions in bc_1 complexes can be reversed and result in quinone reduction at the expense of a reduced electron donor and the transmembrane proton gradient. This process can be stimulated in vesicles

(Miki et al. 1994) and may occur *in vivo* in mitochondria and proteobacteria when the ATP to NAD(P)H ratio is high. Some autotrophic bacteria that rely on electron donors such as iron, sulfur and thiosulfate use this mechanism for carbon fixation and NAD(P)H production (Ingledew 1982).

In chloroplasts and cyanobacteria reversed electron transfer has not been reported but cyclic electron transfer from NAD(P)H (reduced by PSI or RCI) via the cytochrome b_6f complex back to PSI occurs and Heliobacteria rely on this cyclic mode. The electron may travel from NAD(P)H to the quinone pool via complex I or via ferredoxin, FNR and heme c_i to a quinone in the Q_i binding site, resulting in the generation of a quinol. After its release from the Q_i-site this quinol could then be oxidized at the Q_o site of the b_6f complex. This scheme gains support from a recently isolated supercomplex from the green alga *Chlamydomonas reinhardtii*. This complex consists of antennas, PSI, b_6f complex and FNR, and exhibits light-induced cytochrome *b* reduction (Iwai et al. 2010).

E. Inter-monomer Electron Transfer and Interactions Between Q_o and Q_i Sites

Inhibitor and site directed mutagenesis studies have been performed in order to assess possible functional consequence of the dimeric nature of the Rieske/cyt*b* complexes (reviewed in Kramer et al. 2009). Recently, construction of chimeric complexes with a non-functional quinone oxidation side in one monomer and a non-functional quinone reducing site in the other monomer showed that inter-monomer electron transfer via the two hemes b_L (Fig. 8.4, right) can indeed take place (Swierczek et al. 2010), albeit at lower rate than intra-monomer electron transfer.

Influence of Q_i site inhibitors on the position of the Rieske head domain and differential binding of the first and the second molecule of the Q_i site inhibitor Antimycin A have been considered as indications for crosstalk between the quinone binding sites (reviewed in Kramer et al. 2009). Communication between the two monomers may gate inter-monomer *versus* intra-monomer electron transfer towards the Q_i binding site depending on whether it possesses a quinone ready to accept an electron, but this remains to be studied.

IV. Phylogeny and Evolution of Rieske/Cytochrome *b* Complexes

Rieske/cyt*b* complexes are present in most organisms living today (see legend to Fig. 8.3 for details). Phylogenetic analysis showed that the Rieske/cyt*b* complex was already present in the common ancestor (LUCA) of all living beings: the phylogenetic tree of cytochrome *b* (Fig. 8.3) follows the 16S rRNA tree with a few exceptions that indicate lateral gene transfer events in Haloarchaea, *Geobacter* species, Aquificales and possibly Chlorobiaceae. The common origin of all Rieske/cyt*b* complexes is further corroborated by the phylogenetic tree of the Rieske subunit that matched the cytochrome *b* tree and can be rooted by the paralogous Rieske subunit from the arsenite oxidase (Lebrun et al. 2006). A common gene order of the gene coding for the Rieske protein immediately followed by the gene(s) coding for cytochrome *b*, can be observed in all cases with the exception of a few organisms such as plants and some cyanobacteria that dissociated the gene cluster.

The presence of the complex in LUCA indicates that the ancestor of all living beings was relying on chemiosmosis for its energy metabolism. Rieske/cyt*b* complexes use reduced quinones as substrates, an indication that quinones were also present in LUCA. Today, quinones become reduced by complex I, complex II, sulfide oxidase, PSII/RCII and hydrogenase I. Rieske/cyt*b* complexes also need an oxidized electron acceptor to turn over. In organisms living today, this electron carrier provides electrons to oxidases, PSI/RCI, RCII, nitric oxide reductase, nitrite reductase and archaeal nitrate reductase (Fig. 8.1). Among the enzymes working today in bioenergetic reaction chains with Rieske/cyt*b* complexes, we have phylogenetic hints that hydrogenase I (Brugna-Guiral et al. 2003) and NO-reductase

(Ducluzeau et al. 2009) were present in LUCA, in agreement with the fact that H_2 and NO may have been present on the early Earth.

Rieske/cyt*b* complexes seem to have existed in a single copy in LUCA. Today several groups of organisms are known that harbor more than one Rieske/cyt*b* complex. Among them are Archaea that seem to have duplicated the gene coding for the enzyme early in evolution since most known Thermoprotei and Thermoplasmata show both copies. *Acidothiobacillus*, on the other hand, has two bc_1 complexes that originate from a much more recent gene duplication event and may serve forward and reverse electron transfer, respectively. Haloarchaea acquired two Rieske/cyt*b* complexes via lateral gene transfer from different bacterial donors early on. *Geobacter* species seem to have inherited one Rieske/cyt*b* complex vertically and acquired a second one. One species of Planctomycetes and Acidobacteria each are known to encode for three or four Rieske/cyt*b* complexes, respectively. The reasons for the occurrence of multiple copies of the enzyme in a single organism remain to be studied (ten Brink et al. 2013).

Acknowledgment

The authors thank Barbara Schoepp-Cothenet, Wolfgang Nitschke and Daniel Picot for careful reading of the manuscript and helpful discussions. They are grateful to Daniel Picot and David Stroebel for the structure coordinates of the b_6f complex from *Chlamydomonas reinhardtii* in the presence of NQNO. Frauke Baymann and Felix ten Brink are supported by a grant from the French 'Agence nationale de la recherche' (ANR-10-BLAN-1506).

References

Alric J, Pierre Y, Picot D, Lavergne J, Rappaport F (2005) Spectral and redox characterization of the heme c_i of the cytochrome b_6f complex. Proc Natl Acad Sci USA 102:15860–15865

Anemüller S, Schäfer G (1990) Cytochrome aa3 from sulfolobus acidocaldarius. A single subunit, quinol-oxidizing archaebacterial terminal oxidase. Eur J Biochem 191:297–305

Baymann F, Nitschke W (2010) Heliobacterial Rieske/cyt*b* complex. Photosynth Res 104:177–187

Baymann F, Lebrun E, Nitschke W (2004) Mitochondrial cytochrome c_1 is a collapsed di-heme cytochrome. Proc Natl Acad Sci 101:17737–17740

Baymann F, Giusti F, Picot D, Nitschke W (2007) The c_i/b_H-moiety in the cytochrome b_6f complex studied by EPR. A pair of strongly interacting hemes. Proc Natl Acad Sci USA 104:519–524

Brugna M, Nitschke W, Asso M, Guigliarelli B, Lemesle-Meunier D, Schmidt C (1998a) Redox components of cytochrome bc-type enzymes in acidophilic prokaryotes. II. The Rieske protein of phylogenetically distant acidophilic organisms. J Biol Chem 274:16766–16772

Brugna M, Albouy D, Nitschke W (1998b) Diversity of cytochrome bc complexes: example of the Rieske protein in green sulfur bacteria. J Bacteriol 180:3719–3723

Brugna M, Rodgers S, Schricker A, Montoya G, Kazmeier M, Nitschke W, Sinnig I (2000) A spectroscopic method for observing the domain movement of the Rieske iron-sulfur protein. Proc Natl Acad Sci USA 97:2069–2074

Brugna-Guiral M, Tron P, Nitschke W, Stetter KO, Burlat B, Guigliarelli B, Bruschi M, Giudici-Orticoni MT (2003) [NiFe] hydrogenases from the hyperthermophilic bacterium Aquifex aeolicus: properties, function, and phylogenetics. Extremophiles 7:145–157

Crofts AR, Meinhardt SW, Jones KR, Snozzi M (1983) The role of the quinone pool in the cyclic electron-transfer chain of Rhodopseudomonas sphaeroides. Biochim Biophys Acta 723:202–218

Crofts AR, Shinkarev VP, Kolling DRJ, Hong S (2003) The modified Q-cycle explains the apparent mismatch between the kinetics of reduction of cytochrome c_1 and b_H in the bc_1 complex. J Biol Chem 278:36191–36201

Clark RD, Hind G (1983) Isolation of a five-polypeptide cytochrome *b-f* complex from spinach chloroplasts. J Biol Chem 258:10348–10354

Ducluzeau AL, Chenu E, Capowiez L, Baymann F (2008) The Rieske/cytochrome *b* complex of Heliobacteria. Biochim Biophys Acta 1777:1140–1146

Ducluzeau AL, van Lis R, Duval S, Schoepp-Cothenet B, Russell MJ, Nitschke W (2009) Was nitric oxide the first deep electron sink? Trends Biochem Sci 34:9–15

Esser L, Quinn B, Li YF, Zhang M, Elberry M, Yu L, Yu CA, Xia D (2004) Crystallographic studies of quinol oxidation site inhibitors: a modified classification of inhibitors for the cytochrome bc1 complex. J Mol Biol 341:281–302

Glaser E, Dessi P (1999) Integration of the mitochondrial-processing peptidase into the cytochrome bc1 complex in plants. J Bioener Biomemb 31:259–274

Gray KA, Dutton PL, Daldal F (1994) Requirement of Histidine 217 for ubiquinol reductase activity (Q_i site) in the cytochrome bc_1 complex. Biochemistry 33:723–733

Hatefi Y, Haavik AG, Griffith DE (1962) Studies on the electron transfer system. J Biol Chem 237:1681–1685

Hill R (1954) The cytochrome *b* component of chloroplasts. Nature 4428:501–503

Iwai M, Takizawa K, Tokutsu R, Okamuro A, Takahashi Y, Minagawa J (2010) Isolation of the elusive supercomplex that drives cyclic electron flow in photosynthesis. Nature 464:1210–1214

Ingledew WJ (1982) Thiobacillus ferrooxidans. The bioenergetics of an acidophilic chemolithotroph. Biochim Biophys Acta 683:89–117

Iwasaki T, Matsuura K, Oshima T (1995) Resolution of the aerobic respiratory system of the thermoacidophilic Archaeon, *Sulfolobus sp.* Strain 7. J Biol Chem 270:30881–30892

Lebrun E, Santini JM, Brugna M, Ducluzeau AL, Ouchane S, Schoepp-Cothenet B, Baymann F, Nitschke W (2006) The Rieske protein; a case study on the pitfalls of multiple sequence alignments and phylogenetic reconstruction. Mol Biol Evol 23:1180–1191

Komorowski L, Verheyen W, Schäfer G (2002) The Archaeal respiratory supercomplex SoxM from *S. acidocaldarius* combines features of quinol and cytochrome *c* oxidases. Biol Chem 383:1791–1799

Kramer DM, Nitschke W, Cooley JW (2009) The cytochrome bc_1 and related *bc* complexes: Rieske/cytochrome *b* complexes as the functional core of a central electron/proton transfer complex. In: Hunter N, Daldal F, Thurnauer MC, Beatty JT (eds) The purple photosynthetic bacteria. Springer Science+Business Media BV, Dordrecht, pp 451–473

Kurisu G, Zhang H, Smith JL, Cramer WA (2003) Structure of the cytochrome b_6f complex of oxygenic photosynthesis: tuning the cavity. Science 302:1009–1014

Malnoë A, Wollman FA, de Vitry C, Rappaport F (2011) Photosynthetic growth despite a broken Q-cycle. Nature Comm 2:301

Martinez SE, Huang D, Szczepaniak A, Cramer WA, Smith JL (1994) Crystal structure of chloroplast cytochrome f reveals a novel cytochrome fold and unexpected heme ligation. Structure 2:95–105

Martinez-Espinosa RM, Dridge EJ, Bonete MJ, Butt JN, Butler CS, Sargent F, Richardson DJ (2007) Look on the positive side! The orientation, identification and bioenergetics of 'Archaeal' membrane-bound nitrate reductases. FEMS Microbiol Lett 276:129–139

Miki T, Mikin M, Orii Y (1994) Membrane potential-linked reversed electron transfer in the beef heart cytochrome bc_l complex reconstituted into potassium-loaded phospholipid vesicles. J Biol Chem 269:1827–1833

Mitchell P (1976) Possible molecular mechanisms of the protonmotive function of cytochrome systems. J Theor Biol 62:327–367

Moret V, Pinamonti S, Fornasari E (1961) Polarographic study on the redox potential of ubiquinones. Biochim Biophys Acta 54:381–383

Nelson N, Neumann J (1972) Isolation of a cytochrome b_6-f particle from chloroplasts. J Biol Chem 247:1817–1824

Okayama S (1976) Redox potential of plastoquinone A in spinach chloroplasts. Biochim Biophys Acta 440:331–336

Rieske JS, Zaugg WS, Hansen RE (1964) Studies of the electron transfer system. LIX. Distribution of Iron and of the component giving an electron paramagnetic resonance signal at g = 1.90 in subfractions of complex III. J Biol Chem 239:3023–3030

Robertson DE, Prince RC, Bowyer JR, Matsuura K, Dutton L, Ohnishi T (1984) Thermodynamic properties of the semiquinone and its binding site in the ubiquinol-cytochrome *c* (c2) oxidoreductase of respiratory and photosynthetic systems. J Biol Chem 259:1758–1763

Schoepp B, Brugna M, Riedel A, Nitschke W, Kramer D (1999a) The Q_o-site inhibitor DBMIB favours the proximal position of the chloroplast Rieske protein and induces a pK-shift of the redox linked proton. FEBS Lett 450:245–250

Schoepp B, Brugna M, Lebrun E, Nitschke W (1999b) Iron-sulfur centers involved in photosynthetic light reactions. Adv Inorg Chem 47:335–360

Schütz M, Brugna M, Lebrun E, Baymann F, Huber R, Stetter KO, Hauska G, Toci R, Lemesle-Meunier D, Tron P, Schmidt C, Nitschke W (2000) Early evolution of cytochrome *bc*-complexes. J Mol Biol 300:663–676

Schütz M, Schoepp-Cothenet B, Lojou E, Woudstra M, Lexa D, Tron P, Dolla A, Durand MC, Stetter KO, Baymann F (2003) The naphthoquinol oxidizing cytochrome bc_1 complex of the hyperthermophilic Knallgasbacterium *Aquifex aeolicus*: properties and phylogenetic relationships. Biochemistry 42:10800–10808

Sone N, Fujiwara Y (1991) Effects of aeration during growth of *Bacillus stearothermophilus* on proton pumping activity and change of terminal oxidases. J Biochem 110:1016–1021

Sone N, Nagata K, Kojima H, Tajima J, Kodera Y, Kanamaru T, Noguchi S, Sakamoto S (2001) A novel hydrophobic diheme c-type cytochrome. Purification from Corynebacterium glutamicum and analysis of the QcrCBA operon encoding three subunit proteins of a putative cytochrome reductase complex. Biochim Biophys Acta 1503:279–290

Sone N, Fukuda M, Katayama S, Jyoudai A, Syugyou M, Nogguchi S, Sakamoto J (2003) QcrCAB operon of a nocardia-form actinomyces Rhodococcus rhodochorus encodes cytochrome reductase complex with diheme cytochrome *cc* subunit. Biochim Biophys Acta 1557:125–131

Stroebel D, Choquet Y, Popot JL, Picot D (2003) An atypical haem in the cytochrome *b*(6)*f* complex. Nature 426:413–418

Swierczek M, Cieluch E, Sarewicz M, Borek A, Moser CC, Dutton PL, Osyczka A (2010) An electronic bus bar lies in the core of cytochrome bc_1. Science 329:451–454

Tanaka T, Inoue M, Sakamoto J, Sone N (1996) Intra- and inter-complex cross-linking of subunits in the quinol oxidase super-complex from thermophilic *Bacillus* PS3. J Biochem 119:482–486

ten Brink F, Schoepp-Cothenet B, van Lis R, Nitschke W, Baymann F (2013) Multiple Rieske/cyt*b* complexes in a single organism. Biochim Biophys Acta 1827:1392–1406

Wagner GC, Kassner RJ, Kamen MD (1974) Redox potentials of certain Vitamins K: Implications for a role in sulfite reduction by obligately Anaerobic Bacteria. Proc Natl Acad Sci USA 71:253–256

Widger WR, Cramer WA, Herrmann RG, Trebst A (1984) Sequence homology and structural similarity between cytochrome *b* of mitochondrial complex III and the chloroplast b_6-*f* complex: position of the cytochrome *b* hemes in the membrane. Proc Natl Acad Sci 81:674–678

Xia D, Yu CA, Kim H, Xia JZ, Kachurin AM, Zhang L, Yu L, Deisenhofer J (1997) Crystal structure of the cytochrome bc_1 complex from bovine heart mitochondria. Science 277:60–66

Yamashita E, Zhang H, Cramer WA (2007) Structure of the cytochrome b_6f complex: quinone analogue inhibitors as ligands of heme c_n. J Mol Biol 370:39–52

Yu J, Le Brun N (1988) Studies of the cytochrome subunits of menaquinone: cytochrome *c* reductase (*bc-complex*) of *Bacillus subtilis*. Evidence for the covalent attachment of heme to the cytochrome *b* subunit. J Biol Chem 273:8860–8866

Zhang Z, Huang L, Shulmeister VM, Chi YI, Kim KK, Hung LW, Crofts A, Berry E, Kim SH (1998) Electron transfer by domain movement in cytochrome bc_1. Nature 392:677–684

Zhang H, Primak A, Bowman MK, Kramer DM, Cramer WA (2004) Characterization of the high-spin heme x in the cytochrome b_6f complex of oxygenic photosynthesis. Biochemistry 43: 16329–16336

Chapter 9

Quinol Oxidases

Allison E. McDonald*
Department of Biology, Wilfrid Laurier University, Science Building, 75 University Avenue West, Waterloo, ON N2L 3C5, Canada

and

Greg C. Vanlerberghe
Department of Biological Sciences and Department of Cell and Systems Biology, University of Toronto Scarborough, 1265 Military Trail, Toronto, ON M1C1A4, Canada

Summary

Quinol oxidases catalyze the four-electron reduction of oxygen to water using electrons provided by a quinol. Examples of such oxidases can be found in all kingdoms of life and within several unrelated protein families including the heme-copper oxidase family, the cytochrome *bd* family and the di-iron carboxylate family. In prokaryotes, there are examples of quinol oxidases from each of these families. However, only quinol oxidases of the di-iron carboxylate type are found in eukaryotes. These include the mitochondrial-localized alternative oxidase and the plastid-localized plastoquinol terminal oxidase. The quinol oxidases differ in terms of their impact on energy conservation. In general, quinol oxidases may aid in maintaining metabolic homeostasis by providing some additional flexibility in systems coupling energy and carbon metabolism.

*Author for correspondence, e-mail: amcdonald@wlu.ca

M.F. Hohmann-Marriott (ed.), The Structural Basis of Biological Energy Generation, Advances in Photosynthesis and Respiration 39, pp. 167–185, DOI 10.1007/978-94-017-8742-0_9,

I. Introduction

This chapter will focus on quinol oxidases, enzymes that catalyze the four-electron reduction of oxygen to water using electrons provided by a quinol. Examples of such oxidases can be found in all kingdoms of life and within several unrelated protein families (Fig. 9.1). Dioxygen reductases (O_2Red) are *heme-containing* enzymes that catalyze the four-electron reduction of oxygen to water using electrons provided by either a quinol or cytochrome (cyt) *c*. They are found in two unrelated protein families, the heme-copper superfamily (that includes both cyt *c* oxidases and quinol oxidases) and the cyt *bd* oxidase family (quinol oxidases). The heme-copper superfamily can be further divided into three major groups (A–O_2Red, B–O_2Red and C–O_2Red), as well as several minor groups, depending upon specific features of the core subunits. Both the A–O_2Red and B–O_2Red groups contain quinol oxidases, but the C–O_2Red group does not. All the heme-copper superfamily enzymes couple oxygen reduction to a direct pumping of protons across the membrane. This is not the case with cyt *bd* quinol oxidases, although their activity still results in the generation of a proton gradient, as a result of charge separation associated with the quinol oxidation. In addition to the heme-containing quinol oxidases in the above groups, there is another, unrelated class of quinol oxidases that *lack heme* and are members of a diiron carboxylate family of proteins. They also catalyze the four-electron reduction of oxygen to water, but do not act as proton pumps. These include alternative oxidase (AOX) and plastoquinol terminal oxidase (PTOX) (Fig. 9.1).

Abbreviations: AOX – Alternative oxidase; Cyt – Cytochrome; EPR – Electron paramagnetic resonance; FTIR – Fourier transform infrared spectroscopy; O_2Red – Dioxygen reductases; PTOX – Plastoquinol terminal oxidase; ROS – Reactive oxygen species

II. Heme-Copper Quinol Oxidase

These quinol oxidases belong to the larger heme-copper oxidase superfamily that is comprised of eight different oxidase subfamilies (A through H), as well as nitric oxide reductase, all of which have been identified through phylogenetic analyses (Hemp et al. 2007; Gribaldo et al. 2009). Oxidases in this superfamily require multiple hemes and copper co-factors in order to function and contain a binuclear reaction centre that catalyzes the formation of water using four electrons (Schäfer et al. 1996). All oxidases in this superfamily contain one homologous subunit that contains one low-spin heme, one high-spin heme, and coupling between one copper (Cu_B) and the high-spin heme in the binuclear reaction center (Bernroitner et al. 2008; Hemp and Gennis 2008).

Within the heme-copper oxidase superfamily, the quinol oxidases differ from the cyt *c* oxidases in that they utilize reduced quinols as substrate (rather than cyt *c*) and are missing a binuclear copper (Cu_A) site in subunit II (Schäfer et al. 1996; Fig. 9.2). It is hypothesized that in order to access the quinol substrate, these quinol oxidases contain a series of polar residues on the membrane-spanning domain of subunit I that are oriented towards the lipid bilayer. These polar residues are not present in the cyt *c* oxidases (Abramson et al. 2000). It has been proposed that reduced quinols bind to a region of subunit I characterized by having several polar and charged residues (Bernroitner et al. 2008) and that electrons are then transferred to heme *b* and the heme o–Cu_B binuclear centre (Kobayashi et al. 2009; Fig. 9.2). During this process, protons are translocated across the membrane through one or more channels (e.g. D-channel and/or K channel) (Schäfer et al. 1996; Hemp et al. 2007; Kobayashi et al. 2009). It has become clear that the hemes present in these quinol oxidases can vary within and between species (Schäfer et al. 1996; Contreras-Zentella et al. 2003). Several highly conserved histidine residues have been identified as important for quinol binding and the co-ordination of

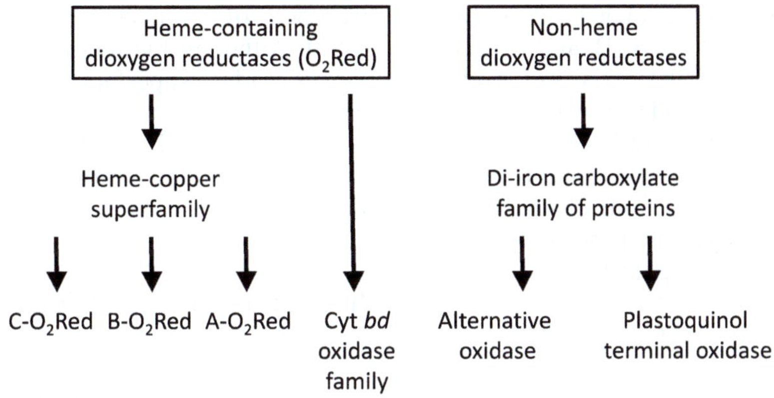

Fig. 9.1. Quinol oxidases are found in both heme-containing protein families (such as the heme-copper superfamily and the cyt *bd* oxidase family) as well as in a di-iron carboxylate family of proteins that lacks heme. The cyt *bd* family, alternative oxidase (AOX) and plastoquinol terminal oxidase (PTOX) are all quinol oxidases, as are some members of the A–O_2Red and B–O_2Red groups of heme-copper enzymes.

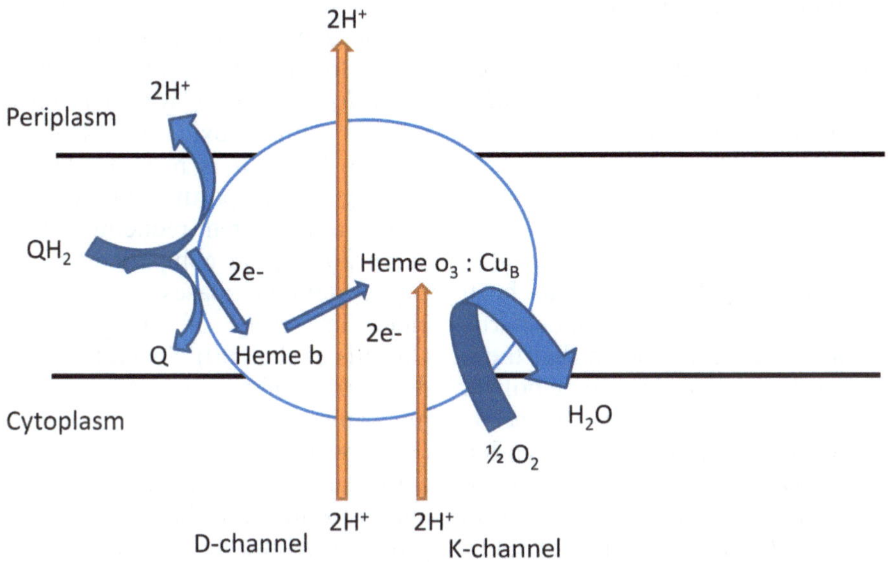

Fig. 9.2. A model for the oxidation of quinols by heme-copper quinol oxidases to produce water and the pumping of protons across the membrane (Modified from Bernroitner et al. (2008), García-Horsman et al. (1994), and Abramson et al. (2000)).

the hemes (Kobayashi et al. 2009) and recently a novel histidine-tyrosine reaction mechanism has been identified in the active site (Hemp et al. 2007). Each quinol oxidase in this superfamily is made up of several protein subunits with the total number varying by species (Schäfer et al. 1996; Hemp et al. 2007). The crystal structure of the cyt bo_3 quinol oxidase from *E. coli* demonstrated that the enzyme is composed of four integral membrane proteins (Abramson et al. 2000).

A. Distribution in Nature

While some members of the heme-copper oxidase superfamily are found in both prokaryotes and eukaryotes (i.e. cyt *c* oxidase), the quinol terminal oxidase members are found only in prokaryotes (Schäfer et al. 1996). Members of the A, B, and C families are found in the Eubacteria and Archaea, but families D through H appear to be unique to Archaea (Hemp and Gennis 2008). Interestingly, one of the quinol oxidases belonging to this superfamily is encoded by the QOX operon (*qoxBAC*) and is very similar in sequence to the operon encoding cyt *c* oxidase (Bernroitner et al. 2008). Usually, it is the absence of the Cu_A site in subunit II that is used to predict that an enzyme is a quinol oxidase rather than a cyt *c* oxidase. In addition, in a fashion similar to cyt *c* oxidase, QOX requires assembly proteins [e.g. surf1c (cyt *c* oxidase) and surf1q (QOX)] (Bundschuh et al. 2008). All nitrogen-fixing cyanobacterial species investigated (see Chap. 14) contain at least one QOX operon (Bernroitner et al. 2008). The most characterized heme-copper oxidase in Archaea (SoxM) is that from *Sulfolobus acidocaldarius,* which is coupled to a bc_1 homolog (Hemp and Gennis 2008).

B. Physiology

Transcription of the QOX genes has been detected in several species of cyanobacteria, and inhibitor studies have shown that these oxidases contribute to energy metabolism (Bernroitner et al. 2008). In *Synechococcus* sp. PCC 7002 the QOX oxidase, encoded by the CtaCIIDIIEII operon, was shown to mediate cyanide-sensitive respiration (Nomura et al. 2006). A function of QOX in establishing an oxygen-free environment may be surmised from the activity of the QOX promoter in *Paracoccus denitrificans.* The promotor activity is three times higher under denitrifying conditions compared to control conditions and its activity is low under aerobic growth conditions (Otten et al. 2001). Scavenging of oxygen may also be the reason for the presence of QOX in all nitrogen-fixing cyanobacteria (see Chap. 14). In *E. coli* the cyt *bo* quinol oxidase is repressed by PdhR, a transcription factor that is sensitive to pyruvate levels; expression of the cyt *bo* oxidase is derepressed in the presence of pyruvate (Ogasawara et al. 2007). Although the members of the heme-copper oxidase superfamily that utilize cyt *c* have been well characterized, especially in mammals (see Chap. 10), a great deal of work remains to be performed on the quinol-using members.

III. Cytochrome *bd* Oxidase

Cyt *bd* oxidase couples quinol oxidation to a four electron reduction of oxygen to water, but does not pump protons across the membrane; rather it generates a proton motive force via transmembrane charge separation (Borisov et al. 2011a; Fig. 9.3). Cyt *bd* oxidase contains one *d* heme and two *b* hemes (b_{558} and b_{595}) (Borisov et al. 2011a; Fig. 9.3), but does not contain copper or non-heme iron (Borisov et al. 2011b). The low-spin heme b_{558} has been hypothesized to play a role in quinol oxidation (Borisov et al. 2011a). Heme *d* is responsible for oxygen binding and in partnership with heme b_{595} forms a dioxygen reducing site (Borisov et al. 2011a). In comparison to the heme-copper quinol oxidases, cyt *bd* is half as efficient in terms of energy conservation (Jasaitis et al. 2000). Several conserved residues proposed to play a role in the catalytic mechanism have recently been summarized (Borisov et al. 2011b). Cyt *bd* oxidase is composed of two integral membrane proteins and is localized to the plasma membrane in prokaryotes and has also been suggested to be present in intracellular (thylakoid) membranes in cyanobacteria (Borisov et al. 2011b). The substrates for cyt *bd* have been reported to be ubiquinol, menaquinol, and perhaps plastoquinol (Borisov et al. 2011b).

A. Distribution in Nature

Cyt *bd* oxidase has been found in prokaryotes (many eubacteria and limited species of archaea), but not in eukaryotes (Baughn and Malamy 2004; Giuffrè et al. 2011). It is hypothesized that the cyt *bd* oxidase arose first in eubacteria and has spread throughout

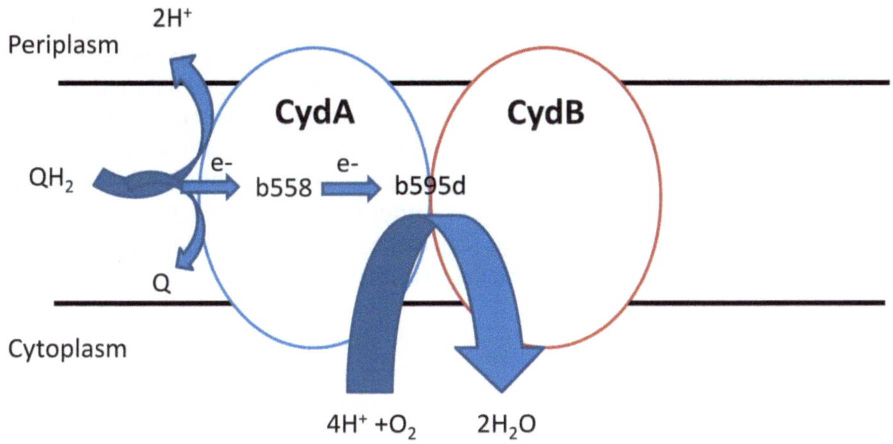

Fig. 9.3. A model for the oxidation of quinols by cyt *bd* quinol oxidases to produce transmembrane charge separation and the production of water (Modified from Zhang et al. (2004)).

the eubacterial and archaeal domains by horizontal gene transfer (Gribaldo et al. 2009). In *E. coli*, two different cyt *bd* operons are present; cydAB (cyt *bd* I) and appBC (cyt *bd* II) (Shepherd et al. 2010) and multiple cyt *bd* family members are found in other prokaryotes (Borisov et al. 2011b). Surprisingly, cyt *bd* oxidase is the only oxidase present in a few prokaryotes (i.e. there is no cyt *c* oxidase) and cyt *bd* is often absent from the genomes of aerobic prokaryotes (Borisov et al. 2011b).

A recent suggestion has been to classify the enzymes based on the size of their Q-loop, a region hypothesized to play a role in quinol binding (i.e. long, group A; short, group B), and their sensitivity to CN (group C) (Borisov et al. 2011b). Due to the fact that cyt *bd* oxidase is only found in prokaryotes, it has been suggested as a possible drug target for diseases caused by pathogenic microbes (Borisov et al. 2011b).

B. Biochemical Regulation

In *E. coli*, cydA encodes subunit I (a 57 kDa polypeptide) and cydB encodes subunit II (43 kDa) of cyt *bd* oxidase (Borisov et al. 2011b). Two additional proteins, CydC and CydD are encoded by a separate operon (CydCD) and are required for cyt *bd* oxidase assembly, but are not themselves part of the complex (Borisov et al. 2011b). These proteins form a heterodimer involved in glutathione transport (Borisov et al. 2011b). In addition to its interactions with its substrates of quinol and oxygen, additional interactions of the protein with other compounds have been proposed to regulate its activity. While the majority of cyt *bd* oxidases are sensitive to the respiratory inhibitor cyanide, a subset of cyt *bd* oxidases contain a region in subunit I that is proposed to be involved in quinol binding (Q-loop), which is shorter than that found in other organisms. It has been suggested that this shorter loop may confer resistance to CN in the cyanide-insensitive oxidases. Furthermore, cyanide-insensitivity has been correlated with the presence of less heme *d* in the enzyme; however this relationship is still controversial (Borisov et al. 2011b). Cyt *bd* oxidase has also been proposed to interact with carbon monoxide and H_2O_2 (Borisov et al. 2011b). The enzyme will also interact and bind to an array of nitrogen containing compounds through the formation of heme-NO adducts. However, unlike cyt *c* oxidase, this reaction appears to be highly and quickly reversible in cyt *bd* oxidases (Borisov et al. 2011b). Cyt *bd* oxidase has been shown to be specifically inhibited by pentachlorophenol (Pils et al. 1997). At the moment, no information on the regulation of cyt *bd* oxidase enzyme activity via post-translational means has been reported.

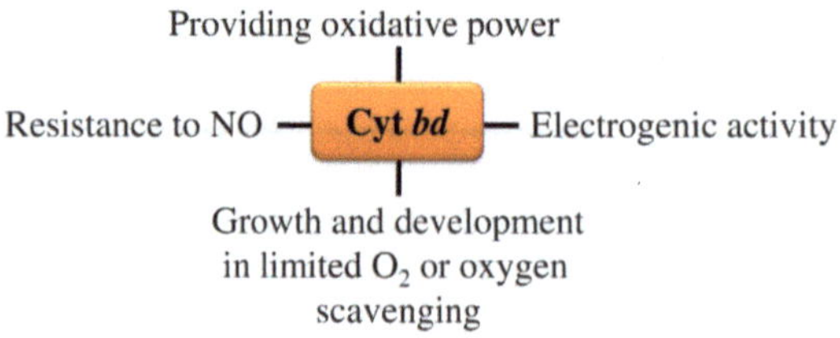

Fig. 9.4. Some emerging themes regarding the functional importance of cyt *bd* oxidase activity. See text for details.

C. Physiology

Several physiological roles have been proposed for cyt *bd* oxidase based on expression studies and an analysis of potential transcription factor binding sites in the upstream regions of genes (Fig. 9.4). Studies in different organisms suggest that cyt *bd* plays an important role in oxygen-depleted environments. Somewhat surprisingly, cyt *bd* oxidase genes have been found in several bacterial groups that are anaerobic (Giuffrè et al. 2011). In *E. coli* cydAB gene expression is controlled by the transcription factors ArcB and Fnr, which are active at low oxygen levels or under anaerobic conditions (Shepherd et al. 2010; Borisov et al. 2011b). This observation may be related to the use of varying terminal electron acceptors for respiration. The transcription of both subunits of cyt *bd* oxidase were upregulated when respiration used tetrachloroethene or nitrate (Peng et al. 2012). In *E. coli* living in the mouse intestine (an anaerobic environment), cyt *bd* oxidase was used during the reduction of fumarate and nitrate, with cells preferring the cyt *bd* oxidase-mediated pathway over nitrate reductase or fumarate reductase (Jones et al. 2011). Cyt *bd* oxidase has also been implicated in scavenging O_2 in order to protect nitrogenase or to allow the growth of anaerobic or microaerobic species (Giuffrè et al. 2011). Mutation of the cyt *bd* oxidase operon of *Azotobacter vinelandii* resulted in the loss of the ability to fix atmospheric nitrogen, suggesting that the oxidase is serving as an oxygen scavenger to protect nitrogenase in this species (Kelly et al. 1990).

A link is proposed to exist between the presence of cyt *bd* oxidase and virulence of bacterial pathogens (Giuffrè et al. 2011). In particular, the expression of cyt *bd* oxidase may convey some resistance to NO or other reactive nitrogen species due to its reversible inhibition characteristics. While cyt *bd* oxidase is reversibly inhibited by NO, as are other heme oxidases, the recovery from inhibition is much faster in cyt *bd* oxidases (Giuffrè et al. 2011).

Cyt *bd* oxidase is also proposed to play a role in cellular redox balance and the regeneration of oxidizing power for various cellular functions. The cyt *bd*-II oxidase does not pump protons and does not generate any proton motive force and has been hypothesized to serve as a means of oxidizing NADH uncoupled from ATP generation (Shepherd et al. 2010). In *E. coli*, results have shown that cyt *bd*-I oxidase likely plays a role in providing oxidative power, which can be utilized by an enzyme in the heme biosynthetic pathway (HemG) (Baughn and Malamy 2004) and systems responsible for the formation of disulfide bonds during protein folding (DsbA–DsbB) (Giuffrè et al. 2011). In a fascinating recent study, cyt *bd* oxidase has been proposed to be electrogenic (i.e. it donates electrons to external electron acceptors in the environment) in cyanobacteria during illumination (Pisciotta et al. 2011). The source of these electrons was found to be photosystem II and the evidence suggests that the PQ pool and cyt *bd* oxidase are responsible for the electrogenic activity. The authors propose that this could serve a protective function during high light conditions, or conditions resulting in photoinhibition. In *E. coli*, the second cyt *bd* oxidase operon (cyt bd-II) is induced by phosphate and carbon limitation, and stationary phase growth via the RpoS, AppY, and ArcA pathways (Shepherd et al. 2010; Borisov et al. 2011b).

IV. Alternative Oxidase and Plastoquinol Terminal Oxidase

Alternative oxidase (AOX) and plastoquinol terminal oxidase (PTOX) are related members of a non-heme diiron carboxylate family of proteins (Berthold and Stenmark 2003).

They couple the oxidation of ubiquinol (in the case of AOX) or plastoquinol (in the case of PTOX) to the four-electron reduction of molecular oxygen to water (Fig. 9.5). The diiron carboxylate proteins are distinguished by an iron-binding motif consisting of six conserved amino acids (four Glu and two His) that coordinate two irons within a structurally conserved four-helix-bundle conformation (Fig. 9.6) (Berthold and Stenmark 2003). Site-directed mutagenesis studies have confirmed the essential nature of these residues in AOX and PTOX as well as identifying other residues (e.g. Tyr residues) essential for activity (Albury et al. 2002; Nakamura et al. 2005; Fu et al. 2005, 2009; Moore and Albury 2008). EPR and FTIR spectroscopy studies of AOX have shown that the active site for the reduction of oxygen to water does indeed comprise a binuclear iron center (Berthold et al. 2002; Moore et al. 2008; Maréchal et al. 2009) and other work has begun to identify residues important for ubiquinol binding (Moore and Albury 2008; Albury et al. 2010). While no three dimensional structure of any membrane-bound diiron carboxylate protein has yet been achieved, there has been a recent report of preliminary analysis of crystals of a trypanosome AOX (Kido et al. 2010).

In eukaryotes, AOX is an interfacial membrane protein, oriented toward the matrix side of the inner mitochondrial membrane. AOX is non-proton pumping and since it bypasses proton-pumping complexes III and IV, electron flow to AOX dramatically reduces the energy yield of respiration (Finnegan et al. 2004). In contrast to the location of AOX in mitochondria, eukaryotic PTOX is an interfacial membrane protein found in plastids. In the chloroplast, PTOX is a thylakoid membrane protein oriented toward the stromal side of the membrane (Lennon et al. 2003; Rumeau et al. 2007).

In plants, AOX is usually encoded by a small nuclear gene family (Considine et al. 2002). In both *Arabidopsis* and tobacco, the expression of a single gene family member is strongly responsive to various stress conditions, such as cold temperature or drought. Other gene family members can display tissue and developmental specificity in their expression. In contrast, it appears that PTOX is encoded by a single gene in plants, that its expression is less responsive to stress (at least in comparison to AOX) and that its activity is critical for normal plastid (chloroplast) development (Carol et al. 1999; Wu et al. 1999; Rosso et al. 2006; see Chap. 23).

A. Distribution in Nature

In extant prokaryotes, AOX appears limited to some proteobacteria (e.g. *Novosphingobium aromaticivorans*, Stenmark and Nordlund 2003), while PTOX appears limited to some oxygen-evolving photosynthetic cyanobacteria (e.g. *Prochlorococcus marinus*, McDonald et al. 2003). It is likely that these proteins entered eukaryotic lineages via the ancient proteobacterial endosymbiont that gave rise to mitochondria and the ancient cyanobacterial endosymbiont that gave rise to chloroplasts (McDonald and Vanlerberghe 2006). Typically, AOX and PTOX share ~25 % protein sequence identity within the core protein delimited by the first and last iron-binding residues (Fig. 9.6). Phylogenetic analyses that include a broad taxonomic range of both prokaryotic and eukaryotic proteins show that AOXs and PTOXs clearly fall into two separate clades, consistent with the hypothesis than an ancestral protein (perhaps an ancient oxygen reductase; Gomes et al. 2001) diversified to AOX and PTOX prior to the primary endosymbiotic events that give rise to mitochondria and chloroplasts (McDonald and Vanlerberghe 2006). Hence, amongst extant eukaryotes, the "aplastidic lineages" (those thought to have never been exposed to the cyanobacterial endosymbiont that gave rise to chloroplasts, including amoebozoa, some protists, fungi, ichthyosporeans, choanoflagellates and animals) include many members containing AOX but no members containing PTOX. On the other hand, "plastidic" eukaryotes (that we define as those lineages that have contained a "chloroplast" at some time during their evolutionary history, and including the Glaucocystophyta, the "green"

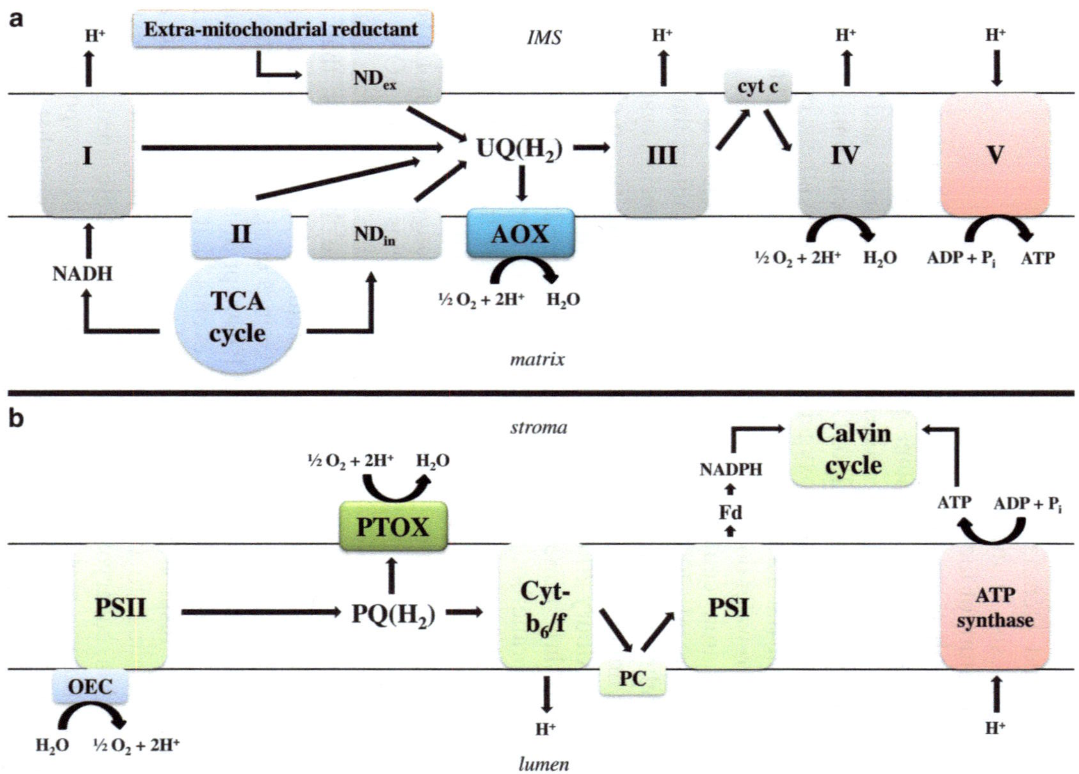

Fig. 9.5. The position of (**a**) alternative oxidase (AOX) in the respiratory electron transport chain, and (**b**) plastoquinol terminal oxidase (PTOX) in the photosynthetic electron transport chain of some eukaryotes. AOX and PTOX each represent a branch in these respective electron transport chains, facilitating electron flow directly from the quinone pool (either UQ[H_2] or PQ[H_2]) to oxygen. I, II, III, IV, V refer to Complexes I (NADH dehydrogenase), II (succinate dehydrogenase), III (cytochrome c reductase), IV (cytochrome *c* oxidase) and V (ATP synthase); cyt *c*, cytochrome *c*; ND$_{ex}$, rotenone-resistant NADH and NAD(P)H utilizing enzymes on the external side of the inner mitochondrial membrane; ND$_{in}$, rotenone-resistant NADH and NAD(P)H utilizing enzymes on the internal side of the inner mitochondrial membrane; *PSI* photosystem I, *PSII* photosystem II, *OEC* oxygen evolving complex, *cyt-b$_6$/f* cytochrome b$_6$/f, *PC* plastocyanin. See text for information regarding the taxonomic distribution of AOX and PTOX amongst eukaryotes. Note that AOX and PTOX also display a limited distribution in prokaryotic respiratory or photosynthetic membranes, respectively (see text for details).

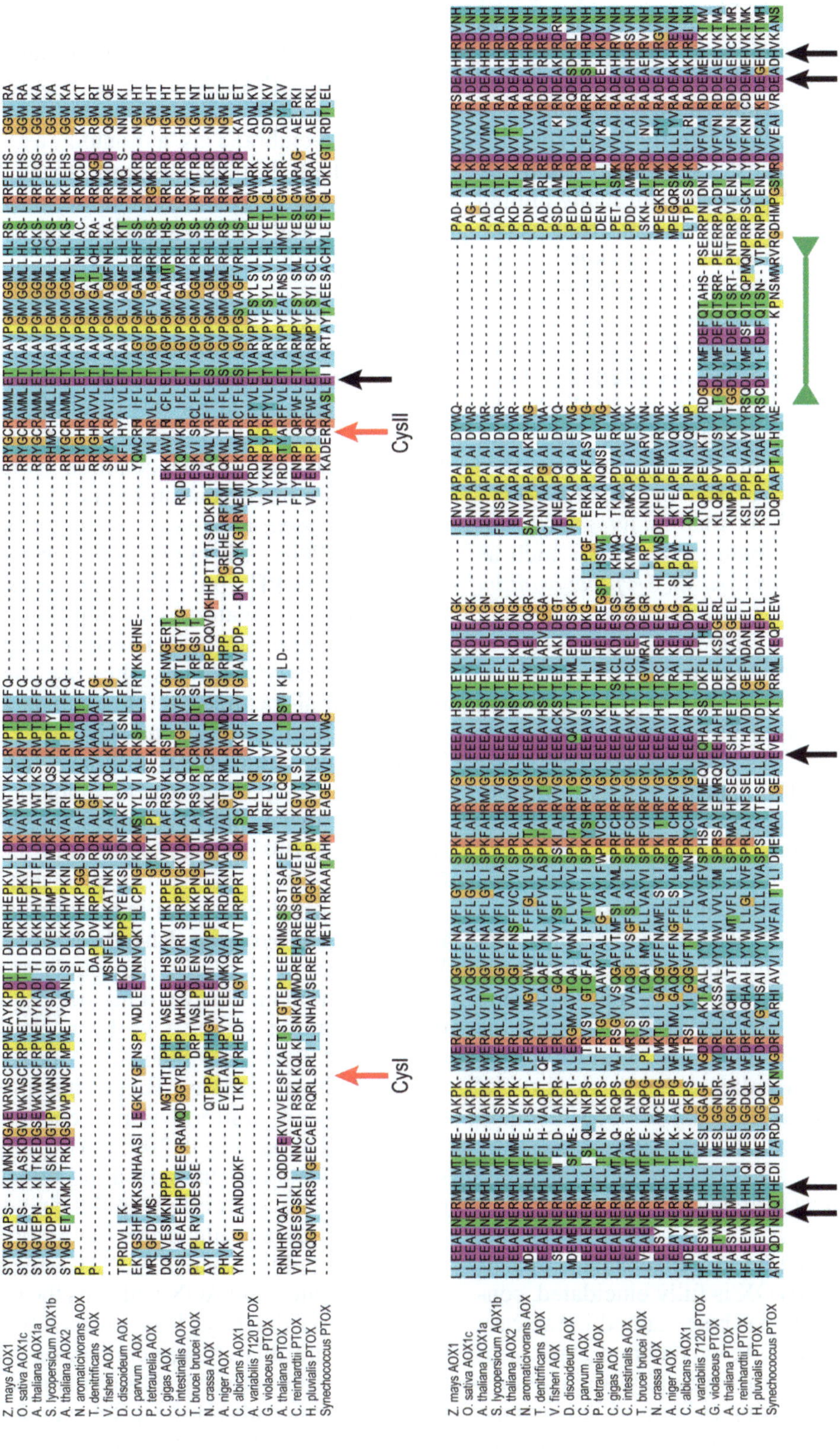

Fig. 9.6. A multiple sequence alignment of the catalytic core of AOX and PTOX proteins from a wide variety of organisms. The location of conserved cysteine residues found in higher plant AOXs are indicated by *red arrows*, while the amino acid residues required for iron binding are indicated by *black arrows*. An insert that is specific to the majority of PTOX sequences is shown by the *green bar*.

(see Chap. 17) lineage and the "red" lineage (see Chap. 16) have many examples (as expected) that contain AOX, PTOX or both (McDonald and Vanlerberghe 2006).

Interestingly, while AOX is ubiquitous in the plant kingdom, it is only more sporadically found in Protista, Fungi and Animalia. For example, while present in many animal phyla, AOX is clearly absent from vertebrates and arthropods (McDonald and Vanlerberghe 2006). These sporadic distributions suggest an evolutionary loss of AOX and PTOX from many eukaryotic lineages while maintenance has occurred in others. The ubiquitous presence of AOX in the Plantae suggests a particularly important functional role in this group. Another striking feature of the organisms found to contain AOX is that many are pathogenic protists or pathogenic fungi such as *Trypanosoma brucei brucei, Cryptosporidium parvum, Phytophthora infestans, Candida albicans* and *Cryptococcus neoformans*. Hence, AOX is often discussed as a potential therapeutic target in treating such human infections (Missall et al. 2004; Chaudhuri et al. 2006).

Thus far, PTOX appears strictly limited to organisms carrying out oxygenic photosynthesis (including cyanobacteria, algae and plants), but has thus far not been found in, for example, bacteria capable of non-oxygenic photosynthesis or plastidic eukaryotes that lack the capacity for photosynthesis. Interestingly, organisms that carry out oxygenic photosynthesis but lack PTOX (such as *Synechocystis* sp. PCC 6803), always have other quinol oxidases (such as cyt *bd* oxidase) instead (Hart et al. 2005; McDonald and Vanlerberghe 2006).

B. Biochemical Regulation

While the structure and mechanism of neither AOX nor PTOX is fully elucidated, considerable insight has been gained in the case of the plant AOX to understand aspects of biochemical control. Typically, such studies have been done using either purified enzyme or isolated mitochondria, and tools such as site-directed mutagenesis (combined with yeast or bacterial expression systems) have been used to identify residues important for biochemical control.

In plants, AOX exists in the inner mitochondrial membrane as a homodimer, the state of which dramatically impacts activity of the enzyme (Fig. 9.7). The dimer may be either non-covalently linked (reduced form) or covalently linked by a regulatory disulfide bond between the two monomers (oxidized form) (Umbach and Siedow 1993). A conserved Cys residue toward the N-terminus (Cys I; Figs. 9.6 and 9.7) and within the matrix is responsible for this disulfide bond (Rhoads et al. 1998; Vanlerberghe et al. 1998). Reduction of the disulfide bond is facilitated by the oxidation of specific TCA cycle substrates and, based upon the substrate specificity, it is hypothesized that specifically NADPH provides the reducing power for this regulatory reduction (Fig. 9.7) (Vanlerberghe et al. 1995). This is in keeping with *in organello* studies showing that a mitochondrial-localized thioredoxin is able to reduce this disulfide bond (Gelhaye et al. 2004). Once reduced, AOX is sensitive to activation by specific α-keto acids, most notably pyruvate, but also others (Fig. 9.7) (Millar et al. 1993; Vanlerberghe et al. 1995; Carré et al. 2011). While it is clear that the exposed sulfhydryls are an important site of interaction for pyruvate (Rhoads et al. 1998; Vanlerberghe et al. 1998), the details of this interaction remain elusive. It has been shown that substitution of the Cys by a charged amino acid (positive or negative) provides increased basal activity, suggesting that a charge-induced conformational change is important (Umbach et al. 2002). Also, recent studies indicate that the activating effect of pyruvate is due to an effect on the enzyme's apparent V_{max}, resulting from the ability of pyruvate to stabilize active AOX (Carré et al. 2011). In some plant AOX isoforms the regulatory Cys residue (Cys I) is replaced by Ser (Fig. 9.6) and this change confers on AOX the ability to be activated by succinate rather than by α-keto acids (Holtzapffel et al. 2003; Grant et al. 2009). The mechanism of this

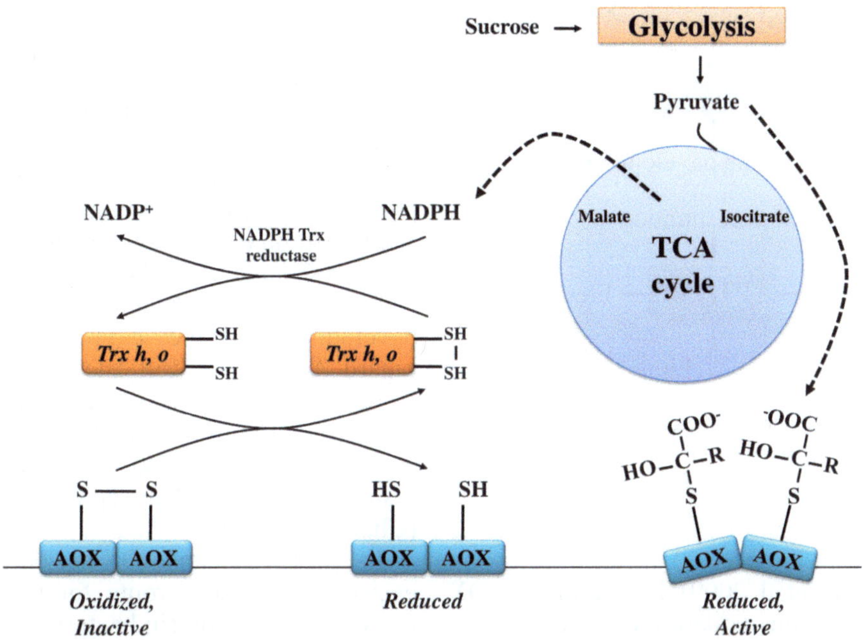

Fig. 9.7. A model for the biochemical control of AOX in plant mitochondria. Electron flow to AOX may be activated in a feed-forward manner by upstream respiratory metabolism. This involves a two-step biochemical activation of the AOX protein (a covalent modification followed by an allosteric activation) in response to the redox (NADPH) and carbon (pyruvate) status of the mitochondrial matrix. *Trx* thioredoxin. See text for details.

activation is poorly understood. Also, a second conserved and N-terminal Cys residue (Cys II; Fig. 9.6), along with other postulated sites may also be important in the activation of at least some AOX isoforms (Crichton et al. 2005; Umbach et al. 2006).

Less is known about the details of biochemical regulation of AOX in prokaryotes, protists, fungi and animals. However, it has been established in several fungi and protists that guanine and adenine mono- and dinucleotides (GMP, AMP, GDP, ADP) act as allosteric activators (Sakajo et al. 1997; Jarmuszkiewicz et al. 2002) and that these groups generally lack the regulatory Cys and activation by α-keto acids that is common in plants (Fig. 9.6) (Umbach and Siedow 2000). Also, a recent study suggests that the allosteric activation of AOX by mono- and dinucleotides in an amoeba can be offset by ATP acting as an allosteric inhibitor (Woyda-Ploszczyca et al. 2009). However, the site(s) of activation and inhibition by purine nucleotides has not been investigated.

At present, no information is available on potential biochemical regulatory properties of prokaryotic or eukaryotic PTOX. It also lacks the regulatory Cys residue found in many plant AOXs (Fig. 9.6).

C. Physiology

The metabolic and physiological roles of AOX and PTOX have been most extensively studied in higher plants. A number of important themes can be highlighted from such studies (Figs. 9.8 and 9.9) and will be briefly introduced here.

Since AOX respiration reduces the otherwise tight coupling between carbon metabolism, mitochondrial electron transport and ATP turnover, it could play a number of general roles in the optimization of respiratory metabolism, as well as the integration of this metabolism with other major metabolic pathways that impact the supply of or demand for carbon skeletons, reducing power and ATP (Finnegan et al. 2004;

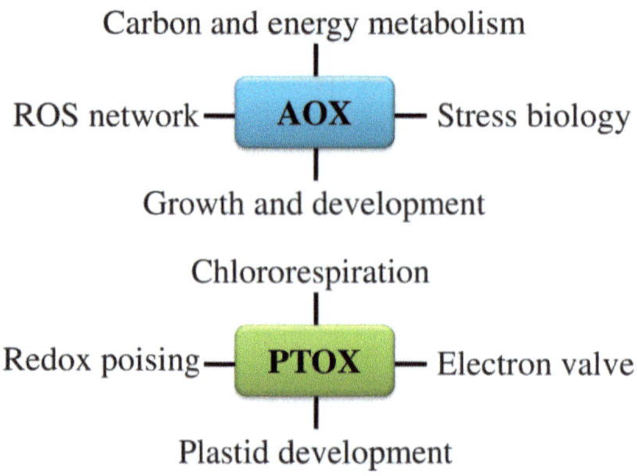

Fig. 9.8. Some emerging themes regarding the functional importance of AOX and PTOX activity in mitochondria and chloroplasts, respectively. See text for details.

Vanlerberghe et al. 2009). For example, AOX may allow for the oxidation of excess reducing power generated by the chloroplast, in this way acting to optimize photosynthetic metabolism (Raghavendra and Padmasree 2003). Changes in AOX activity might also adjust for large differences in reductant demand between processes, for example, the assimilation of ammonium versus nitrate (Elhafez et al. 2006). Also, by acting as a P_i-independent respiratory pathway, it could be of importance when growth is limited by the availability of phosphate (Theodorou and Plaxton 1993; Sieger et al. 2005).

Another central hypothesis is that AOX respiration may act to dampen reactive oxygen species (ROS) generation by the mitochondrial electron transport chain, since its energy-dissipating nature will act to moderate membrane potential (Purvis and Shewfelt 1993). Transgenic plants and green algae with reduced levels of AOX show induced levels of ROS-scavenging enzymes, suggestive of higher rates of ROS production (Amirsadeghi et al. 2006; Mathy et al. 2010).

A notable feature is that the expression of some AOX gene family members is highly responsive to abiotic and biotic stress conditions such as cold, drought, high salt, nutrient limitation, ozone and pathogen infection (Clifton et al. 2006). This might be important to compensate for a damaged cyt pathway (Ederli et al. 2006), to adjust to new metabolic conditions (Giraud et al. 2008), to appropriately modulate growth rate in response to the adverse conditions (Hansen et al. 2002; Sieger et al. 2005), or perhaps even to modulate stress signaling pathways from the mitochondrion (Arnholdt-Schmitt et al. 2006; Vanlerberghe et al. 2009; Wang et al. 2011; Cvetkovska and Vanlerberghe 2012).

Finally, some AOX genes show tissue and development-specific expression. In this regard, AOX has a well-established thermogenic role in the specialized floral structures of a restricted subset of plant species (Watling et al. 2006), but may also have a more broadly distributed role in plant reproductive tissues (Chai et al. 2010).

In other organisms AOX has been hypothesized to provide a route for nitric oxide-resistant respiration (in bacteria; Dunn et al. 2010), for sulfide-resistant respiration (in animals; Hildebrandt and Grieshaber 2008), or to reduce the overflow metabolism (Crabtree effect) that leads to reduced products such as ethanol (in fungi; Vemuri et al. 2007).

PTOX was first discovered as a critical factor in the biosynthesis of carotenoids during early chloroplast development (Carol et al. 1999; Wu et al. 1999). In this role, PTOX activity supplies the oxidized plastoquinone needed by the carotenoid biosynthetic pathway as an electron acceptor (Fig. 9.9b). At this early stage of chloroplast development, the photosynthetic electron transport chain is not yet fully assembled, meaning that downstream electron transport chain components (i.e. PSI) are not yet competent to oxidize plastoquinol. Under these conditions, the plastoquinone pool of an *Arabidopsis* mutant lacking PTOX (called *immutans*) is more highly reduced (Rosso et al. 2009), generating a block in the carotenoid biosynthetic pathway. In the absence of these photoprotective pigments at this critical developmental stage, normal chloroplast biogenesis in *immutans* in response to light is severely disrupted in some leaf sectors, resulting in a variegated leaf phenotype. PTOX similarly has a role in carotenoid biosynthesis in other plastid types (Josse et al. 2000).

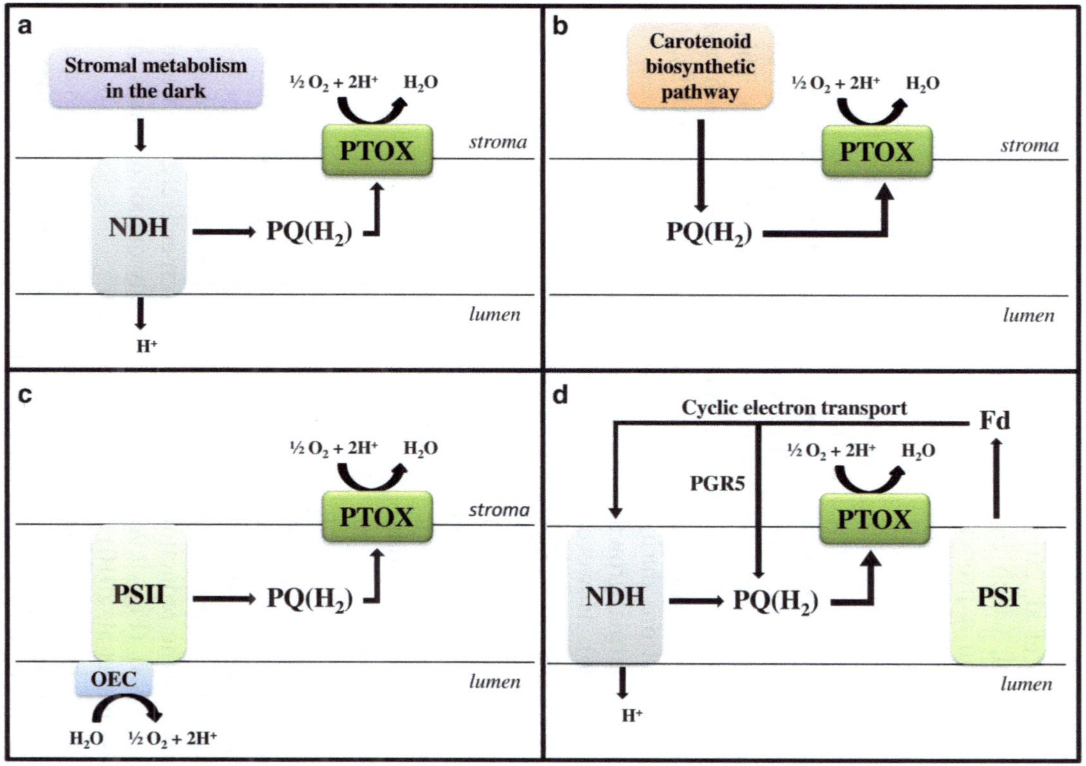

Fig. 9.9. Models of electron flow in the chloroplast involving PTOX. In (**a**) carbon oxidation pathways in the chloroplast during the dark generate reducing equivalents subsequently oxidized by a chloroplast NADH dehydrogenase-like complex embedded in the thylakoid membrane. Electrons are passed to the plastoquinone pool, which is then oxidized by PTOX. This process has been termed chlororespiration. In (**b**) PTOX activity is linked to carotenoid biosynthesis such as during early chloroplast development. A desaturation step in the carotenoid biosynthetic pathway is dependent upon plastoquinone as electron acceptor. The plastoquinol is subsequently oxidized by PTOX. In (**c**) PTOX acts as an electron valve, accepting electrons from PSII in excess of what can be processed by downstream components (i. e. cyt b_6f, PSI). This essentially represents a H_2O-to-H_2O cycle between PSII and PTOX. In (**d**) some electrons being cycled back to the plastoquinone pool from PSI (via a cyclic electron transport pathway that may or may not include NDH) are being passed to PTOX. In this case, a redox poising of the plastoquinone pool by PTOX might be necessary for effective cyclic electron flow.

PTOX is also expressed (be it at generally low levels) in mature chloroplasts and may take part in chloroplast metabolism in both the light and the dark. In the dark, PTOX likely acts in concert with the chloroplast NADH dehydrogenase-like complex (NDH) to oxidize reductant generated by carbon oxidation pathways in the stroma. Stromal reductant oxidized by NDH (probably reduced ferredoxin, Yamamoto et al. 2011) reduces the plastoquinone pool, which can then be oxidized by PTOX, in a process typically referred to as chlororespiration (Fig. 9.9a) (Peltier and Cournac 2002).

In the light, the activity and role of PTOX is much more difficult to evaluate because of multiple potential pathways of plastoquinone oxidation/reduction. In general, overexpression or lack of PTOX in *Arabidopsis* had no obvious effects on photosynthetic electron transport, suggesting that PTOX was not acting as a major sink for electrons during photosynthesis, at least under normal growth conditions (Rosso et al. 2006). However, PTOX expression does increase under conditions when downstream demand for electrons may be reduced (such as cold), suggesting the possibility that PTOX might act as a sink for excess electrons under some conditions (Streb et al. 2005; Quiles 2006; Stepien and Johnson 2009). In green algae and cyanobacteria, it was shown that iron limitation (which increases the proportion of PSII relative to PSI) also increased PTOX, again suggestive that it might act as an electron sink under conditions when there is an imbalance between plastoquinone reduction by PSII and oxidation by PSI (Fig. 9.9c) (Cardol et al. 2008; Grossman et al. 2010).

Interestingly, it has been shown that the leaf variegation in *immutans* is suppressed in plants also defective in cyclic electron transport, suggesting that cyclic-electron transport-dependent PQ reduction is linked to PTOX-dependent PQ oxidation during chloroplast development (Okegawa et al. 2010). Further, the same study showed that the slow growth phenotype of mutant plants defective in cyclic electron transport can be alleviated by *immutans*. These are somewhat surprising results and may indicate that the redox poise of the PQ pool, as determined by an interplay of these oxidizing and reducing pathways is critical to plant growth and development (Fig. 9.9d). Much more study will clearly be needed to understand these links between PTOX, cyclic electron transport, growth and chloroplast biogenesis.

AOX and PTOX have both been suggested to act as electron sinks within their respective organelles. It is well known that reducing power can also flow between these organelles. It is interesting therefore that transgenic plants lacking AOX display increased PTOX expression (Amirsadeghi et al. 2006), while *immutans* displays increased AOX expression (Rosso et al. 2009). This indicates that perhaps the two pathways can compensate for one another's activity, though this is not yet directly demonstrated. Also, in the green alga *Chlamydomonas*, the two pathways have been shown to be regulated in a coordinate manner in response to phosphorus deprivation (Moseley et al. 2006).

V. Conclusion

Based upon taxonomic distribution and protein phylogenies it is clear that quinol oxidases are ancient enzymes (Fig. 9.10). The heme-containing quinol oxidases appear to have arisen and diverged prior to the origin of AOX and PTOX in proteobacteria and cyanobacteria, respectively. AOX and PTOX represent the only quinol oxidases in eukaryotes (Fig. 9.10).

It is likely that quinol oxidases are important for maintaining metabolic homeostasis, for example during periods of environmental stress. When more than one quinol oxidase is present in a given organism, evidence suggests that they may have either distinct or overlapping roles. In general, these enzymes may provide additional metabolic flexibility within systems coupling energy and carbon metabolism.

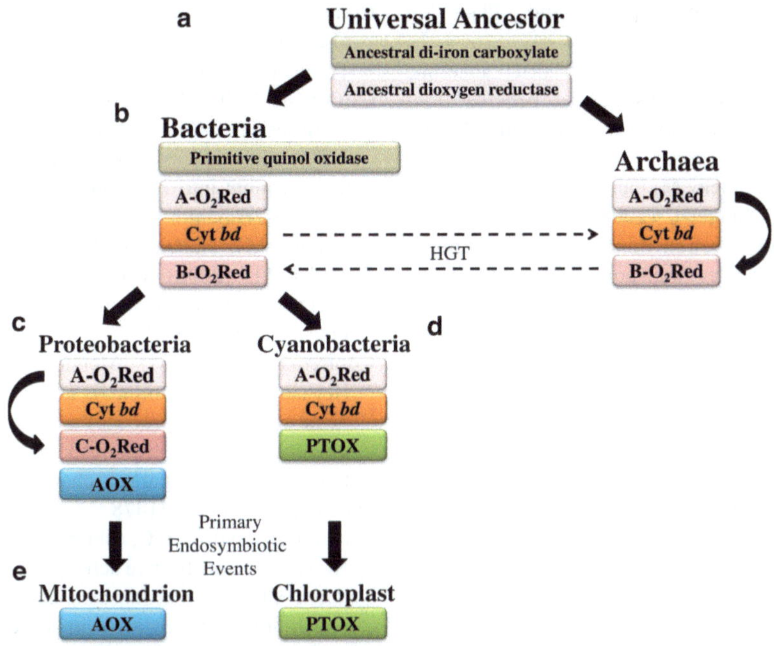

Fig. 9.10. A model for the evolutionary history of quinol oxidases in all kingdoms of life, based on data and ideas presented in Pereira et al. (2001), McDonald and Vanlerberghe (2006), Gribaldo et al. (2009), and Brochier-Armanet et al. (2009). (**a**) It is proposed that a universal ancestor of Bacteria and Archaea contained an ancestral di-iron carboxylate protein and ancestral dioxygen reductase (a primitive A-family O_2Red). (**b**) In the bacterial lineage, the di-iron carboxylate protein evolved into a more efficient primitive quinol oxidase and the A-family dioxygen reductase (A–O_2Red) was inherited vertically. The bacterial lineage also evolved a cyt *bd* oxidase (Cyt *bd*). In the archaeal lineage, the A-family O_2Red was inherited and diverged to produce the B–O_2Red family. It is also proposed that horizontal gene transfer events (HGT) introduced the cyt *bd* into some Archaea and the B–O_2Red into some Bacteria. (**c**) The proteobacteria are proposed to have vertically inherited A–O_2Red and Cyt *bd* and to have evolved alternative oxidase (AOX) from the primitive quinol oxidase. The A–O_2Red family also gave rise to the C–O_2Red family in this group. (**d**) The cyanobacteria are proposed to have vertically inherited the A–O_2Red and Cyt *bd* and to have evolved plastoquinol terminal oxidase (PTOX) from the primitive quinol oxidase. (**e**) AOX and PTOX were introduced into eukaryotic lineages via the primary endosymbiotic events that gave rise to mitochondria and chloroplasts, respectively. See text for further details.

Acknowledgments

Both authors acknowledge financial support from the Natural Sciences and Engineering Research Council of Canada (NSERC) and A.E.M. acknowledges the support of the Faculty of Science, Wilfrid Laurier University.

References

Abramson J, Rijstama S, Larsson G, Jasaitis A, Svensson-Ek M, Laakkonen L, Puustinen A, Iwata S, Wikström M (2000) The structure of the ubiquinol oxidase from *Escherichia coli* and its ubiquinone binding site. Nat Struct Biol 7:910–917

Albury MS, Affourtit C, Crichton PG, Moore AL (2002) Structure of the plant alternative oxidase: site-directed mutagenesis provides new information on the active site and membrane topology. J Biol Chem 277:1190–1194

Albury MS, Elliott C, Moore AL (2010) Ubiquinol-binding site in the alternative oxidase: mutagenesis reveals features important for substrate binding and inhibition. Biochim Biophys Acta 1797:1933–1939

Amirsadeghi S, Robson CA, McDonald AE, Vanlerberghe GC (2006) Changes in plant mitochondrial electron transport alter cellular levels of reactive oxygen species and susceptibility to cell death signaling molecules. Plant Cell Physiol 47:1509–1519

Arnholdt-Schmitt B, Costa JH, de Melo DF (2006) AOX – a functional marker for efficient cell reprogramming under stress? Trends Plant Sci 11:281–287

Baughn A, Malamy M (2004) The strict anaerobe *Bacteroides fragilis* grows in and benefits from nanomolar concentrations of oxygen. Nature 29:441–444

Bernroitner M, Zamocky M, Pairer M, Furtmüller PG, Peschek GA, Obinger C (2008) Heme-copper oxidases and their electron donors in cyanobacterial respiratory electron transport. Chem Biodivers 5:1927–1961

Berthold DA, Stenmark P (2003) Membrane-bound diiron carboxylate proteins. Annu Rev Plant Biol 54:497–517

Berthold DA, Voevodskaya N, Stenmark P, Gräsland A, Nordlund P (2002) EPR studies of the mitochondrial alternative oxidase: evidence for a diiron carboxylate center. J Biol Chem 277:43608–43614

Borisov VB, Forte E, Sarti P, Giuffrè A (2011a) Catalytic intermediates of cytochrome *bd* terminal oxidase at steady-state: ferryl and oxy-ferrous species dominate. Biochim Biophys Acta 1807:503–509

Borisov VB, Gennis RB, Hemp J, Verkhovsky MI (2011b) The cytochrome *bd* respiratory oxygen reductases. Biochim Biophys Acta 1807:1398–1413

Brochier-Armanet C, Talla E, Gribaldo S (2009) The multiple evolutionary histories of dioxygen reductases: implications for the origin and evolution of aerobic respiration. Mol Biol Evol 26:285–297

Bundschuh FA, Hoffmeier K, Ludwig B (2008) Two variants of the assembly factor Surf1 target specific terminal oxidases in *Paracoccus denitrificans*. Biochim Biophys Acta 1777:1336–1343

Cardol P, Bailleul B, Rappaport F, Derelle E, Béal D, Breyton C, Bailey S, Wollman FA, Grossman A, Moreau H, Finazzi G (2008) An original adaptation of photosynthesis in the marine green alga *Ostreococcus*. Proc Natl Acad Sci U S A 105:7881–7886

Carol P, Stevenson D, Bisanz C, Breitenbach J, Sandmann G, Mache R, Coupland G, Kuntz M (1999) Mutations in the *Arabidopsis* gene *immutans* cause a variegated phenotype by inactivating a chloroplast terminal oxidase associated with phytoene desaturation. Plant Cell 11:57–68

Carré JE, Affourtit C, Moore AL (2011) Interaction of purified alternative oxidase from thermogenic *Arum maculatum* with pyruvate. FEBS Lett 585:397–401

Chai T-T, Simmonds D, Day DA, Colmer TD, Finnegan PM (2010) Photosynthetic performance and fertility are repressed in GmAOX2b antisense soybean. Plant Physiol 152:1638–1649

Chaudhuri M, Ott RD, Hill GC (2006) Trypanosome alternative oxidase: from molecule to function. Trends Parasitol 22:484–491

Clifton R, Millar AH, Whelan J (2006) Alternative oxidases in *Arabidopsis*: a comparative analysis of differential expression in the gene family provides new insights into function of non-phosphorylating bypasses. Biochim Biophys Acta 1757:730–741

Considine MJ, Holtzapffel RC, Day DA, Whelan J, Millar AH (2002) Molecular distinction between alternative oxidase from monocots and dicots. Plant Physiol 129:949–953

Contreras-Zentella M, Mendoza G, Membrillo-Hernández J, Escamilla JE (2003) A novel double heme substitution produces a functional bo3 variant of the quinol oxidase aa3 of *Bacillus cereus*. Purification and partial characterization. J Biol Chem 278:31473–31478

Crichton PG, Affourtit C, Albury MS, Carré JE, Moore AL (2005) Constitutive activity of *Sauromatum guttatum* alternative oxidase in *Schizosaccharomyces pombe* implicates residues in addition to conserved cysteines in α-keto acid activation. FEBS Lett 579:331–336

Cvetkovska M, Vanlerberghe GC (2012) Coordination of a mitochondrial superoxide burst during the hypersensitive response to bacterial pathogen in *Nicotiana tabacum*. Plant Cell Environ 35:1121–1136

Dunn AK, Karr EA, Wang Y, Batton AR, Ruby EG, Stabb EV (2010) The alternative oxidase (AOX) gene in *Vibrio fischeri* is controlled by NsrR and upregulated in response to nitric oxide. Mol Microbiol 77:44–55

Ederli L, Morettini R, Borgogni A, Wasternack C, Miersch O, Reale L, Ferranti F, Tosti N, Pasqualini S (2006) Interaction between nitric oxide and ethylene in the induction of alternative oxidase in ozone-treated tobacco plants. Plant Physiol 142:595–608

Elhafez D, Murcha MW, Clifton R, Soole KL, Day DA, Escobar MA, Geisler DA, Rasmusson AG (2006) Reorganization of the alternative pathways of the *Arabidopsis* respiratory chain by nitrogen supply: opposing effects of ammonium and nitrate. Plant J 45:775–788

Finnegan PM, Soole KL, Umbach AL (2004) Alternative mitochondrial electron transport proteins in higher plants. In: Day DA, Millar AH, Whelan J (eds) Plant mitochondria: from gene to function. Kluwer Academic Publishers, Great Britian, pp 163–230

Fu A, Park S, Rodermel S (2005) Sequences required for the activity of PTOX (immutans), a plastid terminal oxidase: in vitro and in planta mutagenesis of

iron-binding sites and a conserved sequence that corresponds to exon 8. J Biol Chem 280: 42489–42496
Fu A, Alura M, Rodermel SR (2009) Conserved active site sequences in *Arabidopsis* plastid terminal oxidase (PTOX). J Biol Chem 284:22625–22632
García-Horsman JA, Barquera B, Rumbley J, Ma J, Gennis RB (1994) The superfamily of heme-copper respiratory oxidases. J Bacteriol 176:5587–5600
Gelhaye E, Rouhier N, Gérard J, Jolivet Y, Gualberto J, Navrot N, Ohlsson P-I, Wingsle G, Hirasawa M, Knaff DB, Wang H, Dizengremel P, Meyer Y, Jacquot J-P (2004) A specific form of thioredoxin h occurs in plant mitochondria and regulates the alternative oxidase. Proc Natl Acad Sci U S A 101:14545–14550
Giraud E, Ho LHM, Clifton R, Carroll A, Estavillo G, Tan Y-F, Howell KA, Ivanova A, Pogson BJ, Millar AH, Whelan J (2008) The absence of alternative oxidase 1a in Arabidopsis results in acute sensitivity to combined light and drought stress. Plant Physiol 147:595–610
Giuffrè A, Borisov VB, Mastronicola D, Sarti P, Forte E (2011) Cytochrome *bd* oxidase and nitric oxide: from reaction mechanisms to bacterial physiology. FEBS Lett 586:622–629
Gomes CM, Le Gall J, Xavier AV, Teixeira M (2001) Could a diiron-containing four-helix-bundle protein have been a primitive oxygen reductase? Chem Bio Chem 2:583–587
Grant N, Onda Y, Kakizaki Y, Ito K, Watling J, Robinson S (2009) Two Cys or not two Cys? That is the question; alternative oxidase in the thermogenic plant sacred lotus. Plant Physiol 150:987–995
Gribaldo S, Talla E, Brochier-Armanet C (2009) Evolution of the haem copper oxidases superfamily: a rooting tale. Trends Biochem Sci 34:375–381
Grossman AR, Mackey KRM, Bailey S (2010) A perspective on photosynthesis in the oligotrophic oceans: hypotheses concerning alternate routes of electron flow. J Phycol 46:629–634
Hansen LD, Church JN, Matheson S, McCarlie VW, Thygerson T, Criddle RS, Smith BN (2002) Kinetics of plant growth and metabolism. Thermochem Acta 388:415–425
Hart SE, Schlarb-Ridley BG, Bendall DS, Howe CJ (2005) Terminal oxidases of cyanobacteria. Biochem Soc Trans 33:832–835
Hemp J, Gennis R (2008) Diversity of the heme-copper superfamily in archaea: insights from genomics and structural modeling. Results Probl Cell Differ 45:1–31
Hemp J, Han H, Roh JH, Kaplan S, Martinez TJ, Gennis RB (2007) Comparative genomics and site-directed mutagenesis support the existence of only one input channel for protons in the C-family (cbb3 oxidase) of heme-copper oxygen reductases. Biochemistry 46:9963–9972
Hildebrandt TM, Grieshaber MK (2008) Redox regulation of mitochondrial sulfide oxidation in the lugworm, *Arenicola marina*. J Exp Biol 211: 2617–2623
Holtzapffel RC, Castelli J, Finnegan PM, Millar AH, Whelan J, Day DA (2003) A tomato alternative oxidase protein with altered regulatory properties. Biochim Biophys Acta 1606:153–162
Jarmuszkiewicz W, Behrendt M, Navet R, Sluse FE (2002) Uncoupling protein and alternative oxidase of *Dictyostelium discoideum*: occurrence, properties and protein expression during vegetative life and starvation-induced early development. FEBS Lett 532:459–464
Jasaitis A, Borisov V, Belevich N, Morgan J, Konstantinov A, Verkhovsky M (2000) Electrogenic reactions of cytochrome *bd*. Biochem 39:13800–13809
Jones SA, Gibson T, Maltby RC, Stewart V, Cohen PS, Conway T (2011) Anaerobic respiration of *Escherichia coli* in the mouse intestine. Infect Immun 79:4218–4226
Josse E-M, Simkin AJ, Gaffe J, Laboure A-M, Kuntz M, Carol P (2000) A plastid terminal oxidase associated with carotenoid desaturation during chromoplast differentiation. Plant Physiol 123:1427–1436
Kelly M, Poole R, Yates M, Kennedy C (1990) Cloning and mutagenesis of genes encoding the cytochrome bd terminal oxidase complex in *Azotobacter vinelandii*: mutants deficient in the cytochrome d complex are unable to fix nitrogen in air. J Bacteriol 172:6010–6019
Kido Y, Maréchal A, Kita K, Moore AL, Rich PR (2009) Three redox states of *Trypanosoma brucei* alternative oxidase identified by infrared spectroscopy and electrochemistry. J Biol Chem 284: 31827–31833
Kido Y, Shiba T, Inaoka DK, Sakamoto K, Nara T, Aoki T, Honma T, Tanaka A, Inoue M, Matsuoka S, Moore A, Harada S, Kita K (2010) Crystallization and preliminary analysis of cyanide-insensitive alternative oxidase from *Trypanosoma brucei brucei*. Acta Crystallogr F66:275–278
Kobayashi K, Tagawa S, Mogi T (2009) Electron transfer processes in subunit I mutants of cytochrome bo quinol oxidase in *Escherichia coli*. Biosci Biotechnol Biochem 73:1599–1603
Lennon AM, Prommeenate P, Nixon PJ (2003) Location, expression and orientation of the putative chlororespiratory enzymes, Ndh and immutans, in higher-plant plastids. Planta 218:254–260

Mathy G, Cardol P, Dinant M, Blomme A, Gérin S, Cloes M, Ghysels B, DePauw E, Leprince P, Remacle C, Sluse-Goffart C, Franck F, Matagne RF, Sluse FE (2010) Proteomic and functional characterization of a *Chlamydomonas reinhardtii* mutant lacking the mitochondrial alternative oxidase 1. J Proteome Res 9:2825–2838

McDonald AE, Vanlerberghe GC (2006) Origins, evolutionary history, and taxonomic distribution of alternative oxidase and plastoquinol terminal oxidase. Comp Biochem Physiol Part D 1:357–364

McDonald AE, Amirsadeghi S, Vanlerberghe GC (2003) Prokaryotic orthogues of mitochondrial alternative oxidase and plastid terminal oxidase. Plant Mol Biol 53:865–876

Millar AH, Wiskich JT, Whelan J, Day DA (1993) Organic acid activation of the alternative oxidase of plant mitochondria. FEBS Lett 329:259–262

Missall TA, Lodge JK, McEwen JE (2004) Mechanisms of resistance to oxidative and nitrosative stress: implications for fungal survival in mammalian hosts. Eukaryot Cell 3:835–846

Moore AL, Albury MS (2008) Further insights into the structure of the alternative oxidase: from plants to parasites. Biochem Soc Trans 36:1022–1026

Moore AL, Carré JE, Affourtit C, Albury MS, Crichton PG, Kita K, Heathcote P (2008) Compelling EPR evidence that the alternative oxidase is a diiron carboxylate protein. Biochim Biophys Acta 1777: 327–330

Moseley JL, Chang C-W, Grossman AR (2006) Genome-based approaches to understanding phosphorus deprivation responses and PSR1 control in *Chlamydomonas reinhardtii*. Eukaryot Cell 5:26–44

Nakamura K, Sakamoto K, Kido Y, Fujimoto Y, Suzuki T, Suzuki M, Yabu Y, Ohta N, Tsuda A, Onuma M, Kita K (2005) Mutational analysis of the *Trypanosoma vivax* alternative oxidase: the $E(X)_6Y$ motif is conserved in both mitochondrial alternative oxidase and plastid terminal oxidase and is indispensable for enzyme activity. Biochem Biophys Res Commun 334:593–600

Nomura CT, Persson S, Shen G, Inoue-Sakamoto K, Bryant DA (2006) Characterization of two cytochrome oxidase operons in the marine cyanobacterium *Synechococcus* sp. PCC 7002: inactivation of ctaDI affects the PS I:PS II ratio. Photosynth Res 87:215–228

Ogasawara H, Ishida Y, Yamada K, Yamamoto K, Ishihama A (2007) PdhR (pyruvate dehydrogenase complex regulator) controls the respiratory electron transport system in *Escherichia coli*. J Bacteriol 189:5534–5541

Okegawa Y, Kobayashi Y, Shikanai T (2010) Physiological links among alternative electron transport pathways that reduce and oxidize plastoquinone in Arabidopsis. Plant J 63:458–468

Otten MF, Stork DM, Reijnders WN, Westerhoff HV, Van Spanning RJ (2001) Regulation of expression of terminal oxidases in *Paracoccus denitrificans*. Eur J Biochem 268:2486–2497

Peltier G, Cournac L (2002) Chlororespiration. Annu Rev Plant Biol 53:523–550

Peng X, Yamamoto S, Vertès AA, Keresztes G, Inatomi K, Inui M, Yukawa H (2012) Global transcriptome analysis of the tetrachloroethene-dechlorinating bacterium *Desulfitobacterium hafniense* Y51 in the presence of various electron donors and terminal electron acceptors. J Ind Microbiol Biotechnol 39(2):255–268

Pereira MM, Santana M, Teixeira M (2001) A novel scenario for the evolution of haem-copper oxygen reductases. Biochim Biophys Acta 1505:185–208

Pils D, Gregor W, Schmetterer G (1997) Evidence for in vivo activity of three distinct respiratory terminal oxidases in the cyanobacterium *Synechocystis* sp. strain PCC6803. FEMS Microbiol Lett 152:83–88

Pisciotta JM, Zou Y, Baskakov IV (2011) Role of the photosynthetic electron transfer chain in electrogenic activity of cyanobacteria. Appl Microbiol Biotechnol 91:377–385

Purvis AC, Shewfelt RL (1993) Does the alternative pathway ameliorate chilling injury in sensitive plant tissues? Physiol Plant 88:712–718

Quiles M (2006) Stimulation of chlororespiration by heat and high light intensity in oat leaves. Plant Cell Environ 29:1463–1470

Raghavendra AS, Padmasree K (2003) Beneficial interactions of mitochondrial metabolism with photosynthetic carbon assimilation. Trends Plant Sci 8:546–553

Rhoads DM, Umbach AL, Sweet CR, Lennon AM, Rauch GS, Siedow JN (1998) Regulation of the cyanide-resistant alternative oxidase of plant mitochondria: identification of the cysteine residue involved in α-keto acid stimulation and intersubunit disulfide bond formation. J Biol Chem 273:30750–30756

Rosso D, Ivanov AG, Fu A, Geisler-Lee J, Hendrickson L, Geisler M, Stewart G, Krol M, Hurry V, Rodermel SR, Maxwell DP, Hüner NPA (2006) Immutans does not act as a stress-induced safety valve in the protection of the photosynthetic apparatus of Arabidopsis during steady-state photosynthesis. Plant Physiol 142:574–585

Rosso D, Bode R, Li W, Krol M, Saccon D, Wang S, Schillaci LA, Rodermel SR, Maxwell DP, Hüner NPA (2009) Photosynthetic redox imbalance governs leaf sectoring in the *Arabidopsis thaliana* variegation mutants *immutans*, *spotty*, *var1*, and *var2*. Plant Cell 21:3473–3492

Rumeau D, Peltier G, Cournac L (2007) Chlororespiration and cyclic electron flow around PSI during photosynthesis and plant stress response. Plant Cell Environ 30:1041–1051

Sakajo S, Minagawa N, Yoshimoto Y (1997) Effects of nucleotides on cyanide-resistant respiratory activity in mitochondria isolated from antimycin a-treated yeast *Hansenula anomala*. Biosci Biotech Biochem 61:397–399

Schäfer G, Purschke W, Schmidt C (1996) On the origin of respiration: electron transport proteins from archaea to man. FEMS Microbiol Rev 18:173–188

Shepherd M, Sanguinetti G, Cook GM, Poole RK (2010) Compensations for diminished terminal oxidase activity in *Escherichia coli*: cytochrome *bd-II-mediated* respiration and glutamate metabolism. J Biol Chem 285:18464–18472

Sieger SM, Kristensen BK, Robson CA, Amirsadeghi S, Eng EWY, Abdel-Mesih A, Møller IM, Vanlerberghe GC (2005) The role of alternative oxidase in modulating carbon use efficiency and growth during macronutrient stress in tobacco cells. J Exp Bot 56:1499–1515

Stenmark P, Nordlund P (2003) A prokaryotic alternative oxidase present in the bacterium *Novosphingobium aromaticivorans*. FEBS Lett 552:189–192

Stepien P, Johnson GN (2009) Contrasting responses of photosynthesis to salt stress in the glycophyte Arabidopsis and the halophyte *Thellungiella*: role of the plastid terminal oxidase as an alternative electron sink. Plant Physiol 149:1154–1165

Streb P, Josse E-M, Gallouët E, Baptist F, Kuntz M, Cornic G (2005) Evidence for alternative electron sinks to photosynthetic carbon assimilation in the high mountain plant species *Ranunculus glacialis*. Plant Cell Environ 28:1123–1135

Theodorou ME, Plaxton WC (1993) Metabolic adaptations of plant respiration to nutritional phosphate deprivation. Plant Physiol 101:339–344

Umbach AL, Siedow JN (1993) Covalent and noncovalent dimers of the cyanide-resistant alternative oxidase protein in higher plant mitochondria and their relationship to enzyme activity. Plant Physiol 103:845–854

Umbach AL, Siedow JN (2000) The cyanide-resistant alternative oxidase from the fungi *Pichia stipis* and *Neurospora crassa* are monomeric and lack regulatory features of the plant enzyme. Arch Biochem Biophys 378:234–245

Umbach AL, Gonzàlez-Meler MA, Sweet CR, Siedow JN (2002) Activation of the plant mitochondrial alternative oxidase: insights from site-directed mutagenesis. Biochim Biophys Acta 1554:118–128

Umbach AL, Ng VS, Siedow JN (2006) Regulation of plant alternative oxidase activity: a tale of two cysteines. Biochim Biophys Acta 1757:135–142

Vanlerberghe GC, Day DA, Wiskich JT, Vanlerberghe AE, McIntosh L (1995) Alternative oxidase activity in tobacco leaf mitochondria: dependence on tricarboxylic acid cycle-mediated redox regulation and pyruvate activation. Plant Physiol 109:353–361

Vanlerberghe GC, McIntosh L, Yip JYH (1998) Molecular localization of a redox-modulated process regulating plant mitochondrial electron transport. Plant Cell 10:1551–1560

Vanlerberghe GC, Cvetkovska M, Wang J (2009) Is the maintenance of homeostatic mitochondrial signaling during stress a physiological role for alternative oxidase. Physiol Plant 137:392–406

Vemuri GN, Eiteman MA, McEwen JE, Olsson L, Nielsen J (2007) Increasing NADH oxidation reduces overflow metabolism in *Saccharomyces cerevisiae*. Proc Natl Acad Sci U S A 104:2402–2407

Wang J, Rajakulendran N, Amirsadeghi S, Vanlerberghe GC (2011) Impact of mitochondrial alternative oxidase on the response of *Nicotiana tabacum* to cold temperature. Physiol Plant 142:339–351

Watling JR, Robinson SA, Seymour RS (2006) Contribution of the alternative pathway to respiration during thermogenesis in flowers of the sacred lotus. Plant Physiol 140:1367–1373

Woyda-Ploszczyca AM, Sluse FE, Jarmuszkiewicz W (2009) Regulation of *Acanthamoeba castellanii* alternative oxidase activity by mutual exclusion of purine nucleotides; ATP's inhibitory effect. Biochim Biophys Acta 1787:264–271

Wu D, Wright DA, Wetzel C, Voytas DF, Rodermel SR (1999) The immutans variegation locus of Arabidopsis defines a mitochondrial alternative oxidase homolog that functions during early chloroplast biogenesis. Plant Cell 11:43–55

Yamamoto H, Peng L, Fukao Y, Shikanai T (2011) An Src homology 3 domain-like fold protein forms a ferredoxin binding site for the chloroplast NADH dehydrogenase-like complex in *Arabidopsis*. Plant Cell 23:1480–1493

Zhang J, Barquera B, Gennis RB (2004) Gene fusions with beta-lactamase show that subunit I of the cytochrome bd quinol oxidase from *E. coli* has nine transmembrane helices with the O_2 reactive site near the periplasmic surface. FEBS Lett 561:58–62

Chapter 10

Probing the Action of Cytochrome c Oxidase

Vangelis Daskalakis and Constantinos Varotsis*
Department of Environmental Science and Technology, Cyprus University of Technology, P.O. Box 50329, 3603 Lemesos, Cyprus

Summary

Density functional theory (DFT) and combined Molecular Mechanics/Quantum Mechanics (MM/QM-MD) calculations have been applied to models of the cytochrome *c* oxidase (CcO) including the Fe–Cu_B binuclear center, where dioxygen is bound and subsequently reduced to water. The properties of several intermediates of the CcO dioxygen reaction have been investigated by theoretical approaches. In this chapter, we investigate the dynamics of the binuclear heme Fe–Cu_B throughout the O_2 catalytic cycle. We are focused on the effects of the protein matrix and proton/water motion exerted on the heme a_3 group. For this, we have built models of CcO, which vary at the heme a_3 environment. This variability is based on hydrogen bonding interactions and amino acid protonation states. Different control points have been identified for the transition from one intermediate to the next. The hydrogen bonding networks in the proximity of heme a_3 area also have consequences for the characteristics of the binuclear center. A theoretical framework for the direct link between an H^+ delivery channel (termed D) and an accumulation of waters, termed '*water pool*' close to the active site, has been achieved at the QM/MM level of theory. Two proton valves (E278 and His403) and an electron/proton coupling site (propionate-A/Asp399) exist in this pathway for the aa_3 CcO from *P. denitrificans*. The ferryl intermediate, produced subsequent to the O–O bond scission, is found to have characteristics highly dependent on the basicity of the proximal His411, in contrast to the hydroxyl intermediate that is sensitive to distal effects.

*Author for correspondence, e-mail: c.varotsis@cut.ac.cy

M.F. Hohmann-Marriott (ed.), The Structural Basis of Biological Energy Generation, Advances in Photosynthesis and Respiration 39, pp. 187–198, DOI 10.1007/978-94-017-8742-0_10,

I. Introduction

Theoretical studies are routinely used to probe complex biological systems in atomistic details. These studies include, among others, Density Functional Theory (DFT), Molecular Mechanics (MM) and Molecular Dynamics (MD), as well as combined Molecular Mechanics/Quantum Mechanics (QM/MM) methodologies. Cytochrome *c* oxidase (CcO), a terminal respiratory enzyme (see also Chap. 9), catalyzes the transfer of electrons from the reduced cytochrome *c* to the molecular oxygen and translocates protons vectorially across the mitochondrial inner membrane. Four redox active metal centers are present in the enzyme: two hemes *a* and three associated copper atoms. In eukaryotes, the electrons coming from the substrate cytochrome *c* enter the homodinuclear copper center, Cu_A. From there, electrons are transferred, via the low-spin heme *a*, to the bi-nuclear center containing a high-spin heme a_3 and a Cu_B complex. The latter two metal sites constitute the active catalytic site, where oxygen is reduced to water by four electrons and four protons (Wikström 1989; Babcock and Wikström 1992; Ferguson-Miller and Babcock 1996; Gennis 1998; Michel 1998a, b). The function and proposed mechanisms of action of CcO enzymes have been reported before (Babcock and Wikström 1992; Ferguson-Miller and Babcock 1996; Michel et al. 1998; Belevich and Verkhovsky 2008). Binding of O_2 to the heme a_3-iron leads to the first oxy intermediate $Fe(II/III)_{a3}$–O_2|Cu_B(I), which subsequently is converted to the peroxy $Fe(III)_{a3}$–O—O–|Cu_B(II) species. Reduction by the third and fourth electrons yields the ferryl $Fe(IV)_{a3}$=O|Cu_B(II)–OH and hydroxyl $Fe(III)_{a3}$–OH|Cu_B(II)–OH intermediates, respectively (Wikström 1989; Babcock and Wikström 1992; Ferguson-Miller and Babcock 1996; Gennis 1998; Michel 1998a, b).

The ferryl intermediate has gained a lot of attention by several research groups (see literature from Babcock, Gennis, Ferguson-Miller, Michel, Varotsis, and Wikström groups), as it presents a key structure produced immediately after the O–O bond cleavage, a step not yet characterized mechanistically in atomistic detail. There was an initial assignment based on the Optical Absorption spectra of the species with a major peak at 580 nm to an oxyferryl (F) heme iron, while the 607 nm species to a ferric heme peroxide (P) structure, and they were named F and P, respectively (Wikström 1981). Later studies attributed an oxo-ferryl character to both species (Varotsis and Babcock 1990; Ogura et al. 1996; Han et al. 2000). The 804/790 cm^{-1} ν(Fe–O) modes in the resonance Raman spectra have been theoretically attributed to the same oxidation level with oxo-ferryl character (P_M) in the CcO dioxygen reaction (Daskalakis et al. 2008). The protonation state of the propionate-A/Asp399 (prop-A/D399) pair seems not to influence significantly the location of the 804/790 cm^{-1} bands but only their intensities. Thus, different protonated/deprotonated states of this pair vary the intensity of one band over the other and their contribution to the experimental spectra. This results in the appearance of one or two prominent positive bands in the difference spectra of isotopically-substituted species such as the ferryl-oxo intermediates in the CcO dioxygen reaction (Daskalakis et al. 2008). Moreover, for the enhancement of the δ(His–Fe=O) ferryl bending mode, we have identified crucial resonances, and linked them to conformational/structural changes during enzymatic turnover, including the Fe–Cu_B distance, which is controlled by

Abbreviations: BLYP – Functional to describe the exchange and correlation part of the electron-electron interaction energy in the theoretical calculations based on Becke's correlation functional; Lee, Yang, Parr exchange terms; CcO – Cytochrome *c* oxidase; DFT – Density functional theory; MD – Molecular dynamics; MM – Molecular mechanics; OPLS – Optimized potentials for liquid simulations force field; QM – Quantum mechanics; riBLYP – Resolution of identity (ri) methodology is an approximation to the computation of two-electron four-center integrals with significantly improved time efficiency; TZVP – Triple-zeta basis sets, TZVP contains a set of d-functions on the heavy atoms and one set of p-functions on hydrogen atoms

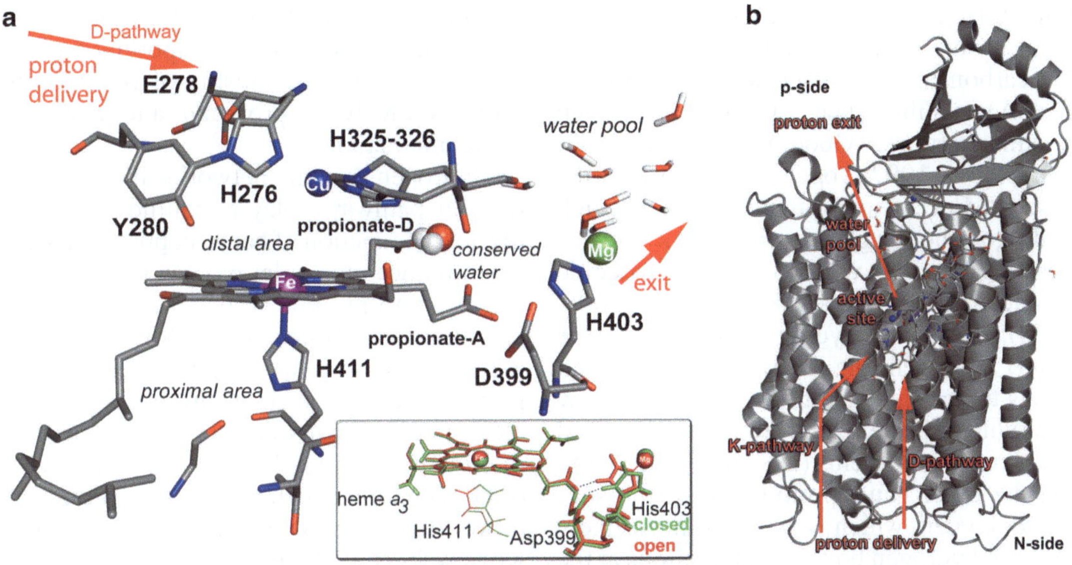

Fig. 10.1. Cytochrome *c* oxidase from *P. denitrificans*. (**a**) Active site with His403 conformations are shown in the inset. Most important surrounding amino acids are also visible. Glu278 is depicted as part of the proton delivery channel termed D-pathway. (**b**) Proton delivery channels (K- and D-pathways), as well as some waters in the water pool can be seen in the structure.

electrostatic interactions in the active site and especially in the area of the heme a_3 propionates (Daskalakis et al. 2011). The position of Cu_B in respect to the heme a_3 plane appears also sensitive to the protonation events in the propionates. In addition, we have proposed that conformational changes in the region of the ring A propionate of heme a_3, due to protonation events, are communicated via the Asp399/His403 pair to Gly386/Thr389 in the proximal area, thereby altering the Hydrogen bonding networks in the region. This, in turn, controls the effect of proximal ligand His411 to the frequency of the ferryl Fe=O bond.

In order to identify the key effects of the protein matrix on the CcO active site, we start by studying several crystal structures of the aa_3–CcO solved to date (Iwata et al. 1995; Tsukihara et al. 1995; Ostermeier et al. 1997; Yoshikawa et al. 1998; Koepke et al. 2009). The fifth axial ligand of heme a_3–Fe is a histidine, termed His proximal. In this proximal area, Gly and Thr residues are also found in the crystal structures of aa_3 from *Paracoccus denitrificans* (Ostermeier et al. 1997) and bovine heart (Koepke et al. 2009), while the ba_3 from *Thermus thermophilus* lacks the Thr residue; instead, a second Gly is located at the proximal site (Soulimane et al. 2000; Hunsicker-Wang et al. 2005). Oxygen binds as the sixth axial ligand, distal to heme a_3 area, between heme a_3 and the Cu_B metal. The latter is coordinated to three His residues. One of the unique properties of the binuclear center, determined by the crystal structures and conserved among different CcOs (Rauhamäki et al. 2006), is the covalent link between a Tyr residue and one of the three His ligands of Cu_B. The cross-linked Tyr forms an hydrogen bonding interaction with the hydroxyethyl-farnesyl side chain of heme a_3. This cross-linked Tyr is also in hydrogen bonding distance to the substrate oxygen in the distal area. Figure 10.1a shows the heme a_3–Fe site of CcO from *P. denitrificans* (Koepke et al. 2009) indicating the *proximal* and *distal* regions, as well as the most important amino acids in the vicinity of the active site.

It has been previously reported that hydrogen bonding networks affect the basicity of the axial to the heme iron ligand to support oxidation states of heme iron greater than III

(Goodin and McRee 1993; Vogel et al. 1999). Cytochrome *c* oxidase, sulfite reductase and the carbon monoxide-sensing heme protein (CooA) exhibit such proposed hydrogen bonding networks (Goodin and McRee 1993; Crane et al. 1997; Vogel et al. 1999). In the case of CcO, a proton movement through hydrogen bonding networks also induces changes in the vibrational spectroscopic characteristics (Daskalakis et al. 2008, 2010). Moreover, hydrogen bonding to the Nδ–H hydrogen (Fig. 10.1a) of the proximal His is the crucial step in understanding how reactivity of the heme-iron is controlled proximally in proteins (Decatur et al. 1999). On the other hand, protonation of the substrate oxygen, which is achieved most likely through hydrogen bonding interactions at the distal site, is crucial for the H_2O production in the CcO/O_2 reaction. Thus, it is of great importance to probe the sites proximal and distal to heme a_3 that are influential during the reduction of O_2 by CcO. Hydrogen bonding networks can be altered by proton and water motion throughout the catalytic cycle of the CcO enzyme. The transfer of a proton (H^+) from E278 (*P. denitrificans* numbering) to an unknown residue above hemes a/a_3 is the first step mediated by CcO that leads to the translocation of protons across the mitochondrial membrane. During the P → F transition, E278 is protonated via the D-pathway. This proton is subsequently transferred to the active site and E278 is re-protonated, while another proton is released to the P-side, as indicated in Fig. 10.1b (Bellevich et al. 2006), as it can be seen in Fig. 10.1b. Several approaches have been used to predict either long distance (QM/MM, MD) or short distance (DFT) effects in models of CcO from proton and electron transfer processes.

II. The Dioxygen Activation Models

Theoretical studies of electron and proton transfer events in the CcO enzyme are a crucial tool for elucidating the interactions between metals, prosthetic groups, ligands and the protein matrix. In addition to insights specifically relevant for the CcO enzymes, these studies also elucidate important aspects into the action mechanisms of other heme proteins involving proximal and distal to heme interactions.

Hydrogen bonding networks, like electron transfer pathways, play a significant role in the double function of heme-copper oxidases: (a) to reduce molecular oxygen to water and (b) to pump protons across the inner mitochondrial membrane. DFT, MD, QM/MM calculations on heme-copper center models provide profound insight into the structural, electronic and dynamic properties of several intermediates throughout the catalytic cycle of O_2 reduction in cell respiration (Blomberg et al. 2000a, b, 2003; Bassan et al. 2006; Kozlowski et al. 2006).

In this chapter we elucidate the heme proximal and distal dynamics and quantify the effects of hydrogen bonds in the regions around the active site for several intermediates in the dioxygen activation catalytic cycle. Experimental quantification would be difficult, as only overall differences in the hydrogen bonding networks could be measured. Furthermore, atomistic details of such reaction dynamics are accurately probed by theoretical calculations.

The crystal structure of aa_3 cytochrome *c* oxidase from *P. denitrificans* (PDB 1AR1) provides the initial x, y, z Cartesian coordinates for the atoms of each model studied. The size of the model depends on the theoretical methodology used. In DFT calculations we restrict our systems to the active site including the first coordination sphere for each metal (Fe_{a3}–Cu_B) and some of the hydrogen bonding networks in the proximal region, while QM/MM-MD calculations encompass the whole protein in a water solvent.

DFT calculations were performed to probe the proximal and distal effects on the CcO heme a_3 active site. For the DFT, ethylic acids can be used in the proximal region, near heme-iron axial ligand, to reproduce the hydrogen bonding network found in the area, thereby substituting the Gly and/or Thr residues. Carbonyl groups of the peptidic bonds in the enzyme, found near the proximal His, are in this way represented by carboxyl groups donating the electronegative atom in the hydrogen bonding network. Ethylic acids are found to be quite convenient small molecules for the study that do not introduce unwanted hydrogen bonds or other interactions in the proximal region. For each structure considered, a full geometry optimization was performed on riBLYP/TZVP level of theory using the *Turbomole V.5-8-0* software package (Turbomole 2005). For metals (iron and copper) an effective core potential (ECP) is applied (termed "ecp-10-mdf" for iron and "ecp-18 arep" for copper atom), as implemented in *Turbomole software package*. Moreover, we have implemented the computational advantage of the RI-J (*Resolution of the Identity*) approximation (Billingsley and Bloor 1971; Eichkorn et al. 1995), as defined in *Turbomole* with the default parameterization. Geometry optimized DFT models of the active site of CcO with relevant vibrational frequencies are presented in Fig. 10.2.

A mixed quantum mechanics/molecular mechanics (QM/MM) approach was used to probe protonation/deprotonation and water motion events near the active site of CcO. QM/MM calculations can seamlessly join together QM and MM representations of a complex system. The fundamental idea is to divide a large, condensed phase system into two regions; the reactive chemical event is contained within the QM region (which includes the active site as in the DFT calculations), while the surrounding condensed phase is modeled via molecular mechanics. The DFT/MM calculations were performed using the QSite module of Schrodinger 2010 Suite of Programs (Schrodinger). B3LYP functional and lacvp* basis set was used for the QM part, while the OPLS2005 force field (FF) was used for the MM part throughout this work.

We prepared initial structures for the QM/MM calculations, by varying the protonation state of propionate-A and Asp399 carboxylic groups. We have built complete models of the solvated CcO protein by adding missing hydrogen atoms and a water buffer of more than 14 Å (T3P FF) between the protein matrix and the boundaries of the simulation box. Crystallographic waters, inside the protein structure, were retained before and after the solvation. An initial minimization/relaxation process was performed by the Desmond Software of Schrodinger Suite 2010, before each QM/MM calculation. The relaxation was based on the default protocol proposed in the Desmond manual. This included stages of solute-restrained minimization, no restrain minimization, NVT simulation of 24 ps, 10 K, NPT simulation of 10 K, 1 atm, NPT simulation of 24 ps, 300 K and 1 atm with non-hydrogen solute atoms restrained and a final NPT 24 ps simulation at 300 K and 1 atm (Schrodinger). For the QM/MM geometries after the minimization/relaxation, we reduced the water buffer significantly by excluding every solvent molecule beyond a distance of 8 Å away from the protein matrix and restrained to their coordinates those waters beyond a 20 Å layer around the heme iron according to their minimized positions. A cutoff distance of 100 Å was also set for the Coulombic interactions. The hydrogen-cap approach was used for the covalent QM-MM boundary region, where hydrogen atoms are used to "cap" the QM-cut/unsaturated bonds. Gaussian charges were implemented to provide the electrostatic contribution from the MM part to the QM Hamiltonian. Figure 10.3 depicts a typical QM/MM system under study.

III. Proton and Water Motion Drives the CcO Dioxygen Reaction

A. The Oxy Intermediate

In the oxy intermediate, DFT calculations show that the O–O bond strength is not at all affected by the proximal hydrogen bonding network (Fig. 10.2, models 5–6/5w). The O–O weakening in this phase of the catalytic cycle is not favored by either the proximal, nor by the distal interactions. So how does the enzyme proceed to the next intermediate?

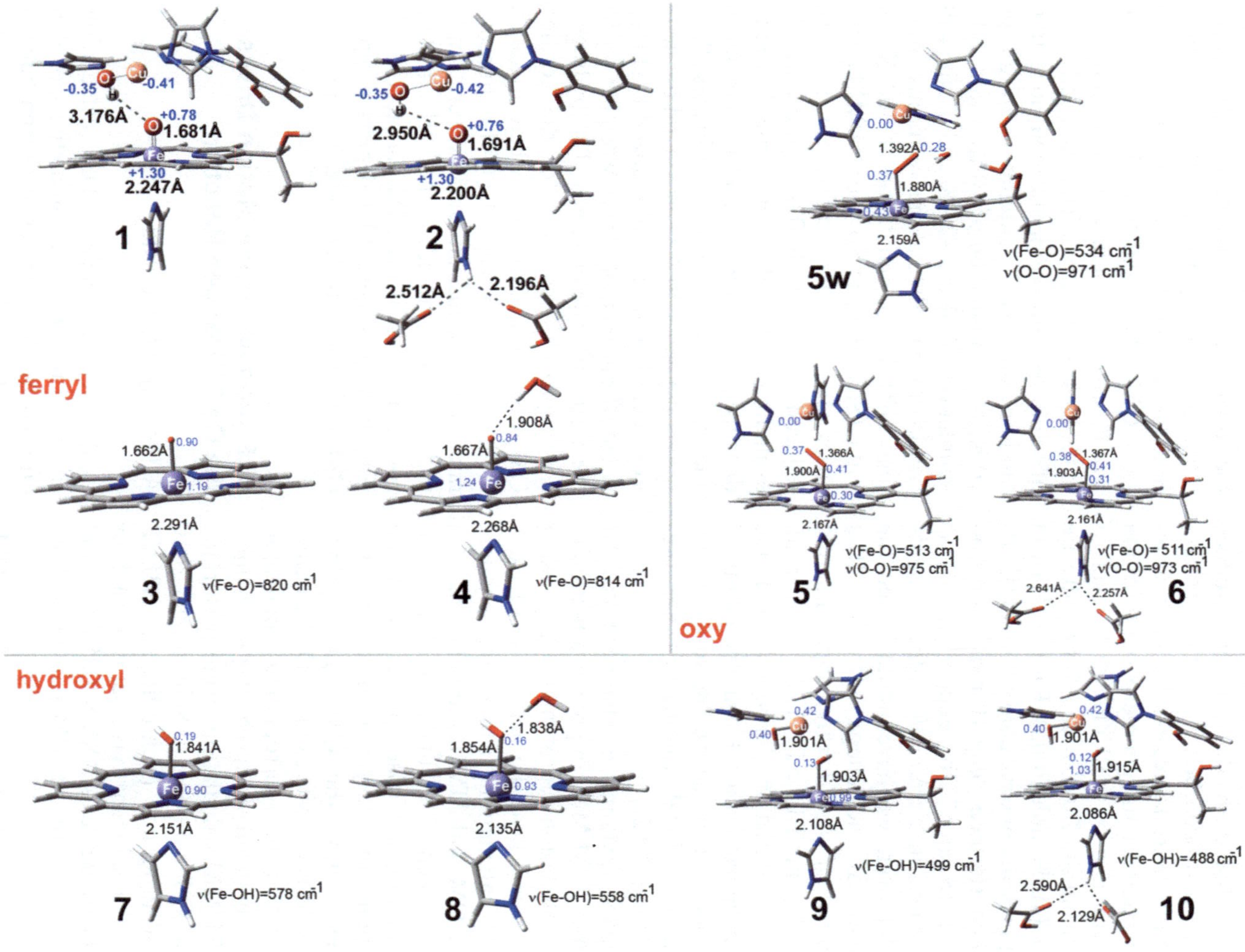

Fig. 10.2. Geometry optimized structures of DFT models of the active site of CcO in different oxidation levels. Selected interatomic distances (imidazole–Fe, Fe–O and O–O) are also shown, as well as selected spin populations in *blue* (Fe, Cu, and O).

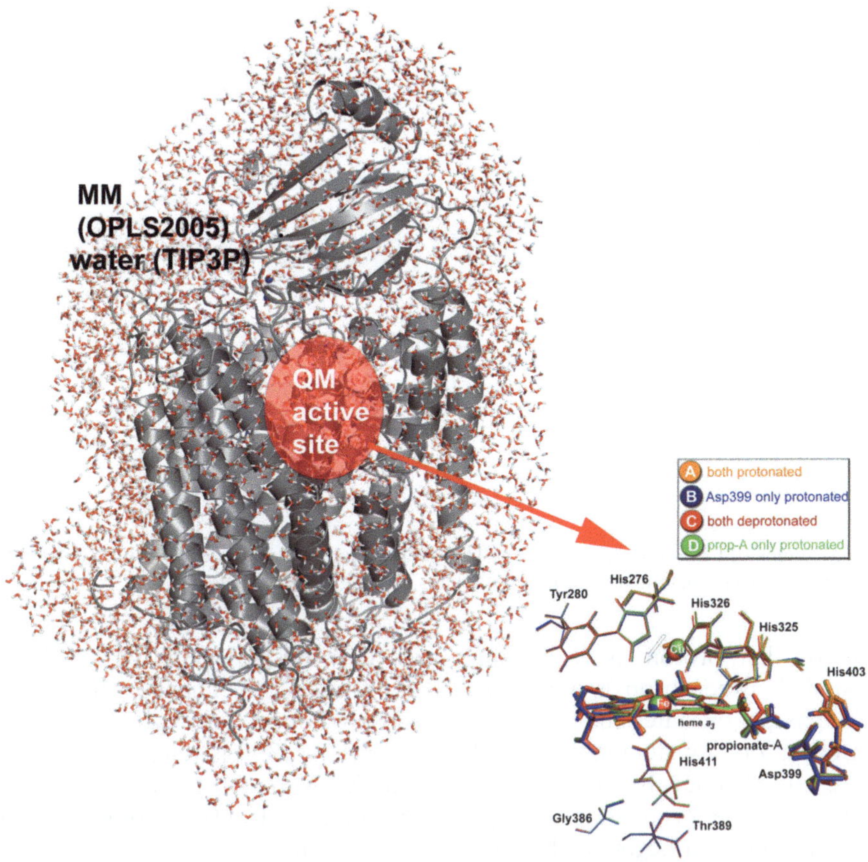

Fig. 10.3. QM/MM model used for the highly elaborate QM/MM calculations, including the whole protein and the water solvent. The active site is represented in QM (DFT) level of theory, while the rest of the protein, as well as the solvent are represented by the OPLS2005/TIP3P Force Fields, respectively. The active site, that is probed at the B3LYP/tzvp* (DFT) level of theory at different protonation states of the Asp399/propionate-A pair, is enlarged on the *bottom right* side of the figure. Cu is optimized at a closer distance to the heme a_3 iron, in the case where the negative charges increase on the Asp399/propionate-A pair.

Waters are possible sites for the accommodation of a proton in the form of transient hydronium (H_3O^+) species, acting as proton carriers. Interestingly, a single water molecule is highly conserved in the crystal structures of bacterial and mammalian CcOs, between the heme a_3 propionates. If such a hydronium molecule interacts with the protein matrix in the area of the heme a_3 propionates, its proton is transferred to the heme a_3 ring A propionate in our QM/MM simulations. This implicates a propionate-D/Arg → prop-A/D399 proton transfer via the conserved water molecule in the cases where a single proton is shared between prop-A and Asp399 (Daskalakis et al. 2011). Figure 10.4 illustrates a schematic mechanism of action involving the early stages of dioxygen activation by CcO. In this model the His403 residue plays the role of a valve, controlling the flow of protons in the region. By substantially changing the pKa values of the prop-A or Asp399, a proton is trapped or released by prop-A. In the closed conformation (strong hydrogen bonding between prop-A and Asp399), the proton is trapped on the prop-A site (Daskalakis et al. 2011). Starting from a structure where a single proton is shared between the latter prop-A/D399 pair (in the oxy intermediate, and the closed His403 conformation), the proton resides 'locked' on the prop-A side. Active site oxidation will trigger

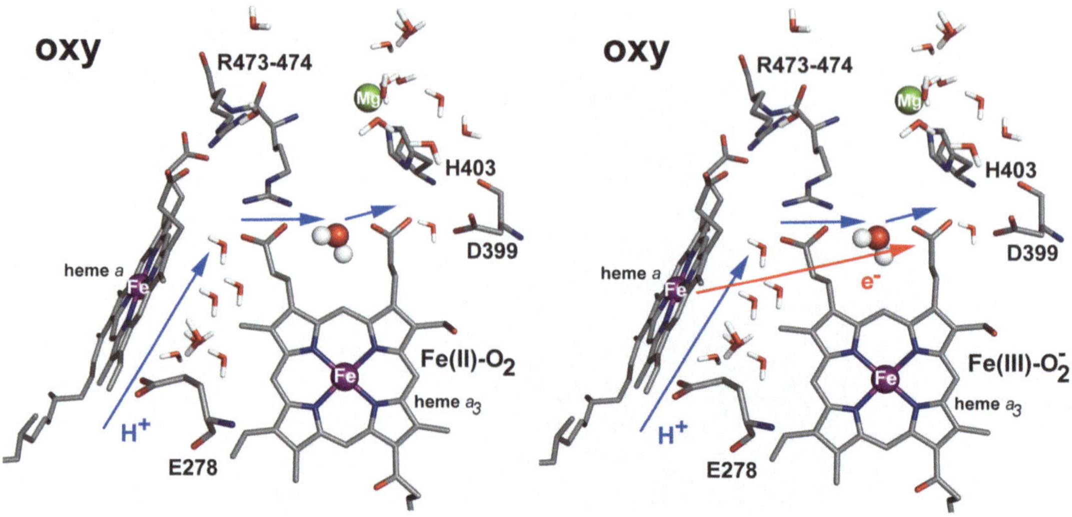

Fig. 10.4. Early stages of CcO dioxygen reaction. The intermediate's oxy are depicted with *arrows* to point the proton (*blue*) and electron (*red*) movement in critical sites around the active site.

the release of that proton as it becomes loosely bound (between prop-A/Asp399) in the rest of the oxidation states and the His403 closed conformation. A water molecule in the area could easily abstract that labile proton. Deprotonation of the prop-A/Asp399 pair increases the presence of negative charges in the area, reallocating Cu_B closer to heme a_3 iron, accommodating also a half spin in prop-A. The heme a_3 Fe(II) is oxidized to Fe(III) upon protonation of the prop-A/Asp399 pair via the D-pathway (Daskalakis et al. 2011). Electron transfer should occur from heme *a*, concurrently to the previous proton transfer, so that the half spin population on prop-A is quenched. This should eventually trigger the further oxidation of the active site to the peroxy and ferryl intermediates. Protonation should change the His403 conformation as discussed previously, thereby trapping the proton on the Asp399 site (His403 open conformation). Another release of a proton to the heme a_3 propionates area via the D-pathway is able to neutralize the prop-A/Asp399 charges and as a consequence His403 residue changes conformation with the protons to become loosely attached to the pair (His403 closed conformation). A water molecule that is leaving the active site can be a possible proton carrier, as well as a trigger for the neutralization of the prop-A/Asp399 pair and the His403 conformational change. In each case, a loosely bound carboxylic proton residing on prop-A/Asp399 is released to the water pool via the Mg that is coordinated to His403 (Schmidt et al. 2003). From the water pool the proton reaches to the solvent immediately, or after His403 changes conformation from open to closed.

B. The Ferryl and Hydroxyl Intermediates

A distal hydrogen bond is the cause of a 20 cm^{-1} shift in (Fe–O) for the hydroxyl intermediate (Fig. 10.2, models 7–8). This distal effect appears significantly enhanced, having twice the amplitude of the proximal effect (Fig. 10.2, models 9–10), partly justifying the proposal of an hydrogen bond in distal region being the cause of the two experimentally observed ν(Fe–O) modes at 450 and 475 cm^{-1} (Han et al. 1989, 2000) for the hydroxyl intermediate.

It is worth noting that the same distal interaction on the ferryl species was calculated to be only 6 cm^{-1} (riBLYP/TZVP level of theory) (Fig. 10.2, models 3–4). The proximal effect on bimetallic (Fe/Cu_B) ferryl complexes accounts for a shift of up to 20 cm^{-1} at the same level of theory (riBLYP/TZVP) and under the same hydrogen bonding network

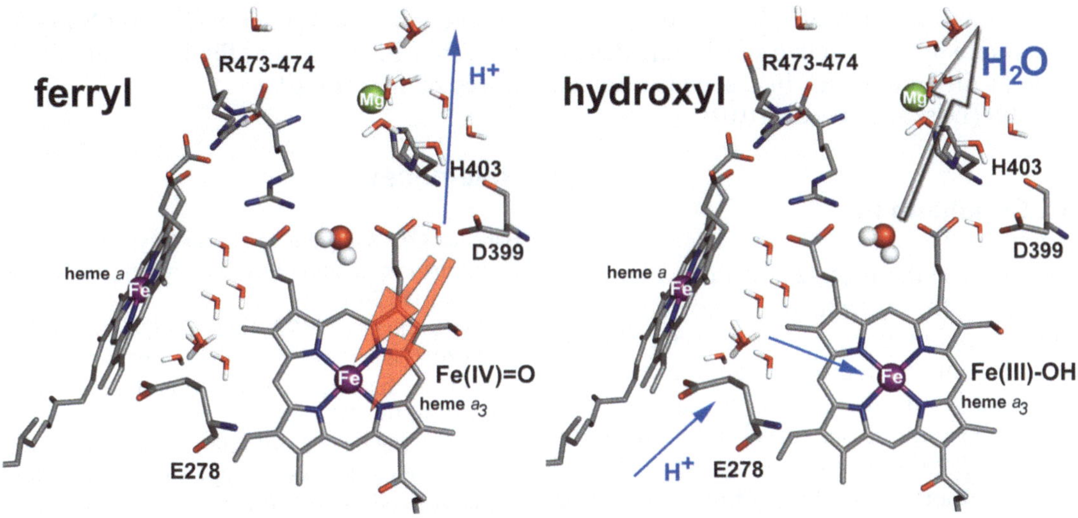

Fig. 10.5. Late stages of CcO dioxygen reaction. The intermediates ferryl and hydroxyl are depicted with *arrows* indicating the proton (*blue*) movement in critical sites around the active site. *Red arrows* are designed to show the effect of the propionate-A site on ferryl heme a_3 Fe=O bond and proximal area characteristics. The *arrow* outlined in *black* depicts the product water movement out of the active site towards the *water pool.*

models (Fig. 10.2, models 1–2) (Pinakoulaki et al. 2013). This pronounced distal effect of the hydroxyl species, is in fact consistent with the advancing of the catalytic cycle. The protonation at the distal site is the key step for the CcO cycle to proceed further, leading to the formation of water molecules at the active site. The presence of H^+ in the distal area should thus immediately affect the hydroxyl Fe(III)–OH moiety and easily induce the transition towards the next step of the catalytic cycle. On the other hand, the presence of H^+ in the distal site at the ferryl intermediate stage does not influence the chemistry of that intermediate. It is proposed that the protonation(s) near the active site (at the heme a_3 propionates area, Fig. 10.3) are communicated to the proximal His (Daskalakis et al. 2008).

Figure 10.5 depicts a schematic view of the last stages of dioxygen activation cycle in CcO. Different protonation states of the prop-A/Asp399 pair control the intensities of the ν(Fe–O) bands (Daskalakis et al. 2008). His403 and Asp399 are connected to Gly386/Thr389 by a short chain of amino acids. Thus, protonation/deprotonation events in the prop-A/Asp399 pair, as well as the His403 conformational changes are communicated via these connecting amino acids to the proximal site of His411. Changes in the hydrogen bonding strengths in the proximal region (Fig. 10.2, models 1–2) are able to control the His411 basicity and thus the ν(Fe–O) modes via resonances (Daskalakis et al. 2010). Glu278 resides at the entrance of the D-pathway. It has been proposed that water molecules are present in the hydrophobic cavity near Glu278, connecting it to the heme a_3 propionate-D and Cu_B (Puustinen et al. 1997; Riistama et al. 1997; Zheng et al. 2003; Olkhova et al. 2004).

In this model, Glu278 has the role of a proton shuttle and/or a valve, mediating the flow of protons towards Cu_B/Fe_{a3} and the propionate-D/Arg salt bridge (Olkhova et al. 2004; Xu and Voth 2005). The propionate-D/Arg salt bridge has been proposed to be in a redox-dependent thermal equilibrium, acting as a gate for a proton or a protonated water molecule (Wikström et al. 2005). As discussed in a previous study (Daskalakis et al. 2011) the Cu_B position is extremely sensitive to the protonation states of the heme a_3 propionates. This thermal equilibrium and proton transient gate should alter the Fe_{a3}–Cu_B distance, which in turn may trigger vibrational resonances in the His–Fe^{IV}=O ferryl moiety of the active site that are detectable as different bands in resonance Raman spectra (Daskalakis et al. 2010).

IV. Conclusions

We propose the presence of three important aspects in the O_2 reduction cycle in the heme a_3 Fe/Cu_B binuclear center of CcO: (1) The effect from the hydrogen bonding networks proximally to heme a_3 is significantly enhanced only at the ferryl oxidation state. (2) The distal to heme a_3 effect is calculated to be important at the hydroxyl oxidation level. (3) The presence of charges in the vicinity of the active site is crucial to the action mechanism of CcO dioxygen reaction, as it controls the Cu_B position, the His403 conformation (aa_3 from *P. denitrificans* numbering) and consequently the release of protons.

In this chapter we link water and proton motion to spectroscopic characteristics of the active site of CcO. A concerted proton (water mediated) and electron motion has been identified in the case of the oxy intermediate. In hydroxyl and oxy species, the proximal region or the axial His basicity induces extremely weakened changes for the Fe–O/O–O bonds and the respective spectroscopic characteristics. We propose that water and proton motion affect the intensities of the ν(Fe–O)/δ(His–Fe=O) vibrational frequencies and alters the His403 conformation to enable the release of waters/protons out of the heme a_3/Cu_B active site.

Acknowledgments

All calculations have been performed either at the Institute of Electronic Structure and Laser of the Foundation for Research and Technology – Hellas (IESL-FORTH), the Barcelona Supercomputing Center (BSC) at the Life Sciences Department, at the Supercomputing Center in Okazaki National Research Institutes and the European GRID infrastructure (EGEE).

References

Babcock GT, Wikström M (1992) Oxygen activation and the conservation of energy in cell respiration. Nature 356:301–309

Bassan A, Blomberg MRA, Borowski T, Siegbahn PEM (2006) Theoretical studies of enzyme mechanisms involving high-valent iron intermediates. J Inorg Biochem 100:727–743

Belevich I, Verkhovsky MI (2008) Molecular mechanism of proton translocation by cytochrome *c* oxidase. Antiox Redox Signal 10:1–30

Bellevich I, Verkhovsky MI, Wikström M (2006) Proton-coupled electron transfer drives the proton pump of cytochrome c oxidase. Nature 440:829–832

Billingsley FP, Bloor JE (1971) Limited Expansion of Diatomic Overlap (LEDO): a near-accurate approximate *ab initio* LCAO MO method. I. Theory and preliminary investigations. J Chem Phys 55:5178–5190

Blomberg MRA, Siegbahn PEM, Babcock GT, Wikström M (2000a) O-O bond splitting mechanism in cytochrome *c* oxidase. J Inorg Biochem 80:261–269

Blomberg MRA, Siegbahn PEM, Babcock GT, Wikström M (2000b) Modeling cytochrome oxidase – a quantum chemical study of the O-O bond cleavage mechanism. J Am Chem Soc 122:12848–12858

Blomberg MRA, Siegbahn PEM, Babcock GT (2003) Metal-bridging mechanism for O-O bond cleavage in cytochrome C oxidase. Inorg Chem 42:5231–5243

Crane BR, Siegel LM, Getzoff ED (1997) Structures of the siroheme- and Fe4S4-containing active center of sulfite reductase in different states of oxidation: heme activation via reduction-gated exogenous ligand exchange. Biochemistry 36:12101–12119

Daskalakis V, Farantos SC, Varotsis C (2008) Assigning vibrational spectra of ferryl-oxo ontermediates of cytochrome c oxidase by periodic orbits and molecular dynamics. J Am Chem Soc 130:12385–12393

Daskalakis V, Farantos SC, Guallar V, Varotsis C (2010) Vibrational resonances and Cu_B displacement controlled by proton motion in cytochrome c oxidase. J Phys Chem B 114:1136–1143

Daskalakis V, Farantos SC, Guallar V, Varotsis C (2011) Regulation of electron and proton transfer by the protein matrix of cytochrome c oxidase. J Phys Chem B 115:3648–3655

Decatur SM, Belcher KL, Rickert PK, Franzen S, Boxer SG (1999) Hydrogen bonding modulates binding of exogenous ligands in a myoglobin proximal cavity mutant. Biochemistry 38:11086–11092

Eichkorn K, Treutler O, Öhm H, Häser M, Ahlrichs R (1995) Auxiliary basis-sets to approximate coulomb potentials. Chem Phys Lett 240:283–289

Ferguson-Miller S, Babcock GT (1996) Heme/copper terminal oxidases. Chem Rev 96:2889–2908

Gennis RB (1998) Multiple proton-conducting pathways in cytochrome oxidase and a proposed role for the active-site tyrosine. Biochim Biophys Acta 1365:241–248

Goodin DB, McRee DE (1993) The Asp-His-iron triad of cytochrome c peroxidase controls the reduction potential electronic structure, and coupling of the tryptophan free radical to the heme. Biochemistry 32:3313–3324

Han S, Ching YC, Rousseau DL (1989) Evidence for a hydroxide intermediate in cytochrome c oxidase. J Biol Chem 264:6604–6607

Han S, Takahashi S, Rousseau DL (2000) Time-dependence of the catalytic intermediates in cytochrome c oxidase. J Biol Chem 275:1910–1919

Hunsicker-Wang LM, Pacoma RL, Chen Y, Fee JA, Stout CD (2005) A novel cryoprotection scheme for enhancing the diffraction of crystals of recombinant cytochrome ba3 oxidase from *Thermus thermophiles*. Acta Crystallogr Sect D 61:340–343

Iwata S, Ostermeier C, Ludwig B, Michel H (1995) Structure at 2.8 Å resolution of cytochrome c oxidase from *Paracoccus denitrificans*. Nature 376: 660–669

Koepke J, Olkhova E, Angerer H, Müller H, Peng G, Michel H (2009) High resolution crystal structure of *Paracoccus denitrificans* cytochrome c oxidase: new insights into the active site and the proton transfer pathways. Biochim Biophys Acta 1787:635–645

Kozlowski PM, Kuta J, Ohta T, Kitagawa T (2006) Resonance Raman enhancement of Fe–IV=O stretch in high-valent iron porphyrins: an insight from TD-DFT calculations. J Inorg Biochem 100: 744–750

Michel H (1998a) The mechanism of proton pumping by cytochrome c oxidase. Proc Natl Acad Sci USA 95:12819–12824

Michel H (1998b) Cytochrome c oxidase: catalytic cycle and mechanisms of proton pumping – a discussion. Biochemistry 38:15129–15140

Michel H, Behr J, Harrenga A, Kannt A (1998) Cytochrome c oxidase: structure and spectroscopy. Ann Rev Biophys Biom Struct 27:329–356

Ogura T, Hirota S, Proshlyakov DA, Shinzawa-Itoh K, Yoshikawa S, Kitagawa T (1996) Time-resolved resonance Raman rvidence for tight coupling between electron transfer and proton pumping of cytochrome c oxidase upon the change from the Fe^{V} oxidation level to the Fe^{IV} oxidation level. J Am Chem Soc 118:5443–5449

Olkhova E, Hutter C, Lill MA, Helms V, Michel H (2004) Dynamic water networks in cytochrome *c* oxidase from *Paracoccus denitrificans* investigated by molecular dynamics simulations. Biophys J 86:1873–1889

Ostermeier C, Harrenga A, Ermler U, Michel H (1997) Structure at 2.7 Å resolution of the *Paracoccus denitrificans* two-subunit cytochrome c oxidase complexed with an antibody FV fragment. Proc Natl Acad Sci USA 94:10547–10553

Pinakoulaki E, Daskalakis V, Ohta T, Richter O-MH, Budiman K, Kitagawa T, Ludwig B, Varotsis C (2013) The structure of two ferryl-oxo intermediates at the same oxidation level in the heme-copper binuclear center of cytochrome c oxidase: the protein effect. J Biol Chem 288:20261–20266

Puustinen A, Bailey JA, Dyer RB, Mecklenburg SL, Wikström M, Woodruff WH (1997) Fourier transform infrared evidence for connectivity between Cu_B and glutamic acid 286 in cytochrome bo_3 from *Escherichia coli*. Biochemistry 36: 13195–13200

Rauhamäki V, Baumann M, Soliymani R, Puustinen A, Wikström M (2006) Identification of a histidine-tyrosine cross-link in the active site of the *cbb*3-type cytochrome *c* oxidase from *Rhodobacter sphaeroides*. Proc Natl Acad Sci USA 103: 16135–16140

Riistama S, Hummer G, Puustinen A, Dyer RB, Woodruff WH, Wikström M (1997) Bound water in the proton translocation mechanism of the haem-copper oxidases. FEBS Lett 414:275–280

Schmidt B, McCracken J, Ferguson-Miller S (2003) A discrete water exit pathway in the membrane protein cytochrome *c* oxidase. Proc Natl Acad Sci USA 100:15539–15542

Schrodinger Suite of Programs (2010) http://www.schrodinger.com

Soulimane T, Buse G, Bourenkov GP, Bartunik HD, Huber R, Than ME (2000) Structure and mechanism of the aberrant ba3-cytochrome c oxidase from *Thermus thermophilus*. EMBO J 19:1766–1776

Tsukihara T, Aoyama H, Yamashita E, Tomizaki T, Yamaguchi H, Shinzawa-Itoh K, Nakashima R, Yaono R, Yoshikawa S (1995) Structure of metal sites of oxidized bovine heart cytochrome c oxidase at 2.8 Å resolution. Science 269:1069–1074

Turbomole V5–8–0 24 Nov. 2005, University of Karlsruhe

Varotsis C, Babcock GT (1990) Appearance of the $\nu(Fe^{IV}=O)$ vibration from a ferryl-oxo intermediate in the cytochrome oxidase/dioxygen reaction. Biochemistry 29:7357–7362

Vogel KM, Spiro TG, Shelver D, Thorsteinsson MV, Roberts GP (1999) Resonance Raman evidence for a novel charge relay activation mechanism of the CO-dependent heme protein transcription factor CooA. Biochemistry 38:2679–2687

Wikström M (1981) Energy-dependent reversal of the cytochrome oxidase reaction. Proc Natl Acad Sci USA 78:4051–4054

Wikström M (1989) Identification of the electron transfers in cytochrome oxidase that are coupled to proton-pumping. Nature 338:776–778

Wikström M, Ribacka C, Molin M, Laakkonen L, Verkhovsky M, Puustinen A (2005) Gating of proton and water transfer in the respiratory enzyme cytochrome *c* oxidase. Proc Natl Acad Sci USA 102:10478–10481

Xu J, Voth GA (2005) Computer simulation of explicit proton translocation in cytochrome c oxidase. Proc Natl Acad Sci USA 102:6795–6800

Yoshikawa S, Shinzawa-Itoh K, Tsukihara T (1998) Crystal structure of bovine heart cytochrome c oxidase at 2.8 A resolution. J Bioenerg Biomembr 30:7–14

Zheng X, Medvedev DM, Swanson J, Stuchebrukhov AA (2003) Computer simulation of water in cytochrome c oxidase. Biochim Biophys Acta 1557:99–107

Chapter 11

Evolution of Structural and Coordination Features Within the Methionine Sulfoxide Reductase B Family

Elena Shumilina, Olena Dobrovolska, and Alexander Dikiy*
Department of Biotechnology, Norwegian University of Science and Technology, N-7491 Trondheim, Norway

Summary

In this review, we summarize the evolution, sequence, structural and coordination peculiarities of proteins belonging to the Methionine Sulfoxide Reductase B family (MsrBs). These proteins represent important redox proteins. MsrBs are found in all kingdoms of life. Whereas prokaryotes have only one type of MsrB, mammals possess three, MsrB1, MsrB2 and MsrB3, distributed in different cellular compartments, and regulated by alternative splicing and specific targeting signals. Structural analysis of mammalian and bacterial MsrBs revealed a well-conserved β-core, and dramatic variability in C-and N-terminus. Mostly, MsrBs contain structural zinc ions coordinated by four cysteines. However, some of MsrBs lack coordinating cysteines and, therefore may not contain zinc ion.

*Author for correspondence, e-mail: alex.dikiy@biotech.ntnu.no

M.F. Hohmann-Marriott (ed.), The Structural Basis of Biological Energy Generation, Advances in Photosynthesis and Respiration 39, pp. 199–215, DOI 10.1007/978-94-017-8742-0_11,

I. Introduction

All amino acids found in proteins are susceptible to oxidation. However, only two of them, the sulfur-containing cysteine and methionine, can be reversibly oxidized. Along with tryptophan these residues are the most susceptible to oxidation (Vogt 1995).

Living organisms developed a defence system against unwanted protein oxidation which includes the reduction of reversibly oxidized amino acids (Cys and Met) as well as the removal of the irreversible oxidation products from the cell. Methionine sulfoxide reductases (Msrs), being a part of this system, can reduce both free and protein-bound methionine sulfoxide back to methionine (Fig. 11.1) (Brot et al. 1981; Brot and Weissbach 1983). Msr genes are found among the most conserved in almost all kingdoms of life (Kryukov et al. 2002; Ezraty et al. 2005).

II. Methionine Residue in Proteins: Its Oxidation and Reduction

The process of methionine (Met) oxidation consists of two steps (Fig. 11.1). During the first step, Met is reversibly oxidized to methionine sulfoxide (MetO). Methionine sulfone is the product of a further, irreversible oxidation of MetO during the second step. Methionine oxidation can be caused either by physiological agents, like superoxide, hydrogen peroxide, hydroxyl radicals (termed as reactive oxygen species – ROS), hypochlorous acid, chloramines and some of the metal ions (as copper and iron). Physiological oxidants are also capable to carry out Met oxidation. These include reactants able to oxidize all amino acids (Brot and Weissbach 1983; Shechter 1986), as well as hydrogen peroxide, chloramine T, N-chloro-succinimide and dimethyl sulfoxide, which oxidize Met rather selectively (Vogt 1995; Levine et al. 1996). It was found that the ability of Met to be oxidized strongly depends on its solvent exposure (Levine et al. 1999). Indeed, surface-exposed residues are more readily oxidized, than the ones hidden inside a protein's interior.

The reversibility of Met oxidation can be seen from different points of view. First, surface-located Met residues, being exposed to ROS attack, reduce the amount of free radicals within a cell, thus, acting as endogenous antioxidants (Levine et al. 1996, 1999). Second, oxidation of some Met residues may lead to loss of enzyme activity, thus Msrs reduction of such residues allows proteins to recover their functions (Brot and Weissbach 2000). Finally, the process of periodic oxidation/reduction of Met in proteins plays a regulatory role, inhibiting or inducing some physiological events. For example, a functional role of the methionine oxidation in regulation of the Shaker voltage dependent K^+ channels was examined (Ciorba et al. 1997, 1999). The results show that Met oxidation and reduction facilitated by MsrA regulate the channel inactivation time course. Another example for Msrs-mediated regulation is the calcium-binding protein – calmodulin. Oxidation of one of the calmodulin Met residues results in a 30-fold decrease in the ability of the protein to bind calcium (Yao et al. 1996). Additional studies have shown that this decrease can be reversed by the addition of MsrA (Sun et al. 1999). It was suggested that the reversible oxidation of specific Met residues in calmodulin may regulate its activity and calcium cell homeostasis (Ciorba et al. 1997; Brot and Weissbach 2000; Hansel et al. 2005).

III. Different Classes of Methionine Sulfoxide Reductases

There are two main classes of Msrs: MsrA, which can reduce protein-bound or free S-epimer of methionine sulfoxide, (Caldwell et al. 1978; Moskovitz et al. 1996, 2000; Kuschel et al. 1999; Sharov et al. 1999; Lee

Abbreviations: Met – Methionine; Msr – Methionine sulfoxide reductase; MsrB – Methionine sulfoxide reductase B; ROS – Reactive oxygen species

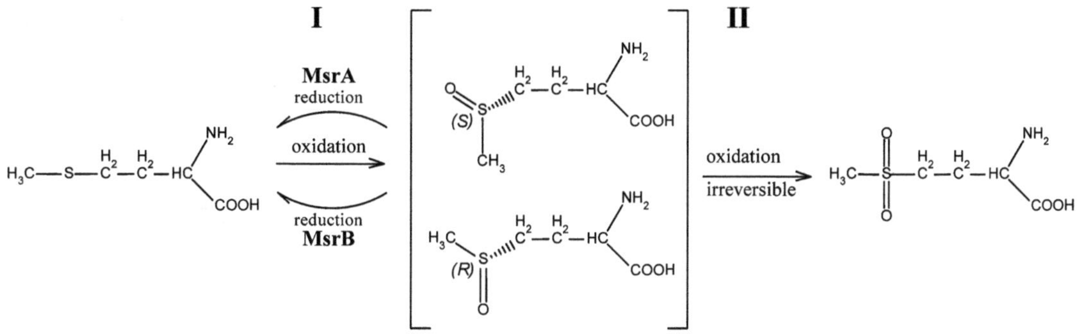

Methionine (Met) Methionine Sulfoxide (MetO) Methionine Sulfone

Fig. 11.1. Mechanism of methionine oxidation. Step I: reversible oxidation of methione results in two enantiomers: S- and R-methionine sulfoxide. Step II: irreversible oxidation of methionine sulfoxide to methionine sulfone.

et al. 2005; Boschi-Muller et al. 2008) and MsrB, which reduce both protein-bound R-epimer of MetO, and, with less efficiency, free R-MetO (Fig. 11.1) (Huang et al. 1999; Bar-Noy and Moskovitz 2002; Hansel et al. 2003; Kim and Gladyshev 2004a, b). Since each Msr class possesses its own type of three-dimensional fold, and significantly distinct, well conserved amino acid sequence, it is possible to suggest the convergent evolution to account for MetO reduction in living organisms (Kauffmann et al. 2002; Lowther et al. 2002; Ezraty et al. 2005; Zhang and Weissbach 2008). MsrA protein sequences, are characterized by the invariable signature motif "GCFWG/C" (Vogt 1995; Lowther et al. 2000; Kumar et al. 2002) and MsrB proteins possess a "RXCXN" motif (where X indicates any amino acid), or in case of MsrB1 – "RXUXF" (Fig. 11.2) (Brot et al. 1981; Kryukov et al. 2002; Lowther et al. 2002). Structures of some Archaea, bacterial and eukaryotic MsrB proteins have previously been determined using either X-ray or NMR techniques (Table 11.1, Fig. 11.3).

An overall structural comparison of the MsrB family was reported earlier (Aachmann et al. 2010, 2011). It was emphasized that the central cores of all proteins in the family are well conserved. However, while bacterial MsrB and mammalian MsrB2 have both α-helix and β-sheets as structural elements, MsrB1 represents a more flexible structure containing only β-sheets (Fig. 11.3). The detailed structural comparison of MsrB proteins is reported in the Sect. VI of this chapter.

Both MsrA and MsrB were found in most genomes (Kryukov et al. 2002); however, their distribution varies greatly between different organisms and kingdoms. Without exception, all studied eukaryotes and cyanobacteria contain msrA and msrB genes (Zhang and Weissbach 2008). In prokaryotes, MsrA and MsrB genes can form two separate transcription units (Ezraty et al. 2005) that are not adjacent on the chromosome. Alternatively, genes can either be transcriptionally (Singh and Moskovitz 2003) or translationally fused and form in the last case the two-domain protein – MsrAB (Olry et al. 2002). In prokaryotes, the Msrs distribution is rather variable: in some thermophile or anaerobic Archaea no Msrs are present. Some organisms possess both classes Msrs, while others only MsrA and no MsrB. However, there are no MsrB-containing organisms, which do not also contain MsrA (Zhang and Weissbach 2008). A possible explanation for found MsrA/MsrB distribution in diverse organisms might be the existence of an enzymatic stereospecific preference for S-epimers during Met oxidation in cells, or, alternatively, a more damaging effect of S–MetO for the cell. Also it is probable, that MsrA and/or MsrB might have other unknown biochemical functions along

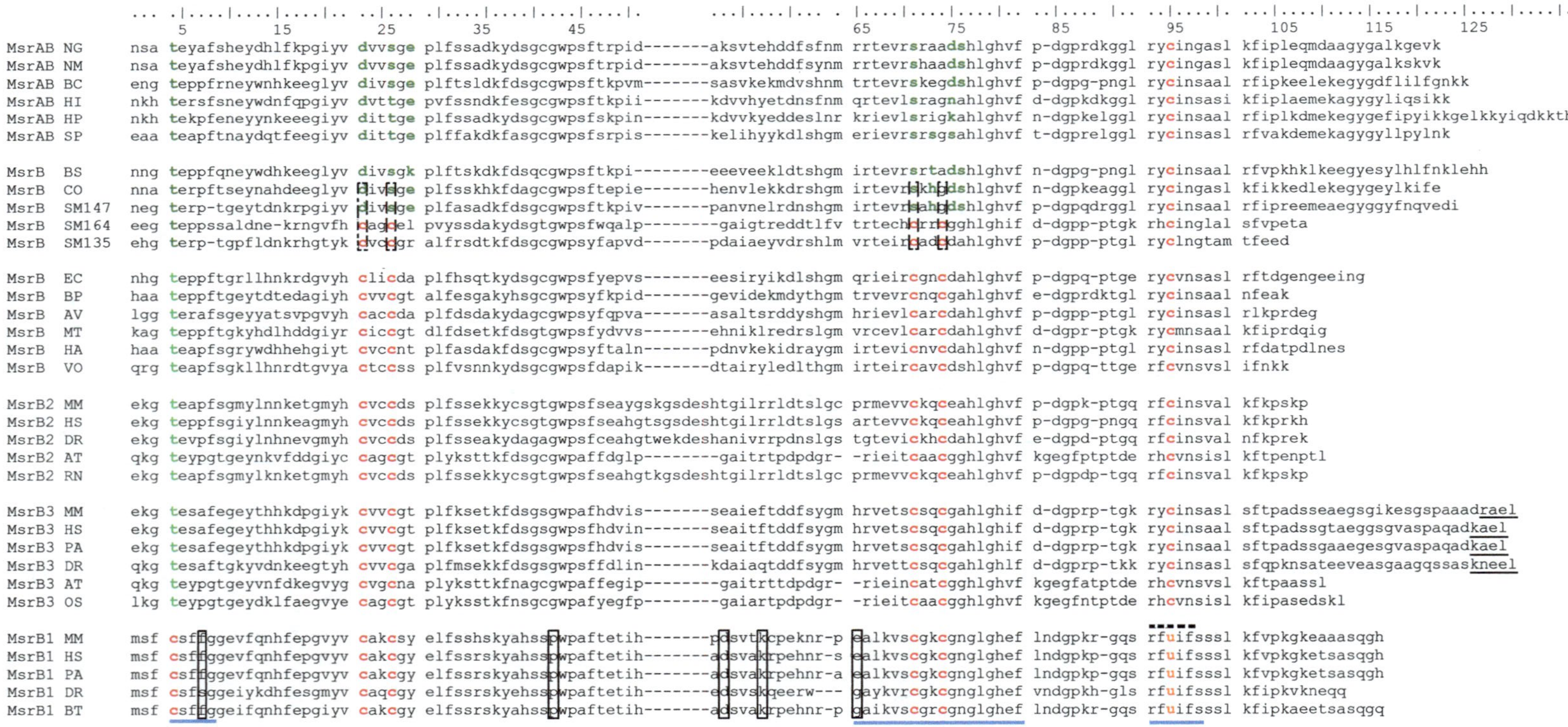

Fig. 11.2. Multiple sequence alignment of proteins belonging to MsrBs family. Numbering follows consensus sequence with mouse MsrB1. In abbreviation the MsrAB only MsrB domain of protein is shown. Conserved catalytic site of MsrBs is indicated with *black dash line*. erTS of MsrB3 underline is indicated in *black*. Amino acids from MsrB1 involved in the initial interacting stage with Trx are indicated by a *blue line* (Dobrovolska et al. 2012). The organisms abbreviations are as follows: *NG Neisseria Gonorrhoeae*, *NM Neisseria meningitides*, *BC Bacillus cereus*, *HI Haemophilus influena*, *HP Helicobacter pylori*, *SP Streptococcus pneumoniae*, *BS Bacillus subtilis*, *CO Clostridium*, *Alkaliphilus oremlandii*, *SM147 Sinorhizobium meliloti* SM11, 147 amino acids, *SM164 Sinorhizobium meliloti* SM11, 164 amino acids, *SM135 Sinorhizobium meliloti* SM11, 135 amino acids, *EC Escherichia coli*, *BP Burkholderia pseudomallei*, *AV Azotobacter vinelandii*, *MT Methanothermobacter thermautotrophicus*, *HA Herminiimonas arsenicoxydans*, *VO Vibrio ordalii*, *MM Mus Musculus*, *HS Homo sapience*, *DR Danio rerio* (zebrafish), *AT Arabidopsis thaliana*, *RN Rattus norvegicus*, *PA Pongo abelii*, *OS Oryza sativa* Japonica Group, *BT Bos taurus*.

Table 11.1. Structural data of methionine sulfoxide reductase B proteins.

Superkingdom/ Phylum	Species	Method	PDB entries	Year	References
MsrB family					
Proteobacteria	*Neisseria gonorrhoeae*	X-ray	1L1D	2002	Lowther et al. (2002)
Proteobacteria	*Burkholderia pseudomallei*	X-ray	3CEZ/3CXK	2008	Gerdts et al. (2008)
Archaea/Euryarchaeota	*Methanothermobacter thermautotrophicus*	NMR	2K8D	2008	unpublished
Bacteria	*Streptococcus pneumoniae*	X-ray	3E0M/3E0O	2009	Kim et al. (2009)
Proteobacteria	*Xanthomonas campestris*	X-ray	3HCI	2009	Ranaivoson et al. (2009)
Proteobacteria	*Neisseria meningitidis*	X-ray	3HCG	2009	Ranaivoson et al. (2009)
Bacteria	*Bacillus subtilis*	NMR	2KZN/1XM0	2012	Lange et al. (2012)
MsrB1 family					
Mammalia	*Mus musculus*, MsrB1	NMR	2KAO/2KV1	2010	Aachmann et al. (2010)
Mammalia	*Homo sapiens*	X-ray	3MAO	2010	unpublished
MsrB2 family					
Mammalia	*Mus musculus*, MsrB2	NMR	2L1U	2011	Aachmann et al. (2011)
MsrB3 family	No reported structures				

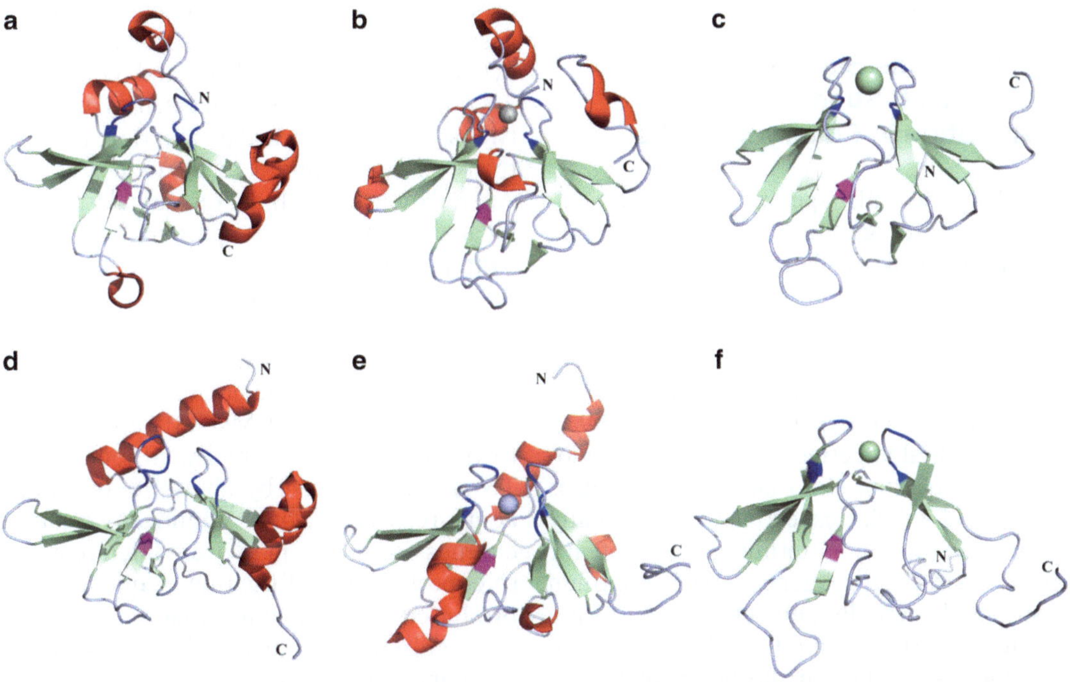

Fig. 11.3. Three-dimensional structures of proteins belonging to MsrB family. (**a**) MsrB *N. Gonorrhoeae*, 3HCH, (Lowther et al. 2002); (**b**) MsrB *X. campestris*, 3HCI, (Ranaivoson et al. 2009); (**c**) MsrB1 *H. sapience*, 3MAO, not pubblished; (**d**) MsrB *B. subtilis*, 1XM0, not pubblished; (**e**) MsrB2 *M. musculus*, 2L1U, (Aachmann et al. 2011); (**f**) MsrB1 *M. musculus*, 2KV1, (Aachmann et al. 2010).

with protein repair, which may account for their unequal distribution in genomes (Zhang and Weissbach 2008).

The number of *msrA* and *msrB* ortholog genes within organisms also varies greatly. For example, *E. coli* contains one copy of *msrA* and *msrB*, *Arabidopsis* harbors five *msrA*s and nine *msrBs*, *Rhizobium meliloti* has three *msrA* and three *msrB* genes, while mammals have one *msrA* and three *msrB* genes, coding for MsrB1, MsrB2 and MsrB3, respectively (Moskovitz et al. 1996; Kryukov et al. 2002; Vougier et al. 2003; Kim and Gladyshev 2004a, b; Hansel et al. 2005; Lee et al. 2009). Gene duplication is a frequent event in genome evolution across all three domains of life. A necessity to have alleles expressed under different conditions might be a possible explanation for the redundancy of *msr* genes in many genomes (Ohno 1970; Zhang 2003, 2012; Ezraty et al. 2005).

IV. Methionine Sulfoxide Reductase B Subcellular Distribution in Eukaryotic Cells

In eukaryotic cells, specific metabolic reactions (defense against ROS attack (Go and Jones 2008)) take place within different intracellular compartments (see Chaps. 21 and 24). For this reason most proteins are allocated to only a single cellular compartment and, thus, need some structural information for their correct targeting (Neupert 1997). Mitochondrial targeting sequences and endoplasmic reticulum (ER) targeting sequences (mTS and erTS, respectively) are usually positioned at the N-terminus of proteins. Nuclear targeting sequences (nTS) are often distributed internally, and peroxisomal TS are located at the C-termini. It has been proposed that the N-terminus of the polypeptide that bears mTS attaches to specific mitochondrial membrane receptors and initiates its translocation before the complete protein is folded (Neupert 1997; Truscott et al. 2001).

Mammals have three MsrBs (MsrB1, MsrB2 and MsrB3) proteins that are localized in different cellular compartments (Table 11.2) (Kim and Gladyshev 2004a, b).

Table 11.2. Subcellular distribution of mammalian MsrBs.

Family	Location	Representatives
MsrB1	Cytosol, Nucleus	MsrB1
MsrB2	Mitochondria	mTS- MsrB2
Rodent MsrB3	ER	erTS-mTS- MsrB3 -RAEL
MsrB3A	ER	MsrB3A -KAEL
MsrB3B	Mitochondria	mTS- MsrB3B -KAEL

MsrB1 is a cytosolic and nuclear protein, MsrB2 is translocated into mitochondria, while MsrB3 occurs in the endoplasmic reticulum and mitochondria (Kim and Gladyshev 2004a, b). Alternative splicing is a conserved mechanism to regulate subcellular distribution of methionine sulfoxide reductases in mammals and other animals (Kim and Gladyshev 2006). The distribution of MsrB3 will be discussed separately below.

MsrB1, the most abundant and active MsrB protein in mammals, is present in cytosol and nucleus (Kim and Gladyshev 2004a, b). The nTS protein consists of one or two short sequences that are rich in positively charged amino acids (e.g. Lys and Arg). Nuclear localization signals can be located anywhere in the protein, but usually form patches or loops on the protein surface. Although MsrB1 does not have clearly predictable nuclear TS, it is rich in positively charged residues, which could serve to transfer the protein into the nucleus.

Both, MsrB2 and MsrB3B are targeted to mitochondria. Normally, mitochondrial TS are represented by about 20–60 amino acid residues with abundant positive charges and frequent hydroxylated residues (Neupert 1997). Targeting sequences are predicted to form amphipathic α-helices in membranes or in membrane-like environments, whereas in aqueous solution they show little structural organization (von Heijne 1986a, b; Roise and Schatz 1988; Lemire et al. 1989; von Heijne 1990). Mouse MsrB2 contains a typical mitochondrial signal with the high proportion of arginine residues at the N-terminus (MARLLRALRGLPLLQAPGRLARG) (Kim and Gladyshev 2004a, b).

MsrB3 should be considered separate from the other MsrBs. In humans and other species (e.g. zebrafish) MsrB3 were found to be translated into both endoplasmic reticulum (MsrB3A) and mitochondria (MsrB3B), due to alternative splicing that produces contrasting signals, for targeting the protein to mitochondria or endoplasmic reticulum (Kim and Gladyshev 2004a, b). It should be noted, that the mitochondrial localization of the human MsrB3B, was not immediately clear because it had contrasting N- and C-terminal signals (Kim and Gladyshev 2004a, b). At the N-terminus a mitochondrial, and at the C-terminus a KDEL-like endoplasmic reticulum translocation signals are present (Fig. 11.2) (Kim and Gladyshev 2004a, b). In contrast, rodents (rat and mouse) have only one form of MsrB3 found only in the endoplasmic reticulum (Kim and Gladyshev 2004a, b). At the same time, mouse MsrB3 has both ER and mitochondrial signal peptides at the N-terminus. The mitochondrial signal is located between the endoplasmic reticulum signal and the common MsrB domain (Table 11.1). The role of the mouse MsrB3 mitochondrial signal, which is functional if placed as an N-terminal sequence, remains unclear and requires further research (Kim and Gladyshev 2004a, b). MsrB3A and rodent MsrB3 are targeted to the endoplasmic reticulum with C-terminus (KDEL-like) ER retention sequence (KAEL for human and RAEL for mouse MsrB) (Kim and Gladyshev 2004a, b).

It follows from the above discussion, that oxidized methionines can be repaired in different cellular compartments in mammals. Mitochondria, the major source of ROS in the cell, have two different MsrBs: MsrB2 and MsrB3B. MsrB2 is the most active at lower concentrations of methionine and is inhibited by high concentrations of the substrate, whereas MsrB3B is most active at concentrations of methionine sulfoxide higher than 1 mM (Kim and Gladyshev 2004a, b). In addition, these two MsrBs show differential tissue expression (Jung et al. 2002). This argumentation can probably explain the occurrence of two MsrBs in mitochondria.

V. Selenocysteine in Methionine Sulfoxide Reductases

Selenocysteine (Sec) is one of the naturally occurring amino acids in proteins. In this residue, which is a cysteine analog, sulfur is replaced by selenium. These two elements have a similar electronegativity [2,58 for sulfur and 2,55 for selenium (Muttenthaler and Alewood 2008)], however, Se is a stronger nucleophile (Huber and Criddle 1967; Pearson et al. 1968; Hondal et al. 2001). Furthermore, the pK_a of selenocysteine is more acidic than that of cysteine (pK_a of Sec is 5.2–5.6 while pK_a value of Cys is around 8.3) (Huber and Criddle 1967; Muttenthaler and Alewood 2008). This means that at physiological pH the Sec residue will be deprotonated (anionic form) and more reactive, while the cysteine residue would still remain protonated. Several oxidoreductases including glutathione peroxidase, thioredoxin reductase and methionine sulfoxide reductase, have Sec in their active sites (Kim and Gladyshev 2005). Due to the higher selenium nucleophilicity, selenoproteins are typically more active than their respective cysteine (Cys) mutants (Metanis et al. 2006). This high catalytic activity has been regarded as a key reason why Sec is used in biological systems (Bock et al. 1991; Bell et al. 1993; Stadtman 1996; Hatfield and Gladyshev 2002). However, the immediate protein environment of Cys or Sec, as well as the overall structure of the protein, can have significant influence on the redox potential of both residues. For example, it has been shown that small changes in the amino acid composition in the region close to the active site can lead to comparable catalytic efficiency of the sulfur homolog of a selenium-dependent enzyme (Gromer et al. 2003). This finding suggests that the requirement for higher activity may not be the unique reason why Sec is used instead of Cys in selenoproteins.

Thus, Sec utilization by living organisms represents a compromise. On the one hand, Sec utilization provides higher activity, a broader range of substrates and of microenvironmental conditions (e.g. pH), in which the enzyme is active. On the other hand, there

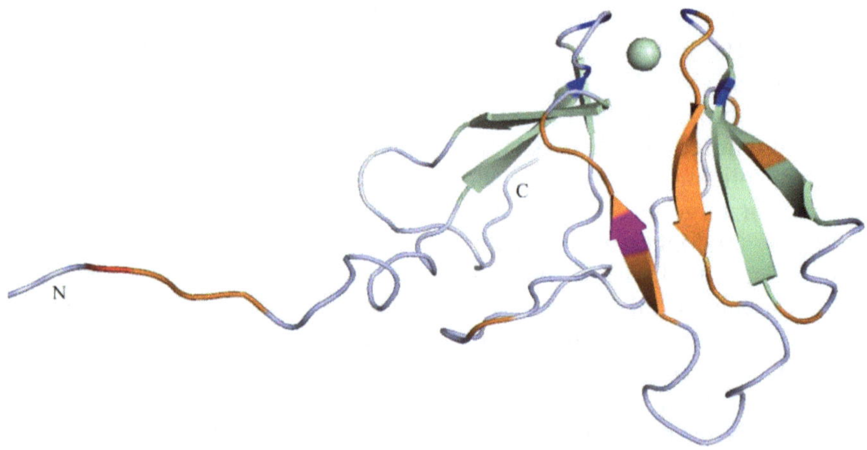

Fig. 11.4. Three-dimensional structure of mouse MsrB1, 2KV1, (Aachmann et al. 2010). Conserved regions of the Se-containing MsrB1 (indicated in *orange*) are involved in the initial stage of Trx interaction (The data taken from Dobrovolska et al. 2012).

are limitations imposed by: (a) dependence on selenium availability, (b) complexity of the Sec insertion system, and (c) correlations of redox potentials of Sec and its substrates (Gromer et al. 2003).

It is worth mentioning, that the replacement of sulfur by selenium may also completely change function of a protein (Kim and Gladyshev 2005). For example, Sec-containing form of subtilisin or glutathione S-transferase, converts these proteins into peroxidases (Bell et al. 1993; Yu et al. 2005).

Seleno-containing MsrBs are found in invertebrates and vertebrates, but not in bacteria or plants (Kryukov et al. 2002; Fomenko et al. 2007). In contrast, selenoprotein forms of MsrA are also found in bacteria and lower eukaryotes, including a unicellular green algae (Novoselov et al. 2002; Kim et al. 2006; Fomenko et al. 2007). Mammals have one selenoprotein MsrB (MsrB1) and two cystein-containing homologs (MsrB2 and MsrB3). MsrB1 contains selenocysteine in the place of the catalytic cysteine residue normally present in other MsrBs (Fig. 11.2) (Kryukov et al. 2002; Kim and Gladyshev 2004a, b). MsrB1 is the most active in the reduction of MetO amongst all members of the MsrB family. The mechanism of a catalytic MetO reduction by Msrs employs sulfenic acid chemistry and S–S/Se–S formation (Boschi-Muller et al. 2000, 2005; Olry et al. 2004; Kim and Gladyshev 2007; Kim and Kim 2008). Thioredoxin (Trx) is generally thought to be a natural reductant for MsrBs (Russel and Model 1986; Lin 1999; Boschi-Muller et al. 2005; Kim and Kim 2008). The Se–S bond is characterized by a lower potential and can be a more challenging target for reduction than the S–S bond (Kim and Gladyshev 2005).

Figure 11.2 illustrates a multiple-sequence alignment of MsrBs, and reveals a set of conserved regions in MsrB1 with respect to other MsrBs. Recently, a simulation of the Trx-MsrB1 interaction dynamics was published (Dobrovolska et al. 2012). NMR titration and molecular dynamics simulations revealed that S5-F7, P42, A66-G72, G75-F82 and R93-S98 amino acids of MsrB1 are involved in the initial interaction with Trx (Fig. 11.2, underlined with blue lines and black rectangle). The sequence alignment indicates that all these groups of amino acids belong to the well-conserved MsrB1 regions. In addition, the amino acid composition of these conserved regions is significantly different from non-selenium MsrBs. The spatial arrangement of these groups of amino acids is shown in Fig. 11.4. It therefore appears that in the course of evolution the amino acid composition of Se-containing MsrB1 has changed, to specifically facilitate the Se–S reduction by Trx. Indeed, selenoprotein MsrB1 is a better substrate for Trx than the Cys-containing MsrB2 and MsrB3 proteins (Kim and Gladyshev 2005).

VI. Methionine Sulfoxide Reductase B Structural Description

All structurally characterized MsrB (both mammalian and bacterial) have a highly similar β-fold consisting of two β-sheets, one with three strands and another with five strands (Table 11.1, Fig. 11.3). However, mammalian MsrB2 protein and bacterial MsrBs additionally have several α-helices at the N-terminal part. A common feature of these α-helices is that they are not rigidly fixed to the β-strand core structure. It has been suggested that these helixes might assist for interaction of MsrBs with different substrates (Aachmann et al. 2010). The latter structural feature is absent in MsrB1 proteins. From the sequence alignment of different MsrBs it appears that the selenoprotein MsrB1 is the smallest and the most compact amongst MsrBs (Lee and Kim 2008; Aachmann et al. 2010). Recent NMR studies reveal that MsrB1 possesses mobile N- and C-terminal tails (Aachmann et al. 2010). The mobility of N-terminus is extremely important, because this tail is involved in catalytic activity of MsrBs. An interesting aspect of the evolution of MsrB superfamily is the dramatic variability with regards to the length and the secondary structure composition of the N- and the C-terminal regions in these proteins. This can be clearly observed from superimposition of MsrB structures (Aachmann et al. 2010): while the β-core is well conserved, the terminal parts of the proteins show significant differences. This feature is further supported by bacterial MsrBs: although their core structures are very similar, their respective N- and C-terminal regions are not superimposable. Therefore, it appears that the evolutionary forces tightly couple the shorter N-terminal region of MsrB1, which is not characterized by any obvious secondary structure.

Comparative structural analysis of mouse MsrB1 and bacterial MsrBs indicates the structural differences between selenoprotein and non-selenoprotein MsrBs and the different mechanisms for the redox reaction catalyzed by MsrBs. These differences may also be deduced from the analysis of MsrBs secondary structure. The presence of the α-helical structures in the N-terminal region of bacterial MsrBs suggests that large structural alterations during catalytic act are unlikely in these enzymes due to rigidity imposed by α-helices. In contrast, the absence of any secondary structure in the N-terminus of Sec-containing MsrB1 does not impose any rigidity constraints. As a result, the catalytic reaction in Sec-containing MsrB1 occurs through the formation of internal selenide-sulfide bridge between the catalytic Sec and the resolving Cys situated on the mobile N-terminus. The reaction between methionine sulfoxide and other MsrBs occurs through formation of a sulfenic acid intermediate and no intramolecular disulfide bridge is formed within this reaction (Kim and Gladyshev 2005).

A structural comparison between MsrB proteins indicates (Fig. 11.3) that mammalian MsrB2 is more similar to bacterial MsrBs rather than to mammalian MsrB1, as it contains three α-helices on the protein exterior at the N-terminal region. Indeed, the proposed catalytic mechanism for mammalian MsrB2 is the same as that proposed for bacterial MsrBs (Aachmann et al. 2011).

VII. Zinc Ion in Methionine Sulfoxide Reductase B

Zinc is one of the most important biological metals (Gladyshev et al. 1996; Bertini et al. 2001; Andreini et al. 2008; Mulkidjanian and Galperin 2009). The ability to predict protein structure through the analysis of DNA sequences allows to categorize different zinc binding sites (Auld 2001a, b). It is possible to distinguish three types of zinc binding sites: *catalytic*, *co-catalytic* and *structural*. For the purpose of this review, we will discuss only the latter two classes here. Structural zinc sites have four protein ligands and no bound water molecule. There is at least one short spacer between coordinating zinc ion residues, generally containing two amino acids (Vallee and Auld 1990; Auld 2001a, b). Cysteines provide the most common, but not the sole ligand in such sites. The second most prevalent ligand is His. 'Zinc fingers' possess such Cys and Cys/His-containing sites. In addition, Asp and Glu residues are also able to bind zinc, even if they perform this task less frequently (Whitlow et al. 1991; Korndorfer et al. 2000; Auld 2001a, b). Multiple Glu/Asp metal binding sites can be too flexible for zinc and lead to weak binding constants due to fast dissociation rates of the zinc ion from such sites (Auld 2001a, b). There is an interesting example of ligand substitution in adenylate kinases belonging to different organisms. The adenylate kinase from *Bacillus stearothermophilus* contains a zinc-binding site composed of four Cys ligands (Berry and Phillips 1998). The adenylate kinase from *Bacillus subtilis* also contains a structural zinc site (Perrier et al. 1994). In the latter case, the fourth Cys has been replaced by an Asp residue in this structural zinc site.

In co-catalytic sites, metals may be important to the overall structure of the protein, as well as the catalytic function. Asp and His predominate as ligands in this type of zinc-binding sites, where the frequency of coordinating residue is maximal for aspartate and histidine, and lower for glutamate (Auld 2001a, b). These sites can also contain unusual zinc ligands such as amide carbonyls provided by Asn, Gln and the peptide backbone; hydroxyl groups from Ser, Thr and Tyr and the amine nitrogen of Lys or the N-terminal amino acid of the protein (Auld 2001a, b). The ligands are often part of a β-sheet or are provided by amino acids one or two residues before or after a β-sheet. A related interesting case was found for the co-catalytic Zn-binding site of β-lactamase. The crystallization of β-lactamase from the *Bacillus cereus* at a pH 5.6 in 0.1 M $ZnSO_4$ in a citrate/cacodylate buffer leads to the protein with one Zn ion (Carfi et al. 1995). However, when the same protein was crystallized in a more basic condition (at pH 7.0 in the presence of 0.5 mM $ZnSO_4$ in Tris buffer, (Fabiane et al. 1998) the resulting protein contained two Zn ions. Thus, the prediction of a Zn binding site based only on the presence of Cys/His ligands cannot give a complete pictures of the zinc distribution in proteins.

Sequence analysis of MsrBs homologs shows the existence of the proteins with two different sets of metal binding sites: the one with the "classical" four Cys ligands for zinc-binding and another with "non-classical" amino acids at the correspondent coordinating positions (Zhang and Gladyshev 2011). Based on this observation, it was suggested that it may be possible to divide zinc-containing proteins into two families. The first one is a strictly Zn-dependent family with the "classical" four cysteines as zinc ligands. The other family normally has Zn ion in metal binding site, but could be expressed in a zinc-independent form where cysteine ligands are substituted by other, "non-classical" residues (Makarova et al. 2001). Thus, for some protein families the same protein may exist either with zinc ion or without. A systematic genome analysis revealed that approximately 20 % of the zinc protein families in the Protein Data Bank (PDB) have a significant number of zinc-independent forms (zinc-dependent form was defined as a homolog, with three or four Cys or His ligands, and a zinc-independent form as a homolog with fewer than three known ligands (Fig. 11.5) (Zhang and Gladyshev 2011). It has been suggested (Panina et al. 2003) that non-zinc-binding paralogs of Zur (a repressor of zinc transport regulation in bacteria) were expressed under zinc-restricted conditions, thereby freeing up some zinc to be used by the essential zinc-binding proteins.

The majority of MsrB proteins have four cysteines that bind Zn ion (Kumar et al. 2002; Olry et al. 2005). They include MsrB1, MsrB2, MsrB3 and the most of bacterial MsrBs (Fig. 11.2). Some of the bacterial

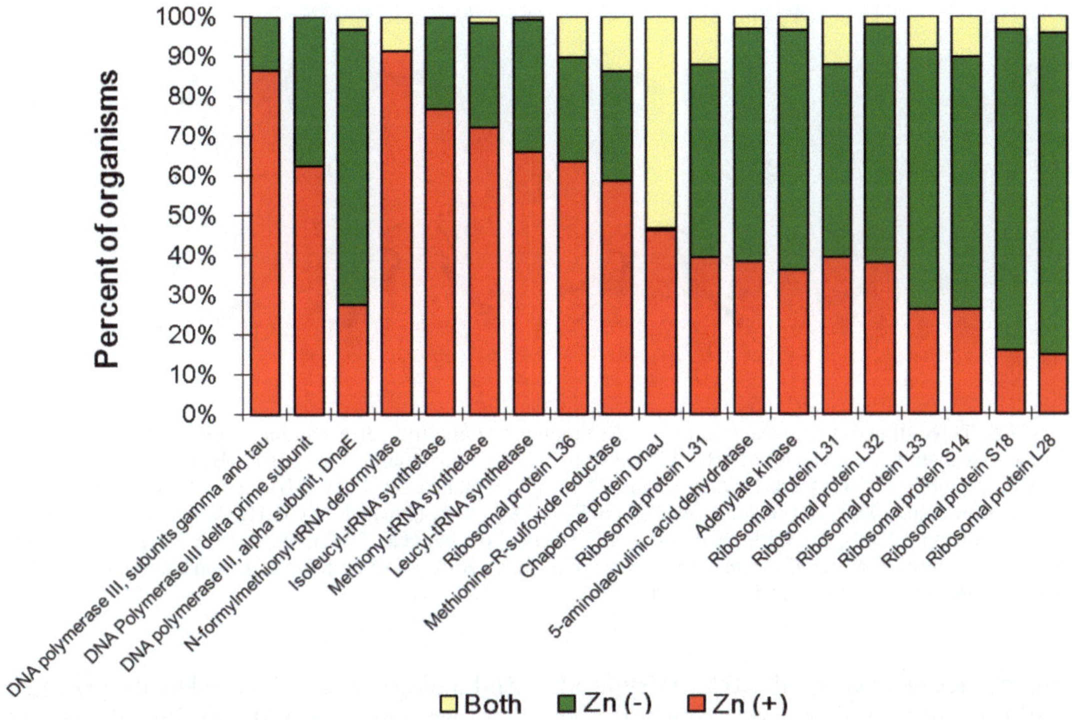

Fig. 11.5. Occurrence of Zn-dependent and Zn-independent forms of representative Zn protein families in bacteria. 'Zn (+)', organisms only containing Zn-dependent form; 'Zn (−)', organisms only containing Zn-independent form; 'both', organisms containing both forms. The figure is taken with permission from (Zhang and Gladyshev 2011).

MsrBs, however, have only Zn-independent forms (e.g. the MsrB domain of MsrAB from *Neisseriaceae*, *Bacillus* sp. and *St. pneumoniae*), some have only Zn-containing forms (MsrB from *B. pseudomallei*, *X. campestris*), and others have both forms (MsrB from *S. meliloti*; *V. cholera*) (unpublished result from (Zhang and Gladyshev 2011). For example, *S. meliloti* has both, the Zn-dependent and Zn-independent forms. In the Zn-independent form, the four cysteines are substituted by Asp, Ser and Gly (Fig. 11.2). While Asp and Ser can hypothetically bind Zn, Gly has not been reported as Zn ligand. However, in the immediate vicinity of Gly highly conserved His, Asp, Glu and Ser are present, and it could be that the latter residues also weakly coordinate the zinc ion. The same "non-classical" ligands are present in MsrB domain of bacterial fused MsrAB.

Several structures of MsrB proteins are presently reported (Table 11.1, Fig. 11.6). Some of them are characterized by the presence of Zn ion, while others lack it. From the structural comparison (Fig. 11.6), it is evident that both Zn-containing and Zn-independent forms have an identical structure of the metal binding site and the corresponding region of non-metal-binding homologs. All MsrBs have two well-conserved unstructured loops between four β-sheets. It should be noted, that the unpublished structure of MsrB1 from *Homo sapiens* has iron ion in the metal binding site, while all other structural features remained unaltered to the rest of the proteins reported shown in Fig. 11.6. In the cases where the Zn ion is absent, it is possible that alternative bonds, and not the coordination of Zn ion, preserve the fold of the region. Studies of the Zn-binding site were carried out by different

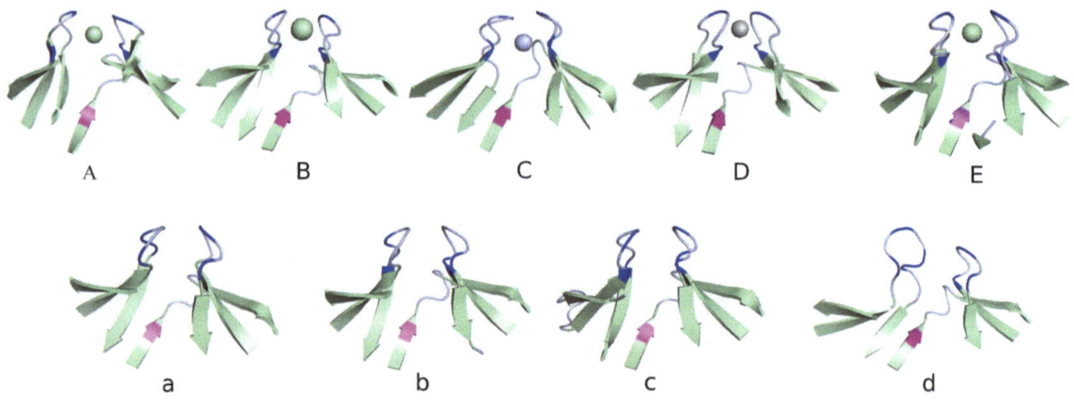

Fig. 11.6. Three-dimensional structure of the Zn-binding site and the corresponding regions of MsrBs family. For the explanation see text. (**A**) MsrB1, *M. musculus*, 2KV1, (Aachmann et al. 2010); (**B**) MsrB1, *H. sapience*, 3MAO, not published; (**C**) MsrB2, *M. musculus*, 2L1U, (Aachmann et al. 2011); (**D**) MsrB, *X. campestris*, 3HCI, (Ranaivoson et al. 2009); (**E**) *B. pseudomallei*, 3CEZ/3CXK, not published; (**a**) MsrB, *N. meningitidis*, 3HCH, (Ranaivoson et al. 2009); (**b**) MsrB, *N. gonorhoeae*, 1L1D, (Lowther et al. 2002); (**c**) MsrB, *S. pneumoniae* 3E0O, (Kim et al. 2009); (**d**) MsrB, *B.subtilis*, 1XM0 (not published). *Blue color*: cysteine and "non classical" ligands in metal binding site; magenta: catalytic cysteine.

laboratories (Kumar et al. 2002; Olry et al. 2005). In these studies it was shown that in MsrB from *Drosophila* and *E.coli*, mutation of Cys at the positions 23, 26, 71 and 74 (numbering corresponds to MsrB1 from *Mus musculus*) to GSGS, DSSS and DSSA results in a loss of Zn. In contrast mutation of DSSS to CCCC results in a more stable, Zn-containing protein. It was suggested, that the role of the metal in Zn-containing MsrBs is to stabilize the core structure and thus, to adopt the active site conformation for efficient reductase activity. However, this suggestion is in apparent contradiction with the fact that the bacterial MsrB, lacking the two CXXC signatures, is as active as the Zn-containing MsrB (Olry et al. 2005). It has been suggested that a more detailed analysis of stabilizing elements for MsrBs, which do not contain the CXXC motif, should be performed.

The analysis of the crystallization conditions for the published structures of MsrBs where the Zn ion is absent ("non-classical" Zn-binding CXXC motif) suggests that the crystallization occurs in mild acidic environments (Lowther et al. 2002; Kim et al. 2009; Ranaivoson et al. 2009). Some of the "non-classical" Zn ligands can be protonated at these conditions. As mentioned above, Zn-binding can be influenced by the pH value. In other words, when the Zn-ligands are protonated, they are not able to coordinate metal ion. In addition, the reported absence of Zn ion for some experimentally characterized MsrB may simply be due to the fact that the metal ion is lost during the protein's isolation and purification, because the coordination between Zn ion and the oxygen containing ligands are weaker with respect to the "classical" cysteine ligands. Thus, zinc ion could escape detection. The structural similarities within metal coordinating loops among all MsrBs would indirectly support this suggestion.

Recently our laboratory demonstrated that Zn ion can be replaced by Co ion (Shumilina et al. 2014) in Cys-containing MsrB1, expressed in *E. coli* in cobalt-containing M9 media. The incorporation of Co ion can be easily detected by both UV-Vis (Co-containing proteins have a characteristic light blue color) and NMR spectroscopies (by the appearance of paramagnetically shifted signals). It is known (Lippard and Berg 1994; Bertini et al. 2001) that Co ion can be used to substitute Zn ion, since both have similar radii and coordination properties. Thus, this method can be used to check and confirm, whether MsrBs without the 'classical' four cysteines do not bind Zn at a more basic pH.

Summarizing, most MsrBs have four cysteines that bind zinc ion. It has been suggested that the metal- binding site plays a structural role in the MsrB family. Some of the organisms have also MsrB proteins, which possess "non-classical" ligands as alternatives to cysteines. The determined structures of these MsrB with "non-classical" ligands possess a conserved structure that might suggest that these proteins still weakly bind metal ion.

VIII. Conclusion

Living organisms have developed efficient and complex defense system against oxidative stress. MsrB proteins play an important role in this system. They can reduce R-isomer of methionine sulfoxide back to methionine. The reversibility of Met oxidation provides the organism with the possibility to reduce intracellular level of ROS, restore the enzymatic function of oxidized protein and inhibit or induce certain cellular events.

MsrBs were found in almost all kingdoms of life. The absence of MsrB in some thermophilic and anaerobic bacteria may indicate that either the MetO reduction at high temperature does not require a catalyst, or MetO is not produced in significant amounts. Whereas prokaryotes have only one type of MsrB, mammals possess three: MsrB1 (the most active mammalian MsrB1 containing catalytic selenocysteine), MsrB2 and MsrB3. The use of Sec in proteins can represent a compromise between a more active enzyme that can work with different substrates at a wider pH range, and a complex genetic mechanism of Se insertion with more challenging Se–S bond reduction. The different eukaryotic MsrBs have peculiar cellular distribution that allow them to be effectively involved in various redox and other catalytic pathways. MsrBs compartmentalization is mostly achieved by alternative splicing, and importing the proteins into organelles by specific signals.

The structure of all catalytic cores of the structurally characterized MsrBs is so far rather similar, containing two β-sheets. In addition, bacterial MsrBs and mammalian MsrB2 are characterized by the presence of N-terminal α-helixes, while mammalian MsrB1 has only β-sheets and very flexible N- and C-termini. This flexibility plays an important functional role leading to differences in catalytic mechanism between MsrB1 and other proteins of MsrB family.

Mostly, MsrBs are Zn-containing proteins. The existence of the metal was predicted by bioinformatics based on the presence of two Zn-binding CXXC motifs. However, some of the bacterial MsrBs have different, well-conserved amino acids at the positions of the four cysteines in mammalian MsrB. Overall, the structures of the metal-binding site in Zn-containing proteins, and the corresponding region in MsrBs without metal, are very similar. Theoretically, there is a possibility that Zn ion can be coordinated by non-cysteine ligands at more alkaline pH and with less affinity. Nevertheless, this speculation requires experimental confirmation.

Thus, in the course of evolution the members of MsrB family developed their amino acids sequences and resulting folds for efficient functioning in different organisms, and different cell compartments with different environmental conditions.

Acknowledgements

AD acknowledges the support from NT Faculty, NTNU. ES acknowledges the NT Faculty, NTNU, for financial support through a post-doctoral fellowship. OD acknowledges the NT Faculty, NTNU, for financial support through a PhD fellowship.

References

Aachmann FL, Sal LS, Kim HY, Marino SM, Gladyshev VN, Dikiy A (2010) Insights into function, catalytic mechanism, and fold evolution of selenoprotein methionine sulfoxide reductase B1 through structural analysis. J Biol Chem 285:33315–33323

Aachmann FL, Kwak GH, Del Conte R, Kim HY, Gladyshev VN, Dikiy A (2011) Structural and biochemical analysis of mammalian methionine sulfoxide reductase B2. Proteins 79:3123–3131

Andreini C, Bertini I, Cavallaro G, Holliday GL, Thornton JM (2008) Metal ions in biological catalysis: from enzyme databases to general principles. J Biol Inorg Chem 13:1205–1218

Auld DS (2001a) Zinc coordination sphere in biochemical zinc sites. Biometals 14:271–313

Auld DS (2001b) Zinc sites in metalloenzymes and related proteins. In: Bertini I, Sigel A, Sigel H (eds) Handbook on metalloproteins. Marcel Dekker, New York, pp 881–959

Bar-Noy S, Moskovitz J (2002) Mouse methionine sulfoxide reductase B: effect of selenocysteine incorporation on its activity and expression of the seleno-containing enzyme in bacterial and mammalian cells. Biochem Biophys Res Commun 297:956–961

Bell IM, Fisher ML, Wu ZP, Hilvert D (1993) Kinetic studies on the peroxidase activity of selenosubtilisin. Biochemistry 32:3754–3762

Berry MB, Phillips GN Jr (1998) Crystal structures of Bacillus stearothermophilus adenylate kinase with bound Ap5A, Mg2+ Ap5A, and Mn2+ Ap5A reveal an intermediate lid position and six coordinate octahedral geometry for bound Mg2+ and Mn2+. Proteins 32:276–288

Bertini I, Sigel A, Sigel H (2001) Handbook on metalloproteins. Marcel Dekker, New York

Bock A, Forchhammer K, Heider J, Baron C (1991) Selenoprotein synthesis: an expansion of the genetic code. Trends Biochem Sci 16:463–467

Boschi-Muller S, Azza S, Sanglier-Cianferani S, Talfournier F, Van Dorsselear A, Branlant G (2000) A sulfenic acid enzyme intermediate is involved in the catalytic mechanism of peptide methionine sulfoxide reductase from *Escherichia coli*. J Biol Chem 275:35908–35913

Boschi-Muller S, Olry A, Antoine M, Branlant G (2005) The enzymology and biochemistry of methionine sulfoxide reductases. Biochim Biophys Acta 1703:231–238

Boschi-Muller S, Gand A, Branlant G (2008) The methionine sulfoxide reductases: catalysis and substrate specificities. Arch Biochem Biophys 474:266–273

Brot N, Weissbach H (1983) Biochemistry and physiological role of methionine sulfoxide residues in proteins. Arch Biochem Biophys 223:271–281

Brot N, Weissbach H (2000) Peptide methionine sulfoxide reductase: biochemistry and physiological role. Biopolymers 55:288–296

Brot N, Weissbach L, Werth J, Weissbach H (1981) Enzymatic reduction of protein-bound methionine sulfoxide. Proc Natl Acad Sci U S A 78:2155–2158

Caldwell P, Luk DC, Weissbach H, Brot N (1978) Oxidation of the methionine residues of *Escherichia coli* ribosomal protein L12 decreases the protein's biological activity. Proc Natl Acad Sci U S A 75:5349–5352

Carfi A, Pares S, Duee E, Galleni M, Duez C, Frere JM, Dideberg O (1995) The 3-D structure of a zinc metallo-beta-lactamase from *Bacillus cereus* reveals a new type of protein fold. EMBO J 14:4914–4921

Ciorba MA, Heinemann SH, Weissbach H, Brot N, Hoshi T (1997) Modulation of potassium channel function by methionine oxidation and reduction. Proc Natl Acad Sci U S A 94:9932–9937

Ciorba MA, Heinemann SH, Weissbach H, Brot N, Hoshi T (1999) Regulation of voltage-dependent K + channels by methionine oxidation: effect of nitric oxide and vitamin C. FEBS Lett 442:48–52

Dobrovolska O, Rychkov G, Shumilina E, Nerinovski K, Schmidt A, Shabalin K, Yakimov A, Dikiy A (2012) Structural insights into interaction between mammalian methionine sulfoxide reductase B1 and thioredoxin. J Biomed Biotechnol 2012:586539

Ezraty B, Aussel L, Barras F (2005) Methionine sulfoxide reductases in prokaryotes. Biochim Biophys Acta 1703:221–229

Fabiane SM, Sohi MK, Wan T, Payne DJ, Bateson JH, Mitchell T, Sutton BJ (1998) Crystal structure of the zinc-dependent beta-lactamase from *Bacillus cereus* at 1.9 A resolution: binuclear active site with features of a mononuclear enzyme. Biochemistry 37:12404–12411

Fomenko DE, Xing W, Adair BM, Thomas DJ, Gladyshev VN (2007) High-throughput identification of catalytic redox-active cysteine residues. Science 315:387–389

Gerdts CJ, Elliott M, Lovell S, Mixon MB, Napuli AJ, Staker BL, Nollert P, Stewart L (2008) The plug-based nanovolume Microcapillary Protein Crystallization System (MPCS). Acta Crystallogr Sect D 64:1116–1122

Gladyshev VN, Jeang KT, Stadtman TC (1996) Selenocysteine, identified as the penultimate C-terminal residue in human T-cell thioredoxin reductase, corresponds to TGA in the human placental gene. Proc Natl Acad Sci U S A 93:6146–6151

Go YM, Jones DP (2008) Redox compartmentalization in eukaryotic cells. Biochim Biophys Acta 1780:1273–1290

Gromer S, Johansson L, Bauer H, Arscott LD, Rauch S, Ballou DP, Williams CH Jr, Schirmer RH, Arner ES (2003) Active sites of thioredoxin reductases: why selenoproteins? Proc Natl Acad Sci U S A 100:12618–12623

Hansel A, Jung S, Hoshi T, Heinemann SH (2003) A second human methionine sulfoxide reductase

(hMSRB2) reducing methionine-R-sulfoxide displays a tissue expression pattern distinct from hMSRB1. Redox Rep 8:384–388
Hansel A, Heinemann SH, Hoshi T (2005) Heterogeneity and function of mammalian MSRs: enzymes for repair, protection and regulation. Biochim Biophys Acta 1703:239–247
Hatfield DL, Gladyshev VN (2002) How selenium has altered our understanding of the genetic code. Mol Cell Biol 22:3565–3576
Hondal RJ, Nilsson BL, Raines RT (2001) Selenocysteine in native chemical ligation and expressed protein ligation. J Am Chem Soc 123:5140–5141
Huang W, Escribano J, Sarfarazi M, Coca-Prados M (1999) Identification, expression and chromosome localization of a human gene encoding a novel protein with similarity to the pilB family of transcriptional factors (pilin) and to bacterial peptide methionine sulfoxide reductases. Gene 233:233–240
Huber RE, Criddle RS (1967) Comparison of the chemical properties of selenocysteine and selenocystine with their sulfur analogs. Arch Biochem Biophys 122:164–173
Jung S, Hansel A, Kasperczyk H, Hoshi T, Heinemann SH (2002) Activity, tissue distribution and site-directed mutagenesis of a human peptide methionine sulfoxide reductase of type B: hCBS1. FEBS Lett 527:91–94
Kauffmann B, Favier F, Olry A, Boschi-Muller S, Carpentier P, Branlant G, Aubry A (2002) Crystallization and preliminary X-ray diffraction studies of the peptide methionine sulfoxide reductase B domain of Neisseria meningitidis PILB. Acta Crystallogr D Biol Crystallogr 58:1467–1469
Kim HY, Gladyshev VN (2004a) Characterization of mouse endoplasmic reticulum methionine-R-sulfoxide reductase. Biochem Biophys Res Commun 320:1277–1283
Kim HY, Gladyshev VN (2004b) Methionine sulfoxide reduction in mammals: characterization of methionine-R-sulfoxide reductases. Mol Biol Cell 15:1055–1064
Kim HY, Gladyshev VN (2005) Different catalytic mechanisms in mammalian selenocysteine- and cysteine-containing methionine-R-sulfoxide reductases. PLoS Biol 3:e375
Kim HY, Gladyshev VN (2006) Alternative first exon splicing regulates subcellular distribution of methionine sulfoxide reductases. BMC Mol Biol 7:11
Kim HY, Gladyshev VN (2007) Methionine sulfoxide reductases: selenoprotein forms and roles in antioxidant protein repair in mammals. Biochem J 407:321–329
Kim HY, Kim JR (2008) Thioredoxin as a reducing agent for mammalian methionine sulfoxide reductases B lacking resolving cysteine. Biochem Biophys Res Commun 371:490–494
Kim HY, Fomenko DE, Yoon YE, Gladyshev VN (2006) Catalytic advantages provided by selenocysteine in methionine-S-sulfoxide reductases. Biochemistry 45:13697–13704
Kim YK, Shin YJ, Lee WH, Kim HY, Hwang KY (2009) Structural and kinetic analysis of an MsrA-MsrB fusion protein from *Streptococcus pneumoniae*. Mol Microbiol 72:699–709
Korndorfer IP, Fessner WD, Matthews BW (2000) The structure of rhamnose isomerase from *Escherichia coli* and its relation with xylose isomerase illustrates a change between inter and intra-subunit complementation during evolution. J Mol Biol 300:917–933
Kryukov GV, Kumar RA, Koc A, Sun Z, Gladyshev VN (2002) Selenoprotein R is a zinc-containing stereo-specific methionine sulfoxide reductase. Proc Natl Acad Sci U S A 99:4245–4250
Kumar RA, Koc A, Cerny RL, Gladyshev VN (2002) Reaction mechanism, evolutionary analysis, and role of zinc in Drosophila methionine-R-sulfoxide reductase. J Biol Chem 277:37527–37535
Kuschel L, Hansel A, Schonherr R, Weissbach H, Brot N, Hoshi T, Heinemann SH (1999) Molecular cloning and functional expression of a human peptide methionine sulfoxide reductase (hMsrA). FEBS Lett 456:17–21
Lange OF, Rossi P, Sgourakis NG, Song Y, Lee HW, Aramini JM, Ertekin A, Xiao R, Acton TB, Montelione GT, Baker D (2012) Determination of solution structures of proteins up to 40 kDa using CS-Rosetta with sparse NMR data from deuterated samples. Proc Natl Acad Sci U S A 109:10873–10878
Lee TH, Kim HY (2008) An anaerobic bacterial MsrB model reveals catalytic mechanisms, advantages, and disadvantages provided by selenocysteine and cysteine in reduction of methionine-R-sulfoxide. Arch Biochem Biophys 478:175–180
Lee BC, Lee YK, Lee HJ, Stadtman ER, Lee KH, Chung N (2005) Cloning and characterization of antioxidant enzyme methionine sulfoxide-S-reductase from *Caenorhabditis elegans*. Arch Biochem Biophys 434:275–281
Lee BC, Dikiy A, Kim HY, Gladyshev VN (2009) Functions and evolution of selenoprotein methionine sulfoxide reductases. Biochim Biophys Acta 1790:1471–1477
Lemire BD, Fankhauser C, Baker A, Schatz G (1989) The mitochondrial targeting function of randomly generated peptide sequences correlates with predicted helical amphiphilicity. J Biol Chem 264:20206–20215

Levine RL, Mosoni L, Berlett BS, Stadtman ER (1996) Methionine residues as endogenous antioxidants in proteins. Proc Natl Acad Sci U S A 93:15036–15040

Levine RL, Berlett BS, Moskovitz J, Mosoni L, Stadtman ER (1999) Methionine residues may protect proteins from critical oxidative damage. Mech Ageing Dev 107:323–332

Lin TY (1999) G33D mutant thioredoxin primarily affects the kinetics of reaction with thioredoxin reductase. Probing the structure of the mutant protein. Biochemistry 38:15508–15513

Lippard SJ, Berg JM (1994) Principles of bioinorganic chemistry. University Science, Mill Valley

Lowther WT, Brot N, Weissbach H, Matthews BW (2000) Structure and mechanism of peptide methionine sulfoxide reductase, an "anti-oxidation" enzyme. Biochemistry 39:13307–13312

Lowther WT, Weissbach H, Etienne F, Brot N, Matthews BW (2002) The mirrored methionine sulfoxide reductases of *Neisseria gonorrhoeae* pilB. Nat Struct Biol 9:348–352

Makarova KS, Ponomarev VA, Koonin EV (2001) Two C or not two C: recurrent disruption of Zn-ribbons, gene duplication, lineage-specific gene loss, and horizontal gene transfer in evolution of bacterial ribosomal proteins. Genome Biol 2 (9):research0033.1–0033.14

Metanis N, Keinan E, Dawson PE (2006) Synthetic seleno-glutaredoxin 3 analogues are highly reducing oxidoreductases with enhanced catalytic efficiency. J Am Chem Soc 128:16684–16691

Moskovitz J, Weissbach H, Brot N (1996) Cloning the expression of a mammalian gene involved in the reduction of methionine sulfoxide residues in proteins. Proc Natl Acad Sci U S A 93:2095–2099

Moskovitz J, Poston JM, Berlett BS, Nosworthy NJ, Szczepanowski R, Stadtman ER (2000) Identification and characterization of a putative active site for peptide methionine sulfoxide reductase (MsrA) and its substrate stereospecificity. J Biol Chem 275:14167–14172

Mulkidjanian AY, Galperin MY (2009) On the origin of life in the zinc world. 2. Validation of the hypothesis on the photosynthesizing zinc sulfide edifices as cradles of life on Earth. Biol Direct 4:27

Muttenthaler M, Alewood PF (2008) Selenopeptide chemistry. J Pept Sci 14:1223–1239

Neupert W (1997) Protein import into mitochondria. Annu Rev Biochem 66:863–917

Novoselov SV, Rao M, Onoshko NV, Zhi H, Kryukov GV, Xiang Y, Weeks DP, Hatfield DL, Gladyshev VN (2002) Selenoproteins and selenocysteine insertion system in the model plant cell system, *Chlamydomonas reinhardtii*. EMBO J 21:3681–3693

Ohno S (1970) Evolution by gene duplication. Springer, Berlin

Olry A, Boschi-Muller S, Marraud M, Sanglier-Cianferani S, Van Dorsselear A, Branlant G (2002) Characterization of the methionine sulfoxide reductase activities of PILB, a probable virulence factor from *Neisseria meningitidis*. J Biol Chem 277:12016–12022

Olry A, Boschi-Muller S, Branlant G (2004) Kinetic characterization of the catalytic mechanism of methionine sulfoxide reductase B from *Neisseria meningitidis*. Biochemistry 43:11616–11622

Olry A, Boschi-Muller S, Yu H, Burnel D, Branlant G (2005) Insights into the role of the metal binding site in methionine-R-sulfoxide reductases B. Protein Sci 14:2828–2837

Panina EM, Mironov AA, Gelfand MS (2003) Comparative genomics of bacterial zinc regulons: enhanced ion transport, pathogenesis, and rearrangement of ribosomal proteins. Proc Natl Acad Sci U S A 100:9912–9917

Pearson RG, Sobel H, Songstad J (1968) Nucleophilic reactivity constants toward methyl iodide and trans-[Pt(Py)2cl2]. J Am Chem Soc 90:319

Perrier V, Surewicz WK, Glaser P, Martineau L, Craescu CT, Fabian H, Mantsch HH, Barzu O, Gilles AM (1994) Zinc chelation and structural stability of adenylate kinase from *Bacillus subtilis*. Biochemistry 33:9960–9967

Ranaivoson FM, Neiers F, Kauffmann B, Boschi-Muller S, Branlant G, Favier F (2009) Methionine sulfoxide reductase B displays a high level of flexibility. J Mol Biol 394:83–93

Roise D, Schatz G (1988) Mitochondrial presequences. J Biol Chem 263:4509–4511

Russel M, Model P (1986) The role of thioredoxin in filamentous phage assembly. Construction, isolation, and characterization of mutant thioredoxins. J Biol Chem 261:14997–15005

Sharov VS, Ferrington DA, Squier TC, Schoneich C (1999) Diastereoselective reduction of protein-bound methionine sulfoxide by methionine sulfoxide reductase. FEBS Lett 455:247–250

Shechter Y (1986) Selective oxidation and reduction of methionine residues in peptides and proteins by oxygen exchange between sulfoxide and sulfide. J Biol Chem 261:66–70

Shumilina E, Dobrovolska O, Del Conte R, Holen HW, Dikiy A (2014) Competitive cobalt for zinc substitution in mammalian methionine sulfoxide reductase B1 overexpressed in E.coli: structural and functional insight. J Biol Inorg Chem 19(1):85–95

Singh VK, Moskovitz J (2003) Multiple methionine sulfoxide reductase genes in *Staphylococcus aureus*: expression of activity and roles in tolerance of oxidative stress. Microbiology 149:2739–2747

Stadtman TC (1996) Selenocysteine. Annu Rev Biochem 65:83–100

Sun H, Gao J, Ferrington DA, Biesiada H, Williams TD, Squier TC (1999) Repair of oxidized calmodulin by methionine sulfoxide reductase restores ability to activate the plasma membrane Ca-ATPase. Biochemistry 38:105–112

Truscott KN, Pfanner N, Voos W (2001) Transport of proteins into mitochondria. Rev Physiol Biochem Pharmacol 143:81–136

Vallee BL, Auld DS (1990) Zinc coordination, function, and structure of zinc enzymes and other proteins. Biochemistry 29:5647–5659

Vogt W (1995) Oxidation of methionyl residues in proteins: tools, targets, and reversal. Free Radic Biol Med 18:93–105

von Heijne G (1986a) Mitochondrial targeting sequences may form amphiphilic helices. EMBO J 5:1335–1342

von Heijne G (1986b) Towards a comparative anatomy of N-terminal topogenic protein sequences. J Mol Biol 189:239–242

von Heijne G (1990) Protein targeting signals. Curr Opin Cell Biol 2:604–608

Vougier S, Mary J, Friguet B (2003) Subcellular localization of methionine sulphoxide reductase A (MsrA): evidence for mitochondrial and cytosolic isoforms in rat liver cells. Biochem J 373:531–537

Whitlow M, Howard AJ, Finzel BC, Poulos TL, Winborne E, Gilliland GL (1991) A metal-mediated hydride shift mechanism for xylose isomerase based on the 1.6 A *Streptomyces rubiginosus* structures with xylitol and D-xylose. Proteins 9:153–173

Yao Y, Yin D, Jas GS, Kuczer K, Williams TD, Schoneich C, Squier TC (1996) Oxidative modification of a carboxyl-terminal vicinal methionine in calmodulin by hydrogen peroxide inhibits calmodulin-dependent activation of the plasma membrane Ca-ATPase. Biochemistry 35:2767–2787

Yu HJ, Liu JQ, Bock A, Li J, Luo GM, Shen JC (2005) Engineering glutathione transferase to a novel glutathione peroxidase mimic with high catalytic efficiency. Incorporation of selenocysteine into a glutathione-binding scaffold using an auxotrophic expression system. J Biol Chem 280:11930–11935

Zhang J (2003) Evolution by gene duplication: an update. Trends Ecol Evol 18:292–298

Zhang J (2012) Genetic redundancies and their evolutionary maintenance. In: Soyer OS (ed) Evolutionary systems biology, vol 751, Advances in experimental medicine and biology. Springer, New York, pp 279–300

Zhang Y, Gladyshev VN (2011) Comparative genomics of trace element dependence in biology. J Biol Chem 286:23623–23629

Zhang XH, Weissbach H (2008) Origin and evolution of the protein-repairing enzymes methionine sulphoxide reductases. Biol Rev Camb Philos Soc 83:249–257

Chapter 12

Respiratory Chain Supercomplexes in Mitochondria

Natalya V. Dudkina*
Department of Biological Sciences, Birkbeck, University of London, Malet Str., London WC1E 7HX, United Kingdom

Egbert J. Boekema
Department of Biophysical Chemistry, GBB, University of Groningen, Nijenborgh 7, 9747AG Groningen, The Netherlands

and

Hans-Peter Braun
Institute of Plant Genetics, University of Hannover, Herrenhäuser Str. 2, D-30419 Hannover, Germany

Summary

Mitochondria have an intricate, heavily folded inner membrane, which is occupied by many copies of the respiratory chain complexes (I, II, III, IV). These complexes, together with the ATP synthase complex (complex V), are responsible for energy production stored as ATP. All five complexes specifically interact and form defined supercomplexes. Electron microscopy has provided structural data describing the interaction between complexes I and III, among I, III and IV and between two ATP synthase monomers in a dimeric form of complex V. Cryo-electron tomography has given new insights how these supercomplexes are arranged within intact mitochondria. The structural data can help to define the functional role of these supercomplexes, in particular for the dimeric ATP synthase complex, which appears to be responsible for the folding of the inner mitochondrial membrane.

*Author for correspondence, e-mail: n.dudkina@mail.cryst.bbk.ac.uk

M.F. Hohmann-Marriott (ed.), The Structural Basis of Biological Energy Generation, Advances in Photosynthesis and Respiration 39, pp. 217–229, DOI 10.1007/978-94-017-8742-0_12,

I. Introduction

Mitochondria are the intracellular organelles responsible for oxygen-dependent respiration in eukaryotes. They accommodate in their inner or cristae membrane the oxidative phosphorylation (OXPHOS) system that consists of five multi-subunit complexes (complexes I–V). The complexes I–IV are oxidoreductases which, with the exception of complex II, couple electron transport with the translocation of protons across the inner mitochondrial membrane (Fig. 12.1). The generated proton motive force is used by ATP synthase (complex V) for ATP synthesis from ADP and phosphate. Considerable knowledge is available about the high-resolution structure of the OXPHOS complexes of fungi and animals based on X-ray crystallography and biochemical studies.

II. Structure and Function of Respiratory Chain Complexes I–V

Complex I or NADH dehydrogenase (also see Chap. 21) is the first and major entrance point of electrons to the mitochondrial respiratory chain. It transfers electrons from NADH molecules to a lipophilic quinone called ubiquinone, coupling electron transport to the translocation of four protons across the membrane. A total of 45 different subunits are known to form part of complex I in animal mitochondria (Carroll et al. 2003) and more recently even 48 subunits were found in the model plant *Arabidopsis thaliana* (Klodmann and Braun 2011). It has a molecular mass of about 1 MDa and is composed of two domains, which run perpendicular to each other and together give it an L-like shape. One domain is embedded in the membrane and contains four proton translocating channels, the other protrudes into the mitochondrial matrix and is responsible for the oxidation of NADH. NADH donates two electrons, which pass via flavine mononucleotide (FMN) to the chain of seven Fe–S clusters and quinone (Sazanov and Hinchliffe 2006). Electron transfer is coupled to conformational changes in the hydrophilic domain close to the membrane, which result in changing conformation of the membrane subunits and pumping protons (Efremov et al. 2010).

Complex II or succinate dehydrogenase (see Chap. 9) represents an alternative entrance point of electrons to the respiratory chain. It transmits electrons from succinate to ubiquinone and directly connects the citric acid cycle to the respiratory chain. It is the smallest complex of the OXPHOS system and consists of two soluble matrix-exposed subunits, which are attached to two hydrophobic membrane proteins (for review, see Horsefield et al. 2004). In plants, four additional subunits form part of this complex. Their functional role in complex II so far is not known (Millar et al. 2004).

Complex III or cytochrome *c* reductase (see Chap. 8) is a functional dimer of about 500 kDa. It transfers electrons from reduced ubiquinone to cytochrome *c* – a small mobile electron carrier associated with the outer surface of the inner membrane. Each monomer is composed of 10 or 11 distinct subunits, three hemes and one Fe–S cluster (Hunte et al. 2000; Berry et al. 2000). About one quarter of the complex is embedded in the inner mitochondrial membrane, a small part protrudes out into the mitochondrial intermembrane space; the larger hydrophilic parts of the "core" proteins core I and core II stick out far into the mitochondrial matrix. The latter proteins are related to the two subunits of the mitochondrial processing peptidase (MPP) and in plants even represent the subunits of this enzyme (Braun et al. 1992).

Complex IV or cytochrome *c* oxidase (see Chap. 10) catalyzes electron transfer from

---

Abbreviations: ADP – Adenosine diphosphate; ATP – Adenosine triphosphate; BN-PAGE – Blue native polyacrylamide gel electrophoresis; EM – Electron microscopy; F_0 – Membrane-embedded part of the ATP synthase; F_1 – Extra-membranous part of the ATP synthase; FMN – Flavine mononucleotide; NADH – Nicotinamide adenine dinucleotide reduced form; OXPHOS – Oxidative phosphorylation; SDS – Sodium dodecyl sulphate

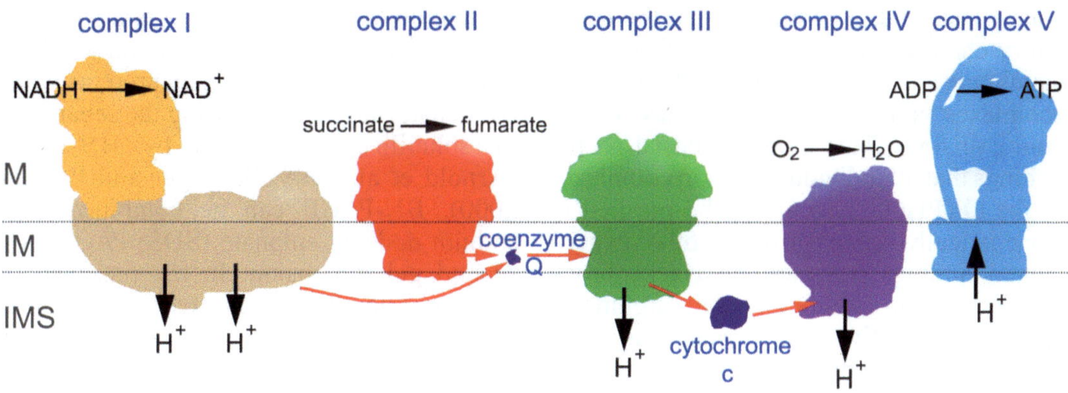

Fig. 12.1. Schematic representation of the mitochondrial respiratory chain. The position of the matrix (*M*), the intermembrane space (*IMS*) and cristae or inner membrane (*IM*) has been indicated.

cytochrome *c* to molecular oxygen thereby reducing the latter to water. The mammalian complex has a molecular mass of about 210 kDa and is composed of 13 subunits, some of which are very hydrophobic (Richter and Ludwig 2003). Four cofactors are attached to the complex (two heme a and two Cu^{2+}). In crystals used to solve the high-resolution structure by X-ray crystallography the bovine complex IV is in a dimeric form (Tsukihara et al. 1996), but the monomeric form was found upon mild detergent solubilization of the inner mitochondrial membrane (Eubel et al. 2003), suggesting that the dimers may not present in the membrane.

Complex V or adenosine triphosphate (ATP) synthase complex (see Chap. 6) uses the chemiosmotic gradient (Mitchell 1961), created by complexes I–IV, to catalyse the formation of ATP by adenosine diphosphate (ADP) phosphorylation at the matrix-exposed side of the inner membrane. It is composed of a water-soluble F_1-headpiece located within the mitochondrial matrix and a hydrophobic F_0-part embedded within the inner mitochondrial membrane, which are linked by a central and a peripheral stalk (Stock et al. 2000). Proton movement across the membrane domain triggers the rotation of the oligomeric ring of c subunits within the F_0 together with the central stalk, which causes conformational changes in the F_1 headpiece resulting in the phosphorylation of ADP. The total molecular mass of complex V lies in the range of 500–600 kDa.

Besides the classical oxidoreductase complexes, some organisms have alternative oxidoreductases. Four distinct alternative NAD(P)H dehydrogenases and one alternative terminal oxidase are described in plants (Møller 2002). All these enzymes participate in electron transport without contributing to the proton gradient across the inner mitochondrial membrane. On the first view this function seems to be a waste of energy. However, in plants the alternative oxidoreductases are believed to form the basis for an "overflow protection mechanism" for the respiratory chain, which is especially important in the presence of high-light conditions. Indeed, experimental evidence has been presented that the mitochondria of plant cells have the task to keep the redox state of the entire cell in a balance (Dutilleul et al. 2003).

III. Supramolecular Organization of the Oxidative Phosphorylation System

For a long time the "fluid-state" model (reviewed in Hackenbrock et al. 1986) of the organization of the OXPHOS system was widely accepted. It postulates that the respiratory chain complexes freely diffuse in the membrane and that electron transfer between complexes takes place on the basis of random collisions. This model is based on the findings that all individual protein complexes of the OXPHOS system can be purified in enzymatically active form, and other experimental findings (reviewed in Hackenbrock et al. 1986). Although the arguments appeared attractive, the model has been abandoned and replaced by the "solid-state model" (Chance and Williams 1955). This model proposes stable interactions between the OXPHOS complexes within larger supercomplexes and is currently supported by different experimental lines of evidence: (1) Supercomplexes can be resolved by blue native polyacrylamide gel electrophoresis (BN-PAGE) (Schägger and Pfeiffer 2000; Eubel et al. 2003). (2) Supercomplexes are active as shown by in-gel activity measurements within blue native gels (Schägger and Pfeiffer 2000; Eubel et al. 2004a). (3) Electron microscopy (EM) structures revealed defined interactions of OXPHOS complexes within isolated respiratory supercomplexes (Dudkina et al. 2005a, b; Minauro-Sanmiguel et al. 2005; Schäfer et al. 2006; Heinemeyer et al. 2007). (4) Flux control experiments (Boumans et al. 1998; Bianchi et al. 2004) confirm that the respiratory chain operates as one functional unit. (5) Point mutations in genes encoding one of the subunits of one OXPHOS complex affect the stability of another complex (Acín-Pérez et al. 2004; Diaz et al. 2006; Suthammarak et al. 2010). (6) Oxygen uptake by isolated mitochondria of potato correlates with the abundance of supercomplexes (Eubel et al. 2004a). (7) Some supercomplexes need cardiolipin for their formation (Zhang et al. 2002; Pfeiffer et al. 2003).

The solid-state model became especially advocated after first evidence for supercomplexes was obtained by BN-PAGE. BN-PAGE is a powerful separation technique based on the solubilization of mitochondrial membranes with non-ionic detergents such as digitonin or dodecyl maltoside, followed by incubation of the generated protein solutions with Coomassie-blue dye. Coomassie-blue introduces negative charges to proteins without denaturing them, allowing the separation of solubilized complexes by BN-PAGE (Arnold et al. 1998; Schägger and Pfeiffer 2000). BN-PAGE can be combined with sodium dodecyl sulphate (SDS) PAGE as a second gel dimension. In the presence of SDS complexes are dissected into separate subunits, which form vertical rows of spots on the resulting 2D gels (Fig. 12.2). Alternatively, first dimension BN-PAGE can be combined with a second BN-PAGE, which is carried out in the presence of a different detergent, allowing the separation of dissection products of supercomplexes or protein complexes. Both 2D gel systems nicely allow to investigate the supramolecular associations of proteins of the OXPHOS system. Sufficiently stable supercomplexes can be structurally analyzed by other methods such as single particle electron microscopy (EM). For this approach, isolated mitochondria are treated with non-ionic detergents and supercomplexes are resolved by sucrose gradient ultracentrifugation (Dudkina et al. 2005a) or electro-eluted from BN-PAGE (Schäfer et al. 2006). Fractions can be directly used for EM analyses and image processing (Dudkina et al. 2005a, b).

By the above described strategy, defined supercomplexes could be identified, which have a $I+III_2$, III_2+IV_{1-2}, $I+III_2+IV_{1-4}$ and V_2 composition. These supercomplexes, comprised of complex I, III and IV, were found in a wide range of organisms (Table 12.1). In other words, there is no specific class of organisms where supercomplexes are highly abundant or fully lacking. However, there are species-depending differences in the stability, which will also be discussed below. Only complex II seems not to form part of any respiratory chain supercomplexes. Although it is involved in the citric acid cycle and electron transfer function towards complex III, its proper function is apparently not crucially depending on the incorporation in a supercomplex. However, Acín-Pérez and colleagues reported the detection of complex II

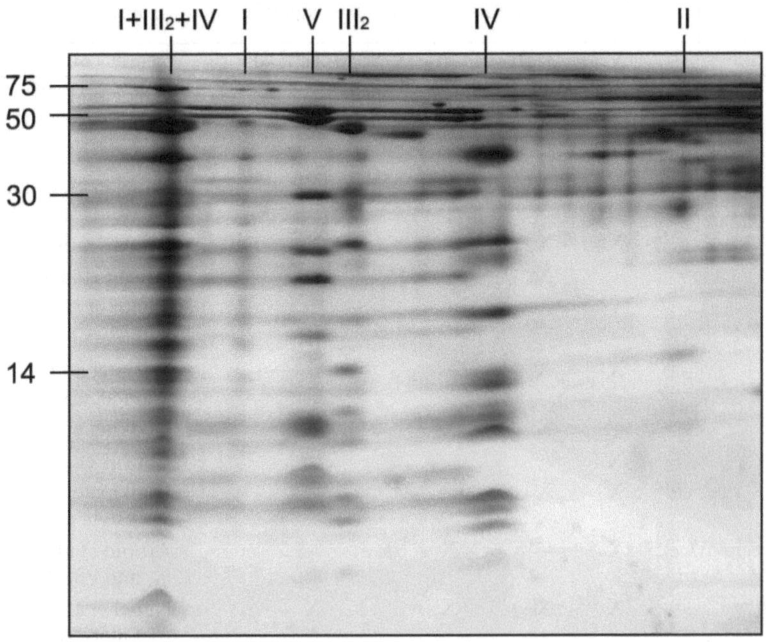

Fig. 12.2. Separation of respiratory chain complexes and supercomplexes from bovine heart by 2D BN/SDS-PAGE. Mitochondrial membranes were solubilized with 25 g of digitonin per g of protein. The molecular weights of standard proteins are given to the left and the identities of the protein complexes above the gel. I+III_2+IV: supercomplex formed by complexes I, dimeric complex III and IV, I: complex I, V: ATP synthase; III_2: dimeric complex III, IV: complex IV, II: complex II.

containing supercomplexes in mouse mitochondria by BN-PAGE (Acín-Pérez et al. 2008). Complexes I, III and IV form a I+III_2+IV_{1-4} supercomplex, the largest unit known to exist in some mitochondrial membranes. It is also called the 'respirasome' because it can autonomously carry out respiration in the presence of the mobile carriers ubiquinone and cytochrome *c*. Although the advantages of the respiratory supercomplexes are obvious, they are considered to co-exist in the membrane with single OXPHOS complexes. On the other hand, treatment of biological membranes with detergent could lead to artificial associations of membrane-protein complexes on the basis of random hydrophobic interaction, but up to now, no clear indications for such artifacts have been found. In contrast, BN-PAGE and electron microscopy analysis demonstrate only defined associations and do not reveal any non-specifically interacting complexes.

The observed interactions between respiratory chain complexes within supramolecular structures should make sense in respect to the known physiological processes. Or in other words, the association of respiratory chain complexes into specific supercomplexes must be functionally advantageous. It was proposed that OXPHOS supercomplexes allow the flow enhancement of electrons between the complexes by reducing the distance of diffusion of the mobile electron carriers ubiquinone and cytochrome *c* or by substrate channeling between complexes (Heinemeyer et al. 2007; Schägger 2001). Supercomplex formation can also be important in preventing excess formation of harmful oxygen radicals (Lenaz and Genova 2009). Besides functional reasons, supercomplex formation is proposed to be necessary for assembly and stability of its individual components. As an example, complex I is necessary for fully assembled

Table 12.1. A list of mitochondrial OXPHOS supercomplexes identified by blue-native PAGE.

Organism	V_2	$I+III_2$	III_2+IV_{1-2}	$I+III_2+IV_{1-4}$	Reference[a]
Arabidopsis thaliana	X	X			Eubel et al. (2003)
Barley		X			Eubel et al. (2003)
Bean		X			Eubel et al. (2003)
Potato	X	X[b]	X	X	Eubel et al. (2003, 2004b); Bultema et al. (2009)
Spinach	X	X	X	X	Krause et al. (2004b)
Tobacco		X			Pineau et al. (2005)
Pea		X			Taylor et al. (2005)
Sunflower				(X)[c]	Sabar et al. (2005)
Zea mays	X	X			Peters et al. (2008); Dudkina et al. (2008)
Asparagus		X	X		Dudkina et al. (2006a)
Chlamydomonas	X				van Lis et al. (2003)
Polytomella	X	X			Dudkina et al. (2005b), Atteia et al. (2003)
Saccharomyces cerevisiae	X	–[d]	X	–[d]	Arnold et al. (1998); Schägger and Pfeiffer (2000)
Yarrowia lipolytica	X	X	X	X	Nübel et al. (2009), Guerrero-Castillo et al. (2009)
Podospora anserina	X	X[e]	X	X[e]	Krause et al. (2004a)
Tetrahymena thermophila	X	X			Balabaskaran et al. (2010)
Bovine	X	X	X	X	Schägger and Pfeiffer (2000)
Homo sapiens		X		X	Schägger et al. (2004)

[a]Reference to the first report on the occurrence of a specific supercomplex

[b]In potato two forms of I+III supercomplexes occur, which have $I+III_2$ and I_2+III_4 composition

[c]In sunflower, a complex IV containing supercomplex of >1,000 kDa was described, which probably has $I+III_2+IV_{1-4}$composition

[d]Complex I is not included in the respiratory chain of *Saccharomyces cerevisiae*, therefore complex-I-containing supercomplexes are absent

[e]In *Podospora anserina* complex I containing supercomplexes were reported to have I_2 and I_2III_2 composition

complex III in human patients with mutations in the complex I subunits NDUFS2 and NDUFS4 (Ugalde et al. 2004) and the absence in complex III results in a dramatic loss of complex I in humans (Blakely et al. 2005). Furthermore, complex IV is required for the assembly of complex I in fibroblasts (Diaz et al. 2006), complex III_2 is essential for the stability of complex I in mouse cells (Acín-Pérez et al. 2004) and the respirasome in *Caenorhabditis elegans* (Suthammarak et al. 2010).

The occurrence of ATP synthase dimers has a special function. Dimerization induces a local curvature of the cristae membrane. Ultrastructural studies were performed on yeast mutants which lack ATP synthase dimers because the dimer-specific e and g subunits are absent. The mitochondrial membranes of these mutant strains exhibit an unusual onion-shape morphology without any membrane foldings (Paumard et al. 2002; Giraud et al. 2002). This strongly suggests that the dimerization of ATP synthases is essential for cristae formation. A recent study on mammalian and algal mitochondria proved the organization of ATP synthases in long rows of dimers (Strauss et al. 2008; Dudkina et al. 2010a). The mitochondrial cristae may act as proton traps and enhance the local proton gradient necessary optimizing in this way their own performance under proton-limited conditions (Strauss et al. 2008).

IV. Respiratory Supercomplexes

A. I+III$_2$, III$_2$+IV$_{1-2}$ and I+III$_2$+IV$_{1-4}$ Supercomplexes

On BN gels, complex I forms a stable association with the complex III dimer of the respiratory chain (Table 12.1, Schägger and Pfeiffer 2000; Eubel et al. 2003). Single-particle EM revealed the lateral association of dimeric complex III with the tip of the membrane part of complex I in *Arabidopsis* (Dudkina et al. 2005a), *Zea mays* (Peters et al. 2008), potato (Bultema et al. 2009) and *Tetrahymena thermophila* (Balabaskaran et al. 2010). The 2-dimensional EM projection maps clearly show the curved shape of the membrane-embedded part of complex I (Dudkina et al. 2005a; Peters et al. 2008; Bultema et al. 2009).

Complex III$_2$ can associate with one or two copies of complex IV as it was described for potato (Bultema et al. 2009), spinach (Krause et al. 2004b), *Asparagus* (Dudkina et al. 2006a), *Saccharomyces cerevisiae* (Heinemeyer et al. 2007) and bovine heart (Schägger and Pfeiffer 2000). Baker's yeast mitochondria lack complex I and in this organism complex III$_2$ exclusively binds to complex IV, forming a very stable III$_2$+IV$_{1-2}$ supercomplex (Heinemeyer et al. 2007). The first detailed EM structure of this supercomplex allowed the generation of a pseudo-atomic 3D model (Heinemeyer et al. 2007), which showed that monomeric cytochrome *c* oxidase complexes are attached to dimeric complex III at two alternate sides with their convex sides facing complex III$_2$. This is the opposite to the side involved in the formation of complex IV dimers as described by X-ray crystallography. Association of the complex III with complex IV depends on the presence of cardiolipin within the inner mitochondrial membrane (Zhang et al. 2002; Pfeiffer et al. 2003). Another EM study on potato respiratory chain revealed the III$_2$-IV$_1$ supercomplex with similar assembly features (Bultema et al. 2009). These particles were also identified in bovine mitochondria but at lower concentration (Schägger and Pfeiffer 2000).

Analyses of bovine mitochondria by BN-PAGE revealed abundant supercomplexes consisting of the OXPHOS complexes I, III$_2$ and IV (Fig. 12.2). Up to four copies of complex IV are present within these supercomplexes, which represent the largest form of OXPHOS units and are termed "respirasomes" (Schägger and Pfeiffer 2000).

The structure obtained by cryo-electron tomography and subtomogram averaging demonstrated that the complex III$_2$ is attached to the middle portion of the membrane arm of complex I while complex IV attaches to the tip of NADH dehydrogenase (Dudkina et al. 2011) (Fig. 12.3). Fitting of X-ray structures clearly showed some space between the complexes I, III$_2$ and IV (Fig. 12.3c). This space is most likely filled with lipids, which could play a stabilising role as it was shown for cardiolipin in the III$_2$+IV$_2$ supercomplex (Pfeiffer et al. 2003). One of the main functions of the respirasome is believed to be the more efficient electron transfer between its single components due to the shortening of travelling distances for ubiquinone and cytochrome *c* (Fig. 12.3d).

B. Dimeric ATP Synthase

Mitochondrial F$_1$F$_0$ ATP synthase is a complex of 600 kDa formed by 15–18 subunits (Stock et al. 2000). The water-soluble F$_1$-part consists of three α and three β catalytic subunits. The F$_1$-part is connected to the membrane-embedded ring-like subunit c oligomer of the F$_0$-part by a central and a peripheral stalk. The F$_0$-part is composed of subunits a (Su 6), A6L (Su 8), e, f, g, the central stalk consists of the γ, δ and ε subunits and the peripheral stalk is made from subunits OSCP (Su 5), b, d, F6 (h) (the nomenclature of subunits in yeast is indicated in brackets, reviewed in Wittig and Schägger 2008). The yeast enzyme has two specific additional subunits i and k, which belong to the transmembrane part. The first biochemical evidence about a dimeric organization of the ATP synthases complex in yeast came from

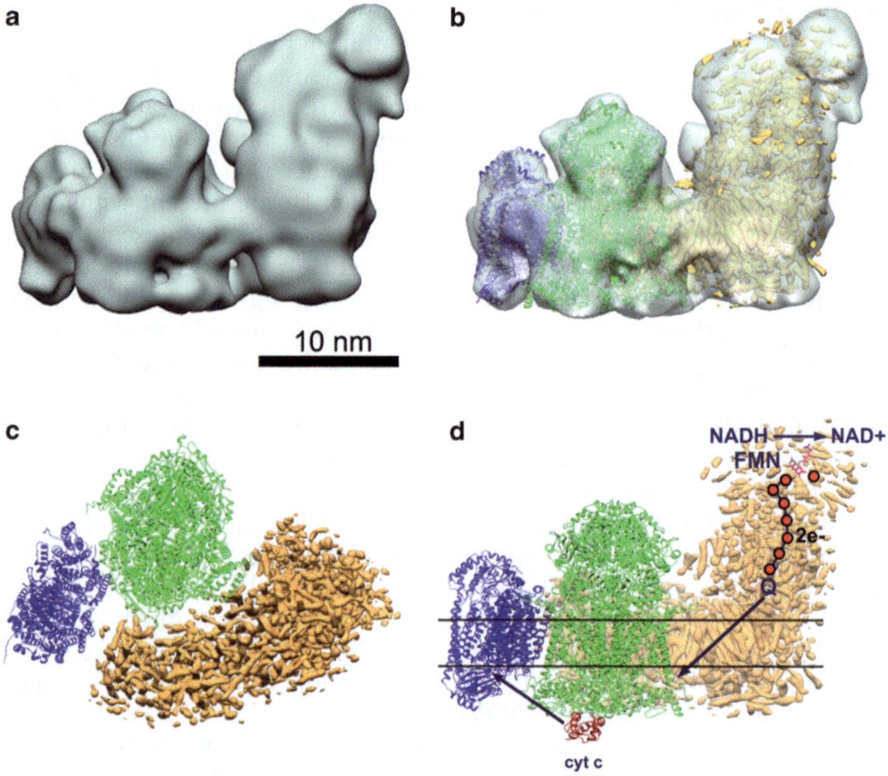

Fig. 12.3. Three-dimensional cryo-electron microscopy density map of bovine I+III_2+IV supercomplex and scheme of electron pathways within the supercomplex. (**a**) Surface representation of the map from aside. (**b**) Fitting of X-ray structures of complexes I, III_2 and IV to the map. (**c**) Pseudo atomic model of respirasome seen from the intermembrane space. (**d**) Scheme of electron flow in the respirasome. In *green*, *purple* and *red* X-ray structures of the bovine dimeric complex III_2 (Iwata et al. 1998), monomeric complex IV (Tsukihara et al. 1996) and cytochrome *c* correspondingly. In *yellow*: the density map of complex I from Yarrowia lipolytica (Hunte et al. 2010). *Horizontal lines* indicate the position of the membrane, *red balls* show the positions of iron-sulphur clusters. *Q* is ubiquinol, *cyt c* is cytochrome *c* (Modified from Dudkina et al. 2011).

the BN-PAGE work of Arnold and colleagues (1998). The 1,200 kDa dimer includes dimer-specific subunits named e, g and k, which were not detected in monomers. Later, dimers were found in beef, *Arabidopsis* and several other organisms (Schägger and Pfeiffer 2000; Eubel et al. 2003). The dimer-specific subunits e and g, as well as the subunits a, b, h and i of the monomer, stabilize the monomer–monomer interface (Wittig and Schägger 2008). Low-resolution 2D structures of dimeric ATP synthase were obtained from bovine mitochondria (Minauro-Sanmiguel et al. 2005), the colorless green alga *Polytomella* (Dudkina et al. 2005b) and *Saccharomyces cerevisiae* (Dudkina et al. 2006b). In all these organisms two monomers associate via the membrane F_0 parts and make an angle of 35–90°. Dimeric ATP synthase from green algae (Fig. 12.4d) seems to be very stable and is represented exclusively by molecules with an angle of 70°.

Although the ATP synthase dimers were first described in the late 1990s (Arnold et al. 1998), earlier work even hints at a higher type of organization as oligomers. The presence of rows of ATP synthase dimers was demonstrated in mitochondria of *Paramecium* by rapid-freeze deep-etch electron microscopy (Allen et al. 1989). BN-PAGE evidence of trimeric and tetrameric organizations of mammalian ATP synthases (Krause et al. 2005)

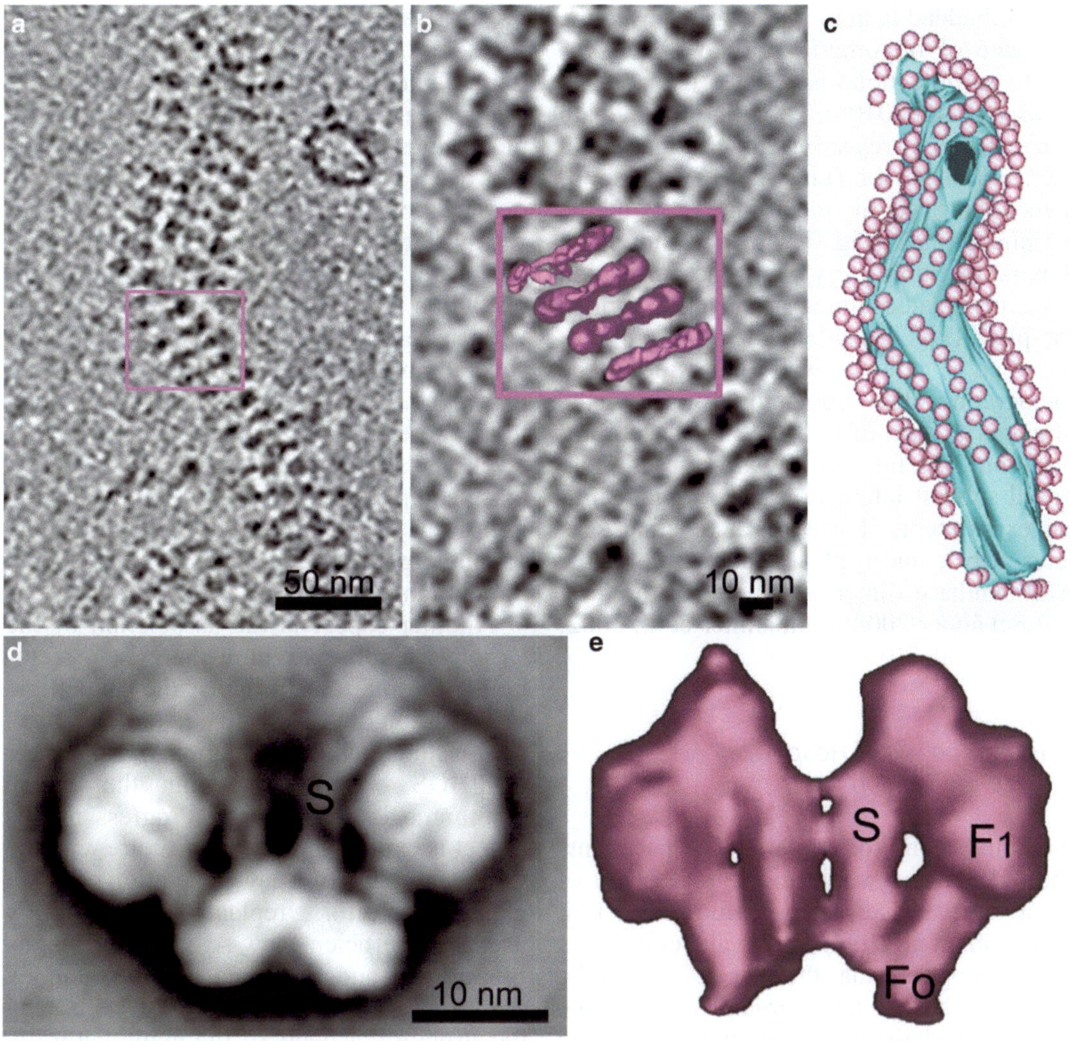

Fig. 12.4. Dimeric ATP synthase. (**a**) Fragment of tubular cristae obtained by cryo-electron tomography of the whole mitochondrion, (**b**) a closer look at the cristae with ATP synthase oligomers in a *top-view* position, corresponding to the *pink square* in the frame (**a**, **c**); surface rendering of the cristae (*blue-green*) from the frame (**a**) in the same orientation, demonstrating the row of dimeric ATP synthases (*pink*), (**d**) average of negatively stained *side-views* of isolated dimeric ATP synthase, (**e**) average of tomographic sub-volumes of ATP synthase dimer seen from a side revealing the F_1 headpiece, the F_0 membrane part and the peripheral stalk (*S*) or stator (Modified from Dudkina et al. 2010a).

and evidence from ultrasectioning and negative staining of osmotically shocked *Polytomella* mitochondria (Dudkina et al. 2006b) support the hypothesis of an oligomeric arrangement of ATP synthases.

Some time ago data based on electron cryo-tomography were presented, which show that the ATP synthases of mammalian mitochondria are arranged in long rows of dimers located at the apex of cristae membranes (Strauss et al. 2008). Sub-tomogram averaging revealed larger angles between monomers than reported for isolated bovine dimeric ATP synthase (Minauro-Sanmiguel et al. 2005). Cryo-electron tomography experiments on intact *Polytomella* mitochondria embedded in amorphous ice showed the presence of oligomeric rows of ATP synthases in mitochondrial cristae (Fig. 12.4a, Dudkina et al. 2010b). The oligomeric chain is composed of repeating dimeric units with a spacing of 12 nm (Fig. 12.4b). The monomers contact via the membrane part and the peripheral stalk (Fig. 12.4e) and individual dimers resemble very well the projection maps of isolated dimers in negative stained samples (Fig. 12.4d). The dimer–dimer interface is not yet well understood. The tomography data convincingly show the oligomeric state of the ATP synthase complex, but are far too low in resolution to reveal subunit interactions. From biochemical experiments it was predicted that subunits a, b, e and g play a major role in the ATP synthase dimerization process (Wittig and Schägger 2008; Soubannier et al. 2002; Velours et al. 2011).

C. Higher Levels of Respiratory Chain Organization

Previous inhibitor titration experiments showed that the whole respiratory chain of yeast behaves like a single functional unit (Boumans et al. 1998). Wittig and colleagues (2006) proposed that the respirasome is the unit of a higher level organization of the respiratory chain, that is structured into long "string"-like structures and interconnected with III_2+IV_4 supercomplexes. Such an organization could provide more effective respiration by making the electron transfer rate through the OXPHOS system more efficient. However, the recent work based on cryo-electron tomography excludes the existence of such strings (Davies et al. 2011). Oligomeric ATP synthases were found along cristae edges of lamellae, while complexes I-IV would be expected on flat surface of cristae (Davies et al. 2011). In contrast, the tubular cristae of the green alga *Polytomella* were fully covered by oligomeric rows of ATP synthases (Dudkina et al. 2010a), suggesting that the other OXPHOS enzymes might be arranged in between.

V. Perspectives

Large amounts of data on the OXPHOS system have been collected with different biochemical and biophysical techniques. However, a complete picture of the protein organization in the cristae membrane is not yet available. Multi-subunit membrane proteins are difficult objects for any high-resolution structural investigation method. Higher resolution information on specific supercomplexes than available now is needed to give a full understanding of subunit interactions and electron transport at the molecular level. The structures of the dimeric ATP synthase complex and the respirasome could be improved by either X-ray crystallography or single particle cryo-electron microscopy/tomography.

Another aspect is the association of the various complexes within the cristae membrane. Despite the fact, that the oligomeric organization of ATP synthases has now been demonstrated, the association of the other OXPHOS complexes, within the cristae membrane is still unclear. Progress in this aspect is expected in the near future with the implementation of a new generation of direct electron detectors, replacing the currently used, but inefficient slow-scan CCD cameras. Finally, our understanding of supercomplex organization will also increase by studying the structural aspects of the many naturally occurring mutant strains with variations in the OXPHOS system.

Acknowledgments

Research in our laboratories is supported by The Netherlands Organisation for Scientific Research (NWO) and the Deutsche Forschungsgemeinschaft (DFG).

References

Acín-Pérez R, Bayona-Bafaluy M, Fernández-Silva P, Moreno-Loshuertos R, Pérez-Martos A, Bruno C, Moraes C, Enríquez J (2004) Respiratory complex

III is required to maintain complex I in mammalian mitochondria. Mol Cell 13:805–815

Acín-Pérez R, Fernández-Silva P, Peleato ML, Pérez-Martos A, Enriquez JA (2008) Respiratory active mitochondrial supercomplexes. Mol Cell 32: 529–539

Allen RD, Schroeder CC, Fok AK (1989) An investigation of mitochondrial inner membranes by rapid-freeze deep-etch techniques. J Cell Biol 108:2233–2240

Arnold I, Pfeiffer K, Neupert W, Stuart RA, Schägger H (1998) Yeast mitochondrial F_1F_0-ATP synthase exists as a dimer: identification of three dimer-specific subunits. EMBO J 17:7170–7178

Atteia A, van Lis R, Mendoza-Hernandez G, Henze K, Martin W, Riveros-Rosas H, Gonzalez-Halphen D (2003) Bifunctional aldehyde/alcohol dehydrogenase (ADHE) in chlorophyte algal mitochondria. Plant Mol Biol 53:175–188

Balabaskaran NP, Dudkina NV, Kane LA, van Eyk JE, Boekema EJ, Mather MW, Vaidya AB (2010) Highly divergent mitochondrial ATP synthase complexes in *Tetrahymena thermophila*. PLoS Biol 8(7):e1000418

Berry EA, Guergova-Kuras M, Huang LS, Crofts AR (2000) Structure and function of cytochrome *bc* complexes. Annu Rev Biochem 69:1005–1075

Bianchi C, Genova ML, Castelli GP, Lenaz G (2004) The mitochondrial respiratory chain is partially organized in a supercomplex assembly: kinetic evidence using flux control analysis. J Biol Chem 279:36562–36569

Blakely EL, Mitchell AL, Fisher N, Meunier B, Nijtmans LG, Schaefer AM, Jackson MJ, Turnbull DM, Taylor RW (2005) A mitochondrial cytochrome *b* mutation causing severe respiration chain enzyme deficiency in humans and yeast. FEBS J 14:3583–3592

Boumans H, Grivell LA, Berden JA (1998) The respiratory chain in yeast behaves as a single functional unit. J Biol Chem 273:4872–4877

Braun HP, Emmermann M, Kruft V, Schmitz UK (1992) The general mitochondrial processing peptidase from potato is an integral part of cytochrome *c* reductase of the respiratory chain. EMBO J 11: 3219–3227

Bultema JB, Braun HP, Boekema EJ, Kouřil R (2009) Megacomplex organization of the oxidative phosphorylation system by structural analysis of respiratory supercomplexes from potato. Biochim Biophys Acta 1767:60–67

Carroll J, Fearnley IM, Shannon RJ, Hirst J, Walker JE (2003) Analysis of the subunit composition of complex I from bovine heart mitochondria. Mol Cell Proteomics 2:117–126

Chance B, Williams GR (1955) A method for the localization of sites for oxidative phosphorylation. Nature 176:250–254

Davies KM, Strauss M, Daum B, Kief JH, Osiewacz HD, Rycovska A, Zickermann V, Kühlbrandt W (2011) Macromolecular organization of ATP synthase and complex I in whole mitochondria. Proc Natl Acad Sci U S A 108:14121–14126

Diaz F, Fukui H, Garcia S, Moraes CT (2006) Cytochrome *c* oxidase is required for the assembly/stability of respiratory complex I in mouse fibroblasts. Mol Cell Biol 26:4872–4881

Dudkina NV, Eubel H, Keegstra W, Boekema EJ, Braun HP (2005a) Structure of a mitochondrial supercomplex formed by respiratory-chain complexes I and III. Proc Natl Acad Sci U S A 102:3225–3229

Dudkina NV, Heinemeyer J, Keegstra W, Boekema EJ, Braun HP (2005b) Structure of dimeric ATP synthase from mitochondria: an angular association of monomers induces the strong curvature of the inner membrane. FEBS Lett 579:5769–5772

Dudkina NV, Heinemeyer J, Sunderhaus S, Boekema EJ, Braun HP (2006a) Respiratory chain supercomplexes in the plant mitochondrial membrane. Trends Plant Sci 11:232–240

Dudkina NV, Sunderhaus S, Braun HP, Boekema EJ (2006b) Characterization of dimeric ATP synthase and cristae membrane ultrastructure from *Saccharomyces* and *Polytomella* mitochondria. FEBS Lett 580:3427–3432

Dudkina NV, Sunderhaus S, Boekema EJ, Braun HP (2008) The higher level of organization of the oxidative phosphorylation system: mitochondrial supercomplexes. J Bioenerg Biomembr 40:419–424

Dudkina NV, Oostergetel GT, Lewejohann D, Braun HP, Boekema EJ (2010a) Row-like organization of ATP synthase in intact mitochondria determined by cryo-electron tomography. Biochim Biophys Acta 1797:272–277

Dudkina NV, Kouřil R, Bultema JB, Boekema EJ (2010b) Imaging of organelles by electron microscopy reveals protein-protein interactions in mitochondria and chloroplasts. FEBS Lett 584:2510–2515

Dudkina NV, Kudryashev M, Stahlberg H, Boekema EJ (2011) Interaction of complexes I, III and IV within the bovine respirasome by single particle cryoelectron tomography. Proc Natl Acad Sci U S A 108:15196–15200

Dutilleul C, Garmier M, Noctor G, Mathieu C, Chétrit P, Foyer CH, De Paepe R (2003) Leaf mitochondria modulate whole cell redox homeostasis, set antioxidant capacity, and determine stress resistance through altered signaling and diurnal regulation. Plant Cell 15:1212–1226

Efremov RG, Baradaran R, Sazanov LA (2010) The architecture of respiratory complex I. Nature 465:441–445

Eubel H, Jänsch L, Braun HP (2003) New insights into the respiratory chain of plant mitochondria supercomplexes and a unique compositions of complex II. Plant Physiol 133:274–286

Eubel H, Heinemeyer J, Sunderhaus S, Braun HP (2004a) Respiratory chain supercomplexes in plant mitochondria. Plant Physiol Biochem 42:937–942

Eubel H, Heinemeyer J, Braun HP (2004b) Identification and characterization of respirasomes in potato mitochondria. Plant Physiol 134:1450–1459

Giraud MF, Paumard P, Soubannier V, Vaillier J, Arselin G, Salin B, Schaeffer J, Brethes D, di Rago P, Velours J (2002) Is there a relationship between the supramolecular organization of the mitochondrial ATP synthase and the formation of cristae? Biochim Biophys Acta 1555:174–180

Guerrero-Castillo S, Vázquez-Acevedo M, González-Halphen D, Uribe-Carvajal S (2009) In *Yarrowia lipolytica* mitochondria, the alternative NADH dehydrogenase interacts specifically with the cytochrome complexes of the classic respiratory pathway. Biochim Biophys Acta 1787:75–85

Hackenbrock CR, Chazotte B, Gupte SS (1986) The random collision model and a critical assessment of diffusion and collision in mitochondrial electron transport. J Bioenerg Biomembr 427(18): 331–368

Heinemeyer J, Braun HP, Boekema EJ, Kouřil R (2007) A structural model of the cytochrome *c* reductase/oxidase supercomplex from yeast mitochondria. J Biol Chem 282:12240–12248

Horsefield R, Iwata S, Byrne B (2004) Complex II from a structural perspective. Curr Protein Pept Sci 5:107–118

Hunte C, Koepke J, Lange C, Rossmanith T, Michel H (2000) Structure at 2.3 Å resolution of the cytochrome bc_1 complex from the yeast *Saccharomyces cerevisiae* co-crystallized with an antibody Fv fragment. Structure 8:669–684

Hunte C, Zickermann V, Brandt U (2010) Functional modules and structural basis of conformational coupling in mitochondrial complex I. Science 329: 448–451

Iwata S, Lee JW, Okada K, Lee JK, Iwata M, Rasmussen B, Link TA, Ramaswamy S, Jap BK (1998) Complete structure of the 11-subunit bovine mitochondrial cytochrome bc_1 complex. Science 281:64–71

Klodmann J, Braun HP (2011) Proteomic approach to characterize mitochondrial complex I from plants. Phytochemistry 72:1071–1080

Krause F, Scheckhuber CQ, Werner A, Rexroth S, Reifschneider NH, Dencher NA, Osiewacz HD (2004a) Supramolecular organization of cytochrome *c* oxidase- and alternative oxidase-dependent respiratory chains in the filamentous fungus *Podospora anserina*. J Biol Chem 279:26453–26461

Krause F, Reifschneider NH, Vocke D, Seelert H, Rexroth S, Dencher NA (2004b) "Respirasome"-like supercomplexes in green leaf mitochondria of spinach. J Biol Chem 279:48369–48375

Krause F, Reifschneider NH, Goto S, Dencher NA (2005) Active oligomeric ATP synthases in mammalian mitochondria. Biochem Biophys Res Commun 329:583–590

Lenaz G, Genova ML (2009) Structural and functional organization of the mitochondrial respiratory chain: a dynamic super-assembly. Int J Biochem Cell Biol 41:1750–1772

Millar AH, Eubel H, Jänsch L, Kruft V, Heazlewood L, Braun HP (2004) Mitochondrial cytochrome *c* oxidase and succinate dehydrogenase contain plant-specific subunits. Plant Mol Biol 56:77–89

Minauro-Sanmiguel F, Wilkens S, García JJ (2005) Structure of dimeric mitochondrial ATP synthase: novel F_0 bridging features and the structural basis of mitochondrial cristae biogenesis. Proc Natl Acad Sci U S A 102:12356–12358

Mitchell P (1961) Coupling of phosphorylation to electron and hydrogen transfer by a chemi-osmotic type of mechanism. Nature 191:144–148

Møller IM (2002) A new dawn for plant mitochondrial NAD(P)H dehydrogenases. Trends Plant Sci 7:235–237

Nübel E, Wittig I, Kerscher S, Brandt U, Schägger H (2009) Two-dimensional native electrophoretic analysis of respiratory supercomplexes from *Yarrowia lipolytica*. Proteomics 9:2408–2418

Paumard P, Vaillier J, Coulary B, Schaeffer J, Soubannier V, Mueller DM, Brethes D, di Rago JP, Velours J (2002) The ATP synthase is involved in generating mitochondrial cristae morphology. EMBO J 21:221–230

Peters K, Dudkina NV, Jänsch L, Braun HP, Boekema EJ (2008) A structural investigation of complex I and I+III_2 supercomplex from *Zea mays* at 11–13 Å resolution: assignment of the carbonic anhydrase domain and evidence for structural heterogeneity within complex I. Biochim Biophys Acta 1777:84–93

Pfeiffer K, Gohil V, Stuart RA, Hunte C, Brandt U, Greenberg ML, Schägger H (2003) Cardiolipin stabilizes respiratory chain supercomplexes. J Biol Chem 278:52873–52880

Pineau B, Mathieu C, Gerard-Hirne C, De Paepe R, Chetrit P (2005) Targeting the NAD7 subunit to mitochondria restores a functional complex I and a wild type phenotype in the *Nicotiana sylvestris* CMS II mutant lacking nad7. J Biol Chem 280:25994–26001

Richter OM, Ludwig B (2003) Cytochrome *c* oxidase–structure, function, and physiology of a redox-driven molecular machine. Rev Physiol Biochem Pharmacol 147:47–74

Sabar M, Balk J, Leaver CJ (2005) Histochemical staining and quantification of plant mitochondrial respiratory chain complexes using blue-native polyacrylamide gel electrophoresis. Plant J 44:893–901

Sazanov LA, Hinchliffe P (2006) Structure of the hydrophilic domain of respiratory complex I from *Thermus thermophilus*. Science 311:1430–1436

Schäfer E, Seelert H, Reifschneider NH, Krause F, Dencher NA, Vonck J (2006) Architecture of active mammalian respiratory chain supercomplexes. J Biol Chem 281:15370–15375

Schägger H (2001) Respiratory chain supercomplexes. IUBMB Life 52:119–128

Schägger H, Pfeiffer K (2000) Supercomplexes in the respiratory chains of yeast and mammalian mitochondria. EMBO J 19:1777–1783

Schägger H, de Coo R, Bauer MF, Hofmann S, Godinot C, Brandt U (2004) Significance of respirasomes for the assembly/stability of human respiratory chain complex I. J Biol Chem 279:36349–36353

Soubannier V, Vaillier J, Paumard P, Coulary B, Schaeffer J, Velours J (2002) In the absence of the first membrane-spanning segment of subunit 4(b), the yeast ATP synthase is functional but does not dimerize or oligomerise. J Biol Chem 277:10739–10745

Stock D, Gibbons C, Arechaga I, Leslie AGW, Walker JE (2000) Rotary mechanism of ATP synthase. Curr Opin Struct Biol 10:672–679

Strauss M, Hofhaus G, Schröder RR, Kühlbrandt W (2008) Dimer ribbons of ATP synthase shape the inner mitochondrial membrane. EMBO J 27:1154–1160

Suthammarak W, Morgan PG, Sedensky MM (2010) Mutations in mitochondrial complex III uniquely affect complex I in *Caenorhabditis elegans*. J Biol Chem 285:40724–40731

Taylor NL, Heazlewood JL, Day DA, Millar AH (2005) Differential impact of environmental stresses on the pea mitochondrial proteome. Mol Cell Proteomics 4:1122–1133

Tsukihara T, Aoyama H, Yamashita E, Tomizaki T, Yamaguchi H, Shinzawa-Itoh K, Nakashima R, Yaono R, Yoshikawa S (1996) The whole structure of the 13-subunit oxidized cytochrome *c* oxidase at 2.8 Å. Science 272:1136–1144

Ugalde C, Janssen RJ, van den Heuvel LP, Smeitink JA, Nijtmans LG (2004) Differences in assembly or stability of complex I and other mitochondrial OXPHOS complexes in inherited complex I deficiency. Hum Mol Genet 13:659–667

van Lis R, Atteia A, Mendoza-Hernandez G, Gonzalez-Halphen D (2003) Identification of novel mitochondrial protein components of *Chlamydomonas reinhardtii*. A proteomic approach. Plant Physiol 132:318–330

Velours J, Stines-Chaumeil C, Habersetzer J, Chaignepain S, Dautant A, Brethes D (2011) Evidence of the proximity of ATP synthase subunits 6 (a) in the innermitochondrial membrane and in the supramolecular forms of Saccharomyces cerevisiae ATP synthase. J Biol Chem 286:35477–35484

Wittig I, Schägger H (2008) Structural organization of mitochondrial ATP synthase. Biochim Biophys Acta 1777:592–598

Wittig I, Carrozzo R, Santorelli FM, Schägger H (2006) Supercomplexes and subcomplexes of mitochondrial oxidative phosphorylation. Biochim Biophys Acta 1757:1066–1072

Zhang M, Mileykovskaya E, Dowhan W (2002) Gluing the respiratory chain together. Cardiolipin is required for supercomplex formation in the inner mitochondrial membrane. J Biol Chem 277:43553–43556

Chapter 13

Energy Conservation in Heliobacteria: Photosynthesis and Central Carbon Metabolism

W. Matthew Sattley*
Division of Natural Sciences, Indiana Wesleyan University, 4201 S. Washington St., Marion, IN 46953-4974, USA

Marie Asao
Department of Microbiology, The Ohio State University, 484 W, 12th Avenue, Columbus, OH 43210-1292, USA

Joseph Kuo-Hsiang Tang
Carlson School of Chemistry and Biochemistry, and Department of Biology, Clark University, 950 Main Street, Worcester, MA 01610, USA

and

Aaron M. Collins
Center for Integrated Nanotechnologies, Los Alamos National Laboratory, P.O. Box 1663, MS K771, Los Alamos, NM 87185-1413, USA

*Author for correspondence, e-mail: matthew.sattley@indwes.edu

M.F. Hohmann-Marriott (ed.), The Structural Basis of Biological Energy Generation, Advances in Photosynthesis and Respiration 39, pp. 231–247, DOI 10.1007/978-94-017-8742-0_13,

Summary

Heliobacteria are a group of anoxygenic phototrophic bacteria that use a unique pigment, bacteriochlorophyll *g*, for photosynthetic energy conversion within a type I homodimeric reaction center. Like their nonphotosynthetic relatives the clostridia, heliobacteria have a gram-positive cell structure and can form heat-resistant endospores. Heliobacteria are also unusual in that they are the only anaerobic anoxygenic phototrophs that lack a mechanism for autotrophic growth. Growth of heliobacteria is therefore dependent upon the presence of usable organic carbon sources and occurs either photoheterotrophically or chemotrophically (via pyruvate fermentation). While knowledge of heliobacterial photosynthesis and physiology has steadily increased since the relatively recent discovery of these phototrophs in the 1980s, high-resolution structural data pertaining to features of the heliobacterial photosynthetic apparatus are not yet available. This chapter summarizes our current understanding of energy conservation in heliobacteria as it relates to central carbon metabolism (in both light and dark conditions), electron transport, and light harvesting and photochemistry within the reaction center.

I. Introduction

A. Cellular Structure, Growth, and Ecology of Heliobacteria

Heliobacteria are a major group of anoxygenic phototrophic bacteria. Unique among all phototrophic bacteria, heliobacteria lack an outer membrane and have a gram-positive cell wall structure. Like species of *Clostridium* and *Bacillus*, heliobacteria are members of the bacterial phylum *Firmicutes* and cells of heliobacteria can differentiate into heat-resistant endospores containing the signature features of these structures, dipicolinic acid and sequestered Ca^{2+} (Ormerod et al. 1990; Madigan 2001; Madigan et al. 2006; Overmann et al. 2006). Also contrary to other phototrophic bacteria, transmission electron micrographs of cells of heliobacteria show no specialized pigment-containing structures (e.g., chromatophores, membrane lamellae, or chlorosomes (see Chap. 5) (Kimble et al. 1995) (Fig. 13.1). Instead, photosynthetic pigments are localized within the cytoplasmic membrane and are incorporated into a comparatively simple type I or FeS-type homodimeric reaction center (RC) (Table 13.1). The pigments themselves are also unique to heliobacteria. The major photosynthetic pigment is bacteriochlorophyll (BChl) *g*, but a small amount of 8^1–OH–chlorophyll (Chl) *a* is also present in the heliobacterial RC (discussed in Sect. II below; Table 13.1) (Fuller et al. 1985; Trost and Blankenship 1989; van de Meent et al. 1991; Amesz 1995a). Carotenoids in heliobacteria vary between species (discussed below). Neutrophilic heliobacteria (pH optimum near 7) contain C_{30} carotenoid pigments that primarily consist of 4,4′-diaponeurosporene, while alkaliphilic species (pH optimum 8–9) contain OH-diaponeurosporene glucoside esters (Takaichi et al. 1997, 2003).

Abbreviations: A_0 – Chlorophyll acting as an electron carrier within a reaction center; A_1 – Quinone acting as an electron carrier within a reaction center; ATP – Adenosine triphosphate; BChl –Bacteriochlorophyll; Chl – Chlorophyll; CoA – Coenzyme A; ED pathway – Entner-Doudoroff pathway; EMP pathway – Embden-Meyerhof-Parnas pathway; FeS-type RC –Reaction center where iron-sulfur clusters are the terminal electron carriers; F_X F_A, F_B – Iron-sulfur electron carrier; GSB – Green sulfur bacteria; KDH – α-ketoglutarate dehydrogenase; OTCA cycle – Oxidative or forward tricarboxylic acid cycle; P_{800} – Primary electron donor in the heliobacterial RC; PEPCK – Phosphoenolpyruvate carboxykinase; PFOR – Pyruvate:ferredoxin oxidoreductase; PSI – Photosystem I; Q-type RC – Reaction center where quinones are the terminal electron acceptors; RC – Reaction center; RTCA cycle – Reductive or reverse tricarboxylic acid cycle; TCA cycle – Tricarboxylic acid cycle

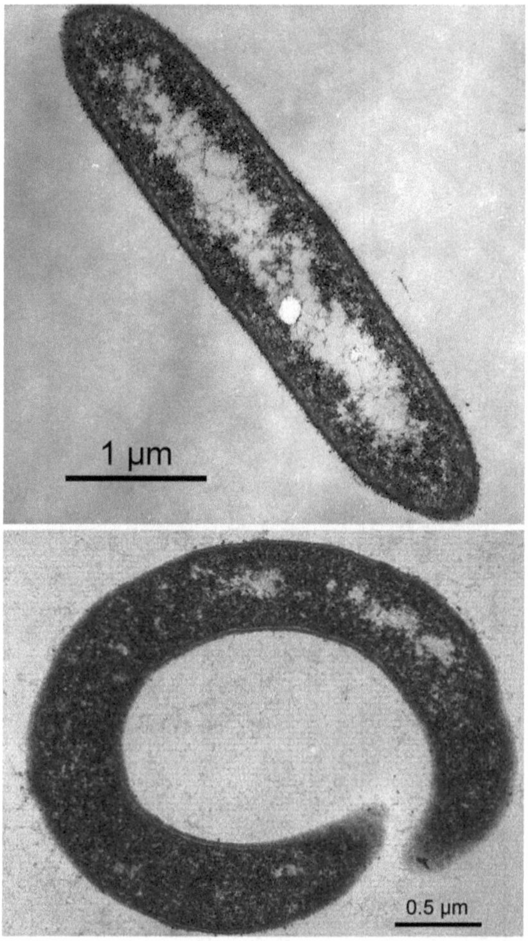

Fig. 13.1. Transmission electron micrographs of heliobacteria. *Top*: *Heliobacterium modesticaldum* (*neutrophilic*). *Bottom*: *Heliorestis convoluta* (*alkaliphilic*). Note the gram-positive cell wall structure and lack of intracytoplasmic photosynthetic membranes or chlorosomes.

In stark contrast to all other anaerobic anoxygenic phototrophic bacteria, heliobacteria are obligate heterotrophs; a limited range of single carbon sources are used for growth. Pyruvate supports good growth of all known species, while lactate and acetate (plus CO_2) are used by most, but not all, heliobacteria. Carbon sources used by only a few species of heliobacteria include ribose, glucose, fructose, casein hydrolysate, malate, ethanol (+CO_2), butyrate (+CO_2), and propionate (+CO_2) (Asao and Madigan 2010). While most neutrophilic heliobacteria can also use yeast extract as a carbon source, alkaliphilic heliobacteria have not shown this property (Asao and Madigan 2010).

Heliobacteria occupy a unique ecological niche in that they are endospore-forming strict anaerobes that primarily inhabit soils (Table 13.1), and in this regard they can be considered "phototrophic clostridia" (Stevenson et al. 1997; Madigan et al. 2006, b; Asao and Madigan 2010). The ability of many heliobacteria to grow chemotrophically (anoxic/dark) by the fermentation of pyruvate to acetate supports this description (Kimble et al. 1994; Pickett et al. 1994). Such a view is also consistent with recent evidence suggesting that the primary pathway of carbon flow during growth of heliobacteria is through an incomplete oxidative tricarboxylic acid (OTCA) cycle (Tang et al. 2010a). The major pathways of central carbon metabolism in heliobacteria are discussed in more detail in Sect. III below.

B. Taxonomy and Phylogeny of Heliobacteria

Heliobacteria are the only phototrophic bacteria to group within the *Firmicutes* (Table 13.1). Cultured species of heliobacteria can be placed into one of two groups based on their growth pH optima. The neutrophilic heliobacteria consist of seven species of the genera *Heliobacterium*, *Heliobacillus*, and *Heliophilum*. The other group consists of alkaliphilic heliobacteria of the genus *Heliorestis* (Fig. 13.2). For a full summary of properties of described species of heliobacteria to date, see Asao and Madigan (2010).

Although the pH relations of heliobacteria are largely consistent with their phylogenetic grouping based on 16S rRNA gene sequences (Fig. 13.2), a newly characterized alkaliphilic heliobacterium, "*Candidatus* Heliomonas lunata," is a novel exception since it is phylogenetically more closely related to neutrophilic heliobacteria than to alkaliphilic species of the genus

Table 13.1. Summary of major properties of different groups of anoxygenic phototrophic bacteria.

Property	Heliobacteria	Green sulfur bacteria	Purple bacteria	Filamentous anoxygenic phototrophs[a]	Phototrophic acidobacteria[b]
Photosynthetic pigments[c]	BChl *g*; 8^1–OH–Chl *a*	BChl *a*, *c*, *d*, *e*; Chl *a*	BChl *a*, *b*; BPhe *a*, *b*	BChl *a*, *c*, *d*; BPhe *a*	BChl *a*, *c*
Photosynthetic membranes	Cytoplasmic membrane	Chlorosomes	Intracellular lamellae	Chlorosomes[d]	Chlorosomes
Reaction center	Type-I homodimeric	Type-I homodimeric	Type-II heterodimeric	Type-II heterodimeric	Type-I homodimeric
Phylogenetic group	Phylum *Firmicutes*	Phylum *Chlorobi*	Phylum *Proteobacteria*	Phylum *Chloroflexi*	Phylum *Acidobacteria*
Habitats	Paddy or volcanic soils, neutral to alkaline hot springs, and soda lakes	Acidic to neutral, high sulfide (anoxic) hot springs and lakes	Fresh or saline, acidic to alkaline waters, often containing sulfide	Neutral to alkaline hot spring mats, temperate fresh and marine waters, and hypersaline mats	Alkaline hot spring mats
Photoautotrophy	No	Yes, reductive TCA cycle	Yes[e], Calvin Cycle	Yes[f], 3–OH–propionate bi-cycle or Calvin cycle	No
Oxygen relationship	Strictly anaerobic	Anaerobic	Facultatively aerobic[g]	Facultatively aerobic[g]	Aerobic

[a]Also called "Green Nonsulfur Bacteria"
[b]Only one cultured representative from this group is known—*Candidatus* "Chloracidobacterium thermophilum"
[c]Excludes carotenoids. Some pigments are present in some, but not all, genera of each group of anoxygenic phototrophic bacteria listed
[d]Absent in species of the genera *Roseiflexus* and *Heliothrix*, which subsequently also lack BChl *c*
[e]Some aerobic purple phototrophs, which do not carry out photosynthesis under anaerobic conditions, are incapable of autotrophic growth
[f]The capacity for autotrophy in some filamentous anoxygenic phototrophs has not yet been determined
[g]Some species are obligate anaerobes

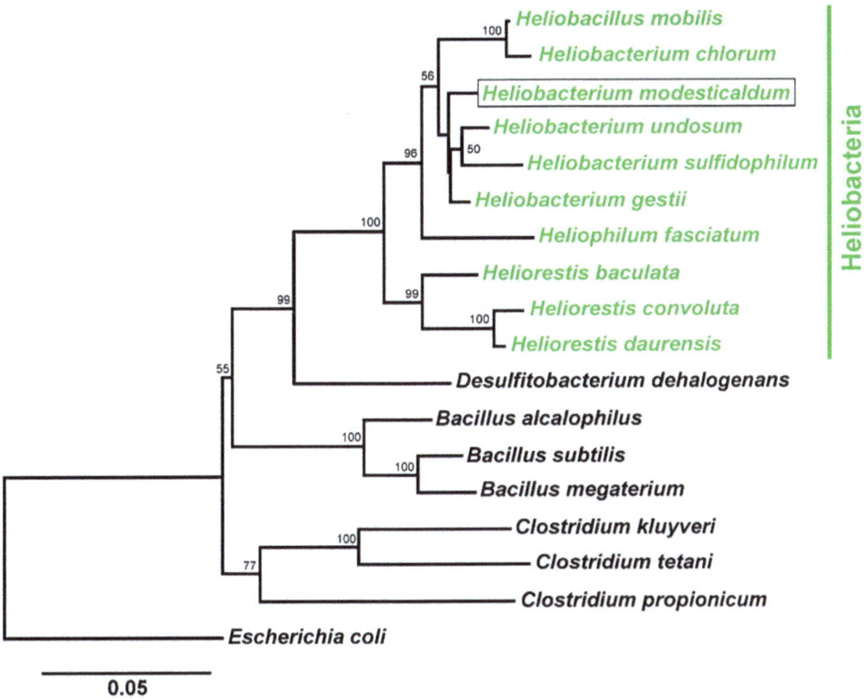

Fig. 13.2. Phylogenetic tree of heliobacteria and other *Firmicutes* based on 1022 nucleotide positions of the 16S rRNA gene. Phototrophic species are shown in *green. Heliobacterium modesticaldum* (*boxed*) has emerged as a model organism for biochemical and genomic studies of heliobacteria. The tree was generated using the neighbor-joining method corrected for multiple substitutions by the F84 algorithm (transition/transversion ratio = 2.0, empirical base frequencies) using PHYLIP version 3.68 (Felsenstein 1989). Bootstrap analysis was conducted on 1,000 replications, and confidence values of ≥50 % are indicated at the nodes. The tree was rooted using *Escherichia coli.* All 16S rRNA gene sequences have been deposited in GenBank with the following accession numbers: *Heliobacillus mobilis* (U14560), *Heliobacterium chlorum* (M11212), *Heliobacterium modesticaldum* (U14559), *Heliobacterium undosum* (AF249679), *Heliobacterim sulfidophilum* (AF249678), *Heliobacterium gestii* (L36198), *Heliophilum fasciatum* (L36197), *Heliorestis baculata* (AF249680), *Heliorestis convoluta* (DQ266255), *Heliorestis daurensis* (AF079102), *Desulfitobacterium dehalogenans* (L28946), *Bacillus alcalophilus* (X76436), *Bacillus subtilis* (AJ276351), *Bacillus megaterium* (X60629), *Clostridium kluyveri* (M59092), *Clostridium tetani* (X74770), *Clostridium propionicum* (X77841), and *Escherichia coli* (J01859).

Heliorestis (data not shown) (Asao et al. 2012). From an energy metabolism standpoint, all alkaliphilic heliobacteria cultured thus far are strictly photoheterotrophic (Bryantseva et al. 1999, 2000; Asao et al. 2006). This is in contrast to neutrophilic heliobacteria, which are capable of growth via pyruvate fermentation in darkness (Kimble et al. 1994).

Complete genome sequences for representative species of both neutrophilic and alkaliphilic heliobacteria (*Heliobacterium* (*Hbt.*) *modesticaldum* and *Heliorestis* (*Hrs.*) *convoluta*, respectively; Fig. 13.1) have now been obtained. Sattley et al. (2008) have provided a description of the former, while at the time of this writing, genomic characterization is still underway for *Hrs. convoluta.* Because of their contrasting natural habitats (hot spring soils versus shoreline soils of a temperate soda lake), important physiological insights may come from comparative analysis of the genomes of these two phototrophs.

II. Photosynthesis and Energy Conservation

A. Photosynthetic Pigments

Photoheterotrophy supports the fastest growth rates in heliobacteria, and it is likely the dominant metabolism in these organisms. Heliobacteria employ bacteriochlorophyll (BChl) *g*, a pigment synthesized only by members of the *Heliobacteriaceae*, to facilitate light-harvesting and primary photochemistry (Table 13.1). BChl *g* is of the bacteriochlorin-type chlorophylls (possessing single bonds between C-7/C-8 and C-17/C-18) and is substituted at the C-3^1 (ring A) and C-8^1 (ring B) positions with vinyl and ethylidine groups, respectively (Fig. 13.3a, left) (Brockmann and Lipinski 1983). Furthermore, the tetrapyrrole ring is esterified at the C-17^3 position with farnesol, a 15-carbon, acyclic alcohol also found in

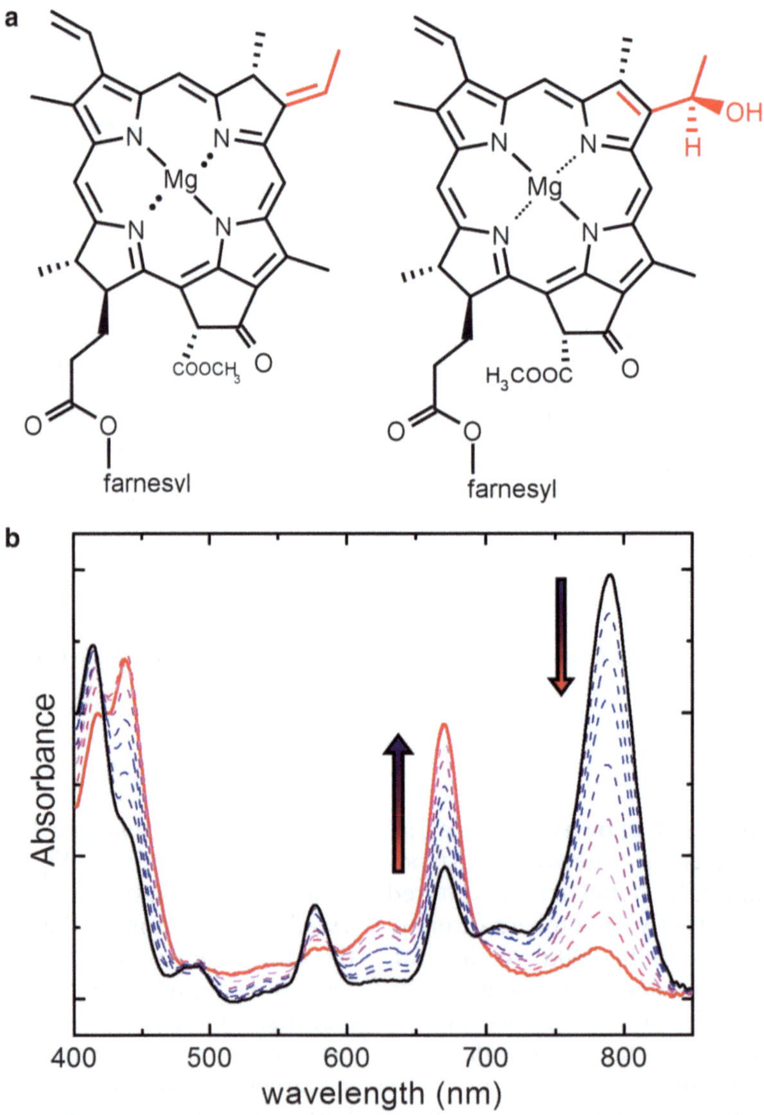

Fig. 13.3. Photosynthetic pigments in heliobacteria. (**a**) Structures of BChl *g* (*left*) and 8^1–OH–Chl *a* (*right*). The subtle differences between the two structures are highlighted in *red*. (**b**) Time-dependent, irreversible photoisomerization of BChl *g* to a spectroscopic equivalent of Chl *a* when cells are exposed to oxygen in the presence of light. The absorbance spectrum of native cells of *Heliobacterium modesticaldum* is shown in *black*.

BChls *c*, *d*, and *e* of green sulfur bacteria (GSB) (Michalski et al. 1987).

Heliobacteria also possess a minor amount of 8^1–OH–chlorophyll (Chl) *a* that is esterified with farnesol and found specifically in the photosynthetic reaction center (RC). The structure and full stereochemistry of this pigment has been unambiguously determined from NMR spectroscopy (van de Meent et al. 1991; Mizoguchi et al. 2005) and are presented in Fig. 13.3a (right). For a comprehensive review of the structure and function of all known Chls and BChls, see Scheer (2006). Heliobacteria are strict anaerobes (Table 13.1), and in the presence of O_2 and light, the C–8^1 ethylidene group of BChl *g* rapidly and irreversibly undergoes intramolecular hydrogen transfer to form 8^1–OH–Chl *a*. A time-course of the photoisomerization of BChl *g* to 8^1–OH–Chl *a* in cells of *Hbt. modesticaldum* is shown in Fig. 13.3b. While this event jeopardizes the viability of the cells, heliobacteria still require a small amount of 8^1–OH–Chl *a* in their photosynthetic RC, which contains approximately 17 BChl *g* molecules per molecule of 8^1–OH–Chl *a* (van de Meent et al. 1991). The in vitro conversion of BChl *g* to 8^1–OH–Chl *a* was demonstrated in the presence of a weak acid and under anoxic conditions (Kobayashi et al. 1998), and this finding may give insight into the biosynthetic pathway of the minor pigment (Mizoguchi et al. 2005).

The extreme O_2 sensitivity of BChl *g* partly explains the limited structural, spectroscopic, and biochemical investigations into the photosynthetic machinery of heliobacteria. However, some advances in these areas have been made. The following sections highlight some of the key structural and spectroscopic properties of the photosynthetic machinery of heliobacteria that allow for efficient energy conversion.

B. Light-Harvesting Antenna

Heliobacteria possess the simplest photosynthetic machinery of all known phototrophs (Sattley et al. 2008). The processes of light harvesting and photochemistry take place within the same complex, which consists of a dimer of the PshA protein that forms the so-called core-complex and resides in the cytoplasmic membrane. Only about 22–40 BChl *g* (per RC) compose the light-harvesting antenna (Trost and Blankenship 1989; Kobayashi et al. 1991a; Heinnickel et al. 2006), and the in vivo absorption of these molecules gives rise to bands around 790, 575, 410 and 368 nm (Fig. 13.3b), with each band referring to a different $\pi \rightarrow \pi^*$ transition (Amesz 1995b). The utilization of a novel BChl gives the heliobacteria a light-absorption niche distinct from other phototrophs (Oelze 1985). For example, algae and green plants absorb light around 650–700 nm (see Chap. 19), and the purple bacteria can absorb light wavelengths of >800 nm. Cooling membranes or purified heliobacterial RCs to cryogenic temperatures reveals that the long wavelength band is split into three distinguishable BChl *g* features absorbing around 778, 793, and 808 nm (Amesz 1995a; Neerken and Amesz 2001); the precise wavelengths of these bands is species dependent. In other words, the ~22–40 BChl *g* that comprise the antenna are not all homogenous in their local environment. Indeed, linear dichroism and fluorescence polarization measurements on membranes from *Hbt. chlorum* have shown that the pigments comprising the 808 nm band are generally oriented in the plane of the membrane, while the shorter wavelength bands form large angles in the membrane plane (van Dorssen et al. 1985).

Measurements at room or cryogenic temperature have demonstrated that excitation energy captured by the different pools of antenna BChl *g* is equilibrated among the other antenna pigments or transferred directly to the RC special pair. While two competing models have been proposed to describe the kinetics of energy transfer in heliobacteria (i.e., the transfer-limited and trap-limited models), the end result is that excitation energy is trapped in a time frame of ~5–30 ps by the RC special pair. For reviews of the contrasting kinetic models, see Lin et al. (1994), Neerken and Amesz (2001), and references therein.

C. Reaction Center Protein Composition

Heliobacteria possess a homodimeric RC (Table 13.1), and this finding was gleaned from initial molecular and biochemical studies. For example, the first sequencing study of the gene encoding the *Heliobacillus* (*Hba.*) *mobilis* RC polypeptide (gene *pshA*) gave no indication of any other RC gene (Liebl et al. 1993), and gel filtration experiments suggested the RC core may be a dimer (van de Meent et al. 1990). Moreover, the predicted protein sequence contained a binding motif for one-half of a [4Fe–4S] cluster, implying a homodimer of PshA would be necessary to bind such a cofactor (Liebl et al. 1993; Trost et al. 1992). The PshA protein from *Hba. mobilis* shows low (<20 %) overall sequence identity to the photosystem I (PSI) core polypeptides PsaA and PsaB found in oxygenic phototrophs or the RC PscA protein from the green sulfur bacterium *Chlorobium limicola* (Fig. 13.4). However, the highest degree of homology

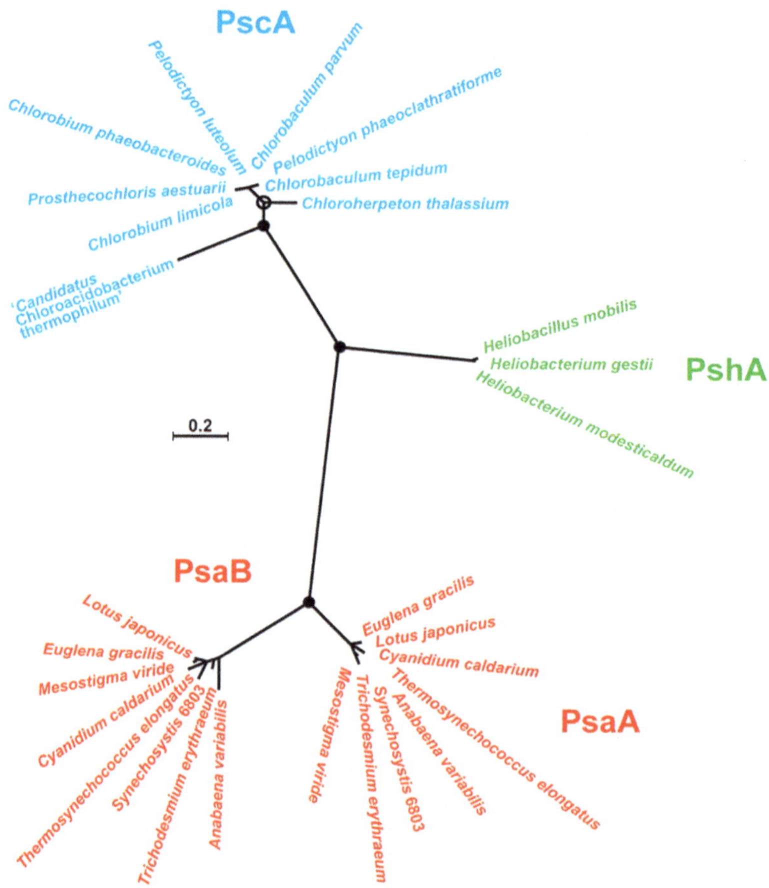

Fig. 13.4. Unrooted neighbor-joining tree of type-I RC proteins based on the conserved 111 amino acid residues at the C terminus. *Colors* highlight RC core polypeptide diversity: *Green*, PshA of heliobacteria; *Blue*, PscA of *green* sulfur bacteria and phototrophic acidobacteria; *Red*, PsaA and PsaB of photosystem I from oxygenic phototrophs. The multiple sequence alignment for the tree was performed using ClustalX 2.0.12 (Larkin et al. 2007), and the resulting alignment was manually edited using Sequence Alignment Editor v2.0a11. The tree was generated using the programs PROTDIST (Jones-Taylor-Thornton matrix) and NEIGHBOR implemented in PHYLIP version 3.68 (Felsenstein 1989). Bootstrap analysis was conducted on 1,000 replications. Bootstrap support of 100 % (●) or >70 % (○) is indicated at a node.

was found in the 12-residue putative F_X binding region, with 75 % and 92 % sequence identity, respectively, for the above-mentioned RC polypeptides. In addition, the tertiary structure of PshA was predicted to have 11 transmembrane helices, a feature conserved among all type I RCs (Liebl et al. 1993; Oh-oka 2007). Despite low sequence identity, functional studies of the heliobacterial RC have clearly demonstrated more similarity to type I (FeS-type) RCs than to type II (Q-type) RCs. For a review of such studies see Amesz (1995a, b), Neerken and Amesz (2001), Heinnickel and Golbeck (2007), and Oh-oka (2007).

D. Electron Transfer Pathways and Reaction Center Cofactors

Photosynthetic energy conversion occurs through an electron transport chain consisting of several components encoded by genes present in the heliobacterial photosynthesis gene cluster (Xiong et al. 1998; Sattley et al. 2008). Electrons from NADH enter the chain through NADH:menaquinone oxidoreductase, where they are transferred via menaquinone to the cytochrome *bc* complex (see Chap. 8) (Fig. 13.5). Electrons donated to the Rieske [2Fe–2S] subunit (PetC) and cytochrome *c* of the complex are then transferred to the membrane-anchored cytochrome c_{553} (PetJ), which then serves as the electron donor to P_{800}, the primary electron donor in the RC (Oh-Oka et al. 2002). An electrochemical gradient is established as electrons are transferred through the membrane-embedded cytochrome *bc* complex. This electrochemical gradient is used to generate phosphoanhydride bonds by the ATP synthase (Sattley et al. 2008) (see Chap. 6).

Electron transfer through the heliobacterial RC appears similar to electron transport in PSI (Brettel 1997): $P_{800} \rightarrow A_0 \rightarrow A_1 \rightarrow F_X \rightarrow F_A \rightarrow F_B$. However, the heliobacterial RC shows features more related to the green sulfur bacterial RC concerning the electron carrier A_1. Because the heliobacterial RC is a homodimer, electron transfer may be bifurcated along the redox centers of each PshA monomer.

A special pair of BChls serves as the primary electron donor in the RC and consists of two molecules of the C-13^1 epimer of BChl *g*, BChl *g*' (Kobayashi et al. 1991a, b). This cofactor, called P_{800} for the maximal wavelength of its reversible photobleaching (Fuller et al. 1985), has an especially reducing redox potential of 225–240 mV (Prince et al. 1985). Nuijs and coworkers first used picosecond flash-induced absorption difference spectroscopy to study primary charge separation in membranes from *Hbt. chlorum*. The authors concluded that, in addition to photooxidation of P_{800}, a spectroscopic feature at 670 nm rapidly formed (Nuijs et al. 1985). Analogous to PSI, the cofactor responsible for this observation is termed A_0 (Ohashi et al. 2010), and it has been identified as 8^1–OH Chl *a* (van de Meent et al. 1991; Mizoguchi et al. 2005). There is one 8^1–OH Chl *a* per BChl *g*'; in other words, each PshA monomer likely binds one of each of these cofactors (Kobayashi et al. 1991a; van de Meent et al. 1991). Using a trap-limited model and taking the number of antenna pigments into account, the rate of primary charge separation was calculated to be 1.2 ps (Lin et al. 1994). The kinetic signal ascribed to A_0 was also found to decay in 500–600 ps (Nuijs et al. 1985; Lin et al. 1994; Brettel et al. 1998), while a transient signature suggestive of F_X from PSI (Ke 1973) was determined to form in that same time frame (Kleinherenbrink et al. 1994; Lin et al. 1995). This cofactor was later discovered to be a [4Fe–4S] cluster by electron paramagnetic resonance (EPR) spectroscopy (Heinnickel et al. 2006; Miyamoto et al. 2006).

In PSI, a phylloquinone cofactor called A_1 is situated between A_0 and F_X and functions as an electron transfer intermediate (Brettel 1997). Heliobacteria appear to use only menaquinone of various tail lengths for cellular processes (Hiraishi 1989), and the RC from *Hba. mobilis* was purified with 1.4 menaquinone-9 per P_{800} (Trost and Blankenship 1989). Additionally, a putative quinone binding site is predicted in the heliobacterial RC based on sequence comparisons

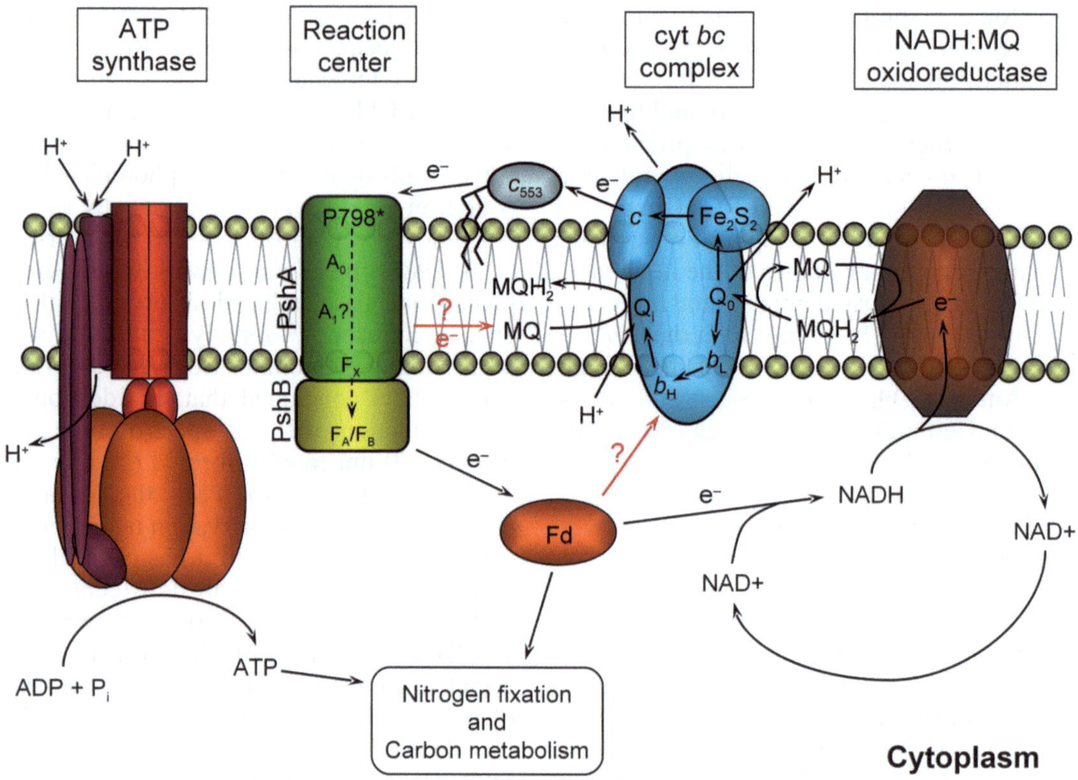

Fig. 13.5. Electron transport and energy conservation in heliobacteria based on genes identified in *Heliobacterium modesticaldum*. Question marks indicate features of electron transfer that have not been confirmed experimentally (Adapted from Sattley et al. 2008).

to PsaA/PsaB from PSI (Miyamoto et al. 2006; Oh-oka 2007), and the photoaccumulation of semiquinone in RC preparations has also been reported (Trost et al. 1992; Miyamoto et al. 2008). However, despite the presence of menaquinone in the heliobacterial RC, evidence for its participation in forward electron transfer remains controversial. For example, transient spectroscopic signatures characteristic of semi-reduced quinone have not been observed on the pico-, nano-, or microsecond time scales (Lin et al. 1994; Brettel et al. 1998). Moreover, ether extraction of all non-covalently bound quinones from membranes of *Hbt. chlorum* did not alter forward or reverse electron transfer kinetics (Kleinherenbrink et al. 1993). It is notable that an A_1 cofactor is also apparently not required for electron transfer in the RC of GSB (Hauska et al. 2001).

Electrons are transferred out of the RC and into the soluble phase by PshB, a dicluster [4Fe–4S] ferredoxin that houses F_A and F_B and is analogous to PsaC of PSI (Nitschke et al. 1990; Heinnickel et al. 2007). Romberger and coworkers (2010) have shown that two PshB orthologs (PshBI and PshBII) are present in *Hb. modesticaldum*. Both of these terminal electron acceptors were found to be loosely bound to the heliobacterial RC, whereas PsaC is bound tightly to PSI (Romberger and Golbeck 2010).

What is the fate of low potential electrons from the RC? Cyclic electron transfer has been reported in *Hba. mobilis* (Kramer et al. 1997), where electrons from the RC were proposed to reduce a cytoplasmic ferredoxin that is used by NADH:menaquinone oxidoreductase to reduce menaquinone. In addition, since all heliobacteria tested show strong nitrogen-fixing capabilities (Asao and Madigan 2010) nitrogenase has also been proposed as a potentially important electron sink (Collins et al. 2010).

III. Central Carbon Metabolism

A. Anaplerotic CO_2 Assimilation

While the central carbon metabolism of heliobacteria is not yet fully understood, several studies have provided significant insight in this area. As obligate photoheterotrophs, heliobacteria require an organic carbon source for growth (Table 13.1). Genomic analysis revealed that while no complete autotrophic CO_2 fixation pathway is encoded in the *Hbt. modesticaldum* genome, most of the genes encoding enzymes of the reductive (reverse) tricarboxylic acid (RTCA) cycle are present (Sattley et al. 2008). Importantly, the genes *aclAB*, which encode ATP citrate lyase (ACL), a key enzyme in the RTCA cycle that splits citrate into acetyl-CoA and oxaloacetate (OAA), were not found in the *Hbt. modesticaldum* genome. The RTCA cycle is used by GSB and some autotrophic chemolithotrophs to assimilate CO_2 and produce acetyl-CoA for growth and metabolism (Evans et al. 1966a, b; Feng et al. 2010; Hügler and Sievert 2011; Sirevåg 1995; Sirevåg and Ormerod 1970; Tang and Blankenship 2010). The missing *aclAB* genes, along with the absence of genes for other autotrophic pathways in the *Hbt. modesticaldum* genome, likely explain the obligately heterotrophic nature of heliobacteria; acetyl-CoA cannot be generated from CO_2 in an incomplete RTCA cycle. This genomic data is consistent with biochemical studies that showed no ACL activity in heliobacteria (Pickett et al. 1994; Tang et al. 2010b).

Considering these results, it seems unlikely that major carbon flow occurs through the (incomplete) RTCA cycle in heliobacteria. Further support of this is provided by comparisons with GSB, which require CO_2 for growth on pyruvate (Tang and Blankenship 2010). If carbon transformations carried out by enzymes of the RTCA cycle represented the major pathway of carbon flow in heliobacteria, one would expect their growth on pyruvate to be enhanced with the addition of CO_2. However, adding sodium bicarbonate to growth media does not enhance pyruvate-supported phototrophic growth of *Hbt. modesticaldum* (Tang et al. 2010b). Furthermore, recent studies using ^{13}C-labeled proteins and bacteriochlorophylls indicate that glutamate is mainly synthesized through a partial oxidative TCA (OTCA) cycle (from citrate to α-ketoglutarate) in heliobacteria (Tang et al. 2010a, b). Similar conclusions were obtained from ^{13}C-NMR studies of different strains of heliobacteria (Pickett et al. 1994).

Recent studies suggest that anaplerotic (nonautotrophic) CO_2-assimilation in *Hbt. modesticaldum* is catalyzed by phosphoenolpyruvate carboxykinase (PEPCK), and this generates ATP via substrate-level phosphorylation during both phototrophic and chemotrophic growth (Fig. 13.6) (Tang et al. 2010a, b). Moreover, pyruvate:ferredoxin oxidoreductase (PFOR) was shown to assimilate CO_2 in *Hbt. modesticaldum* using reduced ferredoxin during phototrophic growth (Fig. 13.6). By comparison, while GSB also assimilate CO_2 through an active anaplerotic pathway, they carry out autotrophic CO_2 assimilation during phototrophic growth using the RTCA cycle (Evans et al. 1966a, b; Feng et al. 2010; Tang and Blankenship 2010).

Central carbon flow in heliobacteria has been shown to occur largely through the OTCA cycle (Tang et al. 2010a). Interestingly, heliobacteria use (*Re*)-citrate synthase rather than the common (*Si*)-citrate synthase to synthesize citrate (Fig. 13.7) (Tang et al. 2010a). Several species of clostridia, nonphototrophic relatives of heliobacteria, and other nonphototrophic anaerobes are also known to use (*Re*)-citrate synthase rather than (*Si*)-citrate synthase to catalyze the aldol condensation of acetyl-CoA and OAA (Fig. 13.7) (Jahn et al. 2007; Li et al. 2007; Pingitore et al. 2007; Tang et al. 2009). Thus, the central carbon metabolism of heliobacteria

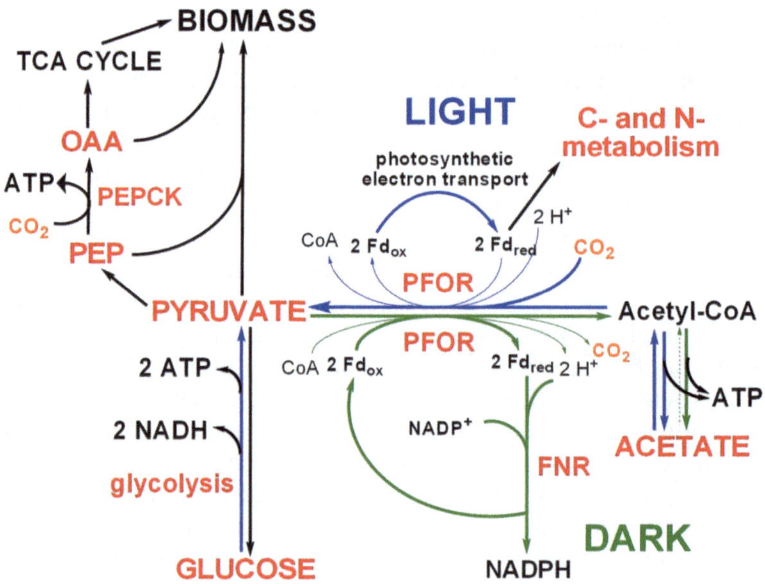

Fig. 13.6. Pyruvate and carbohydrate metabolism of *Heliobacterium modesticaldum* showing active pathways during phototrophic (*blue*) versus chemotrophic (*green*) growth (Adapted from Tang et al. 2010b).

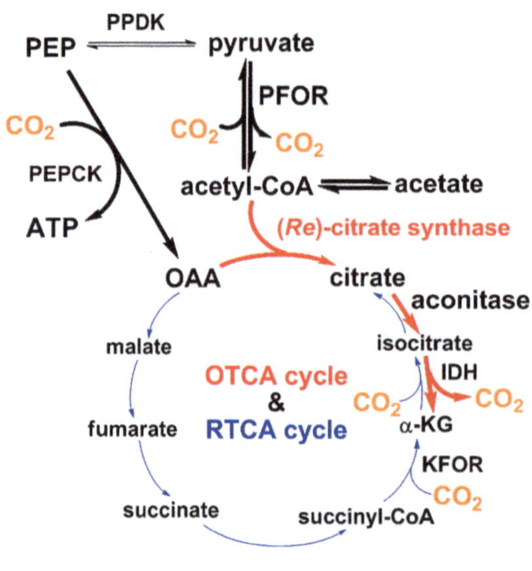

Fig. 13.7. Central carbon metabolism and CO_2 assimilation in heliobacteria. *Heliobacterium modesticaldum* uses a CO_2-anaplerotic pathway to assimilate inorganic carbon during phototrophic growth and has greater carbon flow through the partial oxidative TCA (OTCA) cycle than the reductive TCA (RTCA) cycle. (*Re*)-citrate synthase is used to synthesize citrate in the partial OTCA cycle (Adapted from Tang et al. 2010a).

is more closely related to clostridia than to GSB (Tang et al. 2010a), an observation that is in accordance with phylogenetic associations between these groups of bacteria (Woese et al. 1985). Furthermore, the observation of Tang et al. (2010a) that *Hbt. modesticaldum* has a partial OTCA cycle that progresses from citrate to α-ketoglutarate, a precursor of glutamate, is consistent with the fact that genes encoding α-ketoglutarate dehydrogenase (KDH) were not identified in the *Hbt. modesticaldum* genome (Sattley et al. 2008; Sattley and Blankenship 2010), and no activity of KDH was detected in cell extracts of heliobacteria (Pickett et al. 1994; Tang et al. 2010b). It is of particular interest to note that the major central carbon flow is switched from the RTCA cycle in GSB to the partial OTCA cycle in heliobacteria, with a novel citrate synthase (i.e., (*Re*)-citrate synthase) in heliobacteria substituting for ATP citrate lyase in GSB. Considering that (*Re*)-citrate synthase has also been identified in several species of *Clostridium* but not in

GSB and other phototrophic bacteria, the central carbon metabolism of heliobacteria nicely coincides with the ecological, physiological, and phylogenetic features of other anaerobic, heterotrophic *Firmicutes*.

B. Pyruvate Metabolism

To develop a comprehensive understanding of heliobacterial physiology, it is essential to know how heliobacteria conserve energy via (central) carbon metabolism during chemotrophic growth. As already mentioned, pyruvate is the best carbon source for phototrophic growth of heliobacteria, but it is also the only carbon source to sustain chemotrophic growth of heliobacteria (Kimble et al. 1994, 1995). Thus, pyruvate metabolism is an especially important pathway in heliobacteria. In *Hbt. modesticaldum*, pyruvate is oxidized to acetyl-CoA by PFOR, as pyruvate dehydrogenase is absent (Sattley et al. 2008). The reaction catalyzed by PFOR is reversible during phototrophic growth and irreversible (i.e., CO_2 cannot be assimilated) during chemotrophic growth of *Hbt. modesticaldum* (Fig. 13.6) (Tang et al. 2010b). Furthermore, during chemotrophic growth of heliobacteria, pyruvate fermentation produces acetyl-CoA for lipid biosynthesis (Pickett et al. 1994; Tang et al. 2010b) and citrate for the OTCA cycle (Tang et al. 2010a). Hydrolysis of acetyl-CoA to acetate yields ATP for fermentative growth. Compared with phototrophic growth, many energy- and reducing equivalent-demanding processes, such as nitrogen assimilation and hydrogen production, are either inactive or downregulated during chemotrophic growth of *Hbt. modesticaldum* (Tang et al. 2010b).

C. Carbohydrate Metabolism

The utilization of sugars as carbon sources is rare among cultured heliobacteria. However, D-fructose and D-glucose can support phototrophic growth of *Heliobacterium gestii*, a species isolated from tropical paddy soil (Madigan et al. 2006, Overmann et al. 2006). *Hbt. modesticaldum* is phylogenetically more closely related to *Hbt. gestii* than to other cultured heliobacteria (Asao and Madigan 2010; Bryantseva et al. 1999). All of the genes required for the Embden-Meyerhof-Parnas (EMP) pathway (glycolysis) but not other carbohydrate catabolic pathways (e.g., the Entner-Doudoroff (ED) pathway and the oxidative pentose phosphate pathway) have been annotated in the *Hbt. modesticaldum* genome (Sattley and Blankenship 2010). Consistent with genomic annotation and phylogenetic analyses, recent physiological studies and stable isotope labeling followed by mass spectroscopy analyses have demonstrated that D-glucose, D-fructose and D-ribose can support slow phototrophic growth of *Hbt. modesticaldum* when a trace amount of yeast extract (0.02 % final) is supplied (Tang et al. 2010b). The study shows that when sugars are supplied as carbon sources, *Hbt. modesticaldum* uses the EMP pathway for carbohydrate catabolism and hydrogen (H_2) production (Fig. 13.6).

IV. Final Comments

The heliobacteria are a unique and intriguing group of phototrophic bacteria for several reasons: (1) Although they stain gram-negatively, cells of heliobacteria have a gram-positive cell wall structure and are able to differentiate into resilient endospores; (2) as primarily soil organisms, heliobacteria occupy an ecological niche that is vastly different from the aquatic habitat occupied by other anoxygenic phototrophs (Asao and Madigan 2010); (3) the phylogenetic positioning of heliobacteria within the *Firmicutes* isolates them from other anoxygenic phototrophs but places them closer to the cyanobacteria (see Chap. 14) (based on full-length 16S rRNA gene sequences; Rappé and Giovannoni 2003), which raises interesting questions concerning the origin of oxygenic photosynthesis; (4) heliobacteria carry out photosynthesis by synthesizing and incorporating unique pigments into a reaction center that has perhaps the minimum complexity required for a functional photosynthetic apparatus; (5) and finally, although they are phototrophic, heliobacteria are obligate heterotrophs with no apparent capacity for autotrophic CO_2 fixation. Recent and continued genomic analyses are helping to broaden our understanding of the genetic capacity of heliobacteria, but additional biochemical characterizations, including high-resolution structural studies of the heliobacterial RC, are necessary if we are to gain a thorough understanding of photosynthesis and energy conservation in these anoxygenic phototrophs.

Acknowledgments

We thank Drs. Robert Blankenship and Michael Madigan for helpful comments and discussions during the preparation of the manuscript and Ann Sattley for assistance in preparing some of the figures.

References

Amesz J (1995a) The antenna-reaction center complex of heliobacteria. In: Blankenship RE, Madigan MT, Bauer CE, Amesz J (eds) Anoxygenic photosynthetic bacteria, vol 2. Kluwer Academic Publishers, Dordrecht, pp 687–697, Advances in Photosynthesis and Respiration

Amesz J (1995b) The heliobacteria, a new group of photosynthetic bacteria. J Photochem Photobiol B Biol 30:89–96

Asao M, Madigan MT (2010) Taxonomy, phylogeny, and ecology of the heliobacteria. In: Golbeck J (ed) Photosynthetic phototrophs with homodimeric type-I reaction centers. Photosynth Res 104:103–111

Asao M, Jung DO, Achenbach LA, Madigan MT (2006) *Heliorestis convoluta* sp. nov., a coiled, alkaliphilic heliobacterium from the Wadi El Natroun, Egypt. Extremophiles 10:403–410

Asao M, Takaichi S, Madigan MT (2012) Amino acid-assimilating phototrophic heliobacteria from soda lake environments: *Heliorestis acidaminivorans* sp. nov. and '*Candidatus* Heliomonas lunata'. Extremophiles 16:585–595

Brettel K (1997) Electron transfer and arrangement of the redox cofactors in photosystem I. Biochim Biophys Acta 1318:322–373

Brettel K, Leibl W, Liebl U (1998) Electron transfer in the heliobacterial reaction center: evidence against a quinone-type electron acceptor functioning analogous to A_1 in photosystem I. Biochim Biophys Acta 1363:175–181

Brockmann H, Lipinski A (1983) Bacteriochlorophyll *g*. A new bacteriochlorophyll from *Heliobacterium chlorum*. Arch Microbiol 136:17–19

Bryantseva IA, Gorlenko VM, Kompantseva EI, Achenbach LA, Madigan MT (1999) *Heliorestis daurensis*, gen. nov. sp.nov., an alkaliphilic rod-to-coiled-shaped phototrophic heliobacterium from a siberian soda lake. Arch Microbiol 172:167–174

Bryantseva IA, Gorlenko VM, Kompantseva EI, Tourova TP, Kuznetsov BB, Osipov GA (2000) Alkaliphilic heliobacterium *Heliorestis baculata* sp. nov. and emended description of the genus *Heliorestis*. Arch Microbiol 174:283–291

Collins AM, Redding KE, Blankenship RE (2010) Modulation of fluorescence in Heliobacterium modesticaldum cells. In: Golbeck J (ed) Photosynthetic phototrophs with homodimeric type-I reaction centers. Photosynth Res 104:283–292

Evans MC, Buchanan BB, Arnon DI (1966a) New cyclic process for carbon assimilation by a photosynthetic bacterium. Science 152:673

Evans MC, Buchanan BB, Arnon DI (1966b) A new ferredoxin-dependent carbon reduction cycle in a photosynthetic bacterium. Proc Natl Acad Sci U S A 55:928–934

Felsenstein J (1989) PHYLIP-Phylogeny inference package (Version 3.2). Cladistics 5:164–166

Feng X, Tang KH, Blankenship RE, Tang YJ (2010) Metabolic flux analysis of the mixotrophic metabolisms in the green sulfur bacterium *Chlorobaculum tepidum*. J Biol Chem 285:39544–39550

Fuller RC, Sprague SG, Gest H, Blankenship RE (1985) A unique photosynthetic reaction center from *Heliobacterium chlorum*. FEBS Lett 182: 345–349

Hauska G, Schoedl T, Remigy H, Tsiotis G (2001) The reaction center of green sulfur bacteria. Biochim Biophys Acta 1507:260–277

Heinnickel M, Golbeck J (2007) Heliobacterial photosynthesis. Photosynth Res 92:35–53

Heinnickel M, Agalarov R, Svensen N, Krebs C, Golbeck JH (2006) Identification of F_X in the heliobacterial reaction center as a [4Fe–4S] cluster with an S = 3/2 ground spin state. Biochemistry 45:6756–6764

Heinnickel M, Shen G, Golbeck JH (2007) Identification and characterization of PshB, the dicluster ferredoxin that harbors the terminal electron acceptors F_A and F_B in *Heliobacterium modesticaldum*. Biochemistry 46:2530–2536

Hiraishi A (1989) Occurrence of menaquinone as the sole isoprenoid quinone in the photosynthetic bacterium *Heliobacterium chlorum*. Arch Microbiol 151:378–379

Hügler M, Sievert SM (2011) Beyond the calvin cycle: autotrophic carbon fixation in the ocean. Annu Rev Mar Sci 3:261–289

Jahn U, Huber H, Eisenreich W, Hugler M, Fuchs G (2007) Insights into the autotrophic CO_2 fixation pathway of the archaeon *Ignicoccus hospitalis*: comprehensive analysis of the central carbon metabolism. J Bacteriol 189:4108–4119

Ke B (1973) The primary electron acceptor of photosystem I. Biochim Biophys Acta 301:1–33

Kimble LK, Stevenson AK, Madigan MT (1994) Chemotrophic growth of heliobacteria in darkness. FEMS Microbiol Lett 115:51–55

Kimble LK, Mandelco L, Woese CR, Madigan MT (1995) *Heliobacterium modesticaldum*, sp. nov., a thermophilic heliobacterium of hot springs and volcanic soils. Arch Microbiol 163:259–267

Kleinherenbrink FAM, Ikegami I, Hiraishi A, Otte SCM, Amesz J (1993) Electron transfer in menaquinone-depleted membranes of *Heliobacterium chlorum*. Biochim Biophys Acta 1142:69–73

Kleinherenbrink FAM, Chiou H-C, LoBrutto R, Blankenship RE (1994) Spectroscopic evidence for the presence of an iron-sulfur center similar to F_X of Photosystem I *Heliobacillus mobilis*. Photosynth Res 41:115–123

Kobayashi M, van de Meent EJ, Erkelens C, Amesz J, Ikegami I, Watanabe T (1991a) Bacteriochlorophyll *g* epimer as a possible reaction center component of heliobacteria. Biochim Biophys Acta 1057:89–96

Kobayashi M, Watanabe T, Ikegami I, van de Meent EJ, Amesz J (1991b) Enrichment of bacteriochlorophyll *g*' in membranes of *Heliobacterium chlorum* by ether extraction: unequivocal evidence for its existence in vivo. FEBS Lett 284:129–131

Kobayashi M, Hamano T, Akiyama M, Watanabe T, Inoue K, Oh-oka H, Amesz J, Yamamura M, Kise H (1998) Light-independent isomerization of bacteriochlorophyll *g* to chlorophyll *a* catalyzed by weak acid in vitro. Anal Chim Acta 365:199–203

Kramer DM, Schoepp B, Liebl U, Nitschke W (1997) Cyclic electron transfer in *Heliobacillus mobilis* involving a menaquinol-oxidizing cytochrome *bc* complex and an RCI-type reaction center. Biochemistry 36:4203–4211

Larkin MA, Blackshields G, Brown NP, Chenna R, McGettigan PA, McWilliam H, Valentin F, Wallace IM, Wilm A, Lopez R, Thompson JD, Gibson TJ, Higgins DG (2007) Clustal W and Clustal X version 2.0. Bioinformatics 23:2947–2948

Li F, Hagemeier CH, Seedorf H, Gottschalk G, Thauer RK (2007) *Re*-citrate synthase from *Clostridium kluyveri* is phylogenetically related to homocitrate synthase and isopropylmalate synthase rather than to *Si*-citrate synthase. J Bacteriol 189:4299–4304

Liebl U, Mockensturm-Wilson M, Trost JT, Brune DC, Blankenship RE, Vermaas W (1993) Single core polypeptide in the reaction center of the photosynthetic bacterium *Heliobacillus mobilis*: structural implications and relations to other photosystems. Proc Natl Acad Sci U S A 90:7124–7128

Lin S, Chiou HC, Kleinherenbrink FA, Blankenship RE (1994) Time-resolved spectroscopy of energy and electron transfer processes in the photosynthetic bacterium *Heliobacillus mobilis*. Biophys J 66: 437–445

Lin S, Chiou HC, Blankenship RE (1995) Secondary electron transfer processes in membranes of *Heliobacillus mobilis*. Biochemistry 34:12761–12767

Madigan MT (2001) Heliobacteriaceae. In: Boone DR, Castenholtz RW, Garrity GM (eds) Bergey's manual of systematic bacteriology, vol 1, 2nd edn. Springer, New York, pp 625–630

Madigan MT (2006) The family Heliobacteriaceae. In: Dworkin M, Falkow S, Rosenberg E, Schleifer K-H, Stackebrandt E (eds) The prokaryotes, vol 4, 3rd edn. Springer, New York, pp 951–964

Michalski TJ, Hunt JE, Bowman MK, Smith U, Bardeen K, Gest H, Norris JR, Katz JJ (1987) Bacteriopheophytin *g*: properties and some speculations on a possible primary role for bacteriochlorophylls *b* and *g* in the biosynthesis of chlorophylls. Proc Natl Acad Sci U S A 84:2570–2574

Miyamoto R, Iwaki M, Mino H, Harada J, Itoh S, Oh-oka H (2006) ESR signal of the iron-sulfur center F_X and its function in the homodimeric reaction center of *Heliobacterium modesticaldum*. Biochemistry 45:6306–6316

Miyamoto R, Mino H, Kondo T, Itoh S, Oh-oka H (2008) An electron spin-polarized signal of the P_{800}^+ A_1 $(Q)^-$ state in the homodimeric reaction center core complex of *Heliobacterium modesticaldum*. Biochemistry 47:4386–4393

Mizoguchi T, Oh-oka H, Tamiaki H (2005) Determination of stereochemistry of bacteriochlorophyll g_F and 8^1-hydroxy-chlorophyll a_F from *Heliobacterium modesticaldum*. Photochem Photobiol 81:666–673

Neerken S, Amesz J (2001) The antenna reaction center complex of heliobacteria: composition, energy conversion and electron transfer. Biochim Biophys Acta 1507:278–290

Nitschke W, Setif P, Liebl U, Feiler U, Rutherford AW (1990) Reaction center photochemistry of *Heliobacterium chlorum*. Biochemistry 29: 11079–11088

Nuijs AM, Dorssen RJ, Duysens LNM, Amesz J (1985) Excited states and primary photochemical reactions in the photosynthetic bacterium *Heliobacterium chlorum*. Proc Natl Acad Sci U S A 82:6865–6868

Oelze J (1985) Analysis of bacteriochlorophylls. Meth Microbiol 18:257–284

Ohashi S, Iemura T, Okada N, Itoh S, Furukawa H, Okuda M, Ohnishi-Kameyama M, Ogawa T, Miyashita H, Watanabe T, Itoh S, Oh-oka H, Inoue K, Kobayashi M (2010) An overview on chlorophylls and quinones in the photosystem I-type reaction centers. Photosynth Res 104:305–319

Oh-oka H (2007) Type 1 reaction center of photosynthetic heliobacteria. Photochem Photobiol 83:177–186

Oh-oka H, Iwaki M, Itoh S (2002) Electron donation from membrane-bound cytochrome *c* to the photosynthetic reaction center in whole cells and isolated membranes of *Heliobacterium gestii*. Photosynth Res 71:137–147

Ormerod J, Nesbakken T, Torgerson Y (1990) Phototrophic bacteria that form heat-resistant endospores. In: Baltscheffsky M (ed) Current research in photosynthesis, vol 4. Kluwer, Dordrecht, pp 935–938

Overmann J (2006) The family Chlorobiaceae. In: Dworkin M, Falkow S, Rosenberg E, Schleifer K-H, Stackebrandt E (eds) The prokaryotes, vol 7, 3rd edn. Springer, New York, pp 359–378

Pickett MW, Williamson MP, Kelly DJ (1994) An enzyme and 13C-NMR of carbon metabolism in heliobacteria. Photosynth Res 41:75–88

Pingitore F, Tang Y, Kruppa GH, Keasling JD (2007) Analysis of amino acid isotopomers using FT-ICR MS. Anal Chem 79:2483–2490

Prince RC, Gest H, Blankenship RE (1985) Thermodynamic properties of the photochemical reaction center of *Heliobacterium chlorum*. Biochim Biophys Acta 810:377–384

Rappé MS, Giovannoni SJ (2003) The uncultured microbial majority. Ann Rev Microbiol 57:369–394

Romberger S, Golbeck J (2010) The bound iron–sulfur clusters of Type-I homodimeric reaction centers. Photosynth Res 104:333–346

Romberger S, Castro C, Sun Y, Golbeck J (2010) Identification and characterization of PshBII, a second F_A/F_B-containing polypeptide in the photosynthetic reaction center of *Heliobacterium modesticaldum*. Photosynth Res 104:293–303

Sattley WM, Madigan MT, Swingley WD, Cheung PC, Clocksin KM, Conrad AL, Dejesa LC, Honchak BM, Jung DO, Karbach LE, Kurdoglu A, Lahiri S, Mastrian SD, Page LE, Taylor HL, Wang ZT, Raymond J, Chen M, Blankenship RE, Touchman JW (2008) The genome of *Heliobacterium modesticaldum*, a phototrophic representative of the *Firmicutes* containing the simplest photosynthetic apparatus. J Bacteriol 190:4687–4696

Sattley WM, Blankenship RE (2010) Insights into heliobacterial photosynthesis and physiology from the genome of *Heliobacterium modesticaldum*. Photosynth Res 104:113–122

Scheer H (2006) An overview of chlorophylls and bacteriochlorophylls: biochemistry, biophysics, functions and applications. In: Grimm B, Porra RJ, Rüdiger W, Scheer H (eds) Chlorophylls and bacteriochlorophylls, vol 25. Springer, The Netherlands, pp 1–26, Advances in Photosynthesis and Respiration

Sirevåg R (1995) Carbon metabolism in green bacteria. In: Blankenship RE, Madigan MT, Bauer CE (eds) Anoxygenic photosynthetic bacteria. Kluwer, Amsterdam, pp 871–883

Sirevåg R, Ormerod JG (1970) Carbon dioxide—fixation in photosynthetic green sulfur bacteria. Science 169:186–188

Stevenson AK, Kimble LK, Woese CR, Madigan MT (1997) Characterization of new phototrophic heliobacteria and their habitats. Photosynth Res 53:1–11

Takaichi S, Inoue K, Akaike M, Kobayashi M, Oh-Oka H, Madigan MT (1997) The major carotenoid in all species of heliobacteria is the C_{30} carotenoid 4,4′-diaponeurosporene, not neurosporene. Arch Microbiol 168:277–281

Takaichi S, Oh-Oka H, Maoka T, Jung DO, Madigan MT (2003) Novel carotenoid glucoside esters from alkaliphilic heliobacteria. Arch Microbiol 179:95–100

Tang KH, Blankenship RE (2010) Both forward and reverse TCA cycles operate in green sulfur bacteria. J Biol Chem 285:35848–35854

Tang YJ, Yi S, Zhuang WQ, Zinder SH, Keasling JD, Alvarez-Cohen L (2009) Investigation of carbon metabolism in "*Dehalococcoides ethenogenes*" strain 195 by use of isotopomer and transcriptomic analyses. J Bacteriol 191:5224–5231

Tang KH, Feng X, Zhuang WQ, Alvarez-Cohen L, Blankenship RE, Tang YJ (2010a) Carbon flow of heliobacteria is related more to clostridia than to the green sulfur bacteria. J Biol Chem 285: 35104–35112

Tang KH, Yue H, Blankenship RE (2010b) Energy metabolism of *Heliobacterium modesticaldum* during phototrophic and chemotrophic growth. BMC Microbiol 10:150

Trost JT, Blankenship RE (1989) Isolation of a photoactive photosynthetic reaction center-core antenna complex from *Heliobacillus mobilis*. Biochemistry 28:9898–9904

Trost JT, Brune DC, Blankenship RE (1992) Protein sequences and redox titrations indicate that the electron acceptors in reaction centers from heliobacteria are similar to Photosystem I. Photosynth Res 32:11–22

van de Meent EJ, Kleinherenbrink FAM, Amesz J (1990) Purification and properties of an antenna-reaction center complex from heliobacteria. Biochim Biophys Acta 1015:223–230

van de Meent EJ, Kobayashi M, Erkelens C, van Veelen PA, Amesz J, Watanabe T (1991) Identification of 8^1-hydroxychlorophyll *a* as a functional reaction center pigment in heliobacteria. Biochim Biophys Acta 1058:356–362

van Dorssen RJ, Vasmel H, Amesz J (1985) Antenna organization and energy transfer in membranes of *Heliobacterium chlorum*. Biochim Biophys Acta 809:199–203

Woese CR, Debrunner-Vossbrinck BA, Oyaizu H, Stackebrandt E, Ludwig W (1985) Gram-positive bacteria: possible photosynthetic ancestry. Science 229:762–765

Xiong J, Inoue K, Bauer CE (1998) Tracking molecular evolution of photosynthesis by characterization of a major photosynthesis gene cluster from *Heliobacillus mobilis*. Proc Natl Acad Sci U S A 95:14851–14856

Chapter 14

The Architecture of Cyanobacteria, Archetypes of Microbial Innovation

Claire S. Ting*
Department of Biology, Williams College, 214 Thompson Biology Laboratory, 59 Lab Campus Drive, Williamstown, MA 01267, USA

Summary

With an impressive and unparalleled evolutionary history spanning over two billion years, cyanobacteria have evolved to thrive in a diverse range of habitats and numerically dominate vast regions, such as the open oceans. Impacts of this microbial lineage on our planet have been far-reaching: ancestors of extant cyanobacteria had a pivotal role not only in the establishment of oxygen as a major component of the atmosphere, but in the rise of embryophytic algae and land plants. The evolution of innovative cellular structure and function in response to abiotic and biotic selection pressures has resulted in a striking diversity in cyanobacteria. Given the fact that the relationship between form and function is complex, and that cellular structures are often multifunctional and dynamic, this review uses a comparative approach in understanding how biological functions or strategies impact cellular architecture. Notably, differences existing in cellular architecture among genera, and even between strains of the same genus, often reflect evolutionary innovations that have permitted a group to flourish in a particular environment. This review also addresses how a conserved cellular feature, namely the compartmentalization of key functions, has promoted metabolic flexibility and survival.

*Author for correspondence, e-mail: cting@williams.edu

M.F. Hohmann-Marriott (ed.), The Structural Basis of Biological Energy Generation, Advances in Photosynthesis and Respiration 39, pp. 249–275, DOI 10.1007/978-94-017-8742-0_14,

This compartmentalization involves structures such as the internal membranes, heterocysts and carboxysomes, and has permitted the functioning and integration of diverse, and at times, incompatible, processes in a single cell. With heightened interest in utilizing photosynthetic organisms to address global challenges in food and energy resources, research on this remarkable lineage will undoubtedly continue to inform, as well as inspire.

I. Introduction

The cyanobacterial lineage comprises a relatively ubiquitous and heterogeneous group of photosynthetic prokaryotes, which utilize chlorophyll pigments in photosynthetic reactions that typically evolve oxygen. This lineage has had a profound impact on the biosphere, and its members have been deemed "life's microbial heroes" and "arguably the most important organisms ever to appear on Earth" (Knoll 2003a). Although cyanobacteria exhibit breathtaking diversity, they share several structural features and major physiological strategies. Notably, the majority of cyanobacteria evolved to utilize water as an electron source in photosynthetic reactions that involve two photosystems and result in the evolution of oxygen. This far-reaching innovation in the photosynthetic reaction by an ancestor of extant cyanobacteria, set the stage for the ecological success of this lineage and for major changes in the Earth's biosphere. However, this innovation was not without its provisions, which included optimizing the efficient functioning of two photosystems and the regulation of energy distribution, as well as reducing the impact of oxygen on specific enzymes and processes.

Abbreviations: ATP – Adenosine triphosphate; Chl – Chlorophyll CP43 – Chl-binding protein (apparent molecular mass 43 kDa) associated with PSII; CP47 – Chl-binding protein (apparent molecular mass 47 kDa) associated with PSII; LHCII – Mobile light-harvesting Chl *a/b*-binding antenna complex; Mbp – Million base pairs; Pcb – Prochlorophyte Chl *a/b*-binding antenna protein; PQ – Plastoquinone; PS I – Photosystem I; PS II – Photosystem II; TEM – Transmission electron microscopy

A. Evolutionary History and Impact

Efforts to pinpoint the earliest appearance of cyanobacteria have been controversial. Reports based on characterization of fossils from the Apex Basalt of northwestern Western Australia suggest that cyanobacteria may already have been present during the Archean Era about 3.5 billion years (Gyr) ago (Schopf 1993, 2000; Olson 2006), and hydrocarbon-based analyses of shales from the Pilbara Cratan, Australia suggest the existence of oxygenic cyanobacteria 2.7 Gyr ago (Summons et al. 1999; Brocks et al. 1999, 2003). However, recent micrometre-scale $^{13}C/^{12}C$-based analyses of organic-rich shale also from Pilbara Cratan, Australia refute these latter findings (Rasmussen et al. 2008). These researchers instead suggest that the most robust morphological evidence from the fossil record suggests cyanobacteria were present 2.15 Gyr ago during the early Proterozoic eon (Precambrian) (Hofmann 1976). This supports models linking the evolution of cyanobacteria with increases in atmospheric oxygen levels approximately 2.45–2.32 Gyr ago (Knoll 2003b; Bekker et al. 2004; Holland 2006; Rasmussen et al. 2008).

The evolution of oxygenic photosynthesis and subsequent establishment of oxygen as a major component of the atmosphere were unprecedented events that laid a foundation for the extant biosphere. Moreover, the opportune engulfment of a cyanobacterium by a eukaryotic cell that was the common ancestor of Glaucophytes, Rhodophytes and Viridiplantae set the stage for the rise of embryophytic algae and land plants (Gray 1992; Moreira et al. 2000; Bhattacharya and Medlin 1995; Bhattacharya et al. 2004, 2007; Delwiche 1999; McFadden 2001; Bhattacharya et al. 2007; Reyes-Prieto and Bhattacharya 2007). The rise of land plants, which includes over 250,000 species of angiosperms, in turn had a major impact on the evolution of many other organisms, including insects.

Cyanobacteria have evolved exquisite complexity, as well as flexibility, in their cellular architecture. This complicated the use of morphological characteristics as the basis of taxonomic systems developed in the nineteenth century. During this time, the naming and classification of cyanobacteria fell under the purview of the International Code of Botanical Nomenclature and members of the cyanobacterial lineage were commonly referred to as "blue-green algae." In his treatise published in 1935 on *The Structure and Reproduction of Algae*, Fritsch included cyanobacteria among the 11 classes of Algae, specifically in the Myxophyceae (Cyanophyceae). However, with the subsequent recognition of cyanobacteria as prokaryotes (Stanier and van Niel 1962; Buchanan and Gibbons 1974), Stanier and colleagues submitted a proposal to include cyanobacteria under the International Code of Nomenclature of Bacteria (Stanier et al. 1978). They pointed out that methods used for establishing taxonomic classifications with the Bacterial Code, which relies upon pure cultures of type strains, would be more effective than the Botanical Code that primarily utilizes nonliving botanical materials (Stanier et al. 1978). In the following year, Rippka et al. (1979) published a comprehensive study stemming from their comparative analysis of 178 cyanobacterial strains that proposed dividing cyanobacteria into five sub-groups based on distinct structural and developmental characteristics.

Subsequent molecular phylogenies using 16S rRNA gene sequences have classified cyanobacteria unambiguously among the eubacteria (Woese 1987). Within the cyanobacteria, 16S rRNA and *nifH* gene sequence-based phylogenetic trees suggest that those cyanobacteria capable of cell differentiation and forming heterocysts group in a clade of more recently differentiated lineages and form a monophyletic lineage (Giovannoni et al. 1988; Wilmotte 1994; Turner 1997; Zehr et al. 1997; Fewer et al. 2002; Livaitis 2002; Gugger and Hoffmann 2004). Tree topologies also support multiple evolutionary origins of unicellular and filamentous (non-heterocystous) forms (Giovannoni et al. 1988; Wilmotte 1994). Within this context it is important to mention that cyanobacterial cell morphology can be plastic and that transformations between unicellular and filamentous forms have been documented even within a particular strain (Ingram and van Baalen 1970; Wilmotte and Golubic 1991). 16S rRNA gene sequence analyses have been particularly effective in clarifying the phylogenetic positions of the "prochlorophytes" and have demonstrated that they form distinct lineages within the cyanobacteria (Turner et al. 1989; Urbach et al. 1992). Notably, the three genera (*Prochlorococcus*, *Prochlorothrix hollandica*, *Prochloron*) that were originally grouped as prochlorophytes, have been shown to cluster in separate branches within the cyanobacterial lineage, with *Prochloron* demonstrating closer phylogenetic ties with *Synechocystis* PCC6803 than with either *Prochlorothrix hollandica* or *Prochlorococcus* (Palenik and Haselkorn 1992; Urbach et al. 1992; Wilmotte 1994).

B. Comparative Genomics

The depth and breadth of our understanding of the relationships among members of the cyanobacterial lineage and between cyanobacteria and other organisms have increased due in large part to comparative genomics and related approaches. At present, over 70 cyanobacterial genome sequences are available and over 60 additional cyanobacterial sequencing projects are in progress (http://www.ncbi.nlm.nih.gov/genome/ as of February 2014). Genome sizes differ significantly within the cyanobacterial lineage (Table 14.1), with the most minimal genomes being about five times smaller than the largest (approximately 1.64 Mbp for the cultured *Prochlorococcus* MIT9301 strain and 1.4 Mbp for an uncultured cyanobacterium to about 8.23 Mbp for *Nostoc punctiforme*

Table 14.1. Genome properties of cyanobacteria from diverse habitats.

Environment (originally isolated)	Genus	Strain	Chromosome size (bp)	Genes	%G+C
Marine (open ocean, coastal, intertidal zone)	*Prochlorococcus*	MIT9301	1,641,879	1,962	31.3
		MED4	1,657,990	1,762	30.8
		MIT9515	1,704,176	1,964	30.8
		AS9601	1,669,886	1,965	31.3
		MIT9215	1,738,790	2,054	31.1
		MIT9312	1,709,204	1,856	31.2
		MIT9211	1,688,963	1,900	38.0
		NATL1A	1,864,731	2,250	35.0
		NATL2A	1,842,899	2,228	35.1
		SS120	1,751,080	1,930	36.4
		MIT9313	2,410,873	2,330	50.7
		MIT9303	2,682,675	3,136	50.0
	Synechococcus	WH8102	2,434,428	2,581	59.4
		WH5701	3,043,834	3,401	65.4
		CC9605	2,510,659	2,756	59.2
	Cyanobium	PCC7001	2,832,697	2,817	68.7
	Trichodesmium	IMS101	7,750,108	5,126	34.1
	Crocosphaera	WH8501	6,238,156	5,996	37.1
	Lyngbya	PCC8106	7,037,511	6,185	41.1
	Nodularia	CCY9414	5,316,258	4,904	41.3
	Acaryochloris	MBIC11017	6,503,724	6,396	47.3
	Uncultured	UCYN-A	1,443,806	1,241	31.1
Freshwater	*Cylindrospermopsis*	CS-505	3,879,030	3,501	40.2
	Microcystis	NIES-843	5,842,795	6,364	42.3
	Nostoc (Anabaena)	PCC7120	6,413,771	5,430	41.3
	Raphidiopsis	D9	3,186,511	3,057	40.1
	Synechococcus	PCC7942	2,695,903	2,665	55.5
		PCC6301	2,696,255	2,582	55.5
	Synechocystis	PCC6803	3,573,470	3,229	47.7
Other habitats:					
Salt marsh (Woods Hole)	*Microcoleus*	PCC7420	8,651,623	8,347	45.4
Calcareous rock	*Gloeobacter*	PCC7421	4,659,019	4,482	62
Rice field (Senegal)	*Cyanothece*	PCC7425	5,374,574	5,113	50.8
Rice field (India)	*Cyanothece*	PCC7822	6,091,620	5,664	40.2
Hot spring (mat)	*Synechococcus*	JA-3-3Ab	2,932,766	2,897	60.2
Hot spring (Japan)	*Thermosynechococcus*	BP-1	2,593,857	2,525	53.9
Alkaline salt lake	*Arthrospira platensis*	NIES-39	6,788,435	6,676	44.3
Symbiont plants, fungi	*Anabaena variabilis*	ATCC 29413	6,365,727	5,134	41.4
Azolla endosymbiont	*Nostoc azollae*	0708	5,354,700	5,252	38.4
Gymnosperm cycad symbiont (freshwater)	*Nostoc punctiforme*	PCC73102	8,234,322	6,497	41.4

PCC 73102). With the availability of 12 complete *Prochlorococcus* genomes (Table 14.1), this cyanobacterium has presented one of the most powerful model systems for comparative genomic analyses. Although members of the *Prochlorococcus* lineage exhibit less than 3 % difference in their 16S rDNA sequences, extensive genetic diversity has evolved within this lineage (Rocap et al. 2003). This genetic diversity is reflected even in genes encoding proteins involved in critical cellular functions, such as photosynthesis (Ting et al. 2009). For example, although *Prochlorococcus* strains MIT9313 (2.41 Mbp) and MIT9303 (2.68 Mbp) share 99.21 % sequence identity in their 16S rRNA genes, the chromosomal organization of their genes is not well conserved and about 10 % of the MIT9303 genome (about 298 genes) consists of genes that are not shared with MIT9313 (Ting et al. 2009).

C. Diversity and Innovation

Extant members of the cyanobacterial lineage thrive in a range of habitats, including extreme environments, such as hot springs, oligotrophic waters of the Sargasso Sea, and hot and cold deserts (Table 14.1). In certain habitats, they serve as the major primary producers and carbon sinks, and contributors to biogeochemical cycles. Moreover, members of this lineage occur as symbionts in many eukaryotes, including diatoms, dinoflagellates, lichens, Gunnera (angiosperm), and cycads (gymnosperm). The diversity within the cyanobacterial lineage provides us with snapshots of strategies that have evolved or are in the process of evolving in response to biotic and abiotic conditions and challenges. The selective pressures in each habitat have driven and shaped the evolution of the genome, physiology and cellular architecture, and have given rise to the innovation that characterizes this lineage.

Although considerable morphological diversity has evolved within the cyanobacterial lineage, a conserved property involves the cellular compartmentalization of specific functions, including photosynthesis (intracytoplasmic lamellae), respiration (plasma membrane, intracytoplasmic lamellae), carbon fixation (carboxysomes), and nitrogen fixation (heterocysts). The structures involved in these major functions will be discussed in the next sections. These structures directly affect the spatial organization and partitioning of the proteome. Although cyanobacteria do not possess organelles such as mitochondria and vacuoles, the compartmentalization of cellular functions results in a complexity that directly impacts physiology and energy conversion, and thus, efficiency and survival.

With the goal of effectively encompassing the diversity that has evolved within the lineage, this review will use a comparative approach in presenting key elements of cyanobacterial cell organization and structure. In light of the complex interplay between form and function, this will be done in the context of understanding how biological functions or strategies impact cellular structures, which are dynamic and often serve multiple roles. In particular, this review will address the way in which existing differences in cellular architecture among genera, and even between strains of the same genus, illuminate evolutionary innovations that have permitted cells to thrive in diverse environments. Recent advances in this area in conjunction with those in fields including genomics, proteomics, physiology, and ecology have enabled us to achieve a more integrative understanding of the biology of cyanobacteria.

II. Functional Significance of Cell Size and Shape

Cell sizes are variable within the cyanobacterial lineage, and can be even within a genus. While the smallest cyanobacteria include single cells of *Prochlorococcus* with diameters of about 0.7 um (Chisholm et al. 1988; Lichtle et al. 1995; Ting et al. 2007), some of the largest include filaments of *Oscillatoria princeps*, with individual cell widths of about 55 um (Hoiczyk and Baumeister 1995). Cyanobacteria can occur as single cells, filaments, and/or colonies, and morphological variation is present in each of these categories (Fig. 14.1). While single cells can be spherical, oval, or barrel-shaped, filaments can be open helices or can associate to form fusiform tuft aggregates or "puff" aggregates. Colonies have been

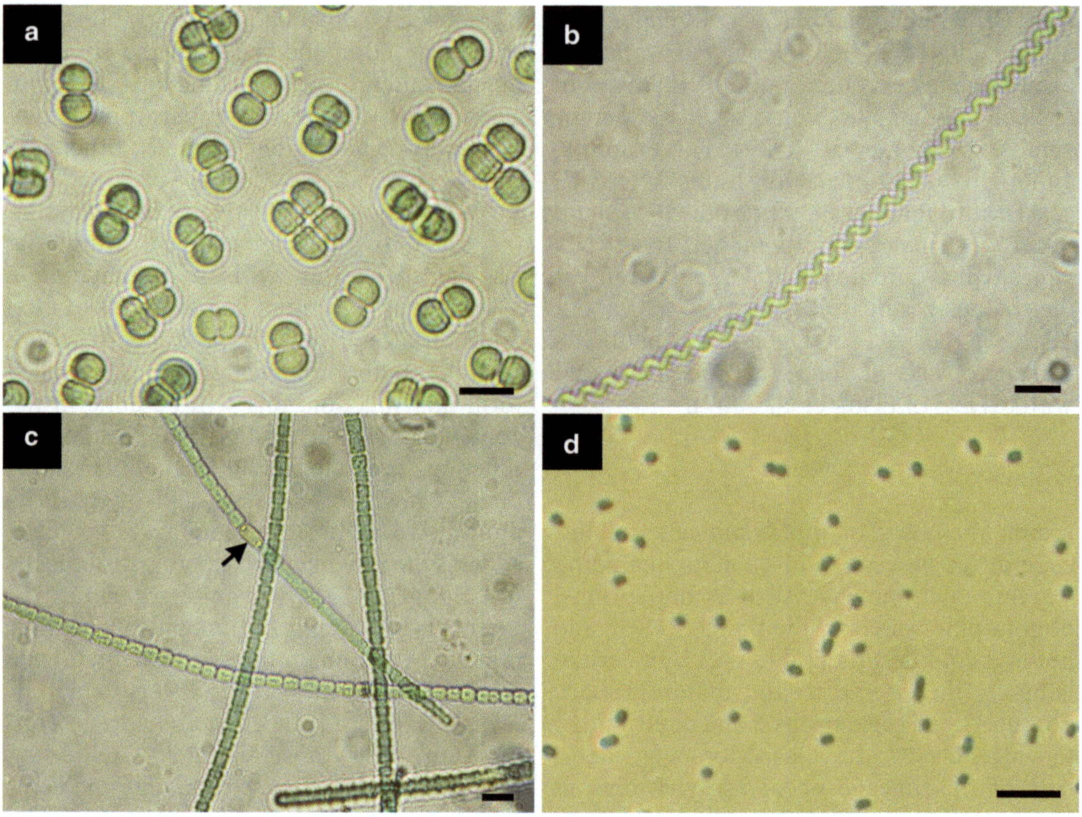

Fig. 14.1. Striking morphological diversity exists within the cyanobacterial lineage. (**a**) *Merismopedia*, scale bar, 8 um. (**b**) *Spirulina*, scale bar, 9 um. (**c**) *Anabaena*, scale bar, 10.5 um., A specialized cell for nitrogen fixation (a heterocyst, *black arrow*) is visible within the filament of vegetative cells. (**d**) *Prochlorococcus* (strain MIT9313), scale bar, 4 um. (Light micrographs by Christie Black and Claire Ting, Williams College).

characterized as spherical ("pearls"), hemispherical, mats, and oncoids. Furthermore, while some cyanobacterial strains exhibit true branching (three different neighboring cells are in contact with the branch point cells (Golubic et al. 1996; Gugger and Hoffman 2004)), others have "false" branching. In some genera, cells are capable of differentiating into heterocysts (cells specialized to conduct nitrogen fixation), akinetes (resting cells that can typically withstand extreme environmental conditions), or hormogonia (motile cell filaments). Traits, such as cell morphology, could ultimately be selective or secondary, or in some cases even superfluous, and distinguishing among these possibilities is not necessarily straightforward (Young 2006). In their work examining how bacterial morphology might correlate with phylogeny, Siefert and Fox (1998) concluded that for the cyanobacteria, there were no clear patterns, except among filamentous heterocystous forms, which Giovannoni et al. (1988) had shown to form a distinct cluster (Siefert and Fox 1998).

Alterations in cell size (volume, biomass) and/or shape can impact many cellular functions, including resource acquisition and light absorption efficiency. For example, smaller cells possess a relatively larger surface area per unit volume, as well as a reduced diffusion boundary layer thickness. These characteristics are likely to facilitate the exchange of solutes (Fogg 1986; Raven 1986, 1998) and could therefore provide a selective advantage in environments where

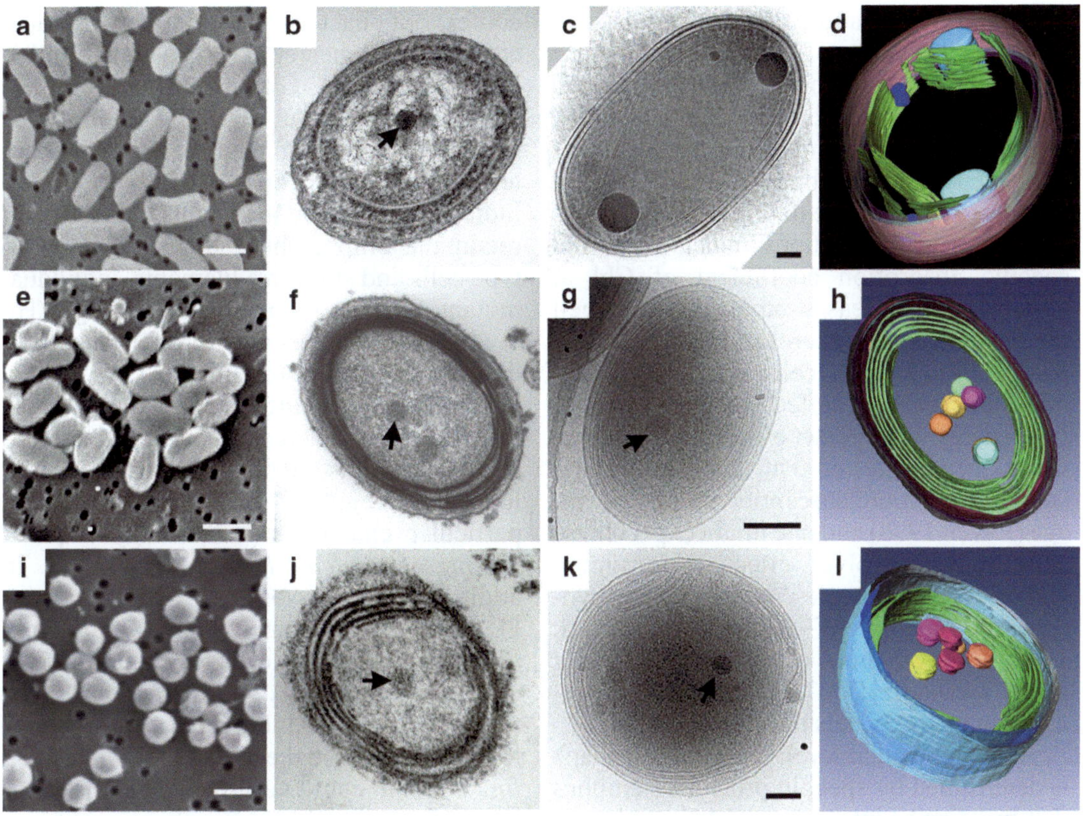

Fig. 14.2. The architecture of marine *Synechococcus* WH8102 (**a–d**), *Prochlorococcus* strain MIT9313 (**e–h**), and *Prochlorococcus* strain MED4 (**i–l**) visualized by scanning electron microscopy (**a**, **e**, **i**), transmission electron microscopy (**b**, **f**, **j**) and cryo-electron microscopy (**c**, **g**, **k**). Cells were preserved by chemical fixation (**a**, **b**, **e**, **f**, **i**, **j**) or by rapid freezing (**c**, **g**, **k**; shown are 1.8 nm tomographic z-slices of frozen-hydrated cells). Three-dimensional reconstructions of cells visualized in a near-native state using cryo-electron microscope tomography are presented in (**d**, **h** and **l**). Note that the intracytoplasmic membranes in *Synechococcus* WH8102 are widely spaced in order to accommodate the light-harvesting phycobilisome complexes (**b**, **c**). In contrast, they are tightly appressed in *Prochlorococcus* strains (**f**, **g**, **j**, **k**). Although the *Prochlorococcus* MED4 cell wall is difficult to preserve using chemical fixation (**j**), it can be visualized in samples preserved by rapid freezing (**k**). Carboxysomes (*black arrow*), or specialized microcompartments surrounded by a polypeptide shell, are visible within cells. Scale bar, 1 um (**a**, **e**), 0.1 um (**c**, **k**), 0.5 um (**i**), 0.25 um (**g**). (**a**, **b**: E. Westly and C. Ting (Williams College); **e**: E. Hambleton and C. Ting (Williams College); **c**, **d**, **g**, **h**, **k**, **l**: C. Ting and S. Sundararaman (Williams College), C. Hsieh, C. Mannella and M. Marko (Wadsworth Center); **f**, **i**, **j**: C. Ting (Williams College)).

nutrients are at low concentrations (oligotrophic regions). Indeed, large sections of the open oceans are oligotrophic, and these areas have been found to be dominated by the relatively small cyanobacteria, *Prochlorococcus* (Figs. 14.1d and 14.2e–l) and *Synechococcus* (Chisholm 1992) (Fig. 14.2a–d). Moreover, with volume being constant, rod-shaped cells have a greater surface area per unit volume compared to spherical cells and this ratio does not change greatly when the length of rod-shaped cells is increased and width is held constant (Young 2006). This fact and the observation that many bacteria form filaments under changing nutrient conditions led Young (2006) to suggest that filament formation could be advantageous with respect to nutrient acquisition.

Furthermore, cell size and shape can affect the efficiency with which light energy is absorbed, and this has been described as the “package effect” (Kirk 1975, 1976, 1994;

Morel and Bricaud 1981; Yentsch and Phinney 1989; Raven 1998). Briefly, because photosynthetic pigments are contained within specific structures (cells, colonies), which can be thought of as "packages", their effectiveness at absorbing light is decreased relative to if they were distributed uniformly. In particular, calculations indicate that the specific absorption coefficient of Chl at 670–680 nm decreases exponentially as cell diameter increases (Morel and Bricaud 1981; Kirk 1994). Smaller cells effectively minimize the "package effect" and can absorb light energy with greater efficiency. However, it has been suggested that this could render smaller cells more susceptible to photodamage (Raven 1998) and one might expect these cells to have evolved effective prevention and/or repair strategies, including the synthesis of "sunscreens" (Garcia-Pichel et al. 1992; Ehling-Schulz et al. 1997; Fleming and Castenholz 2007; Oren and Gunde-Cimerman 2007; Balskus and Walsh 2010).

At the molecular level, bacterial cell shape is determined by specific protein assemblies that function as internal scaffolds and by the synthesis, degradation, and/or deposition of peptidoglycan, a rigid structural component of the cell wall (Young 2003; Cabeen and Jacobs-Wagner 2005; Singh and Montgomery 2011). Proteins such as FtsZ, Mbl, or MreB have a critical role in the construction of the cellular scaffold, which is thought to be a dynamic structure that could also affect the positioning of peptidoglycan-synthesizing enzymes and thus, the deposition of cell wall material (Jones et al. 2001; Carballido-Lopez and Errington 2003; Kroos and Maddock 2003; Young 2003). One model that has been proposed suggests that both the placement of peptidoglycan and the relative stability of the peptidoglycan are important in overall cell shape (Young 2003). In particular, the presence of stable (inert, not regularly recycled) peptidoglocan in key cellular regions, such as the cell poles, could determine the overall shape of bacterial cells (Mobley et al. 1984; De Pedro et al. 2003; Young 2003).

The role of the MreB protein, a bacterial actin, has been studied intensively in the context of bacterial morphology (Wachi et al. 1987; Jones et al. 2001; Young 2003; Figge et al. 2004; Gitai et al. 2004; Cabeen and Jacobs-Wagner 2005). The *mreBCD* gene cluster has been identified in cyanobacteria and research on *Anabaena* PCC7120 indicates that although mutants lacking *mreB* retain the ellipse shape observed in wild-type cells and still form filaments, individual mutant cells are about twice as large (Hu et al. 2007). Interestingly, growth rates, cell division, and the distribution of DNA during division appear unaffected in these mutants (Hu et al. 2007). However, overexpression of *mreBCD* results in filaments composed of long and irregularly shaped cells, and this particular change in morphology requires the presence of *mreB* (Hu et al. 2007). In addition, recent work on *Synechococcus* PCP7942 demonstrated that mutations in *ftn2* and *ftn6*, which leads to an inhibition of cell division, results in filamentous cells; these cells also exhibit increased levels of MreB, and this led the authors to suggest that *ftn2* and *ftn6* could negatively regulate *mreB* expression in wild-type cells (Koksharova et al. 2007).

III. At the Front Line: Role of Cell Envelopes

Maintaining cell shape and stability, mediating the transport of nutrients and metabolites, protection against chemical substances and high/UV light, and cell movement are a few of the potential functions of the cyanobacterial cell envelope (Fig. 14.3) that render it indispensable for survival. In cyanobacteria, the cell wall typically consists of a distinct peptidoglycan layer and an outer membrane, and in many genera, just outside of the cell wall are additional layers composed of (glyco)protein crystalline arrays (S-layers) or carbohydrates. These S-layers could provide additional protection or function in cell adhesion and recognition (Hoiczyk and Hansel 2000), and they have been identified in many genera, including *Synechococcus* (Vaara 1982), *Phormidium*

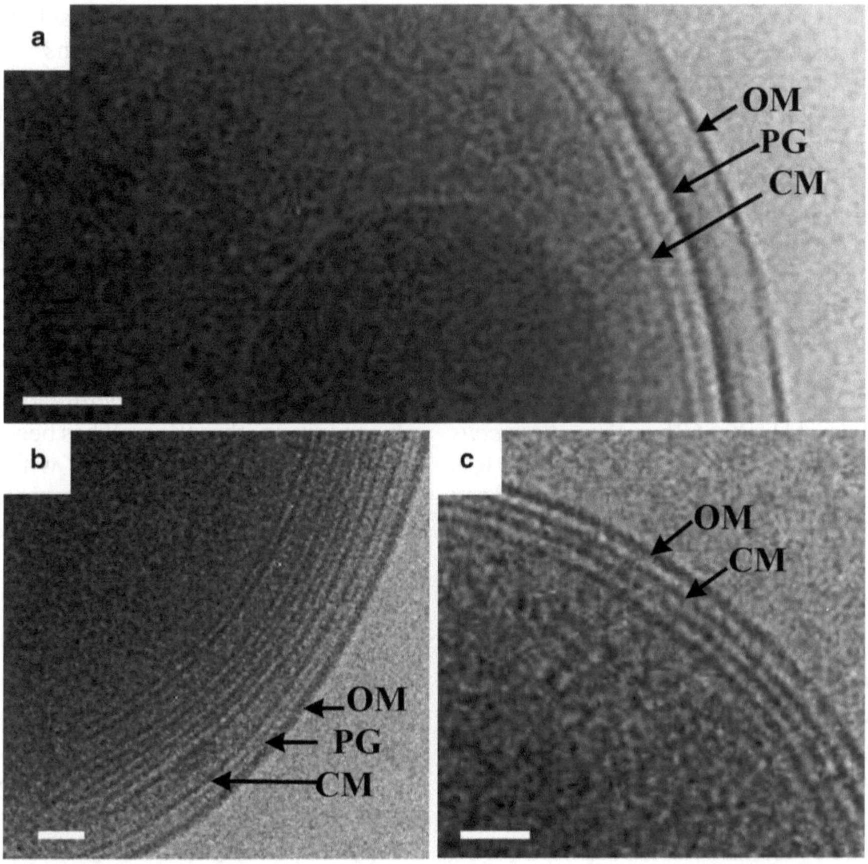

Fig. 14.3. Cell wall architecture can differ significantly between closely related cyanobacteria. Shown are the cell envelopes of marine *Synechococcus* WH8102 and *Prochlorococcus* strains MIT9313 and MED4 as visualized in frozen-hydrated specimens using cryo-electron microscopy (1.8 nm tomographic z-slices are shown). *OM* outer membrane, *P* peptidoglycan layer, *CM* cell membrane. Scale bar, 50 nm (**a**, **c**), 60 nm (**b**). (C. Ting and S. Sundararaman (Williams College), C. Hsieh, C. Mannella and M. Marko (Wadsworth Center), unpublished data).

(Hoiczyk and Baumeister 1995), and *Aphanothece* (Simon 1981). In other cyanobacteria, specific carbohydrates or polysaccharides are deposited directly adjacent to the outer membrane, and these can function in forming a protective sheath in certain strains (Hoiczyk 1998) or could have a role in gliding motility (Hoiczyk and Baumeister 1997). Moreover, oligosaccharide mycosporine amino acids (Scherer et al. 1988; Bohm et al. 1995) or scytonemin pigments (Proteau et al. 1993) have been detected in these sheaths in some cyanobacteria including *Nostoc commune*, and could function as sunscreens in protection against UV radiation.

Associated specifically with the outer membrane are macromolecules and other compounds involved in key physiological processes. These include porins and ATP-binding cassette (ABC) transporters, which facilitate the movement of compounds in and out of the cell. Porins function specifically in the diffusion of small solutes and are thought to consist of monomers of approximately 50–70 kDa in cyanobacteria (Umeda et al. 1996; Hansel et al. 1998; Hansel and Tadros 1998; Hoiczyk and Hansel 2000). In contrast, ABC-transporters facilitate the transport of a range of different molecules, and are thus involved in a variety of functions, including glycolipid transport required for heterocyst formation in *Anabaena* sp. strain PCC 7120 (Fiedler et al. 1998). With the availability of complete cyanobacterial genomes, it has been possible to identify genes encoding putative ABC transporters, which are often abundant in these genomes. For example, in marine *Synechococcus* WH8102, about 60 % of the genes assigned to the transport category are associated with ABC transporter components (Palenik et al. 2003). The outer membrane has also been found to be associated with carotenoids in cyanobacteria, including *Synechococcus leopoliensis* UTEX 625, *Synechococcus elongatus* PCC7942 (*Anacystis nidulans* R2) and *Synechocystis* PCC 6714 (Resch and Gibson 1983; Jurgens and Mantele 1991), and it is likely that these outer membrane cartotenoids function in photoprotection (Clayton 1980; Paerl 1984).

In filamentous cyanobacteria, including *Phormidium uncinatum* strain Baikal (Hoiczyk and Baumeister 1995) and *Anabaena* sp. strain PCC 7120 (Bauer et al. 1995; Flores et al. 2006), the outer membrane is continuous between individual cells of a filament and is not detected in the septum. Although earlier transmission electron microscopy work had suggested this possibility (Ris and Singh 1961), the preservation and visualization of the outer membrane have improved dramatically over the years and have provided robust evidence for this conclusion. It has also been suggested from conventional transmission electron microscopy studies that the outer membrane is continuous between mature heterocysts and vegetative cells (Flores et al. 2006). However, because the junction between these cell types is structurally complex (Fay and Lang 1971; Lang and Fay 1971; Wilcox et al. 1973), additional characterization using techniques such as cryoelectron microscopy would be valuable. In particular, microplasmodesmata have been observed between the vegetative cell and heterocyst in cyanobacteria including *Anabaena cylindrica* (Fay and Lang 1971; Lang and Fay 1971) and *Anabaena catenula* (Wilcox et al. 1973), as well as between vegetative cells of *Anabaena cylindrica* (Giddings and Staehelin 1978), but their function remains unknown. Continuity of the outer membrane between cells of a filament suggests that the periplasm, or the cellular region between the cell membrane and outer membrane, is also continuous. This in turn has important physiological implications for the movement of compounds and for cell-to-cell communication (Flores et al. 2006).

Although the cyanobacterial cell wall most closely resembles that of Gram negative bacteria in overall structure, the peptidoglycan layer within the wall has been reported to share some characteristics with those of Gram positive bacteria. For example, studies on peptidoglycan purified from *Synechocystis* PCC 6714 indicate that the extent (56 %) of tetrapeptide cross-linkages is within the range reported for Gram positive bacteria and that a polysaccharide might be bound to the peptidoglycan, as also reported for some Gram positive bacteria (Jurgens et al. 1983, 1985; Jurgens and Weckesser 1986; Golecki. 1988). The actual thickness of the cyanobacterial peptidoglycan layer can differ significantly between genera (Fig. 14.3), and the measured thickness might also depend on the techniques used for cell preservation and visualization. For example, the cell walls of *Prochlorococcus* strains have been difficult to preserve and visualize with conventional TEM techniques (Fig. 14.2f, j). However, using cryo-electron microscope tomography, where cells are preserved in a near-native state using rapid freezing, we were able to establish that the cell wall peptidoglycan layer of *Prochlorococcus* strain MIT9313 is about 4 nm in thickness (Ting et al. 2007). This is much less than what we observed for the marine cyanobacterium *Synechococcus* WH8102, which possesses a peptidoglycan layer more than two times as thick as that of *Prochlorococcus* MIT9313 (Fig. 14.3). In contrast, in cyanobacteria such as *Oscillatoria* sp. PCC6407, characterized using traditional TEM, the peptidoglycan layer is about 25 nm and in *Oscillatoria princeps* the peptidoglycan layer has been reported to reach approximately 200–650 nm

(Stanier (Cohen-Bazire) 1988; Hoiczyk and Baumeister 1995).

One of the major structural constituents of peptidoglycan is nitrogen. Because of this fact, we previously proposed that the synthesis of a significant peptidoglycan layer could be a major cost to the cell, particularly in environments that contain low concentrations of essential nutrients (oligotrophic regions) (Ting et al. 2007). In particular, we have suggested that cyanobacteria capable of surviving with a reduced peptidoglycan layer could potentially have a competitive advantage, particularly in oligotrophic habitats such as the open oceans. Our work on marine cyanobacteria has revealed that within the *Prochlorococcus* lineage, the MED4 strain lacks a prominent peptidoglycan layer compared to the MIT9313 strain (Fig. 14.3; Ting et al. 2007). MED4 belongs to a large clade of recently differentiated lineages and ecotypes within these lineages have been found to be abundant throughout the water column in subtropical and tropical regions (Johnson et al. 2006). Interestingly, comparative genomic analyses between MED4 and MIT9313 indicate that low sequence identities (35–38 %) exist for predicted peptidoglycan biosynthesis proteins (MurC, MurF, MurG, MurI, MraY), suggesting that the function of these proteins is diverging between these strains (Ting et al. 2007).

IV. The Cellular Energy Matrix: Intracytoplasmic Membrane Systems

The intracytoplasmic lamellae of cyanobacteria consist of a dynamic, interconnected set of membranes that is linked with photosynthetic light-harvesting and electron transfer reactions and to the generation of ATP. In addition, respiratory electron transport components and reactions are associated with these internal membranes and share electron carriers, such the PQ pool (Schmetterer 1994). Notably, research on *Synechocystis* PCC6803 suggests that electrons originating from photosynthetic reactions can potentially enter respiratory electron transport pathways, and *vice versa* (Vermaas et al. 1994; Howitt et al. 1999). Moreover, in some cyanobacteria, processes such as nitrate reduction are associated with these membranes (Flores and Herrero 1994).

Gloeobacter violaceus is currently the only cyanobacterium known to lack an intracytoplasmic lamellar system (Rippka et al. 1974). *Gloeobacter* was first isolated from the surface of limestone rock collected in Switzerland (Table 14.1) and phylogenetic analyses indicate that it is deeply branched within the cyanobacterial lineage (Nelissen et al. 1995; Mimuro et al. 2008). Its unique membrane structure and composition are of interest from both physiological and evolutionary perspectives. Recent studies suggest that its plasma membranes contain distinct domains, one of which is "green" and enriched with Chl, the other of which is "orange" and enriched with carotenoids (Rexroth et al. 2011). Although these domains appear to share a significant fraction of proteins, including PSI and PSII core reaction center polypeptides (PsaA, PsaB, PsbA, PsbD), certain polypeptides were found to be unique to membrane fractions associated with a particular domain, including the cytochrome b_6f (see also Chap. 8) proteins PetB, PetC, and PetD (Rexroth et al. 2011).

The intracytoplasmic membranes of many cyanobacteria are located at the cell periphery and are arranged in bands parallel to the cell membrane (Fig. 14.2). Genera in which this internal membrane architecture has been observed include *Prochloron* (Giddings et al. 1980), *Prochlorococcus* (Chisholm et al. 1988; Ting et al. 2007), *Prochlorothrix* (Burger-Wiersma et al. 1986), *Synechococcus* (Sherman et al. 1994; Nevo et al. 2009), *Acaryochloris* (Miyashita et al. 1996), and *Anacystis* (Gantt and Conti 1969). In other cyanobacteria (*Anabaena*, *Dermocarpa*, *Gleocapsa*, *Pleurocapsa*), the intracytoplasmic membranes have been observed to extend throughout the central cytoplasmic space, clustering in random groups (Kunkel 1982) or to extend from points near the cell periphery to the center in a radial arrangement (*Synechocystis* sp. PCC6803, *Cyanothece* sp. ATCC51142 in Sherman et al. 1994; *Microcoleus chthonoplastes*, Nevo et al. 2009). In cyanobacteria that possess phycobilisomes as their major light-harvesting antenna, the internal membranes are not packed tightly in order to accommodate these large supramolecular complexes located on their cytoplasmic surface (Fig.14.2b–d; Gantt and Conti 1969; Gantt 1994). However, in cyanobacteria such as *Prochlorococcus*, *Prochloron* and *Prochlorothrix*, which do not possess phycobilisomes and instead utilize Chl *a*/*b*-containing antenna complexes that are integral membrane proteins, the intracytoplasmic membranes have been observed to be tightly appressed (Fig. 14.2f–h, j–l; Giddings et al. 1980; Burger-Wiersma et al. 1986; Miller et al. 1988; Ting et al. 2007; Nevo et al. 2009).

In what way might the structure of the internal membrane system reflect function? In particular, can specific physiological and photosynthetic strategies be inferred from membrane organization? In higher plant chloroplasts, it has been proposed that the close packing of chloroplast granal membranes could increase the functional antenna size of PSII, with absorbed light energy transferred between PSII reaction centers not only within a membrane layer, but also between stacked membranes (Boekema et al. 2000; Dekker and Boekema 2005; Mullineaux 2005). Whether this applies to cyanobacteria that have Pcb antennas and tight membrane packing has yet to be established. Furthermore, in higher plants, the stacking of membranes appears to reflect a relatively efficient strategy for acclimating to low light, and increases in both thylakoid stacking and levels of LHCII have been reported under low growth irradiance levels (Anderson 1986; Mullineaux 2005). In *Prochlorococcus*, we have observed an increase in the number of bands of internal membranes when this cyanobacterium is cultured under low irradiance levels. For example, in strain MIT9313, the number of internal membrane bands increased from 2 ± 0.2 at a growth irradiance level 60 umol photons m^{-2} s^{-1}, to 3.8 ± 0.4 at 10 umol photons m^{-2} s^{-1} (C. Ting, unpublished data). These changes in intracytoplasmic membrane content were not accompanied by significant changes in cell size in strain MIT9313 (C. Ting, unpublished data).

In chloroplasts, organization of the photosynthetic reaction centers is linked closely with thylakoid membrane structure, with PSII and PSI located predominantly in the granal and stromal lamellae, respectively. This organization affects major photosynthetic functions, including the transfer of absorbed light energy between the reaction centers, as well as linear and cyclic electron transport reactions (Trissl and Wilhelm 1993; Dekker and Boekema 2005). Notably, freeze-fracture electron microscopy of the internal membranes of *Prochloron* (Giddings et al. 1980) and *Prochlorothrix* (Miller et al. 1988) indicate that similarities exist in the overall distribution of particles associated with the membranes between these cyanobacteria and higher plants/green algae. For example, in *Prochloron*, which lacks phycobilisomes, the size distribution of particles on the exoplasmic fracture face of stacked membranes (EF_s) is similar to that of higher plants and green algae, and suggests that PSII units are also concentrated in these regions; however, the prevalence of smaller size particles indicates that the average PSII unit size might be smaller (Giddings et al. 1980). In *Prochlorothrix*, Miller et al. (1988) also reported that although particle sizes are smaller than observed in higher plants, the general characteristics of the fracture faces are similar to those observed in higher plants; from these analyses, Miller et al. (1988) concluded that the stacked and unstacked regions of the internal membranes of *Prochlorothrix* have different protein compositions.

It is important to point out that in cyanobacteria lacking phycobilisomes (see also Chap. 4) and possessing tightly appressed intracytoplasmic lamellae, both the global structure of the internal membrane system and individual photosynthetic apparatus proteins/supercomplexes, can differ markedly from chloroplasts. For instance, in *Prochlorococcus*, the internal membranes not only form parallel layers near the cell periphery, but are arranged as long bands that extend the length of the cell (Ting et al. 2007). Moreover, in some cellular regions, such as the cell poles in strain MIT9313, the internal membranes appear to be less tightly appressed (Ting et al. 2007). Furthermore, the Pcb light-harvesting complexes of *Prochlorococcus* and other cyanobacteria are structurally and phylogenetically distinct from the major antenna complexes (LHCII) found in higher plants and belong to a family that includes IsiA and CP43 (LaRoche et al. 1996). In both *Prochlorococcus* strain SS120 and *Prochlorothrix hollandica* the Pcb antenna proteins have been shown to form a ring consisting of 18 subunits around PSI trimers (Bibby et al. 2001b; Bumba et al. 2005). With the addition of 270 Chl molecules to each PSI trimer, this ring is expected to significantly increase light-harvesting capabilities (Bibby et al. 2001b). This Pcb-PSI complex is similar to the IsiA-PSI complexes reported in *Synechocystis* PCC6803 and *Synechococcus* PCC7942 when cells are cultured under iron depletion (-Fe), a condition resulting in phycobilisome degradation (Bibby et al. 2001a; Boekema et al. 2001). However, structural variability exists even among *Prochlorococcus* strains, and between *Prochlorococcus* and other Pcb-containing cyanobacteria: Pcb-PSI complexes have been observed in *Prochlorococcus* strain MIT9313 only under –Fe conditions, have not been observed in strain MED4 under normal or –Fe conditions, and have not been observed in *Prochloron didemni* (Bibby et al. 2003a, b). Furthermore specific Pcb proteins associate with PSII dimers to increase the light-harvesting capability of this photosystem in both *Prochlorococcus* and *Prochloron* (Bibby et al. 2003a, b).

The organization of Pcb-PSI and Pcb-PSII complexes within the intracytoplasmic membranes is currently unknown, and it remains to be established if they are uniformly distributed throughout the internal membrane system. Past work on *Synechococcus* PCC7942 suggests that radial asymmetry exists in the distribution of photosynthetic apparatus proteins (Sherman et al. 1994). This cyanobacterium has internal membranes present as parallel bands towards the cell periphery, but also possesses phycobilisomes. The use of TEM and immunocytochemical localization of proteins led Sherman et al. (1994) to propose that PSI complexes are localized primarily in intracytoplasmic membranes closest to the cell membrane. In contrast, polypeptides associated with PSII, including CP43 and CP47, as well as cytochrome b_6f proteins, are distributed throughout the intracytoplasmic membranes (Sherman et al. 1994). However, these authors concluded that the distribution observed in *Synechococcus* PCC7942 is not conserved among all phycobilisome-containing cyanobacteria. In *Synechocystis* PCC6803 and *Cyanothece* ATCC51142, which have internal membranes that extend into the central cytoplasm as described earlier, the distribution of PSI and PSII, as well as other photosynthetic apparatus proteins, appears to be more uniform among all the internal membranes (Sherman et al. 1994).

A particularly striking example of innovation within the cyanobacterial lineage involves the recently discovered *Acarychloris marina* (Miyashita et al. 1996), which contains Chl *d* as a major pigment, and has phycobiliproteins (phycocyanin, allophycocyanin) as well as a Pcb-based light-harvesting antenna (Miyashita et al. 1996; Marquardt et al. 1997; Chen et al. 2005a, b, 2009). Although *Acaryochloris* was isolated originally from colonial ascidians (Miyashita et al. 1996, 2003), *Acaryochloris*-like cyanobacteria have since been found in a variety of habitats, including in a moderately hypersaline lake (Miller et al. 2005), and it has been proposed that they have adapted to ecological niches enriched in far-red wavelengths (Kuhl et al. 2005). Phylogenetic analyses based on 16S rRNA gene analyses indicate *Acaryochloris* forms a distinct clade, separate from other Pcb-containing cyanobacterial lineages (Miyashita et al. 2003; Miller et al. 2005).

Studies using conventional transmission electron microscopy indicate that the *Acarychloris* intracytoplasmic membranes are organized in concentric layers at the cell periphery, and are characterized by distinct regions where membranes are either tightly appressed or well-spaced (Miyashita et al. 1996; Marquardt et al. 2000; Miller et al. 2005). Although phycobilisomes have not been observed in thin sections using conventional transmission electron microscopy (Miyashita et al. 1996; Marquardt et al. 2000), early evidence based on immunocytochemical localization of phycocyanin (Hu et al. 1999) and cell fluorescence (Marquardt et al. 2000) suggested that distinct clusters of biliproteins are associated with the internal membranes. The use of cryo-electron microscopy permitted visualization of distinct phycobiliprotein clusters that are organized as near-crystalline arrays in the cytoplasmic space between layers of intracytoplasmic membranes (Chen et al. 2009). In this work, Chen et al. (2009) observed that phycobiliprotein aggregates are present in membrane regions that are less tightly appressed (maximum 25 nm apart vs 2–3 nm) and are found to be associated with about half of the intracytoplasmic membranes.

In addition to these phycobiliproteins, at least two *pcb* genes (*pcbA*, *pcbC*) have been identified in *Acarychloris* (Chen et al. 2005c) and Pcb proteins have been found to associate with two PSII dimers to form a supercomplex consisting of 16 Pcb subunits (4 Pcb polypeptides associated with each PSII monomer) (Chen et al. 2005b). Furthermore, under conditions of iron limitation, when there are decreased levels of PSI relative to PSII, 18 Pcb subunits have also been found to associate with PSI in a ring-like formation, thus effectively increasing the absorption cross-section of this reaction center (Chen et al. 2005a). Many questions remain to be addressed, including how Pcb-associated supercomplexes are distributed throughout the intracytoplasmic membranes and whether phycobiliproteins and Pcb polypeptides associate with the same PSII dimer.

Significant advances in our understanding of the organization and structure of the intracytoplasmic membrane system derive from three-dimensional reconstructions of cyanobacterial cells (Fig. 14.2). In an early study on the halotolerant *Agmenellum quadruplicatum*, reconstructions from serial thin sections visualized using traditional electron microscopy led Nierzwicki-Bauer et al. (1983) to conclude that the intracytoplasmic membrane system is an interconnected network and exhibits connections at/near the cytoplasmic membrane. The use of electron tomography has provided greater resolution and has revealed that the internal membrane network is connected by the distinct fusion and branching of membranes, and that the membranes can have perforations and fenestrations, which likely permit molecules to move efficiently throughout the membrane system and cytoplasm (Nevo et al. 2007; Ting et al. 2007). Electron tomography of *Cyanothece* 51142 cells that had been subjected to high pressure freezing and freeze substitution revealed a unique spiral organization of the internal membranes, which form an extensive, branched network (Liberton et al. 2011). In *Synechocystis* PCC6803, the use of both conventional TEM and electron tomography on cells that had been cryofixed, freeze-substituted and thick-sectioned led van de Meene et al. (2006) to conclude that the internal membranes ("thylakoids") converged in regions next to the plasma membrane and that these areas included, on occasion, the presence of cylindrical structures with which the internal membranes were associated. These regions, called "thylakoid centers", were first reported in 1982 by Kunkel, and could be important in the organization and biogenesis of the intracytoplasmic membranes. In addition, studies on the near-native, three-dimensional architecture of *Prochlorococcus* have revealed the existence of distinct membrane-lined junctions connecting the lumen of tightly appressed intracytoplasmic membranes (Ting et al. 2007). These channels are approximately 10 nm (length) by 6 nm (diameter), and because they are not abundant, it was suggested that they play a role in maintaining a uniform electrochemical potential across the intracytoplasmic membranes, rather than functioning as major routes for solute diffusion (Ting et al. 2007).

A long-standing question involves the relationship between the intracytoplasmic lamellae and plasma membrane and specifically, whether these major membrane systems are connected. In their study based on three-dimensional reconstructions of cyanobacterial cells, Nierzwicki-Bauer et al. (1983) suggested that the intracytoplasmic membrane system is in contact with the cell membrane at distinct locations. Although several reports have also suggested the existence of these contacts, these associations are characteristically rare and further structural work is necessary in order to unambiguously confirm their existence. Using high-pressure-freezing and serial thin sections, Liberton et al. (2006) constructed three-dimensional models *Synechocystis* PCC 6803 cells, but did not observe well-defined contacts between the intracytoplasmic and cytoplasmic membranes; these authors concluded that these major membrane systems are physically discontinuous. However, in their work on the three-dimensional structure of the same cyanobacterial strain, van de Meene et al. (2006) reported rare contacts between the cytoplasmic and intracytoplasmic membranes and suggested that these might represent transient fusions that permit the exchange of lipids between the two membrane systems. Potential associations between the cytoplasmic and intracytoplasmic membranes could be functionally significant in the biogenesis of protein complexes associated with these membranes. Work by Zak et al. (2001) suggests that early events in the biogenesis of PSI and PSII in *Synechocystis* PCC 6803 occur in the cytoplasmic membranes, and their model postulates that membrane vesicle transport and/or movement through interconnected membrane systems could facilitate biogenesis of the intracytoplasmic membranes.

As the principal sites of energy generating and regulatory processes within a cyanobacterial cell, the intracytoplasmic membranes have been intensely studied. However, much remains to be discovered, particularly with regard to their three-dimensional and functional organization in different cyanobacteria, biogenesis, relationship with the cell membrane, and dynamic responses. Furthermore, the existence of structural and functional heterogeneity between membrane layers, as well as within a single layer, and the composition and interaction of associated proteins are critical to understand. Interactions between potentially intersecting processes, such as linear, cyclic and respiratory electron transport are also important to resolve, as their balance is likely to change under shifting environmental conditions and cellular energy states.

V. Carboxysomes and Cellular Compartmentalization

A prominent feature of cyanobacterial cells are the carboxysomes (see also Chap. 7), polyhedral microcompartments composed of a proteinaceous shell that sequester key enzymes and substrates (Fig. 14.2). The origins, evolution and molecular dynamics of these microcompartments are not well understood. However, the presence of carboxysomes in many chemoautotrophic bacteria (Yeates et al. 2008) and of similar microcompartments in enteric bacteria suggests that these structures are an effective prokaryotic strategy for compartmentalizing major cellular reactions, such as carbon fixation (Badger et al. 2002; Bobik 2006).

In cyanobacteria, carboxysomes function as a component of an efficient system that has evolved to concentrate carbon within the cell and enhance carbon fixation reactions (Fig. 14.4). This system also includes CO_2 and HCO_3^- transporters and carbonic anhydrases that catalyze the intracellular conversion of HCO_3^- to CO_2 (Fig. 14.4). Carboxysomes are grouped into two categories, α-carboxysomes and β-carboxysomes,

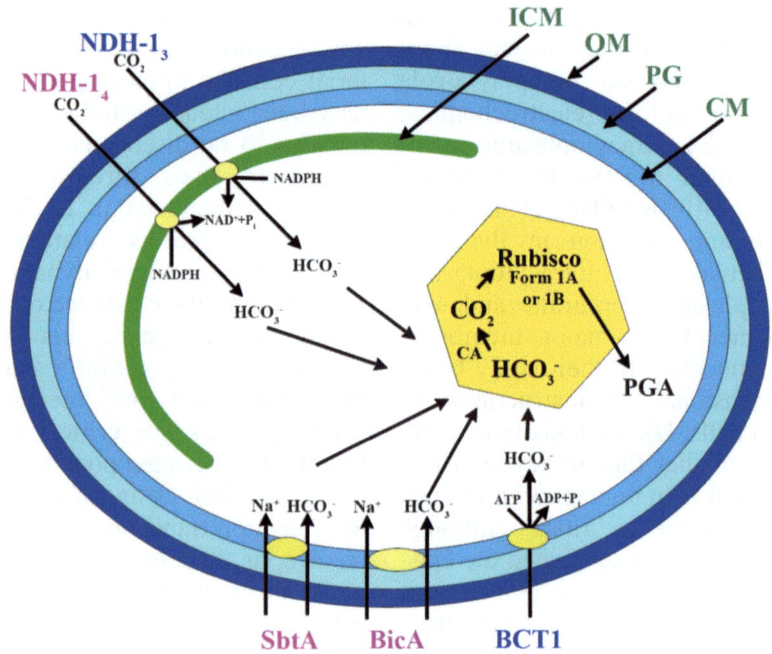

Fig. 14.4. The carbon dioxide-concentrating system of cyanobacteria involves specific transporters (*yellow circle*) for bicarbonate (HCO_3^-)/CO_2 and a microcompartment called the carboxysome (*yellow hexagon*). Within this microcompartment, which is surrounded by a protein shell and contains Rubisco, carbonic anhydrase converts HCO_3^- into CO_2. Transporters whose names are shown in pink have been identified in α- and β-cyanobacteria, and transporters whose names are shown in blue have been identified only in β-cyanobacteria. In cyanobacterial genomes, putative BicA transporters are often annotated as sulfate transporters (Price et al. 2004), and further experimental data is needed to verify their role in bicarbonate transport in specific cyanobacterial strains. *ICM* intracytoplasmic membrane, *OM* outer membrane, *PG* peptidoglycan, *CM* cytoplasmic membrane, *CA* carbonic anhydrase, *PGA* phosphoglycerate (Data in figure adapted from Badger et al. 2006).

based on whether they contain Form 1A Rubisco or Form 1B Rubisco, respectively (Badger et al. 2002). Form 1A Rubisco is commonly found in higher plants and algae, as well as in most chemoautotrophic bacteria, and is composed of eight large and eight small subunits (Kobayashi et al. 1991; Tabita 1999). In contrast, Form 1B Rubisco consists of two large subunits encoded by the *cbbM* gene (Tabita 1988; Shively et al. 1998). Differences also exist in the genes encoding carboxysome shell polypeptides in α-carboxysomes and β-carboxysomes, and the proteins of cyanobacterial α-carboxysomes have been found to share homology with those from several chemoautotrophic bacteria (Cannon et al. 2002; Badger and Price 2003). It has been observed that all marine cyanobacteria characterized to date have α-carboxysomes, and that this carboxysome form might be limited to marine environments (Badger et al. 2006). While α-carboxysomes are also found in chemoautotrophic bacteria, β-carboxysomes appear to be present only in cyanobacteria.

Reports of carboxysome dimensions from different cyanobacteria indicate that they are variable within a genus. For example, among marine cyanobacteria, diameters have been reported to be 114–137 nm (average 123 ± 5 nm) in marine *Synechococcus* WH8102 (Iancu et al. 2007), approximately 90 nm and 130 nm in *Prochlorococcus* strains MED4 and MIT9313, respectively, (Ting et al. 2007), and 250–600 nm in *Cyanothece* 51142 (Liberton et al. 2011). Within cyanobacteria, carboxysomes are typically located in the central cytoplasmic region (Fig. 14.2) and are not interspersed among the intracytoplasmic lamellae (Nierzwicki-Bauer et al. 1983; van de Meene et al. 2006; Liberton et al. 2006, 2011; Ting et al 2007). They often occupy a significant area and can also cluster and/or group closely with polyphosphate bodies (Nierzwicki-Bauer et al. 1983; Ting et al. 2007; Liberton et al. 2011). Reasons for carboxysome clustering have not yet been established, though it has been suggested that this organization could reflect function and facilitate carbon fixation by allowing any CO_2 that is leaked from one carboxysome to potentially reach and be fixed in a neighboring one (Ting et al. 2007). For both *Anabaena* PCC7119 and *Calothrix elenkinii* UAM 225, Orus et al. (2001) reported that increases in the extent of carboxysome clustering depended on environmental growth conditions. However, as Iancu et al. 2010 has pointed out, this clustering could also reflect where the shell polypeptides are synthesized and assembled within the cell. With regard to intracellular location, while Orus et al. (2001) did not observe general changes in carboxysome location in *Anabaena* PCC7119 under different growth conditions, McKay et al. (1993) reported that in *Synechococcus* PCC6301 cells grown under conditions of inorganic carbon limitation, carboxysomes increased in number and were located more towards the cell periphery, immediately adjacent to the intracytoplasmic membranes. It should be pointed out that in studies where carboxysomes are visualized in cyanobacterial cells using conventional transmission electron microscopy, sample preparation steps, such as fixation and dehydration, could potentially affect carboxysome organization and association.

The dynamic nature of microcompartments is supported by a recent study demonstrating that the cytoskeleton plays an important role in carboxysome organization. Savage et al. (2010) successfully visualized carboxysomes in a recent study on *Synechococcus elongatus* PCC7942 cells, which had been fluorescently labeled with probes against the Ccmk4 carboxysome shell protein, as well as the large subunit of Rubisco (RbcL). In these rod-shaped *Synechococcus* cells, Savage et al. (2010) reported that the carboxysomes are positioned along the long axis of the cell. Notably, the deletion of specific cytoskeletal genes, such as *parA*, resulted in a disruption of carboxysome order, as well as in daughter cells that had fewer or no carboxysomes (Savage et al. 2010). These authors concluded that ParA mediates carboxysome cellular organization, as well as carboxysome partitioning during cell division (Savage et al. 2010). The molecular dynamics, nature and impact of the association between carboxysomes and the cytoskeleton are important to characterize further, and also need to be examined in α-carboxysome-containing cyanobacteria and chemoautotrophic bacteria.

With regard to carboxysome structure, recent characterization of shell polypeptides from *Synechocystis* PCC6803 led Kerfeld et al. (2005) to propose that the carboxysome shell has a critical role in regulating the flux of metabolites into and out of the microcompartment. In particular, crystal structures of CcmK2 and CcmK4 indicate that each associates to form a hexamer, which contains a central pore of about 7 Å and 4 Å, respectively (Kerfeld et al. 2005). Notably, Kerfeld et al. (2005) observed that this central pore, as well as the spaces between hexamers, possess a net positive electrostatic potential and could thus facilitate the movement of negatively charged compounds, such as ribulose bis-phosphate, bicarbonate, and phosphoglycerate. Subsequent work by Tanaka et al. (2008) on the crystal structure of shell proteins CcmL and OrfA (CsoS4A) also from *Synechocystis* PCC6803 suggested that these proteins form pentamers and might function as vertices in the intact carboxysome shell. In a separate study by Iancu et al. (2007) on α-carboxysomes from marine *Synechococcus* strain WH8102, the authors concluded that individual carboxysomes are icosahedral, a shape that can be built from hexameric units and has been reported in other bacteria, including *Halothiobacillus neopolitanus* (Schmid et al. 2006; Iancu et al. 2010). Iancu et al. (2007) visualized carboxysomes in a

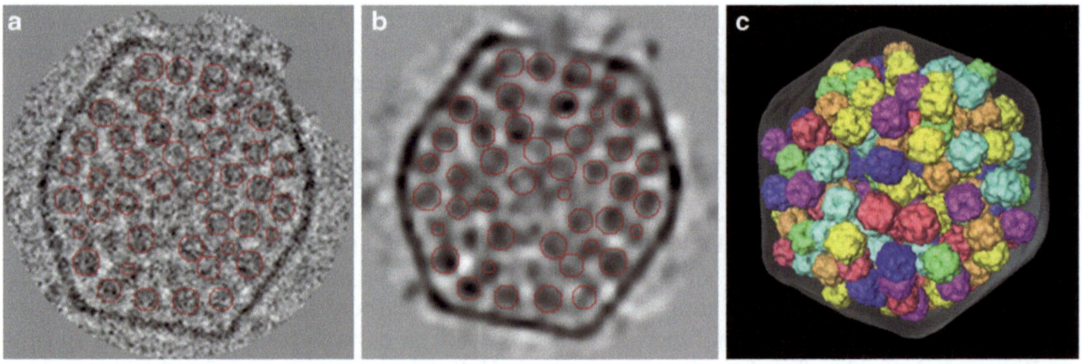

Fig. 14.5. Ribulose 1,5-bisphosphate carboxylase (RuBisCo) identified in a carboxysome purified from marine *Synechococcus* WH8102 and visualized using electron cryotomography. Shown are 6.7 nm slices of a carboxysome (undenoised (**a**), denoised (**b**)) in which template matching followed by customized peak search was used to characterize densities and identify individual RuBisCo proteins (*red circles*). A three-dimensional model of this carboxysome is depicted in (**c**), with the RuBisCo template (shown in different colors) replacing the electron dense structures present in (**a**) and (**b**). As shown (**c**), RuBisCo is often densely packed in carboxysomes, with 232 ± 18 RuBisCo proteins typically organized in three to four concentric layers in each carboxysome (Data and figure adapted from Iancu et al. 2007).

frozen, near-native state using electron cryotomography and found on average approximately 232 ± 18 Rubisco oligomers in each microcompartment (about 27 % of the total carboxysome volume) (Fig. 14.5). These Rubisco molecules are organized in concentric layers and are associated with about three to four carboxysome shell pores (Iancu et al. 2007). With an earlier report of a carbonic anhydrase (formerly CsoS3, presently CsoSCA, catalyzes the conversion of $HCO3^-$ to CO_2) associated with the carboxysome shell of marine *Synechococcus*, as well as of *Prochlorococcus* and *Halothiobacillus neopolitanus* (So et al. 2004), this organization and distribution of Rubisco would appear to maximize the chance of interactions between the enzyme and CO_2.

In terms of the evolution of carboxysomes in cyanobacteria, focus has been primarily on the advantages these microcompartments provide in concentrating CO_2 near Rubisco and thus, enhancing carbon fixation reactions (Badger et al. 2006). In addition, it has been suggested that carboxysomes could serve to protect Rubisco against oxygen (Cannon et al. 2001; Dou et al. 2008), and that microcompartments in general might sequester specialized environments that would minimize the escape of volatile metabolites (Penrod and Roth 2006). I would like to propose that an additional role for carboxysomes could perhaps involve reducing oxidative stress within the internal cellular environment. This might have provided an important selective advantage and thus, could in part have been a driving force in the evolution and retention of these structures in cyanobacteria, as well as other organisms.

A key component in the evolution of the cyanobacterial carbon dioxide-concentrating mechanism involves active inorganic carbon transporters, which permit bicarbonate (HCO_3^-) and CO_2 to accumulate within cells (Badger et al. 2006). In addition, CO_2 can enter cells passively from the surrounding environment (Badger et al. 2006). However, the accumulation of high concentrations of HCO_3^- and CO_2 within the cytoplasm could intensify oxidative stress and compete with cellular mechanisms for the removal of toxic oxygen species. Past work has shown that CO_2 can react with hydrogen peroxide (H_2O_2) to form peroxymonocarbonate (HCO_4^-), which has been demonstrated *in vitro* to oxidize methione and tertiary amines more rapidly than H_2O_2 (Richardson et al. 2003; Balagam and Richardson 2008). Furthermore, HCO_3^-

can react with hydroxyl radicals (HO˙) to form the carbonate radical (CO_3^-), a strong oxidant that can readily oxidize DNA guanine residues and amino acids (Stadtman and Berlett 1991; Shafirovich et al. 2001). In addition, oxidation of amino acids associated with the Fenton reaction has been linked with the presence of HCO_3^-, possibly through a reaction with HO˙ generated by the Fenton reaction (Stadtman and Berlett 1991; Medinas et al. 2007). Data from a recent study on *E. coli* also suggests CO_2 could increase oxidative stress *in vivo* through interactions with reactive oxygen species (Ezraty et al. 2011). In oxygen-evolving photosynthetic organisms, selection for mechanisms that could reduce cellular oxidative stress and damage would certainly provide an advantage. By effectively removing accumulated HCO_3^- and sequestering CO_2, carboxysomes would serve not only to increase the carboxylase activity of Rubisco, but to decrease the chance these compounds might react with different oxygen species and intensify cellular oxidative stress.

VI. Nitrogen Fixation: Lessons in Cyanobacterial Innovation

Many cyanobacteria serve as valuable contributors to global nitrogen fixation and are capable of converting atmospheric nitrogen into biologically useful forms. Because the nitrogenase enzyme catalyzing this reaction is inhibited by oxygen, there has been selective pressure for the evolution of innovative mechanisms to separate nitrogen fixation from oxygen-evolving photosynthetic reactions. Perhaps the best known mechanism involves specialized cells called heterocysts, which are found in some filamentous cyanobacteria (Fig. 14.1c) and will be discussed in greater detail ahead. Unicellular nitrogen-fixing cyanobacteria, such as *Cyanothece*, *Gloeothece* and *Crocosphaera*, rely on the temporal (day vs. night) regulation of photosynthesis and nitrogen fixation reactions that ensures these processes occur at different times (Sherman et al. 1998; Compaore and Stal 2010; Zehr 2011). The nonheterocystous, filamentous marine cyanobacterium, *Trichodesmium*, has evolved a strategy involving both temporal and spatial separation of nitrogen fixation and photosynthetic reactions (Paerl 1994; Bergman et al. 1997; Berman-Frank et al. 2003). Within the trichomes of *Trichodesmium*, although both of these reactions occur during the light period, rates of photosynthetic carbon fixation have been observed to increase in the morning and decrease during the middle of the day, a time when nitrogenase activity reaches a maximum (Berman-Frank et al. 2001). Furthermore, while all cells within a trichome are capable of photosynthesis, individual cells appear to be capable of modulating their photosynthetic activity and increases in photosynthetically inactive cells are correlated with periods of high nitrogen fixation rates (Berman-Frank et al. 2001).

Recent evidence suggests that in many ocean regions, including the South Pacific, Arabian Sea, North Pacific Ocean, and North Atlantic Ocean, a unicellular nitrogen-fixing cyanobacterium (UCYN-A, Table 14.1) exists that lacks the Photosystem II reaction center and associated oxygen-evolving complex (Zehr et al. 2008; Bothe et al. 2010; Moisander et al. 2010; Tripp et al. 2010; Zehr 2011). Genomic analysis of UCYN-A indicates it does possess PSI-associated genes, although it is missing genes encoding proteins involved in the Calvin cycle and tricarboxylic acid (TCA) cycle (Zehr et al. 2008; Tripp et al. 2010). Notably, this cyanobacterium (proposed name *Candidatus* Atelocyanobacterium thalassa) shares a mutualistic association with a prymnesiophyte, which transfers fixed carbon to UCYN-A in exchange for fixed nitrogen (Thompson et al. 2012).

Cyanobacteria that rely on specialized cells, called heterocysts (Fig. 14.1c), for nitrogen fixation have evolved an effective strategy for spatially separating nitrogenase from high concentrations of oxygen generated during photosynthesis (Wolk et al. 1994; Zhang et al. 2006). Heterocysts do contain PSI, though, which can serve as an important source of ATP as well as reducing power, and although polypeptides associated with PSII have been detected, photosynthetic oxygen evolution is not thought to occur (Braun-Howland and Nierzwicki-Bauer 1990; Thiel et al. 1990; Wolk et al. 1994). In an *Anabaena* PCC 7120 filament, one heterocyst typically occurs between 10 and 20 vegetative cells, and is morphologically distinct because of its larger size (Zhang et al. 2006). While the heterocyst supplies vegetative cells with fixed nitrogen, the vegetative cells supply carbohydrates in return (Wolk et al. 1994).

Early electron microscopy studies on the structure of heterocysts revealed that they contain ribosome-like granules, as well as an extensive, branching intracytoplasmic membrane system (Wildon and Mercer 1963; Lang and Fay 1971). Furthermore, three distinct layers were observed to comprise the heterocyst envelope immediately adjacent to the gram negative cell wall: these were described as the laminated (glycolipid), homogeneous (polysaccharide) and loosely organized/fibrous layers of the cell envelope (Lang and Fay 1971; Walsby 2007). In addition to the maintenance of heterocyst structure, the laminated (glycolipid) layer could function as an efficient barrier to oxygen (Wolk et al. 1994; Zhang et al. 2006; Walsby 2007). These early studies concluded that structural characterization of heterocysts is complicated by factors such as stage of differentiation, as well as fixation technique, and that these factors can lead to significant structural differences (Fay and Lang 1971). Subsequent freeze-fracture work by Giddings and Staehelin (1978) revealed the presence of numerous microplasmodesmata (outer diameter <200 Å) connecting adjacent vegetative cells and heterocysts. These microplasmodesmata were observed to be located in a circular region in the middle of the septum. Except for terminal heterocysts, each heterocyst within a filament has a pore at each end, and Walsby (2007) has proposed that these are the major sites for the regulation of gas exchange. Although both N_2 and O_2 could enter via these pore regions, respiratory activity is believed to reduce the concentration of O_2 within heterocysts to levels that do not result in nitrogenase inactivation (Walsby 1985; Wolk et al. 1994).

VII. Cyanobacteria at the Cutting Edge

Cyanobacteria have evolved to thrive in a remarkable range of habitats and at the heart of their ecological success is a stunning diversity interfused with innovation. In just the last three decades, reports of several new cyanobacterial genera have been published, and each of these discoveries has expanded our knowledge of the photosynthetic, physiological and ecological potential of this lineage. Although cyanobacteria are often minimalists, this should not imply a "simplification" of their structure and functions, but instead reflects the evolution of highly complex, yet streamlined, responses to environmental selection pressures, often resulting in increased metabolic efficiency and niche diversification. In particular, the compartmentalization of specific pathways and functions within discrete structures, such as intracytoplasmic membranes, carboxysomes and heterocysts has permitted cellular integration of diverse, and at times, incompatible processes and has promoted metabolic flexibility and dynamism. As we work to solve challenges in food and energy resources this century, it is certainly in our best interest to look closely at Earth's photosynthetic citizens, both in the oceans and on land, for innovative prototypes and potential solutions.

Acknowledgements

This work was supported by the National Science Foundation Award Numbers MCB-0615680 and MCB-0850900.

References

Anderson JM (1986) Photoregulation of the composition, function, and structure of thylakoid memgbranes. Ann Rev Plant Physiol 37:93–136

Badger MR, Price GD (2003) CO_2 concentrating mechanisms in cyanobacteria: molecular components, their diversity and evolution. J Exp Bot 54:609–622

Badger MR, Hanson D, Price GD (2002) Evolution and diversity of CO_2 concentrating mechanisms in cyanobacteria. Funct Plant Biol 29:161–173

Badger MR, Price GD, Long BM, Woodger FJ (2006) The environmental plasticity and ecological genomics of the cyanobacterial CO_2 concentrating mechanism. J Exp Bot 57:249–265

Balagam B, Richardson DE (2008) The mechanism of carbon dioxide catalysis in the hydrogen peroxide N-oxidation of amines. Inorg Chem 47:1173–1178

Balskus EP, Walsh CT (2010) The genetic and molecular basis for sunscreen biosynthesis in cyanobacteria. Science 329:1653–1656

Bauer CC, Buikema WJ, Black K, Haselkorn R (1995) A short-filament mutant of *Anabaena* sp. strain PCC 7120 that fragments in nitrogen-deficient medium. J Bacteriol 177:1520–1526

Bekker A, Holland HD, Wang PL, Rumble D, Stein HJ, Hannah JL, Coetzee LL, Beukes NJ (2004) Dating the rise of atmospheric oxygen. Nature 427:117–120

Bergman B, Gallon JR, Rai AN, Stal LJ (1997) N_2-fixation by non-heterocystous cyanobacteria. FEMS Microbiol Rev 19:139–185

Berman-Frank I, Lundgren P, Chen Y-B, Kupper H, Kolber Z, Bergman B, Falkowski P (2001) Segregation of nitrogen fixation and oxygenic photosynthesis in the marine cyanobacterium *Trichodesmium*. Science 294:1534–1537

Berman-Frank I, Lundgren P, Falkowski P (2003) Nitrogen fixation and photosynthetic oxygen evolution in cyanobacteria. Res Microbiol 154:157–164

Bhattacharya D, Medlin L (1995) The phylogeny of plastids: a review based on comparisons of small subunit ribosomal RNA coding regions. J Phycol 31:489–498

Bhattacharya D, Yoon HS, Hackett JD (2004) Photosynthetic eukaryotes unite: Endosymbiosis connects the dots. Bioessays 26:50–60

Bhattacharya D, Archibald JM, Weber APM, Reyes-Prieto A (2007) How do endosymbionts become organelles? Understanding early events in plastid evolution. Bioessays 29:1239–1246

Bibby TS, Nield J, Barber J (2001a) Iron deficiency induces the formation of an antenna ring around trimeric photosystem I in cyanobacteria. Nature 412:743–745

Bibby TS, Nield J, Partensky F, Barber J (2001b) Antenna ring around photosystem I. Nature 413:590

Bibby TS, Mary I, Nield J, Partensky F, Barber J (2003a) Low-light-adapted *Prochlorococcus* species possess specific antennae for each photosystem. Nature 424:1051–1054

Bibby TS, Nield J, Chen M, Larkum AWD, Barber J (2003b) Structure of a photosystem II supercomplex isolated from *Prochloron didemni* retaining its chlorophyll *a/b* light-harvesting system. Proc Natl Acad Sci U S A 100:9050–9054

Bobik TA (2006) Polyhedral organelles compartmenting bacterial metabolic processes. Appl Microbiol Biotechnol 70:517–525

Boekema EJ, van Breemen JFL, van Roon H, Dekker JP (2000) Arrangement of Photosystem II supercomplexes in crystalline macrodomains within the thylakoid membrane of green plant chloroplasts. J Mol Biol 301:1123–1133

Boekema EJ, Hifney A, Yakushevska AE, Piotrowski M, Keegstra W, Berry S, Michel K-P, Pistorius EK, Kruip J (2001) A giant chlorophyll-protein complex induced by iron-deficiency in cyanobacteria. Nature 412:745–748

Bohm GA, Pfleiderer W, Boger P, Scherer S (1995) Structure of a novel oligosaccharide-mycosporine-amino acid ultraviolet A/B sunscreen pigment from the terrestrial cyanobacterium *Nostoc commune*. J Biol Chem 270:8536–8539

Bothe H, Tripp HJ, Zehr JP (2010) Unicellular cyanobacteria with a new mode of life: the lack of photosynthetic oxygen evolution allows nitrogen fixation to proceed. Arch Microbiol 192:783–790

Braun-Howland EB, Nierzwicki-Bauer SA (1990) Occurrence of the 32-kDa Q_B-binding protein of Photosystem II in vegetative cells, heterocysts and akinetes of *Azolla caroliniana* cyanobionts. Planta 180:361–371

Brocks JJ, Logan GA, Buick R, Summons RE (1999) Archean molecular fossils and the early rise of eukaryotes. Science 285:1033–1036

Brocks JJ, Buick R, Logan GA, Summons RE (2003) Composition and syngeneity of molecular fossils from the 2.78 to 2.45 billion-year-old Mount Bruce Supergroup, Pilbara Craton, Western Australia. Geochim Cosmochim Acta 67:4289–4319

Buchanan RE, Gibbons NE (1974) Bergey's manual of determinative bacteriology, 8th edn. Williams and Wilkins, Co., Baltimore

Bumba L, Prasil O, Vacha F (2005) Antenna ring around trimeric Photosystem I in chlorophyll *b* containing cyanobacterium *Prochlorothrix hollandica*. Biochim Biophys Acta 1708:1–5

Burger-Wiersma T, Veenhuis M, Korthals HJ, Van de Wiel CCM, Mur LR (1986) A new prokaryote containing chlorophylls *a* and *b*. Nature 320:262–264

Cabeen MT, Jacobs-Wagner C (2005) Bacterial cell shape. Nat Rev Microbiol 3:601–610

Cannon GC, Bradburne CE, Aldrich HC, Baker SH, Heinhorst S, Shively JM (2001) Microcompartments in prokaryotes: carboxysomes and related polyhedral. Appl Environ Microbiol 67:5351–5361

Cannon GC, Heinhorst S, Bradburne CE, Shively JM (2002) Carboxysome genomics: a status report. Funct Plant Biol 29:175–182

Carballido-Lopez R, Errington J (2003) The bacterial cytoskeleton: in vivo dynamics of the actin-like protein Mbl of *Bacillus subtilis*. Dev Cell 4:19–28

Chen M, Bibby TS, Nield J, Larkum A, Barber J (2005a) Iron deficiency induces a chlorophyll *d*-binding Pcb antenna system around Photosystem I in *Acaryochloris marina*. Biochim Biophys Acta 1708:367–374

Chen M, Bibby TS, Nield J, Larkum AWD, Barber J (2005b) Structure of a large photosystem II supercomplex from *Acarychloris marina*. FEBS Lett 579:1306–1310

Chen M, Hiller RG, Howe CJ, Larkum AWD (2005c) Unique origin and lateral transfer of prokaryotic Chlorophyll-*b* and Chlorophyll-*d* light-harvesting systems. Mol Biol Evol 22:21–28

Chen M, Floetenmeyer M, Bibby TS (2009) Supramolecular organization of phycobiliproteins in the chlorophyll *d*-containing cyanobacterium *Acaryochloris marina*. FEBS Lett 583:2535–2539

Chisholm SW (1992) Phytoplankton size. In: Falkowski PG, Woodhead AD (eds) Primary productivity and biogeochemical cycles in the sea. Plenum Press, New York, pp 213–237

Chisholm SW, Olson RJ, Zettler ER, Goericke R, Waterbury JB, Welschmeyer NA (1988) A novel free-living prochlorophyte abundant in the oceanic euphotic zone. Nature 334:340–343

Clayton RK (1980) Photosynthesis: physical mechanisms and chemical patterns. Cambridge Unviersity Press, London

Compaore J, Stal LJ (2010) Oxygen and the light-dark cycle of nitrogenase activity in two unicellular cyanobacteria. Environ Microbiol 12:54–62

De Pedro MA, Yong KD, Holtje JV, Schwarz H (2003) Branching in Escherichia coli cells arises from multiple sites of inert peptidoglycan. J Bacteriol 185:1147–1152

Dekker JP, Boekema EJ (2005) Supramolecular organization of thylakoid membrane proteins in green plants. Biochim Biophys Acta 1706:12–39

Delwiche CF (1999) Tracing the thread of plastid diversity through the tapestry of life. Am Nat 154:S164–S177

Dou Z, Heinhorst S, Williams EB, Murin CD, Shively JM, Cannon GC (2008) CO_2 fixation kinetics of *Halothiobacillus neapolitanus* mutant carboxysomes lacking carbonic anhydrase suggest the shell acts as a diffusional barrier for CO_2. J Biol Chem 283:10377–10384

Ehling-Schulz M, Bilger W, Scherer S (1997) UB-B-induced synthesis of photoprotective pigments and extracellular polysaccharides in the terrestrial cyanobacterium *Nostoc commune*. J Bacteriol 179:1940–1945

Ezraty B, Chabalier M, Ducret A, Maisonneuve E, Dukan S (2011) CO_2 exacerbates oxygen toxicity. EMBO Rep 12:321–326

Fay P, Lang NJ (1971) The heterocysts of blue-green algae. I. Ultrastructural integrity after isolation. Proc R Soc Lond B Biol Sci 178:185–192

Fewer D, Friedl T, Budel B (2002) *Chroococcidiopsis* and heterocyst-differentiating cyanobacteria are each other's closest living relatives. Mol Phylogenet Evol 23:82–90

Fieldler G, Arnold M, Hannus S, Maldener I (1998) The DevBCA exporter is essential for envelope formation in heterocysts of the cyanboacterium *Anabaena* sp. strain PCC 7120. Mol Microbiol 27:1193–1202

Figge RM, Divakaruni AV, Gober JW (2004) MreB, the cell shape-determining bacterial actin homologue, co-ordinates cell wall morphogenesis in *Caulobacter crescentus*. Mol Microbiol 51:1321–1332

Fleming ED, Castenholz RW (2007) Effects of periodic desiccation on the synthesis of the UV-screening compound, scytonemin, in cyanobacteria. Environ Microbiol 9:1448–1455

Flores E, Herrero A (1994) Assimilatory nitrogen metabolism and its regulation. In: Bryant DA (ed) The molecular biology of cyanobacteria. Kluwer Academic Publishers, The Netherlands, pp 487–517

Flores E, Herrero A, Wolk CP, Maldener I (2006) Is the periplasm continuous in filamentous multicellular cyanobacteria? Trends Microbiol 14:339–443

Fogg GE (1986) Picoplankton. Proc R Soc Lond B Biol Sci 228:1–30

Fritsch FE (1935) The structure and reproduction of algae. The University Press, Cambridge

Gantt E (1994) Supramolecular membrane organization. In: Bryant DA (ed) The molecular biology of cyanobacteria. Kluwer Academic Publishers, The Netherlands, pp 119–138

Gantt E, Conti SF (1969) Ultrastructure of blue-green algae. J Bacteriol 97:1486–1493

Garcia-Pichel F, Sherry ND, Castenholz RW (1992) Evidence for an ultraviolet sunscreen role of the extracellular pigment scytonemin in the terrestrial cyanobacterium *Chlorogloeopsis* sp. Photochem Photobiol 56:17–23

Giddings TH, Staehelin LA (1978) Plasma membrane architecture of *Anabaena cylindrica*: occurrence of microplasmodesmata and changes associated with heterocyst development and the cell cycle. Cytobiologica 16:235–249

Giddings TH, Withers NW, Staehelin LA (1980) Supramolecular structure of stacked and unstacked regions of the photosynthetic membranes of *Prochloron* sp., a prokaryote. Proc Natl Acad Sci U S A 77:352–356

Giovannoni SF, Turner S, Olsen G, Barns S, Lane DJ, Pace NR (1988) Evolutionary relationships among *Cyanobacteria* and green chloroplasts. J Bacteriol 170:3584–3592

Gitai Z, Dye NA, Shapiro L (2004) An actin-like gene can determine cell polarity in bacteria. Proc Natl Acad Sci U S A 101:8643–8648

Golecki JR (1988) Analysis of the structure and development of bacterial membranes (outer, cytoplasmic and intracytoplasmic membranes). Methods Microbiol 20:61–77

Golubic S, Hernandez-Marine M, Hoffmann L (1996) Developmental aspects of branching in filamentous Cyanophyta/Cyanobacteria. Arch Hydrobiol Suppl 117:303–329

Gray MW (1992) The endosymbiont hypothesis revisited. Int Rev Cytol 141:233–357

Gugger MF, Hoffmann L (2004) Polyphyly of true branching cyanobacteria (Stigonematales). Int J Syst Evol Microbiol 54:349–3357

Hansel A, Tadros MH (1998) Characterization of two pore-forming proteins isolated from the outer membrane of *Synechococcus* PCC 6301. Curr Microbiol 36:321–326

Hansel A, Pattus F, Jurgens UJ, Tadros MH (1998) Cloning and characterization of the genes coding for two porins in the unicellular cyanobacterium *Synechococcus* PCC 6301. Biochim Biophys Acta 1399:31–39

Hofmann HJ (1976) Precambrian microflora, Belcher Islands, Canada: significance and systematics. J Paleontol 50:1040–1073

Hoiczyk E (1998) Structural and biochemical analysis of o the sheath of *Phormidium uncinatum*. J Bacteriol 180:3923–3932

Hoiczyk E, Baumeister W (1995) Envelope structure of four gliding filamentous cyanobacteria. J Bacteriol 177:2387–2395

Hoiczyk E, Baumeister W (1997) Oscillin, an extracellular, Ca+2-binding glycoprotein essential for gliding motility of cyanobacteria. Mol Microbiol 26:699–708

Hoiczyk E, Hansel A (2000) Caynobacterial cell walls: news from an unusual prokaryotic envelope. J Bacteriol 182:1191–1199

Holland HD (2006) The oxygenation of the atmosphere and oceans. Phil Trans R Soc B 361:903–915

Howitt CA, Udall PK, Vermaas WFJ (1999) Type 2 NADH dehydrogenases in the cyanobacterium *Synechocystis* sp. strain PCC6803 are involved in regulation rather than respiration. J Bacteriol 181:3994–4003

Hu Q, Marquardt J, Iwasaki I, Miyashita H, Kurano N, Morschel E, Miyachi S (1999) Molecular structure, localization and function of biliproteins in the chlorophyll *a/d* containing oxygenic photosynthetic prokaryote *Acaryochloris marina*. Biochim Biophys Acta 1412:250–261

Hu B, Yang G, Zhao W, Zhang Y, Zhao J (2007) MreB is important for cell shape but not for chromosome segregation of the filamentous cyanobacterium Anabaena sp. PCC 7120. Mol Microbiol 63:1640–1652

Iancu CV, Ding HJ, Morris DM, Dias DP, Gonzales AD, Martino A, Jensen GJ (2007) The structure of isolated *Synechococcus* strain WH8102 carboxysomes revealed by electron cryotomography. J Mol Biol 372:764–773

Iancu CV, Morris DM, Dou Z, Heinhorst S, Cannon GC, Jensen GJ (2010) Organization, structure, and assembly of alpha-carboxysomes determined by electron cryotomography of intact cells. J Mol Biol 396:105–117

Ingram LO, van Baalen C (1970) Characteristics of a stable, filamentous mutant of a coccoid blue-green alga. J Bacteriol 102:784–789

Johnson ZI, Zinser ER, Coe A, McNulty NP, Woodward EMS, Chisholm SW (2006) Niche partitioning among *Prochlorococcus* ecotypes along ocean-scale environmental gradients. Science 311:1737–1740

Jones LJ, Carballido-Lopz R, Errington J (2001) Control of cell shape in bacteria: helical, actin-like filaments in Bacillus subtilis. Cell 104: 913–922

Jurgens UJ, Mantele W (1991) Orientation of carotenoids in the outer membrane of *Synechocystis* PCC 6714 (Cyanobacteria). Biochim Biophys Acta 1067:208–212

Jurgens UJ, Weckesser J (1986) Polysaccharide covalently linked to the peptidoglycan of the cyanobacterium *Synechocystis* sp. strain PCC6714. J Bacteriol 168:568–573

Jurgens UJ, Drews G, Weckesser J (1983) Primary structure of the peptidoglycan from the unicellular cyanobacterium *Synechocystis* sp. strain PCC 6714. J Bacteriol 154:471–478

Jurgens UJ, Golecki JR, Weckesser J (1985) Characterization of the cell wall of the unicellular cyanobacterium *Synechocystis* PCC 6714. Arch Microbiol 142:168–174

Kerfeld CA, Sawaya MR, Tanaka S, Nguyen CV, Phillips M, Beeby M, Yeates TO (2005) Protein structures forming the shell of primitive bacterial organelles. Science 309:936–938

Kirk JTO (1975) A theoretical analysis of the contribution of algal cells to the attenuation of light within natural waters. II. Spherical cells. New Phytol 75:21–36

Kirk JTO (1976) A theoretical analysis of the contribution of algal cells to the attenuation of light within natural waters. III. Cylindrical and spheroidal cells. New Phytol 77:341–358

Kirk JTO (1994) Light and photosynthesis in aquatic ecosystems, 2nd edn. Cambridge University Press, Cambridge

Knoll AH (2003a) Life on a young planet. The First three billion years of evolution on earth. Princeton University Press, Princeton

Knoll AH (2003b) The geological consequences of evolution. Geobiology 1:3–14

Kobayashi H, Viale AM, Takabe T, Akazawa T, Wada K, Shinozaki K, Kobayashi K, Suiura M (1991) Sequence and expression of genes encoding the large and small subunits of ribulose 1,5-bisphosphate carboxylase/oxygenase from *Chromatium vinosum*. Gene 97:55–62

Koksharova OA, Klint J, Rasmussen U (2007) Comparative proteomics of cell division mutants and wild-type of *Synechococcus* sp. strain PCC 7942. Microbiol 153:2505–2517

Kroos L, Maddock JR (2003) Prokaryotic development: emerging insights. J Bacteriol 185:1128–1146

Kuhl M, Chen M, Ralph PJ, Schreiber U, Larkum AWD (2005) A niche for cyanobacteria containing chlorophyll *d*. Nature 433:820

Kunkel DD (1982) Thylakoid centers: structures associated with the cyanobacterial photosynthetic membrane system. Arch Microbiol 133:97–99

La Roche J, van der Staay GWM, Partensky F, Ducret A, Aebersold R, Li R, Golden SS, Hiller RG, Wrench PM, Larkum AWD, Green BR (1996) Independent evolution of the prochlorophyte and green plant chlorophyll *a*/*b* light harvesting proteins. Proc Natl Acad Sci U S A 93:15244–15248

Lang NJ, Fay P (1971) The heterocysts of blue-green algae. II. Details of ultrastructure. Proc R Soc Lond B Biol Sci 178:193–203

Liberton M, Berg RH, Heuser J, Roth R, Pakrasi HB (2006) Ultrastructure of the membrane systems in the unicellular cyanobacterium *Synechocystis* sp. strain PCC 6803. Protoplasma 227:129–138

Liberton M, Austin JR, Berg RH, Pakrasi HB (2011) Unique thylakoid membrane architecture of a unicellular N2-fixing cyanobacterium revealed by electron tomography. Plant Physiol 155:1656–1666

Lichtle C, Thomas JC, Spilar A (1995) Immunological and ultrastructural characterization of the photosynthetic complexes of the Prochlorophyte *Prochlorococcus* (Oxychlorobacteria). J Phycol 31:934–941

Livaitis MK (2002) A molecular test of cyanobacterial phylogeny: inferences from constraint analyses. Hydrobiological 468:135–145

Marquardt J, Senger H, Miyashita H, Miyachi S, Morschel E (1997) Isolation and characterization of biliprotein aggregates from *Acaryochloris marina*, a *Prochloron*-like prokaryote containing mainly chlorophyll *d*. FEBS Lett 410:428–432

Marquardt J, Morschel E, Rhiel E, Westermann M (2000) Ultrastructure of *Acaryochloris marina*, an oxyphotobacterium containing mainly chlorophyll *d*. Arch Microbiol 174:181–188

McFadden GI (2001) Chloroplast origin and integration. Plant Physiol 125:50–53

McKay RML, Gibbs SP, Espie GS (1993) Effect of dissolved inorganic carbon on the expression of carboxysomes, localization of Rubisco and the mode of inorganic carbon transport in cells of the cyanobacterium *Synechococcus* UTEX 625. Arch Microbiol 159:21–29

Medinas DB, Cerchiaro G, Trindale DF, Augusto O (2007) The carbonate radical and related oxidants derived from bicarbonate buffer. IUBMB Life 59:255–262

Miller KR, Jacob JS, Burger-Wiersma T, Matthijs HCP (1988) Supramolecular structure of the thylakoid membrane of *Prochlorothrix hollandica*: a chlorophyll *b*-containing prokaryote. J Cell Sci 91:577–586

Miller SR, Augustine S, Olson TL, Blankenship RE, Selker J, Wood AM (2005) Discovery of a free-living chlorophyll *d*-producing cyanobacterium with a hybrid proteobacterial/cyanobacterial small-subunit rRNA gene. Proc Natl Acad Sci U S A 102:850–855

Mimuro M, Tomo T, Tsuchiya T (2008) Two unique cyanobacteria lead to a traceable approach of the first appearance of oxygenic photosynthesis. Photosynth Res 97:167–176

Miyashita H, Ikemoto H, Kurano N, Adachi K, Chihara M, Miyachi S (1996) Chlorophyll *d* as a major pigment. Nature 383:402

Miyashita H, Ikemoto H, Kurano N, Miyachi S, Chihara M (2003) *Acarychloris marina* gen. et sp. nov. (cyanobacteria), an oxygenic photosynthetic prokaryote containing Chl *d* as a major pigment. J Phycol 39:1247–1253

Mobley HL, Koch AL, Doyle RJ, Streips UN (1984) Insertion and fate of the cell wall in *Bacillus subtilis*. J Bacteriol 158:169–179

Moisander PH, Beinart RA, Hewson I, White AE, Johnson KS, Carlson CA, Montoya JP, Zehr JP (2010) Unicellular cyanobacterial distributions broaden the oceanic N_2 fixation domain. Science 327:1512–1514

Moreira D, Le Guyader H, Philippe H (2000) The origin of red algae and the evolution of chloroplasts. Nature 405:69–72

Morel A, Bricaud A (1981) Theoretical results concerning light absorption in a discrete medium, and application to specific absorption of phytoplankton. Deep-Sea Res 28:1375–1393

Mullineaux CW (2005) Function and evolution of grana. Trends Plant Sci 10:521–525

Nelissen B, van de Peer Y, Wilmotte A, De Wachter R (1995) An early origin of plastids within the cyanobacterial divergence is suggested by evolutionary trees based on complete 16S rRNA sequences. Mol Biol Evol 12:1166–1173

Nevo R, Charuvi D, Shimoni E, Schwarz R, Kaplan A, Ohad I, Reich Z (2007) Thylakoid membrane perforations and connectivity enable intracellular traffic in cyanobacteria. EMBO J 26:1467–1473

Nevo R, Chuartzman SG, Tsabari O, Reich Z, Charuvi D, Shimoni E (2009) Architecture of thylakoid membrane networks. In: Wada H, Murata N (eds) Lipids in photosynthesis: essential and regulatory functions. Springer Science, Dordrecht, pp 295–328

Nierzwicki-Bauer SA, Balkwill DL, Stevens SE (1983) Three-dimensional ultrastructure of a unicellular cyanobacterium. J Cell Biol 97:713–722

Olson JM (2006) Photosynthesis in the archaen era. Photosynth Res 88:109–117

Oren A, Gunde-Cimerman N (2007) Mycosporines and mycosporine-like amino acids: UV protectants or multipurpose secondary metabolites? FEMS Microbiol Lett 269:1–10

Orus MI, Rodriguez-Buey ML, Marco E, Fernandez-Valiente E (2001) Changes in carboxysome structure and grouping and in photosynthetic affinity for inorganic carbon in *Anabaena* strain PCC 7119 (Cyanophyta) in response to modification of CO_2 and Na+supply. Plant Cell Physiol 42:46–53

Paerl HW (1984) Cyanobacterial carotenoids: their roles in maintaining optimal photosynthetic production among aquatic bloom forming genera. Oecologia 61:143–149

Paerl HW (1994) Spatial segregation of CO_2 fixation in *Trichodesmium* spp.: linkage to N_2 fixation potential. J Phycol 30:790–799

Palenik B, Haselkorn R (1992) Multiple evolutionary origins of prochlorophytes, the chlorophyll *b-containing* prokaryotes. Nature 355:265–267

Palenik B, Brahamsha B, Larimer FW, Land M, Hauser L, Chain P, Lamerdin J, Regala W, Allen EE, McCarren J, Paulsen I, Dufresne A, Partensky F, Webb EA, Waterbury J (2003) The genome of a motile marine *Synechococcus*. Nature 424:1037–1042

Penrod JT, Roth JR (2006) Conserving a volatile metabolite: a role for carboxysome-like organelles in *Salmonella enterica*. J Bacteriol 188:2865–2874

Price GD, Woodger FJ, Badger MR, Howitt SM, Tucker L (2004) Identification of a SulP-type bicarbonate transporter in marine cyanobacteria. Proc Natl Acad Sci U S A 101:18228–18233

Proteau PJ, Gerwick WH, Garcia-Pichel F, Castenholz R (1993) The structure of scytonemin, an ultraviolet sunscreen pigment from the sheaths of cyanobacteria. Experientia 49:825–829

Rasmussen B, Fletcher IR, Brocks JJ, Kilburn MR (2008) Reassessing the first appearance of eukaryotes and cyanobacteria. Nature 455:1101–1104

Raven JA (1986) Physiological consequences of extremely small size for autotrophic organisms in the sea. In: Platt T, Li WKW (eds) Photosynthetic Picoplankton, Can Bull Fish Aquat Sci 214:1–70

Raven JA (1998) The twelfth Tansley Lecture. Small is beautiful: the picophytoplankton. Func Ecol 12:503–513

Resch CM, Gibson J (1983) Isolation of the carotenoid-containing cell wall of three unicellular cyanobacteria. J Bacteriol 155:345–350

Rexroth S, Mullineaux CW, Ellinger D, Sendtko E, Rogner M, Koenig F (2011) The plasma membrane of the cyanboacterium *Gloeobacter violaceus* contains segregated bioenergetic domains. Plant Cell 23:2379–2390

Reyes-Prieto A, Bhattacharya D (2007) Phylogeny of Calvin cycle enzymes supports Plantae monophyly. Mol Phylogenet Evol 45:384–391

Richardson DE, Regino CAS, Yao HR, Johnson JV (2003) Methionine oxidation by peroxymonocarbonate, a reactive oxygen species formed from CO_2/bicarbonate and hydrogen peroxide. Free Radic Biol Med 35:1538–1550

Rippka R, Waterbury J, Cohen-Bazire G (1974) A cyanobacterium which lacks thylakoids. Arch Microbiol 100:419–436

Rippka R, Deruelles J, Waterbury JB, Herdman M, Stanier RY (1979) Generic assignments, strain histories and properties of pure cultures of cyanobacteria. J Gen Microbiol 111:1061

Ris H, Singh RN (1961) Electron microscope studies on blue-green algae. J Biophys Biochem Cytol 9:63–80

Rocap G, Larimer FW, Lamerdin J, Malfatti S, Chain P, Ahlgren NA, Arellano A, Coleman M, Hauser L, Hess WR et al (2003) Genome divergence in two *Prochlorococcus* ecotypes reflects oceaninc niche differentiation. Nature 424:1042–1047

Savage DF, Afonso B, Chen AH, Silver PA (2010) Spatially ordered dynamics of the bacterial carbon fixation machinery. Science 327:1258–1261

Scherer S, Chen TW, Boger P (1988) A new UV-A/B protecting pigment in the terrestrial cyanobacterium *Nostoc commune*. Plant Physiol 88:1055–1057

Schmetterer G (1994) Cyanobacterial respiration. In: Bryant DA (ed) The molecular biology of cyanobacteria. Kluwer Academic Publishers, The Netherlands, pp 409–435

Schmid MF, Paredes AM, Khant HA, Soyer F, Aldrich HC, Chiu W, Shively JM (2006) Structure of the *Halothiobacillus neapolitanus* carboxysomes by cryo-electron tomography. J Mol Biol 364:526–535

Schopf JW (1993) Microfossils of the early archean apex chert: new evidence of the antiquity of life. Science 260:640–646

Schopf JW (2000) The fossil record: tracing the roots of the cyanobacterial lineage. In: Whitton BA, Potts M (eds) The ecology of cyanobacteria. Kluwer Academic Publishers, Dordecht

Shafirovich V, Dourandin A, Huang W, Geacintov NE (2001) The carbonate radical is a site-selective oxidizing agent of guanine in double-stranded oligonucleotides. J Biol Chem 276:24621–24626

Sherman DM, Troyan TA, Sherman LA (1994) Localization of membrane proteins in the cyanobacterium *Synechococcus* sp. PCC7942. Plant Physiol 106:251–262

Sherman LA, Meunier P, Colon-Lopez MS (1998) Diurnal rhythms in metabolism: a day in the life of a unicellular, diazotrophic cyanobacterium. Photosynth Res 58:25–42

Shively JM, Vankeulen G, Meijer WG (1998) Something from nothing – carbon dioxide fixation in chemoautotrophs. Ann Rev Microbiol 52:191–230

Siefert JL, Fox GE (1998) Phylogenetic mapping of bacterial morphology. Microbiology 144:2803–2808

Simon RD (1981) Gliding motility in *Aphanothece halophytica*: analysis of wall proteins in *mot* mutants. J Bacteriol 148:315–321

Singh SP, Montgomery BL (2011) Determining cell shape: adaptive regulation of cyanobacterial cellular differentiation and morphology. Trends Microbiol 19:278–285

So AK-C, Espie GS, Williams EB, Shively JM, Heinhorst S, Cannon GC (2004) A novel evolutionary lineage of carbonic anhydrase (εClass) is a component of the carboxysome shell. J Bacteriol 186:623–630

Stadtman ER, Berlett BS (1991) Fenton chemistry. Amino acid oxidation. J Biol Chem 266: 17201–17211

Stanier RY (1988) Fine structure of cyanobacteria. In: Packer L, Glazer AN (eds) Methods in enzymology. Academic, New York, pp 157–172, Vol 167

Stanier RY, van Niel CB (1962) The concept of a bacterium. Arch Mikrobiol 42:17–35

Stanier RY, Sistrom WR, Hansen TA, Whitton BA, Castenholz RW, Pfenig N, Gorlenko VN, Kondratieva EMN, Eimhjellen KE, Whittenbury R, Gherna RL, Truper HG (1978) Proposal to place the nomenclature of the cyanobacteria (blue-green algae) under the rules of the International Code of Nomenclature of Bacteria. Int J Sys Bacteriol 28:335–336

Summons RE, Jahnke LL, Hope LM, Logan GA (1999) 2-Methylhopanoids as biomarkers for cyanobacterial oxygenic photosynthesis. Nature 400:554–557

Tabita FR (1988) Molecular and cellular regulation of autotrophic carbon dioxide fixation in microorganisms. Microbiol Rev 52:155–189

Tabita FR (1999) Microbial ribulose 1,5-bisphosphate carboxylase/oxygenase: a different perspective. Photosynth Res 60:1–28

Tanaka S, Kerfeld CA, Sawaya MR, Cai F, Heinhorst S, Canon GC, Yeates TO (2008) Atomic-level models of the bacterial carboxysome shell. Science 319:1083–1086

Thiel T, Hartnett T, Pakrasi HB (1990) Examination of Photosystem II in heterocysts of the cyanobacterium *Nostoc* sp. ATCC 29150. In: Murata N (ed) Current research in photosynthesis. Kluwer, Dordrecht, pp 291–294, Vol 1

Thompson AW, Foster RA, Krupke A, Carter BJ, Musat N, Vaulot D, Kuypers MMM, Zehr JP (2012) Unicellular cyanobacterium symbiotic with a single-celled eukaryotic alga. Science 337:1546–1550

Ting CS, Hsieh C, Sundararaman S, Mannella C, Marko M (2007) Cryo-electron tomography reveals the comparative three-dimensional architecture of *Prochlorococcus*, a globally important marine cyanobacterium. J Bacteriol 189:4485–4493

Ting CS, Ramsey ME, Wang YL, Frost AM, Jun E, Durham T (2009) Minimal genomes, maximal productivity: comparative genomics of the photosystem and light-harvesting complexes in the marine cyanobacterium, *Prochlorococcus*. Photosynth Res 101:1–19

Tripp HJ, Bench SR, Turk KA, Foster RA, Desany BA, Niazi F, Affourtit JP, Zehr JP (2010) Metabolic streamlining in an open ocean nitrogen-fixing cyanobacterium. Nature 464:90–94

Trissl H-W, Wilhelm C (1993) Why do thylakoid membranes from higher plants form grana stacks? Trends Biochem Sci 18:415–419

Turner S (1997) Molecular systematics of oxygenic photosynthetic bacteria. Plant Syst Evol 11:14–52

Turner S, Burger-Wiersma T, Giovannoni SJ, Mur LR, Pace NR (1989) The relationship of a prochlorophyte, *Prochlorothrix hollandica*, to green chloroplasts. Nature 337:380–382

Umeda H, Aiba H, Mizuno T (1996) som A, a novel gene that encodes a major outer-membrane protein of *Synechococcus* sp. PCC 7942. Microbiology 142:2121–2128

Urbach E, Robertson DL, Chisholm SW (1992) Multiple evolutionary origins of prochlorophytes within the cyanobacterial radiation. Nature 355: 267–270

Vaara T (1982) The outermost surface structures in chroococcacean cyanobacteria. Can J Microbiol 28:929–941

Van de Meene AML, Hohmann-Marriott MF, Vermaas WFJ, Roberson RW (2006) The three-dimensional structure of the cyanobacterium *Synechocystis* sp. PCC 6803. Arch Microbiol 184:259–270

Vermaas WFJ, Shen G, Styring S (1994) Electrons generated by photosystem II are utilized by an oxidase in the absence of photosystem I in the cyanobacterium *Synechocystis* PCC 6803. FEBS Lett 337:103–108

Wachi M, Doi M, Tamaki S, Park W (1987) Mutant isolation and molecular cloning of *mre* genes, which determine cell shape, sensitivity to mecillinam, and amount of penicillin-binding proteins in *Escherichia coli*. J Bacteriol 169:4935–4940

Walsby AE (1985) The permeability of heterocysts to the gases nitrogen and oxygen. Proc R Soc Lond B 226:345–366

Walsby AE (2007) Cyanobacterial heterocysts: terminal pores proposed as sites of gas exchange. Trends Microbiol 15:340–349

Wilcox M, Mitchison GJ, Smith RJ (1973) Pattern formation in the blue-green alga *Anabaena*. J Cell Sci 13:637–649

Wildon DC, Mercer FV (1963) The ultrastructure of the vegetative cell of blue-green algae. Aust J Biol Sci 16:585–596

Wilmotte A (1994) Molecular evolution and taxonomy of the cyanobacteria. In: Bryant DA (ed) The molecular biology of cyanobacteria. Kluwer Academic Publishers, The Netherlands, pp 1–25

Wilmotte A, Golubic S (1991) Morphological and genetic criteria in the taxonomy of Cyanophyta/Cyanobacteria. Arch Hydrobiol, Suppl 92, Algological Studies 64:1–24

Woese CR (1987) Bacterial evolution. Microbiol Rev 51:221–271

Wolk CP, Ernst A, Elhai J (1994) Heterocyst metabolism and development. In: Bryant DA (ed) The molecular biology of cyanobacteria. Kluwer Academic Publishers, Dordrecht, pp 769–823

Yeates TO, Kerfeld CA, Heinhorst S, Cannon GC (2008) Protein-based organelles in bacteria: Carboxysomes and related microcompartments. Nat Rev Microbiol 6:681–691

Yentsch CS, Phinney DA (1989) A bridge between ocean optics and microbial ecology. Limnol Oceanogr 34:1694–1705

Young KD (2003) Bacterial shape. Mol Microbiol 49:571–580

Young KD (2006) The selective value of bacterial shape. Microbiol Mol Biol Rev 70:660–703

Zak E, Norling B, Maitra R, Huang F, Andersson B, Pakrasi HB (2001) The initial steps of biogenesis of cyanobacterial photosystems occur in plasma membranes. Proc Natl Acad Sci U S A 98:13443–13448

Zehr JP (2011) Nitrogen fixation by marine cyanobacteria. Trends Microbiol 19:162–173

Zehr JP, Mellon MT, Hiorns WD (1997) Phylogeny of cyanobacterial *nifH* genes: evolutionary implications and potential applications to natural assemblages. Microbiology 143:1443–1450

Zehr JP, Bench SR, Carter BJ, Hewson I, Niazi F, Shi T, Tripp HJ, Affourtit JP (2008) Globally distributed uncultivated oceanic N_2-fixing cyanobacteria lack oxygenic Photosystem II. Science 322:1110–1112

Zhang C-C, Laurent S, Sakr S, Peng L, Bedu S (2006) Heterocyst differentiation and pattern formation in cyanobacteria: a chorus of signals. Mol Microbiol 59:367–375

Chapter 15

Respiration and Oxidative Phosphorylation in Mycobacteria

Michael Berney*
Department of Microbiology and Immunology, Albert Einstein College of Medicine, Bronx, NY 10461, USA

and

Gregory M. Cook*
Department of Microbiology and Immunology, University of Otago, 9054 Dunedin, New Zealand

Summary

The genus *Mycobacterium* comprises a group of obligately aerobic bacteria that have adapted to inhabit a wide range of intracellular and extracellular environments. A fundamental feature in this adaptation is the ability of mycobacteria to respire and generate energy for growth or to sustain latency. Mycobacteria harbor multiple primary dehydrogenases to fuel the electron

*Author for correspondence, e-mail: michael.berney@einstein.yu.edu; gregory.cook@otago.ac.nz

M.F. Hohmann-Marriott (ed.), The Structural Basis of Biological Energy Generation, Advances in Photosynthesis and Respiration 39, pp. 277–293, DOI 10.1007/978-94-017-8742-0_15,

transport chain and two terminal respiratory oxidases, an aa_3-type cytochrome *c* oxidase and cytochrome *bd*-type menaquinol oxidase, are present for dioxygen reduction coupled to the generation of a protonmotive force. In mycobacteria, Type II NADH dehydrogenases are favoured over complex I for NADH oxidation and menaquinone acts as the primary conduit between electron-donating and electron-accepting reactions. The molecular mechanisms regulating the expression of the electron transport chain components in mycobacteria remains unknown. Despite being obligate aerobes, mycobacteria have the ability to metabolize in the absence of oxygen and a number of reductases are present to facilitate the turnover of reducing equivalents under these conditions (e.g., nitrate reductase, fumarate reductase). Hydrogenases and ferredoxins are also present in the genomes of mycobacteria suggesting the ability of these bacteria to adapt to an anaerobic-type of metabolism in the absence of oxygen. The exact roles of reductases and hydrogenases is poorly understood. ATP synthesis by the membrane-bound F_1F_O-ATP synthase (see Chap. 6) is essential for growing and non-growing mycobacteria and the enzyme is able to function over a wide range of proton-motive force values (aerobic to hypoxic). Research into mycobacterial respiration and oxidative phosphorylation have been energized by the discovery of a new drug (TMC207) that targets the ATP synthase of mycobacteria, suggesting that inhibitors of respiration and ATP synthesis will provide the next generation of front line drugs to combat tuberculosis and nontuberculous mycobacterial disease.

I. Introduction

Members of the genus mycobacteria are obligate aerobes and therefore are dependent on aerobic respiration to grow and produce energy. Mycobacterial respiration is complex and branched pathways exist for electron transfer from many low potential reductants to oxygen. Unlike most other bacteria, there appears to be little redundancy in the transfer of electrons to dioxygen during growth with only two terminal oxidases present in mycobacteria, an aa_3-type cytochrome *c* oxidase (encoded by *ctaBCDE*) belonging to the heme-copper respiratory oxidase family (see Chap. 10) and cytochrome *bd*-type menaquinol oxidase (*cydABCD*) (see Chap. 9). Mycobacteria regulate the expression of these oxidases in response to oxygen supply, but the molecular mechanisms governing this regulation are unknown. Furthermore, there is no experimental data, other than expression data, to support the proposal that the aa_3-type cytochrome *c* oxidase and cytochrome *bd*-type have markedly different affinities for oxygen in mycobacteria. In the absence of oxygen, mycobacterial growth is inhibited, even if alternative electron acceptors are present (e.g., nitrate, fumarate). Despite growth being inhibited, mycobacteria are able to metabolize exogenous and endogenous energy sources under hypoxia/anaerobiosis. The electron acceptors and mechanisms to recycle reducing equivalents under these conditions are poorly understood. Irrespective of the oxygen concentration, ATP synthesis is obligatorily coupled to

Abbreviations: CCCP – Carbonyl cyanide *m*-chlorophenylhydrazone; CO-DH – Carbon-monoxide dehydrogenase; DCCD – *N,N'*-dicyclohexylcarbodiimide; FAD – Flavin adenine dinucleotide; G3P – Glycerol 3-phosphate; HYD – Hydrogenase; MDH – Malate dehydrogenase; MK – Menaquinone (vitamine K); MKH_2 – Menaquinol; MQO – Malate quinone oxidoreductase; NAD+ – Nicotinamide adenine dinucleotide; NDH – NADH dehydrogenase; NRP – Non-replicating persistence; P5CDH – Pyrroline-5-carboxylate dehydrogenase; PMF – Proton-motive force; PRODH – Proline dehydrogenase; SDH – Succinate dehydrogenase; TCA – Tricarboxylic acid; TMC207 – Tibotec Medicinal Compound 207; $Z\Delta pH$ – Transmembrane pH gradient expressed in mV; $\Delta\Psi$ – Electrical or membrane potential expressed in mV the electron transport chain and the F_1F_o-ATP synthase (see Chap. 6), but the reasons for this remain unexplained. The aim of this chapter is to discuss recent advances in understanding the processes of respiration and oxidative phosphorylation in mycobacteria, with a particular focus on the obligate human pathogen *Mycobacterium tuberculosis*.

II. Energetics of Mycobacterial Growth

In bacteria there is often a trade-off between obtaining the maximum energy yield from a substrate and the maximum flux (rate of ATP production) or growth rate (Pfeiffer et al. 2001). A cell that uses a pathway with a high yield and low rate (e.g., respirers) can produce more ATP from a given amount of substrate compared to a bacterium that produces ATP at a higher rate, but a lower yield (e.g. fermenters). The growth yield of bacterial cultures has been used to estimate the efficiency of energy generation during respiration or fermentation. The energetics of mycobacterial growth has been studied in glycerol-limited continuous culture (Beste et al. 2005). In the slow growing *M. bovis* BCG, at both low (0.01 h^{-1}) and high (0.03 h^{-1}) dilution rates, the cell yield or $Y_{glycerol}$ was 27 g [dry weight] cells/mol glycerol utilized and 33 g [dry weight] cells/mol glycerol utilized respectively (Beste et al. 2005). Under similar growth conditions in continuous culture, the fast growing non-pathogenic saprophyte *M. smegmatis* exhibits a $Y_{glycerol}$ of 30–40 g [dry weight] cells/mol glycerol utilized over the dilution rate range (0.02–0.15 h^{-1}) (Cox and Cook 2007). The glycerol consumption rate for maintenance functions ($m_{glycerol}$) is 0.28 mmoles of glycerol/h/g [dry weight] cells (Cox and Cook 2007), a value that is comparable with *E. coli*. Based on these $Y_{glycerol}$ values, mycobacteria would appear to adopt the first strategy where the cells grow slowly using oxidative phosphorylation to generate large amounts of ATP at a slow rate. Furthermore, this strategy appears to be employed by both, fast growing and slow growing mycobacteria; the cell yield on glycerol is comparable between *M. smegmatis* and *M. bovis* BCG at similar growth rates.

When growing aerobically at an external pH of 7.0, *M. smegmatis* and *M. bovis* BCG generate a protonmotive force (PMF) of approximately −180 mV, which appears to be of a significant magnitude to drive proton-coupled bioenergetic processes (e.g., ATP synthesis, solute transport, etc.) (Rao et al. 2001). Growth of mycobacteria is sensitive to the electrogenic proton translocator carbonyl cyanide *m*-chlorophenylhydrazone (CCCP) demonstrating that a PMF is indispensable for normal growth. Moreover, growth is inhibited by the F-type ATP synthase inhibitor *N,N'*-dicyclohexylcarbodiimide (DCCD), further supporting the role of the PMF in driving ATP synthesis by the membrane-bound F_1F_O-ATP synthase. While proton translocation *via* the respiratory chain generates the PMF during respiration with oxygen as the terminal electron acceptor, it is not clear how the PMF is established in the absence of oxygen under anaerobic growth conditions. Anaerobic bacteria are able to generate a significant PMF (−100 mV) using their membrane-bound F_1F_O-ATP synthase in the ATP hydrolysis direction (Dimroth and Cook 2004). The ATPase activity (proton pumping) of the enzyme is fuelled by ATP produced by substrate level phosphorylation. This mechanism does not appear to operate in mycobacterial cells where the F_1F_O-ATP synthase has been reported to have latent ATPase activity when measured in inverted membrane vesicles (Haagsma et al. 2010; Higashi et al. 1975). Whether the enzyme is also latent in actively growing cells is not known and therefore the potential exists for this enzyme to function as a primary proton pump in the absence of oxygen and a functional respiratory chain to generate the PMF. Rao et al. (2008) have reported that hypoxic, non-replicating *M. tuberculosis* generate a total PMF of −113 mV, −73 mV of electrical potential ($\Delta\Psi$) and −41 mV of $Z\Delta pH$. The addition of thioridazine, a compound that targets NDH-2, results in dissipation of the electrical potential and significant cell death suggesting that NADH is an important electron donor for the generation of the ΔΨ under hypoxic conditions. The addition of R207910 (TMC207), a specific inhibitor of the F_1F_O-ATP synthase was bactericidal against hypoxic, non-replicating *M. tuberculosis*, but had no effect on the ΔΨ (Rao et al. 2008), an observation consistent with the latent ATPase activity of this enzyme (see above).

III. Primary Dehydrogenases

The electron donors and primary dehydrogenases used by mycobacteria to fuel the electron transport chain have largely been inferred from bioinformatics and transcriptional analyses. A summary of the potential electron donors used by mycobacteria is shown in Fig. 15.1.

A. NADH Dehydrogenases I and II

The oxidation of NADH or equivalent by aerobic bacteria is critical for continuous metabolic flux, and in the absence of a fermentative metabolism, NADH oxidation will be carried out primarily by membrane-bound NADH dehydrogenases. NADH dehydrogenase is the first component of the respiratory chain and transfers electrons from NADH to quinones (e.g. ubiquinone or menaquinone). Weinstein et al. have identified genes for two classes of NADH:menaquinone oxidoreductases in the genome of *M. tuberculosis* (Weinstein et al. 2005). NDH-I is encoded by the *nuoABCDEFGHIJKLMN* operon and as this dehydrogenase transfers electrons to menaquinone, it conserves energy by translocating protons across the membrane to generate a PMF (Fig. 15.2). The second class is NDH-II, a non-proton translocating NADH dehydrogenase that does not conserve energy and is present in two copies (*ndh* and *ndhA*) in *M. tuberculosis* (Weinstein et al. 2005). Mutagenesis studies have established that both NDH-1 and *ndhA* are dispensable for growth in vitro (Sassetti et al. 2003). In contrast, the lack of a viable strain with a disrupted or deleted *ndh* suggests that it is essential for growth, and cannot be compensated for by *ndhA* (McAdam et al. 2002; Sassetti et al. 2003; Weinstein et al. 2005). Furthermore, deleterious point mutations in

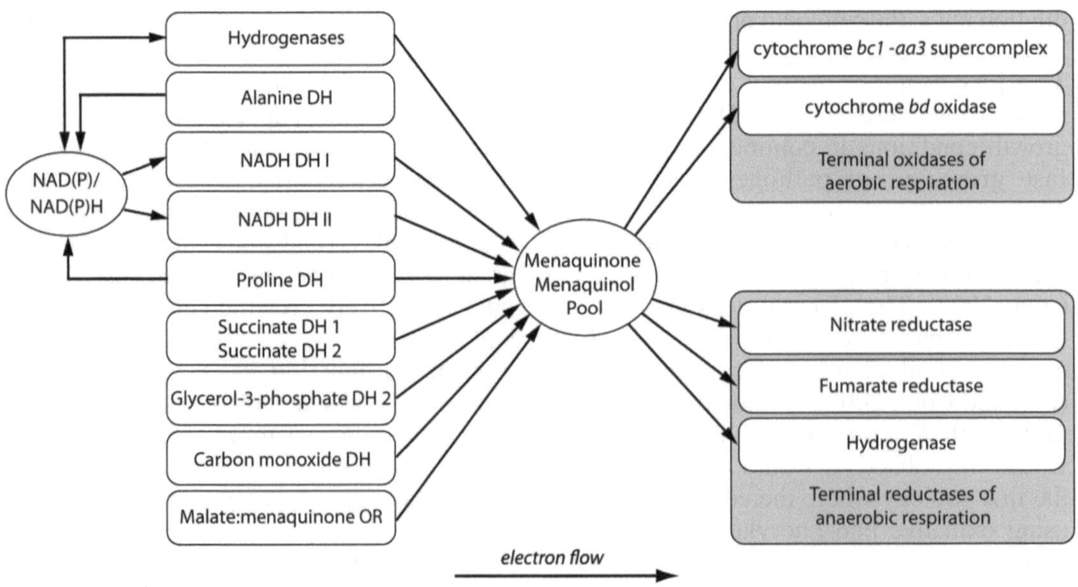

Fig. 15.1. Proposed scheme of enzymes preferentially used in the respiratory chain of Mycobaceria.

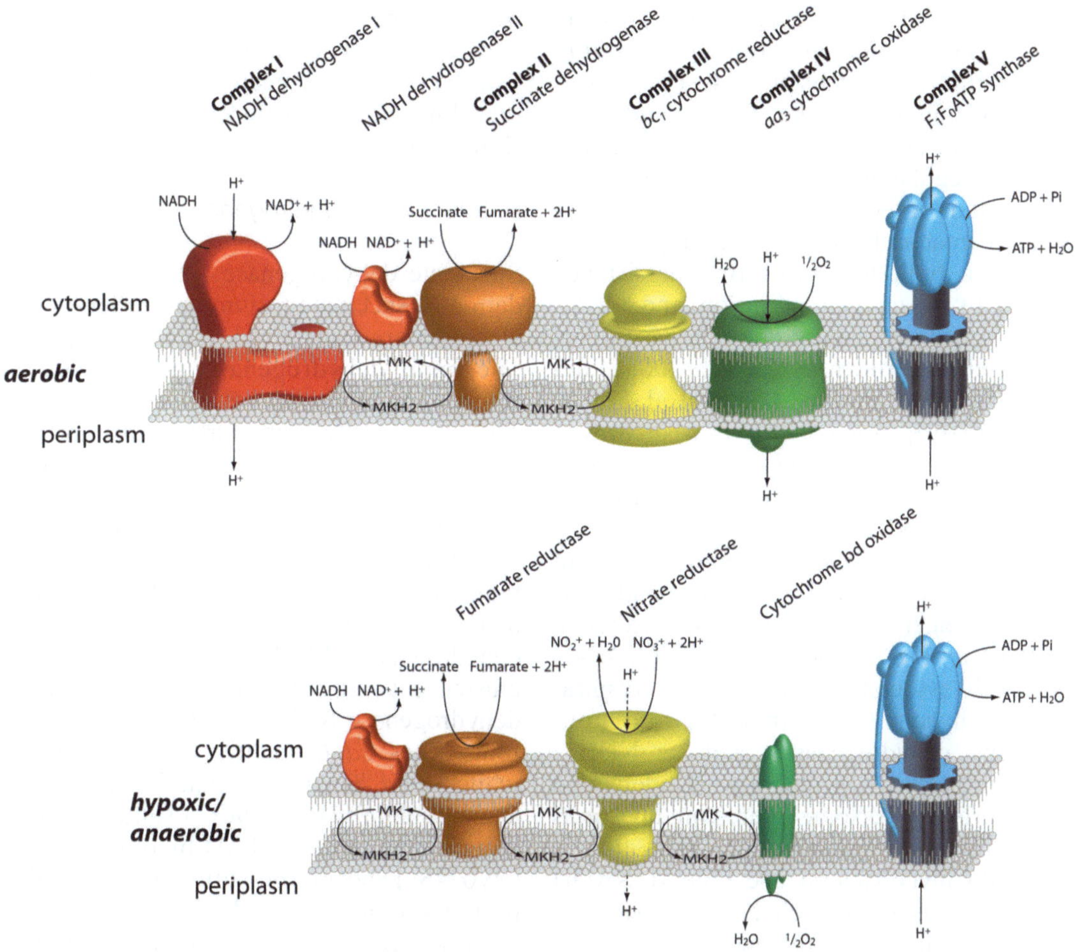

Fig. 15.2. Schematic diagram outlining the electron transport chain components and ATP synthase of Mycobacteria growing under aerobic (*top panel*) and anaerobic conditions (*bottom panel*). *MK* menaquinone, *MKH2* menaquinol.

ndh of *M. smegmatis* are pleiotropic, conferring temperature-sensitive growth arrest and multiple amino acid auxotrophy (Miesel et al. 1998; Vilcheze et al. 2005). The same authors show that some *ndh* mutants also have a 25-fold reduced NADH dehydrogenase activity, implying NDH-II is the primary enzyme responsible for NADH oxidation and it is essential for the viability of *M. smegmatis*.

In *M. tuberculosis*, no energetic role for the non-essential type I NADH:menaquinone oxidoreductases has been established and the persistence of *M. tuberculosis* in an in vitro Wayne model was shown not to be compromised in a *nuo* operon deletion mutant (Rao et al. 2008). Velmurugan et al. (2007) have demonstrated a role for *nuo* (i.e., the NuoG subunit) in the ability of *M. tuberculosis* to inhibit macrophage apoptosis. The *nuo* operon has been lost from the genome of *M. leprae* with a single *nuoN* pseudogene remaining (Cole et al. 2001). *M. smegmatis* also contains genes for a type I NADH: menaquinone oxidoreductases (viz. *nuoA-N*), but enzyme assays for NDH-I activity in *M. smegmatis* failed to detect this enzyme activity suggesting it is not expressed (Miesel et al. 1998). However, increased expression (15-fold) of the *nuo* operon was observed in carbon-limited chemostat in response to a slow down in growth rate (Berney and Cook 2010). Several studies have reported that *nuo* is down-regulated in *M. tuberculosis* during mouse lung infection (Shi et al. 2005), survival in macrophages (Schnappinger et al. 2003), in both non-replicating persistence (NRP)-1 (1 % oxygen saturation) and NRP-2 (0.06 % oxygen saturation) relative to aerated mid-log growth (Shi et al. 2005), and upon starvation in vitro (Betts et al. 2002). The transcription of *ndh* is also down-regulated in *M. tuberculosis* during mouse lung infection, but transcript levels for *ndh* peak during NRP-2 in vitro demonstrating that the pattern of *ndh* regulation is different between in vivo and in vitro conditions (Schnappinger et al. 2003). These data are in contrast to *E. coli* where NuoA-N (NDH-1) is usually associated with anaerobic respiratory pathways (e.g. fumarate) and non-coupling dehydrogenases such as NDH-2 are synthesized aerobically (Unden and Bongaerts 1997). Some interesting questions arise from these observations. The first is "why do mycobacteria use type II NADH dehydrogenases to recycle NADH when they could continue to use the energy conserving and PMF generating NDH-I?" One potential explanation is that because NDH-IIs are non-proton translocating, they will not be impeded by a high PMF, which could ultimately slow down glycolytic flux due to back pressure on the system. This mechanism is akin to a "relief valve" that would allow for a higher metabolic flux and ultimately higher rates of ATP synthesis at the expenses of low energetic efficiency of the respiratory chain. Secondly, why is *ndh* an essential gene when mycobacteria could also use *ndh2* or *nuo*? The fact that *ndh* is essential implies that mycobacteria do not have another mechanism to recycle NADH during normal aerobic growth. Alternatively, this is the only NADH dehydrogenase that is operating under these growth conditions and the activity of this enzyme is essential for maintaining an energized membrane. Compounds that target NDH-II are bactericidal towards hypoxic non-replicating *M. tuberculosis* suggesting that the respiratory chain is essential for the recycling of NADH under these conditions (Rao et al. 2008).

B. Succinate Dehydrogenase

Succinate dehydrogenase catalyzes the oxidation of succinate to fumarate wherein two electrons are transferred to generate quinol. Succinate dehydrogenase is involved in aerobic respiration as well as the TCA cycle. The structurally and functionally related fumarate reductase can catalyze the reverse reaction and is involved in anaerobic respiration. However, succinate dehydrogenase of *E. coli* was also shown to catalyze fumarate reduction under certain conditions. Most mycobacteria harbor a canonical succinate dehydrogenase SDH2 and a second gene cluster encoding for a putative succinate dehydrogenase SDH1. SDH2 is encoded by four genes *sdhC*, *sdhD*, *sdhA* and *sdhB*, while SDH1 consist of a putative membrane anchor subunit (Rv0249c), a subunit resembling SdhA with a FAD-binding domain (Rv0248c) and an iron-sulfur cluster binding protein similar to SdhB (Rv0247c). Upstream of these three genes is a hypothetical protein that is present in all mycobacteria that harbor SDH1. Gene expression data shows that all four genes are expressed in concert (Berney and Cook 2010). Succinate dehydrogenase activity has been measured in *M. tuberculosis* as well as many other mycobacterial species (Tian et al. 2005; Youmans et al. 1956), but it is still unclear which of the two enzymes or if both are responsible for this activity. In fact, the SDH reaction in mycobacteria should have an unfavorable free energy because the redox potential of menaquinone (−80 mV) is lower than that of the succinate to fumarate reaction (+30 mV) (Kana et al. 2009). SDH1 and SDH2 have been shown to be differentially expressed under energy- or oxygen-limiting conditions (Berney and Cook 2010). Under energy-limiting conditions SDH1 was upregulated fourfold while SDH2 was downregulated threefold. Under oxygen-limiting conditions, SDH1 was downregulated up to 30-fold while SDH2 was upregulated twofold. SDH2 responded in concert with other enzymes of the respiratory chain while SDH1 showed opposite expression under all conditions. This argues that the two enzymes have different roles in mycobacterial metabolism. In fact, a recent study indicates that SDH1 or SDH2 could also catalyse fumarate reductase activity in *M. tuberculosis* under anaerobic conditions (Watanabe et al. 2011). These authors show that when *M. tuberculosis* cultures are subjected to anaerobic conditions they start to secrete succinate into the medium indicating fumarate reductase activity. However, a fumarate reductase mutant of *M. tuberculosis* was still accumulating succinate, hence the two putative succinate dehydrogenase clusters could be responsible for this complementation. Furthermore, a recent study suggests that SDH1 is essential for glycerol metabolism, but not for cholesterol metabolism in *M. tuberculosis* (Griffin et al. 2011).

IV. Alternative Electron Donors and Dehydrogenases

Under condition of carbon starvation and slow growth mycobacteria switch to alternative electron donors (Berney and Cook 2010; Beste et al. 2007; Betts 2002).

A. Proline Dehydrogenase

Proline is increasingly being recognized as a critical amino acid in bioenergetics and cellular redox control (Tanner 2008). Proline can be utilized as an electron donor as well as a carbon and nitrogen source. The degradation of proline occurs by means of two enzymes: proline dehydrogenase (PRODH) and pyrroline-5-carboxylate dehydrogenase (P5CDH). These two enzymes catalyze the oxidation of proline to glutamate with four electrons transferred to the respiratory chain (Tanner 2008). In the first step FAD is reduced to $FADH_2$ while in the second step NAD^+ is reduced to NADH. In some bacteria, PRODH and P5CDH are monofunctional enzymes but in the majority of bacterial species they are fused into one protein called proline utilization A flavoenzyme PutA (Menzel and Roth 1981; Tanner 2008). In mycobacteria, PRODH and P5CDH are predicted to be monofunctional enzymes (Tanner 2008). The genes encoding PRODH (*putB*) and P5CDH (*putA*) are expressed as part of an operon with pyrroline-5-carboxylate dehydrogase (Berney et al. 2012). *Mycobacterium smegmatis* can grow on proline as the sole carbon and energy source and it was shown that proline dehydrogenase is an important electron donor under energy-limiting conditions (Berney and Cook 2010) and under hypoxia (Berney et al. 2012). The same authors show that proline metabolism in mycobacteria is regulated by a unique membrane-associated transcriptional regulator called PruC.

B. Hydrogenases

Hydrogenases catalyze the reversible oxidation of molecular hydrogen: $2H^+ + 2e^- \rightarrow H_2$ and play a central role in energy metabolism of bacteria, archaea and eukarya (Vignais et al. 2001). Under physiological conditions, hydrogenases couple H_2 oxidation to respiration (Knallgas reaction) or reduce protons as a way to dispose of surplus reducing equivalents. Four different types of hydrogenases can be found in mycobacteria. All four hydrogenases are annotated to be of the NiFe type. The four hydrogenase complexes each belong to a different group of NiFe-hydrogenases. *M. smegmatis* harbors three of the four hydrogenase complexes (Berney and Cook 2010), Hyd1 aligns closely with Group 2a uptake hydrogenases of the cyanobacteria such as *Nostoc*, indicating it oxidises H_2 (Tamagnini et al. 2007). In contrast, Hyd3 is closely related to the Group 3 cytoplasmic bidirectional hydrogenases. Hyd2 has no whole sequence homology with the four established classes of [NiFe] hydrogenases; instead it is a founding member of the group 5 high-affinity hydrogenases (Constant et al. 2010, 2011). Hyd1, Hyd2 and Hyd3 are all soluble hydrogenases and found in mycobacteria of the slow-growing and fast-growing type, as well as pathogenic and non-pathogenic mycobacteria (Berney and Cook 2010). However, the fourth hydrogenase is only found in pathogenic mycobacteria (including *M. tuberculosis* complex) and seems to be restricted to slow-growers. It shows homology to Group 4 membrane-bound H_2 evolving hydrogenases. It has been shown that *M. smegmatis*, among other mycobacterial species, can oxidize molecular hydrogen in the presence of carbon monoxide, implying that *M. smegmatis* expresses a functional hydrogenase (King 2003). To date no studies have reported on the ability of mycobacteria to produce hydrogen. Gene expression studies suggest that Hyd1 and Hyd2 are used during nutrient starvation as an alternative electron source while Hyd3 and the membrane-bound hydrogenase are more likely to have a function in disposing electrons under anaerobic conditions (Berney and Cook 2010). A knockout mutant of Hyd2 in *M. smegmatis* showed reduced biomass production when grown on complex medium under atmospheric conditions (Berney and Cook 2010) and its homolog in *Streptomyces* sp. was shown to facilitate hydrogen oxidation (Constant et al. 2008, 2010). These data suggest that Hyd2 oxidizes hydrogen at very low concentrations and fits with its purported role as a high-affinity hydrogenase (Constant et al. 2011).

C. Carbon Monoxide Dehydrogenase

Carbon monoxide dehydrogenase (CO-DH) is responsible for the oxidation of CO to carbon dioxide (CO_2) in carboxydobacteria, which grow on CO as a sole source of carbon and energy (Kim and Hegeman 1983). Carboxydobacteria catalyze the oxidation of CO to CO_2 by the following reaction: $CO + H_2O \rightarrow CO_2 + 2H^+ + 2e^-$. Several pathogenic and nonpathogenic mycobacteria including *Mycobacterium tuberculosis* are known to possess CO-DH genes. It has been shown that *M. tuberculosis* H37Ra, which possesses CO-DH activity (Park et al. 2003b), can grow on CO as a sole source of carbon and fuel for energy generation.

D. Glycerol-3-Phosphate Dehydrogenase

Glycerol-3-phosphate dehydrogenase catalyzes the oxidation of glycerol-3-phosphate to dihydroxy-acetone phosphate and reduces quinone in the cytoplasmic membrane (Schryvers et al. 1978). In *E. coli*, glycerol 3-phosphate (G3P) is either used as precursor in the biosynthesis of phospholipids or as a carbon source for energy supply (Boos 1998). *M. tuberculosis* possesses a gene (*glpD*, Rv3302c) with 50 % similarity to the aerobic glycerol-3-phosphate dehydrogenase from *E. coli*. Faster growing mycobacteria like *M. smegmatis* harbor multiple copies of *glpD* on the chromosome. However its role and function in mycobacterial respiration remains unknown.

E. Malate Quinone Oxidoreductase

The membrane associated malate quinone oxidoreductase (MQO, EC 1.1.99.16) oxidizes malate to oxaloacetate and transfers the reducing equivalents to menaquinone (MK) (Molenaar et al. 1998). *M. tuberculosis* harbors a copy of MQO and a cytoplasmic NAD+ -dependent malate dehydrogenase (MDH) (Prasada Reddy et al. 1975). The function and role of MQO in mycobacterial respiration is still unknown. *M. smegmatis* lacks an MDH homologue and it was shown that MQO responded to low oxygen concentration with fourfold increase in gene expression (Berney and Cook 2010).

V. Terminal Electron Acceptors

During aerobic respiration, energy is conserved by the generation of a PMF across a proton-impermeable membrane. The electron transport chain components are membrane-bound and asymmetrically arranged across the membrane to achieve net consumption of protons from the cytoplasm and net release of protons on the outside of the cell. An important part of all electron transport chains are the terminal respiratory oxidases. In order for mycobacteria to utilize oxygen efficiently and obtain the maximum growth yield on a particular carbon and energy source, there must be coordinate regulation of terminal respiratory oxidase expression. For example, in *E. coli* cytochrome *bo* (K_m for oxygen in the micromolar range) and cytochrome *bd* (K_m for oxygen in the nanomolar range) (D'Mello et al. 1995, 1996) are coordinately regulated by the ArcBA system and transcriptional regulator FNR (Cotter et al. 1997). Cytochrome *bo* is synthesized at high oxygen tension (optimal between 15 % and 100 % air saturation) and repressed as the oxygen concentrations decreases (Tseng et al. 1994). This coincides with the induction of cytochrome *bd* at 7 % air saturation, which is turned off (FNR-mediated repression) once the cells enter anaerobiosis (Cotter et al. 1997; Tseng et al. 1994). *E. coli* also uses non-coupling dehydrogenases (NDH-2) during aerobic growth that allow a fast metabolic flux (fast growth rate) and switches to coupling dehydrogenases (NDH-1) during anaerobic growth with fumarate (Unden and Bongaerts 1997). Mycobacteria adopt regulation of oxidase expression to match oxygen supply. Under conditions of low oxygen tension (ca. 1 % air saturation), cytochrome *bd* is induced in *M. smegmatis* as the transition to anaerobiosis is approached (Kana et al. 2001), a value that is tenfold lower than that observed in *E. coli* (ca. 10 % air saturation). In *M. tuberculosis*, cytochrome *bd* is upregulated in the early stages of NRP-1 (i.e., decreasing oxygen) (Voskuil et al. 2004). A strategy that appears to be invoked by mycobacteria is one of down-regulation or a slowing of metabolism as cells enter NRP-1 and NRP-2 (Wayne and Hayes 1996). Transcriptional analysis of *M. tuberculosis* in the macrophage (phagosomal environment) has revealed that NDH-1, menaquinol-cytochrome *c* oxidoreductase and the ATP synthase are all down-regulated when compared to cells growing exponentially, suggesting the reduced need for energy generation during bacteriostasis i.e., the growth state of intraphagosomal *M. tuberculosis* (Schnappinger et al. 2003). Consistent with these observations is the repression of these operons during starvation. In contrast, fumarate reductase, nitrate reductase and NDH-2 are all upregulated under these conditions (Schnappinger et al. 2003). Whilst these proteins do not appear to contribute to increased energy production, it has been suggested that they may play a pivotal role in the recycling of NAD^+ as a result of beta-oxidation of fatty acids (Schnappinger et al. 2003).

A. The aa_3-type Cytochrome c Oxidase

The cytochrome *c* pathway consists of a menaquinol-cytochrome c oxidoreductase termed the bc_1 complex (encoded by the *qcrCAB* operon) and an aa_3-type cytochrome c oxidase (encoded by *ctaBCDE*) belonging to the heme-copper respiratory oxidase family (Boshoff and Barry 2005; Matsoso et al. 2005). The cytochrome *c* oxidase functions as a proton pump and may form a "supercomplex" with menaquinol cytochrome *c* oxidoreductase (Matsoso et al. 2005; Megehee et al. 2006). Support for this hypothesis comes from the observation that such supercomplexes have been reported in other actinomycetes such as *C. glutamicum* (Niebisch and Bott 2003). The bc_1-aa_3 pathway is the major respiratory route in mycobacteria under standard aerobic culturing conditions (Matsoso et al. 2005). Matsoso et al. (2005) have demonstrated that disruption of this pathway in *M. smegmatis* is accompanied by a constitutive upregulation of the cytochrome *bd*-type menaquinol oxidase. In *M. tuberculosis*, the bc_1-aa_3 pathway is essential for growth suggesting an inability of this bacterium to adapt in a manner analogous to *M. smegmatis*. The aa_3 branch is proposed to contain two *ctaD* alleles in *M. smegmatis* versus the one in *M. tuberculosis*, suggesting the existence of alternate isoforms of cytochrome *c* oxidase in *M. smegmatis* (Kana et al. 2001).

B. Cytochrome bd-type Oxidase

M. tuberculosis and other mycobacterial species harbor genes for the cytochrome *bd-type* menaquinol oxidase (*cydAB*) (Kana et al. 2001). The cytochrome *bd* branch is the bioenergetically less efficient branch (non-proton translocating) and is synthesized at low oxygen tensions in mycobacteria (Kana et al. 2001). In addition to cytochrome *bd*, the *M. smegmatis* respiratory chain has been proposed to contain a third possible respiratory branch terminating in the YthAB (*bd-type*) menaquinol oxidase (Kana et al. 2001). The existence of two cytochrome *bd*-type oxidases (I and II) is not unprecedented in bacteria (Poole and Cook 2000), and recent work had reported that cytochrome *bd*-II in *E. coli* is able to generate a PMF with a H^+/e^- ratio of 1.0 (Borisov et al. 2011).

Cytochrome *bd* mutants of *E. coli* have been shown to have a pleiotropic phenotype: they are sensitive to H_2O_2, nitric oxide, temperature, unable to exit from stationary phase and resume aerobic growth at 37 °C, and iron (III) chelators (reviewed in Poole and Cook 2000). A *cydA* mutant of *M. smegmatis* has been generated, but the phenotypes are very subtle compared to *E. coli*. A *M. smegmatis cydA* mutant showed a reduced growth rate at 0.5–1 % air saturation and also a tenfold difference in CFU after 140 h of growth in the presence of cyanide at 21 % air saturation when co-cultured with the wild-type (Kana et al. 2001).

The *cydAB* genes appear to be in an operon with *cydDC* in mycobacterial species, an arrangement similar to that found in *Bacillus*. In *E. coli*, *cydAB* and *cydDC* form two discrete operons and *cydDC* mutants are defective in cytochrome *bd* assembly and the periplasmic space is more oxidized in the mutant versus wild-type (Goldman et al. 1996a, b). In *E. coli*, the CydDC protein (ABC transporter) has been reported to pump glutathione and cysteine into the periplasm to maintain redox homeostasis (Pittman et al. 2005). The role of the *cydDC* genes in mycobacteria is unknown, however, evidence exist that CydDC plays a role during mycobacterial persistance in vivo. A *cydC* mutant of *M. tuberculosis* showed reduced ability to survive the transition from acute to chronic infection in mice (Shi et al. 2005) and Dhar and McKinney have reported that CydC plays a role in mycobacterial persistence in isoniazid-treated mice (Dhar and McKinney 2010).

VI. Alternative Electron Acceptors

In the absence of oxygen, alternative electron acceptors (e.g., nitrate and fumarate) are available for mycobacterial metabolism, but none of these electron acceptors are able to support growth. These potential electron acceptors may play an important role in the disposal of reducing equivalents in the absence of oxygen.

A. Nitrate Reductase

The breakdown of nitric oxide in mammalian tissue by intracellular bacteria provides a source of nitrate that can be used as an alternative electron acceptor. A transport system for the exchange of nitrate and nitrite into and out of the cell are present in the genome of *M. tuberculosis* (e.g., NarK2) (Cole et al. 1998). *M. tuberculosis* contains genes (*narGHJI*) that encode for a putative membrane-bound molybdenum-containing nitrate reductase complex similar to the corresponding *narGHJI* operon of *E. coli*. Moreover, the *narGHJI* operon of *M. tuberculosis* is able to functionally complement a *nar* mutant of *E. coli* to grow on glycerol and reduce nitrate anaerobically (Sohaskey and Wayne 2003). Importantly however, the expression of the *narGHJI* operon in *M. tuberculosis* is not upregulated in response to either hypoxia or stationary phase (Sohaskey and Wayne 2003). Sohaskey and Wayne demonstrate that overexpression of recombinant *M. tuberculosis* nitrate reductase in either *M. tuberculosis* or *M. smegmatis* (low nitrate reductase activity) does not confer the ability of these cells to grow anaerobically i.e. no growth of either species is observed with nitrate anaerobically even though the nitrate reductase activity of whole cells increases (Sohaskey and Wayne 2003). The genome of *M. tuberculosis* also lacks orthologs of transcription regulator FNR which, in combination with NarL, are responsible for the transcriptional activation of the *narGHJI* operon by anaerobic conditions in *E. coli* (Unden and Bongaerts 1997). A putative NarL (*Rv0884c*) has been indentifed in *M. tuberculosis* but the promoter of the *narGHJI* lacks consensus-like binding sites for this regulatory protein. Based on these observations, it is apparent that this enzyme does not support anaerobic growth of mycobacteria and therefore the role of this enzyme in the physiology of mycobacteria is unclear. Given the proposed membrane-bound location of the enzyme and the proton-translocating activity *via* a redox loop of the *E. coli* enzyme, perhaps the primary role of the mycobacterial enzyme is to generate a PMF when the concentration of oxygen is low, and hence its activity increases but not its expression. An alternative role for nitrate reductase may be maintaining the redox balance of the cell during conditions of hypoxia. Sohaskey (2008) has reported that exogenously supplied nitrate has no effect on long-term persistence during gradual oxygen depletion, but played an important role during rapid adaptation to hypoxia (<18 h). This effect required a functional nitrate reductase, suggesting that nitrate reduction may play a role in protecting cells during sudden changes in oxygen concentration leading to disruption of aerobic respiration. Sohaskey (2005) proposes a role for NarK2 in sensing the redox state of the cell such that nitrate is transported into the cell under reducing, but not oxidizing conditions.

B. Fumarate Reductase

Fumarate reductase (encoded by *frdABCD*) is present in *M. tuberculosis* and has been shown to be upregulated during carbon starvation (Betts et al. 2002), oxygen depletion (Bacon et al. 2004) and in macrophages (Schnappinger et al. 2003), suggesting a role for this enzyme in persistence (Fig. 15.2). It has been proposed that fumarate may be an important endogenous electron acceptor for energy production and maintenance of redox balance (oxidation of NADH to NAD+) in hypoxic non-replicating mycobacteria (Rao et al. 2008). Indeed, in a recent study it was shown that *M. tuberculosis* secrets succinate under anaerobic conditions (Watanabe et al. 2011) and 13C flux analysis indicated that a large proportion of the secreted succinate must result from the reduction of fumarate to succinate. However, deletion of *frdA* in *M. tuberculosis* did not stop the accumulation of succinate, indicating that the two putative succinate dehydrogenase operons could compensate or that the secretion of succinate is due to a different metabolic effect. It is interesting to note that only members of the *M. tuberculosis* complex harbor a copy of fumarate reductase and it is missing from all other pathogenic strains like *M. avium paratuberculosis*, *M. marinum*, *M. ulcerans*, *M. leprae*. However, the two putative succinate dehydrogenases are present in all mycobacterial species apart from *M. leprae* where Sdh1 is a degenerate operon (Kana et al. 2009). Therefore, it seems likely that at least one of these putative succinate dehydrogenase could catalyse the reverse reaction under anaerobic conditions. The use of fumarate as an electron acceptor in *E. coli* requires complex I, and expression of the *nuo* operon is stimulated by the presence of fumarate (Unden and Bongaerts 1997). This stands in direct contrast to *M. tuberculosis*, where the *nuo* operon seems to be silent under anaerobic conditions (Shi et al. 2005). Mycobacteria utilize menaquinone/menaquinol as a conduit between electron-donating and -accepting reactions. Menaquinone has a lower midpoint redox potential (E_m=−74 mV) compared to ubiquinone (E_m=+113 mV) and is ideally poised to donate electrons to fumarate during anaerobic conditions (Cecchini et al. 2002).

C. Hydrogenase

The evolution of H_2 is a mechanism commonly employed by anaerobic bacteria to recycle reducing equivalents obtained from anaerobic degradation of organic substrates. However, even strictly aerobic bacteria have been shown to produce H_2 under anaerobic conditions (Kuhn et al. 1984). In the family of *Mycobacteriaceae* two potential hydrogen-evolving hydrogenases are present. One belongs to Group 2 cytoplasmic bidirectional hydrogenases and can be found in slow-growing mycobacteria (e.g., *M. kansasii*) and fast-growing (e.g., *M. smegmatis*) mycobacteria. In *M. smegmatis* this enzyme is termed Hyd3 and has been shown to respond strictly to oxygen-limiting conditions with up to 50-fold increase in gene expression (Berney and Cook 2010). The second type is a yet uncharacterized enzyme that shows homology to Group 4 hydrogenases that are membrane-bound. This putative formate-hydrogen lyase is enocoded in *M. tuberculosis* by the gene cluster (viz. Rv0082 to Rv0087) and is upregulated during infection of human macrophage-like THP-1 cells (Fontan et al. 2008). Moreover, the transcription of the early genes of the *hycP/hycQ* containing operon was shown to be upregulated during anaerobic adaptation in several studies (Bacon et al. 2004; Park et al. 2003a; Sherman et al. 2001; Voskuil et al. 2003, 2004). This gene cluster shows homology to components of hydrogenase 4 and 3 complex of *Escherichia coli*, the latter of which has been shown to catalyze hydrogen evolution at acidic pH (Mnatsakanyan et al. 2004). Recently, it was shown that the operon Rv0081-Rv0087 is positively regulated by the two-component signal transduction systems DosRS-DosT and MprAB and negatively regulated by Rv0081 a member of the ArsR/SmtB family of metal-dependent transcriptional regulators. DosRS-DosT is a redox sensing regulatory system that is important during the adaptation to hypoxic conditions. It is tempting to propose that the acidic, hypoxic and lipid-rich environment in the macrophages might require the expression of a hydrogenase complex in *M. tuberculosis* to help with the recycling of reducing equivalents under these conditions. The requirement for a low potential redox carrier in hydrogen production could be satisfied by reactions that are coupled to ferredoxins and not NADH. In fact when *M. smegmatis* is exposed to hypoxic conditions it switches to NAD^+/NADH-independent enzymes, especially those with ferredoxins as a redox partner. For example, in a recent study it has been shown that *M. tuberculosis* expresses a functional α-ketoglutarate:ferredoxin oxidoreductase (*Rv2454c*/*Rv2455c*) (Baughn et al. 2009). The same enzyme was upregulated in *M. smegmatis* under oxygen-limiting conditions (Berney and Cook 2010). The role of hydrogen in the cycling of reducing equivalents in mycobacteria during hypoxia warrants further investigation.

VII. ATP Synthesis by the F_1F_o ATP Synthase

In *M. tuberculosis* and other mycobacterial species, ATP is synthesized *via* substrate level phosphorylation and oxidative phosphorylation using the membrane-bound F_1F_o-ATP synthases (encoded by the *atpIBEFHAGDC* operon). In *M. tuberculosis*, the *atp* operon is downregulated during growth in macrophages (Schnappinger et al. 2003), the mouse lung and in cells exposed to nitric oxide or hypoxia (Shi et al. 2005). The *atp* operon of *M. bovis* BCG and *M. smegmatis* is down-regulated in response to slow growth rate (Berney and Cook 2010; Beste et al. 2007). When slow growing cells of *M. smegmatis* (70 h doubling time), with low levels of *atp* operon expression, are exposed to hypoxia (0.6 % oxygen saturation), the *atp* operon is upregulated threefold suggesting an important role for this enzyme during adaptation to hypoxia (Berney and Cook 2010).

The F_1F_o-ATP synthase catalyzes ATP synthesis by utilizing the electrochemical gradient of protons to generate ATP from ADP and inorganic phosphate (P_i) and operates under conditions of a high PMF and low intracellular ATP. The enzyme is also capable of working as an ATPase under conditions of high intracellular ATP and an overall low PMF (von Ballmoos et al. 2008). As an ATPase, the enzyme hydrolyzes ATP, while pumping protons from the cytoplasm to the outside of the cell. The ATP synthase of mycobacteria has been studied in detail at a biochemical level in *Mycobacterium phlei* and shown to exhibit latent ATPase activity (Higashi et al. 1975). ATPase activity could be activated by trypsin treatment and magnesium ions, but the mechanism of activation was not elucidated. Recent experiments with inverted membrane vesicles of *M. bovis* BCG and *M. smegmatis* demonstrate latent ATPase activity that could be activated by methanol and the PMF, suggesting regulation by the epsilon subunit and ADP inhibition (Haagsma et al. 2010). The reason for the extreme latency in ATP hydrolysis of the mycobacterial ATP synthase is unknown, but may represent an adaptation to function at low PMF and under hypoxia. Hypoxic non-replicating cells of *M. tuberculosis* generate a PMF in the order of −100 mV and the ATP synthase inhibitor R207910 is bactericidal towards these cells demonstrating that the ATP synthase still continues to function at relatively low PMF (Rao et al. 2008).

The F_1F_o-ATP synthase in *M. tuberculosis* and *M. smegmatis* has been shown to be essential for optimal growth (Sassetti et al. 2003; Tran and Cook 2005). In other bacteria, the F_1F_o-ATP synthase has been shown to be dispensable for growth on fermentable carbon sources (Friedl et al. 1983; Santana et al. 1994), where increased glycolytic flux can compensate for the loss of oxidative phosphorylation. This strategy does not appear to be exploited by *M. smegmatis*: the F_1F_o-ATP synthase is essential for growth even on fermentable substrates, suggesting that ATP production from substrate level phosphorylation alone, despite increased glycolytic flux, may be insufficient to sustain growth of these bacteria (Tran and Cook 2005). This may be due to an extraordinarily high value for the amount of ATP required to synthesize a mycobacterial cell, a possibility that requires further investigation (Cox and Cook 2007). Alternatively, or in conjunction with a high ATP demand for growth, the ATP synthase may be an obligatory requirement for the oxidation of NADH by providing a sink for translocated protons during NADH oxidation coupled to oxygen reduction. Such strict coupling would imply that mycobacteria do not support uncoupled respiration; either they lack a conduit for proton re-entry in the absence of the F_1F_o-ATP synthase or they are unable to adjust the proton permeability of the cytoplasmic membrane to allow a futile cycle of protons to operate. In this context, the cytoplasmic membrane of *M. smegmatis* has been shown to be extremely impermeable to protons (Tran et al. 2005). The ATP synthase of the close mycobacterial phylogenetic relative *Corynebacterium glutamicum* is non-essential for growth on fermentable carbon sources and Δ*atp* mutants of this bacterium show enhanced rates of glucose uptake, oxygen consumption and excretion of pyruvate into the growth medium, suggesting that substrate level phosphorylation alone can sustain growth of this bacterium (Koch-Koerfges et al. 2012).

Several new anti-tubercular compounds have been reported that target oxidative phosphorylation in mycobacteria (Andries et al. 2005; Dhiman et al. 2009; Weinstein et al. 2005). The most promising compounds clinically, the diarylquinolines, have been shown to target the F_1F_o-ATP synthase and inhibit ATP synthesis by the enzyme (Andries et al. 2005; Koul et al. 2007, 2008). Genome sequencing of both *M. tuberculosis* and *M. smegmatis* mutants that are resistant to diarylquinolines (i.e., R207910) revealed that the target of these compounds is the oligomeric *c* ring (encoded by *atpE*) of the enzyme (Andries et al. 2005; Huitric et al. 2007, 2010). The purified *c* ring from *M. smegmatis* binds R207910 with a K_D of 500 nm, and modelling/docking and kinetic studies suggest that R207910 blocks rotary movement of the c-ring during catalysis by mimicking key residues in the proton transfer chain (de Jonge et al. 2007; Haagsma et al. 2011). Further investigations with inverted membrane vesicles of *M. smegmatis* and TMC207 has revealed that TMC207 acts independently of the PMF, and that electrostatic forces play an important role in the interaction of the drug with the ATP synthase (Haagsma et al. 2011).

When mycobacterial cells (growing or non-growing) are treated with TMC207, time-dependent (not dose-dependent) killing is observed (Andries et al. 2005). The mechanism of killing is not clear, but does not involve the dissipation of the membrane potential, which is lethal to all living cells. A dose-dependent decrease in intracellular ATP has been observed when *M. tuberculosis* cells are treated with TMC207 (Koul et al. 2007), but these data do not explain cell death because mycobacterial cells can be depleted of ATP through various deenergization treatments and yet remain viable (Frampton et al. 2012). TMC207 is bactericidal towards most species of mycobacteria (Andries et al. 2005), but is only bacteriostatic against *Mycobacterium avium* (Lounis et al. 2009) and *Mycobacterium smegmatis* (unpublished data). Even when we grew *M. smegmatis* at a doubling time of 70 h in glycerol-limited continuous culture, TMC207 was bacteriostatic (Berney and Cook, unpublished data). The identification of the mechanisms underlying this sensitivity will be important in understanding how TMC207 exerts its inhibitory effects on mycobacterial cells.

Acknowledgments

Research in the authors laboratory is funded by Health Research Council, Lottery Health, Marsden Fund, Royal Society New Zealand and the Maurice Wilkins Centre.

References

Andries K, Verhasselt P, Guillemont J, Göhlmann HW, Neefs JM, Winkler H, Van Gestel J, Timmerman P, Zhu M, Lee E, Williams P, de Chaffoy D, Huitric E, Hoffner S, Cambau E, Truffot-Pernot C, Lounis N, Jarlier V (2005) A diarylquinoline drug active on the ATP synthase of *Mycobacterium tuberculosis*. Science 307:223–227

Bacon J, James BW, Wernisch L, Williams A, Morley KA, Hatch GJ, Mangan JA, Hinds J, Stoker NG, Butcher PD, Marsh PD (2004) The influence of reduced oxygen availability on pathogenicity and gene expression in *Mycobacterium tuberculosis*. Tuberculosis (Edinb) 84:205–217

Baughn AD, Garforth SJ, Vilcheze C, Jacobs WR (2009) An anaerobic-type alpha-ketoglutarate ferredoxin oxidoreductase completes the oxidative tricarboxylic acid cycle of *Mycobacterium tuberculosis*. PloS Pathog 5:e1000662

Berney M, Cook GM (2010) Unique flexibility in energy metabolism allows mycobacteria to combat starvation and hypoxia. PLoS ONE 5:e8614

Berney M, Weimar MR, Heikal A, Cook GM (2012) Regulation of proline metabolism in mycobacteria and its role in carbon metabolism under hypoxia. Mol Microbiol 84:664–681

Beste DJ, Peters J, Hooper T, Avignone-Rossa C, Bushell ME, McFadden J (2005) Compiling a molecular inventory for *Mycobacterium bovis* BCG at two growth rates: evidence for growth rate-mediated regulation of ribosome biosynthesis and lipid metabolism. J Bacteriol 187:1677–1684

Beste DJ, Laing E, Bonde B, Avignone-Rossa C, Bushell ME, McFadden JJ (2007) Transcriptomic analysis identifies growth rate modulation as a component of the adaptation of mycobacteria to survival inside the macrophage. J Bacteriol 189:3969–3976

Betts JC (2002) Transcriptomics and proteomics: tools for the identification of novel drug targets and vaccine candidates for tuberculosis. IUBMB Life 53:239–242

Betts JC, Lukey PT, Robb LC, McAdam RA, Duncan K (2002) Evaluation of a nutrient starvation model of *Mycobacterium tuberculosis* persistence by gene and protein expression profiling. Mol Microbiol 43:717–731

Boos W (1998) Binding protein-dependent ABC transport system for glycerol 3-phosphate of *Escherichia coli*. Methods Enzymol 292:40–51

Borisov VB, Murali R et al (2011) Aerobic respiratory chain of *Escherichia coli* is not allowed to work in fully uncoupled mode. Proc Natl Acad Sci U S A 108(42):17320–17324

Boshoff HI, Barry CE 3rd (2005) Tuberculosis: metabolism and respiration in the absence of growth. Nat Rev Microbiol 3:70–80

Cecchini G, Schroder I, Gunsalus RP, Maklashina E (2002) Succinate dehydrogenase and fumarate reductase from *Escherichia coli*. Biochim Biophys Acta 1553:140–157

Cole ST, Brosch R et al (1998) Deciphering the biology of *Mycobacterium tuberculosis* from the complete genome sequence. Nature 393(6685):537–544

Cole ST, Supply P, Honore N (2001) Repetitive sequences in *Mycobacterium leprae* and their impact on genome plasticity. Lepr Rev 72:449–461

Constant P, Poissant L, Villemur R (2008) Isolation of *Streptomyces* sp. PCB7, the first microorganism demonstrating high-affinity uptake of tropospheric H2. ISME J 2:1066–1076

Constant P, Chowdhury SP, Pratscher J, Conrad R (2010) Streptomycetes contributing to atmospheric molecular hydrogen soil uptake are widespread and encode a putative high-affinity [NiFe]-hydrogenase. Environ Microbiol 12:821–829

Constant P, Chowdhury SP, Hesse L, Pratscher J, Conrad R (2011) Genome data mining and soil survey for the novel group 5 [NiFe]-hydrogenase to explore the diversity and ecological importance of presumptive high-affinity H(2)-oxidizing bacteria. Appl Environ Microbiol 77:6027–6035

Cotter PA, Melville SB, Albrecht JA, Gunsalus RP (1997) Aerobic regulation of cytochrome d oxidase (*cydAB*) operon expression in *Escherichia coli*: roles of Fnr and ArcA in repression and activation. Mol Microbiol 25:605–615

Cox RA, Cook GM (2007) Growth regulation in the mycobacterial cell. Curr Mol Med 7:231–245

D'Mello R, Hill S, Poole RK (1995) The oxygen affinity of cytochrome *bo*' in *Escherichia coli* determined by the deoxygenation of oxyleghemoglobin and oxymyoglobin: Km values for oxygen are in the submicromolar range. J Bacteriol 177:867–870

D'Mello R, Hill S, Poole RK (1996) The cytochrome *bd* quinol oxidase in *Escherichia coli* has an extremely high oxygen affinity and two oxygen-binding haems: implications for regulation of activity in vivo by oxygen inhibition. Microbiology 142:755–763

de Jonge MR, Koymans LH, Guillemont JE, Koul A, Andries K (2007) A computational model of the inhibition of *Mycobacterium tuberculosis* ATPase by a new drug candidate R207910. Proteins 67: 971–980

Dhar N, McKinney JD (2010) *Mycobacterium tuberculosis* persistence mutants identified by screening in isoniazid-treated mice. Proc Natl Acad Sci U S A 107:12275–12280

Dhiman RK, Mahapatra S, Slayden RA, Boyne ME, Lenaerts A, Hinshaw JC, Angala SK, Chatterjee D, Biswas K, Narayanasamy P, Kurosu M, Crick DC (2009) Menaquinone synthesis is critical for maintaining mycobacterial viability during exponential growth and recovery from non-replicating persistence. Mol Microbiol 72:85–97

Dimroth P, Cook GM (2004) Bacterial Na+- or H+ -coupled ATP synthases operating at low electrochemical potential. Adv Microb Physiol 49:175–218

Fontan P, Aris V, Ghanny S, Soteropoulos P, Smith I (2008) Global transcriptional profile of *Mycobacterium tuberculosis* during THP-1 human macrophage infection. Infect Immun 76:717–725

Frampton R, Aggio RB, Villas-Boas SG, Arcus VL, Cook GM (2012) Toxin-antitoxin systems of *Mycobacterium smegmatis* are essential for cell survival. J Biol Chem 287:5340–5356

Friedl P, Hoppe J, Gunsalus RP, Michelsen O, von Meyenburg K, Schairer HU (1983) Membrane integration and function of the three F0 subunits of the ATP synthase of *Escherichia coli* K12. EMBO J 2:99–103

Goldman BS, Gabbert KK, Kranz RG (1996a) Use of heme reporters for studies of cytochrome biosynthesis and heme transport. J Bacteriol 178: 6338–6347

Goldman BS, Gabbert KK, Kranz RG (1996b) The temperature-sensitive growth and survival phenotypes of *Escherichia coli cydDC* and *cydAB* strains are due to deficiencies in cytochrome *bd* and are corrected by exogenous catalase and reducing agents. J Bacteriol 178:6348–6351

Griffin JE, Gawronski JD, Dejesus MA, Ioerger TR, Akerley BJ, Sassetti CM (2011) High-resolution phenotypic profiling defines genes essential for mycobacterial growth and cholesterol catabolism. PLoS Pathog 7:e1002251

Haagsma AC, Driessen NN, Hahn MM, Lill H, Bald D (2010) ATP synthase in slow- and fast-growing mycobacteria is active in ATP synthesis and blocked in ATP hydrolysis direction. FEMS Microbiol Lett 313:68–74

Haagsma AC, Podasca I, Koul A, Andrie K, Guillemont J, Lill H, Bald D (2011) Probing the interaction of the diarylquinoline TMC207 with its target mycobacterial ATP synthase. PLoS ONE 6:e23575

Higashi T, Kalra VK, Lee SH, Bogin E, Brodie AF (1975) Energy-transducing membrane-bound coupling factor-ATPase from *Mycobacterium phlei*. I. Purification, homogeneity, and properties. J Biol Chem 250:6541–6548

Huitric E, Verhasselt P, Andries K, Hoffner SE (2007) In vitro antimycobacterial spectrum of a diarylquinoline ATP synthase inhibitor. Antimicrob Agents Chemother 51:4202–4204

Huitric E, Verhasselt P, Koul A, Andries K, Hoffner S, Andersson DI (2010) Rates and mechanisms of resistance development in *Mycobacterium tuberculosis* to a novel diarylquinoline ATP synthase inhibitor. Antimicrob Agents Chemother 54:1022–1028

Kana BD, Weinstein EA, Avarbock D, Dawes SS, Rubin H, Mizrahi V (2001) Characterization of the *cydAB*-encoded cytochrome *bd* oxidase from *Mycobacterium smegmatis*. J Bacteriol 183: 7076–7086

Kana BD, Machowski EE, Schechter N, Teh J-S, Rubin H, Mizrahi V (2009) Electron transport and respiration in mycobacteria. In: Parish, T, Brown A (eds) Mycobacterium: genomics and molecular biology, Horizon Scientific Press, Poole, UK. pp 35–64

Kim YM, Hegeman GD (1983) Oxidation of carbon monoxide by bacteria. Int Rev Cytol 81:1–32

King GM (2003) Uptake of carbon monoxide and hydrogen at environmentally relevant concentrations by mycobacteria. Appl Environ Microbiol 69:7266–7272

Koch-Koerfges A, Kabus A, Ochrombel I, Marin K, Bott M (2012) Physiology and global gene expression of a *Corynebacterium glutamicum* DeltaF(1) F(O)-ATP synthase mutant devoid of oxidative phosphorylation. Biochim Biophys Acta 1817: 370–380

Koul A, Dendouga N, Vergauwen K, Molenberghs B, Vranckx L, Willebrords R, Ristic Z, Lill H, Dorange I, Guillemont J, Bald D, Andries K (2007) Diarylquinolines target subunit c of mycobacterial ATP synthase. Nat Chem Biol 3:323–324

Koul A, Vranckx L, Dendouga N et al (2008) Diarylquinolines are bactericidal for dormant mycobacteria as a result of disturbed ATP homeostasis. J Biol Chem 283:25273–25280

Kuhn M, Steinbuchel A, Schlegel HG (1984) Hydrogen evolution by strictly aerobic hydrogen bacteria under anaerobic conditions. J Bacteriol 159:633–639

Lounis N, Gevers T, Van den Berg J, Vranckx L, Andries K (2009) ATP synthase inhibition of *Mycobacterium avium* is not bactericidal. Antimicrob Agents Chemother 53:4927–4929

Matsoso LG, Kana BD, Crellin PK, Lea-Smith DJ, Pelosi A, Powell D, Dawes SS, Rubin H, Coppel RL, Mizrahi V (2005) Function of the cytochrome *bc1-aa3* branch of the respiratory network in mycobacteria and network adaptation occurring in response to its disruption. J Bacteriol 187: 6300–6308

McAdam RA, Quan S, Smith DA, Bardarov S, Betts JC, Cook FC, Hooker EU, Lewis AP, Woollard P, Everett MJ, Lukey PT, Bancroft GJ, Jacobs WR Jr, K D Jr (2002) Characterization of a *Mycobacterium tuberculosis* H37Rv transposon library reveals insertions in 351 ORFs and mutants with altered virulence. Microbiology 148:2975–2986

Megehee JA, Hosler JP, Lundrigan MD (2006) Evidence for a cytochrome bcc-aa3 interaction in the respiratory chain of *Mycobacterium smegmatis*. Microbiology 152:823–829

Menzel R, Roth J (1981) Purification of the *putA* gene product. A bifunctional membrane-bound protein from Salmonella typhimurium responsible for the two-step oxidation of proline to glutamate. J Biol Chem 256:9755–9761

Miesel L, Weisbrod TR, Marcinkeviciene JA, Bittman R, Jacobs WR Jr (1998) NADH dehydrogenase defects confer isoniazid resistance and conditional lethality in *Mycobacterium smegmatis*. J Bacteriol 180:2459–2467

Mnatsakanyan N, Bagramyan K, Trchounian A (2004) Hydrogenase 3 but not hydrogenase 4 is major in hydrogen gas production by *Escherichia coli* formate hydrogenlyase at acidic pH and in the presence of external formate. Cell Biochem Biophys 41:357–366

Molenaar D, van der Rest ME, Petrovic S (1998) Biochemical and genetic characterization of the membrane-associated malate dehydrogenase (acceptor) from *Corynebacterium glutamicum*. Eur J Biochem 254:395–403

Niebisch A, Bott M (2003) Purification of a cytochrome *bc-aa3* supercomplex with quinol oxidase activity from *Corynebacterium glutamicum*. Identification of a fourth subunity of cytochrome *aa3* oxidase and mutational analysis of diheme cytochrome *c1*. J Biol Chem 278:4339–4346

Park HD, Guinn KM, Harrell MI, Liao R, Voskuil MI, Tompa M, Schoolnik GK, Sherman DR (2003a) *Rv3133c/dosR* is a transcription factor that mediates the hypoxic response of *Mycobacterium tuberculosis*. Mol Microbiol 48:833–843

Park SW, Hwang EH, Park H, Kim JA, Heo J, Lee KH, Song T, Kim E, Ro YT, Kim SW, Kim YM (2003b) Growth of mycobacteria on carbon monoxide and methanol. J Bacteriol 185:142–147

Pfeiffer T, Schuster S, Bonhoeffer S (2001) Cooperation and competition in the evolution of ATP-producing pathways. Science 292:504–507

Pittman MS, Robinson HC, Poole RK (2005) A bacterial glutathione transporter (*Escherichia coli* CydDC) exports reductant to the periplasm. J Biol Chem 280:32254–32261

Poole RK, Cook GM (2000) Redundancy of aerobic respiratory chains in bacteria? Routes, reasons and regulation. Adv Microb Physiol 43:165–224

Prasada Reddy TL, Suryanarayana Murthy P, Venkitasubramanian TA (1975) Variations in the pathways of malate oxidation and phosphorylation in different species of Mycobacteria. Biochim Biophys Acta 376:210–218

Rao M, Streur TL, Aldwell FE, Cook GM (2001) Intracellular pH regulation by *Mycobacterium smegmatis* and *Mycobacterium bovis* BCG. Microbiology 147:1017–1024

Rao SP, Alonso S, Rand L, Dick T, Pethe K (2008) The protonmotive force is required for maintaining ATP homeostasis and viability of hypoxic, nonreplicating *Mycobacterium tuberculosis*. Proc Natl Acad Sci U S A 105:11945–11950

Santana M, Ionescu MS, Vertes A, Longin R, Kunst F, Danchin A, Glaser P (1994) Bacillus subtilis F0F1 ATPase: DNA sequence of the atp operon and characterization of atp mutants. J Bacteriol 176:6802–6811

Sassetti CM, Boyd DH, Rubin EJ (2003) Genes required for mycobacterial growth defined by high density mutagenesis. Mol Microbiol 48:77–84

Schnappinger D, Ehrt S, Voskuil MI et al (2003) Transcriptional adaptation of *Mycobacterium tuberculosis* within macrophages: insights into the phagosomal environment. J Exp Med 198:693–704

Schryvers A, Lohmeier E, Weiner JH (1978) Chemical and functional properties of the native and reconstituted forms of the membrane-bound, aerobic glycerol-3-phosphate dehydrogenase of *Escherichia coli*. J Biol Chem 253:783–788

Sherman DR, Voskuil M, Schnappinger D, Liao R, Harrell MI, Schoolnik GK (2001) Regulation of the *Mycobacterium tuberculosis* hypoxic response gene encoding alpha -crystallin. Proc Natl Acad Sci U S A 98:7534–7539

Shi L, Sohaskey CD, Kana BD, Dawes S, North RJ, Mizrahi V, Gennaro ML (2005) Changes in energy metabolism of *Mycobacterium tuberculosis* in mouse lung and under in vitro conditions affecting aerobic respiration. Proc Natl Acad Sci U S A 102:15629–15634

Sohaskey CD (2005) Regulation of nitrate reductase activity in *Mycobacterium tuberculosis* by oxygen and nitric oxide. Microbiology 151(Pt 11):3803–3810

Sohaskey CD (2008) Nitrate enhances the survival of *Mycobacterium tuberculosis* during inhibition of respiration. J Bacteriol 190:2981–2986

Sohaskey CD, Wayne LG (2003) Role of *narK2X* and *narGHJI* in hypoxic upregulation of nitrate reduction by *Mycobacterium tuberculosis*. J Bacteriol 185:7247–7256

Tamagnini P, Leitao E et al (2007) Cyanobacterial hydrogenases: diversity, regulation and applications. FEMS Microbiol Rev 31(6):692–720

Tanner JJ (2008) Structural biology of proline catabolism. Amino Acids 35:719–730

Tian J, Bryk R, Itoh M, Suematsu M, Nathan C (2005) Variant tricarboxylic acid cycle in *Mycobacterium tuberculosis*: identification of alpha-ketoglutarate decarboxylase. Proc Natl Acad Sci U S A 102:10670–10675

Tran SL, Cook GM (2005) The F1Fo-ATP synthase of *Mycobacterium smegmatis* is essential for growth. J Bacteriol 187:5023–5028

Tran SL, Rao M, Simmers C, Gebhard S, Olsson K, Cook GM (2005) Mutants of *Mycobacterium smegmatis* unable to grow at acidic pH in the presence of the protonophore carbonyl cyanide m-chlorophenylhydrazone. Microbiology 151:665–672

Tseng CP, Hansen AK, Cotter P, Gunsalus RP (1994) Effect of cell growth rate on expression of the anaerobic respiratory pathway operons *frdABCD*, *dmsABC*, and *narGHJI* of *Escherichia coli*. J Bacteriol 176:6599–6605

Unden G, Bongaerts J (1997) Alternative respiratory pathways of *Escherichia coli*: energetics and transcriptional regulation in response to electron acceptors. Biochim Biophys Acta 1320: 217–234

Velmurugan K, Chen B, Miller JL et al (2007) *Mycobacterium tuberculosis nuoG* is a virulence gene that inhibits apoptosis of infected host cells. PLoS Pathog 3:e110

Vignais PM, Billoud B, Meyer J (2001) Classification and phylogeny of hydrogenases. FEMS Microbiol Rev 25:455–501

Vilcheze C, Weisbrod TR, Chen B, Kremer L, Hazbon MH, Wang F, Alland D, Sacchettini JC, Jacobs WR Jr (2005) Altered NADH/NAD + ratio mediates coresistance to isoniazid and ethionamide in mycobacteria. Antimicrob Agents Chemother 49: 708–720

von Ballmoos C, Cook GM, Dimroth P (2008) Unique rotary ATP synthase and its biological diversity. Annu Rev Biophys 37:43–64

Voskuil MI, Schnappinger D, Visconti KC, Harrell MI, Dolganov GM, Sherman DR, Schoolnik GK (2003) Inhibition of respiration by nitric oxide induces a *Mycobacterium tuberculosis* dormancy program. J Exp Med 198:705–713

Voskuil MI, Visconti KC, Schoolnik GK (2004) *Mycobacterium tuberculosis* gene expression during adaptation to stationary phase and low-oxygen dormancy. Tuberculosis (Edinb) 84: 218–227

Watanabe S, Zimmermann M, Goodwin MB, Sauer U, Barry CE 3rd, Boshoff HI (2011) Fumarate reductase activity maintains an energized membrane in anaerobic *Mycobacterium tuberculosis*. PLoS Pathog 7:e1002287

Wayne LG, Hayes LG (1996) An in vitro model for sequential study of shiftdown of *Mycobacterium tuberculosis* through two stages of nonreplicating persistence. Infect Immun 64: 2062–2069

Weinstein EA, Yano T, Li LS et al (2005) Inhibitors of type II NADH: menaquinone oxidoreductase represent a class of antitubercular drugs. Proc Natl Acad Sci U S A 102:4548–4553

Youmans AS, Millman I, Youmans GP (1956) The oxidation of compounds related to the tricarboxylic acid cycle by whole cells and enzyme preparations of *Mycobacterium tuberculosis* var. Hominis. J Bacteriol 71:565–570

Chapter 16

The Structure and Morphology of Red Algae Chloroplasts

Zenilda L. Bouzon*
Plant Cell Biology Laboratory, Department of Cell Biology, Embryology and Genetics, Federal University of Santa Catarina, Florianópolis, SC, Brazil
Central Laboratory of Electron Microscopy, Federal University of Santa Catarina, 88049-900, CP 476 Florianópolis, SC, Brazil

Carmen Simioni
Post-Graduate Program in Cell Biology and Development, Department of Cell Biology, Embryology and Genetics, Federal University of Santa Catarina, 88049-900, CP 476 Florianópolis, SC, Brazil

and

Eder C. Schmidt
Postdoctoral Research of Postgraduate Program in Cell Biology and Development, Department of Cell Biology, Embryology and Genetics, Federal University of Santa Catarina, 88049-900, CP 476 Florianópolis, SC, Brazil

*Author for correspondence, e-mail: zenilda.bouzon@ufsc.br

M.F. Hohmann-Marriott (ed.), The Structural Basis of Biological Energy Generation, Advances in Photosynthesis and Respiration 39, pp. 295–308, DOI 10.1007/978-94-017-8742-0_16,

Summary

The morphological and structural aspects of red algae chloroplasts are described in this chapter. Structural and physiological features of the red algal chloroplast demonstrate the occurrence of an endosymbiotic event. Structural features of the red algal chloroplast include phycobilisomes, plastoglobuli, genophores, ribosomes, and pyrenoids in some species. The phycobilisomes are photosynthetic macromolecular aggregates of light-harvesting pigment-protein complexes attached to the stromal side of the thylakoid membrane. In the chloroplast stroma, electron-dense lipid droplets (plastoglobuli) are observed between the thylakoids, and in red algae, these structures are interpreted as lipid material. In chloroplasts, the stroma is filled with plastidial ribosomes and a fibrillar region corresponding to a genophore. Red algae can possess a large variety of colors, ranging from purple-red, pinkish-red, red, brown, yellowish to different gradations of green. These colors arise from a variety of photosynthetic pigments, predominantly chlorophyll *a*, with accessory pigments found in the phycobiliproteins (allophycocyanin, phycocyanin and phycoerythrin), as well as different carotenoids (violaxanthin, antheraxanthin, lutein, zeaxanthin, β-cryptoxanthin, α-carotene and β-carotene). Floridean starch is the main storage product synthesized in red algae. The floridean starch is stored in granules, which are localized in the cell cytoplasm – outside of the chloroplast.

I. Introduction

The chloroplasts of red algae provide insight into the evolutionary process of endosymbiosis in that the ultrastructure of the red algae chloroplast is similar to the cellular organization of cyanobacteria.

The origin of the green algal/plant plastids and red algal plastids is primary endosymbiosis, in which cyanobacteria-like cell was engulfment and retained in a host cell. The plastids of both, red algae and green algae, were subsequently transferred by secondary endosymbiotic events, giving rise to plastids that are surrounded by three or four bounding membranes.

The first observations detailing chloroplast structure in red algae came from the pioneering work of Gibbs (1962), who surveyed chloroplast ultrastructure of 13 species. In the *Rhodophyta*, the chloroplasts are limited by a narrow double membrane and contain a granular matrix material made up of dense particles (90–150 Å in diameter) and scattered lipid-like globules. Most noteworthy is the presence of free thylakoids which form compartments and virtually never come into contact with other thylakoids except for occasional inter-connections.

The arrangement of the thylakoid varies between red algae. In *Laurencia,* thylakoids lie in more or less parallel lines across the chloroplast, but there are often single thylakoids, which follow the chloroplast envelope and thus encloses the thylakoids in the chloroplast center. Other red algae have a concentric arrangement of thylakoids, as reported by in *Porphyridium cruentum* (Gantt and Conti 1965). Several authors have studied the size and distribution of intramembranous particles in freeze-fractured thylakoids of red algae (Cole and Sheath 1990). The studies that reveal phycobilisomes as particles attached to the stromal surface of the thylakoids.

Red algal thylakoids contain chlorophyll *a*, carotenoids (violaxanthin, antheraxanthin, lutein, zeaxanthin, β-cryptoxanthin, α-carotene and β-carotene) and with accessory pigments located in the phycobiliproteins (allophycocyanin, phycocyanin and phycoerythrin).

Abbreviations: EM – Electron microscopy; LM – Light microscopy; PAS – Periodic acid-schiff; PBS – Phycobilisome

II. Chloroplast Morphology

The morphology of red algae chloroplasts is highly variable, i.e., from stellate (Fig. 16.1) to discoid to ramifying, highly lobed forms. The chloroplasts can be large and elongated (Fig. 16.2) with the ability to adjust to cellular morphology. Some red algae have one large chloroplast per cell, usually stellate, with a characteristic central pyrenoid, as seen in the genera *Porphyra* (Fig. 16.1) and *Nemalion* (Fig. 16.3), whereas most algae in these groups exhibit numerous small chloroplasts per cell.

The chloroplast envelope of red algae is similar to that of many other photosynthetic organisms, while the internal organization can be distinct form cyanbacteria and green algae and plants. The chloroplast typically has an arrangement of unconnected, parallel, linear thylakoids, a 'periplastidial thylakoid', which follows the contours of the envelope can be present. The number of parallel thylakoids and presence of a 'periplastidal thylakoid' is varies between species and the plastid location in multicellular red algae.

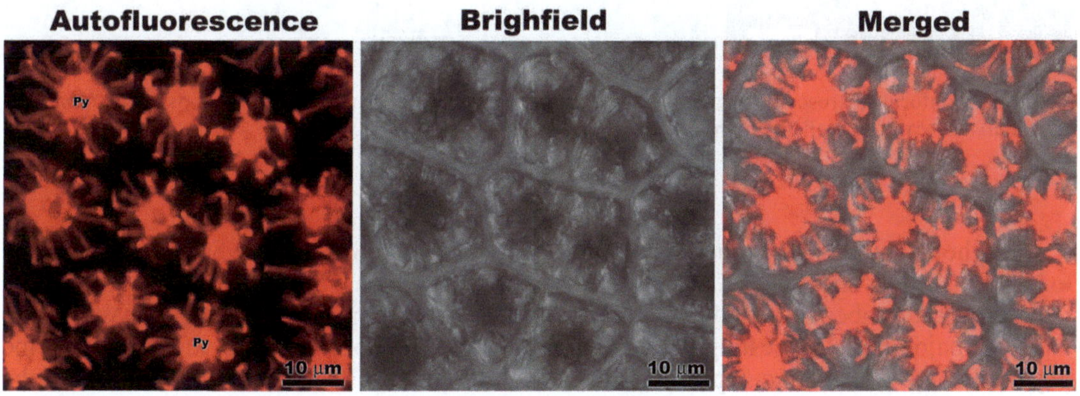

Fig. 16.1. Confocal microscopy images (autofluorescence, bright field and merged composite) of the stellate chloroplast of *Porphyra acanthophora* var. *brasiliensis*. Note the presence of central pyrenoid (Py).

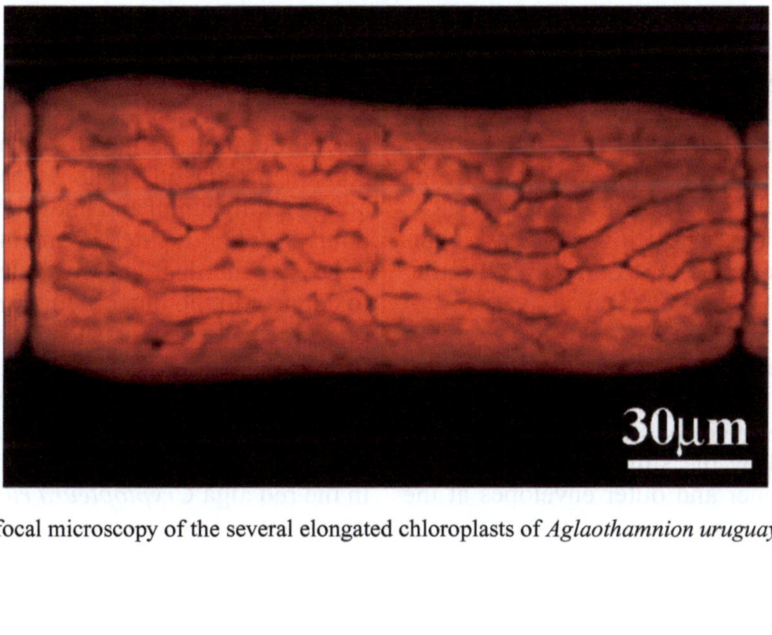

Fig. 16.2. Confocal microscopy of the several elongated chloroplasts of *Aglaothamnion uruguayense*.

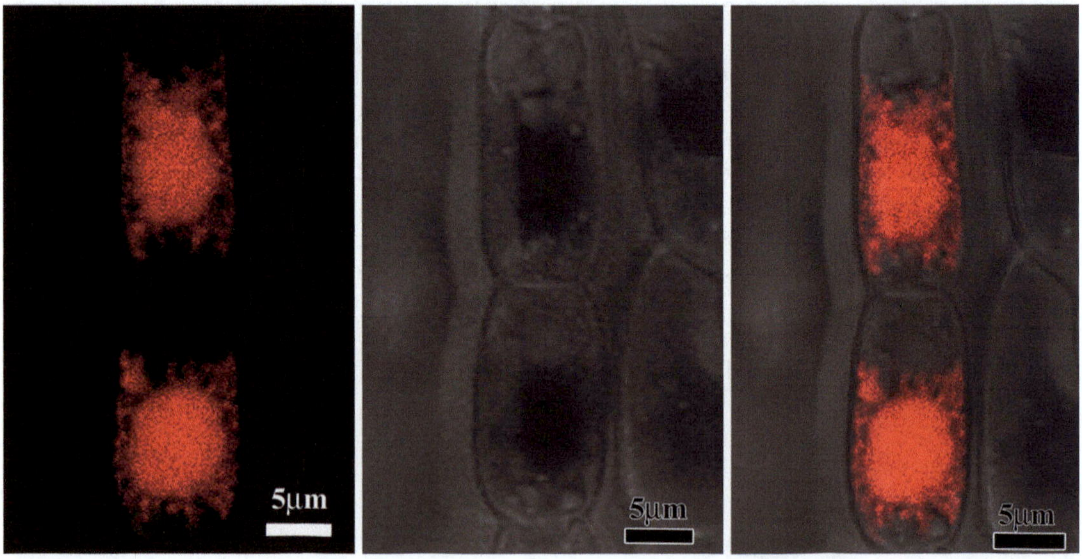

Fig. 16.3. Confocal microscopy images (autofluorescence, bright field and merged composite) of *Nemalion helminthoides* featuring stellate chloroplasts.

The photosynthetic machinery is similar to that found in cyanobacteria. The two photosystems are embedded in the membrane and the major light harvesting system, the phycobilisome (PBS), made up of phycobiliproteins, are in the chloroplast stroma. In *Rhodophyta* the PBS are generally hemispherical, in contrast to cyanobacteria where they are hemidiscoidal.

Red algae possess a sophisticated carbon fixation machinery. Red algal RuBisCo is more closely related to α-proteobacterial origin (type I) than cyanobacterial and green alga and plant RuBisCo.

Some of RuBisCos present in red alga possess the highest known affinity for CO_2. In many red alga species RuBisCo is organized in pyrenoids. Carbon concentrating mechanisms are found in a variety of *Rhodophyta*.

The chloroplasts reproduce by binary fission (Fig. 16.4). Electron microscopic studies established that chloroplast division is performed by the simultaneous constriction of the inner and outer envelopes at the division site. Since the 1980s, electron-dense ring structures at the division site called plastid-dividing (PD) rings have been detected on both the inside and outside of the chloroplast. These observations lead to the suggestion that chloroplast division is achieved by constriction of the ring-like division complex. Recent studies in the unicellular red alga *Cyanidioschyzon merolae*, have identified several components of the division complex using molecular genetics. (Miyagishima 2011) (see Chap. 23).

Undifferentiated chloroplasts, such as those found in callus of young red algae, contain proplastids (Fig. 16.5). Proplastids are generally small and undifferentiated with poorly defined internal membranes, that appear as tubules. These tubules often seem to connect to the inner membrane of the proplastid envelope, but such connections are lost early in development and are apparently absent in mature plastids (Wise and Hooper 2007). Delivopoulos (2003) visualized fission of proplastids and divisions of plastids in the red alga *Cryptopleura ruprechtiana.*

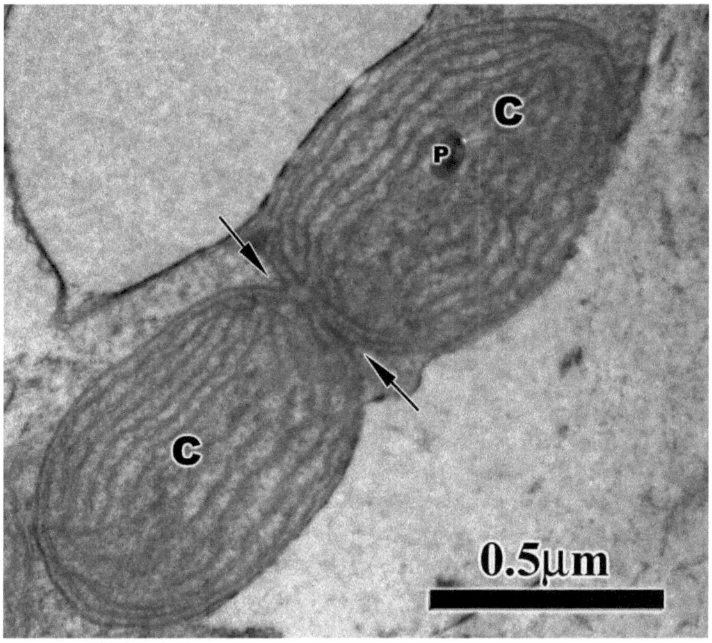

Fig. 16.4. Transmission electron microscopy (TEM) images of Gracilaria caudata. Chloroplast binary fission event is indicated by arrows sister chloroplasts indicated by C. Note the presence of plastoglobuli (P).

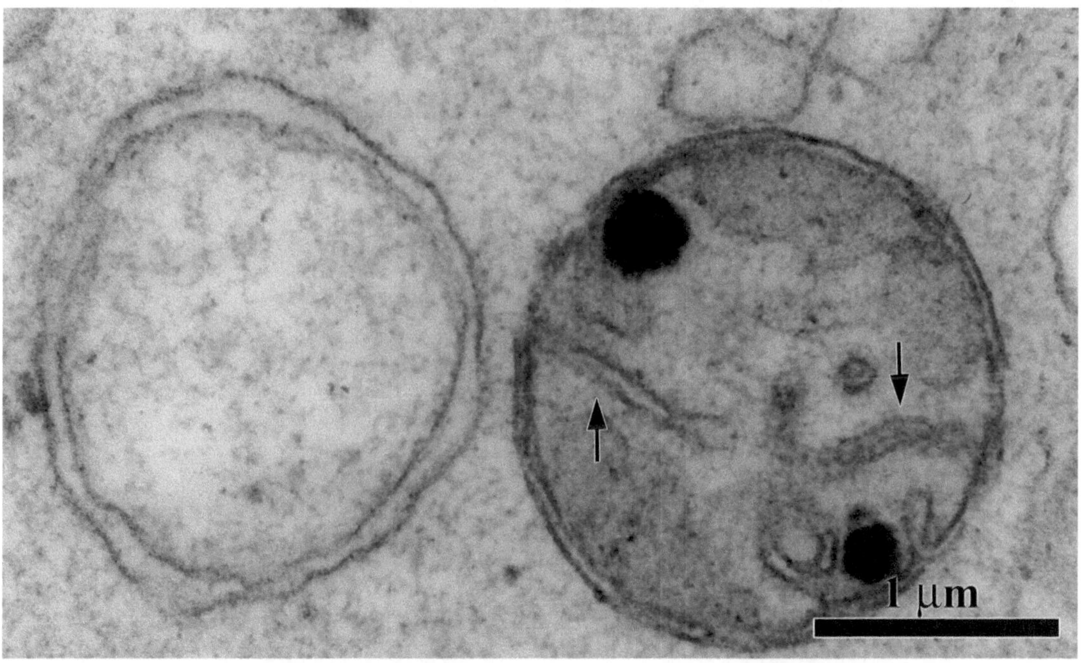

Fig. 16.5. Transmission electron microscopy (TEM) images of *Gracilaria caudata*. Thylakoids in the proplastid are indicated by *arrows*. The inner and the outer chloroplast membrane can be discerned.

III. Phycobilisomes

Phycobilisomes are photosynthetic macromolecular aggregates of light-harvesting pigment-protein complexes attached to the stromal side of the thylakoid membranes of cyanobacteria and red algae (see Chap. 4). The main components of phycobilisomes is an evolutionary related group of hexameric proteins that house phycobilins (e.g., phycoerythrin, phycocyanin) referred to as phycobiliproteins and linker proteins. The phycobilisomes appear as 35 nm granules when phycoerythrin predominates or as discs when phycocyanin predominates (Lee 2008). For example, in *Gracilaria domingensis*, numerous evenly-spaced phycobilisomes are attached on both sides of the outer thylakoids (Fig. 16.6).

When cells are deprived of nitrogen, an ordered degradation of phycobilisomes occurs, involving a progressive degradation of hexamer rod and linker polcoypeptides followed by the core peptides. However, new phycobilisomes are rapidly synthesized upon the addition of nitrogen to the medium. Thus, phycobilisomes are an important source of internal nitrogen and offer red algae an crucial ecological advantage in the open ocean where the supply of nitrogen is limited (Vergara and Niell 1993).

IV. Photosynthetic Pigments

Depending species and on the environment where a red alga grows, the coloration of red algae can vary, ranging from purple-red, pinkish-red, red, brown, yellowish and different gradations of green based on different pigment composition (chlorophyll *a*, phycobiliproteins and carotenoids).

The Rhodophyta family has only chlorophyll *a* as the primary photosynthetic pigment. This pigment has two main absorption bands in vitro, one band in the red light region at 663 nm and the other at 430 nm. The thylakoids contain chlorophyll that is associated with the photosynthetic reaction centers with functions in light harvesting and charge separation reactions; carbon dioxide fixation occurs in the stroma (Lee 2008).

The main accessory pigments in red algae are the phycobiliproteins localized into phycobilisomes on the thylakoid surface.

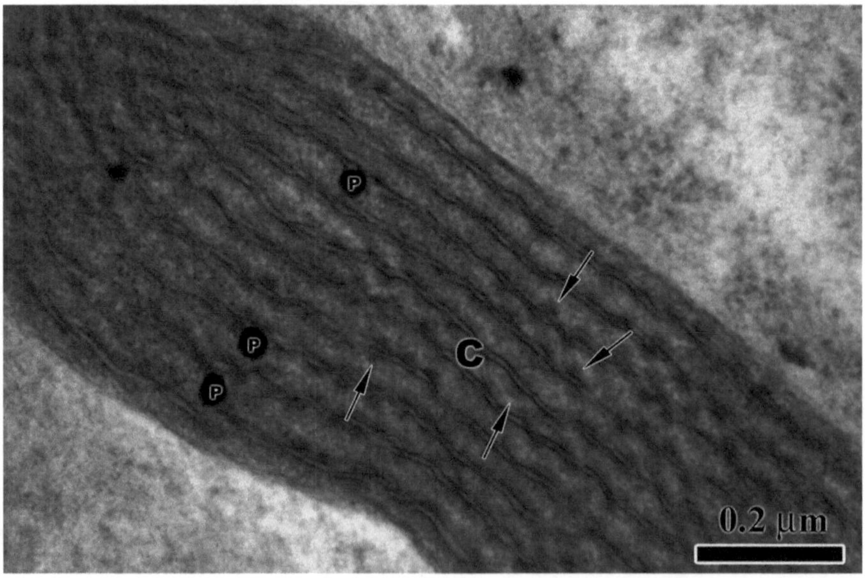

Fig. 16.6. Transmission electron microscopy (TEM) images of *Gracilaria domingensis*. Detail of chloroplast (C) shows the phycobilisomes attached to the outside (cytoplasmic side) of the thylakoid membranes (*arrows*). Note the presence of plastoglobuli (P).

The water-soluble phycobiliproteins, which are comprised of linear tetrapyrrole chromophores covalently bound to apoprotein, fall into three main groups, based on their absorption spectra (Lee 2008): phycoerythrin, phycocyanin and allophycocyanin. Several additional proteins are found within the phycobilisome, and they serve to link the phycobiliproteins together in an ordered fashion and link the phycobilisome to the thylakoid membrane (Zilinskas and Greenwald 1986). The phycobilisome of red algae is composed of an allophycocyanin core and several peripheral rods containing phycocyanin and phycoerythrin (Su et al. 2010). Gantt and Lipschultz (1974) initially reported that phycobilisomes isolated from *Porphyridium cruentum* only consisted of phycobiliproteins with 84 % R- and b-phycoerythrin, 11 % R-phycocyanin and 5 % allophycocyanin.

Other groups of accessory pigments present in red algae are the carotenoids, and their different concentrations are an important factor in determining the color of the algae. Several types of carotenoids in red algae have been described, including violaxanthin, antheraxanthin, lutein, zeaxanthin, β-cryptoxanthin, α-carotene and β-carotene (Schubert et al. 2006).

V. Plastoglobuli

In the chloroplast stroma, electron-dense lipid droplets described as plastoglobuli can be observed between the thylakoids (Fig. 16.6). In red algae, these structure are interpreted as lipid material that function as a lipid reserve. Plastoglobuli were also described by Wetherbee and Wynne (1973) in *Polysiphonia novae-angliae* W. R. Taylor, by Pueschel (1988) in *Hildenbrandia rubra* (Sommerfelt) Meneghini, and by Bouzon (2006) in *Hypnea musciformis*.

Plastoglobuli are considered to passive store of lipids and carotenoids, and their dimensions vary among different species, types and developmental stages of plastids, suggesting a more dynamic function (Bréhélin et al. 2007). Recent publications show that plastoglobuli contain enzymes involved in the metabolism of these secondary metabolites, as well as enzymes of unknown function (Bréhélin et al. 2007). When analyzed by transmission electron microscopy (TEM), algae, such as *Kappaphycus alvarezii* (Schmidt et al. 2009, 2010a, b) (Fig. 16.7), *Gracilaria domingensis* (Schmidt et al. 2010c), *Hypnea musciformis* (Bouzon et al. 2012) (Fig. 16.8) and callus of *Kappaphycus alvarezii* (Zitta 2010) (Fig. 16.9), exposed to different stress factors, e.g. ultraviolet radiation-B and cadmium, an increase in the number of plastoglobuli in the chloroplast is observed. According to Holzinger et al. (2009), when the algae are subjected to stress, including the depletion of nitrogen, the synthesis of lipids is observed. These phenomena is thought to occur, because the metabolic pathways that lead to protein-containing cell structures are suppressed.

VI. Ribosome

In chloroplasts, the stroma is filled with smaller plastidial ribosomes (Fig. 16.10) (Dodge 1973). These ribosomes have the characteristics of 70S ribosomes of bacteria, indicating their endosymbiotic, cyanobacterial heritage. The presence of ribosomes in large numbers suggests a high production of chloroplast proteins to support the chloroplasts metabolic activity.

VII. Genophore

The chloroplasts contain DNA that is localized among the thylakoids, referred to as genophore (Pueschel 1990). Transmission electron microscopy reveals a relatively electron-translucent area containing irregularly arranged fibrils, as observed in species such as *Kappaphycus alvarezii* (Fig. 16.11).

Most chloroplasts contain this DNA in an area of the chloroplast devoid of 70S ribosomes. The DNA is an evolutionary indicator of the cyanobacterium-like organisms

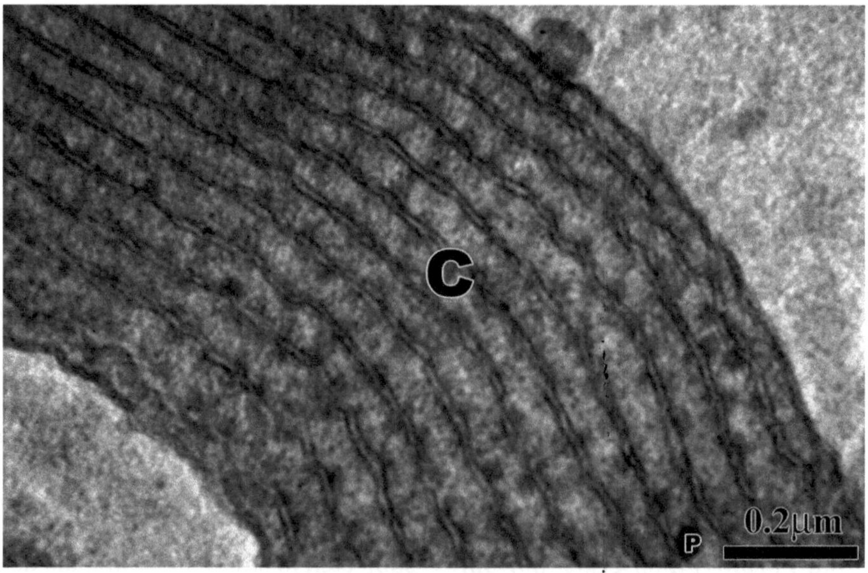

Fig. 16.7. Transmission electron microscopy (TEM) images of *Gracilaria domingensis*. Detail of chloroplast (C) shows the presence of plastoglobulus (P) between thylakoid membranes.

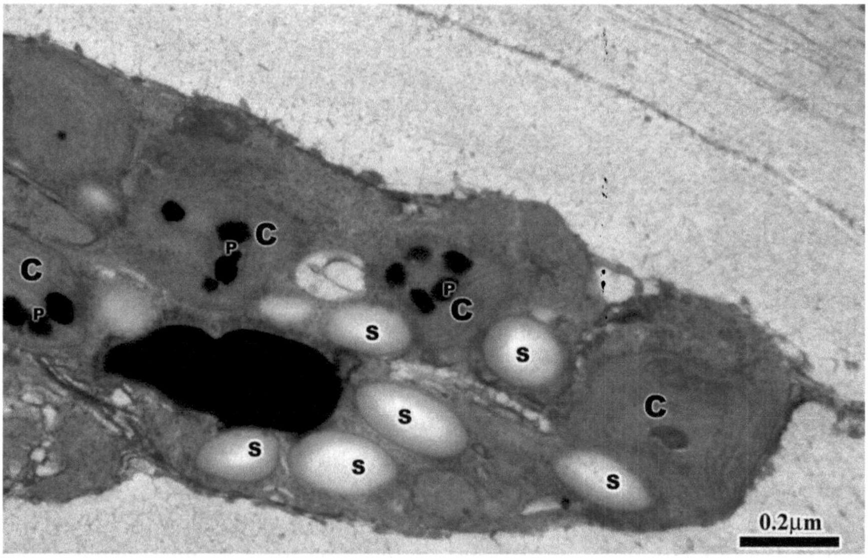

Fig. 16.8. Transmission electron microscopy (TEM) images of *Kappaphycus alvarezii* control (not exposed to UV-B radiation) and treated with ultraviolet radiation-B. Note the disrupted chloroplasts (C) with large quantity of plastoglobuli (P) and some starch grains (S).

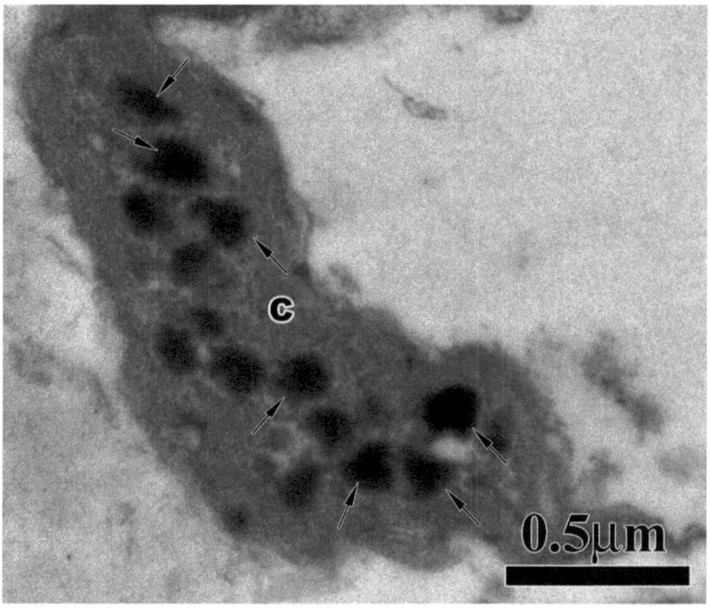

Fig. 16.9. Transmission electron microscopy (TEM) images of *Hypnea musciformis* treated with cadmium. Note the chloroplast with large plastoglobuli volume.

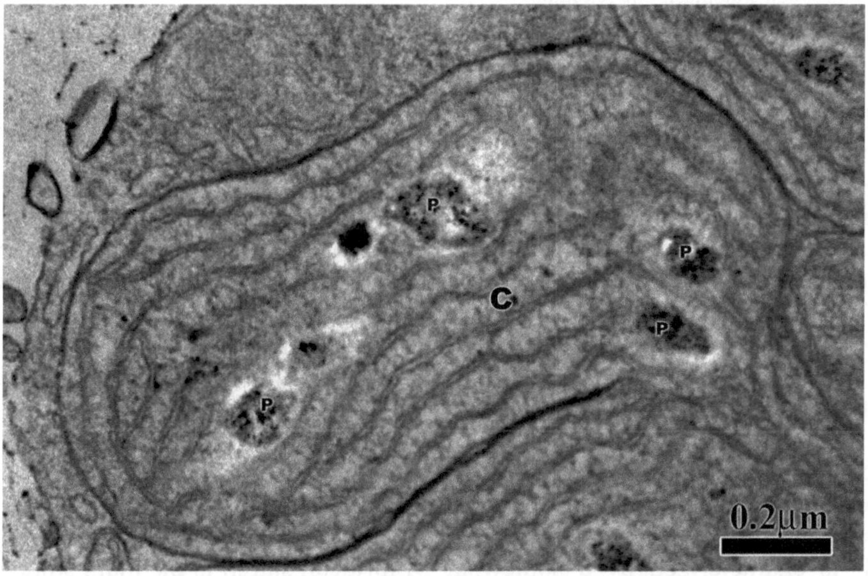

Fig. 16.10. Transmission electron microscopy (TEM) micrographic images of callus of *Kappaphycus alvarezii*. Observe the chloroplast with increase in plastoglobuli volume. Culturing on solid medium can be considered a stress factor that stimulates the production of plastoglobuli, which accumulate metabolites necessary for protection against cellular damage.

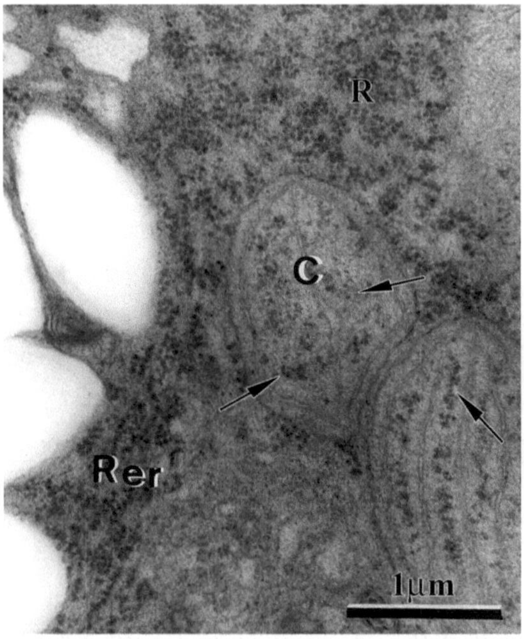

Fig. 16.11. Transmission electron microscopy (TEM) images of *Gracilaria caudata*. Note the plastidial ribosomes (*arrows*), free cytoplasmic ribosomes (R) in the cytoplasm and rough endoplasmatic reticulum (*Rer*).

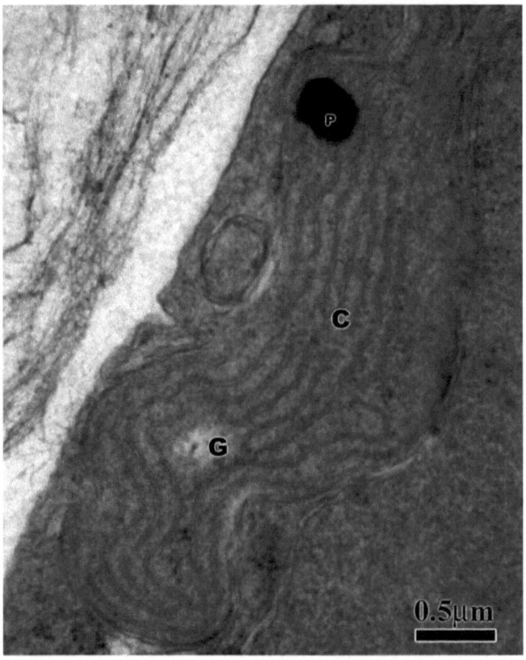

Fig. 16.12. Transmission electron microscopy (TEM) images of *Kappaphycus alvarezii*. Note the presence a fibrillar region corresponding to a genophore (G) in the chloroplast (C), and plastoglobuli (P).

involved in the endosymbiosis event that is the of the chloroplast. The individual DNA microfibrils, which lack basic proteins (histones), are circular and attached to the chloroplast membranes (Lee 2008). DNA is scattered throughout the plastids in *Rhodophyta* (Coleman 1985).

VIII. Pyrenoid

The pyrenoid, which is localized at the center of a rather large chloroplast, is a structure that occurs some primitive red algae (Pueschel 1990). It is characterized by a somewhat undulating membranous structure, probably consisting of one or two thylakoids running through the center of the stroma plastidial. In some red algae, Dodge (1973) found the pyrenoid to be much larger than in previous types and penetrated by a number of chloroplasts. For example, the cells of *Porphyra acanthophora* var. *brasiliensis* (Fig. 16.12) are filled with a single chloroplast with a central pyrenoid, revealing a structure very similar to that of other red algae, having one peripheral thylakoid surrounded by parallel thylakoids. Pyrenoids contain ribulose-1.5-bisphosphate carboxylase/oxygenase (Rubisco) (see Chap. 7), the enzyme that fixes carbon dioxide (Jenks and Gibbs 2000; Nagasato et al. 2003). Consequently, the size of the pyrenoid will vary depending on how much Rubisco is present.

IX. Floridean Starch Granules

Floridean starch is the main storage product of photosynthesis in red algae. Floridean is stored in ‘grains’, which occur in the cytoplasm outside of the chloroplast. The extraplastidic floridean starch synthesis in red algae proceeds via a UDP glucose-selective alpha-glucan synthase, in analogy with the cytosolic pathway of glycogen synthesis in other eukaryotes (Viola et al. 2001). When observed under light microscopy (LM),

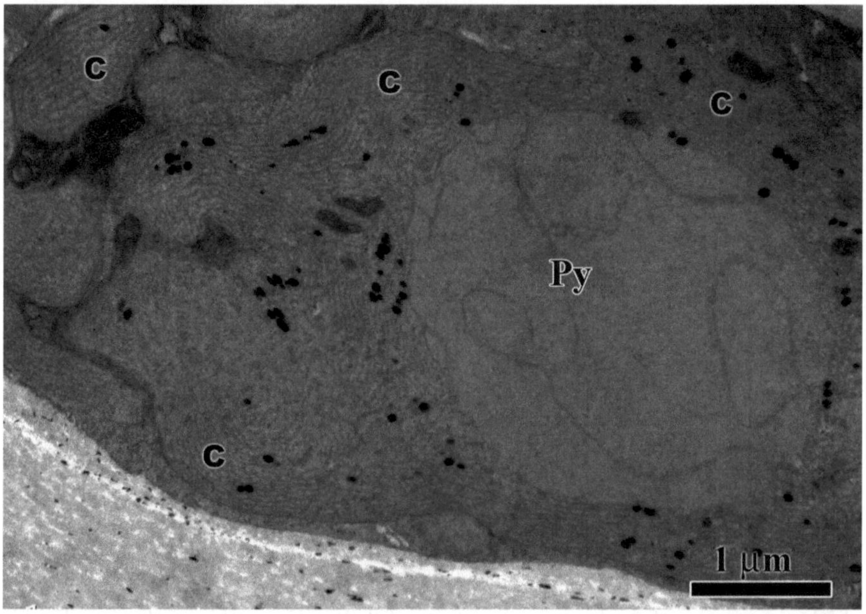

Fig. 16.13. Transmission electron microscopy (TEM) images of *Porphyra acanthophora* var. *brasiliensis*. Note the single chloroplast marked by (C) with central pyrenoid (Py).

floridean starch grains assayed with Periodic Acid-Schiff (PAS), result in a pink structures (Fig. 16.13). This assay is used to identify neutral polysaccharides, such as floridean starch grains, because it requires the presence of 1.2-glycol groups that are oxidized to aldehydes by periodic acid (Trick and Pueschel 1990). The Thiéry test (1967) is also based on the PAS reaction involving the 1.2-glycol groups for images acquired in TEM (Figs. 16.14 and 16.15). During the Thiéry reaction, the thiosemicarbazide reacts with these groups, and their radical *thio* is revealed by the TEM deposits of silver proteinate (Tripodi and De Masi 1975).

X. Perspective

Aspects concerning the export of fixed carbon from the chloroplast and its utilization and storage in the cell cytoplasm remain enigmatic (Linka et al. 2008). In contrast to higher plants, floridean starch is the main carbon storage product in red algae. Also different from the starch granules in plants that are located in the chloroplast, red algal floridean is produced in the cell cytosol, using UDP-Glc as precursor (Patron and Keeling 2005). In photosynthetically active plastids, triose-phosphate is exported via the triose-phosphate translocator (TPT) to sustain soluble and insoluble carbohydrate synthesis (floridoside and floridean starch grains) in the cytosol. Floridoside (2-*O*-glycerol-α-D-galactopyranoside), first isolated from a *Rhodophyceae* (*Rhodymenia palmata)* in 1930 by Colin and Guéguen, is a natural glycerol galactoside found in red algae. Floridoside is believed to play a crucial role in controlling intracellular osmolarity and as a precursor for cell wall biogenesis. These possible functions remain to be evaluated in detail.

Acknowledgments

The authors would like to acknowledge the staff of the Central Laboratory of Electron Microscopy (LCME), Federal University of Santa Catarina, Florianopolis, Santa Catarina,

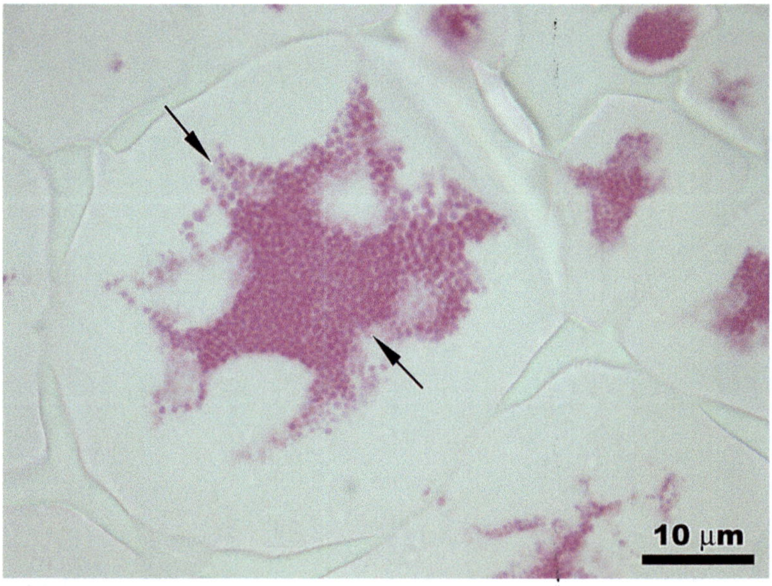

Fig. 16.14. Light microscopy of the transversal sections of thallus *Kappaphycus alvarezii.* Floridean positive starch grains (some indicated by *arrows*) were stained pink by using the Periodic Acid-Schiff assay.

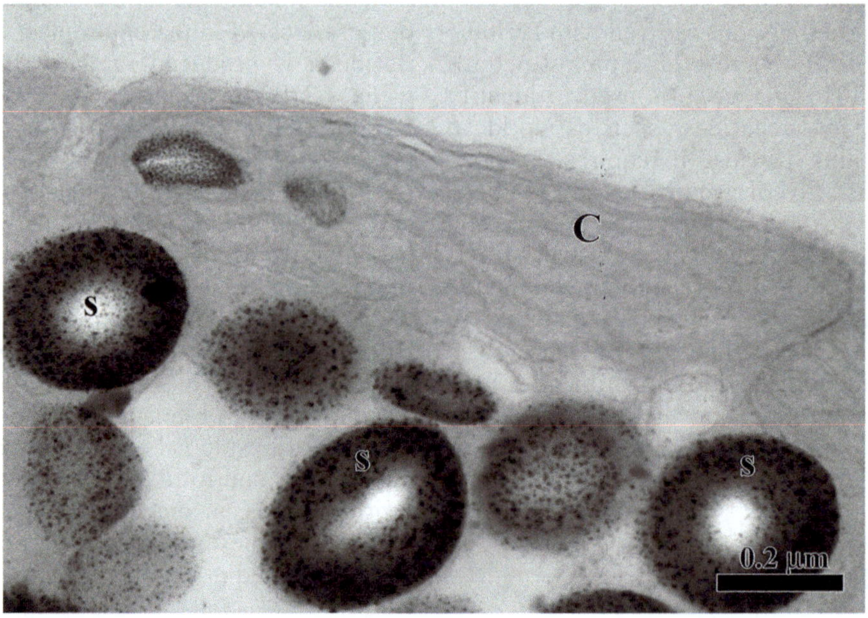

Fig. 16.15. Transmission electron microscopy (TEM) micrographic images of *Kappaphycus alvarezii.* Detail of a cortical cell subjected to Thiéry assay that is used to identify starch grains (*S*). These grains occur in the cytoplasm outside of the chloroplast.

Brazil, for the use of their transmission electron and confocal microscopes. We appreciate the efforts of everyone contributing to the book.

References

Bouzon ZL (2006) Histoquímica e ultra-estrutura da ontogênese dos tetrasporângios de *Hypnea musciformis* (Wulfen) J.V. Lamouroux (Gigartinales, Rhodophyta). Rev Bras Bot 29:229–238

Bouzon ZL, Ferreira EC, Santos R, Scherner F, Horta PA, Maraschin M, Schmidt EC (2012) Influences of cadmium on fine structure and metabolism of *Hypnea musciformis* (Rhodophyta, Gigartinales) cultivated in vitro. Protoplasma 249:637–650

Bréhélin C, Kessler F, Wijk KJ (2007) Plastoglobules: versatile lipoprotein particles in plastids. Trends Plant Sci 12:260–266

Cole KM, Sheath RG (1990) Biology of the red algae. Front cover. Cambridge University Press, Cambridge, Science, 517 pp

Coleman AW (1985) Diversity of plastid DNA configuration among classes of eukaryote algae. J Phycol 21:1–16

Colin H, Guéguen E (1930) The constitution of the principle sugar of *Rhodymenia palmata*. Seances Acad Sci 191:163–164

Delivopoulos SG (2003) Ultrastructure of auxiliary and gonimoblast cells during carposporophyte development in the red alga *Cryptopleura ruprechtiana* (Delesseriaceae, Ceramiales, Rhodophyta). Biol Cell 95:383–392

Dodge JD (1973) The fine structure of algal cells. Academic Press, London, p 261

Gantt E, Contii SF (1965) The ultrastructure of *Porphyridium cruentum*. J Cell Biol 6:365–381

Gantt E, Lipschultz CA (1974) Phycobilisomes of *Porphyridium cruentum*: pigment analysis. Biochemistry 13:2960–2966

Gibbs SP (1962) The ultrastructure of the chloroplasts of algae. J Ultrastruct Res 7:418–435

Holzinger A, Roleda MY, Lütz C (2009) The vegetative arctic freshwater green alga *Zygnema* is insensitive to experimental UV exposure. Micron 40:831–838

Jenks A, Gibbs SP (2000) Immunolocalization and distribution of Form II RUBISCO in the pyrenoid and chloroplast stroma of *Amphidinium carterae* and Form I RUBISCO in the symbiont-derived plastids of *Peridinium foliaceum* (Dinophyceae). J Phycol 36:127–138

Lee RE (2008) Phycology, 4th edn. University Press, Cambridge

Linka ML, Jamai A, Weber PM (2008) Functional characterization of the plastidic phosphate translocator gene family from the thermo-acidophilic red alga *Galdieria sulphuraria* reveals specific adaptations of primary carbon partitioning in green plants and red algae. Plant Physiol 148:1487–1496

Miyagishima S (2011) Mechanism of plastid division: from a bacterium to an organelle. Plant Physiol 155:1533–1544

Nagasato C, Yoshikawa S, Yamashita M, Kawai H, Motomura T (2003) Pyrenoid formation associated with the cell cycle in the brown alga, *Scytosiphon lomentaria* (Scytosiphonales, Phaeophyceae). J Phycol 39:1172–1180

Patron NJ, Keeling PJ (2005) Common evolutionary origin of starch biosynthetic enzymes in green and red algae. J Phycol 41:1131–1141

Pueschel CM (1988) Cell sloughing and chloroplast inclusions in *Hildenbrandia rubra* (Rhodophyta, Hildenbrandiales). Eur J Phycol 23:17–23

Pueschel CM (1990) Cell structure. In: Cole KM, Sheath RG (eds) Biology of the red algae. Cambridge University Press, Cambridge, pp 7–42

Schmidt EC, Scariot LA, Rover T, Bouzon ZL (2009) Changes in ultrastructure and histochemistry of two red macroalgae strains of *Kappaphycus alvarezii* (Rhodophyta, Gigartinales), as a consequence of ultraviolet B radiation exposure. Micron 40:860–869

Schmidt EC, Maraschin M, Bouzon ZL (2010a) Effects of UVB radiation on the carragenophyte *Kappaphycus alvarezii* (Rhodophyta, Gigartinales): changes in ultrastructure, growth, and photosynthetic pigments. Hydrobiologia 649:171–182

Schmidt EC, Nunes BG, Maraschin M, Bouzon ZL (2010b) Effect of ultraviolet-B radiation on growth, photosynthetic pigments, and cell biology of *Kappaphycus alvarezii* (Rhodophyta, Gigartinales) macroalgae brown strain. Photosynthetica 48:161–172

Schmidt EC, Santos R, Horta PA, Maraschin M, Bouzon ZL (2010c) Effects of UVB radiation on the agarophyte *Gracilaria domingensis* (Rhodophyta, Gracilariales): changes in cell organization, growth and photosynthetic performance. Micron 41:919–930

Schubert N, García-Mendoza E, Pacheco-Ruiz I (2006) Carotenoid composition of marine red algae. J Phycol 42:1208–1216

Su HN, Xie BB, Zhang XY, Zhou BC, Zhang YZ (2010) The supramolecular architecture, function, and regulation of thylakoid membranes in red algae: an overview. Photosynth Res 106:73–87

Thiéry JP (1967) Mise en évidence des polysaccharides sur coupes fines en microscopy électronique. J Microsc 6:987–1018
Trick HN, Pueschel CM (1990) Cytochemistry of pit plugs in *Bossiella californica* (Corallinales, Rhodophyta). Phycologia 29:403–409
Tripodi G, De Masi F (1975) Cytological localization of polysaccharidic molecules in some red algae. J Submicrosc Cytol 7:197–209
Vergara JJ, Niell FX (1993) Effects of nitrate avail ability and irradiance on internal nitrogen constituents in *Corallina elongata* (Rhodophyta). J Phycol 29:285–293
Viola R, Nyvall P, Pedersen M (2001) The unique features of starch metabolism in red algae. Proc R Soc Lond B Biol Sci 268:1417–1422
Wetherbee R, Wynne MJ (1973) The fine structure of the nucleus and nuclear associations of developing carposporangia in Polysiphonia novae-angliae (Rhodophyta). J Phycol 9:402–407
Wise RR, Hoober JK (eds) (2007) The structure and function of plastids, vol 23, Advances in photosynthesis and respiration. Springer, The Netherlands
Zilinskas BA, Greenwald LS (1986) Phycobilisome structure and function. Photosynth Res 10:7–35
Zitta CS (2010) Determinação da ploidia de três linhagens de *Kappaphycus alvarezii* (Rhodophyta, Gigartinales) cultivadas em laboratório e análise da ontogênese de calos da linhagem tetrasporofítica marrom. Master dissertation, Federal University of Santa Catarina, Brazil

Chapter 17

Green Algae

Maria Schmidt and Christian Wilhelm*
Department of Plant Physiology, Institute of Biology, University of Leipzig, Johannisallee 23, D-04103 Leipzig, Germany

Summary

Green algae have been treated for a long time as "free living choroplasts" and therefore used as model organisms in photosynthesis research. However, recent progress has provided evidence that they have a paraphyletic origin, resulting in a wide array of different evolutionary lineages. This diversity opens the opportunity to utilise green algae not only for the production of bulk biomass, but also for the extraction of specific biotechnological compounds. This chapter gives an overview of the taxonomic, structural, biochemical, molecular and physiological features of those species which are the most widely used in algal biomass technologies. Based on this description, we suggest how green algal biodiversity and metabolic pathways can be exploited in the future for biological energy generation.

*Author for correspondence, e-mail: cwilhelm@rz.uni-leipzig.de

M.F. Hohmann-Marriott (ed.), The Structural Basis of Biological Energy Generation, Advances in Photosynthesis and Respiration 39, pp. 309–333, DOI 10.1007/978-94-017-8742-0_17,

I. Introduction

First attempts to establish pure algal cultures were made by Pringsheim in 1910 in Halle (Germany). The first published list of cultivated algae was dominated by cyanobacteria and green algae. Especially the genera *Chlorella* and *Scenedesmus* had been used in the early 60s and 70s for algal mass cultures, mainly with the focus on producing proteins for human and animal (fish) nutrition (Stanley and Jones 1976; Söder 1976). Oswald and Golucke (1968) presented in 1960, for the first time, the idea of using green algal mass cultivation to convert solar energy to electric power by anaerobic fermentation of the algal biomass produced with the help of waste water (Li et al. 2011). However, after this first boom, Goldman (1979) summarized the future of outdoor mass cultures to be "limited to small specific applications and not to massive scale projects". The list of green algal species used in biotechnological applications both at the industrial and lab scales is limited to about 20 genera, although many authors have claimed that the huge biodiversity within the green algae waits to be exploited. This list includes the following main genera: *Chlorella*, *Chlamydomonas*, *Scenedesmus*, *Chlorococcum*, *Haematococcus*, *Monoraphidium*, *Desmodesmus*, *Dunaliella*, *Tetraselmis*, *Botryococcus*, *Ostreococcus*, *Muriellopsis*, *Neochloris*, *Parietochloris and Oocystis*. It must be stated that, in the last 10 years, based on molecular phylogenetic analyses green algal systematics have been drastically revised (see below), with the consequence that some species of the same genus have obtained a new position, not only in a different genus or family, but even in a different phylogenetic lineage. The numbers of different species used in biotechnological applications is surprisingly limited, because organisms have to possess key features needed for mass cultivation. First, the cells will be grown in high cell densities and intensive mixing is needed to expose the cells homogenously to the incident light. Both for mixing and for harvest, the cells should be resistant against mechanical stress. Second, the cells should be able to grow quickly with a completely inorganic nutrient supply, otherwise contamination with heterotrophic bacteria, fungi or even eukaryotic protists would disturb the culture growth. Third, species should be single-celled or consist of only a few numbers of cell colonies. Suspensions with cells forming large aggregates are less efficient in light absorption (see below) and need a high mixing velocity to prevent sedimentation. Fourth, cells having a strong potential to adhere to surfaces are also difficult to use in any form of photobioreactor, because this fouling not only reduces light transparency, but the accumulation of dead organic matter is also a source of contamination. Nevertheless, all of these features can be found in all major lineages of green algae. In addition to these more technical cell characteristics, the biochemical composition and special physiological potentials are key parameters for the strain selection in biotechnology. Although Palmucci and Giordano (2011) showed that green algae are much more uniform with respect to their biochemical composition of the cells than Ochrophyta, green algae are important sources for many high valuable products, like carotenoids, lipids, starch and vitamins. Recent work has shown that green algae beside *Chlamydomonas* can be successfully used for genetic transformation (Dawson et al. 1997; Cha et al. 2012), opening the possibility to use green algae as cell factories for pharmaceuticals, drugs or fine chemicals.

However, metabolic engineering is dependent on detailed knowledge of the genes encoding for the enzymes involved in the biochemical pathways of concern. Recent progress in deep phylogeny of green algae has provided surprising evidence that green

Abbreviations: AX – Astaxanthin; CAN – Cantaxanthin; CAR – β-carotene; Chl – Chlorophyll; ITS – Internal transcribe spacer (commonly referring to the ribosomal rRNA operon); LUT – Lutein; Mya – Million years ago; ROS – Reactive oxygen species; VDE – Violaxanthin de-epoxidase; ZEA – Zeaxanthin algae are genetically complex and much more diverse than the "free living chloroplast" of higher plants. Due to their paraphyletic origin, which diverged in different clades very early in evolution (Leliaert et al. 2011), many green algae use genes in their biochemical pathways that are not related to their homologues found in higher plants (Frommolt et al. 2008). Therefore, a detailed understanding of green algal phylogeny is essential to understand the biochemical and physiological differences in the major green algal lineages.

II. Phylogeny of Green Algae

A. Polyphyletic Nature of Chlorophyll b*-Containing Algae*

Algae are a diverse group of unicellular and multicellular organisms, a group which is not based on phylogeny but based only on the definition of similar morphology. Van den Hoek et al. (1996) described them simply as "diverse photosynthetic plants that have neither roots nor leafy shoots" and "lack vascular tissue". Terms such as green algae, red algae (see Chap. 16), brown algae, blue-green algae (see Chap. 14), yellow-green algae etc. illustrate that pigmentation has traditionally been used to group these organism. Since it has become clear that these groups often include non-pigmented organisms and chloroplasts can be transmitted horizontally by endocytobiosis, it has become common to refer to the green, red and glaucocystophyte lineage when it comes to eukaryotic algae. The green lineage can be traced back to a hypothetical ancestral green flagellate dated 750–150 Mya (Leliaert et al. 2011). Since chlorophyll *b* (Chl *b*) was most likely invented only once (Grimm et al. 2006), the green lineage includes all eukaryotic organisms that contain Chl *b*, i.e. the green algae and higher plants (with primary plastids) and all organisms derived from them via secondary or tertiary endocytobiotic events (with complex plastids, surrounded by three or four plastid envelope membranes) (Leliaert et al. 2012, see Fig. 17.1).

The primary plastid bearing organisms with *chlorophyll a* (Chl *a*) and Chl *b* ("Chloroplastida", Adl et al. 2005) are united by additional general characteristics such as grouped thylakoids, characteristic accessory pigments of the α-carotene and β-carotene pathway, pyrenoids (see Chap. 7) that are usually located in the plastid, starch as the main reserve polysaccharide, typically a cellulose cell wall, and motile cell stages with isokont flagellae (van den Hoek et al. 1996). Traditionally, species and higher taxa of green algae were based on available morphological features at the time, especially thallus organization (van den Hoek et al. 1996). The complex and multivariate chloroplast morphology, as well as plastid position and number within a cell, were highly informative features as well (Ettl 1980). The first approaches of algal taxonomy were greatly limited by the resolution of light microscopy. Even though the structure of plant cells was first described in 1665 by Robert Hooke, it took almost another 300 years before the invention of the electron microscope in 1931, which facilitated the incorporation of structures in the nanometer range into the morphological description and thereby aiding in differentiating unicellular species. Up until then, and even afterwards, phylogenetic approaches were mostly based on similarities in cytological morphology. Especially for small green algae, simple morphologies were not character-rich and led to the impression that many species have a cosmopolitan distribution (Fawley et al. 2004). In the case of the genus *Chlorella*, the additional inclusion of physiological/biochemical properties such as hydrogenase activity, secondary carotenoid production, nitrate reduction, vitamin requirements, and extreme pH, salt and temperature tolerance ultimately led to the assignment of 166 strains (based on faint morphological characteristics) to only 17 species (Kessler 1992). In his work on the physiological and biochemical contributions to taxonomy of the genus *Chlorella*, Kessler also defined other properties such as the use

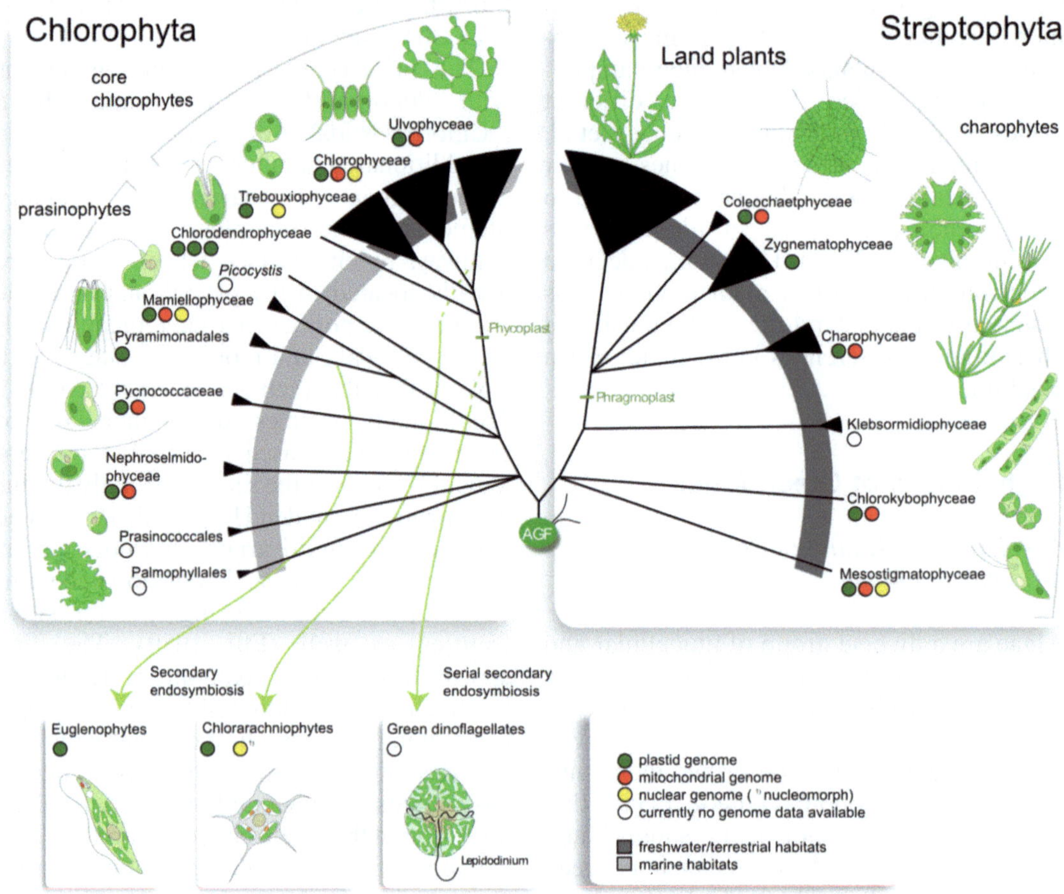

Fig. 17.1. Phylogentic relationships of mayor clades of the green lineage, including primary (Chlorophyta and Streptophyta) and complex (chlorarachniophytes, euglenophytes, green dinoflagellates) plastid-bearing taxa. Width of terminal nodes corresponds to the known species diversity, not genetic diversity within a taxon. *Shaded areas* indicate dominant habitats (*dark grey* – freshwater/terrestrial, *light grey* – marine) of higher taxa. *AGF* Ancestral green flagellate. Information on available genomic data for green algae summarized from http://www.ncbi.nlm.nih.gov/genome (Modified from Leliaert et al. 2012).

of sugars for heterotrophic growth in the dark and organic nitrogen sources (Kessler and Czygan 1970) that were less reliable for predicting species affiliations. Also, these observations were not useful for the determination of evolutionary relatedness, especially at higher levels, and were also not transferable to other problematic taxa.

Consequently, after the introduction of molecular phylogenies based on sequence comparison of marker genes like *rbc*L, 18S rDNA, 28S rDNA and ITS, it became clear that systematics had been greatly mislead by convergent evolution towards a reduced morpholgy, e.g. for the genus *Chlorella* this led to the discovery that some species even belong to a different algae class (Huss et al. 1999; Lewis and McCourt 2004). According to the most recent molecular phylogentic analyses integrating composite data from multiple studies and divergence estimates based on both, fossil records and molecular clock estimates, the green lineage can be split into two main branches, the Chlorophyta and Streptophyta (Leliaert et al. 2011; and therein), both of which contain 'green algae'.

A third group of marine deep-water, palmelloid algae, the Palmophyllales, are recognized as an ancient diverging lineage with yet to be determined closest affiliation (Zechman et al. 2010).

As clearly shown in Fig. 17.1, charophytes represent a polyphyletic taxon of unicellular and filamentous green algae, uniting the diverse algae that share a common origin with the land plants and are usually found in freshwater and terrestrial habitats. These algae display the highest diversity among the unicellular or filamentous Zygnematophyceae, with beautifully symmetrical cells (Gontcharov 2008), and the macroscopic Charophyceae, used in studies on their complex and specialized life cycle in relation to higher plants (Niklas and Kutschera 2010). It has also become clear that the predominantly marine prasinophytes are a polyphyletic taxon as well. They are single-celled flagellates covered with organic scales, for example the species-rich taxa Mamiellophyceae, Pyramimonadales and Nephroselmidophyceae, but other morphologies without flagella or scales have been described as well (for characteristics of the major prasinophytic lineages see Leliaert et al. 2011). A wide array of α- and β-carotene-derived carotenoids was described in this group (Egeland et al. 1997). Even though species numbers are relatively low, members of this class can dominate the marine, eukaryotic picoplankton biomass (Vaulot et al. 2008). The core chlorophytes, as the name suggests, are the most diverse "crown group" of the Chlorophyta, with the mostly freshwater-inhabiting unicelluar Trebouxiophyceae and Chlorophyceae, and also the marine Ulvophyceae, with their well-known macroalgae representatives (green seaweeds). Phylogenetic relationships among these three classes are difficult to discern because of their rapid radiation (Falkowski and Knoll 2007).

The complex plastid-bearing organisms of the green lineage are chlorarachniophytes, photosynthetic euglenids and the "green" dinoflagellates and evolved through secondary (or serial secondary) endocytobiosis of three different green algae taxa (Fig. 17.1) by three different host taxa of Rhizaria, dinoflagellates and Excavata, respectively (for reviews of the fate of plastids, see Keeling (2010) and Elias and Archibald (2009)). In total, complex green plastid-bearing organisms comprise no more than about 1,200 species to date (Guiry and Guiry 2012) and for reasons mentioned in the introduction, they are not suitable for use in biotechnological applications. They are, however, valuable model organisms for the advancement of evolutionary concepts of photosynthetic eukaryotes, such as the analysis of complex protein and metabolite targeting through plastid membranes and gene transfer from organelles to nucleus. A comparison of these organism groups reveals similar evolutionary trends and key inventions in complex plastid-bearing eukaryotes of the red and green lineage (e.g. nuclear genome reduction, Archibald and Lane 2009; plastid protein import, Agrawal and Striepen 2010). Further insights into phylogenetic relationships of these taxa can be found in Kim et al. (2010), Gile et al. (2010) and Matsumoto et al. (2011).

Even though the green algae are by definition paraphyletic (including all descendants of the ancestral green flagellate, except for higher plants, Fig. 17.1), there is also an underlying polyphyletic nature (multiple origin) in Chl *b*-containing algae. One is the obvious transfer of green plastids into other eukaryotic hosts (see complex plastids). A second polyphyly is discovered on a smaller scale by looking at taxa with special adaptations like cryophilic green algae. They are by no means of one descent (monophyletic) and temperate strains of different taxa can be found as close relatives to arctic and antarctic strains (De Wever et al. 2009). Where historical descriptions and definitions no longer conform to molecular data available nowadays, major revisions of green algal taxonomy are under way (Bock et al. 2011; Nakada et al. 2008; Pröschold et al. 2001). It remains to be seen if, by the end of these reclassifications, a better understanding of the phylogeographic distribution of green algal ecotypes is formed.

Overall, the identification and phylogeny of species, inferred from morphology or molecular characters, is important for unambiguous scientific communication and the ability to compare traits of interest.

B. Consequences for Biosynthetic Pathways

It is tempting to deduce biochemical and physiological properties from the molecular phylogentic relationships shown above. However, our knowledge remains fragmented, with only a limited number of fully investigated strains or species. It is important to keep in mind that different lineages of green algae have adapted to diverse habitats worldwide. Aerophytic algae have developed distinct mechanisms like thick cell walls, mucilage and osmotic metabolites to cope with drought and high salinity, and can be found among Klebsormidiophyceae (Mikhailyuk et al. 2008), Chlorophyceae (*Chlorococcum*, Klochkova et al. 2006) and Trebouxiophyceae (Darienko et al. 2010) species. Psycrophilic green algae (snow algae, ice algae), frequently exposed to extreme high light under low temperature and low nutrient conditions, have similarly developed coping strategies involving the accumulation of different pigments in extraplastidial vesicles. They are thought to confer photoprotection or advantages during reactive oxygen species (ROS) removal in the Chlorophyceae (*Chlamydomonas nivalis*, Remias et al. 2005) and Zygnematophyceae (*Mesotaenium berggrenii*, Remias et al. 2012). Freshwater and terrestrial green algae (mostly Chlorophyceae, Trebouxiophyceae and charophytes) in general are often exposed to rapid and drastic changes or extreme conditions of irradiance, temperature, nutrient supply, salinity and water availability and, therefore, have had to cope with one or an array of these factors during speciation. This has led to the conquest of diverse habitats under functionally similar adaptations, but with different genetic backgrounds in the Chlorophyta and Streptophyta, respectively. Currently, species names have to be considered with caution and the best identification is by strain designation that can be unambiguously linked to molecular data.

Two main consequences can be deduced for biosynthetic pathways from observations made above. First, green algae have developed different strategies to solve similar problems (abiotic stressors) via the modification of metabolic pathways and changes in structural genes. Therefore, a multitude of chemical compounds and structures that serve similar functions (e.g. antioxidants in habitats with high irradiance) can be expected to have evolved, so that one has the advantage of choosing the organism, which solves the problem in a way most suitable for biotechnological applications. Second, in related species with similar genetic backgrounds, the basic concepts of physiological properties should be similar as well, even though they might be adapted to differing environmental stressors. This is only true for genes under neutral evolution (Kimura 1968). Genes under positive or negative selective pressure have different nucleotide substitution rates compared to closely related species and can vary greatly. Knowing one species would still lead to a better understanding of the related one, which is the basic concept behind the elaborate research on model organisms. Points of interest regarding model organisms with respect to biosynthetic pathways are metabolite and protein targeting to different compartments (Patron and Waller 2007) and pathway regulation on the genetic and physiological level.

The width of branch tips in Fig. 17.1 corresponds to the species richness found in individual classes of green algae and should not be confused with genetic diversity. It has been shown that the overall genome-wide variability among single-celled algae is greater than can be expected by their marker gene differences used for species delineation (Piganeau et al. 2011), for example in the Mamiellophyceae (Worden et al. 2009).

For biotechnology, an understanding of phylogentic relationships is also essential, with a focus on genetic engineering. Orthologs (members of one gene family in different species) and paralogs (duplicate members of a gene family in one species) of genes can be identified, traced along phylogenetic trees and even introduced into organisms of interest, but only if codon usage information is available, e.g. for *Chlamydomonas* (Genkov et al. 2010). A thorough genetic understanding of the interrelationship of metabolic pathways and their regulatory principles establishes a basis for modifications via gene knockdown, knockout or expression modulation approaches (e.g. via RNA interference, insertional mutagenesis, homologous recombination; see Harris (2008) for a comprehensive summary of these methods in *Chlamydomonas reinhardtii*). It must be noted that even if the genes encoding a given protein have very high similarity between species belonging to a clade of high phylogenetic distance, the gene expression control by the corresponding transcription factors can be completely different. This must be taken into account for metabolic engineering. Figure 17.1 displays information on currently available genomic data of model organisms of the different green algal classes. Figure 17.2 serves as a reference for discussed green algal model organisms and their phylogenetic relationships. Overall, knowledge of phylogenetic relationships is necessary for understanding the relevance of model organisms with regard to metabolic pathway possibilities and regulation.

III. Model Organisms

The selection of an organisms for biotechnological applications is dependent on the technological approach to convert light into chemical products. If bulk biomass is produced for feedstock or for hydrothermal carbonization (Heilmann et al. 2011), the organism of choice should show robust and fast growth rates. In this case species screening should focus on traits like salt, temperature, light and mechanical stress tolerance, which are the most crucial production parameters for outdoor cultures (Torzillo et al. 2003). If the biomass is the source for specific high-valuable products, those species will be chosen which contain: (i) the biochemical component in intrinsic high amount, and (ii) are easy to handle in the refinement process (e.g. extractability). The production of the component of interest can then be improved

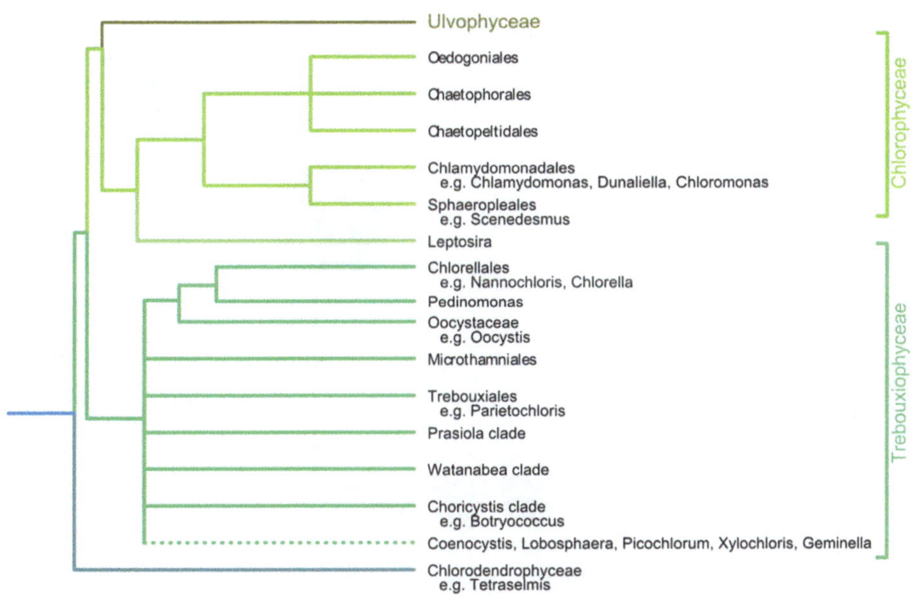

Fig. 17.2. Phylogenetic relationships of core chlorophyte families with examples of genera mentioned in the text (Modified from Leliaert et al. 2012).

by defining the optimal culturing conditions and by metabolic engineering. Therefore, the species should be genetically transformable and the genomes should be sequenced and annotated. Metabolic engineering does not only concern directing metabolic pathways into the product of interest, but also key features for the refinement process to improve the product yield. Therefore, there is no ideal model organism that can be used for all biotechnological applications, but a set of major "players" which are described below.

A. Chlamydomonas

The genus *Chlamydomonas* contains many species with unclear phylogenetic relationships. The physiological features of the different species are very diverse depending on their genetic and ecological origins. Most species are isolated from fresh water habitats, but also psychrophilic snow or ice-acclimated, acidophilic (Langner et al. 2009) or high salt-resistant species (Pocock et al. 2004) are described in the literature. The reader is referred to a thorough revision of the genus by Pröschold et al. (2001). In 1887 Goroschankin described the life cycle of *Chlamydomonas* and recognized the sequence of diploid cells (vegetative zygote, stable diploids and zygospores) and haploid progeny, which were formed during meiosis from the zygospore (see Fig. 17.3). From these tetrads, gametogenesis results in differentiated haploid gametes. The free-living haploid gametes can form pairs and consequently fuse via plasmogamie, karyogamy and plastid fusion to form a diploid zygote. In homothallic forms of *Chlamydomonas* the gametes are morphologically identical and

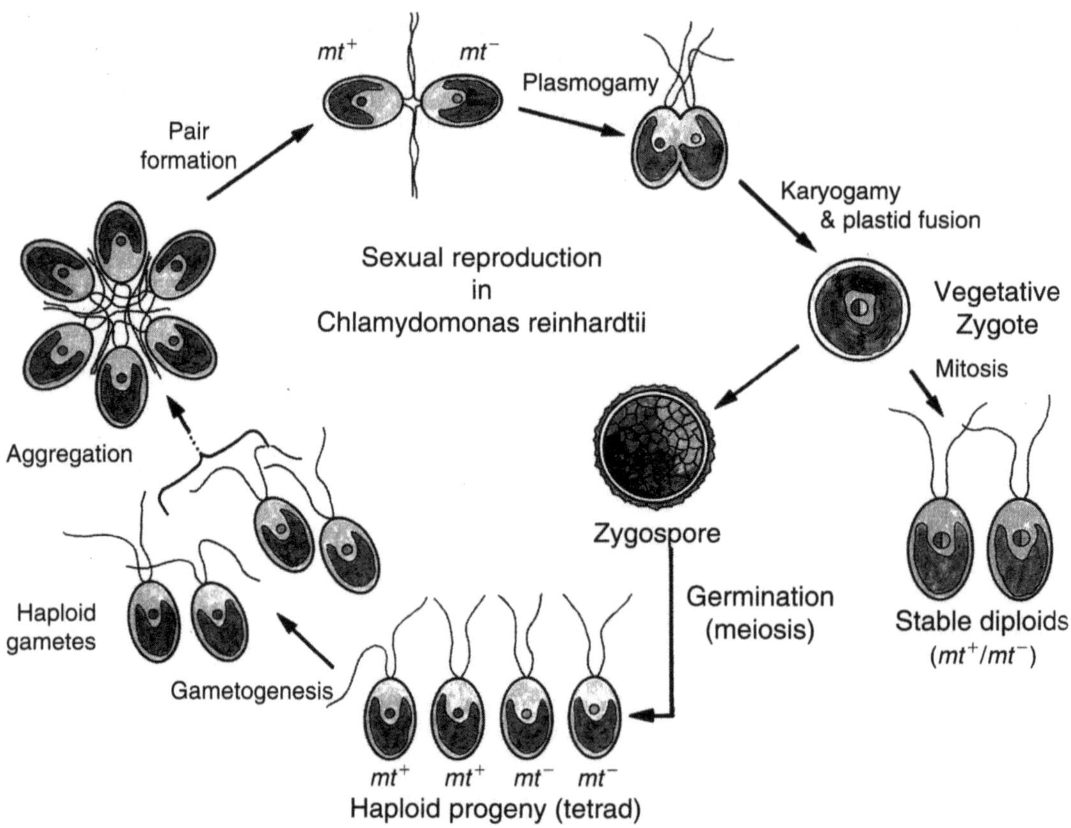

Fig. 17.3. Life cycle of *Chlamydomonas reinhardtii* (Taken from Harris 2008).

are physiologically different termed as "+" and "–" forms. Different stages of the cells during this life cycle can be isolated, because flagellated cells are positively phototactic and can be separated from immotile stages simply by light. In the early 1950s, mutants were described and enabled the analysis of the Mendelian heredity of alleles. Sager (1954) discovered that nitrogen starvation can be used to induce sexual reproduction. Using different antibiotic-resistant mutants, she discovered the non-Mendelian inheritance. Ebersold (1962) presented the first nuclear genetic map for *C. reinhardtii*, which attracted many researchers to study this alga. Therefore, 50 years later *C. reinhardtii* is the best understood eukaryotic photosynthetic organism to date. We recommend that the reader refer to the so-called "*Chlamydomonas* Source Book" edited by E. Harris (2008). Three volumes contain up-to-date information about the cellular structures, metabolic processes and cell behavior, including an extended collection of strains, mutants and experimental recipes. Another important resource for information and material is the "Chlamydomonas Resource Center" which is available on the web at http://www.chlamy.org/.

Chlamydomonas is mainly used in laboratory applications but far less at the industrial scale. To the authors' knowledge, there is no industrial unit in operation using *Chlamydomonas*. In synchronous cultures at 35 °C, with a light intensity of about 400 µE/(m^2 s) Lien and Knutsen (1979) achieved a 4 h doubling time. However, under photobioreactor conditions, Janssen et al. (2000) estimated a maximum specific growth rate of about 0.48 cell divisions per day. Despite these highly stable growth rates, the usage of *Chlamydomonas* in algal farms is limited because it tends to fouling and its biochemical composition is not favorable for most biotechnological applications. The most important bulk products from green algal biotechnology are pigments, lipids and high molecular carbohydrates. *C. nivalis* is known to accumulate ketocarotenoids e.g. astaxanthin (AX) as globules in the cytoplasma (Remias et al. 2005). The chemical composition of the cell wall in *C. reinhardtii* is unusual for green algae, because its protein content is 50 % or more. Instead of the typical cellulosic fibers, the cell wall is mechanically stabilized by hydroxyprolin-rich glycoproteins having some similarities to the extensin of higher plants (Ferris et al. 2001). The ability to adhere to surfaces might be due to the excretion of lytic enzymes which are necessary for the gamete fusion process or due to the release of the sugar components of the fibrous glycoprotein. The prolin-rich protein content is neither favorable for the use of *C. reinhardtii* in aquaculture, nor for biofuel production, where the major carbon pools of the cells should be either starch/cellulose or lipid. Langner et al. (2009) have reported that, under low light conditions, the dry weight of *C reinhardtii* consists of 9 % carbohydrates, 69 % proteins and 17 % lipids.

The genome of *C. reinhardtii* has been published by Merchant et al. (2007), and many protocols for genetic transformation have been published (Kindle 1990), including insertial mutagenesis (Dent et al. 2005) and RNAi (Rohr et al. 2004) . Although transgenic lines can be obtained relatively easily, it is still difficult to express the inserted genes with high frequency (Schroda et al. 2000). Recent progress has been reported by Neupert et al (2009) to induce high transcription levels of the foreign DNA, and by Potvin and Zhang (2010) to improve the efficiency of the production of recombinant proteins in *C. reinhardtii*.

Figure 17.4 shows an electron micrograph of a typical interphase cell of *C. reinhardtii*. The cup-shaped chloroplast accounts for approximately 70 % of the cell volume and contains large starch granules. In some species, starch is stored in a so-called pyrenoid, where the enzymes of the carbon fixation cycle are concentrated. The thylakoids run in parallel stacks of variable numbers of membranes. The number of membranes per band is increased in low light. In contrast to higher plants or mosses, the chloroplast thylakoids are not differentiated into grana membranes

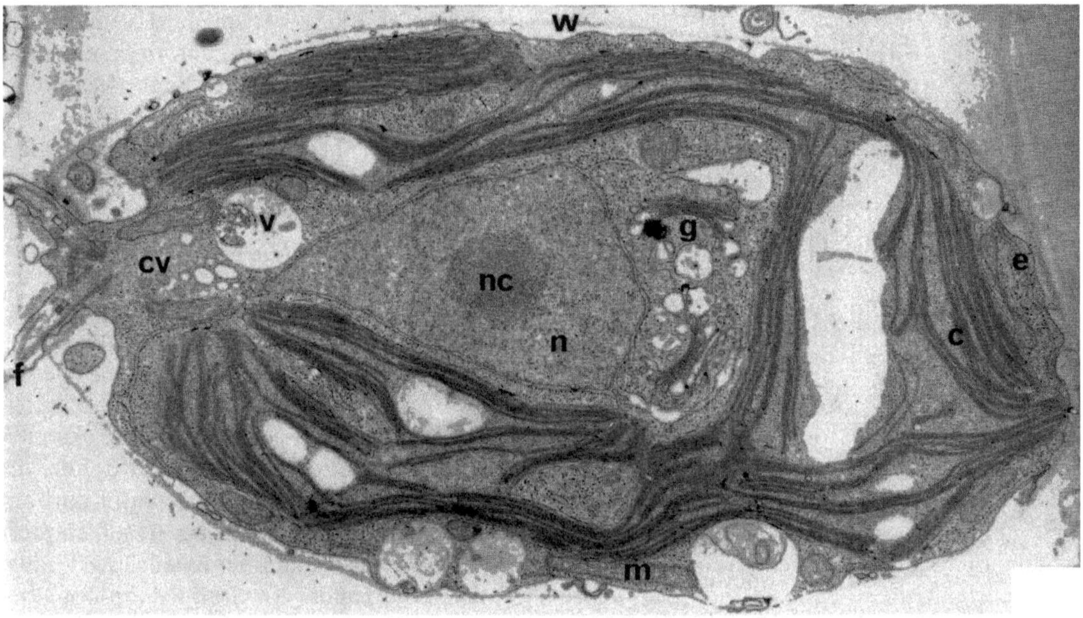

Fig. 17.4. Electron micrograph of an interphase *Chlamydomonas reinhardtii* cell (Taken from Pickett-Heaps 1975).

and stroma membranes. This has led to a less pronounced separation of both photosystems compared to grana-type chloroplasts. The function of grana formation has been discussed as a strategy for a balanced energy distribution of both photosystems (PS) (Trissl and Wilhelm 1993) which can be further optimized by so-called state 1-state 2 transitions where light-harvesting complex (LHC) II units move between both photosystems depending on the redox state of the plastoquinone pool. The weak lateral heterogeneity in photosystem distribution in *Chlamydomonas* thylakoids is compensated by unusual strong state 1-state 2 transitions. Under anaerobic conditions up to 80 % of the absorbed light energy (Forti and Caldiroli 2005) can be redistributed to PSI, whereas in higher plants this ratio is far lower (at most 20 %). The central role of chloroplast kinases for the fine- tuning of the photosynthetic electron transport and the regulation of gene expression in response to changing environmental factors has been recognized for the first time using mutants of *C. reinhardtii* defective in STN 7 and SNT 8 kinases (Reiland et al. 2011). The occurrence of a respiratory electron pathway in the chloroplast, the so-called "chlororespiration" was also detected for the first time in *C. reinhardtii* by Bennoun (1982). Since that time, many alternative electron pathways in the chloroplast of *C. reinhardtii* have been identified, which facilitate the adjustment of the NADPH/ATP ratio to changing conditions of demand. As a freshwater organism, *C. reinhardtii* is exposed to high light under conditions of nutrient limitation, especially under carbon limitation when the pH in the medium rises. Therefore, C. reinhardtii has developed very efficient carbon concentrating mechanisms (CCM) that consume ATP but not reductants (Giordano et al. 2005). Under these conditions, the demand for ATP is increased. Additional ATP is generated by increasing the translocation of protons into the thylakoid lumen trough alternative electron pathways. The most prominent contribution to this extra ATP is made by the photoreduction of oxygen at PSI by the water-water cycle (Asada 2000). In summary, although *C. reinhardtii* is the best understood green alga, it possesses many special genetic, biochemical and physiological features, which may limit its role as a typical model organism.

B. Chlorella

The genus *Chlorella* is characterized as a solitary coccoid green alga with a spherical cell wall that does not possess a mucilagineous surface and has an obligatory asexual reproduction by autospores. Tamiya (1966) showed that autospores and autospore mother cells can be easily separated by selective sedimentation, and applying an appropriate light–dark cycle synchronizes the cell divisions in such a way that at any moment during culture all cells are in identical developmental stages. Wilhelm and Wild (1984) have shown how photosynthetic and respiratory activity change drastically during the cell cycle. Therefore, *Chlorella* cultures with different developmental stages can differ with respect to their physiological activity, not only due to external factors, but also by developmental composition.

Figure 17.5 shows an electron micrograph of an autospore, which shows a cell wall with a thin trilaminar outer wall that is highly resistant to chemical degradation. The polymers that make up the outer cell wall have been characterized by Derenne et al. (1992). Their study showed that these polymers resist a 30 % KOH/phosphoric acid treatment and are preserved during fossilization. Therefore, these polymers can provide information about the biological origin of the sedimentary carbon, providing an important marker for the detection of new fossil oil sources. These polymers provide also high resistance against mechanical stress. Recently a protocol for genetic transformation has been presented (Coll 2006).

Chlorella has the ability to grow on completely inorganic media with high growth rates (Tamiya 1966), making *Chlorella* a suitable platform for producing bulk biomass in algal biotechnology. The number of different *Chlorella* species is not known, because the taxon is still under revision. In the Algaebase 24 species are listed, with most of these species isolated from fresh water or soils, and also several marine species (e.g. *C. marina*). *C. salina* has optimum growth rates under sea water conditions of 3 % salinity and pH 9. Therefore *C. salina* and *C. marina* are cultivated for

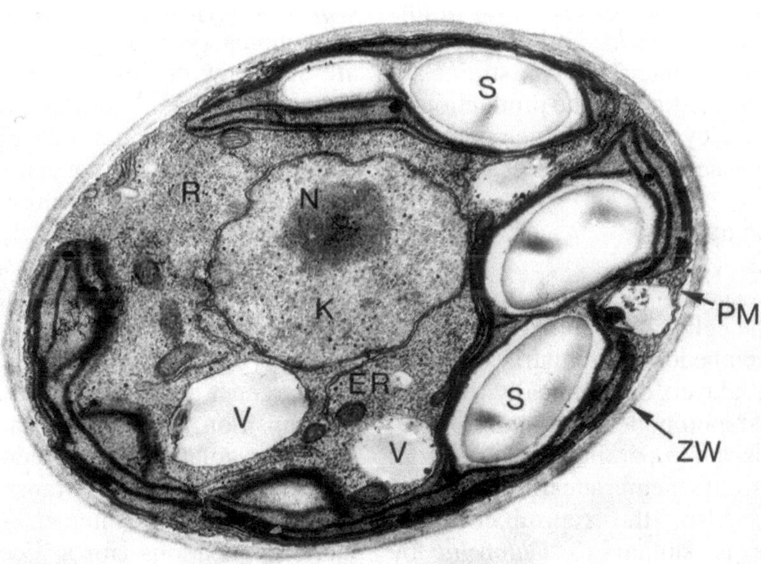

Fig. 17.5. Electron micrograph of *Chlorella fusca* grown under white light conditions (Taken from Wilhelm et al. 1985).

feeding marine zooplankton in fish aquaculture (Jayasankar and Ramamoorthy 1993). In the world's largest microalgae production plant operating as a closed photobioreactor facility, 40 t per year of *Chlorella vulgaris* are produced in glass tubes with a combined length of 500 km for the food additive "Algomed" (for details see the website: www.algomed.de). The productivity of *Chlorella* cultures can be improved by the addition of organic carbon (sugars, organic acids). Interestingly, mixotrophy does not strongly impose a change in the biochemical composition of the cells. Therefore mixotrophic *Chlorella* cultivation is under investigation for the production of lipids, using waste water as a source of organic carbon. A second application of *Chlorella* in waste water treatment uses its potential to remove pollutants. Here, immobilized *Chlorella* biofilms appear as a good solution (de Bashan and Bashan 2010). Some *Chlorella* strains are capable of producing lipids equivalent to 50 % of their dry weight and are used for bio-diesel production in photobioreactors (Feng et al. 2011). Sialve et al. (2009) have suggested anaerobic digestion to convert algal biomass into hydrogen or methane as a promising pathway for second generation biofuels. Lakaniemi et al. (2011) have tested the efficiency of this process using *Chlorella* and *Dunaliella* cultures. Finally, some *Chlorella* strains are also under investigation for the production of hydrogen using cycles of oxygenic photosynthesis and anaerobiosis (Song et al. 2011).

C. *Scenedesmus*

Scendesmus shares many features with *Chlorella*. The cell wall is composed of cellulose fibers embedded in a matrix of hemicelluloses with layers of chemically resistant material called sporopollenin. As in *Chlorella*, the thylakoids run in parallel stacks without clear grana/stroma membrane differentiation (Fig. 17.6). Also, the reproduction of *Scenedesmus* is similar to *Chlorella* by means of autospores, forming cells, which are resistant against damage by turbulence, one of the major reasons for its use in microalgal biotechnology. Finally, the macromolecular composition of *Scenedesmus* resembles that of *Chlorella*. The major morphological difference is that *Scenedesmus* forms aggregates of two up to 16 cells (see scanning electron micrograph in Fig. 17.6). Interestingly, the formation of aggregates is stimulated in the presence of grazers (Lurling 2001), pointing towards a potential function of aggregation in defence of predators. Taxonomic revision in the last few years has revealed that some species formerly classified as *Chlorella* belong to the genus *Scenedesmus* (e.g. *Chlorella fusca*). Some *Scenedesmus* species e.g. *S. almeriensis* can produce large amounts of lutein and zeaxanthin (Granado-Lorencio et al. 2009). Another biotechnological application of *S. almeriensis* is the detoxification of constituents in waste water (Bayramoglu and Arica 2011).

D. *Haematococcus*

Light microscopy reveals the four major stages of the *Haematococcus* cell cycle (Kobayashi 2003) (Fig. 17.7). Biflagellated motile cells perform photosynthesis (Fig. 17.7a); green palmelloid cells occur under replete nutrient conditions (Fig. 17.7b); palmelloids occur under high light and nitrogen depletion showing a slight accumulation of orange carotenoids (Fig. 17.7c); red aplanospores are shown with the corresponding absorption spectrum (Collins et al. 2011) (Fig. 17.7d). The latter accumulate the keto-carotenoid AX up to concentrations of 4 % of cell dry weight. AX is stored in lipid vesicles outside the chloroplast and is assumed to help the cell to resist reactive oxygen species produced under stress conditions and to have a function as an energy reservoir aiding survival (Lemoine and Schoefs 2010). Steinbrenner and Linden (2013) showed that the transition from the "green" to the "red" state is regulated by the redox state of the photosynthetic electron transport chain. AX is of high economic interest, because it is has many applications (for a recent review see Guedes et al. (2011). Traditionally, AX is used as a food colorant (e.g. in fish cultivation), and as a food additive due to its beneficial

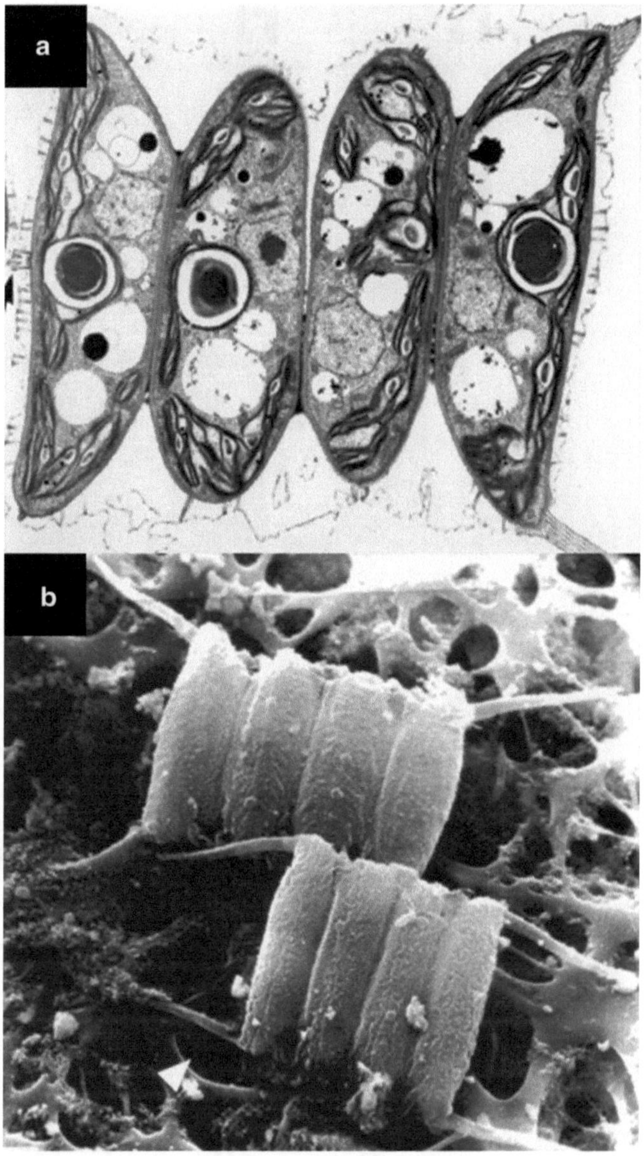

Fig. 17.6. Transmission electron micrograph (*upper part*) and scanning electron micrograph (*lower part*) of *Scenedesmus acutus* (Taken from Pickett-Heaps 1975).

effects as an antioxidant (Lorenz and Cysewski 2000). Recently, AX was shown as a potential therapeutic agent in cardiovascular diseases (Fassett and Coombes 2011). The physiological regulation of AX biosynthesis is still not understood. Since AX is accumulated only under stress conditions, when the cells do not grow any more, the efficiency of AX production is relatively low. Therefore, a better understanding of the process would lead to major improvement in biotechnological productivity. It is being speculated that β-carotene, which is produced via the mevalonic acid pathway (Bouvier et al. 2005) in the chloroplast, is transported across the envelope membranes, and then converted to AX and accumulates in vesicles (Grünewald et al. 2001). However, recently Collins et al. (2011) provided some evidence that, in the lipid vesicles, β-carotene and AX are co-localized and the late biosynthetic steps occur at the vesicle-cytoplasma interface.

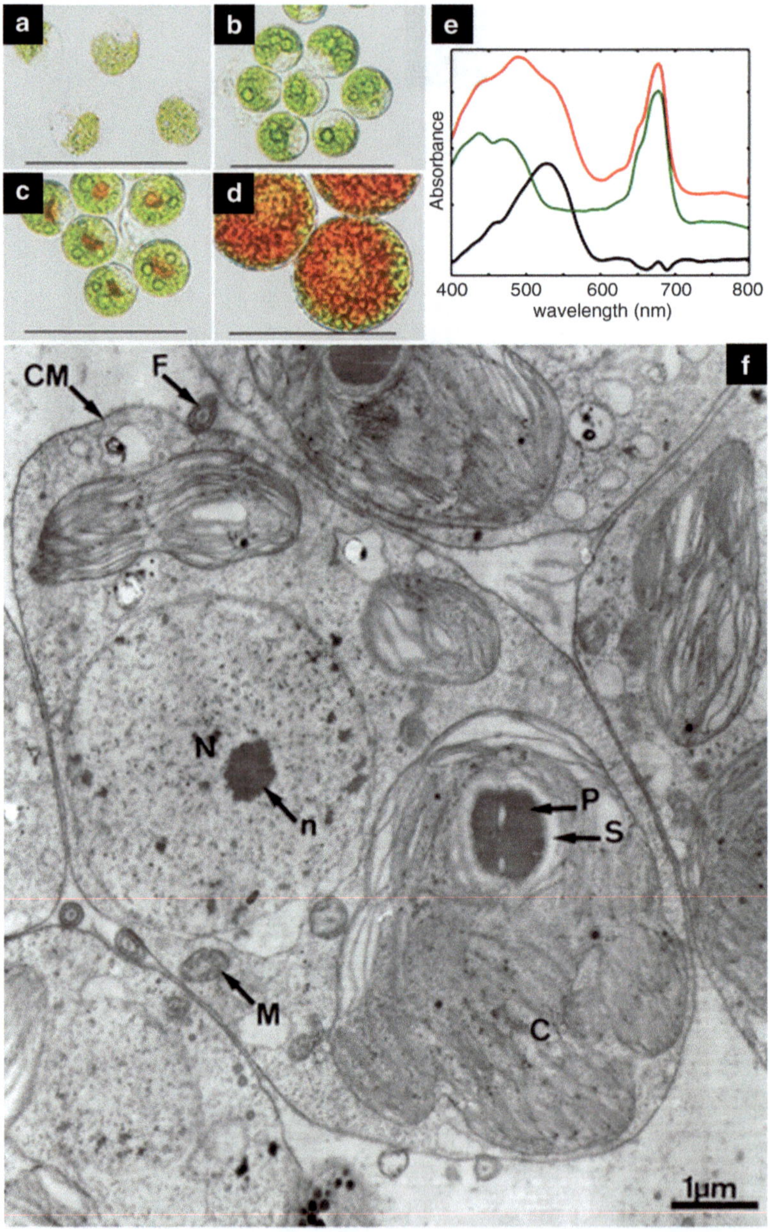

Fig. 17.7. Microscopic picture of the four major cell stages of *Haematococcus* (**a**) biflagellated motile cells performing photosynthesis, (**b**) *green* palmelloid cells under replete nutrient conditions, (**c**) palmelloids under high light and nitrogen depletion showing slight accumulation of *orange* carotenoids, and (**d**) *red* aplanospores with the corresponding absorption spectrum. Light microsopic images (Taken from Collins et al. 2011, electron micrograph taken from Triki et al. 1997).

E. *Ostreococcus*

Ostreococcus tauri (Mamiellophyceae) is known as the smallest eukaryotic phototroph described so far (Chretiennot-Dinet et al. 1995) with a genome size of only 13 MB (Grimsley et al. 2010) and a cell diameter smaller than 1 μ. The marine cells contain only one chloroplast, which is very similar to that of *Mantoniella squamata* with respect

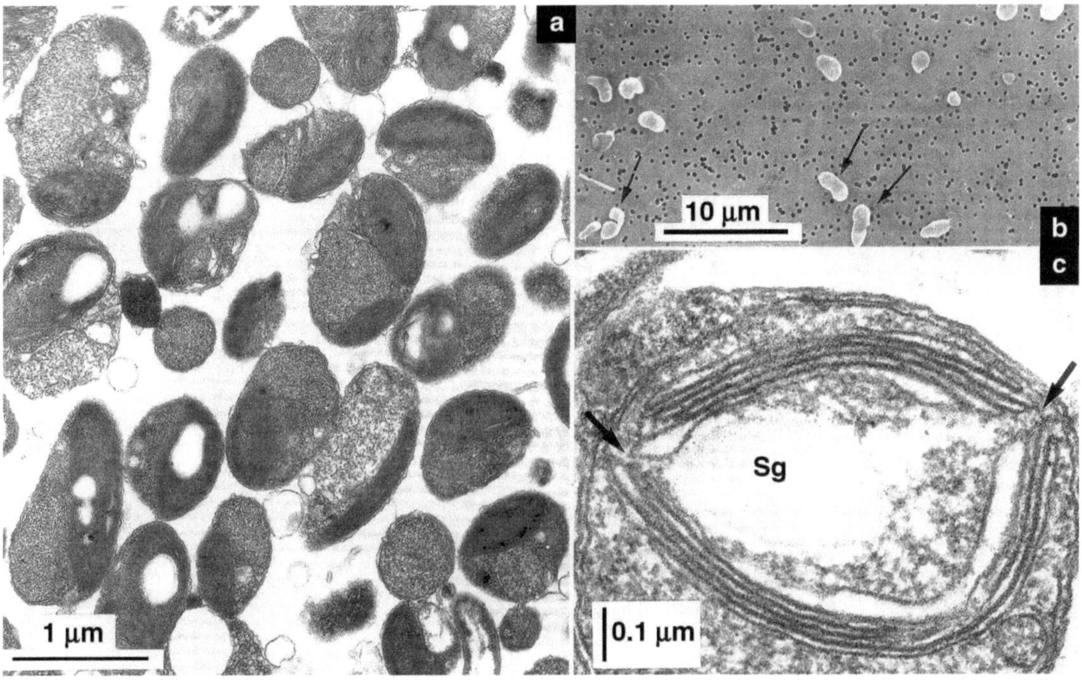

Fig. 17.8. Ostreococcus tauri. Electron micrograph showing the cell size of 1 μm or below (*left side*). Light microscopic picture (*right upper part*), and a section through the chloroplast (*lower part*) showing that the thylakoid lamellae run unstacked in parallel (Taken from Chretiennot-Dinet et al 1995).

to pigmentation and thylakoid organization. Instead of grana/stroma differentiated lamellae, all membranes run non-stacked through the chloroplast (see Fig. 17.8), sometimes forming twins or triplets without having contact zones (Krämer et al. 1988; Chretiennot-Dinet et al. 1995), which can be observed in the partition zones in the grana. The light harvesting apparatus is characterized by the cooperation of three different chlorophyll types Chl *a*, Chl *b* and Mg-Divinylphaeoporphyrine, which has structural similarities to Chl c_1. The carotenoid composition is also unusual, because it is composed of seven different xanthophylls dominated by the brownish colored prasinoxanthin (Wilhelm and Lenartz-Weiler 1987). The cells are protected against excessive light by an unusual violaxanthin/antheraxanthin cycle (Goss et al. 1998). The last step from antheraxanthin to zeaxanthin is blocked because the enzyme VDE does not catalyze the second de-epoxidation step fast enough compared to the back-reaction of the epoxidase. Therefore, the cells accumulate antheraxanthin under high light, but the resulting NPQ is high enough for efficient photoprotection. This shows that the second de-epoxidation step, which is obligatory in higher plants and green algae is functionally not necessary (Goss 2003). The biomass of *O. tauri* is of biotechnological importance, because of its high content in highly unsaturated fatty acids and anionic lipids. Due to its small genome size and its thin cell wall, genetic transformation is easy to accomplish and opens the possibilities of systems biology and metabolic engineering (Sorokina et al. 2011).

F. *Botryococcus*

Botryococcus braunii is a cosmopolitan green colonial microalga that has attracted the interest of biotechnologists because it contains hydrocarbons in large amounts (see Fig. 17.9 adopted from Casadevall et al. 1985).

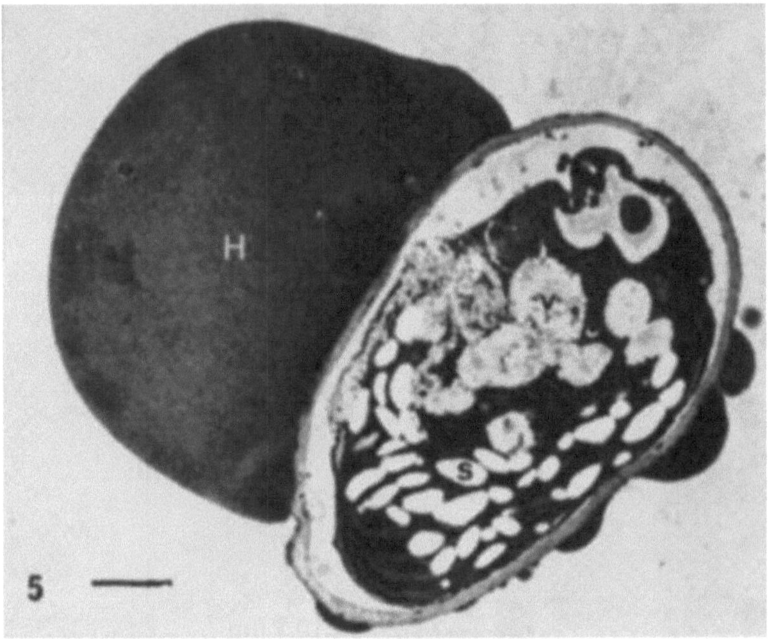

Fig. 17.9. Botryococcus braunii . Cell at the end of the exponental phase. Starch is much more abundant in the chloroplast. A voluminous hydrocarbon droplet, as large as a cell is observed in the trilamellar structure of the cell wall (Adapted from Casadevall et al. 1985).

The hydrocarbons are stored in the three lamellar structures in the cell wall. These are formed by hydrophobic sporopollenin layers, high-molecular weight polymers derived from lipid biosynthesis. The hydrocarbons consist of n-alkadienes and trienes, of triterpenoids with a multiplicity of modifications; the most important are botryococcenes and methylated squalenes. Finally, the cells accumulate tetraterpenoid-like lycopadiene. For a review concerning lipids in *Botryococcus* species, see Metzger and Largeau (2005) and Banerjee et al. (2002).

Brown and Knights (1969) have shown for the first time that *B. braunii* exists in three different physiological forms: A green photosynthetically active form has the highest growth performance and contains a complex mixture of hydrocarbons of the general formula C_nH_{2n}-2(4), with up to 60 % of the dry weight (a), a brownish resting stage accumulates botryococcene up to 90 % of its dry weight (b), large green cells having lost the capability to form colonies, grow very slowly and show only very little synthesis of hydrocarbons restricted to the class of tetraterpenoids (c). These cell stages have also been designated as 'race'. These races are now identified as different strains which differ in their genome sizes, nevertheless the term 'race' has been maintained: Race A for the green cells, race B for the brownish resting stages and race L, which produce only the tetraterpenoids lycopadiene. Phylogenetic analysis of sister groups of *B. braunii* have been performed to explore the different potential for hydrocarbon biosynthesis (Weiss et al. 2011).

It is important to note that the n-alkadienes are formed from fatty acids, whereas the squalene-type hydrocarbons are formed from the isoprenoid pathway (Metzger et al. 2005). Since n-alkadienes can be easily extracted from the cells and transesterification is not needed for biodiesel production, the n-alkadiene biosynthesis in *Botryococcus* has been intensively studied. Another reason for the great biotechnological interest is the fact that the biomass can be used for both commercial strategies: (a) To produce highly valuable products; (b) To use the residual carbon for biofuel production. *B. braunii* delivers the optimal biomass to reach both goals. However, the major problem with *B. braunii* is the slow growth rate. Many attempts have been made to improve growth performance. Recently, Ashokkumar and Rengasamy (2011) have reported laboratory cultures and growth in raceway ponds showing growth rates between 0.2 and 0.3 cell divisions per day, which means that the doubling time is between three and five days. Even under these "optimized" conditions, the hydrocarbon content dropped down to 11 % and 13 %, respectively. Therefore, many researchers follow another approach: To identify the genes coding for those proteins, which catalyze the synthesis of the hydrocarbons, and to transfer those genes to other autotrophic cells, which show a better growth performance. Surprisingly, it became obvious that *B. braunii* does not use a squalen synthase. The squalen synthase, known to be present in other eukaryotes, catalyzes a two step reaction without intermediate release. However, in *B. braunii* two enzymes are required for squalen synthesis (Niehaus et al. 2011). This provides evidence that the triterpene pathway in *B. braunii* might be unique, and metabolic engineering in other organisms might face unexpected problems concerning enzyme regulation.

IV. Special Physiological Traits of Green Algae

A. Carotenoids

Green algae contain both α- and β-carotene-derived xanthophylls, in contrast to heterokontophytes, e.g. diatoms, which can produce only β-type xanthophylls (see Fig. 17.10).

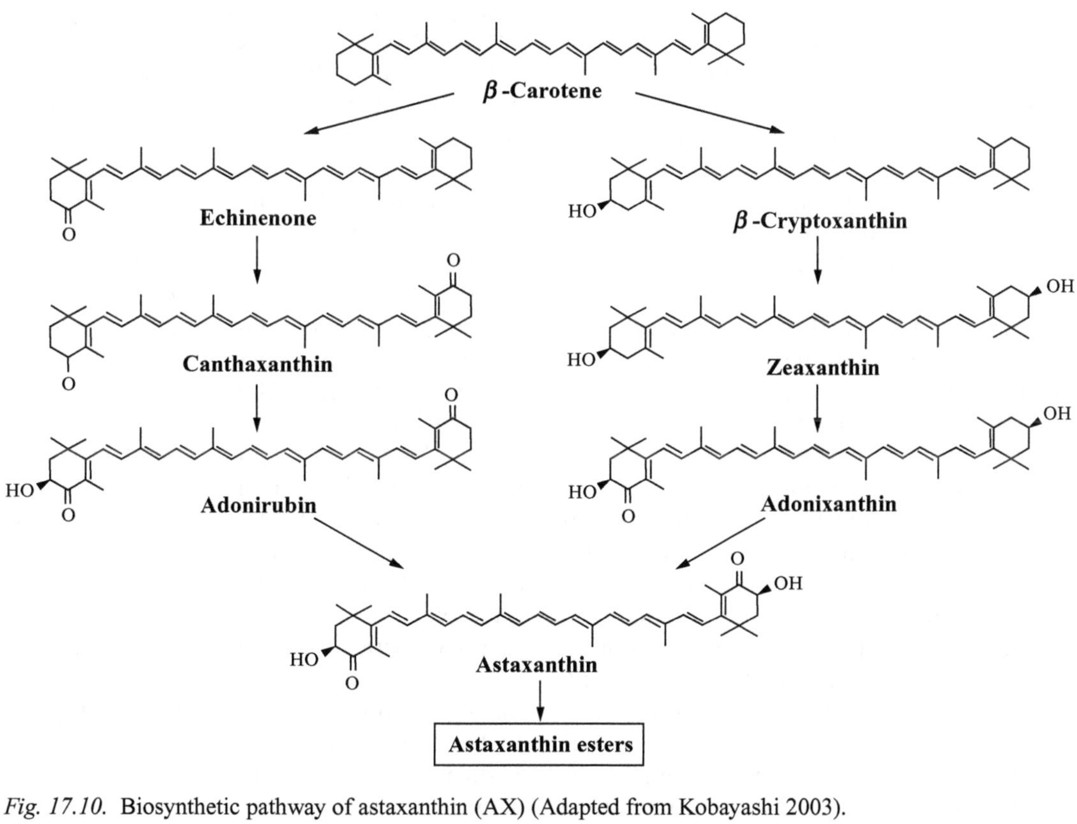

Fig. 17.10. Biosynthetic pathway of astaxanthin (AX) (Adapted from Kobayashi 2003).

The number of alga-derived carotenoids, produced at the industrial scale, is restricted to β-carotene, its derivatives astaxanthin (AX) and zeaxanthin. The α-type xanthophyll lutein is not yet produced from green algae on a large scale, because it can be obtained at lower prices from flowers (e.g. *Tagetes*). The ongoing industrial interest in β-carotene, luteine and zeaxanthin is mainly due to their features as antioxidants.

AX is the most important industrial product from green algae, although AX can also be produced biotechnologically by the red yeast *Xanthophyllomyces denrorhous* (formerly designated as *Phaffia rhodozyma*), which can be metabolically engineered by standard molecular techniques (Visser et al. 2003). A third biotechnological source for AX is the marine bacterium *Paracoccus haeundaensis* (Lee and Kim 2006). The reason that phototrophic green algae are the favorite production system is the lower production price due to the inexpensive inorganic nutrients. AX is commercially produced from *Haematococcus*, in open ponds (Australia), as well as in photobioreactors (Israel). The biosynthetic pathway for AX synthesis branches from β-carotene by the action of β-carotene ketolase, which oxidizes the carbon atoms in the position 4 and 4′ to form canthaxanthin with echinenon as an intermediate (Fan et al. 1995). The carotenoid hydroxilase (Lotan and Hirschberg 1995) hydroxilates the carbon atoms in positions 3 and 3′, leading to the final product AX. Steinbrenner and Sandmann (2006) showed that the transformation of *H. pluvialis* with a phytoene desaturase can accelerate AX biosynthesis.

Recently, it has been discovered that in the green alga *Chlamydomonas nivalis* AX exists as diglucoside diesters (Rezanka et al. 2008), forming "hydrophobic-hydrophilic-hydrophobic" structures. These types of structures are known from "thermozeaxanthins" and "thermocryptoxanthins" found in thermophilic bacteria with a function in stabilizing the membrane at extreme temperatures (Yokoyama et al. 1996). In contrast to *Haematococcus*, the snow alga *C. nivalis* lives at high altitudes and is exposed to extremely low temperatures during the night, and high UV radiation at full sunlight (Remias et al. 2005). These AX-based "hydrophobic-hydrophilic-hydrophobic" structures form small lipid droplets in the cytosol and reduce the risk of ice crystal formation. Therefore, together with the antioxidant capacity, these complexes may be of biotechnological interest as themo-protectants in many applications, e.g. in molecular medicine.

B. Hydrogen

About 30 species/strains of green algae are known to possess the pathway of anaerobic oxygenic photosynthesis. This pathway is induced by anaerobic conditions, when respiration consumes more oxygen than the water splitting apparatus of PSII generates. Under anaerobic conditions, the FeFe-hydrogenase transfers PSI-activated electrons via ferredoxin to protons, thereby generating hydrogen (Melis 2007). This represents, in principle, a CO_2-free biofuel technology. Since the hydrogenase is extremely sensitive to oxygen, hydrogen production occurs in the light only if PSII is inactivated. This can be done either by the addition of PSII inhibitors or by Sulfur-starvation (Melis et al. 2000). However, the inactivation of PSII eliminates water-derived electrons as the source of electrons for hydrogen production. Without an alternative electron source, hydrogen production cannot continue for a long period. As a special case for these green algae, the chloroplast also contains an active fermentative pathway which degrades starch in the absence of oxygen. This fermentation pathway delivers electrons, which are transferred to the plastoquinone pool by the activity of the NADPH-PQ-oxidoreductase, which is also involved in chlororespiration (Bennoun 2002). Therefore, continuous hydrogen production in green algae is only possible in three phases. In the first phase, the oxygenic photosynthetic apparatus is fully active to produce biomass. In the second step, this biomass accumulates starch in the chloroplast. In the last phase, PSII is inactivated under S-starvation, and starch-derived electrons are activated by PSI to generate hydrogen under anaerobic conditions in the light. Green alga-mediated hydrogen production is a very active area of research and the reader is referred to recent review papers (Rupprecht 2009; Kruse and Hankamer 2010).

C. Excretion Products

Some algal excretion products are extremely valuable biotechnological products. Socio-economic studies on the feasibility of algal-derived biofuels have shown that – in addition to biomass production – harvesting and the subsequent refinement are crucial for future industrial applications (Davis et al. 2011). A critical parameter for the evaluation of the ecological feasibility of algal biofuel production is the ratio of the green house gas emissions (GHE) per MJ usable energy. In order to be competitive with fossil energy carriers, harvest and refinement of biofuels need significant improvement (Weinberg et al. 2012). The idea of "milking algae" (Hejazi and Wijffels 2001) is based on the concept that the cells use photosynthetic carbon assimilation to produce biotechnological products. These are excreted and can be recovered from the cell suspension by sophisticated extraction procedures, preventing costly harvesting procedures and, to some extent, also the refinement process. This idea was the basis for research of potential excretion products from algae. In addition to the lipid excretion by *Botryococcus braunii*, excretion of carbohydrates is also of major interest. Another option is the excretion of gaseous products in addition to hydrogen. One such candidate that has been demonstrated in cyanobacteria is isoprene. Large amounts of isoprene are emitted by trees, e.g. oak or poplar trees (Sharkey and Yeh 2001). Isoprene is produced in these plants by the mevalonate pathway, which is located in the chloroplast to synthesize carotenoids. The insertion and activation of the isoprene synthase gene has been shown to transform a cyanobacterium into an isoprene-emitting cell. However, the isoprene production rates that can currently be achieved are low in cyanobacteria. Perhaps metabolic engineering of green alga for isoprene production will prove more successful.

D. Extremophiles

Extremophile phototrophs are found mainly in taxonomic groups different from green algae, e.g. cyanobacteria, which contain mainly thermophilic, acidophilic and basophilic strains, or diatoms, where many species can grow at extremely low temperatures. Known as ice algae, some green algae can also resist harsh environmental conditions. *C. nivalis* grows at extremely low temperatures, resisting relatively high UV-radiation, which reflects its adaptation to its natural environment, glaciers, or *Dunaliella parva*, which can tolerate extreme saline environments. *Haematoccus* sp. is known to be very insensitive to UV-radiation. In the genus *Chlamydomonas acidophila* isolated from an acidic mining area in Germany can actively grow at pH 2.6 with a growth performance similar to non-acidicdophilic *Chlamydomonas* species (Langner et al. 2009). Extremophiles might be of special biotechnological interest because their high physiological performance under extreme conditions reduces the risk of contamination by other organisms.

V. Concepts of Bioenergy Conversion

A. Antenna Size Engineering

Under full sunlight, the total photosynthetic capacity is normally limited by the capacity of 'dark reactions', i.e. carbon incorporation. In the model of Farquhar et al. (1980), it is assumed that the carboxylation rate is either Rubisco-limited at low CO_2 or RuBP regeneration-limited at high CO_2. This limitation leads to a severe decrease in the photosynthetic quantum yield at high light intensity. Cells have developed several mechanisms to dissipate the excessive energy, which has been reviewed recently by Wilhelm and Selmar (2011), and the costs of high light-induced photohinibition have been recently calculated by Raven (2011). Most algae become light-saturated at light intensities above 150–200 μE / (m² * s), which is equal to about 10 % of the intensity of full sunlight. Wilhelm and Selmar (2011) have shown that about 80 % of the yearly PAR radiation at a latitude of 50° is above the inclination point, the light intensity where photosynthesis starts to be saturated. However, under natural light conditions, during 60 % of the year these cells are exposed to dark or limiting light conditions. To address the limited conversion efficiency at high light intensities, it has been proposed that decreasing the absorptivity of the cells will improve photosynthetic performance at high light intensity (for review see Melis 2009). A promising strategy for reducing the absorptivity of the cells is to down-regulate the light-harvesting chlorophyll complexes, which bind about 50 % of the total amount of Chl *a* and about 90 % of Chl *b*. Recently, Beckmann et al. (2009) demonstrated that this can be successfully done by the expression of a LHC-specific translation repressor protein. Mussgnug et al. (2001) have provided evidence that engineering the light harvesting capacity will have a favorable effect on algal mass cultivation, because the suspension can contain a higher biomass per volume, while the mixing energy can be reduced. Although complete energy balances of mutants with truncated antennas are still not available, this approach appears very promising for improved algal productivity.

B. Metabolic Engineering

Metabolic engineering of green microalgae is motivated by many different expectations. In the context of biofuel production, increased oil content can decrease the costs for harvest, extraction and refinery (Weinberg et al. 2012). The combination of highly valuable products e.g., carotenoids, polyunsaturated lipids as byproducts of biofuel production can improve the commercial value and help to make microalgal production feasible for industrial use. Finally, species with optimal biomass composition will be used on a large scale only if "technical" requirements are also met. For instance, the cells should be robust enough to withstand shearing forces during mixing, the cell wall should be easy to break and not covered by a sticky surface, which leads to fouling in the photobioreactor. Although intensive research is under way on screening activities to find new strains, which combine many of these features, there is clear evidence that no species will be optimal for all requirements. Therefore, the most promising approach is to start from cells with physiological and technical robust key features and to modulate the biochemical pathways by metabolic engineering (Gressel 2008). For instance, modified phytoene desaturase gene expression can induce herbicide resistance, with the consequence that open pond cultivation can be protected from contaminants (Gressel et al. 2010). Since ketocarotenoids are of special interest as high value products from green algae, many attempts have been made to engineer the carotenoid pathway accordingly (e.g. León et al. 2007). Recently, several attempts have been published to use the chloroplast of *Chlamydomonas* for the synthesis of recombinant proteins (Rasala et al. 2010; Coragliotti et al. 2011). The analysis of the genes involved in lipid biosynthesis is presently at the state of gene identification, and progress in metabolic engineering can be expected in the near future. To the authors' knowledge, there is at present no GMO alga in industrial use. Schuhmann et al. (2012) point out that an improved lipid content in microalgae based on metabolic engineering will be strongly supported by the recent progress in system biology and metabolic mapping.

Acknowledgements

We would like to acknowledge the German Funding Agency (DFG) for financial support for Maria Schmidt and the German Ministry of Science and Education (BMBF) for support in developing new concepts of algal based biofuels production (VIP 16 V0001)

References

Adl SM, Simpson AGB, Farmer MA, Andersen RA, Anderson OR, Barta JR, Bowser SS, Brugerolle G, Fensome RA, Fredericq S, James TY, Karpov S, Kugrens P, Krug J, Lane CE, Lewis LA, Lodge J, Lynn DH, Mann DG, McCourt RM, Mendoza L, Moestrup Ø, Mozley-Standrige SE, Nerad TA, Shearer CA, Smirnov AV, Spiegel FW, Taylor MFJR (2005) The new higher level classification of eukaryotes with emphasis on the taxonomy of protists. J Euk Microbiol 52:399–451

Agrawal S, Striepen B (2010) More membranes, more proteins: complex protein import mechanisms into secondary plastids. Protist 161:672–687

Archibald JM, Lane CE (2009) Going, going, not quite gone: nucleomorphs as a case study in nuclear genome reduction. J Heredity 100:582

Asada K (2000) The water-water cycle as alternative photon and electron sinks. Philos Trans R Soc Lond B Biol Sci 355:1419–1431

Ashokkumar V, Rengasamy R (2011) Mass culture of *Botryococcus braunii* under open raceway pond for biofuel production. Bioresource Techn 104:394–399

Banerjee A, Sharma R, Chisti Y, Banerjee UC (2002) *Botryococcus braunii*: a renewable source of hydrocarbons and other chemicals. Crit Rev Biotechn 22:245–279

Bayramoglu G, Arica MY (2011) Preparation of a composite biosorbent using *Scenedesmus quadricauda* biomass and alginate/polyvinyl alcohol for removal of Cu(II) and Cd(II) ions: isotherms, kinetics, and thermodynamic studies. Wat Air Soil Poll 221:391–403

Beckmann J, Lehr F, Finazzi G, Hankamer B, Posten C, Wobbe L, Kruse O (2009) Improvement of light to biomass conversion by de-regulation of light-harvesting protein translation in *Chlamydomonas reinhardtii*. J Biotechn 142:70–77

Bennoun P (1982) Evidence for a respiratory chain in the chloroplast. Proc Natl Acad Sci USA 79:4352–4356

Bennoun P (2002) The present model of chlororespiration. Photosynth Res 73:273–277

Brown AC, Knights BA (1969) Hydrocarbon content and its relationship to physiological state in the green alga *Botryococcus braunii*. Phytochem 8:543–547

Bock C, Pröschold T, Krienitz L (2011) Updating the genus *Dictyosphaerium* and description of *Mucidosphaerium* gen. nov. (*Trebouxiophyceae*) based on morphological and molecular data. J Phycol 47:638–652

Bouvier F, Rahier A, Camara B (2005) Biogenesis, molecular regulation and function of plant isoprenoids. Progr Lipid Res 44:357–492

de Bashan LE, Bashan Y (2010) Immobilized microalgae for removing pollutants: review of practical aspects. Biores Techn 101:1611–1627

Casadevall D, Dif D, Largeau C, Gudin C, Chaumont D, Desanti O (1985) Studies on batch and continuous cultures of *Botryococcus braunii*: hydrocarbon production in relation to physiological state, cell ultrastructure and phosphate nutrition. Biotechnol Bioeng 27:286–295

Cha TS, Yee W, Aziz A (2012) Assessment of factors affecting Agrobacterium-mediated genetic transformation of the unicellular green alga, *Chlorella vulgaris*. World J Microbiol Biotechnol 28:1771–1779

Coll JM (2006) Methodologies for transferring DNA into eukaryotic microalgae. S J Agricult Res 4:316–330

Collins AM, Jones HDT, Han D, Hu Q, Hu Q, Beechem TE, Timlin JA (2011) Carotenoid distribution in living cells of *Haematococcus pluvialis* (*Chlorophyceae*). PLoS ONE 6(9):e24302

Coragliotti AT, Beligni MV, Franklin SE, Mayfield SP (2011) Molecular factors affecting the accumulation of recombinant proteins in the *Chlamydomonas reinhardtii* chloroplast. Mol Biotechnol 48:60–75

Chretiennot-Dinet MJ, Courties C, Vaquer A, Neveux J, Claustre H, Lautier J, Machado MC (1995) A new marine picoeukaryote – *Ostreococcus tauri* gen et sp. nov. (*Chlorophyta*, *Prasinophyceae*). Phycologia 34:285–292

Darienko T, Gustavs L, Mudimu O, Menendez CR, Schumann R, Karsten U, Friedl T, Proeschold T (2010) *Chloroidium*, a common terrestrial coccoid green alga previously assigned to *Chlorella* (*Trebouxiophyceae*, *Chlorophyta*). Eur J Phycol 45:79–95

Davis R, Aden A, Pienkos PT (2011) Techno-economic analysis of autotrophic microalgae for fuel production. Appl Energy 88:3524–3531

Dawson HN, Burlingame R, Cannons AC (1997) Stable transformation of *Chlorella*: rescue of nitrate reductase-deficient mutants with the nitrate reductase gene. Curr Microbiol 35:356–362

Dent RM, Haglund CM, Chin BL, Kobayashi MC, Niyogi KK (2005) Functional genomics of eukaryotic photosynthesis using insertional mutagenesis of Chlamydomonas reinhardtii. Plant Physiol 137:545–556

Derenne S, Largeau C, Berkaloff C, Rousseau B, Wilhelm C, Hatcher PG (1992) Non-hydrolysable macromolecular constituents from outer walls of *Chlorella fusca* and *Nanochlorum eukaryotum*. Phytochemistry 31:1923–1929

De Wever A, Leliaert F, Verleyen E, Vanormelingen P, van der Gucht K, Hodgson DA, Sabbe K, Vyverman W (2009) Hidden levels of phylodiversity in Antarctic green algae: further evidence for the existence of glacial refugia. Proc Biol Sci 276:3591–3599

Ebersold WT (1962) Biochemical genetics. In: Lewin RA (ed) Biochemistry and physiology of algae. Academic Press, New York, pp 731–739

Egeland ES, Guillard RRL, Liaaen-Jensen S (1997) Additional carotenoid prototype representatives and a general chemosystematic evaluation of carotenoids in *Prasinophyceae* (*Chlorophyta*). Phytochem 44:1087–1097

Elias M, Archibald JM (2009) Sizing up the genomic footprint of endosymbiosis. BioEssays 31:1273–1279

Ettl H (1980) Grundriss der allgemeinen Algologie. VEB Gustav Fischer Verlag, Jena

Falkowski PG, Knoll AH (2007) Evolution of primary producers in the sea. Academic, Waltham

Fan L, Vonshak A, Gabbay R, Hirschberg J, Cohen Z, Boussiba S (1995) The biosynthetic pathway of astaxanthin in a green alga *Haematococcus pluvialis* as indicated by inhibition with dephenylamine. Plant Cell Physiol 36:1519–1524

Farquhar GD, von Caemmerer S, Berry JA (1980) A biochemical model of photosynthetic CO2 assimilation in leaves of C3 species. Planta 149:78–90

Fassett RG, Coombes JS (2011) Astaxanthin: a potential therapeutic agent in cardiovascular disease. Marine Drugs 9:447–465

Fawley MW, Fawley KP, Buchheim MA (2004) Molecular diversity among communities of freshwater microchlorophytes. Microbial Ecol 48:489–499

Feng P, Deng Z, Hu Z, Lu F (2011) Lipid accumulation and growth of Chlorella zofingiensis in flat plate photobioreactors outdoors. Bioresour Technol 102:10577–10584

Ferris PJ, Woessner JP, Waffenschmidt S, Kilz S, Drees J, Goodenough U (2001) Glycosylated polyprolin II rods with kinks as a structural motif in plant hydroyprolin-rich glycoproteins. Biochemistry 40:2978–2987

Forti G, Caldiroli G (2005) State transitions in *Chlamydomonas reinhardtii*. The role of the Mehler reaction in state 2-to state 1 transitions. Plant Physiol 137:492–499

Frommolt R, Werner S, Paulsen H, Goss R, Wilhelm C, Zauner S, Maier U, Grossman A, Bhattacharya D, Lohr M (2008) Ancient recruitment by Chromists of green algal genes encoding enzymes for carotenoid biosynthesis. Mol Biol Evol 25:2653–2667

Genkov T, Meyer M, Griffiths H, Spreitzer RJ (2010) Functional hybrid rubisco enzymes with plant small subunits and algal large subunits. J Biol Chem 285:19833–19841

Gile GH, Stern RF, James ER, Keeling PJ (2010) DNA barcoding of chlorarachniophytes using nucleomorph ITS sequences. J Phycol 46:743–750

Giordano M, Beardall J, Raven JA (2005) CO2 concentrating mechanisms in algae: mechanisms, environmental modulation and evolution. Annu Rev Plant Biol 56:99–131

Goldman JC (1979) Outdoor algal mass cultures – applications. Water Res 13:1–19

Gontcharov AA (2008) Phylogeny and classification of *Zygnematophyceae* (*Streptophyta*): current state of affairs. Fottea 8:87–104

Goss R (2003) Substrate specificity of the violaxanthin de-epoxidase of the primitive green alga *Mantoniella squamata* (*Prasinophyceae*). Planta 217:801–812

Goss R, Böhme K, Wilhelm C (1998) The xanthophyll cycle of *Mantoniella squamata* converts violaxanthin into antheraxanthin but not to zeaxanthin. Planta 205:613–621

Granado-Lorencio F, Herrero-Barbudo C, Acien-Fernandez G (2009) In vitro bioaccesibility of lutein and zeaxanthin from the microalgae *Scenedesmus almeriensis*. Food Chem 114:747–752

Gressel J (2008) Transgenics are imperative for biofuel crops. Plant Sci 174:246–263

Gressel J, Chen O, Einbinder S, Eisenstadt S, Schatz D, Schlesinger A (2010) Transgenically domesticating marine micro-algae for biofuel and feed uses: no competition with crops for land and water. J Biotechn 150:16

Grimsley N, Péquin B, Bachy C, Moreau H, Piganeau G (2010) Cryptic sex in the smallest eukaryotic marine alga. Mol Biol Evol 27:47–54

Grimm B, Porra RJ, Rüdiger W, Scheer H (eds) (2006) Chlorophylls and bacteriochlorophylls, 1st edn. Springer, Dordrecht

Grünewald K, Hirschberg J, Hagen C (2001) Ketocarotenoid biosynthesis outside of plastids in the unicellular green alga *Haematococcus pluvialis*. J Biol Chem 276:6023–6029

Guedes AC, Maras HM, Malcata FX (2011) Microalgae as sources of high added-value compounds – a brief review of recent work. Biotechnol Prog 27:597–613

Guiry MD, Guiry GM (2012) AlgaeBase [Online]. World-wide electronic publication, National University of Ireland, Galway http://www.algaebase.org. Accessed 14 Mar 2014

Harris EH (2008) *Chlamydomonas* in the laboratory. In: Harris EH (ed) The *Chlamydomonas* sourcebook: introduction to *Chlamydomonas* and its laboratory use. Academic, Waltham, pp 241–302

Hejazi MA, Wijffels H (2001) Milking of microalgae. Trends Biotechnol 22:189–194

Heilmann S, Jader LR, Harned LA (2011) Hydrothermal carbonization of microalgae II. Fatty acid, char, and algal nutrient products. Appl Energy 88:3286–3290

Huss VAR, Frank C, Hartmann EC, Hirmer M, Kloboucek A, Seidel BM, Wenzeler P, Kessler E (1999) Biochemical taxonomy and molecular phylogeny of the genus *Chlorella sensu lato* (*Chlorophyta*). J Phycol 35:587–598

Janssen M, de Bresser L, Baijens T, Tramper J, Mur LR, Snel JFH, Wijffesl RH (2000) Scale-up aspects of photobioreactors: effects of mixing induced light/dark cycles. J Appl Phycol 12:225–237

Jayasankar R, Ramamoorthy N (1993) Some observations on the growth of *Chlorella salina*. Seaweed Res Utiln 16:139–144

Keeling PJ (2010) The endosymbiotic origin, diversification and fate of plastids. Phil Trans R Soc B Biol Sci 365:729–748

Kessler E (1992) *Chlorella* biochemische Taxonomie einer für Forschung und Biotechnologie wichtigen Gattung einzelliger Grünalgen. Naturwissenschaften 79:260–265

Kessler E, Czygan F-C (1970) Physiologische und biochemische Beiträge zur Taxonomie der Gattung Chlorella. Arch Mikrobiol 70:211–216, 1970

Kim JI, Shin W, Triemer RE (2010) Multigene analyses of photosynthetic euglenoids and new family. Phacaceae (Euglenales) J Phycol 46:1278–1287

Kimura M (1968) Evolutionary rate at the molecular level. Nature 217:624–626

Kindle KL (1990) High-frequency nuclear transformation of *Chlamydomonas reinhardtii*. Proc Natl Acad Sci U S A 87:1228–1232

Kobayashi M (2003) Astaxanthin biosynthesis enhanced by reactive oxygen species in the green alga *Haematococcus pluvialis*. Biotechnol Bioproc Eng 8:322–330

Klochkova T, Kang S, Cho G, Pueschel C, West J, Kim G (2006) Biology of a terrestrial green alga, *Chlorococcum* sp (*Chlorococcales*, *Chlorophyta*), collected from the Miruksazi stupa in Korea. Phycologia 45:349–358

Krämer P, Wilhelm C, Wild A, Mörschel E, Riehl E (1988) Ultrastructure and freeze fracture studies of the thylakoids of Mantoniella squamata (Prasinophyceae). Protoplasma 147:170–177

Kruse O, Hankamer B (2010) Microalgal hydrogen production. Curr Opin Biotechnol 21:238–243

Lakaniemi AM, Hulatt CJ, Thomas DN, Tuovinen OH, Puhakka JA (2011) Biogenic hydrogen and methane production from *Chlorella vulgaris* and *Dunaliella tertiolecta* biomass. Biotechnol Biofuels 4:34

Langner U, Jakob T, Stehfest K, Wilhelm C (2009) A complete energy balance for *Chlamydomonas reinhardtii* and *Chlamydomonas acidophila* under neutral and extremely acidic growth conditions. Plant Cell Environ 32:250–258

Lee JH, Kim YT (2006) Cloning and characterization of the astaxanthin biosynthesis gene cluster from the marine bacterium *Paracoccus haeundaensis*. Gene 370:86–95

Leliaert F, Verbruggen H, Zechman F (2011) Into the deep: new discoveries at the base of green plant phylogeny. Bioessays 33:683–692

Leliaert F, Smith DR, Moreau H, Herron MD, Verbruggen H, Delwiche CF, De Clerck O (2012) Phylogeny and molecular evolution of the green algae. Crit Rev Plant Sci 31:1–46

Lemoine Y, Schoefs B (2010) Secondary ketocarotenoid astaxanthin biosynthesis in algae: a multifunctional response to stress. Photosynth Res 106:155–177

Leon R, Couso I, Fernández E (2007) Metabolic engineering of ketocarotenoids biosynthesis in the unicellular microalga *Chlamydomonas reinhardtii*. J Biotechnol 130:143–152

Lewis LA, McCourt RM (2004) Green algae and the origin of land plants. Am J Bot 91:1535–1556

Li Y, Zhou W, Hu B, Min M, Chen P, Ruan RR (2011) Integration of algae cultivation as biodiesel production feedstock with municipal wastewater treatment: strains screening and significance evaluation of environmental factors. Bioresour Technol 102:10861–10867

Lien T, Knutsen T (1979) Synchronous growth of *Chlamydomonas reinhardtii* (*Chlorophyceae*): a review of optimal conditions. J Phycol 15:191–200

Lorenz RT, Cysewski RG (2000) Commercial potential for *Haematococcus* microalgae as a natural source of astaxanthin. Trends Biotechnol 18:160–167

Lotan T, Hirschberg J (1995) Cloning and expressing in *Escherichia coli* the gene encoding β-C-4-oxygenase that converts β-carotene to the ketocarotenoid canthaxanthin in *Haematococcus pluvialis*. FEBS Lett 364:125–128

Lurling M (2001) Grazing associated infochemicals induce colony formation in the green alga *Sencedesmus*. Protist 152:7–16

Matsumoto T, Shinozaki F, Chikuni T, Yabuki A, Takishita K, Kawachi M, Nakayama T, Inouye I, Hashimoto T, Inagaki Y (2011) Green-colored plastids in the dinoflagellate genus *Lepidodinium* are of core chlorophyte origin. Protist 162:268–276

Melis A (2007) Photosynthetic H_2 metabolism in *Chlamydomonas reinhardtii* (unicellular green algae). Plant 226:1075–1086

Melis A (2009) Solar energy conversion efficiencies in photosynthesis: minimizing the chlorophyll antennae to maximize efficiency. Plant Sci 177:272–280

Melis A, Zhang L, Forestier M, Ghirardi ML, Seibert M (2000) Sustained photobiological hydrogen gas production upon reversible inactivation of oxygen evolution in the green alga *Chlamydomonas reinhardtii*. Plant Physiol 122:127–135

Merchant SS, Prochnik SE, Vallon O et al (2007) The Chlamydomonas genome reveals the evolution of key animal and plant functions. Science 318:245–251

Metzger P, Largeau C (2005) *Botryococcus braunii*: a rich source for hydrocarbons and related ether lipids. Appl Microbiol Biotechnol 66:486–496

Mikhailyuk TI, Sluiman HJ, Massalski A, Mudimu O, Demchenko EM, Kondratyuk SY, Friedl T (2008) New streptophyte green algae from terrestrial habitats and an assessment of the genus Interfilum (*Klebsormidiophyceae, Streptophyta*). J Phycol 44:1586–1603

Mussgnug JH, Thomas-Hall S, Rupprecht J, Foo A, Klassen V, McDowall A, Schenk PM, Kruse O, Hankamer B (2001) Engineering photosynthetic light capture: impacts on improved solar energy to biomass conversion. Plant Biotech J 5:802–814

Nakada T, Nozaki H, Pröschold T (2008) Molecular phylogeny, ultrastructure, and taxonomic revision of *Chlorogonium* (*Chlorophyta*): emendation of *Chlorogonium* and description of *Gungnir* gen. nov. and *Rusalka* gen. nov. J Phycol 44:751–760

Niehaus TD, Okad S, Devarenne TP, Watt DS, Sviripa V, Chapell J (2011) Identification of unique mechanisms for triperterpene biosynthesis in *Botryococcus braunii*. Proc Natl Acad Sci USA 108:12260–126265

Niklas KJ, Kutschera U (2010) The evolution of the land plant life cycle. New Phytol 185:27–41

Oswald W, Golucke C (eds) (1968) Algae, man and the environment. Syracuse University Press, Syracuse

Palmucci M, Giordano M (2011) Ecological and evolutionary implications of carbon allocation in marine phytoplankton as a function of nitrogen availability. A Fourier Transform Infrared spectroscopy approach. J Phycol 47:313–323

Patron NJ, Waller RF (2007) Transit peptide diversity and divergence: a global analysis of plastid targeting signals. BioEssays 29:1048–1058

Pickett-Heaps JD (ed) (1975) Green algae, structure, reproduction and evolution in selected genera. Sinauer Associates, Inc, Sunderland

Piganeau G, Eyre-Walker A, Grimsley N, Moreau H (2011) How and why DNA barcodes underestimate the diversity of microbial eukaryotes. PLoS ONE 6:e16342

Pocock T, Lachance MA, Proschold T, Priscu C, Kim SS, Huner NPA (2004) Identification of a psychrophilic green alga from Lake Bonney Antarctica: *Chlamydomonas raudensis* Ettl. (UWO 241) *Chlorophyceae*. J Phycol 40:1138–1148

Potvin G, Zhang Z (2010) Strategies for high-level recombinant protein expression in transgenic microalgae: a Review. Biotechnol Adv 28:910–918

Pröschold T, Marin B, Schlosser UG, Melkonian M (2001) Molecular phylogeny and taxonomic revision of *Chlamydomonas* (*Chlorophyta*). I. Emendation of *Chlamydomonas* Ehrenberg and *Chloromonas* Gobi, and description of *Oogamochlamys* gen. nov. and *Lobochlamys* gen. Nov. Protist 152:265–300

Rasala BA, Muto M, Lee PA, Jager M, Cardos R, Behnke CA, Kirk P, Hokanson CA, Crea R, Mendez MS (2010) Production of therapeutic proteins in algae, analysis of expression of seven human proteins in the chloroplast of *Chlamydomonas reinhardtii*. Plant Biotechn J 8:719–733

Raven JA (2011) The cost of photoinhibition. Physiol Plant 142:87–104

Reiland S, Finazzi G, Endler A et al (2011) Comparative phosphoproteome profiling reveals a function of the STN 8 kinase in fine-tuning of cyclic electron flow. Proc Natl Acad Sci U S A 108:12955–12960

Remias D, Lütz-Meindl U, Lütz C (2005) Photosynthesis, pigments and ultrastructure of the alpine snow alga *Chlamydomonas nivalis*. Eur J Phycol 40:259–268

Remias D, Schwaiger S, Aigner S, Leya T, Stuppner H, Lütz C (2012) Characterization of an UV- and VIS-absorbing, purpurogallin-derived secondary pigment new to algae and highly abundant in *Mesotaenium berggrenii* (*Zygnematophyceae, Chlorophyta*), an extremophyte living on glaciers. FEMS Microbiol Ecol 79:638–648

Rezanka T, Nedbalova L, Sigler K, Cepak V (2008) Identification of astaxanthin di glucoside diesters from snow alga Chlamydomonas nivalis by liquid chromatogrphy-atmospheric pressure chimica ionization mass spectrometry. Phytochem 69:479–490

Rohr J, Sarkar N, Balenger S, Jeong BR, Cerutti H (2004) Tandem inverted repeat system for selection of effective transgenic RNAi strains in Chlamydomonas. Plant J 40:611–621

Rupprecht J (2009) From systems biology to fuel *Chlamydomonas reinhardtii* as a model for a systems biology approach to improve biohydrogen production. J Biotechn 142:10–20

Sager R (1954) Mendelian and non-Mendelian inheritance of streptomycin resistance in *Chlamydomonas reinhardtii*. Proc Natl Acad Sci U S A 40:356–363

Schroda M, Blöcker D, Beck CF (2000) The HSP70A promoter as a tool for the improved expression of transgenes in Chlamydomonas. Plant J 21:121–131

Schuhmann H, Lim DKY, Schenk PM (2012) Perspectives on metabolic engineering for increased lipid contents in microalgae. Biofuels 3:71–86

Sharkey TD, Yeh S (2001) Isoprene emission form plans. Annu Rev Plant Physiol Mol Biol 52:407–436

Sialve B, Bernet N, Bernard O (2009) Anaerobic digestion of microalgae as a necessary step to make microalgal biodiesel sustainable. Biotechnol Adv 27:409–416

Söder C (1976) The use of microalgae in nutrition. Naturwissenschaften 63:131–138

Song W, Rashid N, Choi W, Lee K (2011) Biohydrogen production by immobilized *Chlorella* sp. using cycles of oxygenic photosynthesis and anaerobiosis. Bioresour Technol 102:8676–8681

Sorokina O, Corellou F, Dauvillée D, Sorokin A, Goryanin I, Ball S, Bouget F-Y, Millar AJ (2011) Microarray data can predict diurnal changes of starch content in the picoalga Ostreococcus. BMC System Biol 5:36

Stanley JG, Jones JB (1976) Feeding algae to fish. Aquaculture 7:219–223

Steinbrenner J, Linden H (2013) Licht induction of carotenoid biosynthesis genes in the green algae *Haematococcus pluvialis*: regulation by photosynthetic redox control. Plant Mol Biol 52:343–356

Steinbrenner JA, Sandmann G (2006) Transformation of the green alga *Haematococcus pluvialis* with a phytoene desaturase for accelerated astaxanthin biosynthesis. Appl Environ Microbiol 72:7477–7484

Tamiya H (1966) Synchronous cultures of algae. Annu Rev Plant Physiol 17:1–26

Torzillo G, Pushparaj B, Masojidek J (2003) Biological constraints in algal biotechnology. Biotechnol Bioproc Engineer 8:338–348

Triki A, Maillard P, Gudin C (1997) Gametogenesis in *Haematococcus pluvialis*. F. (*Volvocales, Chlorophyta*). Phycologia 36:190–194

Trissl HW, Wilhelm C (1993) Why do thylakoid membranes from higher plants form grana stacks? Trends Biochem Sci 18:415–419

Visser H, van Ooyen AJ, Verodes JC (2003) Metabolic engineering of the astaxanthin-biosynthetic pathway of *Xanthophyllomyces dendrorhous*. FEMS Yeast Res 4:221–231

van den Hoek C, Mann D, Jahns HM (eds) (1996) Algae: an introduction to phycology. Cambridge University Press, Cambridge

Vaulot D, Eikrem W, Viprey M, Moreau H (2008) The diversity of small eukaryotic phytoplankton (≤3 μm) in marine ecosystems. FEMS Microbiol Rev 32:795–820

Weinberg J, Kaltschmitt M, Wilhelm C (2012) Biofuels from microalgae – an environmental analysis. Biomass Convers Biorefinery 2:179–194

Weiss TL, Johnston JS, Fjijsawa K, Okada S, Devarenne TP (2011) Genome size and phylogenetic analysis of the A and L races of *Botryococcus braunii*. J Appl Phycol 23:833–839

Wilhelm C, Wild A (1984) The variability of the photosynthetic unit in *Chlorella*. II. The effect of light intensity and cell development on photosynthesis, P-700 and cytochrome *f* in homocontinuous and synchronous cultures of *Chlorella*. J Plant Physiol 115:125–135

Wilhelm C, Lenartz-Weiler I (1987) Energy transfer and pigment composition in three chlorophyll *b* containing light-harvesting complexes isolated from *Mantoniella squamata*, *Chlorella fusca* and *Sinapis alba*. Photosynth Res 13:101–111

Wilhelm C, Krämer P, Wild A (1985) Effect of different light qualities on the ultrastructure, thylakoid membrane composition and assimilation metabolism in *Chlorella fusca*. Physiol Plant 64:359–364

Wilhelm C, Selmar D (2011) Energy dissipation is an essential mechanism to sustain the viability of plants: the physiological limits of improved photosynthesis. J Plant Physiol 168:79–87

Worden AZ, Lee J-H, Mock T, Rouzé P, Simmons MP (2009) Green evolution and dynamic adaptations revealed by genomes of the marine picoeukaryotes micromonas. Science 324:268–272

Yokoyama A, Shizuri Y, Hoshino T, Sandmann G (1996) Thermocryptoxanthins: novel intermediates in the carotenoid bio-synthetic pathway of *Thermus thermophilus*. Arch Microbiol 165:342–345

Zechman FW, Verbruggen H, Leliaert F, Ashworth M, Buchheim MA, Fawley MW, Spalding H, Pueschel CM, Buchheim JA, Verghese B, Hanisak MD (2010) An unrecognized ancient lineage of green plants persists in deep marine waters. J Phycol 46:1288–1295

Chapter 18

Carbon Fixation in Diatoms

Yusuke Matsuda*
Department of Bioscience, Research Center for Environmental Bioscience, Kwansei Gakuin University, Sanda, Hyogo 669-1337, Japan

and

Peter G. Kroth
Fachbereich Biologie, Universität Konstanz, D-78457 Konstanz, Germany

Summary

Diatoms are unicellular photoautotrophic algae and very successful primary producers in the oceans. Their high primary productivity is probably sustained by their high adaptability and a uniquely arranged metabolism. Diatom belongs to the Chromista, a large eukaryotic group, which has evolved by multiple endosymbiotic steps. As a result, diatoms possess a plastids with four membranes together with complicated translocation systems to transport proteins

*Author for correspondence, e-mail: yusuke@kwansei.ac.jp

M.F. Hohmann-Marriott (ed.), The Structural Basis of Biological Energy Generation, Advances in Photosynthesis and Respiration 39, pp. 335–362, DOI 10.1007/978-94-017-8742-0_18,

and metabolites including inorganic substances into and out of the plastids. In addition to the occurrence of potential plasma-membrane transporters, there are numerous carbonic anhydrases (CAs) within the matrix of the layered plastidic membranes, strongly suggesting large interconversion activity between CO_2 and HCO_3^- within the chloroplast envelope as a part of a CO_2-concentrating mechanism (CCM). In diatoms also the Calvin cycle and its adjacent metabolism reveal unique characteristics as, for instance, ribulose-1,5-bisphosphate carboxylase/oxygenase (RubisCO) activase, the plastidic sedoheptulose-1,7-bisphosphatase (SBPase), and the plastidic oxidative pentose phosphate pathway (OPP) are absent. Furthermore, the Calvin cycle metabolism in diatoms is not under the strict redox control by the thioredoxin (Trx) system. Instead, a CO_2-supplying system in the pyrenoid shows CA activities which are probably regulated by chloroplastic Trxs. Pyrenoidal CAs are also regulated on the transcriptional level by CO_2 concentrations via cAMP as a second messenger, suggesting an intense control system of CO_2 acquisition in response to CO_2 availability. The photorespiratory carbon oxidation cycle (PCOC) is the major pathway to recycle phosphoglycolate in diatoms although this process might not be involved in recycling of 3-phosphoglycerate but instead produces glycine and serine. In this review we focus on recent experimental data together with supportive genome information of CO_2 acquisition and fixation systems primarily in two marine diatoms, *Phaeodactylum tricornutum* and *Thalassiosira pseudonana*.

I. Introduction

Diatoms have a rather complicated evolutionary history including multiple endosymbiotic events, which have involved unknown host cells as well as cyanobacteria, a red alga and possibly a green alga (Moustafa et al. 2009). As a result of the cellular reorganization following secondary endosymbiosis, diatoms possess four layers of chloroplast envelope membranes, with the outermost being linked to the ER-nuclear envelope network, and a highly redundant metabolic system cooperating across organelles (Kroth et al. 2008; Prihoda et al. 2012). Marine diatoms are major primary producers responsible for up to one fifth of global carbon fixation (Tréguer et al. 1995; Falkowski et al. 2000) and their mechanisms for CO_2 fixation have important implications for biogeochemistry. Like cyanobacteria and green algae, diatoms are believed to possess CO_2-concentrating mechanisms (CCMs), which accumulate dissolved inorganic carbon (DIC) by pumping DIC biophysically from the surrounding media. This stored carbon is then mobilized by internal carbonic anhydrases (CAs), which are located in multiple intracellular compartments (Tachibana et al. 2011). The operation of CCMs is probably crucial in the ocean environment because of its high alkalinity and salinity, which change the equilibrium of DIC species and result in reduced rates of CO_2 formation via HCO_3^- breakdown and dehydration of H_2CO_3 (Goyet and Poisson 1989; Matsuda et al. 2001). This physical environment negatively effects the low affinity system of the major CO_2-fixing enzyme, ribulose-1,5-bisphosphate carboxylase/oxygenase (RubisCO) in diatoms. A part of the CCM in diatoms is thought to be

Abbreviations: AC – Adenylyl cyclase; CA – Carbonic anhydrase; CCM – CO_2-concentrating mechanism; DIC – Dissolved-inorganic carbon; FBA – Fructose-1,6-bisphosphate aldolase; FBPase – Fructose-1,6-bisphosphatase; Fd – Ferredoxin; HGT – Horizontal gene transfer; LCI – Low-CO_2-inducible proteins; NTT – Nucleotide translocators; OPP – Oxidative pentose phosphate pathway; PCOC – Photorespiratory carbon oxidation cycle; PEPCK – Phospho*enol* pyruvate carboxykinase; PPDK – Pyruvate orthophosphate dikinase; RubisCO – Ribulose-1,5-bisphosphate carboxylase/oxygenase; SLC – Solute carrier; Trx – Thioredoxin linked to C_4 type organic acid accumulation in some diatom strains (Reinfelder et al. 2004). This biochemical CCM would be enabled by a redundant organization of (de-) carboxylase enzymes presumably inherited from multiple symbiosis partners and by horizontal gene transfer (HGT) from bacteria. However, molecular details of these biophysical and biochemical CCMs are so far almost totally unknown in diatoms.

The recent annotation of diatom genomes revealed that a number of metabolic pathways involved in carbon metabolism and photosynthesis might be duplicated and/or shifted to different intracellular locations (Bowler et al. 2008; Kroth et al. 2008; Allen et al. 2012). Diatoms possess a large number of isoenzymes that according to phylogenetic analyses originate from different sources including the cyanobacterial primary endosymbiont, the primary and secondary host cells and from HGT from bacteria (Kilian and Kroth 2004; Moustafa et al. 2008; Allen et al. 2011). Consequently, diatoms seem to have developed an unusual metabolic cooperation across organelles and redundant duplications of pathways (Kroth et al. 2008; Prihoda et al. 2012). Cooperation on the metabolic level requires the exchange of substrates between the organelles. This strongly suggests the presence of a number of specific channels and transporters in the membrane systems of mitochondria, chloroplasts, and peroxysomes. In diatoms this exchange is especially complicated considering the additional chloroplast envelope membranes (see below).

There are thousands of unique genes in the diatom genomes, whose functions are yet completely unclear, and which may be involved in unknown processes of diatom carbon metabolism. Genetic tools to study molecular aspects of diatom physiology/biochemistry are so far limited, and there is no established system to control ploidy and the sexual reproduction cycle. Hence predictions of the functions of diatom genes so far largely rely on bio-informatics (Kroth et al. 2008; Fabris et al. 2012). However, recent technological improvements in genetic transformation (Zaslavskaia et al. 2000; Poulsen et al. 2006) and gene silencing (De Riso et al. 2009; Sakaguchi et al. 2011; Lavaud et al. 2012) of diatoms open the gate for functional identifications of genes by reverse genetics. In this chapter, we discuss molecular aspects of the carbon acquisition system, mechanisms of carbon fixation, and the respective carbon metabolism in diatoms, which includes information predicted from the genome analyses, focusing on data with direct experimental support. Although carbon metabolism is strongly connected to photosynthetic electron transport, we have excluded this topic, as it goes far beyond the scope of this review.

II. Structure of Diatom Cells in Relation to Their Evolutionary History

Diatom cells possess a silica-based cell wall. The main parts of the cell wall (termed valve) have highly symmetrical shapes and nanopores (termed areolae) within, whose structures must be imprinted in the genome as they are highly specific for individual species. Due to such chemically stable and geometrically shaped cell walls, diatoms have readily become well identifiable fossils and thus their emergence in the history of the earth is relatively clear. The oldest diatom fossils from the Jurassic are about 180 million years old (Kooistra et al. 2007), thus it is thought that diatoms have evolved in the Metazoic, most probably as marine species with radially symmetrical valves. This oldest type of diatoms are called Centrales or centric diatoms (Sims et al. 2006). Pennate diatoms (Pennales, bilaterally symmetrical) separated from the centrics in the late Cretaceous (Sims et al. 2006). After the Oligocene, diatom populations have expanded and diversified vigorously to about a 100,000 species, which also is their estimated number today (Norton et al. 1996; Prihoda et al. 2012).

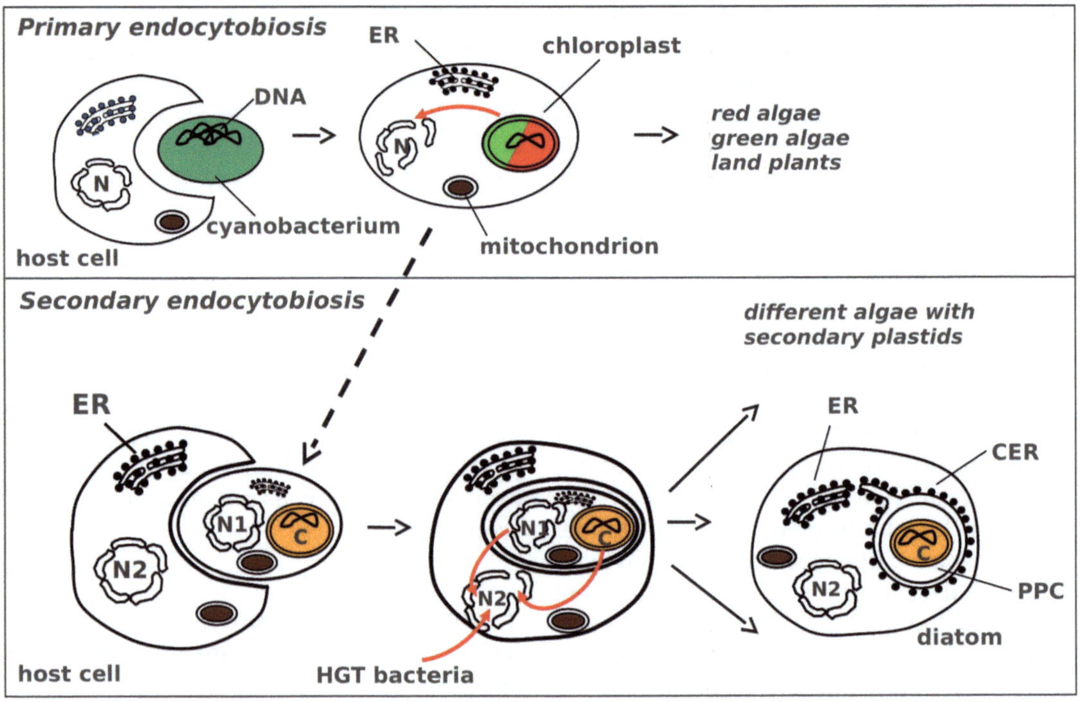

Fig. 18.1. Cellular processes involved in primary and secondary endosymbiotic events. *Top*: In the primary endocytobiosis event a unicellular cyanobacterium apparently has been taken up by a eukaryotic cell into a food vacuole. After the endosymbiont persisted in this vacuole, most of the genes of the endosymbiont had been transferred to the nucleus of the host cell. The eukaryotic algal ancestor developed into the different lineages of rhodophytes, glaucophytes and chlorophytes. *Bottom*: In a secondary endocytobiotic step, eukaryotic ancestors of either green or red algae were taken up by eukaryotic cells. Here only one possible development is shown including the reduction of redundant components of the endosymbiont-like nucleus, mitochondria, ER and further components. In diatoms, a later fusion of the food vacuole with the ER system of the secondary host cell allowed the utilization of the CER for cotranslational import of proteins into the plastids that possess four surrounding membranes. *ER* endoplasmic reticulum, *CER* chloroplast endoplasmic reticulum *N* nucleus, *HGT* horizontal gene transfer, *PPC* periplastidic compartment; *red arrows*: gene transfer processes.

A. Evolution of Diatoms Based Upon Their Genome Structure

As described above, diatoms are among the youngest groups of eukaryotic photoautotrophs. They belong to a large eukaryotic supergroup informally called the chromalveolates, comprising the chromista and the alveolates (Cavalier-Smith 1999). Within this group, there are a number of important marine primary producers, like the chromist algae (brown algae, diatoms, haptophytes and cryptomonads) and some plastid-bearing alveolates (dinoflagellates) (Cavalier-Smith 2000). Chromalveolates (and other groups such as excavates and rhizaria) arose by secondary endosymbioses. From the comparative analyses of multiple genes encoded on the genome of the chlorophyll *c*-containing plastids, it is thought that a diverse group of secondary endosymbionts arose either by a single or up to five independent events of phagocytosis of a red alga probably by a biciliate protozoan about 1,200 million years ago (Delwiche and Palmer 1997; Cavalier-Smith 1999; Yoon et al. 2002). The food vacuole, created in the host cell upon phagocytosis of the red alga, in turn became a stable peripheral vacuole. In the case of the evolution of chromista, this peripheral vacuole has fused with the nuclear envelope, placing the acquired chloroplast in the rough ER lumen (Cavalier-Smith 2000) (Fig. 18.1). This way chromist algae, including diatoms, have

Table 18.1. Size of diatom genomes.

Species name	Genome size	Number of ORF	Reference
Thalassiosira pseudonana	32 Mb	~11,000	Armbrust et al. (2004)
Phaeodactylum tricornutum	27 Mb	~10,000	Bowler et al. (2008)
Fragilariopsis cylindrus	80 Mb	~27,000	JGI[a]
Pseudo-nitzschia multiseries CLN-47	218 Mb	?	JGI[b]

[a]http://genome.jgi-psf.org/Fracy1/Fracy1.home.html
[b]http://genome.jgi-psf.org/Psemu1/Psemu1.home.html

established a very stable nucleus-chloroplast association (Fig. 18.1), which, howver, may be physically constraining the behavior of these two housekeeping organelles.

The first diatom nuclear genome that was fully sequenced belongs to the multipolar centric *Thalassiosira pseudonana* (Armbrust et al. 2004), followed by the nuclear genome of the raphid pennate diatom *Phaeodactylum tricornutum* (Bowler et al. 2008). Plastid and mitochondrial genomes of these model diatoms were also sequenced and compared to other genomes (Oudot-Le Secq et al. 2007; Oudot-Le Secq and Green 2011). The size of the nuclear genomes and the number of genes encoded are summarized in Table 18.1. Recent analyses of the genomes have revealed numerous footprints of an enrichment of foreign nuclear genes via serial endosymbioses, and HGT from proteobacteria as well as from Chlamydiae (Bowler et al. 2008; Moustafa et al. 2009). Of particular interest is that, in spite of possessing plastids of red algal origin, the diatoms *P. tricornutum* and *T. pseudonana* have a large number of nuclear genes that apparently are more related to green algae than to red algae (Moustafa et al. 2009). About 2,500 green or red type genes were identified in the nuclear genomes of these diatoms with about 70 % apparently being green-type genes, which accounts for 16 % of the nuclear genome, while only 5 % of the nuclear genome could be categorized to red-type genes (Moustafa et al. 2009; Prihoda et al. 2012). It is thus suggested that an ancestor of modern green algae might have been the first eukaryotic symbiosis partner, which later was replaced by an ancestor of red alga (Moustafa et al. 2009; Prihoda et al. 2012), however, there is still a controversy about the statistical significance of the genomic data (Deschamps and Moreira 2012). Apparently most genes for the photosynthetic apparatus were transferred from the initial green plastid to the nucleus of the host cell and kept even after the secondary red plastid was established, (Moustafa et al. 2009). These serial secondary endosymbioses together with the subsequent HGT events from bacteria have genetically enriched the nuclear genome of diatoms (Bowler et al. 2008; Moustafa et al. 2009; Qiu et al. 2013) (Fig. 18.1).

B. Chloroplast Structures in Diatoms

As described above, diatom chloroplasts are in a firm physical association with the nucleus linked by an ER network (Fig. 18.1). The outermost membrane of the chloroplast is termed chloroplast ER membrane (CER). In electron micrographs this membrane is often studded with ribosomes similarly as the rough ER (Fig. 18.1). The space between the CER membrane and the next envelope membrane is called CER lumen and probably is continuous with the ER lumen. The space between the second and the third membrane is called periplastidal compartment (PPC) and probably represents the strongly reduced cytosol of the secondary endosymbiont. The two inner membranes apparently are homologous to the two envelope membranes of red algal plastids (Fig. 18.1). The thylakoids in diatoms are generally stacked in pairs of three without grana, while directly below the innermost membrane there is a stack of three thylakoids surrounding most of the plastid (girdle lamella) (Fig. 18.1). In the central part of the chloroplast, there is a large undefined proteinaceous body called pyrenoid, which usually is penetrated

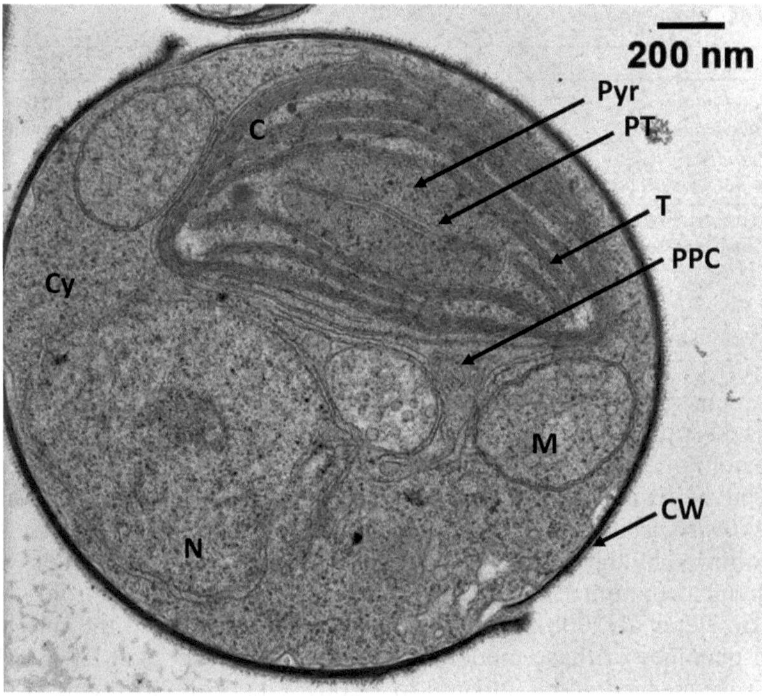

Fig. 18.2. Electron micrograph of a cross-section of *P. tricornutum* showing the plastid (C) including the pyrenoid (Pyr), the internal thylakoid membranes (PT), the regular thylakoid membranes (T) and the periplastidal compartment (PPC). *M* mitochondria, *N* Nucleus, *Cy* cytosol, *CW* cell wall (The images are shown by courtesy of A. Gruber and J. Hentschel, Universität Konstanz).

by a single line of thylakoid membranes (Fig. 18.2). Recent improvements of molecular-labeling techniques using fluorescent proteins like GFP have revealed reticular or vesicular structures that might be located in the CER (Kilian and Kroth 2005). This structure denoted as blob-like structure (BLS) is often observed when periplastidal preproteins or stromal preproteins with modified presequences fused to GFP are expressed in *P. tricornutum*. Thus this structure probably represents an accumulation of proteins in a certain area of the periplastidic space; however, structural details of BLS are unknown yet.

C. Traffic of Small and Large Molecules Across Chloroplast Membranes

Biomembranes represent unique barriers for the translocation of smaller molecules like metabolites or larger molecules like polypeptide chains. Membrane transporters are a large, structurally and functionally heterogeneous group of proteins, consisting of carriers and channels that span lipid bilayers. Chan et al. (2011, 2012) have compiled a large number of putative translocators by studying diatom genomes. However, the functionality of these proteins still has to be confirmed experimentally. Studying nucleotide import translocators, Ast et al. (2009) discovered the presence of six different nucleotide translocators (NTT) in *P. tricornutum* (and eight in *T. pseudonana*), which fall into three classes: some do not possess a presequence, others possess a signal and a transit peptide domain, while the third group additionally shows an "ASAFAP" motif (Gruber et al. 2007), leading to the suggestion that these translocators are integrated in different plastid envelope membranes. Two translocators have been studied more thoroughly, NTT1 and NTT2. NTT1 is more similar to plastid translocators of land plant and algal plastid ATP/ADP exchangers while NTT2 is related to nucleotide importers from parasitic endocytic bacteria. Interestingly the diatom NTT1 shows properties of an H^+-driven ATP translocator, and NTT2 of a nucleotide exchanger (Ast et al. 2009), indicating that the original functional properties of these translocators have been completely altered in diatoms. The apparent capability of diatoms to acquire and combine different translocator systems in order to overcome the additional barriers represented by the four plastid envelope membranes is also reflected in the plastid protein import machinery. Targeting of proteins into diatom plastids has been studied intensely in the recent years. Analyses of nuclear-encoded plastid proteins revealed the presence of an N-terminal ER targeting signal peptide fused to a transit peptide domain known from land plant plastids (Kroth 2002). The presequence structure of proteins targeted to the stroma additionally possesses an "ASAFAP" motif at the interface of both domains. Interestingly the penylalanine residue in this motif appears to be essential for a complete import (Kilian and Kroth 2005). Although the details of import pathway have not been solved yet completely, there is evidence that four different import machineries are involved in this process: (i) in a first step the protein is targeted cotranslationally across the CER membrane using the ER Sec system (Lang et al. 1998), (ii) in the next step a duplicated endosymbiont-specific ER-associated degradation system for lumenal proteins (ERAD-L) may be involved (Sommer et al. 2007), (iii) the next membrane may be traversed by a translocator similar to bacterial Omp85 and plastidic Toc75 proteins (Bullmann et al. 2010), while (iv) for the innermost membrane a protein complex related to plastidic Tic complexes may be involved (Vugrinec, Gruber, Kroth, unpublished). Once the proteins have entered the stroma, they can be sorted to further compartments like the innermost of the four membranes surrounding the plastids, the thylakoid membrane or the thylakoid lumen. Similar to land plant plastids, different import pathways are known for targeting proteins into diatom thylakoids (Lang and Kroth 2001; Chaal et al. 2003).

III. CO_2-Concentrating Mechanisms in Cyanobacteria, Green Algae, and Marine Diatoms

The entry of CO_2 into plant cells is a limiting factor of CO_2 acquisition for a large majority of autotrophic organisms from chemolithotrophic bacteria to higher plants. In higher plants, CO_2 permeates across mesophyll cells through specific channels that belong to the family of aquaporins (Uehlein et al. 2003). The acquisition of CO_2 in water is further hampered, as the concentration of CO_2 dissolved in water is very low (less than 15 μM at the current atmospheric pCO_2 at 20 °C) and diffusion constants of dissolved molecules are 10^4 times smaller compared to those of the gaseous state (Badger et al. 1998). This problem is more serious in marine environments because of the highly alkaline/saline seawater, in which CO_2 formation from HCO_3^- is significantly slower (1/4~1/8) compared to that in freshwater (Matsuda et al. 2001). Aquatic photoautotrophs thus experience difficulties in acquiring CO_2 from surrounding media and it has been demonstrated that they generally are able to take up both CO_2 and HCO_3^- to accumulate dissolved-inorganic carbon (DIC) intracellularly, a process which also is termed biophysical CO_2-concentrating mechanism (CCM) (Badger et al. 1998; Falkowski and Raven 2007; Hopkinson et al. 2011).

A. Lessons from Molecular Studies in Cyanobacteria and Green Algae

Molecular details of the CCMs have been studied in cyanobacteria, where a whole set of CCM components, has been elucidated that fully account for the cyanobacterial CCM. In freshwater β-cyanobacteria, HCO_3^- is directly taken up by three plasma-membrane HCO_3^- transporters, a bicarbonate transporter 1 (BCT1), a sodium bicarbonate transporter A (SbtA), and a bicarbonate transporter A (BicA). Permeated CO_2 is also converted into HCO_3^- by the NADPH-dehydrogenase-CO_2-hydration protein (Ndh-Chp) complexes at the thylakoid membrane and perhaps at the cytoplasmic membrane (Omata et al. 1999; Maeda et al. 2002; Shibata et al. 2002; Price et al. 2004). Accumulated HCO_3^-, a leakage resistant form of dissolved-inorganic carbon (DIC), is then converted into CO_2 only when HCO_3^- passes across the shell of the carboxysome, an icosahedronic crystal body in the cell (Kerfeld et al. 2005) (see Chap. 7). This process is catalyzed by a carbonic anhydrase (CA), a component of the carboxysome shell (So et al. 2002), and released CO_2 is fixed by RubisCO within the carboxysome. The carboxysomes of similar structural characteristics are also detected in chemolithotrophs like *Halothiobacillus neapolitanus* (Dou et al. 2008). Cyanobacterial cells (see Chap. 14) at full operation of the CCM behave like a leakage-proof HCO_3^- chamber with an "efficient CO_2-formation and fixation reactor", which is the carboxysome.

The cyanobacterial system is believed to have evolved after the primary endosymbiosis (Badger et al. 2002). Algal cells have developed a different-type of CCM based upon the pyrenoid. The presence of a pyrenoid does not necessarily mean that the cells possess a CCM (Morita et al. 1998; Ratti et al. 2007; Genkov et al. 2010), but there is the possibility that the pyrenoid functions as a focal point of the CCM. There is some evidence for biochemical functions of the pyrenoid in microalgae. In the freshwater chlorophyte *Chlamydomonas reinhardtii*, a thylakoid-lumenal CA (CAH3), and a pyrenoidal component denoted Low-CO_2-inducible proteins LCIB and LCIC have been demonstrated to provide CO_2 supply to RubisCO under moderately limiting CO_2 conditions (Spalding et al. 1983; Yamano et al. 2010). CAH3 presumably forms CO_2 from HCO_3^- utilizing the acidity of the thylakoid lumen, while the LCIB/C complex is speculated to recapture the CO_2 leaking out of the stroma, forcing it back to the pyrenoid (Yamano et al. 2010; Duanmu and Spalding 2011). A plasma-membrane transporter, LCI1, has been demonstrated to be a HCO_3^- channel which also facilitates the CO_2 acquisition of *C. reinhardtii* under CO_2 limitation (Ohnishi et al. 2010). The function of extracellular CAs as a part of the green algal CCM has been discussed controversially (see Ynalvez et al. 2008).

B. Transports of Inorganic Carbon across the Plasma Membrane and the Chloroplast Envelopes

Both, freshwater and marine diatoms, have been shown to actively take up HCO_3^- and/or CO_2 (Patel and Merrett 1986; Colman and Rotatore 1995; Rotatore et al. 1995; Johnston and Raven 1996; Mitchell and Beardall 1996; Korb et al. 1997; Burkhardt et al. 2001; Rost et al. 2003; Trimborn et al. 2008). The list in Table 18.2 (redrawn from Matsuda et al. 2011) shows the respective results in a number of publications, which all employed relatively direct and reliable measures of DIC accumulations: (silicon-oil centrifugation: SOC) or steady-state DIC flux (membrane-inlet mass spectrometry: MIMS; isotopic disequilibrium technique: IDT). Accordingly, it is clear that these diatom species can take up both CO_2 and HCO_3^- from the surrounding media with significant species-dependent preferences of the respective DIC molecules. Probably, HCO_3^- consumers have adapted to use abundant HCO_3^- in seawater, albeit direct uptake of HCO_3^- can be energetically expensive. Conversely, in marine photoautotrophs, HCO_3^- uptake has been reported to be mediated by an external CA form, which might be a less energy consuming strategy for the use of external HCO_3^-. However, so far there is little molecular evidence, and proteins involved in DIC uptake and the mechanisms of DIC flux regulations within the diatom cells are not known.

A recent study has revealed that plasma membrane-type transporters homologous to the mammalian solute-carrier (SLC) family work as an initial entry gate of DIC in marine diatoms in a low CO_2 environment

Table 18.2. Dissolved inorganic carbon species taken up by diatoms (Redrawn from Matsuda et al. 2011).

Species	pH at measurement	DIC Transport		Method	Reference
		CO_2	HCO_3^-		
Stellarima stellaris	7.9–8.4	No	Yes	MIMS	Trimborn et al. (2008)
Pseudo-nitzschia multiseries	7.9–8.4	Yes	Yes	MIMS	Trimborn et al. (2008)
Nitzschia navis-varingica	7.9–8.4	Yes	No	MIMS	Trimborn et al. (2008)
Nitzschia frigida	7.5	Yes[a]		SOC	Mitchell and Beardall (1996)
Phaeodactylum tricornutum	8.0	Yes		SOC	Patel and Merrett (1986)
	7.5	Yes		SOC	Colman and Rotatore (1995)
	8.0	Yes	Yes	MIMS	Burkhardt et al. (2001)
	7.5	Yes		SOC	Colman and Rotatore (1995)
	8.0	Yes		SOC	Johnston and Raven (1996)
Cyclotella sp.	7.5	Yes		SOC	Colman and Rotatore (1995)
	7.5	Yes	–	MIMS	Rotatore et al. (1995)
Thalassiosira weissflogii	8.0	Yes	Yes	MIMS	Burkhardt et al. (2001)
Skeletonema costatum	8.0	Yes	Yes	IDT	Korb et al. (1997)
	8.0	Yes	Yes	MIMS	Rost et al. (2003)
Ditylum brightwellii	8.0	Yes	Yes	IDT	Korb et al. (1997)
Chaetoceros calcitrans	8.0	Yes	Yes	IDT	Korb et al. (1997)

MIMS membrane inlet mass spectrometry, *SOC* silicon-oil centrifugation, *ITD* isotopic disequilibrium technique
[a]Ci species could not be differentiated

(Nakajima et al. 2013). One subfamily of the diatom SLC, PtSLC4-2, was shown to transport specifically HCO_3^- in the presence of relatively high concentrations of sodium ions with a saturation point of about 100 mM Na^+ (Nakajima et al. 2013; Hopkinson 2013). This protein, together with other two closely related putative transporters PtSLC4-1 and PtSLC4-4 are induced specifically under CO_2-limited conditions, indicating that several transporters engage in the enhancement of HCO_3^- influx into the cell and perhaps into the plastid under low CO_2 conditions (Nakajima et al. 2013; Hopkinson 2013). Accumulated DIC might pass across the chloroplast membranes by specific transporters (Hopkinson 2013) and CAs densely localized in the matrixes of the four-layered chloroplast membranes very likely control the efficiency of permeation (Tachibana et al. 2011; Samukawa et al. 2014). It is interesting that diatom's SLC4 family transporters cluster phylogenetically with members of the human SLC4 family with reasonably high bootstrap values at the transmembrane region (Nakajima et al. 2013), while HCO_3^- transporters already identified in cyanobacteria and *C. reinhardtii* do not share homology with the respective proteins in diatoms. This strongly suggests that eukaryotic algae perhaps have acquired HCO_3^- transporters independently from various ancestral eukaryotic hosts. Indeed, it has been suggested that diatom transporters share a common origin with those in human cells (Nakajima et al. 2013).

C. Convergent Evolution of Carbonic Anhydrases

There are six subtypes of CAs so far found in living organisms and they are categorized from α to ζ (Tripp et al. 2001; So et al. 2004; Lane et al. 2005). The α-CAs are found predominantly in animals but also occur in bacteria, higher plants and green algae, the β-CAs are known to be abundant in plants, green algae, eubacteria and archaea, and the γ-CAs may be the most ancient form protein related to CAs. They have evolved long before the α class and are predominantly found in bacteria, although their activity is so far confirmed only in very limited candidates (Tripp et al. 2001). The eukaryotic algae studied so far possess α- to γ- CAs

(Satoh et al. 2001), with some exceptions such as the red alga *Cyanidioschyzon merolae*, which lacks α-CA genes (Matsuzaki et al. 2004). Interestingly, among diatoms, β-type CAs have only been found so far in the pennate diatom *P. tricornutum*, but there is no homologous gene in the genome of the centric diatom *T. pseudonana* (Montsant et al. 2005; Tachibana et al. 2011) or the polar-sea species, *Fragilariopsis cylindrus*, (http://genome.jgi-psf.org/Fracy1/Fracy1.home.html).

The δ- and ζ-type CAs are unique in diatoms and were first discovered in *Thalassiossira weissflogii*, denoted as TWCA1 and CDCA1 (Roberts et al. 1997; Lane and Morel 2000; Lane et al. 2005; Xu et al. 2008). These two CA types are also present in the centric diatom *T. pseudonana*, while a δ-type CA was found in the haptophyte *Emiliania huxleyi* (Soto et al. 2006) and the dinoflagellate *Lingulodinium polyedrum* (Lapointe et al. 2008). These two "marine CAs" were identified as *bona fide* CAs (Xu et al. 2008; Viparelli et al. 2010; Alterio et al. 2012; Lee et al. 2013). The crystal structure of the ζ-type CA has revealed a striking similarity of the active center structure to that of β-CA (Xu et al. 2008), while the active center of δ-CA is reported to be similar to that of α-CA (Cox et al. 2000). Indeed, it was recently reported that δ-CA in *T. weissflogii* possess an esterase activity, a feature unique to α-CAs (Lee et al. 2013).

The ε-CA was first identified in the chemolithotrophic bacterium *H. neapolitanus* as CsoS3/CsoSCA, representing a component of the carboxysomal shell. Genes homologous to *csoS3* were also found in cyanobacteria and chemolithotrophic bacteria (So et al. 2004). Since the crystal structure of ε-CA is very similar to β-type CAs, despite its unique primary amino acid sequence, this protein was later recategorized into β-type CA (Sawaya et al. 2006). CsoS3/CsoSCA is essential for the operation of the CCM in *H. neapolitanus* because of its role in dehydrating HCO_3^- at the carboxysomal shell in order to supply CO_2 to RubisCO within the carboxysome (Dou et al. 2008).

The discovery of these novel lineages of convergent evolution revealed a strong variation between CAs that have evolved from an unexpectedly wide variety of origins and functions in marine autotrophs. Moreover, the cofactor of diatom's ζ-CAs can be substituted by Cd when Zn is not available (Xu et al. 2008), strongly suggesting their competence in a metal depleted environment. However, the function of numerous CAs in diatoms are still speculative and need to be identified at the molecular level.

D. Importance of Carbonic Anhydrases in Algal CO_2-Concentrating Mechanisms

CAs catalyze the reversible reaction of CO_2 hydration and thus their physiological function deeply relies on the local pH of the individual cellular location of the enzymes. Recent investigations revealed that RubisCO and β-type CAs (denoted as PtCA1 and PtCA2) are located in the pyrenoid of *P. tricornutum* and that these CAs are induced remarkably at low CO_2 concentrations. The biochemical composition of the pyrenoid in algae is not well known, but similarities have been demonstrated among a limited number of proteins localized in the pyrenoids of eukaryotic algae. CA in the areas of pyrenoids are likely responsible for efficient formation of CO_2 at the proximity of RubisCO in the pyrenoid. This process is probably affected by the redox condition in the stroma via thioredoxins (Trxs) in response to the light intensities and the oxygen concentrations, in order to ensure optimal energy flow and carbon fixation under changing environmental conditions (described in the following section).

The occurrence and the function of external CAs in microalgae have been discussed controversially. It has been postulated that CA activity in the periplasmic space may enhance CO_2 consumption in alkaline environments, where the major carbon source is HCO_3^- . However, silencing of the external CA CAH1 in *C. reinhardtii* did not alter cells to a high CO_2-requiring phenotype (Van and Spalding 1999), although the possibility still remains that the other external CA, CAH8 may complement the function of CAH1 (Ynalvez et al. 2008). In marine photoautotrophs, the importance of external CAs has been stressed as well as that of acidification at the periplasmic unstirred layer, a space surrounding cells where cellular activities of H^+ and OH^- pump are the predominant factor to determine pH (Milligan and Morel 2002). As aforementioned, high salinity and alkalinity of seawater tend to remove CO_2 from cell exterior which may force cells to invest energy in pumping HCO_3^-, while CO_2 uptake via external CA is energetically less expensive. In fact, in the dinoflagellates *Prorocentrum micans* and *Lingulodinium polyedrum*, external CAs have been shown to be essential (Nimer et al. 1999; Lapointe et al. 2008). However, it is also known that external CAs are absent in some marine microalgae (Danson et al. 2004). In marine diatoms, the occurrence of external CAs can be inconsistent even in a single species (John-Mckay and Colman 1997).

In contrast, intracellular CAs are likely an essential component of the CCM and of photosynthesis under CO_2 limitations. For example, knock-out of the internal β-CA gene *icfA* (later renamed as *ccaA*) in cyanobacteria resulted in a high CO_2-requiring phenotype (Fukuzawa et al. 1992). Similarly, a mutation of *Cah3* caused a high CO_2-requiring phenotype in *C. reinhardtii* (Funke et al. 1997). The respective CAs have been localized in the carboxysome shell in cyanobacteria (Long et al. 2007) and the thylakoid lumen in *C. reinhardtii* (Karlsson et al. 1998). They are believed to supply CO_2 from accumulated HCO_3^- at the cellular proximity of RubisCO (Funke et al. 1997; Raven 1997; Van and Spalding 1999). In diatoms, internal CAs have been demonstrated to be essential on the physiological level. In *P. tricornutum*, the highly permeable CA inhibitor ethoxyzolamide (EZA) severely suppressed the operation of high-affinity photosynthesis, whereas the weakly permeable inhibitor acetazolamide (AZA) had little effect (Satoh et al. 2001), strongly suggesting that the role of external CA is rather small compared to that of internal CAs.

Table 18.3. Carbonic anhydrases and their localization in *P. tricornutum*.

Name	Class	Location
CA-I	α	PPC
CA-II	α	PPC
CA-III	α	CER
CA-IV (PtCA1)	β	Pyrenoid
CA-V (PtCA2)	β	Pyrenoid
CA-VI	α	CER
CA-VII	α	CER
CA-VIII	γ	Mitochondria
CA-IX	γ	N.D.
CA-X	ζ	N.D.

Data are summarized from Tachibana et al. (2011)
PPC periplastidal compartment, *CER* chloroplastic endoplasmic reticulum, *N.D.* not detected

E. Localizations and Functions of Diatom Carbonic Anhydrases

As summarized in Table 18.3, the individual location of diatom CAs has been extensively investigated in the pennate marine diatom, *P. tricornutum* (Tanaka et al. 2005; Kitao et al. 2008; Kitao and Matsuda 2009; Tachibana et al. 2011) and the centric diatom, *T. pseudonana* (Tachibana et al. 2011). There are genes encoding 5 α-, 2 β-, and 2 γ-type CAs in the genome of *P. tricornutum* and these subclasses are located in specific organelles depending on the subtype (Tachibana et al. 2011). All putative α-CAs were located in the four-layered chloroplast envelopes, two β-CAs were within the pyrenoid, and one putative γ-CA, CA-VIII, was found in the mitochondria (Tanaka et al. 2005; Kitao et al. 2008; Tachibana et al. 2011). The remaining γ-CA candidate, CA-IX most likely is also located in the mitochondria (Nawary and Matsuda, unpublished). A total of 13 CA candidate genes have been identified in the genome of *T. pseudonana* (encoding 3 α-, 5 γ-, 4 δ-, 1 ζ-type enzymes). Nine CAs out of these 13 candidates have recently been localized. That is, α-CA1 was localized in the stroma; γ-CA1, 3, and 4 were in the mitochondria; γ-CA2 was in the cytosol; δ-CA1, 2, and 3 were respectively localized in the periplasmic space, mitochondria, and periplastidal compartment; ζ-CA1 was localized at the periplasmic space (Tachibana et al. 2011; Samukawa et al. 2014). However, the localization of the other CAs awaits further studies. Nonetheless, the CA localization data in *T. pseudonana* displayed a significant diversity in CA localization in diatoms, strongly suggesting the diversity of the mode of DIC flux control in diatoms (Samukawa et al. 2014).

In *P. tricornutum*, five α-CAs that are located in the chloroplast envelope system, are transcriptionally active and are expressed independently of the CO_2 concentrations (Tachibana et al. 2011). This suggests that these CAs may work constantly to control the permeation of DIC from the cytoplasm to the stroma and *vice versa*. This finding may be strongly related to the occurrence of DIC channels in the envelope membranes and potentially different pH values in the compartment between the plastid envelope membranes, although the details are yet clear. Two pyrenoidal β-type CAs, PtCA1 and PtCA2, are exclusive CO_2-responsive CAs in *P. tricornutum*, being de-repressed under low CO_2 conditions (Harada and Matsuda 2005; Harada et al. 2005; Tachibana et al. 2011), in which the pyrenoid would constitute a strong driving force to direct CO_2 to RubisCO for fixation. During such a physiological state – where CO_2 is efficiently directed to RubisCO under CO_2 limitations – the input of DIC from the bulk medium should also be ensured by specific transporters (as described above).

The putative γ-CAs in the mitochondria studied so far are not related to photosynthetic carbon fixation, and the activity of γ-CA candidate is not confirmed so far. Their function is thus so far obscure, but their mitochondrial localization strongly suggests that some of these proteins are involved in anaplerotic processes to provide HCO_3^- to phosphoenol pyruvate or pyruvate in the mitochondria (Giordano et al. 2003).

F. Is There a C_4 Metabolism in Diatoms?

A C_4-type biochemical CCM was shown to be a part of the diatom CCM in the relatively large marine centric diatom, *Thalassiosira weissflogii* (Reinfelder et al. 2000, 2004; Roberts et al. 2007b; McGinn and Morel 2008). However, radio tracer experiments for photosynthetic products revealed in another centric species, *T. pseudonana*, the absence of a transient increase in C_4 metabolites prior to the increment of 3-PGA, indicating that this centric diatom fixes CO_2 by a C_3-based metabolism. It is thought that the occurrence of C_4-type photosynthesis might be limited to particular species (Roberts et al. 2007b). Haimovich-Dayan et al. (2013) recently investigated gene silencing of the pyruvate orthophosphate dikinase (PPDK) gene in *P. tricornutum* cells and proposed that the C_4 metabolism may not increase net CO_2 fixation but rather helps the cells to dissipate excess light energy and to maintain pH homeostasis. Genes encoding carboxylating and decarboxylating enzymes exist in the genomes of *P. tricornutum* and *T. pseudonana*. *In silico* analyses of both *P. tricornutum* and *T. pseudonana* genome databases indicate that the decarboxylating enzymes phosphoenol pyruvate carboxykinase (PEPCK) and malic enzyme do not possess plastid targeting sequences (Kroth et al. 2008), and in fact, a recent localization study has shown the lack in the decarboxylation enzyme in the chloroplast of *T. pseudonana* (Tanaka et al. 2014). Moreover, there is no evidence so far for the presence of malate and/or oxaloacetate transporters in the plastid membranes (Kroth et al. 2008). Supply of oxaloacetate to intact chloroplast isolated from the diatom *Odontella sinensis* did not stimulate photosynthetic oxygen evolution (Wittpoth et al. 1998), also suggesting the potential import of oxaloacetate into the plastids of diatoms. It is interesting that C_4-type diatoms like *T. weissflogii* still actively take up both CO_2 and HCO_3^- at equal rates (Table 18.2) (Burkhardt et al. 2001). One explanation could be that C_4-type diatoms might also employ active DIC uptakes, which consequently could provide a part of CO_2 directly from the developed DIC pool to the Calvin cycle, constituting a hybrid-type, i.e., a biophysical and a biochemical CCM (McGinn and Morel 2008).

The evolutionary driving force in diatoms to acquire a biochemical CCM in addition to the existing biophysical CCM is unknown. One of the possible explanations could be the extreme diversification of diatom cell sizes that might force the development of a redundant CCM system to overcome diffusion resistance of pooled DIC molecules (Matsuda et al. 2011). Alternatively (or simultaneously), it is also possible that both, biochemical and biophysical CCMs, may contribute to dissipation of light energy by consuming ATP. Of particular interest is that the biophysical CCM as well as the C_4 system could be an efficient consumption system of ATP by cycling CO_2 through uptake and leakage across the cytoplasmic membrane when pooled DIC is not fixed efficiently and light excitation pressure continuously closes the photosystems (Tchernov et al. 1997).

G. Transcriptional Regulation of CO_2-Concentrating Mechanism Components

Algal CO_2 acquisition systems are regulated at the transcriptional level via CO_2 concentrations. This phenomenon raises the intriguing question how algal cells sense the concentration of environmental CO_2. Furthermore it is unclear how such control mechanisms within the algal CCM may influence net photosynthesis at increasing environmental CO_2 concentrations, which are expected to occur in the future due to human activity.

Physiological aspects of the CO_2 response of CCMs have been reported in cyanobacteria and green algae since the early 1980s (Marcus et al. 1983; Mayo et al. 1986; Dionisio-Sese et al. 1990; Sueltemeyer et al. 1991; Matsuda and Colman 1995a, b). In summary, two systems have been proposed: (i) In cyanobacteria, the critical controlling factor for overall CCM activity is the CO_2/O_2 ratio (Marcus et al. 1983) and the total amount of DIC in the medium (Mayo et al. 1986), but not the pH or individual DIC species (Mayo et al. 1986). This strongly suggests the participation of a photorespiratory pathway in transmitting the CO_2 signal to gene expression in cyanobacteria (Marcus et al. 1983; Mayo et al. 1986). (ii) On the other hand, CCMs in the green algae *Chlorella ellipsoidea* and *C. reinhardtii* respond to $CO_{2(aq)}$ in the medium rather than other DIC species or total DIC (Matsuda and Colman 1995b; Bozzo and Colman 2000). Furthermore, neither the internal DIC concentration nor the light intensity was correlated with the CCM expression levels in *C. ellipsoidea*, suggesting that a CO_2 response mechanism in eukaryotic algae does not necessarily depend on metabolite feedback, but also includes more direct sensing mechanisms (Matsuda and Colman 1995b).

The metabolite signaling hypothesis in cyanobacteria is supported indeed, by the molecular evidence that a transcription factor of *Synechocystis* sp. PCC 6803, CmpR, which is the LysR family inducing factor of the HCO_3^- transporter operon *cmpABCD*, can bind directly to the promoter region of the *cmpABCD* operon and that binding is significantly stimulated in the presence of a physiological concentrations of 2-phosphoglycolate, the first product of oxygenase reaction by RubisCO (Omata et al. 2001; Nishimura et al. 2008). In eukaryotic algae, on the other hand, most of the molecular work on transcriptional controls of putative CCM components has been done with the green alga *C. reinhardtii*. The periplasmic CA gene *Cah1* in *C. reinhardtii* is a well known CO_2-responsive gene and regulation of the promoter region of *Cah1* are governed by tandemly aligned enhancer and silencer regions on the promoter (Kucho et al. 1999). A zinc finger protein, CCM1/CIA5 was found to control the up-regulation of most of the low-CO_2-inducible genes including *Cah1* and thus is considered to be a master regulator for CO_2-responsive gene expressions (Fukuzawa et al. 2001; Xiang et al. 2001; Miura et al. 2002, 2004; Wang et al. 2005; Kohinata et al. 2008; Yamano et al. 2008). However, the key factor for capturing the CO_2 signal in algae is unknown.

Similarly to green algae, the major determinant of the extent of CCM expression in *P. tricornutum* likely is the $CO_{2(aq)}$ concentration in the medium rather than the pH value

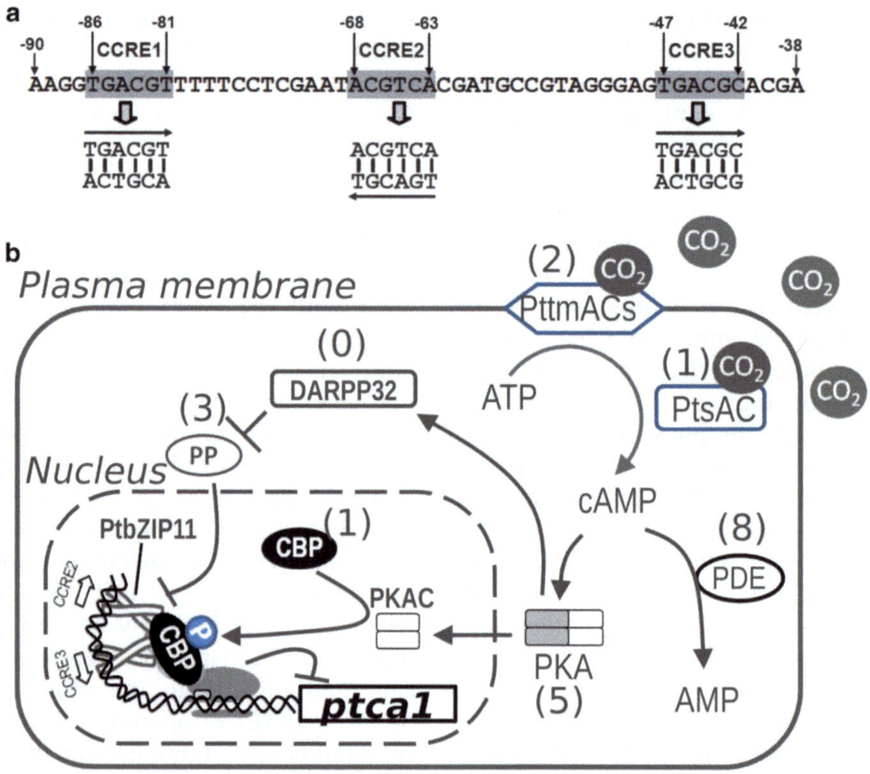

Fig. 18.3. The CO_2 responsive *ptca1* promoter and the putative CO_2 signaling pathway. (**a**) Structure of the core-regulatory region of the ptca1 promoter with 3 CCRE elements (Redrawn from Ohno et al. 2012). (**b**) Putative CO_2 signaling pathway predicted based upon the mammalian cAMP signaling cascade. *PDE* cAMP phosphodiesterase, *PKA* protein kinase A, *PKAC* C subunit of PKA, *PP* protein phosphatase, *CBP* CREB (cAMP binding protein) binding protein, *DARPP* dopamine-and cAMP-regulated phosphoprotein. *Parentheses* indicate the number of candidate genes in the *P. tricornutum* genome. So far, there is no known candidate for a functional analogue of DARPP. The model depicts the signaling route under high CO_2 conditions, which may activate tmACs and/or sACs. PtbZIP11 binds to CCREs to form a repressor complex on the *ptca1* promoter.

or other DIC species (Matsuda et al. 2001). The number of known typical CO_2 responsive molecules is still low in marine diatoms and the aforementioned PtCA1 and 2 enzymes have been the best model for molecular studies on the mechanisms of CO_2 responses. The genes *ptca1* and *ptca2* are known to be markedly de-repressed at the transcriptional level in response to a decrease in [CO_2] (Satoh et al. 2001; Harada et al. 2005, 2006). The β-glucuronidase (GUS) reporter assay of the promoter region of *ptca1* (P*ptca1*) revealed that a relatively short sequence of up to 100 bps upstream the transcription-start site plays a key role in CO_2-responsive transcription, and that this region is rich in animal-type cAMP-response elements (Harada et al. 2006). Indeed, the ability of the cAMP analogue dibutyryl cAMP (dbcAMP) to mimic high CO_2 conditions and to repress transcription of *ptca*s (Harada et al. 2006) disappears by impairing the P*ptca1* sequences (Harada et al. 2006). Further investigations have revealed that the critical CO_2-responsive *cis*-element of P*ptca1* contains three analogous sequences TGACGT/C which are tandemly aligned in an invert direction within the core P*ptca1* sequence. These were denoted CO_2/cAMP responsive elements (CCRE) 1, 2, and 3 (Ohno et al. 2012) (Fig. 18.3). The CCREs are the typical targets of mammalian ATF6-type basic-ZIP-transcription factors such as cAMP-responsive-element-binding protein

(CREB). Therefore the genome of *P. tricornutum* was screened for genes encoding potential ATF6-type-bZIP proteins based upon the well conserved DNA binding site of ATF6 with an adjacent leucine zipper motif (Ohno et al. 2012). As a result, one of the eight candidate bZIP proteins, denoted PtbZIP11, was found to bind to CCREs specifically (Ohno et al. 2012) (Fig. 18.3). These investigations clearly revealed that diatoms use cAMP signaling, which is perhaps similar to the mammalian system, to respond to ambient CO_2 concentrations (Fig. 18.3). In fact, most of the genes involved in the typical mammalian-type cAMP signaling pathways have been identified in diatom EST databases (Fig. 18.3).

Adenylyl cyclase (AC) is known to serve as a sensor for inorganic carbon signals in a wide variety of living organisms from cyanobacteria, fungi, to mammals, except for higher plants (Chen et al. 2000; Klengel et al. 2005; Matsuda et al. 2011). In rat testis, the activity of soluble adenylyl cyclase (sAC) is modulated by HCO_3^- and pH, and this sAC is closely related to bacterial-type sACs (Chen et al. 2000). In cyanobacteria, indeed, the activities of sAC, CyaB1 (*Anabaena* sp.) and Cya1 (*Synechocystis* sp. PCC6803) were shown to be regulated by inorganic carbon (most probably by CO_2) (Hammer et al. 2006). Cyclic AMP in cyanobacteria has been related to blue-light responses, the mating system, and cell division, but so far not to CCM regulation (Matsuda et al. 2011). In diatoms, there are also putative CO_2-sensing ACs (one for soluble ACs and two for transmembrane-type ACs, tmAC1 and tmAC2 in *P. tricornutum*; there are two candidate tmACs in *T. pseudonana*), which were denoted PtsAC, PttmAC1, and PttmAC2, respectively, in *P. tricornutum*, and TptmAC1 and TptmAC2 in *T. pseudonana* (Matsuda et al. 2011). These ACs were suggested to be a part of CO_2-sensing system in *P. tricornutum*, and in fact, a significant stimulation of AC activity in total lysates of *P. tricornutum* by DIC was observed, although the function of each AC yet has to be identified (Matsuda et al. 2011). CCM regulation in *P. tricornutum* is highly depending on light as well as on CO_2 concentrations (Harada et al. 2005), strongly suggesting the occurrence of a crosstalk between the CO_2-responsive cAMP-signaling pathway and light.

IV. Delivery Systems of CO_2 to RubisCO and CO_2 Fixation

The final step of the CCM perhaps cooperates deeply with the initial step of the Calvin cycle. The CO_2-fixing enzyme, RubisCO in diatoms belongs to the Form I (red-type) enzyme (Badger et al. 1998); the activating mechanism has yet to be elucidated. RubisCO as well as CAs have been localized in diatom pyrenoids (Jenks and Gibbs 2000; Tachibana et al. 2011), strongly suggesting that the CO_2 concentration is elevated in close proximity of RubisCO by the formation of CO_2 from HCO_3^- catalyzed by CAs (Hopkinson et al. 2011). However, detailed mechanisms of CO_2 supply to the Calvin cycle and regulations of the CCM and the Calvin cycle still have to be elucidated in diatoms.

A. Biochemical Components of the Pyrenoid

Pyrenoids have been studied on the morphological level, but their biochemical structure and function are still largely unknown. The morphology of the pyrenoid can be extremely diverse, but there is similarity in biochemical components associated with the pyrenoid in different species. In the green alga *C. reinhardtii* most of RubisCO protein is located in the pyrenoid (Lacoste-Royal and Gibbs 1987; Kuchitsu et al. 1988; Borkhsenious et al. 1998). Similarly, pyrenoids of *Chlorella pyrenoidosa*, *Porphyridium cruentum*, *Euglena gracilis*, and the *Anthocerotae* ferns were identified as RubisCO aggregates (Mckay and Gibbs 1989, 1990; Osafune et al. 1990; Vaughn et al. 1990). Besides RubisCO, nitrate reductase (NR), RubisCO activase, and the LCIB/C complex were also localized in the pyrenoid of *C. reinhardtii* (Lopez-Ruiz et al. 1985; McKay et al. 1991; Yamano et al. 2010). In *C. reinhardtii*, CAH3 was also found in the lumen of the pyrenoid-penetrating thylakoid (Karlsson et al. 1998). The function of the LCIB/C complex is not clear yet, but impairment of LCIB results in a lethal phenotype under ambient CO_2, while the cells survive under reduced CO_2 concentrations. This indicates its vital function under moderate CO_2 starvation. It further indicates that another high-affinity system may participate under severer CO_2 starvation conditions (Wang and Spalding 2006). LCIB homologues were also found in many eukaryotic algae including diatoms (Yamano et al. 2010), suggesting that this protein is conserved in pyrenoids across species. It has also been reported earlier that ribulose-5-phosphate isomerase and ribulose-5-phosphate kinase activities were detected in the pyrenoidal fraction of the giant green alga *Eremospheaera viridis* (Holdsworth 1971), suggesting that a part of reactions in the Calvin cycle may occur in the pyrenoid.

In *P. tricornutum*, the structure of the pyrenoid has been reported as a lens-shaped body residing in the central part of the stroma. Two layered thylakoid membranes are known to penetrate in the center of the pyrenoid (Fig. 18.2) (Jenks and Gibbs 2000). RubisCO was localized in the pyrenoid of *P. tricornutum* (Jenks and Gibbs 2000), together with two β-type CAs (PtCA1 and PtCA2) and two plastidic fructose-1,6-bisphosphate aldolases (FBAs) (Allen et al. 2012). PtCA1 and 2 are the first CAs to be found in the pyrenoid, although this localization already has been proposed earlier in algae with CCMs (Pronina and Semenenko 1984; Kuchitsu et al. 1991). PtCA1 and 2 are known to form large clumped particles within the stroma (Tanaka et al. 2005; Kitao et al. 2008) and this particle formation was demonstrated to be driven by an amphipathic helix at the C-terminal ends (Kitao and Matsuda 2009). PtCAs are not uniformly distributed within the pyrenoid. Their location appears to be limited to the central part of the lens-shaped pyrenoid (Tachibana et al. 2011), suggesting a layered structure within the pyrenoid (Matsuda et al. 2011). It should be noted that the LCIB/C hetero-hexamer apparently surrounds the pyrenoid structure in *C. reinhardtii* (Yamano et al. 2010), presumably shielding the pyrenoid from CO_2 leakage or recapturing leaking CO_2 from the pyrenoid in cooperation with the stromal CAH6 (Mitra et al. 2004; Yamano et al. 2010). The layered structure of the pyrenoid in *P. tricornutum* may also be important to avoid leakage of CO_2. Co-localization of fructose bisphosphatases, FBAC1 and FBAC5 with PtCA1 in the pyrenoid (Allen et al. 2012) implies that the initial part of CO_2 input and the first step to hexose output of the Calvin cycle may occur in the pyrenoid. Interestingly FBA is one of the primary rate limiting factors of the Calvin cycle in higher plants and also PtCA1 seems to play a major role in supplying substrate to RubisCO. This strongly suggests that diatom pyrenoids play a central role in controlling carbon metabolism within and around the Calvin cycle.

B. Redox Regulation of CO_2 Acquisition and Fixation Systems in the Stroma

Redox states play an important role in plastids of higher plants regulating photosynthetic carbon reduction in the Calvin cycle, nitrogen metabolism, fatty acid biosynthesis, carbohydrate storage and translation (Meyer et al. 2009). Chloroplasts of algae and land plants generally possess the ferredoxin (Fd) / thioredoxin (Trx) system (Buchanan and Balmer 2005). The *Arabidopsis* genome contains at least 12 Trx isoforms, which are grouped into seven subfamilies denoted Trx *f, h, m, o, x, y* and *z* based on their primary structures (Mestres-Ortega and Meyer 1999; Lemaire and Miginiac-Maslow 2004; Arsova et al. 2010). Of these, Trxs *f, m, x, y* and *z* are known to be typical chloroplast proteins (Schuerman and Jacquot 2000; Collin et al. 2003; Lemaire et al. 2003; Arsova et al. 2010). In chloroplasts of higher plants and green algae, the function of Trx seems to be very extensive, directly targeting more than 300 proteins, including numerous plastidic factors of unknown function (Motohashi et al. 2001; Balmer et al. 2003; Lemaire et al. 2004). In contrast, the function of plastidic Trxs in diatoms seems to be rather limited.

Eight genes encoding a set of Trxs have been identified in the genome of *P. tricornutum*, and at least three Trx candidates (*f*, *m*, and *y*) are located in the chloroplast together with a ferredoxin-thioredoxin reductase (FTR) gene, essential for Trx reduction via the electron transport chain (Weber et al. 2009). However, none of the several key enzymes in the Calvin cycle, which are under redox regulation in higher plants, appear to be regulated by Trxs in diatoms (Liaud et al. 2000; Michels et al. 2005; Kroth et al. 2008; Weber et al. 2009), with the so far only exception of the fructose-1,6-bisphosphatase (FBPase) and the phosphoglycerate kinase (Michels et al. 2005; Bosco et al. 2012). It is thus likely that the plastidic Trx systems in diatoms and in other chromist algae have undergone a unique functional specialization.

Activities of CCMs in marine diatoms are controlled by light and by environmental CO_2 concentrations (Colman and Rotatore 1995; Johnston and Raven 1996; Matsuda et al. 2001, 2002), strongly indicating the importance of CCM regulation in response to the availability of light and CO_2. As aforementioned, PtCA1 and PtCA2 in *P. tricornutum* presumably take a vital role in supplying CO_2 to RubisCO in the pyrenoid, thus controlling the initial reaction of the Calvin cycle. A recent study has demonstrated that both PtCA1 and PtCA2 are targets of plastidic Trx*m* and Trx*f* of *P. tricornutum*, and are activated via reduction by DTT (Kikutani et al. 2012). The presence of these Trxs significantly stimulated the efficiency of the reductive activation of PtCAs (Kikutani et al. 2012). Interestingly, PtCA1 requires Trx to be stimulated to its maximum activity, while PtCA2 only revealed stimulation of the efficacy of reductive activation (Kikutani et al. 2012). In *P. tricornutum* Trx targets two Cys residues (105 and 166 in PtCA1; 102 and 163 in PtCA2; relative to the mature N-termini) and disulfide formation of these two Cys residues is specifically regulated by oxygen concentrations above atmospheric levels (Kikutani et al. 2012). Interestingly, these Cys residues are not conserved in plastidic β-CAs in green algae, red algae, and higher plants (Kikutani et al. 2012). It was shown that the disulfide bond (such as between Cys 105 and 166 in PtCA1) is critical for regulation of PtCA activity, being a unique characteristic in β-CAs in chromists, opisthokonts, and bacteria and suggesting that this particular redox moiety was obtained via an HGT event (Kikutani et al. 2012). The redox potentials of PtCA1 and PtCA2 were determined to be around −370 mV (unpublished data), a value which allows reduction exclusively by the reduction side of the photosystem I (PSI) via Fd (Kikutani et al. 2012). A disulfide is formed between these Cys residues specifically by molecular oxygen above atmospheric levels to inactivate PtCAs, and none of other oxidants which commonly occur in the chloroplast (such as oxidized glutathione, dehydro-ascorbate, and H_2O_2), may substitute this function of molecular oxygen (Kikutani et al. 2012). These results strongly suggest that the CO_2 acquisition system, and not the fixation system, is the target of redox control via Trxs in the diatom plastid, and that the function of the pyrenoid is under the control of light-driven reduction mediated by Trx and molecular oxygen generated in PSII (Fig. 18.4). This implies that activities of PSI and PSII compete for control of the redox state of pyrenoid-forming CAs. Thus, oxygen concentrations above the atmospheric level efficiently inactivate PtCAs by forming a disulfide across the active center (Kikutani et al. 2012), but electrons from the reducing side of the PSI regularly cleave this disulfide, regulating the activity of PtCAs in response to oxygen concentration and light intensity. This strongly suggests that the function of the pyrenoidal CAs is fine-tuned on the post-translational level reflecting the balance of energy distribution between PSI and PSII.

V. Carbon Metabolism Relating to Photosynthesis and Respiration

The Calvin cycle is the central reaction in the plastid and produces hexoses, consuming ATP and NADPH derived from the light

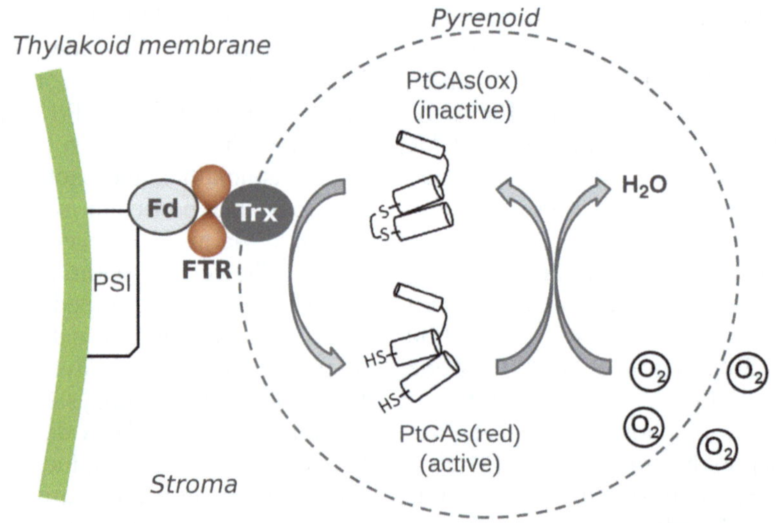

Fig. 18.4. A schematic drawing of the redox regulation of affecting structure and activity of PtCAs via chloroplastic Trx. Electrons derived from the reducing (stromal) side of photosystem I (PSI) are transferred to the ferredoxin (Fd) – thioredoxin (Trx) reductase complex, which constitute an electron transfer metabolon with Fd and Trx. Reduced Trx then reduces the target protein (PtCAs in the pyrenoid), cleaving the intramolecular disulfide bond to activate PtCAs. The reduced active form of PtCAs is oxidized by molecular oxygen above atmospheric levels which originates from active PSII.

reactions of the photosystems. Carbon dioxide is the major sink of light-driven ATP production and reducing power. However, amino acid synthesis including nitrogen and sulfur assimilation, photorespiration, and any other ATP/NADPH consuming process can also be an important sink for light energy. Moreover, the Calvin cycle is an essential process converting the length of carbon skeletons between C_3-C_7, thereby interfacing with several other biosynthesis pathways such as glycolysis, pentose phosphate pathway, a part of gluconeogenesis, as well as fatty acid and amino acid biosynthesis. The reactions of the Calvin cycle and the adjacent carbon metabolism have therefore to be highly regulated to avoid short circuits of energy and substrate consumption. In diatoms, the essential regulatory enzymes of these processes are redundant and their relations and regulations are still largely unclear.

A. RubisCO and the Calvin Cycle

The main pathway for photosynthetic carbon fixation in photosynthetic organisms is the Calvin cycle formally termed reductive pentose phosphate pathway. The RubisCO catalyses the first step in carbon fixation and consists of two subunits, the small and the large subunit. In contrast to land plants, both RubisCO subunits are encoded on the plastid genome in *P. tricornutum* and *T. pseudonana* (Oudot-Le Secq et al. 2007). It is noteworthy that, despite several nuclear-encoded genes are related to those in green algae, there is no RubisCO activase encoded in the genomes of *P. tricornutum* and *T. pseudonana* (Kroth et al. 2008), suggesting that diatom formI RubisCO is under the control of unknown activation factors similar to that of red algae. A recent study demonstrated that an ATPase-associated various cellular activities (AAA^+) type protein, CbbX, is the activase of red-type formI RubisCO in the purple bacterium, *Rhodobacter sphaeroides* (Mueller-Cajar et al. 2011). Interestingly, *P. tricornutum* and *T. pseudonana* possess putative CbbX genes in both plastidic and nuclear genomes (Chishiro, Kikutani, Matsuda, unpublished). The function of these gene products await to be studied.

All further Calvin cycle enzymes are encoded by the nuclear genomes. The two subsequent enzymatic steps after caryboxylation, the reduction to carbohydrates, are catalysed by a plastid targeted phosphoglycerate kinase (PGK) and a glyceraldehyde-3-phosphate dehydrogenase (GAPDH). In these two cases, only single plastidic isoenzymes have been identified in *P. tricornutum* and *T. pseudonana*. Interestingly, for the following reactions, the regeneration of the CO_2 acceptor ribulose-1,5-bisphosphate, multiple nuclear-encoded and plastid-targeted isoforms of enzymes are found: for instance in *P. tricornutum* there are three fructose-1, 6-bisphosphate aldolase (FBA) isoforms (two in *T. pseudonana*), four fructose-1, 6-bisphosphatase (FBPase) isoforms (two in *T. pseudonana*), two triosephosphate isomerase (TPI) isoforms (one in *T. pseudonana*) and two transaldolase (TAL) isoforms (three in *T. pseudonana*) (Gruber et al. 2009). This multitude of isoenzymes may be a result of endosymbiotic processes (see above); however, the reason why the cells contain different isoenzymes in the same compartment remains unclear. This is especially interesting in the case of the FBPases, as two of the isoenzymes are redox-regulated, while the other two enzymes apparently are not (Gruber et al. 2009). One possible explanation could be the formation of regulatory hetero-oligomeric complexes. The other plastidic Calvin cycle enzymes ribose-5-phosphate isomerases (RPI), ribulose-phosphate epimerases (RPE) and transketolases (TKL), again, only appear as single enzymes. Interestingly, although a gene for a sedoheptulose-1,7-bisphosphatase (SBPase) is present in the diatom genomes, the respective protein does not possess a plastidic presequence, indicating that it is a cytosolic enzyme, which is supported by experimental evidence (Gruber et al. 2009). It remains yet unclear what the actual function of the cytosolic SBPase of diatoms could be. So far cytosolic SBPases are unknown in photosynthetic eukaryotes, but present in some non-photosynthetic organisms like ciliates, kinetoplastids and ascomycetes (Rogers and Keeling 2004; Teich et al. 2007). Possibly, the cytosolic diatom SBPase is a part of a modified oxidative pentose phosphate pathway (OPP), which in diatoms has been removed completely from the chloroplast and which is situated in the cytosol (Gruber et al. 2009).

Oxidative pentose phosphate pathways can be found in plants and green algae in the cytosol as well as in the plastids. These pathways are mainly used for the generation of NADPH and ribose-5-phosphate and erythrose-4-phosphate as substrates for anabolic pathways. In isolated plastids of the diatom *Odontella sinensis*, no glucose-6-phosphate dehydrogenase (GPDH) and 6-phosphogluconate dehydrogenase (PGDH) activity was detectable (Michels et al. 2005), and only cytosolic isoforms of GPDH and PGDH were identified in other diatoms (Kroth et al. 2008), leading to the suggestion that the OPP in diatoms might be generally restricted to the cytosol (Wilhelm et al. 2006). Rather unexpected was also the finding of a PGDH in the periplastidic space of *P. tricornutum* (Gruber et al. 2009). During the reaction from gluconate-6-P to ribulose-5-P, the GPDH releases CO_2 and thus could potentially increase the CO_2 concentration in the periplastidic space. It is unclear yet, whether this reaction might be involved in the CCM in diatoms.

B. Other CO_2-Fixing Enzymes in Diatoms

Annotation of the diatom genomes revealed a number of surprises as several genes have been identified of which the deduced proteins may be involved in CO_2 fixation (Kroth et al. 2008). For instance two PEP carboxylases, which in C_4 plants are involved in pre-fixation of CO_2, have been identified in the genome of *P. tricornutum*. One of these PEP carboxylases is located in the mitochondria, while the second enzyme apparently is located in the periplastidic space of the plastids (Ewe, Gruber, Kroth, unpublished). Two pyruvate carboxylases, enzymes that usually are involved in gluconeogenesis, also have been discovered. One of them is located in the mitochondria, the other in the plastids. All of these enzymes could be involved in a biochemical CCM, however, most of them are separated from the RubisCO by several membranes, thus fixed carbon would have to be transported actively across several membranes. Furthermore no plastidic decarboxylase has yet been identified that may allow a release of the bound CO_2 in close proximity to RubisCO. Reinfelder et al. (2000) found that phospho-*enol* pyruvate carboxykinase (PEPCK) activity co-localizes with RubisCO activity in isolated plastid-enriched fractions of *T. weissflogii* and concluded that decarboxylation occurred within the plastids, however, all decarboxylases identified so far are not located in the plastids: the PEPCK and two malic enzymes are located in the mitochondria. This of course does not exclude the possibility of occurrences of a plastidic decarboxylase, which simply might not be identified yet, due to lacking sequence similarity.

C. Photorespiration

In addition to its carboxylating activity, the RubisCO may also function as an oxygenase by fixing molecular oxygen at the same active site as CO_2, thus splitting the substrate ribulose-1,5-bisphosphate into 2-phosphoglycolate (2PG) and 3-PGA. This oxygenase reaction is the first step of the photorespiratory pathway and is known to be an essential process dissipating light excitation energy in the photosystems when CO_2 supply is limited and/or under high light irradiation (Kozaki and Takeba 1996). In plants with C_4 metabolism and in microalgae possessing CCMs, photorespiration does not occur even under atmospheric CO_2 concentrations or under high light conditions, where photorespiration is unavoidable in C_3-type plants. In diatoms, there are a few reports describing the occurrence of photorespiration under ambient conditions (Parker et al. 2004; Roberts et al. 2007a), suggesting that the diatom photorespiration operates even under the operation of the CCM. There seems to be a set of photorespiratory genes present in the genome of diatoms according to the DiatomCyc database (see Fabris et al. 2012; http://akongo.psb.ugent.be/), suggesting that the photorespiratory carbon oxidation cycle (PCOC) is the major pathway to recycle phosphoglycolate in diatoms (Armbrust et al. 2004; Bowler et al. 2008). In cyanobacteria, glycolate oxidation is found to be catalyzed by bacterial-type of glycolate dehydrogenase (GDH) instead of the plant-type glycolate oxidase (GOX) (Eisenhut et al. 2006). This oxygen-independent glycolate oxidation is also found in several groups of eukaryotic algae (like *Chlorophyceae*, *Prasinophyceae*, *Cryptophyceae*, and *Bacillariophyceae*) in contrast to the plant-type, oxygen-dependent pathway in *Chrysophyceae*, *Eustigmatophyceae*, *Raphidophyceae*, *Xanthophyceae*, and *Rhodophyceae* (Suzuki et al. 1991). Interestingly, some diatoms apparently possess two genes for both GOX and GDH (Kroth et al. 2008). A set of these enzymes revealed typical peroxisome-targeting signals at their C-termini, while the other enzymes seem to be equipped with mitochondrial targeting sequences, suggesting that glycolate oxidation may occur in these two organelles (Kroth et al. 2008).

A genome-based study by Kroth et al. (2008) suggested that PCOC occurs mainly in the mitochondria and glycolate is also integrated into the glyoxylate pathway via malate synthase in the peroxisome. Another alternative pathway of glycolate metabolism is the tartronic semialdehyde pathway. The existence of this pathway in diatoms (Kroth et al. 2008), originaly identified in cyanobacteria (Eisenhut et al. 2006), is suggested by precedential biochemical studies with *T. pseudonana* and *Cylindrotheca fusiformis* (Paul and Volcani 1974, 1976). It is also pointed out that so far no gene for the glycerate kinase (GK), an enzyme of the last step of PCOC, which forms 3-PGA, has yet been identified (Kroth et al. 2008). This strongly suggests that PCOC in diatoms would not result in recycling 3-PGA for fixation by the Calvin cycle, but rather supplying glycine and serine (Kroth et al. 2008).

Acknowledgements

This work was supported by Grant-in-Aid for Scientific Research B (grant no. 24310015 to Y. M.), by Grant-in-Aid for Challenging Exploratory Research (grant no. 24651119 to Y. M.) from the Japan Society for the Promotion of Science (JSPS), by MEXT-Supported Program for the Strategic Research Foundation at Private Universities (2010–2014), by the Program for Research on Halophilic Organism of the Salt Science Research Foundation (grant no. 06B02 to Y. M.), and by the Steel Industry Foundation for the Advancement of Environmental Protection Technology to Y. M. PGK is grateful for financial support by the German Research Foundation (DFG), grant KR1661/7-1, the German Israeli Foundation (GIF), the University of Konstanz, and is thankful to A. Gruber and J. Hentschel for providing an unpublished electron micrograph.

References

Allen AE, Dupont CL, Obornik M, Horák A, Nunes-Nesi A, McCrow JP, Zheng H, Johnson DA, Hu H, Fernie AR, Bowler C (2011) Evolution and metabolic significance of the urea cycle in photosynthetic diatoms. Nature 473:203–207

Allen AE, Moustafa A, Montsant A, Eckert A, Kroth PG, Bowler C (2012) Evolution and functional diversification of fructose bisphosphate aldolase genes in photosynthetic marine diatoms. Mol Biol Evol 29:367–379

Alterio V, Langella E, Viparelli F, Vullo D, Ascione G, Dathan NA, Morel FMM, Supuran CT, De Simone G, Monti SM (2012) Structural and inhibition insights into carbonic anhydrase CDCA1 from the marine diatom *Thalassiosira weissflogii*. Biochimie 94:1232–1241

Armbrust EV, Berges JA, Bowler C, Green BR, Martinez D, Putnam NH, Zhou S, Allen AE, Apt KE, Bechner M, Brzezinski MA, Chaal BK, Chiovitti A, Davis AK, Demarest MS, Detter JC, Glavina T, Goodstein D, Hadi MZ, Hellsten U, Hildebrand M, Jenkins BD, Jurka J, Kapitonov VV, Kröger N, Lau WW, Lane TW, Larimer FW, Lippmeier JC, Lucas S, Medina M, Montsant A, Obornik M, Parker MS, Palenik B, Pazour GJ, Richardson PM, Rynearson TA, Saito MA, Schwartz DC, Thamatrakoln K, Valentin K, Vardi A, Wilkerson FP, Rokhsar DS (2004) The genome of the diatom *Thalassiosira pseudonana*: ecology, evolution, and metabolism. Science 306:79–86

Arsova B, Hoja U, Wimmelbacher M, Greiner E, Üstün S, Melzer M, Petersen K, Lein W, Börnke F (2010) Plastidial thioredoxin *z* interacts with two fructokinase-like proteins in a thiol-dependent manner: evidence for an essential role in chloroplast development in *Arabidopsis* and *Nicotiana benthamiana*. Plant Cell 22:1498–1515

Ast M, Gruber A, Schmitz-Esser S, Neuhaus HE, Kroth PG, Horn M, Haferkamp I (2009) Diatom plastids depend on nucleotide import from the cytosol. Proc Natl Acad Sci U S A 106:3621–3626

Badger MR, Andrews TJ, Whitney SM, Ludwig M, Yellowlees DC, Leggat W, Price GD (1998) The diversity and coevolution of RubisCO, plastids, pyrenoids, and chloroplast-based CO_2-concentrating mechanisms in algae. Can J Bot 76:1052–1071

Badger MR, Hanson D, Price GD (2002) Evolution and diversity of CO_2-concentrating mechanisms in cyanobacteria. Funct Plant Biol 29:183–194

Balmer Y, Koller A, del Val G, Manieri W, Schürmann P, Buchanan BB (2003) Proteomics gives insight into the regulatory function of chloroplast thioredoxins. Proc Natl Acad Sci U S A 100:370–375

Borkhsenious ON, Mason CB, Moroney JV (1998) The intracellular localization of ribulose-1,5-bisphosphate carboxylase/oxygenase in *Chlamydomonas reinhardtii*. Plant Physiol 116:1585–1591

Bosco MB, Aleanzi MC, Iglesias AA (2012) Plastidic phosphoglycerate kinase from *Phaeodactylum tricornutum*: on the critical role of cysteine residues for the enzyme function. Protist 163:188–203

Bowler C, Allen AE, Badger JH, Grimwood J, Jabbari K, Kuo A, Maheswari U, Martens C, Maumus F, Otillar RP, Rayko E, Salamov A, Vandepoele K, Beszteri B, Gruber A, Heijde M, Katinka M, Mock T, Valentin K, Verret F, Berges JA, Brownlee C, Cadoret JP, Chiovitti A, Choi CJ, Coesel S, De Martino A, Detter JC, Durkin C, Falciatore A, Fournet J, Haruta M, Huysman MJ, Jenkins BD, Jiroutova K, Jorgensen RE, Joubert Y, Kaplan A, Kröger N, Kroth PG, La Roche J, Lindquist E, Lommer M, Martin-Jezequel V, Lopez PJ, Lucas S, Mangogna M, McGinnis K, Medlin LK, Montsant A, Oudot-Le Secq MP, Napoli C, Obornik M, Parker MS, Petit JL, Porcel BM, Poulsen N, Robison M, Rychlewski L, Rynearson TA, Schmutz J, Shapiro H, Siaut M, Stanley M, Sussman MR, Taylor AR, Vardi A, von Dassow P, Vyverman W, Willis A, Wyrwicz LS, Rokhsar DS, Weissenbach J, Armbrust EV, Green BR, Van de Peer Y, Grigoriev IV (2008) The *Phaeodactylum genome* reveals the evolutionary history of diatom genomes. Nature 456:239–244

Bozzo GG, Colman B (2000) The induction of inorganic carbon transport and external carbonic anhydrase in *Chlamydomonas reinhardtii* is regulated by external CO_2 concentration. Plant Cell Environ 23:1137–1144

Buchanan BB, Balmer Y (2005) Redox regulation: a broadening horizon. Annu Rev Plant Biol 56:187–220

Bullmann L, Haarmann R, Mirus O, Bredemeier R, Hempel F, Maier UG, Schleiff E (2010) Filling the gap, evolutionarily conserved Omp85 in plastids of chromalveolates. J Biol Chem 285:6848–6856

Burkhardt S, Amoroso G, Riebesell U, Sültemeyer D (2001) CO_2 and HCO_3^- uptake in marine diatoms acclimated to different CO_2 concentrations. Limnol Oceanogr 46:1378–1391

Cavalier-Smith T (1999) Principles of protein and lipid targeting in secondary symbiogenesis: euglenoid, dinoflagellate, and sporozoan plastid origins and the eukaryote family tree. J Eukaryot Microbiol 46:347–366

Cavalier-Smith T (2000) Membrane heredity and early chloroplast evolution. Trends Plant Sci 5:174–182

Chaal BK, Ishida K, Green BR (2003) A thylakoidal processing peptidase from the heterokont alga *Heterosigma akashiwo*. Plant Mol Biol 52:463–472

Chan CX, Reyes-Prieto A, Bhattacharya D (2011) Red and green algal origin of diatom membrane transporter: insights into environmental adaptation and cell evolution. PLoS One 6:e29138

Chan CX, Bhattacharya D, Reyes-Prieto A (2012) Endosymbiotic and horizontal gene transfer in microbial eukaryotes: impacts on cell evolution and the tree of life. Mob Genet Elem 2:101–105

Chen Y, Cann MJ, Litvin TN, Iourgenko V, Sinclair ML, Levin LR, Buck J (2000) Soluble adenylyl cyclase as an evolutionarily conserved bicarbonate sensor. Science 289:625–628

Collin V, Issakidis-Bourguet E, Marchand C, Hirasawa M, Lancelin JM, Knaff DB, Miginiac-Maslow M (2003) The *Arabidop* plastidial thioredoxins: new functions and new insights into specificity. J Biol Chem 278:23747–23752

Colman B, Rotatore C (1995) Photosynthetic inorganic carbon uptake and accumulation in two marine diatoms. Plant Cell Environ 18:919–924

Cox EH, McLendon GL, Morel FMM, Lane TW, Prince RC, Pickering IJ, George GN (2000) The active site structure of *Thalassiosira weissflogii* carbonic anhydrase 1. Biochemistry 39:12128–12130

Danson JS, Huertas IE, Colman B (2004) Source of inorganic carbon for photosynthesis in two marine dinoflagellates. J Phycol 40:285–292

De Riso V, Raniello R, Maumus F, Rogato A, Bowler C, Falciatore A (2009) Gene silencing in the marine diatom *Phaeodactylum tricornutum*. Nucl Acids Res 37:e96

Delwiche CF, Palmer JD (1997) The origin of plastids and their spread via secondary symbiosis. Plant Syst Evol 11:53–86

Deschamps P, Moreira D (2012) Reevaluating the green contribution to diatom genomes. Genome Biol Evol 4:683–688

Dionisio-Sese ML, Fukuzawa H, Miyachi S (1990) Light-induced carbonic anhydrase expression in *Chlamydomonas reinhardtii*. Plant Physiol 94:1103–1110

Dou Z, Heinhorst S, Williams EB, Murin CD, Shively JM, Cannon GC (2008) CO_2 fixation kinetics of *Halothiobacillus neapolitanus* mutant carboxysomes lacking carbonic anhydrase suggest the shell acts as a diffusional barrier for CO_2. J Biol Chem 283:10377–10384

Duanmu D, Spalding MH (2011) Insertional suppressors of *Chlamydomonas reinhardtii* that restore growth of air-dier *lcib* mutants in low CO_2. Photosynth Res 109:123–132

Eisenhut M, Kahlon S, Hasse D, Ewald R, Lieman-Hurwitz J, Ogawa T, Ruth W, Bauwe H, Kaplan A, Hagemann M (2006) The plant-like C2 glycolate cycle and the bacterial like glycerate pathway cooperate in phophoglycolate metabolism in cyanobacteria. Plant Physiol 142:333–342

Fabris M, Matthijs M, Rombauts S, Vyverman W, Goossens A, Baart GJE (2012) The metabolic blueprint of *Phaeodactylum tricornutum* reveals a eukaryotic Entner–Doudoroff glycolytic pathway. Plant J 70:1004–1014

Falkowski PG, Raven JA (2007) Aquatic photosynthesis, 2nd edn. Princeton University Press, Princeton

Falkowski P, Scholes RJ, Boyle E, Canadell J, Canfield D, Elser J, Gruber N, Hibbard K, Hoegberg P, Linder S, Mackenzie FT, III Moore B, Pedersen T, Rosenthal Y, Seitzinger S, Smetacek V, Steffen W (2000) The global carbon cycle: a test of our knowledge of Earth as a system. Science 290:291–296

Fukuzawa H, Suzuki E, Komukai Y, Miyachi S (1992) A gene homologous to chloroplast carbonic anhydrase (*icfA*) is essential to photosynthetic carbon dioxide fixation by *Synechococcus* PCC7942. Proc Natl Acad Sci U S A 89:4437–4441

Fukuzawa H, Miura K, Ishizaki K, Kucho KI, Saito T, Kohinata T, Ohyama K (2001) *Ccm1*, a regulatory gene controlling the induction of a carbon-concentrating mechanism in *Chlamydomonas reinhardtii* by sensing CO_2 availability. Proc Natl Acad Sci U S A 98:5347–5352

Funke RP, Kovar JL, Weeks DP (1997) Intracellular carbonic anhydrase is essential to photosynthesis in *Chlamydomonas reinhardtii* at atmospheric levels of

CO_2. Demonstration via genomic complementation of the high-CO_2-requiring mutant ca-1. Plant Physiol 114:237–244

Genkov T, Meyer M, Griffiths H, Spreitzer RJ (2010) Functional hybrid RubisCO enzymes with plant small subunits and algal large subunits: engineered rbcS cDNA for expression in *Chlamydomonas*. J Biol Chem 285:19833–19841

Giordano M, Norici A, Forssen M, Eriksson M, Raven JA (2003) An anaplerotic role for mitochondrial carbonic anhydrase in *Chlamydomonas reinhardtii*. Plant Physiol 132:2126–2134

Goyet C, Poisson A (1989) New determination of carbonic acid dissociation constants in seawater as a function of temperature and salinity. Deep-Sea Res 36:1635–1654

Gruber A, Vugrinec S, Hempel F, Gould SB, Maier UG, Kroth PG (2007) Protein targeting into complex diatom plastids depends on the signal peptide's cleavage site within the bipartite presequence. Plant Mol Biol 64:519–530

Gruber A, Weber T, Bártulos CR, Vugrinec S, Kroth PG (2009) Intracellular distribution of the reductive and oxidative pentase phosphate pathways in two diatoms. J Basic Microbiol 49:58–72

Haimovich-Dayan M, Garfinkel N, Ewe D, Marcus Y, Gruber A, Wagner H, Kroth PG, Kaplan A (2013) The role of C_4 metabolism in the marine diatom *Phaeodactylum tricornutum*. New Phytol 197: 177–185

Hammer A, Hodgson DR, Cann MJ (2006) Regulation of prokaryotic adenylyl cyclases by CO_2. Biochem J 396:215–218

Harada H, Matsuda Y (2005) Identification and characterization of a new carbonic anhydrase in the marine diatom *Phaeodactylum tricornutum*. Can J Bot 83:909–916

Harada H, Nakatsuma D, Ishida M, Matsuda Y (2005) Regulation of the expression of intracellular β-carbonic anhydrase in response to CO_2 and light in the marine diatom *Phaeodactylum tricornutum*. Plant Physiol 139:1041–1050

Harada H, Nakajima K, Sakaue K, Matsuda Y (2006) CO_2 sensing at ocean surface mediated by cAMP in a marine diatom. Plant Physiol 142:1318–1328

Holdsworth RH (1971) The isolation and partial characterization of the pyrenoid protein of *Eremosphaera viridis*. J Cell Biol 51:499–513

Hopkinson BM (2013) A chloroplast pump model for the CO_2 concentrating mechanism in the diatom *Phaeodactylum tricornutum*. Photosynth Res. doi:10.1007/s11120-013-9954-7

Hopkinson BM, Dupont CL, Allen AE, Morel FMM (2011) Efficiency of the CO_2-concentrating mechanism of diatoms. Proc Natl Acad Sci USA 108:3830–3837

Jenks A, Gibbs SP (2000) Immunolocalization and distribution of form II RubisCO in the pyrenoid and chloroplast stroma of *Amphidinium carterae* and form I RubisCO in the symbiont-derived plastids of *Perinidium foliaceum* (Dinophyceae). J Phycol 36:127–138

John-Mckay ME, Colman B (1997) Variation in the occurrence of external carbonic anhydrase among strains of the marine diatom *Phaeodactylum tricornutum* (Bacillariophyceae). J Phycol 33:988–990

Johnston AM, Raven JA (1996) Inorganic carbon accumulation by the marine diatom *Phaeodactylum tricornutum*. Eur J Phycol 31:285–290

Karlsson J, Clarke AK, Chen ZY, Hugghins SY, Park YI, Husic HD, Moroney JV, Samuelsson G (1998) A novel α-type carbonic anhydrase associated with the thylakoid membrane in *Chlamydomonas reinhardtii* is required for growth at ambient CO_2. EMBO J 17:1208–1216

Kerfeld CA, Sawaya MR, Tanaka S, Nguyen CV, Phillips M, Beeby M, Yeates TO (2005) Protein structures forming the shell of primitive bacterial organelles. Science 309:936–938

Kikutani S, Tanaka R, Yamazaki Y, Hara S, Hisabori T, Kroth PG, Matsuda Y (2012) Redox regulation of carbonic anhydrases via thioredoxin in the chloroplast of the marine diatom *Phaeodactylum tricornutum*. J Biol Chem 287:20689–20700

Kilian O, Kroth PG (2004) Presequence acquisition during secondary endocytobiosis and the possible role of introns. J Mol Evol 58:712–721

Kilian O, Kroth PG (2005) Identification and characterization of a new conserved motif within the presequence of proteins targeted into complex diatom plastids. Plant J 41:175–183

Kitao Y, Matsuda Y (2009) Formation of macromolecular complexes of carbonic anhydrases in the chloroplast of a marine diatom by the action of the C-terminal helix. Biochem J 419:681–688

Kitao Y, Harada H, Matsuda Y (2008) Localization and targeting mechanisms of two chloroplastic β-carbonic anhydrases in the marine diatom *Phaeodactylum tricornutum*. Physiol Plant 133:68–77

Klengel T, Liang WJ, Chaloupka J, Ruoff C, Schröppel K, Naglik JR, Eckert SE, Mogensen EG, Haynes K, Tuite MF, Levin LR, Buck J, Mühlschlegel FA (2005) Fungal adenylyl cyclase integrates CO_2 sensing with cAMP signaling and virulence. Curr Biol 15:2021–2026

Kohinata T, Nishino H, Fukuzawa H (2008) Significance of zinc in a regulatory protein, CCM1, which regulates the carbon-concentrating mechanism in *Chlamydomonas reinhardtii*. Plant Cell Physiol 49:273–283

Kooistra WHCF, Gersonde R, Medlin LK, Mann DG (2007) The origin and evolution of the diatoms: their adaptation to a planktonic existence. In: Falkowski PG, Knoll AH (eds) Evolution of primary producers in the sea. Academic Press, Burlington, pp 207–249

Korb RE, Saville PJ, Johnston AM, Raven JA (1997) Sources of inorganic carbon for photosynthesis by three species of marine diatom. J Phycol 33:433–440

Kozaki A, Takeba G (1996) Photoinhibition protects C_3 plants from photooxidation. Nature 384:557–560

Kroth PG (2002) Protein transport into secondary plastids and the evolution of primary and secondary plastids. Int Rev Cytol 221:191–255

Kroth PG, Chiovitti A, Gruber A, Martin-Jezequel V, Mock T, Parker MS, Stanley MS, Kaplan A, Caron L, Weber T, Maheswari U, Armbrust EV, Bowler C (2008) A model for carbohydrate metabolism in the diatom *Phaeodactylum tricornutum* deduced from comparative whole genome analysis. PLoS One 3:e1426

Kuchitsu K, Tsuzuki M, Miyachi S (1988) Characterization of the pyrenoid isolated from unicellular green alga *Chlamydomonas reinhardtii*: particulate from RubisCO protein. Protoplasma 144:17–24

Kuchitsu K, Tsuzuki M, Miyachi S (1991) Polypeptide composition and enzyme activities of the pyrenoid and its regulation by CO_2 concentration in unicellular green algae. Can J Bot 69:1062–1069

Kucho K, Ohyama K, Fukuzawa H (1999) CO_2-responsive transcriptional regulation of CAH1 encoding carbonic anhydrase is mediated by enhancerand silencer regions in *Chlamydomonas reinhardtii*. Plant Physiol 121:1329–1337

Lacoste-Royal G, Gibbs SP (1987) Immunocytochemical localization of ribulose-1,5-bisphosphate carboxylase in the pyrenoid and thylakoid region of the chloroplast of *Chlamydomonas reinhardtii*. Plant Physiol 83:602–606

Lane TW, Morel FMM (2000) A biological function for cadmium in marine diatoms. Proc Natl Acad Sci U S A 97:4627–4631

Lane TW, Saito MA, George GN, Pickering IJ, Prince RC, Morel FMM (2005) A cadmium enzyme from marine diatom. Nature 435:42

Lang M, Kroth PG (2001) Diatom fucoxanthin chlorophyll *a*/*c*-binding protein (FCP) and land plant light-harvesting proteins use a similar pathway for thylakoid membrane insertion. J Biol Chem 276:7985–7991

Lang M, Apt KE, Kroth PG (1998) Protein transport into "complex" diatom plastids utilizes two different targeting signals. J Biol Chem 273:30973–30978

Lapointe M, Mackenzie TDB, Morse D (2008) An external δ-carbonic anhydrase in a free-living marine dinoflagellate may circumvent diffusion-limited carbon acquisition. Plant Physiol 147:1427–1436

Lavaud J, Materna AC, Sturm S, Vugrinec S, Kroth PG (2012) Silencing of the violaxanthin de-epoxidase gene in the diatom *Phaeodactylum tricornutum* reduces diatoxanthin synthesis and non-photochemical quenching. PLoS One 7:e36806

Lee RBY, Smith JAC, Rickaby REM (2013) Cloning, expression and characterization of the δ-carbonic anhydrase of *Thalassiosira weissflogii* (Bacillariophyceae). J Phycol 49:170–177

Lemaire SD, Miginiac-Maslow M (2004) The thioredoxin superfamily in *Chlamydomonas reinhardtii*. Photosynth Res 82:203–220

Lemaire SD, Collin V, Keryer E, Quesada A, Miginiac-Masalow M (2003) Characterization of thioredoxin y, a new type of thioredoxin identified in the genome of *Chlamydomonas reinhardtii*. FEBS Lett 543:87–92

Lemaire SD, Guillon B, Le Marechal P, Keryer E, Miginiac-Maslow M, Decottignies P (2004) New thioredoxin targets in the unicellular photosynthetic eukaryote *Chlamydomonas reinhardtii*. Proc Natl Acad Sci U S A 101:7475–7480

Liaud MF, Lichtlé C, Apt K, Martin W, Cerff R (2000) Compartment-specific isoforms of TPI and GAPDH are imported into diatom mitochondria as a fusion protein: evidence in favor of a mitochondrial origin of the eukaryotic glycolytic pathway. Mol Biol Evol 17:213–223

Long BM, Badger MR, Whitney SM, Price GD (2007) Analysis of carboxysomes from *Synechococcus* PCC7942 reveals multiuple RubisCO complexes with carboxysomal proteins CcmM and CcaA. J Biol Chem 282:29323–29335

Lopez-Ruiz A, Verbelen JP, Roldan JM, Diez J (1985) Nitrate reductase of green algae is located in the pyrenoid. Plant Physiol 79:1006–1010

Maeda S, Badger MR, Price GD (2002) Novel gene products associated with NdhD3/D4-containing NDH-1 complexes are involved in photosynthetic CO_2 hydration in the cyanobacterium *Synechococcus* sp. PCC7942. Mol Microbiol 43:425–436

Marcus Y, Harel E, Kaplan A (1983) Adaptation of the cyanobacterium *Anabaena variabilis* to low CO_2 concentration in their environment. Plant Physiol 71:208–210

Matsuda Y, Colman B (1995a) Induction of CO_2 and bicarbonate transport in green alga *Chlorella ellipsoidea*. Time course of induction of two systems. Plant Physiol 108:247–252

Matsuda Y, Colman B (1995b) Induction of CO_2 and bicarbonate transport in green alga *Chlorella ellip-*

soidea. Evidence for induction in response to external CO_2 concentration. Plant Physiol 108:253–260

Matsuda Y, Hara T, Colman B (2001) Regulation of the induction of bicarbonate uptake by dissolved CO_2 in the marine diatom *Phaeodactylum tricornutum*. Plant Cell Environ 24:611–620

Matsuda Y, Satoh K, Harada H, Satoh D, Hiraoka Y, Hara T (2002) Regulation of the expressions of HCO_3^- uptake and intracellular carbonic anhydrase in response to CO_2 concentrating in the marine diatom *Phaeodactylum* sp. Funct Plant Biol 29:279–287

Matsuda Y, Nakajima K, Tachibana M (2011) Recent progresses on the genetic basis of the regulation of CO_2 acquisition systems in response to CO_2 concentration. Photosynth Res 109:191–203

Matsuzaki M, Misumi O, Shin-I T, Maruyama S, Takahara M, Miyagishima S, Mori T, Nishida K, Yagisawa F, Nishida K, Yoshida Y, Nishimura Y, Nakao S, Kobayashi T, Momoyama Y, Higashiyama T, Minoda A, Sano M, Nomoto H, Oishi K, Hayashi H, Ohta F, Nishizaka S, Haga S, Miura S, Morishita T, Kabeya Y, Terasawa K, Suzuki Y, Ishii Y, Asakawa S, Takano H, Ohta N, Kuroiwa H, Tanaka K, Shimizu N, Sugano S, Sato N, Nozaki H, Ogasawara N, Kohara Y, Kuroiwa T (2004) Genome sequence of the ultrasmall unicellular red alga *Cyanidioschyzon merolae*10D. Nature 428:653–657

Mayo WP, Williams TG, Birch DG, Turpin DH (1986) Photosynthetic adaptation by *Synechococcus leopoliensis* in response to exogenous dissolved inorganic carbon. Plant Physiol 80:1038–1040

McGinn PJ, Morel FMM (2008) Expression and inhibition of the carboxylating and decarboxylating enzymes in the photosynthetic C_4 pathway of marine diatoms. Plant Physiol 146:300–309

McKay RML, Gibbs SP (1989) Immunocytochemical localization of ribulose-1,5-bisphosphate carboxylase/oxygenase in light-limited and light-saturated cells of *Chlorella pyrenoidosa*. Protoplasma 149:31–37

McKay RML, Gibbs SP (1990) Phycoerythrin is absent from the pyrenoid of *Porphyridium cruentum*: photosynthetic implications. Planta 180:249–256

McKay RML, Gibbs SP, Vaughn KC (1991) RubisCO activase is present in the pyrenoid of green algae. Protoplasma 162:38–45

Mestres-Ortega D, Meyer Y (1999) The *Arabidopsis thaliana* genome encodes at least four thioredoxins m and a new prokaryotic-like thioredoxin. Gene 240:307–316

Meyer Y, Buchanan BB, Vignols F, Reichheld JP (2009) Thioredoxins and glutaredoxins: unifying elements in redox biology. Annu Rev Genet 43:335–367

Michels AK, Wedel N, Kroth PG (2005) Diatom plastids possess a phosphoribulokinase with an altered regulation and no oxidative pentose phosphate pathway. Plant Physiol 137:911–920

Milligan AJ, Morel FMM (2002) A proton buffering role for silica in diatoms. Science 297:1848–1850

Mitchell C, Beardall J (1996) Inorganic carbon uptake by an Antarctic sea-ice diatom, *Nitzschia frigida*. Polar Biol 16:95–99

Mitra M, Lato SM, Ynalvez RA, Xiao Y, Moroney JV (2004) Identification of a new chloroplast carbonic anhydrase in *Chlamydomonas reinhardtii*. Plant Physiol 135:173–182

Miura K, Kohinata T, Yoshioka S, Ohyama K, Fukuzawa H (2002) Regulation of a carbon concentrating mechanism through CCM1 in *Chlamydomonas reinhardtii*. Funct Plant Biol 29:211–219

Miura K, Yamano T, Yoshioka S, Kohinata T, Inoue Y, Taniguchi F, Asamizu E, Nakamura Y, Tabata S, Yamato KT, Ohyama K, Fukuzawa H (2004) Expression profiling-based identification of CO_2-responsive genes regulated by CCM1 controlling a carbon-concentrating mechanism in *Chlamydomonas reinhardtii*. Plant Physiol 135: 1595–1607

Montsant A, Jabbari K, Maheswari U, Bowler C (2005) Comparative genomics of the pennate diatom *Phaeodactylum tricornutum*. Plant Physiol 137:500–513

Morita E, Abe T, Tsuzuki M, Fujiwara S, Sato N, Hirata A, Sonoike K, Nozaki H (1998) Presence of the CO_2-concentraitng mechanism in some species of the pyrenoid-less free-living algal genus *Chloromonas* (Volvocales, Chlorophyta). Planta 204:269–276

Motohashi K, Kondoh A, Stumpp MT, Hisabori T (2001) Comprehensive survey of proteins targeted by chloroplast thioredoxin. Proc Natl Acad Sci U S A 98:11224–11229

Moustafa A, Reyes-Prieto A, Bhattacharya D (2008) Chlamydiae has contributed at least 55 genes to Plantae with predominantly plastids functions. PLoS One 3:e2205

Moustafa A, Beszteri B, Maier UG, Bowler C, Valentin K, Bhattacharya D (2009) Genomic footprints of a cryptic plastid endosymbiosis in diatoms. Science 324:1724–1726

Mueller-Cajar O, Stotz M, Wendler P, Hartl FU, Bracher A, Hayer-Hartl M (2011) Structure and function of the AAA^+ protein CbbX, a red-type RubisCO activase. Nature 479:194–199

Nakajima K, Tanaka A, Matsuda Y (2013) SLC4 family transporters in a marine diatom directly pump bicarbonate from seawater. Proc Natl Acad Sci USA 110(5):1767–1772

Nimer NA, Brownlee C, Merrett MJ (1999) Extracellular carbonic anhydrase facilitates carbon dioxide availability for photosynthesis in the marine dinoflagellate *Prorocentrum micans*. Plant Physiol 120:105–112

Nishimura T, Takahashi Y, Yamaguchi O, Suzuki H, Maeda S, Omata T (2008) Mechanism of low CO_2-induced activation of the cmp bicarbonate transporter operon by a LysR family protein in the cyanobacterium *Synechococcus elongatus* strain PCC 7942. Mol Microbiol 68:98–109

Norton TA, Melkonian M, Anderson RA (1996) Algal biodiversity. Phycologia 35:308–326

Ohnishi N, Mukherjee B, Tsujikawa T, Yanase M, Nakano H, Moroney JV, Fukuzawa H (2010) Expression of a low CO_2-inducible protein, LCI1, increases inorganic carbon uptake in the green alga *Chlamydomonas reinhardtii*. Plant Cell 22:3105–3117

Ohno N, Inoue T, Yamashiki R, Nakajima K, Kitahara Y, Ishibashi M, Matsuda Y (2012) CO_2-cAMP-responsive cis-elements targeted by a transcription factor with CREB/ATF-like basic zipper domain in the marine diatom *Phaeodactylum tricornutum*. Plant Physiol 158:499–513

Omata T, Price GD, Badger MR, Okamura M, Gohta S, Ogawa T (1999) Identification of an ATP-binding cassette transporter involved in bicarbonate uptake in the cyanobacterium *Synechococcus* sp. strain PCC 7942. Proc Natl Acad Sci U S A 96:13571–13576

Omata T, Gohta S, Takahashi Y, Harano Y, Maeda S (2001) Involvement of a CbbR homolog in low CO_2-induced activation of the bicarbonate transporter operon in cyanobacteria. J Bacteriol 183:1891–1898

Osafune T, Yokota A, Sumida S, Hase E (1990) Immunogold localization of ribulose-1, 5-bisphosphate carboxylase with reference to pyrenoid morphology in chloroplasts of synchronized *Euglena gracilis* cells. Plant Physiol 92:802–808

Oudot-Le Secq MP, Green BR (2011) Complex repeat structures and novel features in the mitochondrial genomes of the diatoms *Phaeodactylum tricornutum* and *Thalassiosira pseudonana*. Gene 476:20–26

Oudot-Le Secq MP, Grimwood J, Shapiro H, Armbrust EV, Bowler C, Green BV (2007) Chloroplast genomes of the diatoms *Phaeodactylum tricornutum* and *Thalassiosira pseudonana*: comparison with other plastid genomes of the red lineage. Mol Genet Genomics 277:427–439

Parker MS, Armbrust EV, Piovia-Scott J, Keil RG (2004) Induction of photorespiration by light in the centric diatom *Thalassiosira weissflogii* (Bacillariophyceae): molecular characterization and physiological consequences. J Phycol 40:557–567

Patel BN, Merrett MJ (1986) Inorganic-carbon uptake by the marine diatom *Phaeodactylum tricornutum*. Planta 169:222–227

Paul JS, Volcani BE (1974) Photorespiration in diatoms I. The oxidation of glycolic acid in *Thalassiosira pseudonana*. Arch Microbiol 101:115–120

Paul JS, Volcani BE (1976) Photorespiration in diatoms IV. Two pathways of glycolate metabolism in synchronized cultures of *Cylindrotheca fuciformis*. Arch Microbiol 110:247–252

Poulsen N, Chesley PM, Kröger N (2006) Molecular genetic manipulation of the diatom *Thalassiosira pseudonana* (Bacillariophyceae). J Phycol 42:1059–1065

Price GD, Woodger FJ, Badger MR, Howitt SM, Tucker L (2004) Identification of a SulP-type bicarbonate transporter in marine cyanobacteria. Proc Natl Acad Sci U S A 101:18228–18233

Prihoda J, Tanaka A, de Paula WBM, Allen JF, Tirichine L, Bowler C (2012) Chloroplast-mitochondria cross-talk in diatoms. J Exp Bot 63:1543–1557

Pronina NA, Semenenko VE (1984) Localization of membrane bound and soluble forms of carbonic anhydrase in the *Chlorella* cell. Fiziol Rast (Moscow) 31:241–251

Qiu H, Yoon HS, Bhattacharya D (2013) Algal endosymbionts as vectors of horizontal gene transfer in photosynthetic eukaryotes. Front Plant Sci 19:366

Ratti S, Giordano M, Morse D (2007) CO_2-concentrating mechanisms of the potentially toxic dinoflagellate *Protoceratium reticulatum* (Dinophyceae, Gonyaulacales). J Phycol 43:693–701

Raven JA (1997) CO_2-concentrating mechanisms: a direct role for thylakoid lumen acidification? Plant Cell Environ 20:147–154

Reinfelder JR, Kraepiel AML, Morel FMM (2000) Unicellular C_4 photosynthesis in a marine diatom. Nature 407:996–999

Reinfelder JR, Milligan AJ, Morel FMM (2004) The Role of the C_4 pathway in carbon accumulation and fixation in a marine diatom. Plant Physiol 135:2106–2111

Roberts SB, Lane TW, Morel FMM (1997) Carbonic anhydrase in the marine diatom *Thalassiosira weissflogii* (Bacillariophyceae). J Phycol 33:845–850

Roberts K, Granum E, Leegood RC, Raven JA (2007a) Carbon acquisition by diatoms. Photosynth Res 93:79–88

Roberts K, Granum E, Leegood RC, Raven JA (2007b) C_3 and C_4 pathways of photosynthetic carbon assimilation in marine diatoms are under genetic, not environmental, control. Plant Physiol 145:230–235

Rogers M, Keeling PJ (2004) Lateral transfer and recompartmentalization of Calvin cycle enzymes of plants and algae. J Mol Evol 58:367–375

Rost B, Riebesell U, Burkhardt S (2003) Carbon acquisition of bloom-forming marine phytoplankton. Limnol Oceanogr 48:55–67

Rotatore C, Colman B, Kuzuma M (1995) The active uptake of carbon dioxide by the marine diatom *Phaeodactylum triconrutum* and *Cyclotella* sp. Plant Cell Environ 18:913–918

Sakaguchi T, Nakajima K, Matsuda Y (2011) Identification of the UMP stnthase gene by establishment of uracil auxotrophic mutants and the phenotypic complementation system in the marine diatom *Phaeodactylum tricornutum*. Plant Physiol 156:78–89

Samukawa M, Shen C, Hopkinson BM, Matsuda Y (2014) Localization of putative carbonic anhydrases in the marine diatom, *Thalassiosira pseudonana*. Photosynth Res. doi:10.1007/s11120-014-9967-x

Satoh D, HiraokaY, Colman B, Matsuda Y (2001) Physiological and molecular biological characterization of intracellular carbonic anhydrase from the marine diatom *Phaeodactylum tricornutum*. Plant Physiol 126:1459–1470

Sawaya MR, Cannon GC, Heinhorst S, Tanaka S, Williams EB, Yeates TO, Kerfeld CA (2006) The structure of β-carbonic anhydrase from the carboxysomal shell reveals a distinct subclass with one active site for the price of two. J Biol Chem 281:7546–7455

Schuerman P, Jacquot JP (2000) Plant thioredoxin systems revisited. Annu Rev Plant Physiol Plant Mol Biol 51:371–400

Shibata M, Katoh H, Sonoda M, Ohkawa H, Shimoyama M, Fukuzawa H, Kaplan A, Ogawa T (2002) Genes essential to sodium-dependent bicarbonate transport in cyanobacteria: function and phylogenetic analysis. J Biol Chem 277:18658–18664

Sims PA, Mann DG, Medlin LK (2006) Evolution of the diatoms: insights from fossil, biological and molecular data. Phycologia 45:361–402

So AK, Cot SSW, Espie GS (2002) Characterization of the C-terminal extension of carboxysomal carbonic anhydrase from *Synechocystis* sp PCC6803. Funct Plant Biol 29:183–194

So AK, Espie GS, Williams EB, Shively JM, Heinhorst S, Cannon GC (2004) A novel evolutionary lineage of carbonic anhydrase (ε class) is a component of the carboxysome shell. J Bacteriol 186:623–630

Sommer MS, Gould SB, Lehmann P, Gruber A, Przyborski JM, Maier UG (2007) Der1-mediated preprotein import into the periplastid compartoment of chromalveolates? Mol Biol Evol 24:918–928

Soto AR, Zheng H, Shoemaker D, Rodriguez J, Read BA, Wahlund TM (2006) Identification and preliminary characterization of two cDNAs encoding unique carbonic anhydrases from the marine alga *Emiliania huxleyi*. Appl Environ Microbiol 72:5500–5511

Spalding MH, Spreitzer RJ, Ogren WL (1983) Carbonic anhydrase-deficient mutant of *Chlamydomonas reinhardtii* requires elevated carbon-dioxide concentration for photoautotrophic growth. Plant Physiol 73:268–272

Sueltemeyer DF, Fock HP, Canvin DT (1991) Active uptake of inorganic carbon by *Chlamydomonas reinhardtii*: evidence for simultaneous transport of HCO_3^- and CO_2 and characterization of active CO_2 transport. Can J Bot 69:995–1002

Suzuki K, Iwamoto K, Yokoyama S, Ikawa T (1991) Glycolate-oxidizing enzymes in algae. J Phycol 27:492–498

Tachibana M, Allen AE, Kikutani S, Endo Y, Bowler C, Matsuda Y (2011) Localization of putative carbonic anhydrases in two marine diatoms, *Phaeodactylum tricornutum* and *Thalassiosira pseudonana*. Photosynth Res 109:205–221

Tanaka Y, Nakatsuma D, Harada H, Ishida M, Matsuda Y (2005) Localization of soluble β-carbonic anhydrase in the marine diatom *Phaeodactylum tricornutum*. Sorting to the chloroplast and cluster formation on the girdle lamellae. Plant Physiol 138:207–217

Tanaka R, Kikutani S, Mahardika A, Matsuda Y (2014) Localization of enzymes relating to C_4 organic acid metabolisms in the marine diatom, *Thalassiosira pseudonana*. Photosynth Res. doi:10.1007/s11120-014-9968-9

Tchernov D, Hassidim M, Luz B, Sukenik A, Reinhold L, Kaplan A (1997) Sustained net CO_2 evolution during photosynthesis by marine microorganism. Curr Biol 7:723–728

Teich R, Zauner S, Baurain D, Brinkmann H, Petersen J (2007) Origin and distribution of Calvin cycle fructose and sedoheptulase bisphosphatases in plantae and complex algae: a single secondary origin of complex red plastids and subsequent propagation via tertiary endosymbioses. Protist 158:263–276

Tréguer P, Nelson DM, Bennekom AJ, DeMaster DJ, Leynaert A, Quéquiner B (1995) The silica balance in the world ocean: a reestimate. Science 268: 375–379

Trimborn S, Lundholm N, Thomas S, Richter KU, Krock B, Hansen PJ, Rost B (2008) Inorganic carbon acquisition in potentially toxic and non-toxic diatoms: the effect of pH-induced changes in seawater carbonate chemistry. Physiol Plant 133:92–105

Tripp BC, Smith K, Ferry JG (2001) Carbonic anhydrase: new insights for an ancient enzyme. J Biol Chem 276:48615–48618

Uehlein N, Lovisolo C, Siefritz F, Kaldenhoff R (2003) The tobacco aquaporin NtAQP1 is a membrane CO_2 pore with physiological functions. Nature 425:734–737

Van K, Spalding MH (1999) Periplasmic carbonic anhydrase structural gene (*Cah1*) mutant in *Chlamydomonas reinhardtii*. Plant Physiol 120:757–764

Vaughn KC, Campbell EO, Hasegawa J, Owen HA, Renzaglia KS (1990) The pyrenoid is the site of ribulose 1,5-bisphosphate carboxylase/oxygenase accumulation in the hornwort (Bryophyta: Anthocerotae) chloroplast. Protoplasma 156:117–129

Viparelli F, Monti SM, De Simone G, Innocenti A, Scozzafava A, Xu Y, Morel FMM, Supuran CT (2010) Inhibition of the R1 fragment of the cadmium-containing ζ-class carbonic anhydrase from the diatom *Thalassiosira weissflogii* with anions. Bioorg Med Chem Lett 20:4745–4748

Wang Y, Spalding MH (2006) An inorganic carbon transport system responsible for acclimation specific to air levels of CO_2 in *Chlamydomonas reinhardtii*. Proc Natl Acad Sci U S A 103:10110–10115

Wang Y, Sun ZH, Horken KM, Im CS, Xiang Y, Grossman AR, Weeks DP (2005) Analyses of CIA5, the master regulator of the carbon-concentrating mechanism in *Chlamydomonas reinhardtii*, and its control of gene expression. Can J Bot 83:765–779

Weber T, Gruber A, Kroth PG (2009) The presence and localization of thioredoxins in diatoms, unicellular algae of secondary endosymbiotic origin. Mol Plant 2:468–477

Wilhelm C, Büchel C, Fisahn J, Goss R, Jakob T, Laroch J, Lavaud J, Lohr M, Riebesell U, Stehfest K, Valentin K, Kroth PG (2006) The regulation of carbon and nutrient assimilation in diatoms is significantly different from green algae. Protist 157:91–124

Wittpoth C, Kroth PG, Weyrauch K, Kowallik KV, Strotmann H (1998) Functional characterization of isolated plastids from two marine diatoms. Planta 206:79–85

Xiang YB, Zhang J, Weeks DP (2001) The *Cia5* gene controls formation of the carbon concentrating mechanism in *Chlamydomonas reinhardtii*. Proc Natl Acad Sci U S A 98:5341–5346

Xu Y, Feng L, Jeffrey PD, Morel FMM (2008) Structure and metal exchange in the cadmium carbonic anhydrase of marine diatoms. Nature 452:56–61

Yamano T, Miura K, Fukuzawa H (2008) Expression analysis of genes associated with the induction of the carbon-concentrating mechanism in *Chlamydomonas reinhardtii*. Plant Physiol 147:340–354

Yamano T, Tsujikawa T, Hatano K, Ozawa SI, Takahashi Y, Fukuzawa H (2010) Light and low-CO_2 dependent LCIB-LCIC complex localization in the chloroplast supports the carbon-concentrating mechanism in *Chlamydomonas reinhardtii*. Plant Cell Physiol 51:1453–1468

Ynalvez RA, Xiano Y, Ward AS, Cunnusamy K, Moroney JV (2008) Identification and characterization of two closely related β-carbonic anhydrases from *Chlamydomonas reinhardtii*. Physiol Plant 133: 15–26

Yoon HS, Hackett JD, Pinto G, Bhattacharya D (2002) The single, ancient origin of chromist plastids. Proc Natl Acad Sci U S A 99:15507–15512

Zaslavskaia LA, Lippmeier JC, Kroth PG, Grossman AR, Apt KE (2000) Transformation of the diatom *Phaeodactylum tricornutum* (Bacillariophyceae) with a variety of selectable marker and reporter genes. J Phycol 36:379–386

Chapter 19

Leaf: Light Capture in the Photosynthetic Organ

Thomas C. Vogelmann*
Department of Plant Biology, University of Vermont, Burlington, VT 05405, USA

and

Holly L. Gorton
Department of Biology, St. Mary's College of Maryland, St. Mary's City, MD 20686, USA

Summary

Terrestrial photosynthesis fixes about half the carbon on the planet and most of that photosynthesis occurs in chloroplasts in the leaves of higher plants. Leaves protect the chloroplasts, distribute them for light interception, and provide enough surface area interfacing with the atmosphere to facilitate maximum carbon dioxide uptake. There is a balancing act between light levels and carbon dioxide supply: insufficient quantities of either one limit the amount of photosynthesis that a leaf can conduct. When leaves develop under low light they are usually thin because light harvesting limits the amount of carbon that can be fixed and photosynthesis is concentrated within a few cell layers. When leaves develop under high light they are usually thick and light absorption is distributed over many cell layers, greatly increasing the amount of carbon that can be fixed per unit leaf area. In nature, light is rarely

*Author for correspondence, e-mail: thomas.vogelmann@uvm.edu

M.F. Hohmann-Marriott (ed.), The Structural Basis of Biological Energy Generation, Advances in Photosynthesis and Respiration 39, pp. 363–377, DOI 10.1007/978-94-017-8742-0_19,

constant and leaves are often exposed to too little or too much light. This chapter describes adaptations at the level of the leaf that control the amount of light that is absorbed by the photosynthetic tissues. Some of these adaptations are anatomical and the epidermis is the first optical boundary that can play a key role in controlling entry of light into leaves. The anatomy of the photosynthetic tissues and the directional quality of the ambient light also interact to determine the light absorption profile within the tissues, which sets an energetic boundary on the amount of photosynthesis that can be conducted within the individual cell layers. Other adaptations provide for screening of excess light by pigments and fine-tuning of light absorption through chloroplast movement. Additional adaptations occur at the biochemical and whole-plant level to balance light absorption with carbon fixation and this chapter concentrates on the intermediate level: the leaf.

I. Introduction

Light absorption is the first step for photosynthesis and most leaves absorb about 80 % of the light within the photosynthetically active region of the spectrum (PAR, 400–700 nm; Fig. 19.1). The amount of light that a leaf absorbs sets an energetic boundary on how much carbon can be fixed and assimilated. Under low light, where light is limiting to photosynthesis, processes associated with light capture and utilization may be confined to a few cell layers. At the other extreme, under conditions of high light, leaves tend to be much thicker and photosynthesis may be distributed over numerous cell layers leading to higher photosynthetic performance on a leaf-area basis (reviewed by Evans 1999; Evans et al. 2004; Terashima et al. 2009; Terashima et al. 2011). Given that light is highly variable in the natural environment, leaves frequently receive either insufficient light or too much light, and they have developed numerous adaptations to control the amount of light available for photosynthesis. These adaptations are implemented at levels ranging from biochemical reactions to whole-plant organization. This chapter describes some of the adaptations at the level of the leaf (Fig. 19.2).

Abbreviations: FR – Far red; θ_i – Angle of incidence; *n* – Refractive index; PAR – Photosynthetically active region of the spectrum; UV – Ultraviolet

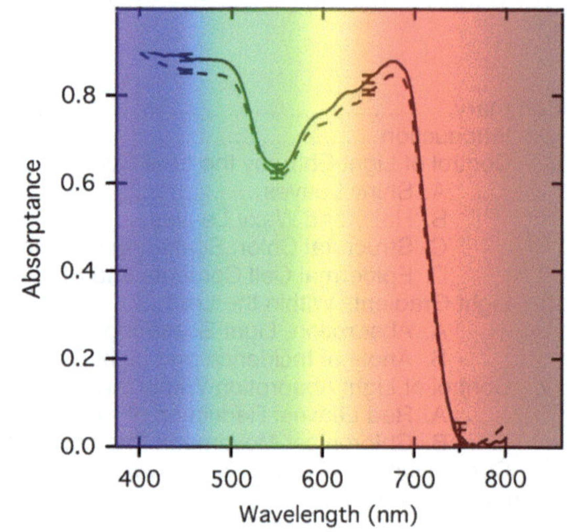

Fig. 19.1. Light absorption by a *Helianthus annuus* leaf within the photosynthetically active region of the spectrum (400–700 nm). The amount of light that is absorbed sets an upper boundary on the amount of carbon that can be assimilated by the leaf. *Solid lines* show absorptance under direct light and *broken lines* diffuse light (After Gorton et al. 2010).

II. Control of Light Entry by the Leaf Surface

The first opportunity for a leaf to influence the extent of light absorption is when photons strike the leaf surface, the cuticle. Few visible photons are absorbed by cuticular waxes, but they can either be accepted and penetrate into the leaf, or rejected through reflectance. The proportion of light taking each of these paths depends on the refractive

Fig. 19.2. Leaves that illustrate optical phenomena described in the chapter. (**a**) A *green leaf* from Norway maple (*Acer platanoides*) shows mainly diffuse *green* reflectance originating in the mesophyll tissue. (**b**) A *shiny leaf* from coffee (*Coffea arabica*) has high specular reflectance from the leaf surface. (**c**) Mullein (*Verbascum thapsus*) has high diffuse reflectance of *white light* because of its densely hairy surface. (**d**) Wax on a leaf from *blue gum* (*Eucalyptus globulus*) causes a *blue tint*. Wax was removed from the lower half of the adaxial surface shown here revealing diffuse, *green* reflectance from the leaf interior. (**e**) Leaves from the tropical fern *Microsorum thailandicum* may be *green* or may show conspicuous *blue* iridescence, depending on growth conditions. In both cases the leaves have high specular reflection of *white light* from the shiny adaxial surface. (**f**) Young leaves of cinnamon (*Cinnamomum verum*) show *red* color from anthocyanin. (**g**) Chloroplast movement causes a striking pattern on a leaf from *Alocasia brisbanensis*. A mask cut in the sunburst shape was placed over the leaf while it was irradiated. Chloroplasts in the shaded area remained along periclinal walls, while those in the irradiated portion of the leaf moved to anticlinal walls causing mutual chloroplast shading, lower absorptance, and a lighter color.

index (n) of the surface and on the angle of incidence (θ_i) of the light beam relative to the surface (Vogelmann 1993). Light is reflected when it encounters a change in n, for example from air ($n=1.0$) to the plant cuticle, which has a much higher n. Estimates of cuticular n vary. For example, the surface of leaf hairs of soybean has a refractive index between 1.47 and 1.49 (Woolley 1975), and theoretical work indicates that epicuticular waxes have refractive index varying from 1.493 at 700 nm to 1.540 at 400 nm (Vanderbilt and Grant 1985). It is noteworthy that the transition from air to cuticle is the largest change in n a photon encounters. Cell walls have $n=1.42$ (Gausman et al. 1974; Woolley 1975), and though the refractive indices of cytoplasm and vacuolar sap vary, the minimum n for cell contents would be that of water, $n=1.33$. Light is reflected as it moves from cuticle to cell wall to cytoplasm to vacuole, but reflectance is low because the change in refractive index at each interface is low.

Changing the epicuticular wax and its refractive index may seem like a possible control point for light absorption, but differences in refractive indices among waxes are small, so changing the wax actually exerts little control over the amount of light rejected through reflection. For an incident beam normal (at a right angle) to the leaf surface, reflectance ranges from 3.6 % with $n=1.47$ to 4.5 % with $n=1.54$. In contrast, θ_i markedly affects reflectance. For a leaf surface with $n=1.5$, reflectance increases from 4.0 to 8.9 % as θ_i increases from 0° (normal to leaf surface) to 60° (Fig. 19.3a). As θ_i approaches 90°, reflectance approaches 100 %. Reflectance of totally diffuse light coming equally from all angles would be 8.2 %. Plants can control θ_i as well as the amount of light striking the leaf by altering leaf angle, for example by solar tracking (diaheliotropism) or solar avoidance (paraheliotropism; Koller 1990). Leaf-surface structure, for example whether it is shiny, hairy, or waxy, can play a role in determining reflectance as well.

A. Shiny Leaves

A smooth, shiny looking, cuticular surface (Fig. 19.2b) will have greater mirror-like reflectance, termed specular reflectance, than a rough one. Specular reflectance is largely independent of wavelength because the refractive index of wax only changes a few percent between 400 and 700 nm, and light is reflected before it can interact with plant pigments located in either epidermis or mesophyll. In addition, specularly reflected light is partially polarized. The amount of specular reflectance observed and degree of polarization depend on viewing angle, and are maximal when the viewing angle equals $-\theta_i$ (Woolley 1971). Human observers detect an increase in the amount of specular reflectance as a decrease in color saturation, but we cannot detect the increase in polarization without instrumentation. Differences among plant species in the amount of specular reflectance, and hence in the degree of polarization of reflected light, can be exploited in remote sensing (Grant 1987).

The cuticular layer and epidermal cell wall absorb very little visible light because they usually do not contain colored pigments. However, in some species both cuticle and wall can absorb UV radiation, as can epicuticular waxes (Pfündel et al. 2006).

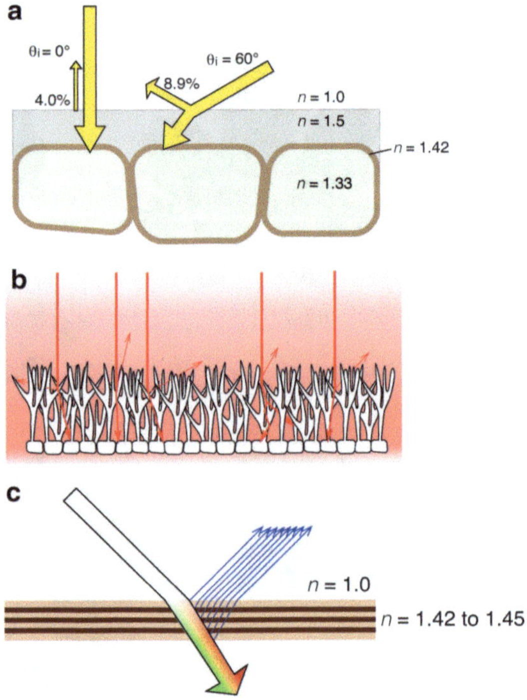

Fig. 19.3. Physical interaction of light with plant surfaces. (**a**) Reflectance requires a change in refractive index (n), as between air and cuticle, shown here in *grey*. Percent reflectance increases with angle of incidence (θ_i). (**b**) Trichomes can increase the number of air/cell interfaces and hence increase diffuse reflectance from a leaf surface. Rays suggest multiple reflections and refraction of light in a complex network of trichomes. Light entering the leaf and reflected light both become more diffuse because of these interactions. (**c**) Iridescence caused by multiple *thin layers*. In this example, only *blue light* reflected from each interface undergoes positive interference and would be observed as iridescence. The wavelength of iridescence depends on the θ_i as well as the thickness and refractive index of the layers. Refractive indices shown for the *wall layers* are estimates. With *blue* iridescence, in contrast to *blue* pigment color, other wavelengths of light are transmitted, not absorbed.

B. Hairy and Waxy Leaves

Reflectance of white light increases when surface structures, such as hairs (Fig. 19.2c) or epicuticular wax (Fig. 19.2d), increase light scattering (Sinclair and Thomas 1970; Ehleringer 1981; Fig. 19.3b). The presence of wax and hairs can more than double reflectance at 680 nm, and wax can increase reflectance of UV radiation at 330 nm more than fivefold (Holmes and Keiller 2002). Reflective, hairy leaves are common in desert plants, where increased reflectance reduces the excessive load of radiation and heat. Reflectance from a hairy leaf surface is generally independent of wavelength in the visible region of the spectrum because little visible light is absorbed by the hairs, that therefore appear white. For example, edelweiss (*Leontopodium nivale* subsp. *alpinum*), the iconic plant of the Alps, has strikingly white, wooly bracts surrounding the flowers. The bracts are densely covered with hollow, transparent, scattering hairs that reflect about 50 % of radiation between 450 and 900 nm (Vigneron et al. 2005). Without the hairs, the bracts reflect only about 5 % of green light and less for red and blue. However, edelweiss also demonstrates that there can be limits to the wavelength neutrality of light scattering by surface hairs. Below 400 nm, reflectance from the bracts drops dramatically to a few percent (Vigneron et al. 2005) suggesting UV absorption by the hollow leaf hairs. In fact, the edelweiss leaf hairs have nanostructure that may enhance UV absorption (Vigneron et al. 2005), and the hairs also may contain UV-absorbing pigments.

C. Structural Color: Scattering and Iridescence

We generally think of leaf color as depending on pigments, especially chlorophylls, carotenoids, and anthocyanins. More rarely, however, leaf color can be structural rather than pigmentary, using processes such as interference and scattering rather than absorption. Structural color cannot be extracted as pigment-based color can. For example, the microstructure of epicuticular waxes on glaucous leaves can cause Tyndall scattering; shorter wavelengths are preferentially scattered back to the observer's eye, causing, the characteristic bluish grey-green color of blue spruce needles (*Picea pugnans*) (Clark and Lister 1975) or blue gum leaves (*Eucalyptus globulus*) (Fig. 19.2d). One can remove the blue cast of the leaves by dissolving the wax (Fig. 19.2d), but the blue color is not found in the solvent because the scattering structure has been destroyed. A bluish cast to glaucous leaves is common but subtle. More rarely, leaves, flowers, and fruits can appear obviously and strikingly blue because of another type of structural color, iridescence (Lee 2009; Glover and Whitney 2010).

Iridescence is eye catching (Fig. 19.2e). Iridescent colors can be quite intense, and a hallmark of iridescence is that both the magnitude of the iridescence and its wavelength depend on viewing angle. Iridescence is most commonly produced either by interference by multiple thin layers or by diffraction from a grating consisting of parallel grooves at an appropriate spacing.

Perhaps the most familiar example of iridescence caused by interference is that produced by a single thin film, for example a soap bubble. Many brilliant colors play across the surface of the bubble, and what the viewer sees depends on minor variations in the thickness of the bubble and on viewing angle. There are reflections from both the front and back surfaces of the film where there is a transition in n first from air to the aqueous medium of the soap film, and then back to air. One must have the appropriate viewing angle to see the reflected light. In addition, because one beam has traveled farther than the other, through the film and back, the two beams have an altered phase relationship to one another. If they are in phase, they interfere constructively, causing intense, visible color. If the reflected beams are 180° out of phase, they interfere destructively and will not be visible. Multiple thin layers can produce even stronger iridescence than a single thin film (Fig. 19.3c).

An example of iridescence caused by a diffraction grating is the rainbow of colors produced by reflection from a compact disc, with its spiral of closely-spaced data tracks. As light hits the tracks, it is split into its component wavelengths because each wavelength has a different diffraction angle. The diffraction results in the spectrum of colors one can see reflected from the disc.

In plants, iridescence occurs in fruits and flowers and leaves, and both interference and diffraction can be involved. Blue iridescence of the fruits of the Australian tropical understory trees *Elaeocarpus angustifolius* (Lee 1991) and *Delarbrea michieana* (Lee 2000) is caused by structures called iridisomes located beneath the outer cell walls of the epidermis. The iridosomes contain many thin layers of cellulose of an appropriate thickness to produce blue iridescence, assuming *n* appropriate for cellulose (Lee 2000). Blue iridescence may provide animals that disperse the seeds with a clear visual signal that persists even during fruit senescence, and unlike a blue color caused by a pigment, it allows red light to penetrate to the fruit interior, where it can drive photosynthesis as the fruit ripens (Lee 1991).

Iridescence in flower petals, including *Tulipa*, *Hibiscus trionum*, and *Mentzelia lindleyi*, has only recently been described and is caused by diffraction from parallel cuticular ridges on elongated epidermal cells (Whitney et al. 2009b). The iridescence might be hidden to human observers if it is masked by underlying pigmented cells, but it can contribute to the suite of visual cues that attract pollinators (Whitney et al. 2009a, b). Such a pollination strategy may be common; more than half of angiosperm species have cuticular ridges on their flower petals, and thus far iridescence has been found in flowers from ten angiosperm families (Whitney et al. 2009b).

Plants with iridescent leaves typically grow in deep shade, in the understory of tropical rainforests, and in all cases described to date the iridescence is attributable to multiple thin layers. This mechanism was first described for two species of *Selaginella*, *S. wildenowii* from Southeast Asia and *S. uncinata* from South China (Hébant and Lee 1984). Leaves that developed in the dim shade of a foliar canopy or in dim artificial light (12 μmol m^{-2} s^{-1}) with enhanced FR had strong blue iridescence. Occurrence of the iridescence was correlated with development of two thin lamellae in the outermost walls of leaf epidermal cells, and the thickness of these layers was consistent with production of iridescence through interference. More recent work has confirmed the existence of those two cell-wall lamellae, and a combination of spectroscopic measurements and modeling supports the connection between the lamellae and the observed iridescence (Thomas et al. 2010). The spectroscopic measurements showed that the specular reflectance peak corresponding to blue iridescence shifted to shorter wavelength with increasing θ_i, as expected for iridescence caused by multiple thin layers. The model suggests that very modest differences in *n* between the thin layers (from $n=1.44$ to $n=1.45$) would be adequate to cause blue iridescence and also to explain the observed shift in wavelength of the blue peak with viewing angle. The wavelength shift observed was more modest than predicted by the model, presumably because of structural irregularities in the leaf surface such as domed epidermal cells. With these irregularities, the thin layers responsible for iridescence become quasi-ordered rather than strictly ordered, and the blue iridescence peak becomes broader, somewhat less intense and less angle-dependent. The angle-invariance of the apparent blue iridescence of the *Morpho rhetenor* butterfly is similarly attributed to quasi-ordered arrays (Vukusic et al. 1999; Kinoshita et al. 2002; Kinoshita and Yoshioka 2005; Doucet and Meadows 2009).

Leaf iridescence from *Selaginella* is perhaps the most thoroughly characterized, but we know something of the mechanism of leaf iridescence in other species as well. Iridescence in the ferns *Danaea nodosa* (Graham et al. 1993), *Diplazium tomentosum*, and *Lindsaea lucida* (Gould and Lee 1996) is attributed to the large number of layers in their outermost cell walls, each with a slightly different orientation of cellulose microfibrils, forming a helicoidal pattern of cellulose orientation through the stack of layers. In the fern *Trichomanes elegans* (Graham et al. 1993) and in angiosperms *Phyllagathis rotundifolia* and *Begonia pavonina* (Gould and Lee 1996) iridescence may be caused by specialized, flattened plastids, termed iridoplasts, in the adaxial epidermis. The iridoplasts have unusual thylakoid stacks with spacing that could generate iridescence. *Selaginella erythropus* has single, large, cup-shaped plastids in its adaxial epidermal cells that appear to have two zones, an upper zone with thylakoid structure similar to iridoplasts, and a lower zone with more typical chloroplast thylakoid structure (Sheue et al. 2007) (see Chap. 17).

It seems likely that flower and fruit iridescence have a role in attracting pollinators and dispersers, but the function of leaf iridescence is not as clear. It is possible that persistent structural iridescence might protect leaves against herbivory, perhaps by disguising them. However, the restricted occurrence of iridescent leaves to deep understory shade, combined with the variety of mechanisms that have evolved to produce iridescence, suggests an advantage related to the understory environment. One possibility is that the thin layers of the cell wall might act as an anti-reflective coating, enhancing entry of photosynthetically active red light, an adaptation to the low-light environment (Lee and Lowry 1975). However, enhanced capture of red light would be coupled to losing equally-important blue light, and the iridescent structures do not seem to act as an anti-reflective coating (Thomas et al. 2010).

A second challenging feature of the understory environment that might relate to leaf iridescence is the occurrence of sunflecks, sudden and intense irradiation that occurs as direct sunlight penetrates through small spaces between the overhanging leaves. Sunflecks can provide much of the photosynthetically active radiation an understory plant receives in a day (Chazdon 1986; Pfitsch and Pearcy 1989), but they can also cause photodamage in leaves adapted to deep shade (Powles and Björkman 1981; Le Gouallec et al. 1990). Blue iridescence might allow plants to reject blue light during a sunfleck to reduce photodamage (Lee 2009). One attractive feature of this hypothesis is that sunflecks differ from the deep shade not just in irradiance but also in the light's directional quality; while light in the deep shade is diffuse, light during sunflecks is mainly direct, collimated light. Directional quality relates to iridescence because iridescence depends on angle and should be greatly reduced under diffuse light. Thus iridescence might be a protective mechanism that is automatically engaged during the potentially damaging, intense, collimated light of a sunfleck, but that is disengaged in the diffuse, dim light of the canopy shade, allowing more of the blue light to enter the leaf to drive photosynthesis. The magnitude of control would be small, however. The eye-catching blue iridescence represents reflectance at the optimal specular reflectance angle on the order of 20 % (Thomas et al. 2010), but iridescence increases total reflectance in the blue by only a few percent (Hébant and Lee 1984).

D. Epidermal Cell Contents and Shape

We have previously discussed epidermal modifications such as hairs, epicuticular wax, and thin layers that produce structural colors. The epidermal cells can further affect both wavelength and directionality of light penetrating to the leaf interior through their vacuolar pigmentation and cell shape. The epidermal cells of flowers often contain anthocyanins that are responsible for flower color, but leaf epidermal cells more often contain UV-absorbing, colorless pigments such as flavonoids and sinapates. Synthesis of these screening pigments is induced by UV exposure, and their role in reducing UV damage is well established (Caldwell et al. 1983; Sheahan 1996; Cockell and Knowland 1999; Pfündel et al. 2006).

Cell shape plays a role because epidermal cells are rarely flat; the outermost wall is generally convex, and can be papillose or even conical. Those curved surfaces, which have refractive index higher than air, can act as lenses, focusing incident light. Strongly papillose epidermal cells in flowers can focus light on the pigmented vacuole, increase absorption, and make floral colors appear more saturated (Gorton and Vogelmann 1996). Similarly, the papillose shape of some leaf epidermal cells may enhance UV absorption by vacuolar pigments like flavonoids or sinapates. Plants with papillose leaf epidermal cells often live in dense shade. In some cases, they may focus light on chloroplasts that are clustered in a focal spot beneath each epidermal cell (Bone et al. 1985), perhaps enhancing photosynthesis in low light, but this idea has not been tested experimentally. Another suggestion relates to the diffuse nature of the light environment of the understory: papillose epidermal cells might reduce specular reflectance and aid in capturing low-angle light (Lee 2009). In the completely diffuse light typical of deep shade, a leaf loses 8.2 % of the incoming light by specular reflectance, and the ability of a papillose leaf to capture this light might be beneficial in such a very low-light environment. However, there is little experimental support for this hypothesis (Brodersen and Vogelmann 2007).

III. Light Gradients Within the Leaf

A. Absorption, Light Scattering and Leaf Anatomy

As discussed above, the leaf surface can alter the amount of light penetrating into the leaf and available for photosynthesis. Further changes occur within the leaf mesophyll, where light is attenuated by scattering and absorption. Light scattering in leaves occurs primarily by reflection from the intercellular air spaces and, within the visible spectrum, the amount of scattering is relatively independent of wavelength (Merzlyak et al. 2009). Absorption occurs primarily by chlorophylls, carotenoids, flavonoids and related pigments that alter the spectral quality within the various leaf tissues. Given that a typical leaf absorbs 80 % of the light that strikes it, it stands to reason that the internal light gradients are relatively steep and there are dramatic changes in spectral quality as one progresses from the top to the bottom of the leaf, creating a highly differentiated light environment for chloroplasts at different depths within the tissues.

The amount of light absorption in leaves depends upon pigment composition, concentration, and wavelength-specific absorption (extinction coefficient). In addition, the photosynthetic pigments are contained within chloroplasts that move within the cells, thereby controlling how much light they intercept and absorb (see below). On the other hand, light scattering depends upon the characteristics of tissue anatomy, primarily cell shape and the volume and shape of the intercellular air spaces. The steepest light gradients in leaves are found within the blue, owing to the particularly strong absorption bands of carotenoids and chlorophylls, whereas green light penetrates deeper because chlorophyll absorbs minimally in this region (Vogelmann and Han 2000; Vogelmann and Evans 2002; Evans and Vogelmann 2003; Brodersen and Vogelmann 2010; Figs. 19.4 and 19.5).

It is possible to estimate the contributions of absorption vs. scattering by infiltrating the intercellular air spaces with water, which eliminates much of the light scattering within the tissues. In spinach, this treatment increases only slightly the amount of blue light that penetrates through the palisade tissue, indicating the relative importance of absorption in this region of the PAR for these chloroplast-packed, columnar cells (Vogelmann and Han 2000; Vogelmann and Evans 2002). The effect of infiltration is greater in the spongy mesophyll tissue, pointing to a greater contribution of light scattering within that tissue (Fig. 19.5) where the cells are more irregular in shape than the tubular cells in the palisade tissue. These effects of infiltration are more pronounced for green than blue light in both tissues, indicating the importance of light scattering effects in this spectral region. The differences in shapes of the light gradients in the palisade vs. spongy mesophyll tissue illustrates the important contributions of

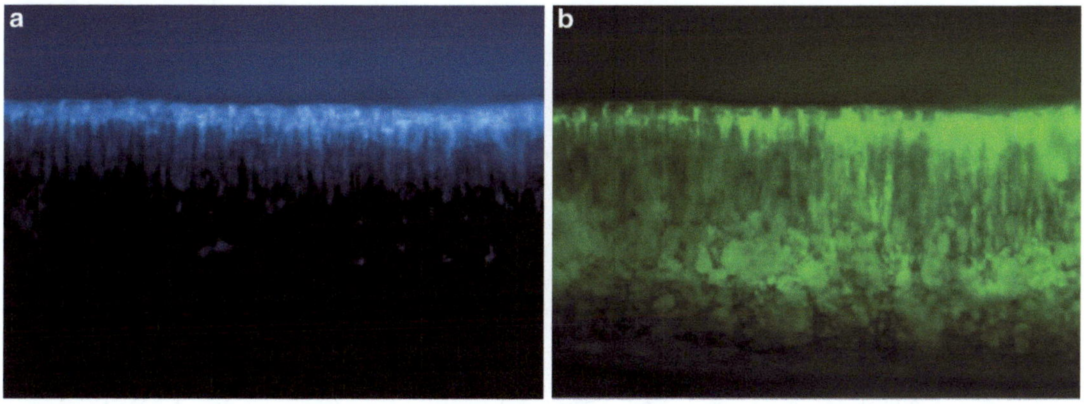

Fig. 19.4. False-color light micrographs of absorption profiles visualized from chlorophyll fluorescence in an *Antirrhinum majus* leaf irradiated with direct light (**a**) in the *blue* (488 nm) and (**b**) *green* (532 nm). *Blue light* penetrates less far than green because it is absorbed more intensely by the photosynthetic pigments. The profile of absorbed *red light* (not shown) is intermediate between that of the *blue* and *green* (From Brodersen and Vogelmann 2010).

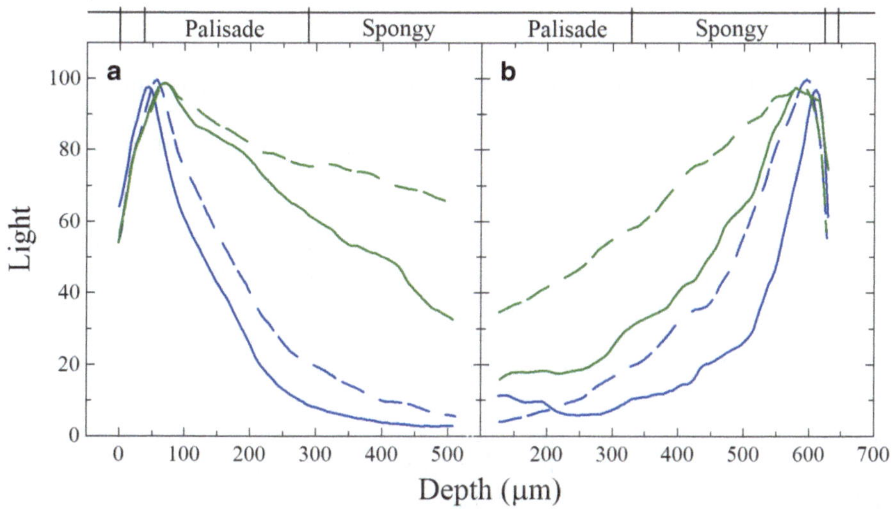

Fig. 19.5. Absorption profiles in spinach leaves irradiated on their upper (adaxial, **a**) and lower (abaxial, **b**) surface with *blue* (*blue lines*, 450 nm) and *green* (*green lines*, 550 nm) monochromatic light. Profiles are shown before (*solid lines*) and after (*broken lines*) the intercellular air space were infiltrated with water to remove most of the internal light scattering (From Vogelmann and Evans 2002).

tissue anatomy in determining both the shape and steepness of light gradients within leaves.

B. Angle of Incidence and Light Directional Quality

In nature, light varies widely in direction, and in directional quality. On sunny days most of the light consists of parallel rays, which, depending upon leaf orientation, can intersect the leaf from different directions. About 85 % of sunlight on clear days is direct whereas the remainder consists of diffuse light that results from scattering by the atmosphere. The presence of clouds, aerosols and plant canopies can increase the amount of diffuse light, so that in

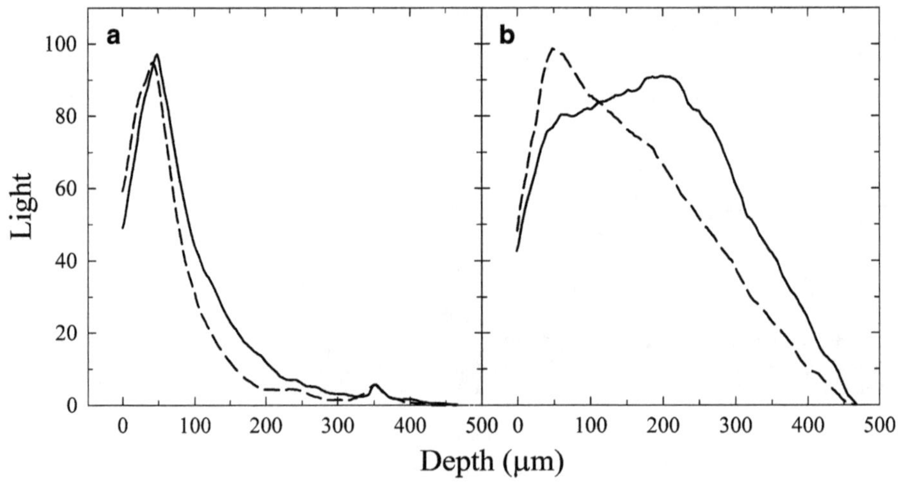

Fig. 19.6. Light absorption profiles in leaves of *Antirrhinum majus* irradiated at different angles of incidence. The adaxial surface was irradiated with *blue* (**a**, 488 nm) or *green* (**b**, 532 nm) light at 0° (normal to the leaf surface, *solid lines*), or 60° (*broken lines*) (From Brodersen and Vogelmann 2010).

nature there is a broad continuum in the directional quality of light, from mostly direct to highly diffuse.

As described previously (Fig. 19.3a) the direction of travel of light into the leaf is influenced by the angle of incidence and it stands to reason that light gradients should be affected by light direction and directional quality. Indeed, light gradients become steeper when direct light changes from normal (90°) to 60° (Fig. 19.6). The largest changes are observed in the green part of the spectrum and interestingly, the shape of the profile is also changed, suggesting that both light-piping and light scattering contribute to creation of the absorption profile within the palisade tissue. Similar effects are observed when gradients are measured in leaves irradiated with direct vs. diffuse light, where the gradients become steeper under conditions of diffuse light (Brodersen and Vogelmann 2010).

IV. Control of Light Absorption Within the Leaf

We have seen that gradients of light absorption within leaves are influenced both by light absorption and by scattering. Scattering is mainly controlled by the size and shape of cells and air spaces, and is not easily altered in mature leaves, but absorption can be adjusted more easily.

A. Red Leaves: Reduction of Internal Light by Screening Pigments

After light enters a leaf, what mechanisms other than anatomy can control the amount of light and its distribution within the photosynthetic tissues? When plants are exposed to excess light, subjected to nutrient or a variety of other stresses, they commonly synthesize red anthocyanins, which, in contrast to UV screening pigments that humans do not see, can be quite noticeable. Anthocyanins are found in the central vacuoles of the epidermis, leaf mesophyll tissues and in stem tissues (Gould et al. 2010). Less commonly, other pigments such as betalains, proanthocyanidins and lignins also create red coloration (Davies 2004). Anthocyanins are often found in young (Fig. 19.2f) and senescing leaves but usually not during the intervening period when the leaf is fully photosynthetically functional (van den Berg et al. 2009). The early and late stages of the lifetime of a leaf are a time when the light-harvesting system is not fully coupled to the carbon-reduction cycle, developmental stages when the absorbed light energy must be dissipated so it does not give rise to damaging oxygen free radicals. Following through with this line of thinking, a leading hypothesis for the presence of red coloration in leaves and stems is that the pigments may serve as light screens to provide photoprotection to the chloroplasts (reviewed by Steyn et al. 2002; Hatier and Gould 2008; Archetti et al. 2009; Gould et al. 2010). Other adaptive functions have also been suggested: the pigments may act as antioxidants (Gould and Lister 2005; Hernandez et al. 2008) or they may provide a visual deterrent to ward off herbivores (Döring et al. 2009).

Given the optical properties of leaves (Fig. 19.1), are the spectral properties of red anthocyanins ideal for a sunscreen role? Anthocyanins typically have a broad absorption band that extends from 400 to 600 nm with an absorption maximum in the green at 520–540 nm that is 2–3 times higher than absorption in the blue. Foliar anthocyanins usually absorb 20–30 % of the light within the green spectral range (Feild et al. 2001), which seems small compared to the amount that sunlight can fluctuate within the environment. This, in combination with the fact that photosynthesis saturates in many plants at irradiances well below full sunlight, leaves one wondering if this level of screening is sufficient to confer adaptive advantage. Blue light is more effective than other wavelengths in causing photoinhibition, so would a pigment with strong absorption in the blue make a better sunscreen? If the physiological strategy were to provide photoprotection primarily to chloroplasts in the palisade tissue, where most of the blue light is absorbed for photosynthesis (Sun et al. 1998), then placing a yellow pigment near the irradiated surface would be more effective than a red pigment. However, green light penetrates deeply into leaves and it drives photosynthesis in all the photosynthetic cell layers with a relatively high quantum yield (Nishio 2000; Terashima et al. 2009). If the goal is to throttle back electron transport in all the cell layers, then placing a red pigment near the irradiated surface would accomplish that goal and anthocyanins seem ideally suited for that purpose.

What evidence supports the idea that anthocyanins serve as light screens for the photosynthetic system? Most of the evidence comes from measurements of photosystem II quantum efficiency, where significant levels of photoprotection were found in leaves containing anthocyanins. However, this finding is not universal (Manetas 2006). Measuring light gradients within leaves to assess screening within the photosynthetic tissues also gives varying results. In some leaves anthocyanins altered the internal light gradients and provided screening to the spongy mesophyll tissue (Gould et al. 2002), whereas in other leaves they did not appear to affect the amount of light absorbed by this tissue (van den Berg et al. 2009).

B. Chloroplast Movement: Photoprotection Through Shading?

Each photosynthetic cell within a leaf contains dozens of chloroplasts that occupy the band of cytoplasm between the central vacuole and cell wall. Their position is not fixed; they move rapidly in response to the irradiance that strikes the leaf surface. Under low light, chloroplasts align along the surface (periclinal) cell walls, whereas under high light they move to the sidewalls (anticlinal), which increases the amount of light that is transmitted in leaves and correspondingly decreases leaf absorptance; this phenomenon is taxonomically widespread among plants (Evans et al. 2004). Chloroplast movement can lead to striking visual changes. For example, placing a mask on a leaf and irradiating it with high light for 20 min can create an image (Fig. 19.2). Regions of the leaf where chloroplasts are positioned along the anticlinal cell walls (low light) appear green, whereas when chloroplasts are positioned along the periclinal walls (high light) the leaf appears less green. The largest optical changes are found in leaves of shade plants where changes in transmittance up to 15 %

have been measured. More typical optical changes are in the range of 6 % for shade leaves and 2 % for sun leaves (Davis et al. 2011). Noninvasive measurements of plants in the field have shown that chloroplasts are in motion most of the time, rarely achieving the extreme anticlinal or periclinal positions (Williams et al. 2003).

The appearance of leaves and the characteristics of their internal light gradients are determined by light scattering and light absorption. Calculating the extent to which these optical parameters are affected by chloroplast movement shows that most of the changes in leaf optical properties are caused by changes in absorption and that chloroplasts have little influence on the light-scattering properties of leaf tissues (Davis et al. 2011).

So why do chloroplasts move? It may by that movement maximizes photosynthesis, where under low light chloroplast positioning along the periclinal cell walls maximizes light absorption for photosynthesis. Similarly, under high light, movement to the anticlinal walls allows more light to penetrate to chloroplasts located deeper within the leaf, resulting in an increase in the overall photosynthetic rate of the leaf. This differential positioning of chloroplasts would therefore increase photosynthetic performance in both cases. While this reasoning appears logical, there is little experimental evidence to support this hypothesis. Another idea is that movement may minimize the distance that CO_2 travels from the intercellular air spaces to the chloroplast, especially when chloroplasts are in the high-light position adjacent to the bulk of intercellular air spaces in the palisade tissue (Evans 1999). Measuring oxygen diffusion times as a proxy for CO_2 migration in leaves of *Alocasia* did not support the notion that chloroplast movement modulates intracellular gas exchange (Gorton et al. 2003). Additional gas-exchange studies with *Arabidopsis thaliana* showed that chloroplast movement to the high-light position had a limiting effect on photosynthesis (Tholen et al. 2008). A third hypothesis is that chloroplast movement and orientation protect against photoinhibition. Chloroplasts are disk-shaped and, similar to the principles of a venetian blind, changing their orientation such that the face vs. the edge intercepts the incident light will significantly affect the amount of light that they absorb. Even though chloroplast movement does not change leaf absorptance very much, movement and orientation have significant effects on the internal light gradient. Movement from the low-light to the high-light position would allow more blue light to penetrate further into the leaf (Evans et al. 2004). The gradient in red light would be expected to change similarly, but the gradient in green light should not change very much. The potential effects of this spectral tuning on photosynthesis remain to be determined, but overall movement has been shown to protect chloroplasts against photoinhibition caused by high light (Kasahara et al. 2002). More research is warranted to more fully evaluate the photosynthetic consequences of chloroplast movement in leaves.

V. Conclusions

Leaves are capable of controlling the amount and distribution of light available for photosynthesis in many ways, both at the leaf surface and within the mesophyll tissue. Some of these control mechanisms can have large effects, for example the 50 % reflectance from the hairy bracts surrounding edelweiss flowers. Most, however, are more subtle. Some, such as anthocyanin synthesis or chloroplast movement, can be rapidly engaged as needed, while others, such as scattering in the mesophyll are structural and difficult to adjust over short time periods in mature leaves. These leaf-level mechanisms work together and in concert with whole-plant mechanisms, such as leaf orientation, and biochemical mechanisms, such as the xanthophyll cycle.

Acknowledgments

We thank William E. Williams for his insightful comments and suggestions on the manuscript.

References

Archetti M, Döring TF, Hagen SB, Hughes NM, Leather SR, Lee DW, Lev-Yadun S, Manetas Y, Ougham HJ, Schaberg PG, Thomas H (2009) Unravelling the evolution of autumn colours: an interdisciplinary approach. Trends Ecol Evol 24:166–173

Bone RA, Lee DW, Norman JM (1985) Epidermal cells functioning as lenses in leaves of tropical rainforest plants. App Opt 24:1408–1412

Brodersen CR, Vogelmann TC (2007) Do epidermal lens cells facilitate the absorptance of diffuse light? Am J Bot 94:1061–1066

Brodersen CR, Vogelmann TC (2010) Do changes in light direction affect absorption profiles? Funct Plant Biol 37:403–412

Caldwell MM, Robberecht R, Flint SD (1983) Internal filters: prospects for UV-acclimation in higher plants. Physiol Plant 58:445–450

Chazdon RL (1986) Light variation and carbon gain in rain forest understorey palms. J Ecol 74: 995–1012

Clark J, Lister G (1975) Photosynthetic action spectra of trees. II. The relationship of cuticle structure to the visible and ultraviolet spectral properties of needles of four coniferous species. Plant Physiol 55:407–413

Cockell CS, Knowland J (1999) Ultraviolet radiation screening compounds. Biol Rev 74:311–345

Davies KM (2004) Plant pigments and their manipulation. Blackwell Publishing, Oxford

Davis PA, Caylor S, Whippo CW, Hangarter RP (2011) Changes in leaf optical properties associated with light-dependent chloroplast movements. Plant Cell Environ 34:2047–2059

Döring TF, Archetti M, Hardie J (2009) Autumn leaves seen through herbivore eyes. Proc R Soc B 276: 121–127

Doucet S, Meadows MG (2009) Iridescence: a functional perspective. JRS Interface 6:S115–S132

Ehleringer J (1981) Leaf absorptances of Mohave and Sonoran Desert plants. Oecologia 49:36–370

Evans JR (1999) Leaf anatomy enables more equal access to light and CO_2 between chloroplasts. New Phytol 143:93–104

Evans JR, Vogelmann TC (2003) Profiles of ^{14}C fixation through spinach leaves in relation to light absorption and photosynthetic capacity. Plant Cell Environ 26:547–560

Evans JR, Vogelmann TC, Williams WE, Gorton HL (2004) Chloroplast to leaf. In: Smith WK, Vogelmann TC, Critchley C (eds) (2005) Photosynthetic adaptation: chloroplast to landscape. Ecological studies, vol 178. Springer, New York, pp 15–41

Feild TS, Lee DW, Holbrook NM (2001) Why leaves turn red in autumn. The role of anthocyanins in senescing leaves of red-osier dogwood. Plant Physiol 127:566–574

Gausman HW, Allen WA, Escobar DE (1974) Refractive index of plant cell walls. App Opt 13:109–111

Glover BJ, Whitney HM (2010) Sturctural colour and iridescence in plants: the poorly studied relations of pigment colour. Ann Bot 105:505–511

Gorton HL, Vogelmann TC (1996) Effects of epidermal cell shape and pigmentation on optical properties of *Antirrhinum* petals at visible and ultraviolet wavelengths. Plant Physiol 112:879–888

Gorton HL, Herbert SK, Vogelmann TC (2003) Photoacoustic analysis indicates that chloroplast movement does not alter liquid-phase CO_2 diffusion in leaves of *Alocasia brisbanensis*. Plant Physiol 132:1529–1539

Gorton HL, Brodersen CR, Williams WE, Vogelmann TC (2010) Measurement of the optical properties of leaves under diffuse light. Photochem Photobiol 86:1076–1083

Gould KS, Lee DW (1996) Physical and ultrastructural basis of blue leaf iridescence in four Malaysian understory plants. Am J Bot 83:45–50

Gould KS, Lister C (2005) Flavonoid functions in plants. In: Andersen ØM, Markham KR (eds) Flavonoids: chemistry, biochemistry, and applications. CRC, Boca Raton, pp 397–441

Gould KS, Vogelmann TC, Han T, Clearwater MJ (2002) Profiles of photosynthesis within red and green leaves of *Quintinia serrata* A. Cunn. Physiol Plant 116:127–133

Gould KS, Dudle DA, Neufeld HS (2010) Why some stems are red: cauline anthocyanins shield photosystem II against high light stress. J Exp Bot 61:2707–2717

Graham RM, Lee DW, Norstog K (1993) Physical and ultrastructoral basis of blue leaf iridescence in two neotropical ferns. Amer J Bot 80:198–203

Grant L (1987) Diffuse and specular characteristics of leaf reflectance. Rem Sens Environ 22:309–322

Hatier J-HB, Gould KS (2008) Anthocyanin function in vegetative organs. In: Gould KS, Davies K, Winefield C (eds) Anthocyanins: biosynthesis, functions, and application. Springer, New York, pp 1–19

Hébant C, Lee DW (1984) Ultrastructural basis and developmental control of blue iridescence in *Selaginella* leaves. Amer J Bot 71:216–219

Hernandez I, Alegre L, Breusegem FV, Munne-Bosch S (2008) How relevant are flavonoids as antioxidants in plants? Trends Plant Sci 14:125–132

Holmes MG, Keiller DR (2002) Effects of pubescence and waxes on the reflectance of leaves in the ultraviolet and photosynthetic bands: a comparison of a range of species. Plant Cell Environ 25:85–93

Kasahara M, Kagawa T, Oikawa K, Suetsugu N, Miyao M, Wada M (2002) Chloroplast avoidance movement reduces photodamage in plants. Nature 420:829–832

Kinoshita S, Yoshioka S (2005) Structural colors in nature: the role of regularity and irregularity in the structure. Chem Phys Chem 6:1442–1459

Kinoshita S, Yoshioka S, Kawagoe K (2002) Mechanisms of structural colour in the *Morpho* butterfly: cooperation of regularity and irregularity in an iridescent scale. Proc R Soc London B Biol Sci 269:1417–1421

Koller D (1990) Light-driven leaf movements. Plant Cell Environ 13:615–632

Le Gouallec J, Cornic G, Blanc P (1990) Relations between sunfleck sequences and photoinhibition of photosynthesis in a tropical rain forest understory herb. Amer J Bot 77:999–1006

Lee DW (1991) Ultrastructural basis and function of iridescent blue colour of fruits in *Elaeocarpus*. Nature 349:260–262

Lee DW (2000) Structural fruit coloration in *Delarbrea michieana* (Araliaceae). Int J Plant Sci 161:297–300

Lee DW (2009) Plant tissue optics: micro- and nanostructures. In: Martin-Palma RJ, Lakhtakia A (eds) Biomimetics and bioinspiration. SPIE, San Diego, pp 740104-1–740104-11

Lee DW, Lowry JB (1975) The physical basis and ecological significance of iridescence in blue plants. Nature 254:50–51

Manetas Y (2006) Why some leaves are anthocyanic and why most anthocyanic leaves are red? Flora 201:163–177

Merzlyak MN, Chivkunova OB, Zhigalova TV, Naqvi KR (2009) Light absorption by isolated chloroplasts and leaves: effects of scattering and 'packing'. Photosyn Res 102:31–41

Nishio JN (2000) Why are plants green? Evolution of the higher plant photosynthetic pigment complement. Plant Cell Environ 23:539–548

Pfitsch WA, Pearcy RW (1989) Daily carbon gain by *Adenocaulon bicolor* (Asteraceae), a redwood forest understory herb, in relation to its light environment. Oecologia 80:465–470

Pfündel EE, Agati G, Cerovic ZG (2006) Optical properties of plant surfaces. In: Riederer M, Müller C (eds) Annual plant reviews. Blackwell, Oxford, pp 216–249

Powles SB, Björkman O (1981) Leaf movement in the shade species *Oxalis oregana*. II. Role in protection against injury by intense light. Carnegie Inst Wash Yearb 80:63–66

Sheahan JJ (1996) Sinapate esters provide greater UV-B attenuation than flavonoids in *Arabidopsis thaliana* (Brassicaceae). Amer J Bot 83: 679–686

Sheue CR, Sarafis V, Kiew R, Liu HY, Salino A, Kuo-Huang LL, UYang YP, Tsai CC, Lin CH, Yong JWH, Ku MSB (2007) Bizonoplast, a unique chloroplast in the epidermal cells of microphylls in the shade plant *Selaginella erythropus* (Selaginellaceae). Amer J Bot 94:1922–1929

Sinclair R, Thomas DA (1970) Optical properties of leaves of some species in arid South Australia. Aust J Bot 18:261–273

Steyn WJ, Wand SE, Holcroft DM, Jacobs G (2002) Anthocyanins in vegetative tissues: a proposed unified function in photoprotection. New Phytol 155:349–361

Sun J, Nishio JN, Vogelmann TC (1998) Green light drives CO_2 fixation deep within leaves. Plant Cell Physiol 39:1020–1026

Terashima I, Fujita T, Inoue T, Chow WS, Oguchi R (2009) Green light drives leaf photosynthesis more efficiently than red light in strong white light: revisiting the enigmatic question of why leaves are green. Plant Cell Physiol 50:684–697

Terashima I, Hanba YT, Tholen D, Niinemets U (2011) Leaf functional anatomy in relation to photosynthesis. Plant Physiol 155:108–116

Tholen D, Boom C, Noguchi KO, Ueda S, Katase T, Terashima I (2008) The chloroplast avoidance response decreases internal conductance to CO_2 diffusion in *Arabidopsis thaliana* leaves. Plant Cell Environ 31:1688–1700

Thomas KR, Kolle M, Whitney HM, Glover BJ, Steiner U (2010) Function of blue iridescence in tropical understorey plants. JRS Interface 7:1699–1707

van den Berg AK, Vogelmann TC, Perkins TD (2009) Anthocyanin influence on light absorption within juvenile and senescing sugar maple leaves – do anthocyanins function as photoprotective visible light screens? Funct Pl Biol 36:793–800

Vanderbilt VC, Grant L (1985) Plant canopy specular reflectance model. IEEE Trans Geosci Rem Sens 23:722–730

Vigneron JP, Rassart M, Vértesy Z, Kertész K, Sarrazin M, Biró L, Ertz D, Lousse V (2005) Optical structure and function of the white filamentary hair covering the edelweiss bracts. Phys Rev E 71

Vogelmann TC (1993) Plant tissue optics. Annu Rev Plant Physiol Plant Mol Biol 44:231–251

Vogelmann TC, Evans JR (2002) Profiles of light absorption and chlorophyll within spinach leaves from chlorophyll fluorescence. Plant Cell Environ 25:1313–1323

Vogelmann TC, Han T (2000) Measurement of profiles of absorbed light within spinach leaves. Plant Cell Environ 23:1303–1312

Vukusic P, Sambles JR, Lawrence CR, Wootton RJ (1999) Quantified interference and diffraction in single *Morpho* butterfly scales. Proc R Soc Lon B 266:1403–1411

Whitney HM, Kolle M, Alvarez-Fernandez R, Steiner U, Glover BJ (2009a) Contributions of iridescence to floral patterning. Comm Int Biol 2:230–232

Whitney HM, Kolle M, Andrew P, Chittka L, Steiner U, Glover BJ (2009b) Floral iridescence, produced by diffractive optics, acts as a cue for animal pollinators. Science 323:130–133

Williams WE, Gorton HL, Witiak SM (2003) Chloroplast movements in the field. Plant Cell Environ 26:2005–2014

Woolley JT (1971) Reflectance and transmittance of light by leaves. Plant Physiol 47:656–662

Woolley JT (1975) Refractive index of soybean leaf cell walls. Plant Physiol 55:172–174

Chapter 20

Lichen Photosynthesis. Scaling from the Cellular to the Organism Level

Miloš Barták*

Faculty of Science, Department of Experimental Biology, Masaryk University, Kamenice 5, 62500 Brno, Czech Republic

Summary

Photosynthesis in lichens is intimately linked to the photosynthetic capacities of the photobiont, *i.e.* autotrophic algae and cyanobacteria, that form the lichen association together with a fungal partner. Lichen photosynthesis in nature is also affected by a complex mixture of internal and external factors.

Intrathalline locatization of photobiont cells, structure of photobiont layer, functional photobiont-mycobiont interlink, and physico-chemical properties of the fungal part of thallus are considered important internal characteristics affecting photosynthesis and utilization of photosynthetic products in lichens. In this chapter, a brief introduction into the anatomy and morphology is provided from a view point of function. Special attention is given to cellular

*Author for correspondence, e-mail: mbartak@sci.muni.cz

M.F. Hohmann-Marriott (ed.), The Structural Basis of Biological Energy Generation, Advances in Photosynthesis and Respiration 39, pp. 379–400, DOI 10.1007/978-94-017-8742-0_20,

structure of photobionts, and especially to the chloroplast of unicellular alga *Trebouxia*, the most abundant symbiotic alga in lichen association. Since lichens are typical poikilohydric organism with no active control of their hydration status, the photosynthetic responses of lichens to full, partial and severely limited water supply are described. In addition the protective mechanisms activated during thallus desiccation are discussed. Several aspects of lichen photosynthesis including light-response curves, photoinhibition, activation of photoprotective mechanisms and reactive oxygen species-induced changes in the amount and activity of antioxidative substances are reviewed. Lichens can photosynthesize over a wide temperature range, including subzero temperature. The photobiont also exhibits response depending on nitrogen availability and exposure to heavy metals.

I. Introduction

Photosynthesis is a key metabolic function of lichens. Therefore a understanding of mechanisms and limitations of photosynthesis in this symbiotic organisms is desirable. Since many lichens are considered extremophilic organisms, that exhibit great tolerance against a variety of stress factors, lichen photosynthesis in response to several of these stress factors has been investigated in detail. Defining and characterizing the range of conditions that lichen can tolerate continuous to be an active area of research. The photosynthetic machinery of lichens responds to these challenges with a number of biophysical and biochemical protective mechanisms. Although some of these protective mechanisms, may have well-characterized homologous counterparts in plants, many mechanisms in lichens are not yet fully understood. This lack of knowledge has given generations of lichenologist, plant physiologist and ecophysiologist the motivation to study lichen photosynthesis under natural and controlled conditions. In recent years, molecular, proteomic and metabolomic approaches are increasingly employed to study lichen photosynthesis. In addition miniaturized and automated measuring devices are now available that enable detailed long-term studies of photosynthesis in the field. This chapter gives an overview of our current knowledge of lichen photosynthesis, with an emphasis on mechanisms that have been revealed under stress conditions.

Abbreviations: APX – Ascorbate peroxidase; CA – Carbonic anhydrase; CAT – Catalase; CCM – Carbon concentrating mechanism; DEPS – Deepoxidation state of xantophyll-cycle pigments; DW – Dry weight; Fo – Background chlorophyll fluorescence; F_m – Maximal fluorescence yield; F_v – Variable chlorophyll fluorescence; F_v/F_m – Potential quantum yield of photochemical processes in photosystem II; FW – Fresh weight; GSH – Reduced glutathione; GSSG – Oxidised glutathione; NPQ – Non-photochemical quenching; PAR – Photosynthetically active radiation; Pn – Net photosynthesis; Pn_{max} – Maximum rate of net photosynthesis; PSII – Photosystem II; ϕ_{PSII} – Effective quantum yield of photochemical processes in photosystem II; qE – Energy dependendt quenching; qI – Photoinhibitory quenching; qT – State 1-state 2 transition quenching; ROS – Reactive oxygen species; RuBisCo – Ribulose bis phosphate carboxylase oxygenase; RWC – Relative water contents; SOD – Superoxid dismutase; UV-A – Ultraviolet radion (B); UV-B – Ultraviolet radion (B); WP – Water potential

II. Lichen Anatomy and Morphology

A. Symbionts and Thallus Forms

Liches are symbiotic organisms consisting of at least two partners: A photobiont and mycobiont that form the structures of a lichenized association, *i.e.* the lichen thallus. The mycobiont, *i.e.* fungal partner (in most cases Ascomycetes) forms the majority of the lichen thallus biomas. The mycobiont is with only a few exceptions, responsible for thallus morphology and growth form. The photobiont, *i.e.* photosynthetizing partner, is an alga or cyanobacterium. The relation

Fig. 20.1. Placopsis contortuplicata is a lichen that has two photosynthesizing partners located in a thallus. In the central *orange* part (cephalodium), there is a nitrogen-fixing cyanobacterium (*Nostoc* sp.). Marginal *white* thallus parts have an algal photobiont (Photo M. Barták).

between these two partners is considered a mutualistic symbiosis.

While bipartite lichens consist of one mycobiont and one photobiont species, tripartite lichens have two photobionts in a single mycobiont. Tripartite lichens have typically an alga (see Chap. 17) and cyanobacteria (see Chap. 14) in association with the mycobiont. Areas where these two photobionts are located have anatomical differences in the majority of cases – see e.g. *Placopsis contortuplicata* (Fig. 20.1).

Regarding thallus morphology, lichens are divided into three main and several additional morphological groups. In this chapter, only the main ones are treated: crustose liches, foliose lichens, and fruticose lichens (for more details see e.g. Büdel and Scheidegger 1996).

Crustose lichens are tightly attached to the substrate (upper surface of stones and rocks) with their lower surface (see Fig. 20.2). A typical feature that distinguishes crustose lichens from the foliose and fruticose lichens is that crustose lichens cannot be removed from the surface without a lot of force and destruction of the thallus structure.

Foliose lichens possess generally a flat leaf-like thallus, which is attached to the surface only partially by special structures: rhizines, cilia, tomentum or umbilicus (see Fig. 20.3). Foliose lichens exhibit considerable difference in coloration and surface structure between the upper and lower thallus surface. The thallus of foliose lichens is often divided into lobes that show various degree of branching. The size of the lobes can vary from several milimeters to more than 10 cm. Among foliose lichens, two subgroups can be distinguished: laciniate and umbilicate. While laciniate thalli typically exhibit numerous lobes arranged radially or forming overlapping structures, umbilicate lichens create circular thalli consisting of one unbranched lobe.

Lichens with fruticose thallus morphology typically form three-dimensional branching structures. These structures may be arranged upward (shrub-like structures) or form hanging branched structures oriented downward from a basal holdfast attached to the substrate (typically tree branches or stems). Fruticose lichens form a cluster of free-standing, branching tubes, which are usually round in cross section (see Fig. 20.4). Some fruticose lichens, however, may have flattened branches such as e.g. *Ramalina* sp.

B. Intrathalline Location of Photobionts and Their Photosynthetic Properties

More than 40 algal and cyanobacterial genera have been described that participate as photobionts in lichen association (Ahmadjian 1993). The most frequent algal photobionts include species of the *Trebouxia*, *Trentepohlia*, *Coccomyxa*, and *Dictyochloropsis* clades. In cyanobacterial lichens, *Nostoc*, *Scytonema*, *Stigonema*, *Gloeocapsa*, and *Calothrix* are the most common genera. Recently, thanks to advanced molecular biology-based taxonomic approaches, a majority of species previously assigned to the genus *Trebouxia* have been reclassified as *Asterochloris* (Peksa and Škaloud 2010). In this chapter, however, the term *Trebouxia* is used even for those newly reclassified algal

Fig. 20.2. Rhizocarpon geographicus is a typical lichen with crustose morphology showing *black* and *yellow* thalli parts (*lower panel*). Crustose lichens are tightly attached to stone or rock surfaces (*upper panel*) (Photos M. Barták).

species. For majority of lichens, symbiotic algae and/or cyanobacteria are located beneath the upper cortex layer that is formed by densely packed fungal hyphae (see Figs. 20.4 and 20.5).

Since more than 50 % of all known lichens possess the unicellular alga *Trebouxia* in their thalli, anatomical and physiological properties are briefly overviewed here. More than 40 species of *Trebouxia* isolated from lichen thalli are currently described. *Trebouxia* species are defined using shape of chloroplast (see Fig. 20.6), as well as pyrenoid (see Chap. 7) and pyrenoglobuli configuration as critical features (Peksa and Škaloud 2011). The pyrenoid is a large central part of the chloroplast, rich in proteins and lipids. It contains the photosynthetic enzyme RuBisCo and is the place of starch synthesis. Within a pyrenoid, there are several pyrenoglobuli that are lipid-rich storage inclusions (Ahmadjian 2001). The pyrenoids contained in *Trebouxia* cells also possess a carbon-concentrating mechanism that can enhance photosynthetic performance (Smith and Griffiths 1998).

Carbon concentrating mechanism (CCM) is a term that refers to the ability of pyrenoid-containing algal chloroplasts, as well as carboxysome-containing cyanobacteria to

Fig. 20.3. Foliose thalli of *Umbilicaria antarctica*, a macrolichen, are abundant in maritime Antarctica. In the dry state, the *upper side* of the thallus exhibits a pale *brown color* (*left*), while the *lower side* of the thallus is *black* (*right*). There is only one point that connects the thallus to a surface (typically rock); this umbilicus is located in the *center* of the thallus (*right*) (Photo M. Barták).

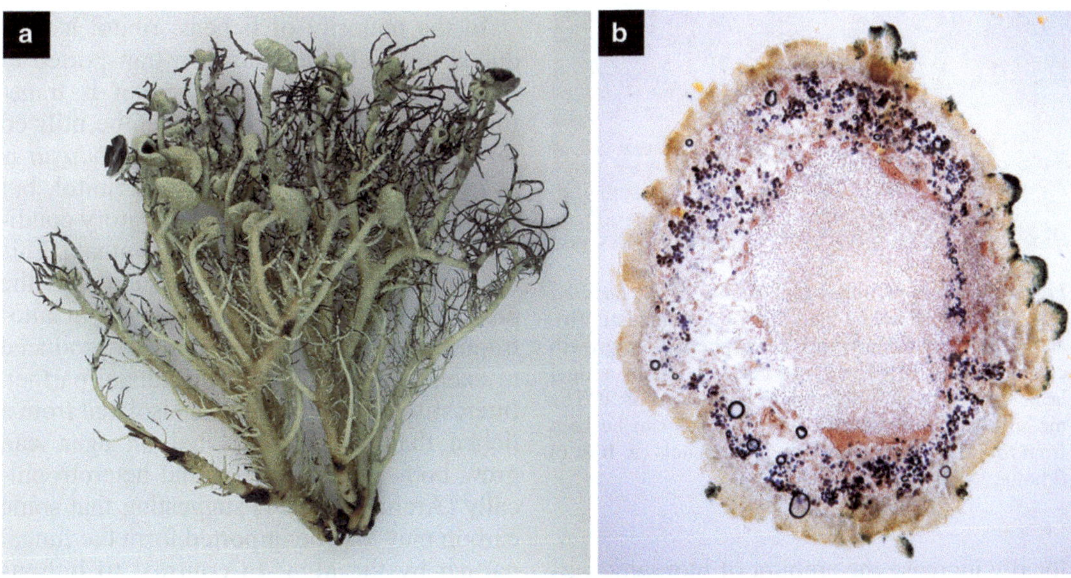

Fig. 20.4. Cluster of fruticose thalli (*left*) and cross section of the middle part of a single thallus of *Usnea aurantiaco-atra* (*right*) showing typical anatomical structures. From the thallus periphery to the thallus *center*, the outer cortex (*brown*), medulla with symbiotic algae (individual *dark points*) forming *round-shaped* layer, and a cell-free internal space (*light red*) can be distinguished (Photos M. Barták (*left*), J. Rotkovská (*right*)).

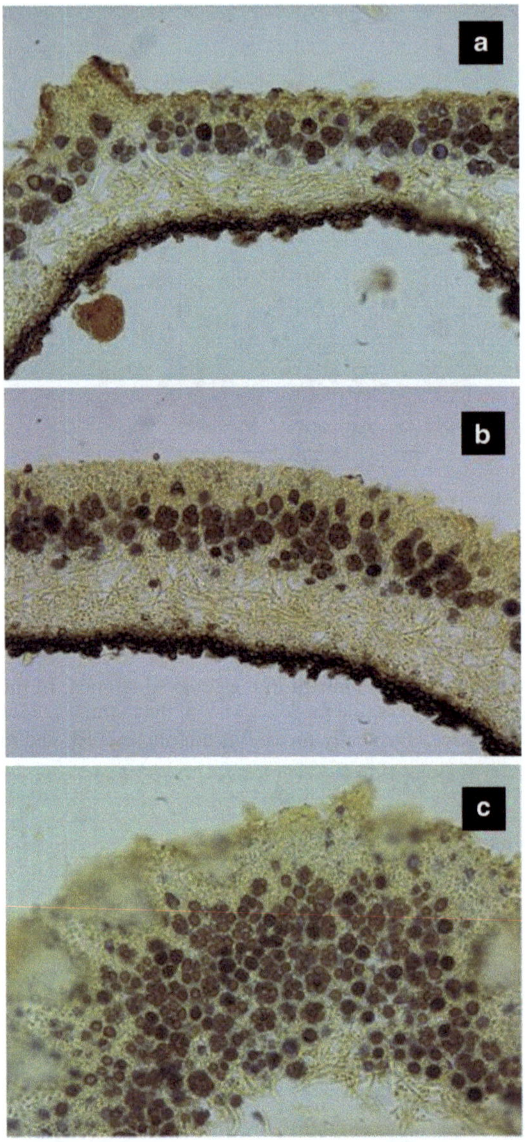

Fig. 20.5. Cross sections of the foliose lichen *Lasallia pustulata* that show the marginal (**a**), intemediate (**b**), and central thallus part, with hemispherical outgrowth called pustulus (**c**). The *upper* cortex contains a algal layer with different thickness in panels **a**, **b** & **c**. The medulla, and *lower* cortex with rhizines can be seen from *top* to the *bottom* at individual panels (**a**, **b** & **c**) (Photos J. Dubová).

locally increase the amount of inorganic carbon that can so be used to efficiently perform carbon fixation. Lichens use CCM during favourable conditions, when the thallus is hydrated and light is sufficiently available. Studies made on isolated photobionts (e.g. Palmqvist and Badger 1996) indicate a key role of carbonic anhydrase (CA) in such lichens. It has been shown for *Trebouxia* sp. and *Coccomyxa* sp. that CA is located both intra- and extracellularly. Species-specific differences in total CA content have been determined by Palmqvist and coworkers (1997).

Polyols (*i.e.* sugar alcohols) are photosynthetic products of symbiotic algae. In lichens, polyols (mainly ribitol, arabitol, mannitol) have several physiological roles (see below). Natural levels of polyols in lichens vary within 1.4–8.8 mg/g dry weight (ribitol) and 0.4–29.0 mg/g (mannitol) between species and physiological state – for review see Hájek et al. (2009). The polyols can make up between 2 and 10 % of the thallus dry weight (Palmqvist 2000). The stored polyols are utilized for supplying the mycobiont with carbon, as energy-storage compound within algae, and as a cryoprotectant. In agreement with these functions is that the highest content of polyols is typically found in late summer (Sundberg et al. 1999).

In the majority of lichens, ribitol is produced in algal cells and then transported to fungal part of a lichen, where it is transformed to mannitol and can then be utilized for metabolic processes. When *Trebouxia* is in lichen association it produces ribitol, but when cultured axenically at laboratory conditions, sucrose is the main photosynthate. This observation may indicate that sucrose is the preferred carbon storage component in autotrophic conditions, while ribitol is produced to exchange carbon with the fungal partner. Interestingly *Trebouxia*, when isolated from a lichen thallus and cultivated on agar, can grow both autotrophically and heterotrophically (Archibald 1977) suggesting that some carbon may also be imported form the fungal partner by the alga. In contrast to lichens in association with alga, cyanolichens utilize glucose as the main transport compound to the mycobiont.

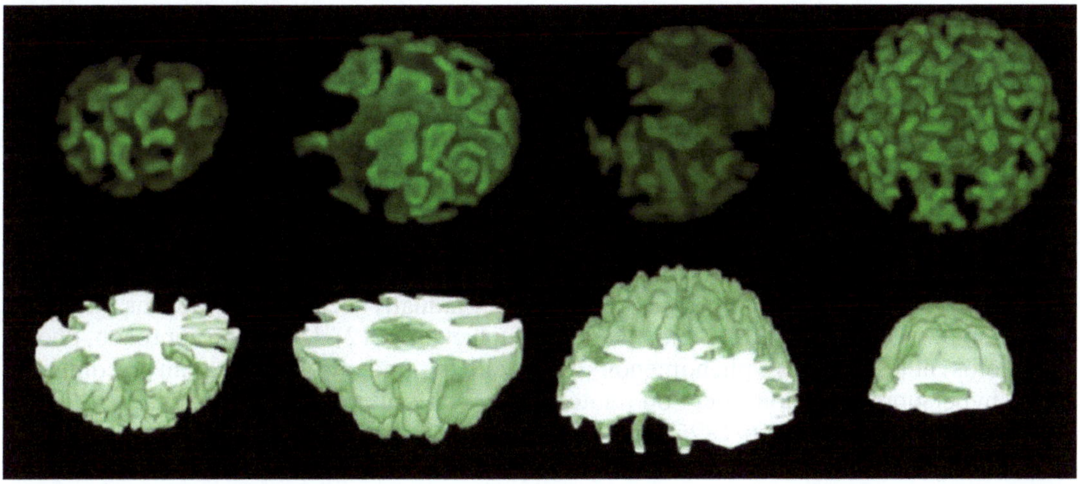

Fig. 20.6. Chloroplast shapes of *Trebouxia/Asterichloris* species. *Upper row* from *left* to *right*: *A. glomerata*, *A. glomerata*, *A. irregularis*, *Aastreochloris sp. Lower row* from *left* to *right*: cross sections of *Asterochloris* sp. (Provided by P. Škaloud, Charles University, Prague).

III. Dependence of Photosynthesis on Physical Factors

A. Hydration-Dependent Photosynthesis

Due to their poikilohydric nature, lichens can respond to the changes in the thallus water content rapidly. While dehydrated lichens are virtually metabolically inactive, increases of the thallus water content leads to gradual re-activation of physiological processes (including photosynthesis and respiration) – Palmqvist and Sundberg 2000. The majority of lichens exhibit dry thalli that are more contracted and possess a different spatial organization compared to their wet state. Extensive packing of thalli layers decreases the volume occupied by the lichen in the dry state. On a cell level, symbiotic algae and cyanobacteria as well as the fungal hyphae are a lot reduced in volume. The cells of the algal photobionts are far from the round shape observed in wet lichens, but are typically collapsed into star-shape structures and densly packed in-between mycobiont cells (de los Rios et al. 1999).

Dessication also has consequences for the ultrastructure in severely desiccated cells of photobionts. For example the pyrenoglobuli may be re-distributed in a pyrenoid and pyrenoglobuli-free pyrenoid matrix areas may be observed in pyrenoid cross section (Ascaso et al. 1988). Moreover, pyrenoglobuli density may be altered according to severity and rapidity of drying. Due to cytorhisis, *i.e.* loss of intracellular water resulting in cell collapse (protoplast, however remains attached to cell wall), the cell wall of photobionts is much less resistant to desiccation-induced changes than the thick cell wall of mycobionts (Scheidegger et al. 1995).

During rehydration, both photo- and mycobiont cells undergo volumetric changes accompanied with gradual re-activation of physiological processes including photosynthesis (see e.g. Sancho and Kappen 1989). When rewetted from the dry state, lichen thalli show an almost linear increase in photosynthesis and respiration with time of hydration (Green et al. 2011). Duration of rehydration and the rate of restoration of photosynthetic processes are, however, species-specific and also dependent on the ecophysiology of the lichen in the preceding period (*i.e.* number of hydration/dehydration cycles, prevailing temperature, the length of period spent dehydrated etc.). At optimum hydration, *i.e.* when the full cellular turgor of photosyntesizing symbiont is reached, the lichens exhibit maximum photosynthetic rate. The optimum water content in a lichen thallus varies amongst species. When the content is above the optimum, surface or extracellular water represents a physical barrier for CO_2 transfer from the air to a photobiont. Over hydration therefore leads to the limitation of CO_2 supply and inhibits the rate of photosynthesis. This phenomenon is called hypersuprasaturation effect on photosynthesis (Lange and Green 1996) and was observed in several lichen species such as e.g. *Ramalina maciformis* Lange (1980), *Xanthoria calcicola* and *Lecanora muralis* Lange (2002), and *Teloschistes capensis* (Lange et al. 2006).

It is well established that, lichens can perform close-to-maximum or even maximum photosynthesis at hydration levels below full thallus hydration. Several studies that applied gasometric measurements of lichen photosynthesis reported such high photosynthesis rates in partially dehydrated thalli (see below). However, such photosynthetic studies expressed water status of thalli as relative water content (RWC, %, Eq. 20.1).

$$RWC = \frac{(FWa - DW)*100}{(FWw - DW)} \qquad (20.1)$$

where FWa is fresh weight of actually hydrated lichen thallus, FWw is fresh weight of fully hydrated lichen thallus, and DW is dry weight of fully dehydrated thallus. In majority of lichens, maximum photosynthetic rates are reached within the RWC range of 60–90 %. However, species-specific differences in ability to hold water in their thalli (per unit of dry weight) exist. Therefore gasometric measurements of photosynthesis are difficult to compare between lichen species, especially when these measurements were obtained in the field.

In contrast to gasometric measurements, the hydration-response curves of photosynthesis measured by chlorophyll fluorescence can be related directly to water potential (WP). In this approach, WP of lichen thallus is measured directly in a chamber of a chilled-mirror dewpoint meter. Immediately after the WP is evaluated, the lichen sample is exposed to light and effective quantum yield of photosystem II (ϕ_{PSII}) is measured using a saturation pulse method. In this way, simultaneous measurements of water status and photosynthetic activity can be performed on gradually desiccating lichens (see e.g. Jupa et al. 2012). Using the approach, photosynthesis under water stress can be compared between poikilohydrous lichens (Fig. 20.7) and homoihydrous higher plants.

While higher plants can hardly perform positive photosynthetis below −2.5 MPa, the majority of lichens do not show any significant decrease within the range of WP −10.0 to 0 MPa (M. Barták, unpublished data). With further dehydration of the lichen thalli, photosynthesis, assessed by effective quantum yield of PSII (ϕ_{PSII}), declines in a curvilinear manner towards critical WP at which $\phi_{PSII}=0$. For the majority of lichens, critical WP is about −30.0 MPa (e.g. *Xanthoria elegans* – Barták et al. 2005), however, some species of the genus *Umbilicaria* have critical WP of as low as −40.0 MPa (Fig. 20.7). The gradual inhibition of photosynthetic processes with ongoing thallus dehydration leads to the activation of protective mechanisms that involves mainly interconversion of xanthophyll-cycle pigments (Calatayud et al. 1997). Such changes are manifested in an dehydration-dependent increase of non-photochemical quenching (NPQ). The rate of NPQ increase is higher when the WP decreases from 0 to −15 MPa than at more pronounced thalli dehydration within the WP range of −30 to −15 MPa (*Lasallia pustulata*, Moudrá 2009).

Dehydration-induced inhibition of photosynthetic processes may be altered by the rapidity of desiccation and light conditions during desiccation. Generally, fast dehydration leads to more pronounced stress to PSII. Therefore, after fast dehydration, a slower and incomplete recovery is observed. This behavior is also observed in the isolated photobiont *Trebouxia erici* (Gasulla et al. 2009).

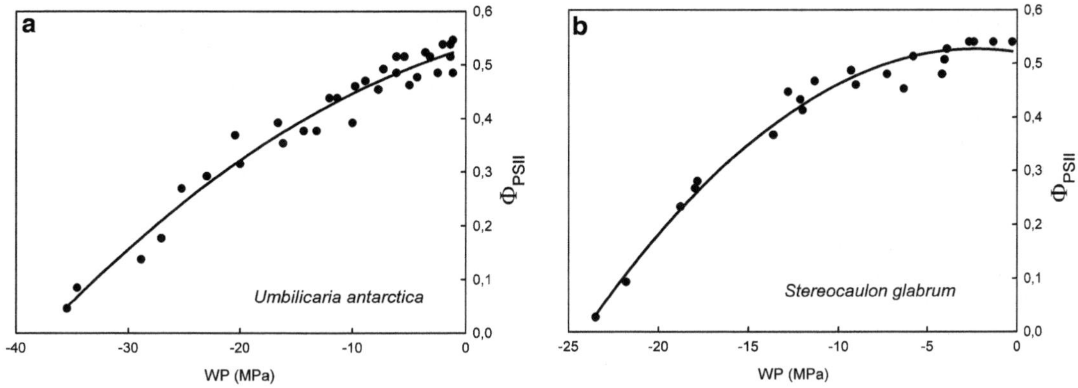

Fig. 20.7. Effective quantum yield of PSII (ϕ_{PSII}) in response to dehydration. With gradual dehydration of the lichen thallus, no substantial decrease of ϕ_{PSII} is between a water potential (WP) of 0 (full hydration) to −12 MPa. Further thalli dehydration (WP below −12 MPa) leads to a rapid decrease of ϕ_{PSII}. The critical WP, at which full inhibition of ϕ_{PSII} occurs is typically about −25 MPa (see *Stereocaulon glabrum – right*). However, several species may exhibit higher tolerance of ϕ_{PSII} against thallus dehydration (see *Umbilicaria antarctica – left*).

Stuctural and functional properties of the lichen thallus may affect dehydration-induced inhibition of photosynthesis as well. Kosugi et al. (2009) showed that an association of the photobionts with the mycobionts in lichen thalli increases tolerance to photoinhibition under drying conditions, compared to isolated photobiont.

A detailed analysis of non-photochemical quenching components (qE, qT, qI) reveals that they differ between lichen thalli passing fast and slow dehydration. Apart from well-known protective mechanisms activated during dehydration of lichen thalli, *i.e.* light-dependent xanthophyll cycle pigments conversion, and state 1–2 transition, there is another effective quencher in the photosynthetic apparatus of PSII of desiccating lichens. It is not related to zeaxanthin formation and independent on light. According to Heber et al. (2006), it may involve conformational changes in pigment-protein complexes in PSII. In this way, highly-effective dissipation centers are formed in which excitation energy is trapped (Heber et al. 2010). Recently, the presence of a strong quencher located in the PSII core and/or PSII antennas has been reported from *Parmelia sulcata* desiccating on light (Veerman et al. 2007), however its chemical structure is not known.

B. Light-Dependent Photosynthesis

1. Light Response Curves

Optimally-hydrated, lichens respond to light in a similar manner to higher plants and light response curves of photosynthesis can be determined in both systems and differences evaluated. At low light intensities of photosynthetically active radiation (PAR), lichens exhibit linear increase in net photosynthetic rate (Pn) with increasing light followed by a curvilinear Pn-PAR relationship until constant photosynthesis (Pn_{max}) in saturating light is reached.

The great majority of lichens is adapted to low light and thus reach typical light-saturated photosynthesis in the PAR range of 100–400 μmol m^{-2} s^{-1} (Demmig-Adams et al. 1990a). However, species from sun-exposed habitats exist that exhibit light-saturated photosynthesis at 1,100 μmol photons m^{-2} s^{-1} (Green et al. 1997). In general, lichens have much lower maximum (light-saturated) photosynthetic rates expressed per area or dry weight unit of thallus than higher plant leaves. This is because, photobionts (alga, cyanobacterium or both) constitute only a small proportion to the lichen thalli, while the bulk is made up of the fungal partner. Typically, maximum rate of net photosynthesis (Pn_{max}) does not exceed 6 μmol (CO_2) m^{-2} s^{-1} as shown e.g. for cyanobacterial lichen *Collema tenax* (Lange et al. 1998), algal species *Xanthoria elegans* (Barták et al. 2005), and *Lecanora muralis* (Lange and Green 2008). For *Lecanora muralis* a Pn_{max} as high as 7.5 was reported for optimally hydrated thalli when measured under controlled laboratory conditions. Many lichens, however, have much lower Pn_{max} ranging 1.0–2.0 μmol (CO_2) m^{-2} s^{-1} as shown for *Lobaria pulmonaria* and *Platismatia glauca* (Sundberg et al. 1997).

2. Photoinhibition

Photoinhibition of photosynthesis frequently occurs in hydrated lichens exposed to high light intensities. Such high-exposure to lichen can occur in wet open habitats such as rocks and light-exposed clear cuts or forest edges. It is well established that the sensitivity to photoinhibition is higher in the lichens possessing cyanobacterial photobionts than those having algal partners (Demmig-Adams et al. 1990b). When photoinhibited, cyanobacterial lichens do not show fast return of photosynthetic parameters to pre-photoinhibition parameters during dark recovery. This in contrast to lichens with algal photobiont, which exhibit recovery to initial values typically within tens of minutes.

When the photosynthetic apparatus of the lichen photobionts is exposed to high light doses, reactive oxygen species (ROS) are formed that are destructive to the photosynthetic machinery (see Chap. 11), especially photosystem II. ROS may cause damage or destruction of pigment-protein complexes and inhibition of photochemical processes. To avoid or reduce the negative consequences of ROS formation, lichen photobionts activate several photoprotective mechanisms that are similar to higher plants.

a. Zeaxanthin Formation

Several laboratory-based studies with high light-treated lichens demonstrated zeaxanthin formation in algal lichens (e.g. Vráblíková et al. 2005). Zeaxantin is involved in effective thermal energy dissipation from overenergized photosystem II of algal photobionts (Demmig-Adams et al. 1990a), *i.e.* high light-induced conversion of violaxanthin to zeaxanthin (DEPS value). The recovery to pre-photoinhition violaxanthin to zeaxanthin ratios is relatively fast ranging from 3 to 10 h (Barták et al. 2008). While the zeaxanthin formation is well established in algal photobionts, cyanobacterial photobionts do not possess such mechanism.

b. Involvement of Glutathione

Glutathione is low-weight thiol that has the function of a antioxidant and plays a role in the photoprotection of lichens. Phytochelatines, another antioxidant is synthesized from glutathione. The amount of intrathaline glutathione ranges typically in lichens between 1.2 and 3.3 μmol/g dry weight. Glutathione levels respond not only to accual light conditions and functioning of photosynthetic apparatus of algal and fungal symbionts during light-induced oxidative stress, but also possess seasonal variation as well as spatial variation within the thallus (Kranner and Grill 1996).

The amount of glutathion Increases during lichen dehydration (Kranner and Birtić 2005; Tretiach et al. 2012), while both increases and decreases can be induced by heavy metals and toxic compound (Mrak et al. 2010).

If exposed to strong, brief high light treatment, reduced glutathione (GSH) is converted to oxidized glutathione (GSSG) and the amount of total glutathione (GSHt) decreases in lichens as algal cells are increasingly photoinhibited. Severe photoinhibtion leads to degradation of GSH (down to 12–30 % of GSHt before light treatment Štěpigová et al. 2007) to glutamyl-cystein. Glutamyl-cystein in turn can be used to synthesis reduced glutathione during recovery from high light treatment. An increase of glutathione content by long-term high light treatment is reported for higher plants (Burritt and MacKenzie 2003), but has never been demonstrated for lichens.

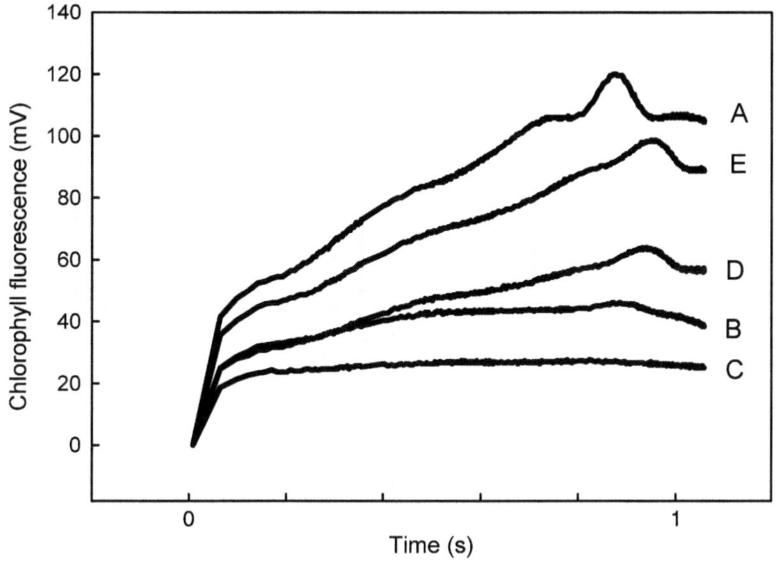

Fig. 20.8. Photoinhibition and recovery in *Usnea antarctica* assessed by fast chlorophyll fluorescence induction curves (OJIP). The fluorescence kinetics are caused by a *short light* pulse applied to a *dark-adapted* lichen thallus. Photoinhibition is characterized by a decrease in chlorophyll fluorescence yield. Partial inhibition (*B*) and full inhibition (*C*) can be compared to the uninhibited control (*A*). After photoinhibition treatment, a 30 or 60 min dark recovery period leads to partial (*D*) and almost complete (*E*) restoration of PSII fluorescence characteristics. This fast recovery indicates high resistance of PSII in *U. antarctica* to short-term high light treatment.

The extent of photoinhibition and recovery can be sensitively monitored in a lichen thallus by several chlorophyll fluorescence techniques and the derived chlorophyll fluorescence parameters. In chloroplasts, photoinhibition can lead to a gradual inactivation and photodestruction of chlorophyll molecules. This leads to a decrease in chlorophyll fluorescence (variable fluorescence – F_v) and efficiency of absorbed energy transfer through PSII and PSI. Figures 20.8 and 20.9 show photoinhibition-induced changes in the shape of fast and slow chlorophyll fluorescence induction kinetics in wet lichen thalli exposed to high doses of photosynthetically active radiation.

c. Antioxidative Enzymes

In symbiotic lichen algae, antioxidative enzymes are involved in cell protection when ROS are formed (del Hoyo et al. 2011). Several studies reported altered levels of catalase (CAT), superoxide dismutase (SOD) and ascorbate peroxidase (APX) during dehydration stress (e.g. Mayaba and Beckett 2001; Gasulla et al. 2009), heavy metal-induced stress (Weissman et al. 2006), and exposure to SO_2 (Kong et al. 1999). It appears very likely that CAT, SOD and APX would be also involved in protecting lichens against ROS formed during high-light stress in hydrated lichens, but experimental efficidence is lacking.

3. *Effects of UV-B Radiation*

Lichens are generally resistant against high doses of UV radiation, especially when dehydrated and thus physiologically inactive. Some algal lichens, however, show limited resistance (Unal and Uyanikgil 2011). Experiments conducted so far led to the conclusion that UV enhancement typically lead to an increase of UV-absorbing pigments and compounds such as phenolics

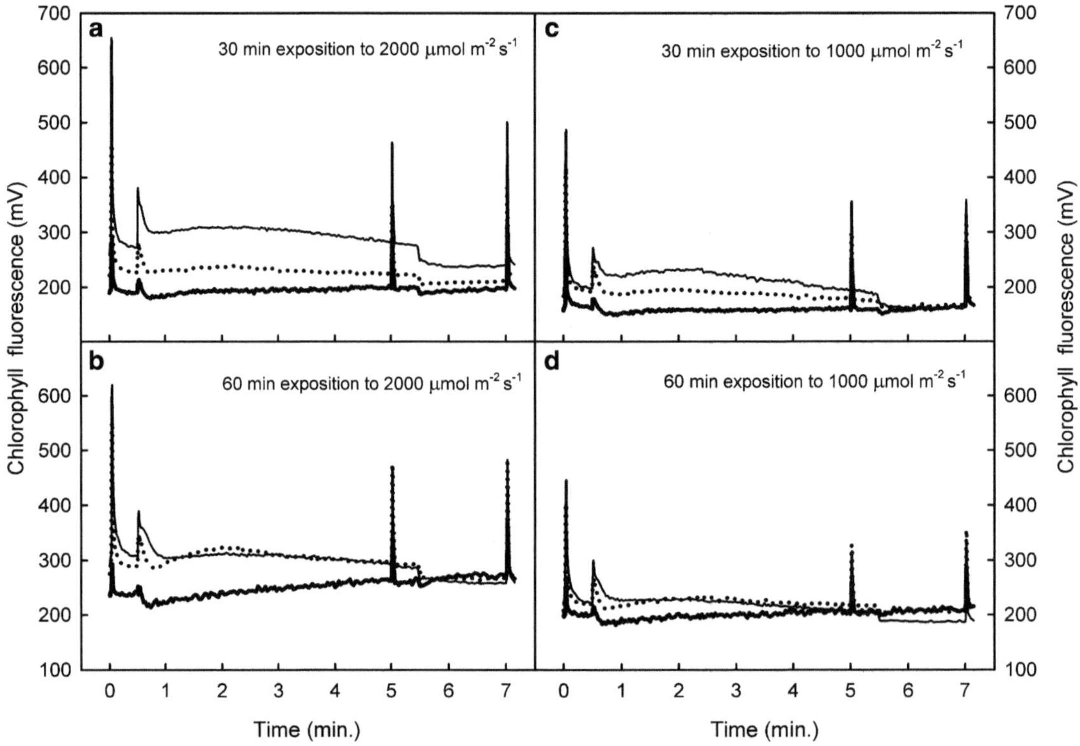

Fig. 20.9. Slow chlorophyll fluorescence kinetics and maximum fluorescence induced by saturation light pulses applied to liches in *dark-adapted* and *light-adapted* states. Untreated *Usnea antarctica* (*solid line*) is exposed to a 30/60 min high *light treatment* (*solid bold line*) and consequent dark recovery (*dashed line*). The recovery in the *dark* is faster when thalli are exposed for longer periods (60 min) indicating the involvement of photoprotective and regulatory process in chloroplasts (Reprinted from Barták et al. 2012 with the permission of the Masaryk University Press).

(Buffoni Hall et al. 2002), usnic acid (Bjerke et al. 2002), rhizocarpic acid (Rubio et al. 2002), parietin (Solhaug and Gauslaa 2004). However, these protective substances cannot fully prevent DNA damage and cell death (Unal and Uyanikgil 2011). UV screening compounds are typically located in the upper cortex of lichen thallus and thus reduce the amount of harmful radiation reaching the algal layer.

Photosynthesis in most lichens appears not to be negatively effected by UV radiation. For example, Lud et al. (2001) reported no change of fluorometric parameters in *Turgidosculum complicatulum* exposed to extra UV-B radiation in Antarctica. Lichens from shade habitats, however, may exhibit some limited sensitivity of supplemental UV and resulting species-specific effects on photosynthetic rate and biomass production (Larsson et al. 2009). For a majority of lichens, exposure to increased quantities of UV-B radiation leads to increased synthesis of photoprotective, UV-absorbing compounds, such as phenolics (Buffoni Hall et al. 2002) parietin, melanic compounds (Solhaug and Gauslaa 2004; Solhaug et al. 2003) and usnic acid (McEvoy et al. 2006). In lichen thalli – in the upper cortex layer, in particular – these compounds protect photosynthetic pigments located in a photobiont cells in underlying layer.

C. Temperature-Dependence of Photosynthesis

Similarly to higher plants, the rate of photosynthesis and production in lichens is also temperature-dependent including aquatic lichens (Davis et al. 2002). Therefore, temperature response curves can be measured and interspecific differences in temperature tolerance can be evaluated. Due to poikilohydric character of the lichen thallus, however, an increase in thallus temperature is accompanied with gradual loss of water. In natural conditions, temperature-induced change in photosynthetic rate can therefore not be distinguished from dehydration-dependent changes. Several studies evaluated photosynthetic temperature-response curves in lichens that were maintained optimally hydrated in laboratory studies. For many lichens, the temperature optimum for photosynthesis varies between 10 and 22 °C. Individual temperature optima depend on lichen adaptation/acclimation to particular growing habitats (tropical, mountainous, polar etc.), as well as other physical environmental factors including light intensity. If a lichen is exposed to a lower than optimum light level, a decrease in optimum temperature for photosynthesis is observed (Reiter et al. 2008). Many lichens are capable of maintaining high photosynthetic rates (of about 80 % of maximum photosynthesis) within a temperature range of 5–20 °C (Hájek et al. 2001). The critical maximum temperature for lichen photosynthesis is about 40 °C for both tropical and high-mountain lichen species (Lange et al. 2004; Hájek et al. 2001). However this parameter (maximum temperature) has mostly theoretical value, as such high temperature will usually also cause rapid dessication of the lichen in most environments. At 40 °C, wet and fully photosynthetically active thalli can hardly be found in the field.

The only exception are tropical lichens. At temperature above optimum, absolute rate of lichen respiration increases as well as proportion of respiration to photosynthesis. This is attributed mainly to the respiration of mycobiont which forms substantial part of overall thallus CO_2 release. Short-term high temperature stress in hydrated lichens leads to the loss of effectivity of photosynthetic processes in photosystem II and, if the capacity of protective mechanisms is low, to the injury of PSII structure in photosynthetizing symbiont (e.g. destruction of the Mn-cluster – Oukarroum et al. 2012), as well as partial disintegration of thylakoid membrane structure.

Contrastingly to higher plants, majority of lichens can perform photosynthesis at temperature far below zero even under snow cover (Kappen 1993, Pannewitz et al. 2006). This capability is facilitated by the presence of molecules in lichen thallus, such as e.g. polyols that prevent freezing. The critical temperature for lichen photosynthesis can drop to values about −20 °C as has been demonstrated by gasometric (Kappen et al. 1996) and fluorometric studies (Barták et al. 2007) in lichens from polar and alpine regions. In many field and laboratory studies, photosynthetic parameters, such as e.g. F_v/F_m and ϕ_{PSII} are used as indicators of cryoresistance of lichens and their ability to survive long-term periods at freezing temperature (in dry state) without significant damage to photosynthetic apparatus. However, freezing experiments with lichen species from cold habitats showed that wet thalli may perform substantial photosynthesis at below zero temperature (e.g. *Usnea antarctica* – see Fig. 20.10). During rapid cooling, both F_v/F_m and ϕ_{PSII} of lichen thallus decline, in a S-curve type manner. While F_v/F_m is more or less constant within the range of 20 to −10 °C, ϕ_{PSII} tends to decline more rapidly indicating gradual temperature-dependent inhibition of Photosystem II.

The algal photobionts are remarkable resilient to freezing. Even when shock-frozen in liquid nitrogen, lichen symbiotic algae of the genus *Trebouxia* survive to a certain extent (typically 30 % of algal population, Hájek et al. 2012) and are able to restore their photosynthetic performance during subsequent cultivation.

IV. Important Chemical Factors Affecting Photosynthesis

A. Availability of Nutrients

The typical lichen habitat only provides poor access to nutrients. This limitation also effects photosynthetic performance. Consequently,

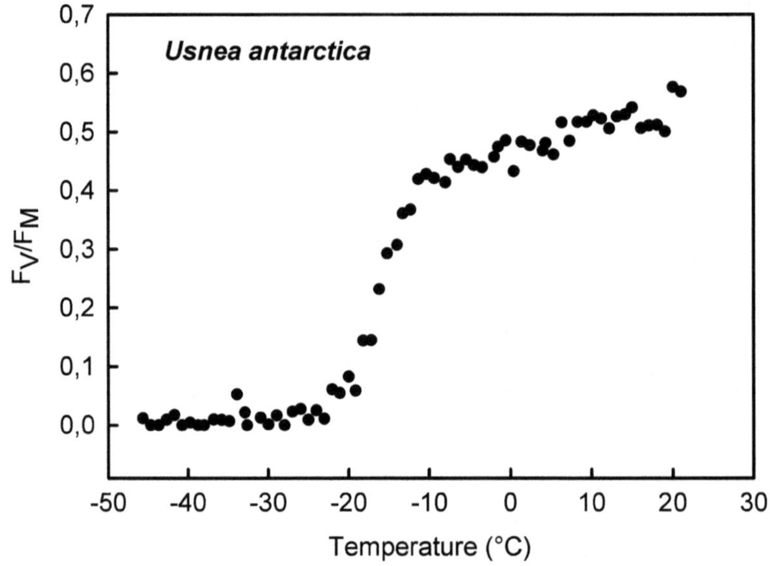

Fig. 20.10. Temperature-dependent decrease of potential quantum yield of PSII (F_v/F_m – potential yield of PSII photosynthesis) recorded during linear cooling of *Usnea antarctica* thallus from +20 to −30 °C (cooling rate 0.5 °C/ min). Photochemical processes in PSII show slight decrease in the temperature range (−12 to +20 °C), dramatic decrease within the range of −12 to −20 °C. The critical point for photosynthesis (assessed by chlorophyll fluorescence) is about −22 °C.

lichens can increase photosynthetic rate when nutrients – nitrogen in particular – are available (see e.g. Davis et al. 2000). A study by Palmqvist et al. (1998) demonstrated that lichens with higher nitrogen content exhibit higher gross photosynthetic rate, but lower photosynthetic nitrogen use efficiency. It is well established that when nitrogen is available to a lichen, larger amount of nitrogen is invested in both, the algal and fungal partner. The nitrogen partiniting between the algal and fungal partner is likely species specific (Palmqvist et al. 2002). Nitrogen fertilization experiments show that additional nitrogen is used to synthesize proteins, chlorophyll and ribitol, the photosynthetic product of algal photobiont that can be supplied to the fungal partner (Dahlman et al. 2003).

B. Heavy Metals

With few exceptions of heavy metal-tolerant species (e.g. *Xanthoria parietina*), lichens are considered good indicators of toxic environmental pollution (Beckett and Brown 1983). It is well established that lichen abundance and distribution decreases with heavy presence of heavy metals. This susceptibility is a result of lichens' disposition to accumulate heavy metals. Heavy metal accumulation is thought to lead to ROS formation, and consequently to lipid peroxidation and damages to proteins and nucleic acids.

On the level of the photobiont, a decrease in the integrity of chlorophyll molecules that can in extreme cases also lead to a severe decrease in of chlorophyll content, that limits and inhibits photosynthetic processes (Bačkor and Fahsett 2008; Bačkor and Loppi 2009). Interestingly evidence from laboratory studies suggest that heavy metal-sensitive and – tolerant *Trebouxia* strains exist.

While there is amply evidence from laboratory experiments, field studies that investigate the effect of heavy metals on photosynthesis are less frequent. Garty et al. (2002) reported a decrease of capacity of photosynthetic processes in PSII (F_v/F_m) in epilithic fruticose lichen *Ramalina maciformis* exposed to a mixture of heavy metal ions from industrial pollution. The decrease in photosynthetic performance is attributed to the degradation of chlorophylls, leading to a decrease in variable chlorophyll fluorescence. Therefore, heavy metal exposure in lichens may be assessed as an decrease in minimum (F_o) fluorescence, variable chlorophyll fluorescence (F_v) and chlorophyll fluorescence parameters derived from those signals (Kupper et al. 1998).

The extent of heavy metal-induced inhibition of photosynthetic processes differs among lichen species. Several species, especially those grown near copper mines, exhibit increased tolerance to heavy metals. Such organisms have increased content of phytochelatins, glutathione (Bačkor et al. 2009) compared to similar organisms living away from pollution sources.

V. Lichen Photosynthesis in the Field

A. Overview of Typical Environments

Over last several decades, numerous short- and long-term studies have been conducted that involved measurements of lichen photosynthesis in the field. In addition to the pioneering studies of O. Lange and his co-workers, who started photosynthesis measurements of lichens in semi-desert ecosystems (Lange et al. 1970), a wide range ecosystems and biotopes has been investigated (see below). Initially these studies have utilized infra-red gas (CO_2) analysers, that were later complemented by a variety of chlorophyll fluorescence measurements and environmental sensors.

Detailed field studies can assess the diurnal courses of lichen photosynthesis simultaneously with microclimatological and environmental parameters. Such approaches enables the identification of the key factors that limit photosynthesis in particular location (see e.g. Lange 2002), as well as primary productivity of lichens (e.g. Green and Lange 1991; Uchida et al. 2006).

In tropical rain forests, net photosynthetic rate of algal and cyanobacterial lichens is generally high, but also show season-dependent variation (Lange et al. 2004). During moist period (*i.e.* overcast days), light is limiting, while during the periods with high photosynthetically active radiation, thalli dehydration is limiting lichen photosynthesis.

Water limitation is the key factor that determins lichen photosynthesis in deserts and semideserts, where lichens are components of soil crusts. It has been shown for a variety of lichen species from the Negev desert (Lange et al. 1977; Palmer and Friedmann 1990) and Namib desert (Lange et al. 1994, 2006) that aerial moisture is a crucial factor for photosynthesis and the survival of crust-forming lichens. Nocturnal fog and/or early morning dew hydrates lichen so that they are able to perform positive net photosynthesis for a short period of time after sunrise.

Several long-term studies on lichen photosynthesis in mountainous ecosystems have been conducted (Reiter and Turk 2000; Reiter et al. 2008). These studies addressed interaction of water availability and temperature and their effect on photosynthetic performance of *Xanthoria elegans*, *Umbilicaria cylindrica*, and *Brodoa atrofusca*, respectively. Continuous field measurements of net photosynthesis and respiration revealed that in the particular alpine ecosystem, the species were photosynthetically active only 16–25 % of the observation time. The most limiting factor for photosynthesis was dehydration of thalli. Interspecific comparison showed that *X. elegans* is the most photosynthetically active among the three species.

Similar long-term studies were carried out in polar regions, particularly in Antarctica. Using automation the effective quantum yield (ϕ_{PSII}) of lichen photosynthesis was continously evaluated. Measurements can be recorded over weeks (see Fig. 20.11) and even several seasons (Schroeter et al. 2011). These measurements clearly show the limitation of lichen photosynthesis by thallus dehydration, freezing of thalli and unavailability of light during the polar night.

While continuous long-term field measurements of lichen photosynthesis provide

Fig. 20.11. Field installation of a fluorometer for long-term measurements of effective quantum yield of photosynthetic processes in *Xanthoria elegans* (James Ross Island, Antarctica) (Photo M. Barták).

unparalleled insights in the ecophysiology of lichens, field studies that acquire data in week or month intervals can also be very informative. Such studies can reveal general trends in acclimation of lichen photosyntheis and provide additional data that require direct interactions with the sample (e.g. quantification of photosynthetic pigments) (Piccotto and Tretiach 2010).

B. Special Environments

In addition to ecosystems on Earth, lichen photosynthesis has been also investigated in extraterrestrial space. Here lichen photosynthesis may serve as a marker for surviving conditions experienced in space (Gomez et al. 2012). Within the last decade, several preflight tests have performed to test the resistance of lichens to freezing temperature, vacuum and UV radiation using chlorophyll fluorescence characteristics in *Rhizocarpon geographicum* and *Xanthoria elegans*.

Short-term exposure of lichens to extraterrestrial (orbital) conditions was evaluated in the year 2005 within the FOTON_M2 spacecraft mission using the BIOPAN facility. Based on these studies, a long-term (18 months) lichen exposure experiment was performed within the EXPOSE-E experiment onboard the International Space Station (ISS) in 2008/2009. These experiments showed that the experimental lichen species survive open space conditions in a dry state and restore their photosynthetic activity with only minor signs of damage when rewetted under laboratory conditions (de la Torre et al. 2010).

VI. Methods for Assessing Lichen Photosynthesis

A wide array of methods is available to study photosynthesis in lichens and their isolated photobionts. Assessment of photosynthetic CO_2 fixation by measuring gas exchange is frequently employed in the field and laboratory. Due to the small dimensions of lichen thalli, necessity of maintaining a constant hydration state, impossibility of deattaching crustose lichen thalli from stone surfaces, these gasometric measurements typically require specifically-modified measuring chambers (e.g. clap-cuvette – Lange et al. 1986, 2007). In spite of these difficulties, many studies demonstrate the acquisition of gas exchange measurements to characterize lichen photosynthesis.

Measurements of chlorophyll fluorescence parameters are also widely used in lichen photosynthesis studies. Fluorometers are relatively robust and cost-effective instruments. Apart of generally well-established indicators of potential (F_v/F_m) and actual (ϕ_{PSII}) effectivity of PSII processes, there are several parameters used in the evaluation of stress in lichen photosynthesis, such as e.g. non-photochemical quenching (NPQ, qN) and its components related to the involvement of energy-dependent (qE), state one-state two transition (qT) and photoinhibitory (qI) processes in PSII. Such approach enables to determine negative effects of particular stress factors in chloroplast of a lichen photobiont. Almost all chlorophyll fluorescence measuring techniques are used in lichens to address specific aspects of lichen photosynthesis. Fast chlorophyll fluorescence transients (OJIPs) were exploited to characterize absorbed light energy transformation in PSII and photochemical photosynthetic machinery. At OJIP curves measured on symbiotic lichen alga *Trebouxia*, Ilík et al. (2006) reported unusual dip that differs from evidence from higher plant. It was attributed to fast reoxidation of electron carriers in chloroplastic thylakoid membranes caused by activation of ferredoxinNADP + oxidoreductase or Mehlerperoxidase. Oukarroum et al. (2012) used OJIPs to evaluate heat stress effect in PSII of optimally hydrated thallus of *Parmelia tiliacea*. The study focused on high temperature-induced occurrence of K and M peaks at the OJIP transient. K and M values are indicative for heat stress effects in PSII. Recently, visualization techniques are increasingly used in many fluorometric studies (e.g. Barták et al. 2000) enabling to distinguish intrathalline differences in lichen photosynthetic performance (see Fig. 20.12). Tens of chlorophyll fluorescence parameters are available to evaluate the heterogenetity of photosynthesis caused over lichen thalli by patterned distribution of photosynthetizing photobionts (Jensen et Siebke 1997) photoinhibitory (Barták et al. 2006), temperature and osmotic stress (Hájek et al. 2006). Recently, development of automated and even robotic fluorometers that are used in many applications in higher plants is very promising and future application chlorophyll fluorescence imaging in ecophysiological studies of lichen photosynthesis might be expected.

Oxymetric methods in lichen photosynthesis research have only been employed recently and only to a limited extent. In principle, the method is based on a Clark-type oxygen electrode housed in a chamber that interfaces with the lichen. Typically these instruments are used to assess photosynthetic O_2 evolution in response to light, temperature and hydration status of lichens (Charles 2011; Aubert et al. 2007). Due to the small chamber volumes of these instruments a constant hydration state of lichen thallus can be maintained for longer periods than during CO_2 exchange measurements.

Another application of oxymetric methods is the measurement of oxygen exchange in algal/cyanobacterial cultures of photobionts isolated from lichen thalli and cultivated in liquid media. However the isolated photobionts cultivated in cultures differ in their photosynthetic performance from their natural behaviour of their lichenized forms. Studies that contrast the performance of mycobiont and photobiont collective with the isolated photobiont can be used to characterize features that arise from interactions within lichens.

Carbon isotopes are also used in the evaluation of lichen photosynthesis. Stable isotope (^{13}C) discrimination is typically used to distinguish between photosynthesis and respiration and also between the presence or absence of carbon concentration mechanisms (Máguas et al. 1993). Radioactive carbon isotope (^{14}C) is used for identification of photosynthetic products of lichen photobionts and secondary metabolites such as sugars, polyols and lichen acids (Eisenreich et al. 2011). Recently, the approach has been used as an effective tool to assess lichen metabolomics in *Cladonia portentosa* (Freitag et al. 2012).

Acknowledgments

I am very thankful to my colleagues from the Laboratory of Photosynthetic Processes (Masaryk University, Brno) who have collaborated with me in many field- and laboratory-based studies focussed on ecophysiology of lichen photosynthesis within last decade. Their particular help during the preparation of manuscript of this chapter is also acknowledged.

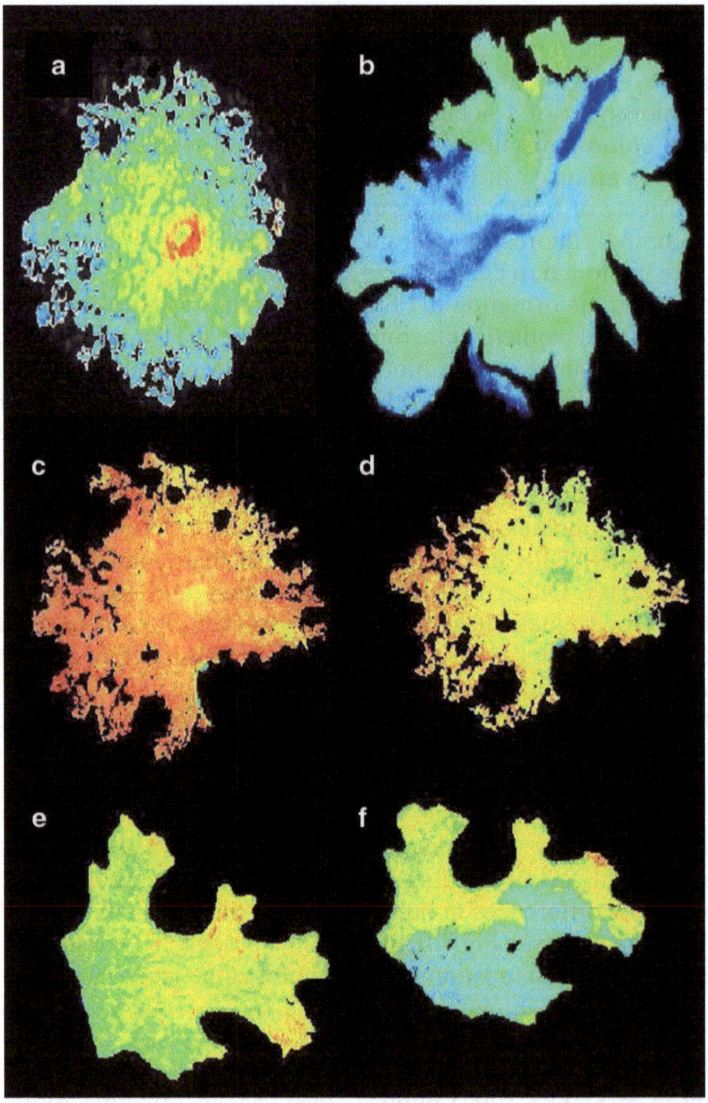

Fig. 20.12. Chlorophyll fluorescence imaging visualizes several physiological and photosynthetic processes in lichens. (**a**) Image of the chlorophyll fluorescence signal (intensity indicated by color) reveals higher amount of chlorophyll (*red*) and less developed *dark pigmentation* (*yellow*) of the central (young and growing) thallus part of *Lasallia pustulata*: Older and more *dark-pigmented* marginal parts show low chlorophyll fluorescence (*green* and *blue*). (**b**) Intrathalline heterogeneity of effective quantum yield of photosynthetic processes (ϕ_{PSII}) in *Umbilicaria antarctica*. The *blue* line across the thallus area indicates inhibited photosynthesis due to mechanical stress caused by repeated dehydration-dependent movements. In response to desiccation, the thallus bends along by the *blue* line. (**c, d**) Distribution of photoinhibition (assessed by F_v/F_m) in a fully-hydrated *Lasallia pustulata* thallus induced by high light treatment. The reduction of F_v/F_m from pre-photoinhibory state ($F_v/F_m = 0.648$) to high light state ($F_v/F_m = 0.530$) is indicated by color shift form orange (see **c**) to *yellow* and *green* (see **d**). (**e, f**) Photoinhibition caused by high light (650 μmol m^{-2} s^{-1} for 2 d) in the dry thallus of foliose lichen *Lobaria pulmonaria*. Maximum photosynthetic activity (ϕ_{PSII}) is observed close to the tips of thallus lobes in the untreated control (*red* and *yellow* in panel **e**). In the desiccated state, a part of thallus bends and so provides a shield against incident radiation. As a consequence, one part of thallus exhibits very limited photoinhibition (indicated in *yellow* in panel **f**) while the rest of thallus (unshielded by overlaped part) is severely photoninhibited (*blue*).

References

Ahmadjian V (1993) The lichen symbiosis. Wiley, Chichester, 250p

Ahmadjian V (2001) *Trebouxia*: reflections on a perplexing and controversial lichen photobiont. In: Seckbach J (ed) Symbiosis. Mechanisms and model systems. Kluwer Academic Publishers, Dordrecht, pp 373–383

Archibald PA (1977) Physiological characteristics of *Trebouxia* (Chlorophyceae, Chlorococcales) and *Pseudotrebouxia* (Chlorophyceae, Chlorosarcinales). Phycologia 16:295–300

Ascaso C, Brown HD, Rapsch S (1988) The effect of desiccation on pyrenoid structure in the oceanic lichen *Parmelia laevigata*. Lichenologist 20:31–39

Aubert S, Juge C, Boisson AM, Gout E, Bligny R (2007) Metabolic processes sustaining the reviviscence of lichen *Xanthoria elegans* (Link) in high mountain environments. Planta 226:1287–1297

Bačkor M, Fahselt D (2008) Lichen photobionts and metal toxicity. Symbiosis 46:1–10

Bačkor M, Loppi S (2009) Interactions of lichens with heavy metals. Biol Plant 53:214–222

Bačkor M, Kováčik J, Dzubaj A, Bačkorová M (2009) Physiological comparison of copper toxicity in the lichens *Peltigera rufescens* (Weis) Humb. and *Cladina arbuscula* subsp. mitis (Sandst.) Ruoss. Plant Growth Regul 58:279–286

Barták M, Hájek J, Gloser J (2000) Heterogeneity of chlorophyll fluorescence over thalli of several foliose macrolichens exposed to adverse environmental factors: interspecific differences as related to thallus hydration and high irradiance. Photosynthetica 38:531–537

Barták M, Gloser J, Hájek J (2005) Visualized photosynthetic characteristics of the lichen *Xanthoria elegans* related to daily courses of light, temperature and hydration: a field study from Galindez Island, maritime Antarctica. Lichenologist 37:433–443

Barták M, Solhaug KA, Vráblíková H, Gauslaa Y (2006) Curling during desiccation protects the foliose lichen *Lobaria pulmonaria* against photoinhibition. Oecologia 149:553–560

Barták M a, Váczi P, Hájek J, Smykla J (2007) Low-temperature limitation of primary photosynthetic processes in Antarctic lichens *Umbilicaria antarctica* and *Xanthoria elegans*. Polar Biol 31:47–51

Barták M, Vráblíková-Cempírková H, Štepigová J, Hájek J, Váczi P, Večeřová K (2008) Duration of irradiation rather than quantity and frequency of high irradiance inhibits photosynthetic processes in the lichen *Lasallia pustulata*. Photosynthetica 46:161–169

Barták M, Hájek J, Očenášová P (2012) Photoinhibition of photosynthesis in Antarctic lichen *Usnea antarctica*. I. Light intensity- and light duration-dependent changes in functioning of photosystem II. Czech Polar Rep 2:42–51

Beckett RP, Brown DH (1983) Natural and experimentally-induced zinc and copper resistance in the lichen genus *Peltigera*. Ann Bot 52:43–50

Bjerke JW, Lerfall K, Elvebakk A (2002) Effects of ultraviolet radiation and PAR on the content of usnic and divaricatic acids in two arctic-alpine lichens. Photochem Photobiol Sci 1:678–685

Büdel B, Scheidegger C (1996) Thallus morphology and anatomy. In: Nash TH (ed) Lichen biology. Cambridge University Press, Cambridge, pp 37–64

Buffoni Hall RS, Bornman JF, Bjorn LO (2002) UV-induced changes in pigment content and light penetration in the fruticose lichen *Cladonia arbuscula* ssp. Mitis. J Photochem Photobiol B 66:13–20

Burritt DJ, Mackenzie S (2003) Antioxidant metabolism during acclimation of *Begonia* × *erythrophylla* to high light levels. Ann Bot 91:783–794

Calatayud A, Deltoro VI, Barreno E, del Valle-Tascon S (1997) Changes in in vivo chlorophyll fluorescence quenching in lichen thalli as a function of water content and suggestion of zeaxanthin-associated photoprotection. Physiol Plant 101:93–102

Charles HC (2011) The physiological response of sub-Arctic lichens to their abiotic environment. Masters thesis, Durham

Dahlman L, Persson J, Näsholm T, Palmqvist K (2003) Carbon and nitrogen distribution in the green algal lichens *Hypogymnia physodes* and *Platismatia glauca* in relation to nutrient supply. Planta 217:41–48

Davis WC, Gries C, Nash TH III (2000) The ecophysiological response of the aquatic lichen *Hydrothyria venosa* to nitrate in terms of weight and photosynthesis over long periods of time. Biblio Lichenol 75:201–208

Davis WC, Gries C, Nash TH III (2002) The influence of temperature on the weight and net photosynthesis of the aquatic lichen *Peltigera hydrothyria* over long periods of time. Biblio Lichenol 83:233–242

de la Torre R, Sancho LG, Horneck G, de los Ríos A, Wierzchos J, Olsson-Francis K, Cockell CS, Rettberg P, Berger T, de Vera J-PP, Ott S, Frías JM, Melendi PG, Lucas MM, Reina M, Pintado A, Demets R (2010) Survival of lichens and bacteria exposed to outer space conditions – results of the Lithopanspermia experiments. Icarus 208:735–748

de los Ríos A, Ascaso C, Wierzchos J (1999) Study of lichens with different state of hydration by the combination of low temperature scanning electron and confocal laser scanning microscopies. Internat Microbiol 2:251–257

del Hoyo A, Álvarez R, Del Campo EM, Gasulla F, Barreno E, Casano LM (2011) Oxidative stress induces distinct physiological responses in the two *Trebouxia* phycobionts of the lichen *Ramalina farinacea*. Ann Bot Lond 107:109–118

Demmig-Adams B, Adams WW III, Czygan F-C, Schreiber U, Lange OL (1990a) Differences in the capacity for radiationless energy dissipation in the photochemical apparatus of green and blue-green algal lichens associated with differences in carotenoid composition. Planta 180:582–589

Demmig-Adams B, Máguas C, Adams WW, Meyer A, Kilian E, Lange OL (1990b) Effect of high light on the efficiency of photochemical energy conversion in a variety of lichen species with green and blue-green phycobionts. Planta 180:400–409

Eisenreich W, Knispel N, Beck A (2011) Advanced methods for the study of the chemistry and the metabolism of lichens. Pytochem Rev 10:445–456

Freitag S, Feldmann J, Raab A, Crittenden PD, Hogan EJ, Squier HA, Boyd KG, Thain S (2012) Metabolite profile shifts in the heathland lichen *Cladonia portentosa* in response to N deposition reveal novel biomarkers. Physiol Plant 146:160–172

Garty J, Tamir O, Cohen Y, Lehr H, Goren AI (2002) Changes in the potential quantum yield of photosystem II and the integrity of cell membranes relative to the elemental content of the epilithic desert lichen *Ramalina maciformis*. Environ Toxicol Chem 21:848–858

Gasulla F, de Nova PG, Esteban-Carrasco A, Zapata JM, Barreno E, Guéra A (2009) Dehydration rate and time of desiccation affect recovery of the lichen alga *Trebouxia erici*: alternative and classical protective mechanisms. Planta 231:195–208

Gomez F, Barták M, Bell EM (2012) Extreme environments on earth as analogues for life on other planets: astrobiology. In: Bell EM (ed) Life at extremes. Environments, organisms and strategies for survival. CABI, Walingford, pp 522–536

Green TGA, Lange OL (1991) Ecophysiological adaptations of the lichen genera *Pseudocyphellaria* and *Sticta* to South temperate rainforests. Lichenologist 23:267–282

Green TGA, Büdel B, Meyer A, Zellner H, Lange OL (1997) Temperate rainforest lichens in New Zealand: light response of photosynthesis. NZ J Bot 35:493–504

Green TGA, Sancho LG, Pintado A (2011) Ecophysiology of desiccation/rehydration cycles in mosses and lichens. In: Luttge U et al (eds) Plant desiccation tolerance, ecological studies 215, Part 2. Springer, Berlin, pp 89–120

Hájek J, Barták M, Gloser J (2001) Effects of thallus temperature and hydration on photosynthetic parameters of *Cetraria islandica* from contrasting habitats. Photosynthetica 39:427–435

Hájek J, Barták M, Dubová J (2006) Inhibition of photosynthetic processes in foliose lichens induced by temperature and osmotic stress. Biol Plant 50: 624–634

Hájek J, Vaczi P, Barták M, Smejkal L, Lipavská H (2009) Cryoproective role of ribitol in *Xanthoparmelia somloensis*. Biol Plant 53:677–684

Hájek J, Váczi P, Barták M, Jahnová L (2012) Interspecific differences in cryoresistance of lichen symbiotic algae of genus *Trebouxia* assessed by cell viability and chlorophyll fluorescence. Cryobiology 64:215–222

Heber U, Bilger W, Shuvalov VA (2006) Thermal energy dissipation in reaction centres and in the antenna of photosystem II protects desiccated poikilohydric mosses against photo-oxidation. J Exp Bot 57:2993–3006

Heber U, Bilger W, Türk R, Lange OL (2010) Photoprotection of reaction centres in photosynthetic organisms: mechanisms of thermal energy dissipation in desiccated thali of the lichen *Lobaria pulmonaria*. New Phytol 185:459–470

Ilík P, Schansker B, Kotabová E, Váczi P, Strasser RJ, Barták M (2006) A dip in the chlorophyll fluorescence induction at 0.2–2 s in Trebouxia-possessing lichens reflects a fast reoxidation of photosystem I. A comparison with higher plants. Biochim Biophys Acta 1757:12–20

Jensen M, Siebke K (1997) Fluorescence imaging of lichens in the macro scale. Symbiosis 23:183–195

Jupa R, Hájek J, Hazdrová J, Barták M (2012) Interspecific differences in photosynthetic efficiency and spectral reflectance in two *Umbilicaria* species from Svalbard during controlled desiccation. Czech Polar Rep 2:31–41

Kappen L (1993) Plant activity under snow and ice, with particular reference to lichens. Arctic 46:297–302

Kappen L, Schroeter B, Scheidegger C, Sommerkorn M, Hestmark G (1996) Cold resistance and metabolic activity of lichens bellow 0° C. Adv Space Res 18:119–128

Kong FX, Hu W, Chao SY, Sang WL, Wang LS (1999) Physiological responses of the lichen *Xanthoparmelia mexicana* to oxidative stress of SO_2. Environ Exper Bot 42:201–209

Kosugi M, Arita M, Shizuma R, Moriyama Y, Kashino Y, Koike H, Satoh K (2009) Responses to desiccation stress in lichens are different from those in their photobionts. Plant Cell Physiol 50:879–888

Kranner I, Birtić S (2005) A modulating role for antioxidants in desiccation tolerance. Integrat Comparat Biol 45:734–740
Kranner I, Grill D (1996) Significance of thiol-disulphide exchange in resting stages of plant development. Bot Acta 109:8–14
Küpper H, Küpper F, Spiller M (1998) In situ detection of heavy metal substituted chlorophylls in water plants. Photosynth Res 58:123–133
Lange OL (1980) Moisture content and CO_2 exchange of lichens. Oecologia 45:82–87
Lange OL (2002) Photosynthetic productivity of the epilithic lichen *Lecanora muralis*: long-term field monitoring of CO2 exchange and its physiological interpretation - I. Dependence of photosynthesis on water content, light, temperature and CO2 concentration from laboratory measurements. Flora 197:233–249
Lange OL, Green TGA (1996) High thallus water content severely limits photosynthetic carbon gain of central European epilithic lichens under natural condition. Oecologia 19:111–118
Lange OL, Green TGA (2008) Diel and seasonal courses of ambient carbon dioxide concentration and their effect on productivity of the epilithic lichen *Lecanora muralis* in a temperate, suburban habitat. Lichenologist 40:449–462
Lange OL, Schulze ED, Koch W (1970) Experimentell-ökologische Untersuchungen an Flechten der Negev-Wüste.II.CO_2-Gaswechsel und Wasserhaushalt von Krusten- und Blattflechten am natürlichen Standort während der sommerlichen Trockenperiode. Flora 159:525–538
Lange OL, Geiger IL, Schulze E-D (1977) Ecophysiological investigations on lichens of the Negev desert. Oecologia 28:247–259
Lange OL, Kilian E, Ziegler H (1986) Water vapor uptake and photosynthesis of lichens: performance differences in species with green and blue-green algae as phycobionts. Oecologia 71:104–110
Lange OL, Meyer A, Zellner H, Heber U (1994) Photosynthesis and water relations of lichen soil crusts: field measurements in the coastal fog zone of the Namib desert. Funct Ecol 8:253–264
Lange OL, Belnap J, Reichenberger H (1998) Photosynthesis of the cyanobacterial soil-crust lichen *Collema tenax* from arid lands in Southern Utah, USA: role of water content on light and temperature responses of CO_2 exchange. Funct Ecol 12:195–202
Lange OL, Budel B, Meyer A, Zellner H, Zotz G (2004) Lichen carbon gain under tropical conditions: water relations and CO_2 exchange of *Lobariaceae* species of a lower montane rainforest in Panama. Lichenologist 36:329–342
Lange OL, Green TGA, Melzer B, Meyer A, Zellner H (2006) Water relations and CO_2 exchange of the terrestrial lichen *Teloschistes capensis* in the Namib fog desert: measurements during two seasons in the field and under controlled conditions. Flora 201:268–280
Lange OL, Reichenberger H, Walz H (2007) Continuous monitoring of CO_2 exchange of lichens in the field: short-term enclosure with an automatically operating cuvette. Lichenologist 29:259–274
Larsson P, Večeřová K, Cempírková H, Solhaug KA, Gauslaa Y (2009) Does UV-B influence biomass growth in lichens deficient in sun-screening pigments? Environ Exp Bot 67:215–221
Lud C, Huiskes A, Ott S (2001) Morphological evidence for the symbiotic character of *Turgidosculum complicatulum* Kohlm. & Kohlm. (= *Mastodia tesselata* Hook.f. & Harvey). Symbiosis 31:141–151
Máguas C, Griffiths H, Ehleringer J, Serôdio J (1993) Characterization of photobiont associations in lichens using carbon isotope discrimination techniques. In: Ehleringer J, Hall T, Farquhar G (eds) Physiological ecology series: perspectives on carbon and water relations from stable isotopes, vol 11. Academic, London, pp 201–212
Mayaba N, Beckett RP (2001) The effect of desiccation on the activities of antioxidant enzymes in lichens from habitats of contrasting water status. Symbiosis 31:113–121
McEvoy M, Nybakken L, Solhaug KA, Gauslaa Y (2006) UV triggers the synthesis of the widely distributed secondary compound usnic acid. Mycol Prog 5:221–229
Moudrá A (2009) Activation of non-photochemical quenching mechanisms of absorbed light energy in lichen thalli exposed to desiccation stress. Master thesis, Masaryk University
Mrak T, Jeran Z, Batič F et al (2010) Arsenic accumulation and thiol status in lichens exposed to As(V) in controlled conditions. Biometals 23:207–219
Oukarroum A, Strasser RJ, Schansker G (2012) Heat stress and the photosynthetic electron transport chain of the lichen *Parmelina tiliacea* (Hoffm.) Ach. in the dry and the wet state: differences and similarities with the heat stress response of higher plants. Photosynth Res 111:303–314
Palmer RJ, Friedmann EI (1990) Water relations and photosynthesis in the cryptoendolithic microbial habitat of hot and cold deserts. Microb Ecol 19: 111–118
Palmqvist K (2000) Carbon economy of lichens. New Phytol 148:11–36
Palmqvist K, Badger MR (1996) Carbonic anhydrase(s) associated with lichens: *in vivo* activities, possible locations and putative roles. New Phytol 132:627–639
Palmqvist K, Sundberg B (2000) Light use efficiency of dry matter gain in five macro-lichens: relative

impact of microclimate conditions and species-specific traits. Plant Cell Environ 23:1–14

Palmqvist K, de los Rios A, Ascaso C, Samuelsson G (1997) Photosynthetic carbon acquisition in the lichen photobionts *Coccomyxa* and *Trebouxia* (Chlorophyta). Oecologia 101:67–76

Palmqvist K, Campbell D, Ekblad A, Johansson H (1998) Photosynthetic capacity in relation to nitrogen content and its partitioning in lichens with different photobionts. Plant Cell Environ 21:361–372

Palmqvist K, Dahlman L, Valladares F, Tehler A, Sancho LG, Mattsson J-E (2002) CO_2 exchange and thallus nitrogen across 75 contrasting lichen associations from different climate zones. Oecologia 133:295–306

Pannewitz S, Green TGA, Schlensog M, Seppelt R, Sancho LG, Schroeter B (2006) Photosynthetic performance of *Xanthoria mawsonii* C. W. Dodge in coastal habitats, Ross Sea region, continental Antarctica. Lichenologist 38:67–81

Peksa O, Škaloud M (2010) Evolutionary inferences based on ITS rDNA and actin sequences reveal extensive diversity of the common lichen alga *Asterochloris* (*Trebouxiophyceae*, *Chlorophyta*). Mol Phyl Evol 54:36–46

Peksa O, Škaloud M (2011) Do photobionts influence the ecology of lichens? A case study of environmental preferences in symbiotic green alga *Asterochloris* (Trebouxiophyceae). Mol Ecol 20:3936–3948

Piccotto M, Tretiach M (2010) Photosynthesis in chlorolichens: the influence of the habitat light regime. Ital J Plant Res 123:763–775

Reiter R, Türk R (2000) Investigations on the CO_2 exchange of lichens in the alpine belt. II. Comparative patterns of net the CO_2 exchange in *Cetraria islandica* and *Flavocetraria nivalis*. Phyton Ann Rei Botanicae 40:161–177

Reiter R, Höftberger M, Green TGA, Türk R (2008) Photosynthesis of lichens from lichen-dominated communities in the alpine/ nival belt of the Alps – II: laboratory and field measurements of CO_2 exchange and water relations. Flora 203:34–46

Rubio C, Fernández E, Hidalgo ME, Quilhot W (2002) Effects of solar UV-B radiation in the accumulation of rhizocarpic acid in a lichen species from alpine zones of Chile. Bol de la Soc Chil de Quím 47:1–10

Sancho LG, Kappen L (1989) Photosynthesis and water relations and the role of anatomy in *Umbilicariaceae* (lichenes) from Central Spain. Oecologia 81:473–480

Scheidegger C, Schroeter B, Frey B (1995) Structural and functional processes during water vapour uptake and desiccation in selected lichens with green algal photobionts. Planta 197:399–409

Schroeter B, Green TGA, Pannewitz S, Schlensog M, Sancho LG (2011) Summer variability, winter dormancy: lichen activity over 3 years at Botany Bay, 77°S latitude, continental Antarctica. Polar Biol 34:13–22

Smith EC, Griffiths H (1998) Intraspecific variation in photosynthetic responses of trebouxioid lichens with reference to the activity of a carbon-concentrating mechanism. Oecologia 113:360–369

Solhaug K-A, Gauslaa Y (2004) Photosynthates stimulate the UV-B induced fungal anthraquinone synthesis in the foliose lichen *Xanthoria parietina*. Plant Cell Environ 27:167–176

Solhaug KA, Gauslaa Y, Nybakken L, Bilger W (2003) UV-induction of sunscreening pigments in lichens. New Phytol 158:91–100

Štěpigová J, Vráblíková H, Lang J, Večeřová K, Barták M (2007) Glutathione and *zeaxanthin* formation during high light stress in foliose lichens. Plant Soil Environ 53:340–344

Sundberg B, Campbell D, Palmqvist K (1997) Predicting CO_2 gain and photosynthetic light acclimation from fluorescence yield and quenching in cyano-lichens. Planta 201:138–145

Sundberg B, Ekblad A, Näsholm T, Palmqvist K (1999) Lichen respiration in relation to active time, temperature, nitrogen and ergosterol concentrations. Funct Ecol 13:119–125

Tretiach M, Pavanetto S, Pittao E, Sanità di Toppi L, Piccotto M (2012) Water availability modifies tolerance to photo-oxidative pollutants in transplants of the lichen *Flavoparmelia caperata*. Oecologia 168:589–599

Uchida M, Nakatsubo T, Kanda H, Koizumi H (2006) Estimation of the annual primary production of the lichen *Cetrariella delisei* in a glacier foreland in the high arctic, Ny-Ålesund, Svalbard. Polar Res 25:39–49

Unal D, Uyanikgil (2011) UV-B induces cell death in the lichen *Physcia semipinnata* (J.F.Gmel). Turk J Biol 35:137–145

Veerman J, Vasil'ev S, Paton GD, Ramanauskas J, Bruce D (2007) Photoprotection in the Lichen *Parmelia sulcata*: the origins of desiccation-induced fluorescence quenching. Plant Physiol 145:997–1005

Vráblíková H, Barták M, Wonisch A (2005) Changes in glutathione and xanthophyll cycle pigments in high light-stressed lichens *Umbilicaria antarctica and Lasallia pustulata*. J Photochem Photobiol B 79:35–41

Weissman L, Fraiberg M, Shine L, Garty J, Hochman A (2006) Responses of antioxidants in the lichen *Ramalina lacera* may serve as an early-warning bioindicator system for the detection of air pollution stress. FEMS Microbiol Ecol 58:41–53

Chapter 21

Electron Transport in the Mitochondrial Respiratory Chain

Maria Luisa Genova*
Dipartimento di Scienze Biomediche e Neuromotorie, Università di Bologna, Via Irnerio 48, 40126 Bologna, Italy

Summary

The metabolic capacity of the eukaryotic cell to convert free energy contained in nutrients into ATP is a process accomplished by a multi-step system: the mitochondrial respiratory chain. This chain involves a series of electron-transferring enzymes and redox co-factors, whose biochemical characterization is the collective result of more than 50 years of scientists' endeavors. The current knowledge describes in detail the structure and function of the individual proton-translocating "core" complexes of the respiratory chain (Complex I, III, IV). However, a holistic approach to the study of electrons transport from NAD-dependent substrates to oxygen has recently directed our attention to the existence of specific albeit dynamic interactions between the respiratory complexes. In this context, the respiratory complexes are envisaged to be either in form of highly ordered assemblies (i.e. supercomplexes) or as individual enzymes randomly distributed in the mitochondrial membrane. Either model of organization has functional consequences, which can be discussed in terms of the structural stability of the protein complexes and the kinetic efficiency of inter-complex electron transfer. Available experimental evidence suggests that Complex I and Complex III behave as assembled supercomplexes (ubiquinone-channeling) or as individual enzymes (ubiquinone-pool), depending on the lipid environment of the membrane. On the contrary, a

*Author for correspondence, e-mail: marialuisa.genova@unibo.it

M.F. Hohmann-Marriott (ed.), The Structural Basis of Biological Energy Generation, Advances in Photosynthesis and Respiration 39, pp. 401–417, DOI 10.1007/978-94-017-8742-0_21,

strict association of Complexes III and Complex IV is not required for electron transfer via cytochrome *c*, although there are supercomplexes in bovine heart mitochondria, known as the respirasomes, that also include some molecules of Complex IV. Our recent experimental results demonstrate that the disruption of the supercomplex I_1–III_2 enhances the propensity of Complex I to generate the superoxide anion; we propose that any primary source of oxidative stress in mitochondria may perpetuate generation of reactive oxygen species by a vicious cycle involving supercomplex dissociation as a major determinant.

I. Introduction

The recent findings in the study of electron transfer and energy transformation in mitochondria are the logical consequence, as we observe in many other sciences, of a continuous development of knowledge, through more than 50 years. In this view, the reader may appreciate an historical introduction that emphasizes some retrospective landmark studies on the respiratory chain prior to a description of more recent scientific endeavors leading to advances in understanding of mitochondrial respiration.

Following this line of thought, we will first point out that it was recognized early on that the metabolic capacity of the cell to convert free energy from nutrients into ATP is a process accomplished by a multi-step system involving molecular oxygen. Researchers soon realized the very complex nature of respiration and fundamental differences of this process from fermentation and glycolysis, which were also investigated at the time.

In the first edition of his textbook about bioenergetics, Albert L. Lehninger (1965) describes the molecular basis of energy-conversion in mitochondria by schematically drawing a flow chart (Fig. 21.1 top) that consists of "a series of cytochromes, which are electron-transferring enzyme molecules containing deeply colored active groups… comprised of porphyrin and iron". In this scheme, electrons can enter via one of two flavoproteins (FP) that accepts electrons either from reduced nicotinamide adenine dinucleotide (now conventionally referred to as NADH, formerly known by the name "reduced diphosphopyridine nucleotide" and abbreviated as DPNH), or from succinate.

The electron donors are arranged in a thermodynamic series of decreasing "electron pressure" where only the last cytochrome, called the "respiratory enzyme" or cytochrome oxidase, gives up its electron directly to molecular oxygen. Four such electron transfers are required to form water (Chance and Williams 1956). Since it was known that the components of the respiratory chain are located in mitochondria in simple ratios to each other (Criddle et al. 1962; Green and Wharton 1963) and cytochromes can carry only one electron at a time whereas FP can carry two at a time, the flow of electrons down the chain was written out allowing each cytochrome to react twice in a series of five sequential oxidation-reduction reactions connected by a common substrate intermediate. However, Lehninger's chart (Lehninger 1965) did not show an important detail of the respiratory chain that a number of investigators had already suggested (Racker 1965): electrons donated from NAD- and FAD-dependent dehydrogenases funnel into a common acceptor, ubiquinone (Morton 1958; Crane et al. 1957), which in turn feeds into the cytochrome system (Lester and Fleischer 1961; Hatefi et al. 1962b; Green and Tzagoloff 1966; Lenaz et al. 1968).

Abbreviations: BN-PAGE – Blue native polyacrylamide gel electrophoresis; $C_{I\text{–}IV}$ – Metabolic flux control coefficient of the corresponding respiratory complex; CoQ – Coenzyme Q ubiquinone; EPR – Electron paramagnetic resonance; ETF – Electron transfer flavoprotein; FP – Flavoprotein; $O_2^{-\cdot}$ – Superoxide anion; OXPHOS – Oxidative phosphorylation system; PL – Phospholipids; ROS – Reactive oxygen species; SDS – Sodium dodecyl sulfate

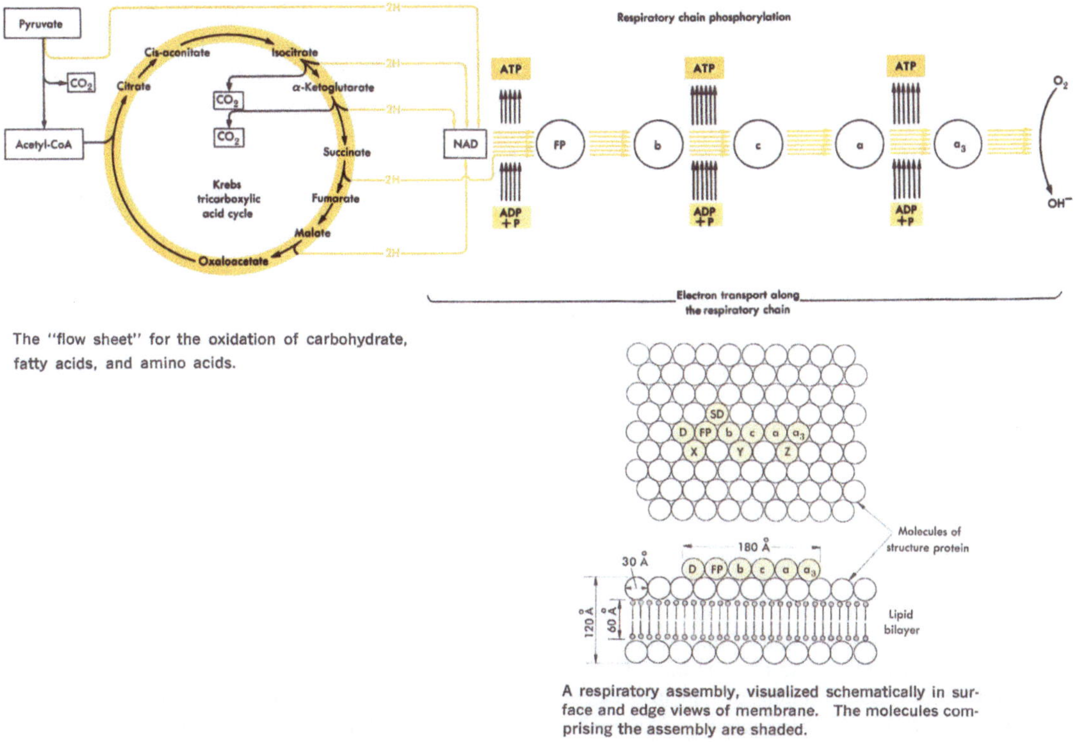

Fig. 21.1. Scheme of the respiratory chain as originally drawn by Lehninger (1965). (*Top*) Flow chart of electron transfer; (*Bottom*) Respiratory assembly. *D* reduced diphosphopyridine nucleotide, *FP* flavoprotein, *SD* succinate dehydrogenase, *b, c, a,* a_3 designate the cytochromes, *X, Y, Z* designate hypothetical high-energy intermediates (see text for details).

It is worth noting that those early investigators presented the electron carriers of the respiratory chain as geometrically fixed assemblies, located next to each other in the exact sequence in which they interact and intimately connected with the lipid components of the mitochondrial membrane (see also Sect. III). These assignments were derived from available staining methods using electron microscopy, which demonstrated the existence of clusters of cytochromes and flavoproteins studded in the inner membrane at regular intervals about 200 Å apart (Fig. 21.1 bottom). These studies also revealed for the first time that the high molecular weight enzymes responsible for the formation of ATP project from the inner surface of the inner mitochondrial membrane (Fernandez-Moran 1963).

In Lehninger's scheme, three segments of the chain contribute a relative large amount of free energy that was suggested to serve for donating high-energy phosphate groups to ATP generation (Chance and Williams 1955). Although the sequence of the cytochromes in the respiratory chain was known in principle, Lehninger (1965) had stated that "many important details remain to be established with more certainty" for a clear description of the bioenergetic role of respiration. Indeed, a lot of effort in the following years was spent in the vain search of the hypothesized high-energy intermediate compounds that could convey energy and phosphate to the ADP molecule (chemical coupling hypothesis).

By the time that the second edition of Lehninger's textbook was published (Lehninger 1971), interesting progress was made in understanding the molecular organization of the respiratory chain. The kinetic properties of the respiratory protein complexes

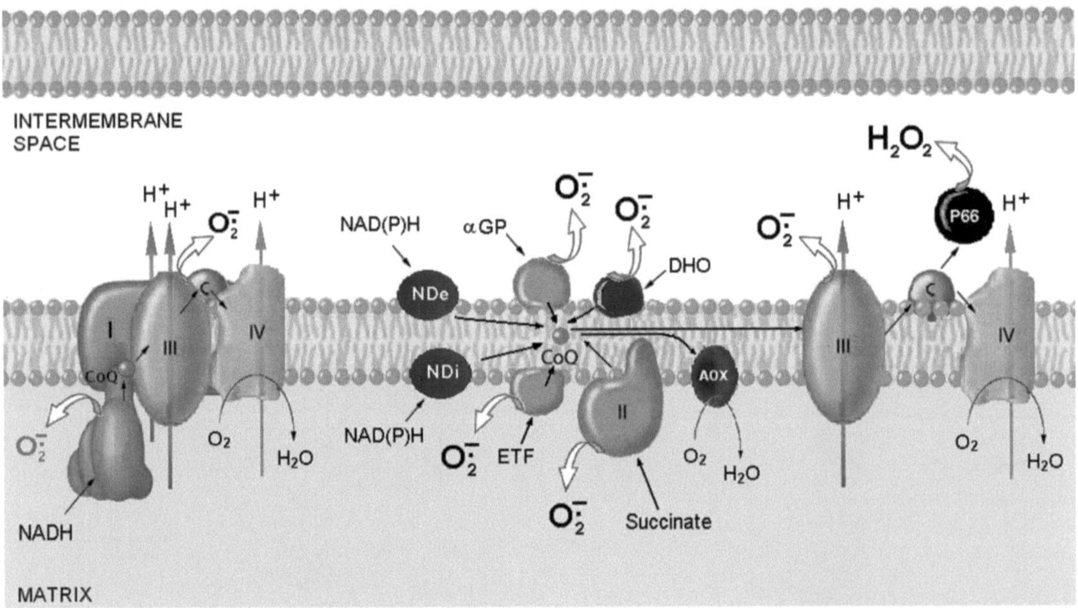

Fig. 21.2. A schematic drawing of the respiratory chain depicting the protein complexes and their substrates in relation to the inner mitochondrial membrane. Complex I is depicted as a component of the $I_1III_2IV_1$ supercomplex; whereas Complex III and Complex IV are also shown in their free form. The *white arrows* represent the major sites of ROS generation. *I* NADH-ubiquinone oxidoreductase, *II* succinate-ubiquinone oxidoreductase, *III* ubiquinol-cytochrome c oxidoreductase, *IV* cytochrome oxidase, ND_i *and* ND_e internal and external alternative NAD(P)H dehydrogenases, *AOX* alternative oxidase, *αGP* glycerol-3-phosphate, *ETF* electron transfer flavoprotein, *DHO* dihydroorotate, *CoQ* Coenzyme Q ubiquinone, *C* cytochrome *c*, *P66* cytochrome p_{66} (see text for details).

were unraveled by systematic studies in purified segments of the respiratory chain (Griffiths and Wharton 1961; Hatefi et al. 1962a; Baum et al. 1967). However, despite years of hard work, the exact mechanism, by which ATP is generated during electron transport, was still debated. It became clearer and clearer, however, that the intactness of the inner membrane is an essential element in the process of oxidative phosphorylation of ADP. Furthermore, it was emphasized that H^+ ions may be pumped out of the mitochondrial matrix during electron transport (Mitchell and Moyle 1967). Those observations suggested possible mechanisms for the conversion of redox energy into phosphate-bond energy (Hinkle et al. 1972; Ragan and Hinkle 1975; Leung and Hinkle 1975) other than the previous 'chemical coupling model': alternative thinking culminated in the 'chemiosmotic hypothesis' proposed by Peter Mitchell (1975) and the demonstration that the proton gradient produced across the inner mitochondrial membrane concomitantly with the electron transport along the respiratory chain is the immediate driving force for ATP formation.

Since then, the characteristics of the respiratory chain from a wide variety of sources have been enriched by growing evidence describing in detail the structure and function of the proton-translocating "core" complexes (Complex I, III, IV). Several auxiliary enzymes that reduce ubiquinone (Coenzyme Q, CoQ), bypassing Complex I, or that deliver electrons directly from reduced ubiquinone (ubiquinol) to oxygen bypassing Complex III and IV were also described (Fig. 21.2). All those auxiliary enzymes are characterized by the lack of energy-conserving mechanisms of proton translocation (see Chap. 9) Lenaz and Genova 2010 for a review).

The present chapter will discuss available evidence concerning the supramolecular organization of the respiratory enzymes that are envisaged to be either in the form of highly ordered assemblies (i.e. supercomplexes), or as individual enzyme units that are randomly distributed in the mitochondrial membrane. We will characterize the functional consequences of inter-complex electron transfer in both of these models of organization.

II. Electron Transport and Proton Translocation in Mitochondrial Membrane Systems

A. Redox Centers

Several redox centers that are associated with enzyme complexes allow electron transfer in the respiratory chain that follows a gradient in midpoint potential (E_m). The electrons donated by NADH at $E_m = -320$ mV are accepted by FMN in Complex I and then transferred to a series of iron-sulphur clusters (Fe-S centers) (Ohnishi et al. 1998). NADH-ubiquinone oxidoreductases from different sources have different numbers of Fe-S centers, most of which share the same midpoint potential (Dutton et al. 1998). Recently, the arrangement of the prosthetic groups in the hydrophilic domain of Complex I from *Thermus thermophilus* has been determined by x-ray crystallography, showing a linear chain of seven conserved clusters (i.e. N3, N1b, N4, N5, N6a, N6b, N2), whereas two additional clusters do not participate in the main pathway and have been proposed to represent an evolutionary remnant (cluster N7) and a possible anti-oxidant center (cluster N1a) (Hinchliffe and Sazanov 2005; Sazanov and Hinchliffe 2006). N2, which is a Fe_4-S_4-type center, has the most positive E_m (between −50 and −150 mV) and is considered to be the direct electron donor to ubiquinone, being most likely located along the interface between the peripheral and the membrane arm of Complex I where the ubiquinone head group could be located, as suggested by Brandt et al. (2003). The mechanism of ubiquinone reduction by Complex I is particularly intriguing, because more than one bound quinone species has been assigned to the enzyme (Magnitsky et al. 2002); a tentative two-step scheme was drawn by Fato et al. (2009) where a Fe-S cluster located upstream of N2 may act as a switch for electron delivery.

Most certainly, ubiquinone molecules bound to Complex I coexist with a mobile pool of molecules in the inner mitochondrial membrane (ubiquinone-pool). This pool is of crucial importance for various ubiquinone-reducing dehydrogenases to connect with Complex III (see also Sect. III). As schematically depicted in Fig. 21.2, among those dehydrogenases is Complex II, which conveys electrons to ubiquinone through one covalently linked FAD, three Fe-S centers acting in sequence, and a single *b*-type heme with a very positive E_m (Yankovskaya et al. 2003; Horsefield et al. 2006). Moreover, some rotenone-insensitive NADH dehydrogenases (Yagi et al. 2001) and other auxiliary enzymes involved in different metabolic pathways (i.e. glycerol-3-phosphate dehydrogenase, ETF-ubiquinone oxidoreductase, choline dehydrogenase, dihydroorotate dehydrogenase, malate dehydrogenase) can donate electrons through their prosthetic groups into the ubiquinone-pool at an E_m around zero. Electrons move from ubiquinol via the Fe-S center and to cytochrome c_1 and involvement of *b*-cytochromes in Complex III (Covian and Trumpower 2008; Cramer et al. 2011; and Chap. 8) to cytochrome *c* ($E_m = +230$ mV).

Electrons transported by cytochrome *c* are donated to Complex IV, where electrons transfered to oxygen via the Cu_A center (which acts as a single-electron receptor), heme a, heme a_3/Cu_B center, and heme a3 (cf. Belevich and Verkhovsky 2008 for extensive review). However, in some mitochondrial systems, like plant mitochondria, alternative oxidases (see Chap. 9) are able to reduce molecular oxygen by directly accepting electrons from ubiquinol (Krab 1995).

B. Protonic Coupling

Electron transport along the respiratory chain is directly coupled to the translocation of protons from the mitochondrial matrix (H^+_{in}) to the intermembrane space. This H^+-pumping is operated by Complex I, III and IV. The change in redox energy mediated by the three sites is far from equal, with the smallest at Complex III (250 mV), the next larger at Complex I (360 mV) and the largest at Complex IV (470 mV).

The coupling of oxidation-reduction reactions to the deprotonation-protonation of ubiquinol/ubiquinone within the cytochrome bc_1 complex is central to the mechanism of proton translocation at site-2, as described in the ubiquinone-cycle model proposed by Mitchell (1975) and in subsequent discussion of his model (for a review, see Cramer et al. 2011). Different energy coupling mechanisms have been proposed for the other two sites. In particular, the x-ray crystallographic structure of bovine cytochrome *c* oxidase shows possible pathways extending across the enzyme where cooperative linkage between the heme a/Cu_A redox center and nearby acid/base protolytic residues is envisaged to result in the uptake and vectorial translocation of protons from the matrix surface (N space) toward the cytosolic surface (P space) in association with the oxygen reduction chemistry at the binuclear center (heme a_3/Cu_B) (Brzezinski and Gennis 2008; Papa et al. 2006a; Belevich et al. 2007). Based on the emerging structures of the bacterial (Efremov et al. 2010) and mitochondrial (Hunte et al. 2010) Complex I, a hypothetical mechanism for its redox-driven proton pumping proposed that two conformational strokes, generated during turnover by the stabilization of semiquinone/quinol molecules in the peripheral arm of the enzyme, can drive two membrane integral pump modules connected by a long helical "transmission element". The overall proposed pumping cycle is reminiscent to the function of a coupling rod of a stem engine (Efremov and Sazanov 2011). However, the actual mechanism how redox chemistry and proton pumping are coupled remains elusive. Very recent insights into the molecular architecture of Complex I from *Thermus thermophilus* (Baradaran et al. 2013) provide a complete picture of all the components of this giant proton pump and suggest that a long chain of charged residues, which pass through the middle of the membrane integral P-module of the enzyme, connects the ubiquinone-binding pocket to four putative pump sites consisting of separate proton-input and -output channels (Brandt 2013). This corroborates an earlier proposal that energy transmission within Complex I may occur through electrostatic coupling (Euro et al. 2008). Future functional testing of the predictions in the models proposed by Brandt and coworkers and by Efremov and coworkers will be of critical importance.

Protons from the intermembrane space (H^+_{out}) are mainly utilized to drive the synthesis of ATP molecules from ADP and inorganic phosphate by the F_0F_1-ATP synthase of the mitochondrial inner membrane. Proton transport stoichiometries have had a turbulent history, but $H^+_{in}/2e^-$ ratios of 4; 2 and 4, respectively for the three proton-translocating complexes, and values of $H^+_{out}/ATP=4.3$ (i.e. including energy consumption for nucleotide- and Pi-transporters) are generally accepted today. As a consequence, the overall P/O ratio of oxidative phosphorylation is about 2.5 when NADH-linked substrates are oxidized through Complex I whereas the electron donation from the flavin dehydrogenases, which reduce ubiquinone without pumping protons result in P/O values close to 1.5 (Hinkle 2005).

The controlled dissipation of the proton-motive force of the H^+ gradient through reverse proton transloaction mediated by the uncoupling proteins of the inner mitochondrial membrane (Ricquier 2005), and slips in the proton pumping of respiratory complexes (Kadenbach 2003; Papa et al. 2006b) can contribute to preventing excessive and harmful metabolic production of reactive oxygen species (ROS) under conditions of high reducing power (see Sect. IV).

III. Overall Organization of Classic and Alternative Redox Complexes

A. Working in the Chain Gang or Free?

The supramolecular organization of the respiratory chain was a major research subject in the 1970s through the 1980s, culminating in the 'random collision model', of electron transfer (Hackenbrock et al. 1986) accepted by the majority of the investigators at the time. Evidence in favor of the random collision model derived from direct observation that: (i) the integral proteins of the inner mitochondrial membrane are randomly distributed in the bilayer (Höchli and Hackenbrock 1976; Sowers and Hackenbrock 1981) and phospholipid dilution of the respiratory proteins slows electron transfer (Schneider et al. 1982); (ii) according to the analysis of Kröger and Klingenberg (1973), both ubiquinone and cytochrome *c* behave kinetically as homogeneous pools, which follow saturation kinetics.

On the other hand, circumstantial evidence for the existence of highly aggregated protein assemblies in the native membrane was already present in early studies on the isolation of the respiratory enzymes. Studies that obtained purified enzyme fractions comprising both Complex I and Complex III (Hatefi et al. 1961) or Complex II plus Complex III (Yu and Yu 1980) are indicative of higher level aggregation. We believe that the investigators who first discovered the mitochondrial respiratory complexes already had the key for demonstrating the supercomplex organization in their hands, but were unaware of it. In addition, a close inspection of the fractured faces in the freeze-fracture electron micrographs (originally taken as a demonstration of long-range random distribution of the mitochondrial particles (Höchli and Hackenbrock 1976)) reveals the apposition of small clusters, and also hints at a possible association of the integral membrane proteins.

A clear demonstration of such protein associations came during the past decade from investigations that reported multicomplex units in yeast and mammalian mitochondria. A distinct structure was observed for all these supercomplexes, supporting the idea of highly ordered associations of respiratory supercomplexes and discarding most doubts that the postulated structures arose through artificial interactions. In particular, polyacrylamide gel electrophoresis under non-denaturating conditions (BN-PAGE (see Chap. 12) in digitonin-solubilized mitochondria of *Saccharomyces cerevisiae*, which possesses no Complex I, revealed two bands with apparent masses of ~750 and 1,000 kDa containing the subunits of complexes III and IV. Similar interactions of supercomplexes were investigated in bovine heart mitochondria: Complex I–III interactions were apparent from the presence of Complex I in the form of the I_1III_2 supercomplex that was also found further assembled into larger supercomplexes (respirasomes) comprising different copy numbers of Complex IV ($I_1III_2IV_{1-4}$). Only 14–16 % of total Complex I was found in free form in the presence of digitonin (Schägger and Pfeiffer 2001). Therefore it seems likely that all Complex I is bound to Complex III in physiological conditions (i.e. in the absence of detergents).

Current literature indicates an average ratio of oxidative phosphorylation complexes I:III:IV equal to 1.1:3:6.7 (Lenaz and Genova 2010). Therefore it appears plausible that approximately one-third of total Complex III in bovine mitochondria is not bound to monomeric Complex I. The fraction of Complex IV in free form represents >85 % of total cytochrome oxidase in mitochondria. Strong evidence has been accumulating that complexes I, III, and IV are organized as stoichiometric supercomplexes also in plants and in fungi (Eubel et al. 2004; Krause et al. 2004a, b; Marques et al. 2007). Interestingly, in contrast to the situation in mammals, complexes III and IV are not essential for assembly/stability of Complex I in the ascomycote fungus *Podospora anserina.* The function of the arrangement of the respiratory complexes in *Podospora anserina* is rather an open question, since electrons can be released from dimeric Complex I to the so-called alternative oxidase, while Complex III is mostly kinetically inactive in this fungus (Krause et al. 2006; Maas et al. 2009).

It is worth noting that, in the prevailing opinion, Complex II acts as an independent enzyme with no specific association with other OXPHOS complexes, although the work of Acín-Pérez et al. (2008) showed Complex II-containing supercomplexes in mouse liver mitochondria. Moreover, neither the alternative NAD(P)H dehydrogenases nor the extra terminal oxidases were found associated with the major respiratory complexes (Sunderhaus et al. 2010), except in *Yarrowia lipolytica* (Ascomycote fungus) where the dehydrogenase NDH2e is reported to form part of a large supercomplex (Guerrero-Castillo et al. 2009).

A precise docking of the existing atomic x-ray structures of individual OXPHOS complexes into the cryo-electron microscopy map of supercomplexes from a variety of sources has resulted in pseudo-atomic models of their three-dimensional structure (Schäfer et al. 2007; Dudkina et al. 2011; Althoff et al. 2011; Heinemeyer et al. 2007; Mileykovskaya et al. 2012). It has been also proposed (Bultema et al. 2009; Strauss et al. 2008) that the OXPHOS complexes may assemble into even higher types of organization, forming row-like megacomplexes composed by supercomplexes as building blocks, with important implication for the morphology of the inner mitochondrial membrane. The characterization of the supercomplexes by biochemical functional analysis is incomplete, so a functional role of the supramolecular assemblies cannot be assigned with certainty.

Most evidence for a structural coordination of OXPHOS complexes has been obtained by indirect observations indicating deviations from "pool behavior" of electron transfer reactions involving ubiquinone, and from studies directly aimed to demonstrate substrate channeling by metabolic flux control analysis of the electron transfer in mitochondrial membranes, whereas the study of the kinetic properties of isolated supercomplexes is still in its infancy. In fact, the first functional demonstration of the existence of supercomplexes was provided by kinetic analysis of the pool function of ubiquinone and cytochrome *c* in mitochondria from *Saccharomyces cerevisiae* (Boumans et al. 1998). The finding that those mitochondria did not follow pool behavior unless treated with chaotropic agents was considered a peculiarity of this organism, because pool behavior was still widely accepted at that time. Later on, our studies by Metabolic Control Analysis (Bianchi et al. 2004) indicated the existence of functional supercomplexes in bovine heart submitochondrial particles. A strong kinetic evidence of a functionally relevant association between Complex I and Complex III is provided by the high metabolic control of both complexes over NADH oxidation (control coefficients equal 1.06 and 0.99, respectively), whereas Complex IV appears to be randomly distributed, as indicated by its low control coefficient ($C_{IV}=0.26$). Moreover, Complex II is fully rate-limiting for succinate oxidation ($C_{II}=0.88$, $C_{III}=0.34$, $C_{IV}=0.20$), clearly indicating the absence of substrate channeling toward Complex III and Complex IV. In permeabilised mitochondria from freshly harvested potato tubers, analogous inhibitor titration experiments indicate that Complexes III and IV are both involved in the formation of a supercomplex assembly that also comprises Complex I, whereas the alternative dehydrogenases, as well as the molecules of Complex II, act as independent structures within the inner mitochondrial membrane (Genova et al. 2008).

B. Solid-State or Fluid-State Model: Kinetic Advantages

Today, a solid-state model of the respiratory chain, based on specific, albeit dynamic interactions between individual respiratory components, is well consolidated (Wittig and Schägger 2009; Lenaz and Genova 2007, 2009a, b, 2010), although its possible relation with a random diffusion model of electron transfer is not completely clarified.

Acin-Perez and coworkers (2008) propose a "Plasticity model" where both types of organization are possible and functional, depending on different mitochondrial systems and on particular physiological states. The dynamic character of the supercomplexes is compatible with the factors affecting the association of their subunit components: some factors may act in a short or medium time scale, such as the membrane potential (Piccoli et al. 2006; Dalmonte et al. 2009) or protein phosphorylation/dephosphorylation (Rosca et al. 2011), whereas other factors may be operative on longer time scales, such as changes in phospholipid composition (for a review, see Lenaz and Genova 2012). However, a recent study (Muster et al. 2010) suggests that supercomplex dissociation, at least in a random fashion, may not be a fast event.

Supercomplex association confers several new properties to the respiratory chain with respect to the non-associated respiratory complexes. From a kinetic point of view, the most obvious is substrate channeling, specifically addressing ubiquinone and cytochrome *c* to interact directly with the partner enzymes, thereby replacing the random diffusion characteristics expected for non-associated components. In that case, inter-complex electron transfer becomes indistinguishable from intra-complex electron transfer, so that the so-called mobile intermediates, predicted to exhibit substrate-like behavior in the classic view of the random collision model (Hackenbrock et al. 1986), would rather be buried at the interface between two consecutive complexes. Intra-protein electron transfer is typically limited by tunneling through the insulating protein medium between the interacting redox centers. This electron tunneling is reasonably well described by a simple exponential decay depending on distance, so that the maximal distances that allow for physiological electron transfer should not exceed 13–14 Å (Moser et al. 2005). Inter-protein electron tunneling obeys the same exponential rate dependence on distance as intra-protein electron transfer; however, in addition small-scale constrained diffusive motions are sometimes necessary to bring redox centers within the 14 Å tunneling limit.

Ideally, in order to describe the mechanism of electron transfer within supercomplexes formed by apposition of individual respiratory complexes connected by potentially mobile cofactors, we should have a detailed knowledge of the molecular structure of the interacting sites. The recent work by Althoff and colleagues (2011) demonstrates that only few points of direct contact are allowed between the three complexes in the mammalian supercomplex $I_1III_2IV_1$, because the average distances exceed 2 nm. Moreover, the same authors indicate that at 19 Å resolution the membrane-embedded part of the supercomplex shows intermediate values of density between that of soluble protein and the hydrophobic membrane interior, thus suggesting that the supercomplex is held together at least partly by lipid–protein interactions. Likely, in Althoff's model, a gap of about 13 nm between Complex I and the proximal Complex III monomer is filled with membrane lipid that would facilitate restricted diffusion (microdiffusion) of ubiquinol between the ubiquinone-binding sites of the two enzyme partners.

Data obtained by us in proteoliposomes enriched with mitochondrial Complex I and Complex III in the presence of ubiquinone 10 (Genova et al. 2008) demonstrate that the respiratory complexes behave as individual enzymes (ubiquinone-pool behavior) or as assembled supercomplexes, depending on the distance induced between the intramembrane particles by phospholipid dilution. In particular, pool behavior is not effective in proteoliposomes at 1:1 protein:lipid ratio, where Complex I and Complex III are within a short range (<50 nm) resembling the mean nearest neighbor distance between respiratory complexes in mitochondria. In this experimental condition, NADH-cytochrome *c* reductase activity is indeed higher than predicted by the pool equation (Table 21.1) and BN-PAGE analysis confirms that Complex I and Complex III are mostly

Table 21.1. Experimental and calculated values of NADH-cytochrome *c* reductase in a mitochondrial protein fraction diluted with different levels of phospholipids.

Protein/PL[a]	Distance[b]	NADH-cytochrome *c*	V_{obs} (calculated)[c]
(w:w)	(nm)	(μmol/min/mg protein)	
1:1	18	0.472±0.091 (7)	0.210
1:30	97	0.273±0.028 (4)	0.236

Values in brackets indicate multiple experiments
[a]A protein fraction enriched in Complex I and Complex III from bovine heart mitochondria was fused with phospholipids (PL) and Coenzyme Q_{10} by cholate dilution (Lenaz et al. 1999)
[b]The theoretical distances between Complex I and Complex III were calculated according to Hackenbrock et al. (1986)
[c]NADH-cytochrome *c* reductase activity as calculated from the pool equation of Kröger and Klingenberg (1973) using experimental values of NADH-CoQ reductase and ubiquinol-cytochrome *c* reductase activity as Vred and Vox, respectively (Data reprinted from Genova et al., Copyright (2008), with permission from Elsevier)

assembled in the form of supercomplex I_1III_2 (Lenaz et al. 2010).

Therefore, the function of the mobile pool of ubiquinone molecules that most certainly exist in the inner mitochondrial membrane has to be reconsidered in view of the presence of supercomplexes. As discussed in detail by Lenaz and colleagues (Lenaz 2001; Lenaz and Genova 2009a), although electron transfer between Complex I and Complex III may not be best described by the pool equation of Kröger and Klingenberg (1973) and it is likely that substrate channeling of ubiquinone takes place, the pool equation still seems to represent the best description of the mechanism of electron transfer from Complex II and other flavin-linked dehydrogenases to Complex III (or from any dehydrogenase, including Complex I, to the alternative oxidase when present). In addition, the ubiquinone pool participates in a reverse electron transfer from Complex II to Complex I (energy-dependent reduction of NAD by succinate).

Moreover, the concept of a ubiquinone pool remains crucial in evaluating ubiquinone channeling from Complex I to Complex III. In fact, we may conclude that this pool is in equilibrium with protein-bound ubiquinone molecules. In addition the ubiquinone molecules that facilitate electron transfer within the supercomplex I–III may well be in dissociation equilibrium with the ubiquinone pool. Therefore, at steady state, the amount of the supercomplex-associated quinone would be dependent on the size of the pool. This proposition conforms with the abovementioned experimental observations of the kinetic behavior in the presence of respiratory supercomplexes, provided that the dissociation rate constants of bound CoQ is considerably slower than the rate constants of inter-complex electron transfer via the same bound quinone molecules; however, our hypothesis should be explored by further experimental analysis.

Concerning cytochrome *c*, the model of the supercomplex $I_1III_2IV_1$ by Althoff and colleagues (2011) suggests that a shallow cavity lined by negative charges on the exterior membrane surface near the Cu atoms in Complex IV directly faces the similarly shaped binding sites of cytochrome *c* on Complex III. This arrangement would make it possible for the small, globular, partly positive cytochrome *c* molecules to pass efficiently from one complex to the other along a trajectory of approximately 11 nm. Indeed, residual amounts of cytochrome *c* remaining bound throughout the purification procedure of this supercomplex suggest at least partial occupancy of specific binding sites in the supercomplex (Althoff et al. 2011). This is in accordance with some observations from our laboratory also showing the presence of bound cytochrome *c* by two dimensional BN/SDS-PAGE analysis and Western blot immuno-detection in digitonin-solubilised respirasomes from potato tuber mitochondria (Lenaz et al. 2010). However, despite the presence of some molecules of Complex IV in the respirasome of bovine heart mitochondria, the kinetic evaluation of NADH oxidase activity by Metabolic Control Analysis (Bianchi et al. 2004) suggests that cytochrome *c* obeys pool behavior. We hypothesize that such behavior may be due to rapid exchange of bound cytochrome *c* and freely diffusing-molecules of cytochrome *c*.

The presence of a large excess of free cytochrome oxidase, together with the loss of cytochrome *c* bound to supercomplexes, as assessed by 2D BN/SDS-PAGE in isolated bovine submitochondrial particles (Lenaz et al. 2010), are in line with our hypothesis. This strongly implies that a strict association of Complexes III and IV is not required for efficient electron transfer by cytochrome *c*.

Very recently, Trouillard et al. (2011) have developed another kinetic approach to address the issue of the functional relevance of the supercomplex III–IV in intact cells of *Saccharomyces cerevisiae*; they also conclude that cytochrome *c* is neither trapped within supercomplexes nor does it encounter particular restriction to diffusion, and that oxidation of cytochrome *c* by a given cytochrome oxidase is a random process.

Likewise, we cannot exclude that the presence of Complex IV in the supercomplex assembly may provide kinetic advantages other than substrate channeling of cytochrome *c*, by modifying the conformation of the partner complexes in order to enhance their catalytic activity, as suggested by the study of Schäfer et al. (2006). This study indicated higher Complex I and Complex III activities in the bovine supercomplex comprising cytochrome oxidase ($I_1III_2IV_1$) compared to the supercomplex devoid of the terminal oxidase (I_1III_2).

IV. By-Products of Aerobiosis: Generation of Reactive Oxygen Species by the Respiratory Chain

Generation of ROS (see also Chap. 24) within mitochondria is closely associated to the primary function of these organelles; namely the supply of electrons to the enzymes of the respiratory chain in order to drive the vectorial transfer of protons into the mitochondrial intermembrane space. This process inadvertently results in leaking of electrons from reduced cofactors onto molecular oxygen and hence ROS generation under normal circumstances. Murphy (2009) has carefully analyzed the thermodynamics of the production of the superoxide anion ($O_2^{-}\cdot$) in mitochondria. Oxygen can theoretically undergo one-electron reduction to generate superoxide at every enzymes of the respiratory chain (see Fig. 21.2). Indeed, Complex I and Complex III have been identified as robust sources of superoxide anion radicals, whereas cytochrome oxidase, which catalyzes the tetravalent reduction of dioxygen without the release of partially reduced molecular species, represents an exception (Wikström 2012).

In Complex III, $O_2^{-}\cdot$ is thought to arise from the reaction of oxygen with a component located at the Q_o site, the quinone binding site on the outer or cytosolic face of the protein (Boveris and Cadenas 1975; Turrens et al. 1985; Muller et al. 2002). Conditions that favor the formation of a semiquinone in the Q_o site should yield the highest rates of superoxide production. Consequently, one might suppose that a fully reduced ubiquinone pool in the presence of antimycin A, an inhibitor that blocks the normal egress of electrons from the system through Q_i site (van den Berg et al. 1979), would lead to the production of superoxide. Remarkably, however, conditions of substrate limitation that allow a partial oxidation rather than a full reduction of the ubiquinone pool (e.g. when oxidation of the substrate succinate is experimentally restricted by the presence of the competitive inhibitor malonate) result in significantly higher rates of superoxide production by Complex III. Mechanistic studies (Dröse and Brandt 2008; Quinlan et al. 2011) of the cytochrome bc_1 complex have clarified that the electron is transferred onto oxygen in a reverse reaction from reduced cytochrome b_{566} via ubiquinone, rather than during the forward ubiquinone-cycle. It was also observed that the production of $O_2^{-}\cdot$ in the antimycin-inhibited enzyme can be suppressed by the protonmotive force generated by the hydrolytic activity of the F_1F_0-ATPase. In this condition a more equal distribution of electrons between cytochrome b_{566} (on the cytosolic side of the mitochondrial inner membrane) and cytochrome b_{562} (on the matrix side) is achieved. It is noteworthy that the controlled state of respiration (state 4) might ensure a relatively reduced ubiquinone-pool and limit the possibility of electron backflow from cytochrome b_{566} to ubiquinone and the generation of the semiquinone. The observation that the $O_2^{-}\cdot$ production by Complex III is negligible in the absence of respiratory chain inhibitors has lead some authors to argue that this phenomenon is likely to be of limited physiological relevance (Adam-Vizi and Chinopoulos 2006). However, it is worth to note that the modulation of the ROS production by modulation of the redox state of the ubiquinone-pool may contribute significantly to the control of mitochondrial redox signaling.

By contrast, the production of $O_2^{-}\cdot$ by Complex I is certainly more relevant from a metabolic point of view (Lenaz 2012), although the precise mechanism that promotes the one-electron reduction of oxygen at Complex I still requires further clarification. The identification of the oxygen-reducing site has been the subject of extensive investigation and several prosthetic groups in the enzyme (FMN, ubisemiquinone, and iron sulphur cluster N2) have been suggested to be the electron donors to oxygen (Lenaz et al. 2006; Fato et al. 2009). In isolated Complex I, fully reduced FMN is considered the major electron donor to oxygen (Galkin and Brandt 2005; Kussmaul and Hirst 2006; Esterházy et al. 2008). However, the effect of specific inhibitors acting downstream of the iron sulfur clusters in Complex I (Fato et al. 2009) suggests the possibility of a second site for oxygen reduction in the enzyme. In fact, EPR studies in bovine heart submitochondrial particles in the presence of NADH (Fato et al. 2009) show that rotenone but not stigmatellin enhances ROS production and quenches the semiquinone radical signal, whereas the FeS cluster N2 is kept reduced in presence of rotenone but is oxidized in presence of stigmatellin. Overall these results are compatible with the idea that N2 can be the source of enhanced production of $O_2^{-}\cdot$ in submitochondrial particles. Ohnishi et al. (2010) presented a new hypothesis that the generation of superoxide in a purified preparation of Complex I reflects a dynamic balance between the flavosemiquinone and the ubiquinone semiquinone. Superoxide generation from the flavosemiquinone would be increased, if electron transfer was inhibited under pathological conditions. The identification of the ubiquinone semiquinone rather than N2 as the electron donor to oxygen is, however, in contrast with the findings reported above, that no superoxide is produced in the presence of stigmatellin (Fato et al. 2009). Moreover, studies in ubiquinone-depleted and reconstituted mitochondria indicate that endogenous ubiquinone is not required for superoxide generation (Genova et al. 2001). The site of oxygen reduction is even less well defined in reverse electron transfer when $O_2^{-}\cdot$ production by Complex I is maximal (Murphy 2009).

The hypothesis that the tight organization of Complex I into supramolecular structures may hide autooxidable prosthetic groups, thereby preventing the reaction with oxygen, suggests that a dynamic assembly of the respiratory supercomplexes can also provide a regulation mechanism for $O_2^{-}\cdot$ production (Fig. 21.3). Very recently, we have directly addressed this issue and produced experimental evidence under conditions in which Complex I is either arranged as a component of the supercomplex I_1III_2 or dissociated as an individual enzyme (Maranzana et al. 2013). Our results show that disruption of the supercomplex I_1III_2 strongly enhances the ability of Complex I to generate ROS both in reconstituted proteoliposomes and in bovine heart mitochondrial membranes. It is easy to predict the implications of these findings in human diseases and in aging, where oxidative stress plays a major etiologic and pathogenic role. We propose that oxidative stress in mitochondria, generated by any primary source, may perpetuate ROS generation by a vicious

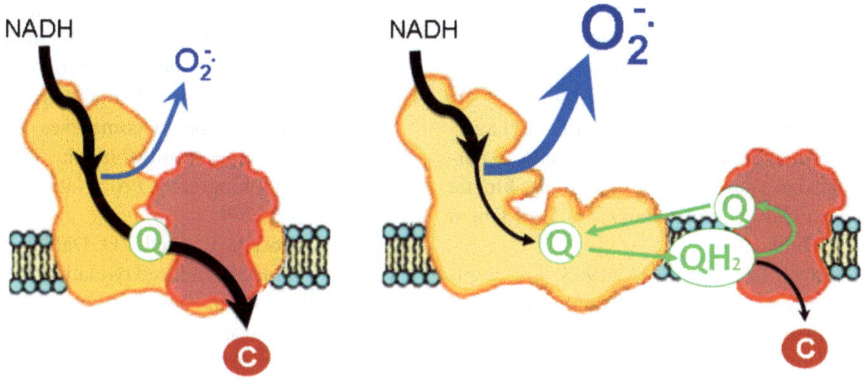

Fig. 21.3. Scheme of electron flux in two models of integrated NADH-cytochrome *c* reductase activity in the respiratory chain. (*Left*) In a supercomplex I_1III_2, any step in the obligatory pathway is regarded as a component of a single enzyme unit where the overall transfer of electrons to the final acceptor is optimized by channeling of the intermediate substrates. Protein-protein interactions modulate the reactivity of the intra-complex redox centers with dioxygen thereby preventing excessive ROS generation from Complex I (*yellow*). (*Right*) In a linear pathway composed of individual Complex I (*yellow*) and Complex III (*red*), a result of supercomplex disaggregation, the overall electron transfer to cytochrome *c* is mediated by the pool behavior of ubiquinone. The efficiency of electron transport is modified by the absence of protein-protein interactions, and $O_2^{-\cdot}$ production by Complex I is strongly enhanced.

cycle involving supercomplex dissociation as a major determinant.

Finally, it is worth mentioning that mitochondria from different tissues may vary conspicuously in their capacity to produce ROS depending on the utilized substrates (Kwong and Sohal 1998), and depending on the animal species and age. As a matter of fact, in addition to the major sites of superoxide formation in the respiratory chain, further sites in the inner mitochondrial membrane may have importance and physiological relevance (see Fig. 21.2). For example, a significant proportion of ROS may be released from several other ubiquinone-reducing enzymes (e.g. Complex II, glycerol-3-phosphate dehydrogenase, dihydroorotate dehydrogenase, ETF-dehydrogenase), whose contribution is still poorly characterized (see Lenaz and Genova 2010; Lenaz 2012, for extended review).

Acknowledgements

I would like to express my gratitude to my scientific mentor, Professor Giorgio Lenaz (University of Bologna, Italy), to whom I am indebted for the critical review of my work. I am also sincerely grateful to him for his valued example of attitude in science and for the generous helpful advices and stimulating ideas that he has shared with me during all these years.

References

Acín-Pérez R, Fernández-Silva P, Peleato ML, Pérez-Martos A, Enriquez JA (2008) Respiratory active mitochondrial supercomplexes. Mol Cell 32:529–539

Adam-Vizi V, Chinopoulos C (2006) Bioenergetics and the formation of mitochondrial reactive oxygen species. Trends Pharmacol Sci 27:639–645

Althoff T, Mills DJ, Popot JL, Kühlbrandt W (2011) Arrangement of electron transport chain components in bovine mitochondrial supercomplex I(1) III(2)IV(1). EMBO J 30:4652–4664

Baradaran R, Berrisford JM, Minhas GS, Sazanov LA (2013) Crystal structure of the entire respiratory complex I. Nature 494:443–448

Baum H, Silman HI, Rieske HS, Lipton SH (1967) On the composition and structural organization of complex 3 of the mitochondrial electron transfer chain. J Biol Chem 242:4876–4887

Belevich I, Verkhovsky MI (2008) Molecular mechanism of proton translocation by cytochrome *c* oxidase. Antioxid Redox Signal 10:1–29

Belevich I, Bloch DA, Belevich N, Wikström M, Verkhovsky MI (2007) Exploring the proton pump mechanism of cytochrome c oxidase in real time. Proc Natl Acad Sci U S A 104:2685–2690

Bianchi C, Genova ML, Parenti Castelli G, Lenaz G (2004) The mitochondrial respiratory chain is partially organized in a supercomplex assembly: kinetic evidence using flux control analysis. J Biol Chem 279:36562–36569

Boumans H, Grivell LA, Berden JA (1998) The respiratory chain in yeast behaves as a single functional unit. J Biol Chem 273:4872–4877

Boveris A, Cadenas E (1975) Mitochondrial production of superoxide anions and its relationship to the antimycin insensitive respiration. FEBS Lett 54:311–314

Brandt U (2013) Inside view of a giant proton pump. Angew Chem Int Ed Engl 52:7358–7360

Brandt U, Kerscher S, Drose S, Zwicker K, Zickermann V (2003) Proton pumping by NADH: ubiquinone oxidoreductase. A redox driven conformational change mechanism? FEBS Lett 545:9–17

Brzezinski P, Gennis RB (2008) Cytochrome c oxidase: exciting progress and remaining mysteries. J Bioenerg Biomembr 40:521–531

Bultema JB, Braun HP, Boekema EJ, Kouril R (2009) Megacomplex organization of the oxidative phosphorylation system by structural analysis of respiratory supercomplexes from potato. Biochim Biophys Acta 1787:60–67

Chance B, Williams GR (1955) A method for the localization of sites for oxidative phosphorylation. Nature 176:250–254

Chance B, Williams GR (1956) The respiratory chain and oxidative phosphorylation. Adv Enzymol Relat Subj Biochem 17:65–134

Covian R, Trumpower BL (2008) The dimeric structure of the cytochrome *bc*(1) complex prevents center P inhibition by reverse reactions at center N. Biochim Biophys Acta 1777:1044–1052

Cramer WA, Hasan SS, Yamashita E (2011) The Q cycle of cytochrome *bc* complexes: a structure perspective. Biochim Biophys Acta 1807:788–802

Crane FL, Hatefi Y, Lester RL, Widmer C (1957) Isolation of a quinone from beef heart mitochondria. Biochim Biophys Acta 25:220–221

Criddle RS, Bock RM, Green DE, Tisdale H (1962) Physical characteristics of proteins of the electron transfer system and interpretation of the structure of the mitochondrion. Biochemistry 1:827–842

Dalmonte ME, Forte E, Genova ML, Giuffrè A, Sarti P, Lenaz G (2009) Control of respiration by cytochrome c oxidase in intact cells: role of the membrane potential. J Biol Chem 284:32331–32335

Dröse S, Brandt U (2008) The mechanism of mitochondrial superoxide production by the cytochrome bc1 complex. J Biol Chem 283:21649–21654

Dudkina NV, Kudryashev M, Stahlberg H, Boekema EJ (2011) Interaction of complexes I, III, and IV within the bovine respirasome by single particle cryoelectron tomography. Proc Natl Acad Sci U S A 108:15196–15200

Dutton PL, Moser CC, Sled VD, Daldal F, Ohnishi T (1998) A reductant-induced oxidation mechanism for complex I. Biochim Biophys Acta 1364:245–257

Efremov RG, Sazanov LA (2011) Respiratory complex I: 'steam engine' of the cell? Curr Opin Struct Biol 21:532–540

Efremov RG, Baradaran R, Sazanov LA (2010) The architecture of respiratory complex I. Nature 465:441–445

Esterházy D, King MS, Yakovlev G, Hirst J (2008) Production of reactive oxygen species by complex I (NADH: ubiquinone oxidoreductase) from Escherichia coli and comparison to the enzyme from mitochondria. Biochemistry 47:3964–3971

Eubel H, Heinemeyer J, Sunderhaus S, Braun HP (2004) Respiratory chain supercomplexes in plant mitochondria. Plant Physiol Biochem 42:937–942

Euro L, Belevich G, Verkhovsky MI, Wikström M, Verkhovskaya M (2008) Conserved lysine residues of the membrane subunit NuoM are involved in energy conversion by the proton-pumping NADH: ubiquinone oxidoreductase (Complex I). Biochim Biophys Acta Bioenerg 1777:1166–1172

Fato R, Bergamini C, Bortolus M, Maniero AL, Leoni S, Ohnishi T, Lenaz G (2009) Differential effects of complex I inhibitors on production of reactive oxygen species. Biochim Biophys Acta 1787: 384–392

Fernandez-Moran H (1963) Subunit organization of mitochondrial membranes. Science 140:381

Galkin A, Brandt U (2005) Superoxide radical formation by pure complex I (NADH:ubiquinone oxidoreductase) from *Yarrowia lipolytica*. J Biol Chem 280:30129–30135

Genova ML, Ventura B, Giuliano G, Bovina C, Formiggini G, Parenti CG, Lenaz G (2001) The site of production of superoxide radical in mitochondrial complex I is not a bound ubisemiquinone but presumably iron-sulfur cluster N2. FEBS Lett 505:364–368

Genova ML, Baracca A, Biondi A, Casalena G, Faccioli M, Falasca AI, Formiggini G, Sgarbi G, Solaini G, Lenaz G (2008) Is supercomplex organization of the respiratory chain required for optimal electron transfer activity? Biochim Biophys Acta 1777:740–746

Green DE, Tzagoloff A (1966) The mitochondrial electron transfer chain. Arch Biochem Biophys 116:293–304

Green DE, Wharton DC (1963) Stoichiometry of the fixed oxidation-reduction components of the electron transfer chain of beef heart mitochondria. Biochem Z 338:335–348

Griffiths DE, Wharton DC (1961) Studies of the electron transport system. XXXV. Purification and properties of cytochrome oxidase. J Biol Chem 236:1850–1856

Guerrero-Castillo S, Vázquez-Acevedo M, González-Halphen D, Uribe-Carvajal S (2009) In *Yarrowia lipolytica* mitochondria, the alternative NADH dehydrogenase interacts specifically with the cytochrome complexes of the classic respiratory pathway. Biochim Biophys Acta 1787:75–85

Hackenbrock CR, Chazotte B, Gupte SS (1986) The random collision model and a critical assessment of diffusion and collision in mitochondrial electron transport. J Bioenerg Biomembr 18:331–368

Hatefi Y, Haavik AG, Jurtshuk P (1961) Studies on the electron transport system. XXX. DPNH-cytochrome *c* reductase I. Biochim Biophys Acta 52:106–118

Hatefi Y, Haavik AG, Griffiths DE (1962a) Studies on the electron transfer system. XL. Preparation and properties of mitochondrial DPNH-coenzyme Q reductase. J Biol Chem 237:1676–1680

Hatefi Y, Haavik AG, Fowler LR, Griffiths DE (1962b) Studies on the electron transfer system, XLII. Reconstitution of the electron transfer system. J Biol Chem 237:2661–2669

Heinemeyer J, Braun HP, Boekema EJ, Kouril R (2007) A structural model of the cytochrome *c* reductase/oxidase supercomplex from yeast mitochondria. J Biol Chem 282:12240–12248

Hinchliffe P, Sazanov LA (2005) Organization of iron-sulfur clusters in respiratory complex I. Science 309:71–74

Hinkle PC (2005) P/O ratios of mitochondrial oxidative phosphorylation. Biochim Biophys Acta 1706:1–11

Hinkle PC, Kim JJ, Racker E (1972) Ion transport and respiratory control in vesicles formed from cytochrome oxidase and phospholipids. J Biol Chem 247:1338–1339

Höchli MK, Hackenbrock CR (1976) Fluidity in mitochondrial membranes: thermotropic lateral translational motion of intramembrane particles. Proc Natl Acad Sci U S A 73:1636–1640

Horsefield R, Yankovskaya V, Sexton G, Whittingham W, Shiomi K, Omura S, Byrne B, Cecchini G, Iwata S (2006) Structural and computational analysis of the quinonebinding site of complex II (succinate-ubiquinone oxidoreductase): a mechanism of electron transfer and proton conduction during ubiquinone reduction. J Biol Chem 281:7309–7316

Hunte C, Zickermann V, Brandt U (2010) Functional modules and structural basis of conformational coupling in mitochondrial complex I. Science 329:448–451

Kadenbach B (2003) Intrinsic and extrinsic uncoupling of oxidative phosphorylation. Biochim Biophys Acta 1604:77–94

Krab K (1995) Kinetic and regulatory aspects of the function of the alternative oxidase in plant respiration. J Bioenerg Biomembr 27:387–396

Krause F, Reifschneider NH, Vocke D, Seelert H, Rexroth S, Dencher NA (2004a) "Respirasome"-like supercomplexes in green leaf mitochondria of spinach. J Biol Chem 279:48369–48375

Krause F, Scheckhuber CQ, Werner A, Rexroth S, Reifschneider NH, Dencher NA, Osiewacz HD (2004b) Supramolecular organization of cytochrome *c* oxidase- and alternative oxidase-dependent respiratory chains in the filamentous fungus *Podospora anserina*. J Biol Chem 279:26453–26461

Krause F, Scheckhuber CQ, Werner A, Rexroth S, Reifschneider NH, Dencher NA, Osiewacz HD (2006) OXPHOS Supercomplexes: respiration and life-span control in the aging model *Podospora anserina*. Ann NY Acad Sci 1067:106–115

Kröger A, Klingenberg M (1973) The kinetics of the redox reactions of ubiquinone related to the electron-transport activity in the respiratory chain. Eur J Biochem 34:358–368

Kussmaul L, Hirst J (2006) The mechanism of superoxide production by NADH:ubiquinone oxidoreductase (complex I) from bovine heart mitochondria. Proc Natl Acad Sci U S A 103:7607–7612

Kwong LK, Sohal RS (1998) Substrate and site specificity of hydrogen peroxide generation in mouse mitochondria. Arch Biochem Biophys 350:118–126

Lehninger AL (1965) Bionergetics: the molecular basis of biological energy transformations. WA Benjamin Inc, New York

Lehninger AL (1971) Bionergetics: the molecular basis of biological energy transformations, 2nd edn. WA Benjamin Inc, Menlo Park

Lenaz G (2001) A critical appraisal of the mitochondrial coenzyme Q pool. FEBS Lett 509:151–155

Lenaz G (2012) Mitochondria and reactive oxygen species. Which role in physiology and pathology? Adv Exp Med Biol 942:93–136

Lenaz G, Genova ML (2007) Kinetics of integrated electron transfer in the mitochondrial respiratory chain: random collisions vs. solid state electron channeling. Am J Physiol Cell Physiol 292:C1221–C1239

Lenaz G, Genova ML (2009a) Mobility and function of Coenzyme Q (ubiquinone) in the mitochondrial respiratory chain. Biochim Biophys Acta 1787:563–573

Lenaz G, Genova ML (2009b) Structural and functional organization of the mitochondrial respiratory chain: a dynamic super-assembly. Int J Biochem Cell Biol 41:1750–1772

Lenaz G, Genova ML (2010) Structure and organization of mitochondrial respiratory complexes: a new understanding of an old subject. Antioxid Redox Signal 12:961–1008

Lenaz G, Genova ML (2012) Supramolecular organisation of the mitochondrial respiratory chain: a new challenge for the mechanism and control of oxidative phosphorylation. Adv Exp Med Biol 748:107–144

Lenaz G, Daves GD Jr, Folkers K (1968) Organic structural specificity and sites of coenzyme Q in succinoxidase and DPNH-oxidase systems. Arch Biochem Biophys 123:539–550

Lenaz G, Fato R, Di Bernardo S, Jarreta D, Costa A, Genova ML, Parenti Castelli G (1999) Localization and mobility of coenzyme Q in lipid bilayers and membranes. Biofactors 9:87–93

Lenaz G, Baracca A, Fato R, Genova ML, Solaini G (2006) New insights into structure and function of mitochondria and their role in aging and disease. Antioxid Redox Signal 8:417–437

Lenaz G, Baracca A, Barbero G, Bergamini C, Dalmonte ME, Del Sole M, Faccioli M, Falasca A, Fato R, Genova ML, Sgarbi G, Solaini G (2010) Mitochondrial respiratory chain super-complex I–III in physiology and pathology. Biochim Biophys Acta 1797:633–640

Lester RL, Fleischer S (1961) Studies on the electron-transport system. 27. The respiratory activity of acetone extracted beef-heart mitochondria: role of coenzyme Q and other lipids. Biochim Biophys Acta 47:358–377

Leung KH, Hinkle PC (1975) Reconstitution of ion transport and respiratory control in vesicles formed from reduced coenzyme Q-cytochrome *c* reductase and phospholipids. J Biol Chem 250:8467–8471

Maas MF, Krause F, Dencher NA, Sainsard-Chanet A (2009) Respiratory complexes III and IV are not essential for the assembly/stability of complex I in fungi. J Mol Biol 387:259–269

Magnitsky S, Toulokhonova L, Yano T, Sled VD, Hagerhall C, Grivennikova VG, Burbaev DS, Vinogradov AD, Ohnishi T (2002) EPR characterization of ubisemiquinones and iron-sulfur cluster N2, central components of the energy coupling in the NADH-ubiquinone oxidoreductase (complex I) in situ. J Bioenerg Biomembr 34:193–208

Maranzana E, Barbero G, Falasca AI, Lenaz G, Genova ML (2013) Mitochondrial respiratory supercomplex association limits production of reactive oxygen species from complex I. Antioxid Redox Signal [Jun 28. Epub ahead of print]

Marques I, Dencher NA, Videira A, Krause F (2007) Supramolecular organization of the respiratory chain in *Neurospora crassa* mitochondria. Eukaryot Cell 6:2391–2405

Mileykovskaya E, Penczek PA, Fang J, Mallampalli VK, Sparagna GC, Dowhan W (2012) Arrangement of the respiratory chain complexes in *Saccharomyces cerevisiae* supercomplex III2IV2 revealed by single particle cryo-electron microscopy. J Biol Chem 287:23095–23103

Mitchell P (1975) Protonmotive redox mechanism of the cytochrome b-c_1 complex in the respiratory chain: protonmotive ubiquinone cycle. FEBS Lett 56:1–6

Mitchell P, Moyle J (1967) Respiration-driven proton translocation in rat liver mitochondria. Biochem J 105:1147–1162

Morton RA (1958) Ubiquinone. Nature 182(4652): 1764–1767

Moser CC, Page CC, Dutton PL (2005) Tunneling in PSII. Photochem Photobiol Sci 4:933–939

Muller F, Crofts AR, Kramer DM (2002) Multiple Q-cycle bypass reactions at the Q_o site of the cytochrome bc_1 complex. Biochemistry 41:7866–7874

Murphy MP (2009) How mitochondria produce reactive oxygen species. Biochem J 417:1–13

Muster B, Kohl W, Wittig I, Strecker V, Joos F, Haase W, Bereiter-Hahn J, Busch K (2010) Respiratory chain complexes in dynamic mitochondria display a patchy distribution in life cells. PLoS One 5:e11910

Ohnishi T, Sled VD, Yano T, Yagi T, Burbaev DS, Vinogradov AD (1998) Structure-function studies of iron-sulfur clusters and semiquinones in the NADH-Q oxidoreductase segment of the respiratory chain. Biochim Biophys Acta 1365:301–308

Ohnishi ST, Shinzawa-Itoh K, Ohta K, Yoshikawa S, Ohnishi T (2010) New insights into the superoxide generation sites in bovine heart NADH-ubiquinone oxidoreductase (complex I): the significance of protein-associated ubiquinone and the dynamic shifting of generation sites between semiflavin and semiquinone radicals. Biochim Biophys Acta 1797:1901–1909

Papa S, Capitanio G, Martino PL (2006a) Concerted involvement of cooperative proton-electron linkage and water production in the proton pump of cytochrome *c* oxidase. Biochim Biophys Acta 1757:1133–1143

Papa S, Lorusso M, Di Paola M (2006b) Operativity and flexibility of the protonmotive activity of mitochondrial respiratory chain. Biochim Biophys Acta 1757:428–436

Piccoli C, Scrima R, Boffoli D, Capitanio N (2006) Control by cytochrome *c* oxidase of the cellular

oxidative phosphorylation system depends on the mitochondrial energy state. Biochem J 396:573–583
Quinlan CL, Gerencser AA, Treberg JR, Brand MD (2011) The mechanism of superoxide production by the antimycin-inhibited mitochondrial Q-cycle. J Biol Chem 286:31361–31372
Racker E (1965) Mechanisms in bioenergetics. Academic, New York
Ragan CI, Hinkle PC (1975) Ion transport and respiratory control in vesicles formed from reduced nicotinamide adenine dinucleotide coenzyme Q reductase and phospholipids. J Biol Chem 250:8472–8476
Ricquier D (2005) Respiration uncoupling and metabolism in the control of energy expenditure. Proc Nutr Soc 64:47–52
Rosca M, Minkler P, Hoppel CL (2011) Cardiac mitochondria in heart failure: normal cardiolipin profile and increased threonine phosphorylation of complex IV. Biochim Biophys Acta 1807:1373–1382
Sazanov LA, Hinchliffe P (2006) Structure of the hydrophilic domain of respiratory complex I from *Thermus thermophilus*. Science 311:1430–1436
Schäfer E, Seelert H, Reifschneider NH, Krause F, Dencher NA, Vonck J (2006) Architecture of active mammalian respiratory chain supercomplexes. J Biol Chem 281:15370–15375
Schäfer E, Dencher NA, Vonck J, Parcej DN (2007) Three-dimensional structure of the respiratory chain supercomplex I1III2IV1 from bovine heart mitochondria. Biochemistry 44:12579–12585
Schägger H, Pfeiffer K (2001) The ratio of oxidative phosphorylation complexes I–V in bovine heart mitochondria and the composition of respiratory chain supercomplexes. J Biol Chem 276:37861–37867
Schneider H, Lemasters JJ, Hackenbrock CR (1982) Lateral diffusion of ubiquinone during electron transfer in phospholipid- and ubiquinone-enriched mitochondrial membranes. J Biol Chem 257:10789–10793
Sowers E, Hackenbrock CR (1981) Rate of lateral diffusion of intramembrane particles: measurement by electrophoretic displacement and rerandomization. Proc Natl Acad Sci U S A 78:6246–6250
Strauss M, Hofhaus G, Schröder RR, Kühlbrandt W (2008) Dimer ribbons of ATP synthase shape the inner mitochondrial membrane. EMBO J 27:1154–1160
Sunderhaus S, Klodmann J, Lenz C, Braun HP (2010) Supramolecular structure of the OXPHOS system in highly thermogenic tissue of Arum maculatum. Plant Physiol Biochem 48:265–272
Trouillard M, Meunier B, Rappaport F (2011) Questioning the functional relevance of mitochondrial supercomplexes by time-resolved analysis of the respiratory chain. Proc Natl Acad Sci U S A 108:e1027–e1034
Turrens JF, Alexandre A, Lehninger AL (1985) Ubisemiquinone is the electron donor for superoxide formation by complex III of heart mitochondria. Arch Biochem Biophys 237:408–414
van den Berg WH, Prince RC, Bashford CL, Takamiya KI, Bonner WD Jr, Dutton PL (1979) Electron and proton transport in the ubiquinone cytochrome b-c2 oxidoreductase of *Rhodopseudomonas sphaeroides*. Patterns of binding and inhibition by antimycin. J Biol Chem 254:8594–8604
Wikström M (2012) Active site intermediates in the reduction of O(2) by cytochrome oxidase, and their derivatives. Biochim Biophys Acta 1817:468–475
Wittig I, Schägger H (2009) Supramolecular organization of ATP synthase and respiratory chain in mitochondrial membranes. Biochim Biophys Acta 1787:672–680
Yagi T, Seo BB, Di Bernardo S, Nakamaru-Ogiso E, Kao MC, Matsuno-Yagi A (2001) NADH dehydrogenases: from basic science to biomedicine. Bioenerg Biomembr 33:233–242
Yankovskaya V, Horsefield R, Törnroth S, Luna-Chavez C, Miyoshi H, Léger C, Byrne B, Cecchini G, Iwata S (2003) Architecture of succinate dehydrogenase and reactive oxygen species generation. Science 299:700–704
Yu A, Yu L (1980) Resolution and reconstitution of succinate-cytochrome *c* reductase: preparations and properties of high purity succinate dehydrogenase and ubiquinolcytochrome *c* reductase. Biochim Biophys Acta 591:409–420

Chapter 22

The Hydrogenosome

Marlene Benchimol*
Laboratório de Ultraestrutura Celular, Universidade Santa Úrsula, Rio de Janeiro, Brazil

Summary

Some organisms lack conventional mitochondria and instead contain divergent mitochondrial-related organelles, called hydrogenosomes, which are double-membrane bound organelles and produce molecular hydrogen. Phylogenetic and biochemical analyses of hydrogenosomes indicated a common origin with mitochondria. Hydrogenosomes are spherical or slightly elongated organelles found in non-mitochondrial organisms such as some protists and fungi which live in anaerobic or microaerophilic environments. The most-studied hydrogenosomes are those in the human pathogen, *Trichomonas vaginalis*. Hydrogenosomes are polyphyletic and have arisen independently in several eukaryotic lineages. Like mitochondria hydrogenosomes produce ATP, participate in the metabolism of pyruvate formed during glycolysis, incorporate calcium, import proteins post-translationally and divide in the same way. However, they differ from mitochondria by the absence of genetic material, at least in trichomonas, lack a respiratory chain and cytochromes, absence of the F_0–F_1 ATPase

*Author for correspondence, e-mail: marlenebenchimol@gmail.com

M.F. Hohmann-Marriott (ed.), The Structural Basis of Biological Energy Generation, Advances in Photosynthesis and Respiration 39, pp. 419–433, DOI 10.1007/978-94-017-8742-0_22,

(see Chap. 6), absence of the tricarboxylic acid cycle, lack of oxidative phosphorylation and absence of cristae. ATP is generated by the Trichomonas hydrogenosome by substrate level phosphorylation involving acetyl CoA released by the decarboxylation of pyruvate. Hydrogenosomes are considered an excellent drug target. The sequencing of the *T. vaginalis* genome allowed bioinformatic identification of putative hydrogenosomal proteins through screening for the conserved N-terminal presequence motif. The *Trichomonas vaginalis* hydrogenosome proteome is highly reduced relative to mitochondria. Of the 569 proteins in the hydrogenosomes proteome, ~30 % are proteins with important functions such as amino acid and energy metabolism, Fe–S cluster assembly, flavin-mediated catalysis, oxygen stress response, membrane translocation, chaperonin functions, proteolytic processing and ATP hydrolysis.

I. Introduction

During the evolutionary process, eukaryotic microorganisms appeared that possess special cytoplasmic organelles related to mitochondria (Figs. 22.1, 22.2, 22.3, 22.4, 22.5, 22.6, 22.7, 22.8, 22.9 and 22.10). The hydrogenosome is a crucial organelle for some organisms that inhabit oxygen-poor environments, protists such as some free-living ciliates, unicellular parasites, termite flagellates and rumen fungi and rumen ciliates (Figs. 22.3 and 22.7). These organisms lack conventional mitochondria (Figs. 22.1 and 22.2). Hydrogenosomes are polyphylogenetic and have arisen independently in several eukaryotic lineages (Embley and Hirt 1998).

The most-studied hydrogenosomes are those in the human pathogen, *Trichomonas vaginalis* (Figs. 22.2 and 22.4b), for which the complete genome sequence is available (Carlton et al. 2007). *Tritrichomonas foetus* and *Trichomonas vaginalis* (Figs. 22.1 and 22.2) are flagellated parasitic protists that inhabit the urogenital tract of cattle and humans, respectively. They often induce reproductive failure and are a cause of significant economic losses and social and health disturbances throughout the world. In addition to their medical relevance, these organisms have also been the subject of extensive investigation due to the presence of hydrogenosomes (Figs. 22.1, 22.2 and 22.4), an energy-producing organelle that raises several questions concerning mitochondria evolution and the origin of the eukaryotes. These parasites, as well other organisms such as flagellates living in the termite gut (Fig. 22.3) and rumen, have a special mode of respiration, since they produces molecular hydrogen and ATP by oxidizing pyruvate or malate under anaerobic conditions (Müller 1993) (Fig. 22.8). Biochemical investigations revealed the functional significance of hydrogenosomes, which showed molecular hydrogen production as a metabolic end product. Consequently, they were named hydrogenosomes by Lindmark and Müller, in 1973.

Phylogenetic and biochemical analyses of hydrogenosomes indicate a common origin with mitochondria. It is proposed that in the course of the mitochondria to hydrogenosome transition, aspects of typical mitochondrial energy metabolism were lost, including the classic pyruvate dehydrogenase complex, the citric acid cycle and the elaborate membrane-associated respiratory chain. In hydrogenosomes, the oxidative decarboxylation of pyruvate is coupled to ATP synthesis and linked to ferredoxin-mediated electron transport (Fig. 22.8).

Until today, hydrogenosomes have not been found in multicellular animals or plants, or in other anaerobic protists, such as amoebas and giardias.

Abbreviations: PFOR – Pyruvate: ferredoxin oxidoreductase

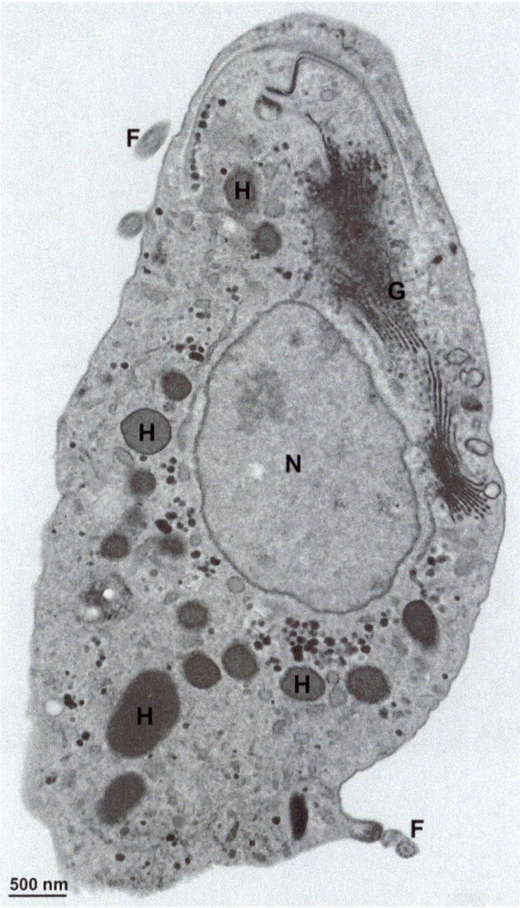

Fig. 22.1. Thin section of *Tritrichomonas foetus* as seen in transmission electron microscopy showing a pear-shaped cell. *F* flagellum, *G* Golgi complex, *H*, hydrogenosomes, *N* nucleus. Bar, 500 nm.

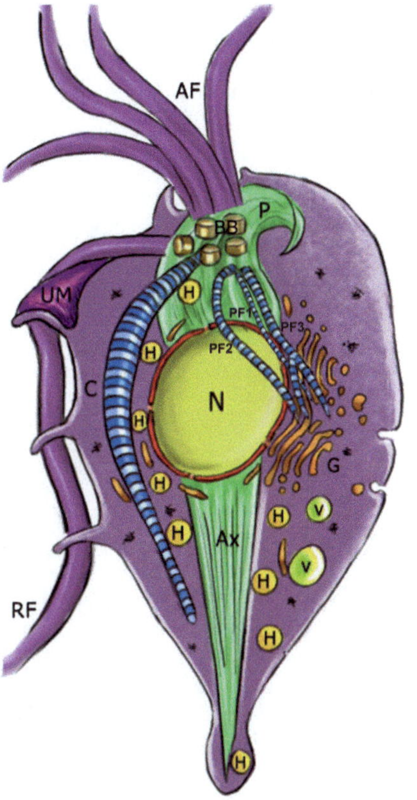

Fig. 22.2. Schematic diagram of a whole view of *Trichomonas vaginalis* showing the main cell structures. *AF* anterior flagella, *Ax* axostyle, *BB* basal body, *C* costa, *G* Golgi, *H* hydrogenosomes, *N* nucleus, *P* pelta, *PF* parabasal filament, *PF1* parabasal filament number 1, *PF2* parabasal filament number 2, *PF3* parabasal filament number 3, *RF* recurrent flagellum, *UM* undulating membrane, *V* vacuole.

II. Hydrogenosome Origin

Nowadays hydrogenosomes are considered modified mitochondria, in the evolutionary sense that they descend from one and the same eubacterial endosymbiont, but they differ from the mitochondria in several aspects.

Mitochondria and hydrogenosomes have these features in common:

- Are enveloped by two membranes (Figs. 22.4 and 22.7) (Benchimol and De Souza 1983)
- Divide autonomously by fission (Fig. 22.6) (Benchimol et al. 1996b)
- Import proteins post-translationally (Johnson et al. 1993)
- Produce ATP (Fig. 22.8) (Lindmark and Müller 1973)
- Contain a beta-succinyl-coenzyme A synthetase, a soluble hydrogenosomal protein with an amino-terminal sequence that resembles mitochondrial presequences (Lahti et al. 1992)
- Contain a membrane targeting pathway, a member of the mitochondrial carrier family (Dyall et al. 2000)
- Can divide at any phase of the cell cycle (Benchimol and Engelke 2003)

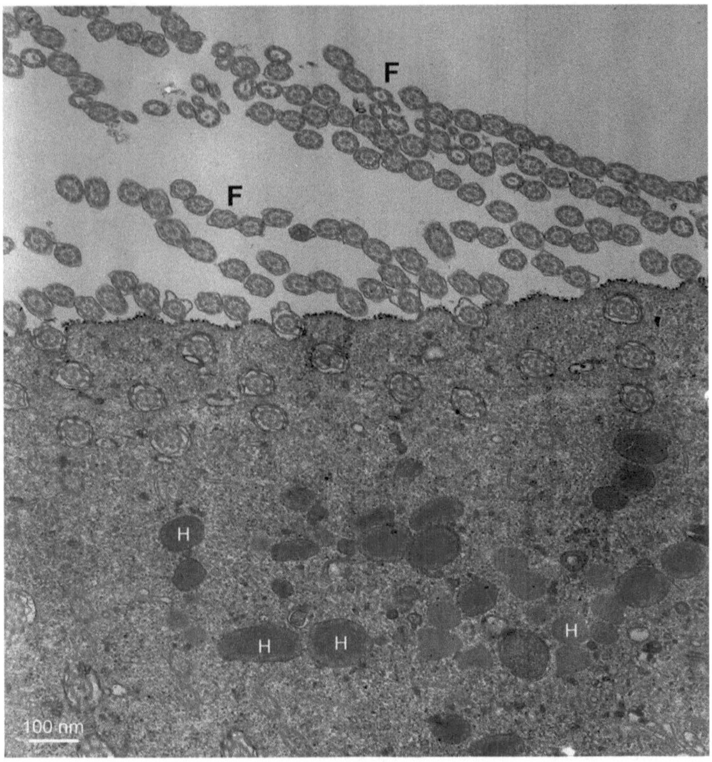

Fig. 22.3. Flagellate protist from termite gut. The hydrogenosomes (H) are seen in clusters and in large numbers; *F* flagella. Bar, 100 nm.

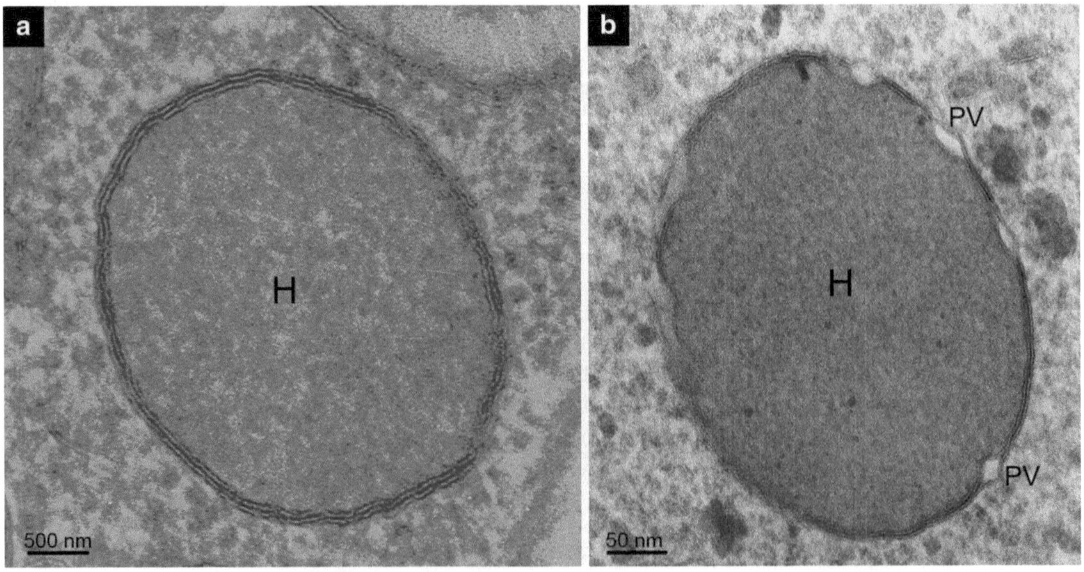

Fig. 22.4. Routine preparation of hydrogenosomes (H) from (**a**) *Tritrichomonas foetus* and (**b**) *Trichomonas vaginalis*. Note that hydrogenosomes are spherical, enveloped by a double membrane. Figure (**b**) shows a hydrogenosome with several peripheral vesicles (PV) characteristic of *T. vaginalis*. Bars, 50 nm.

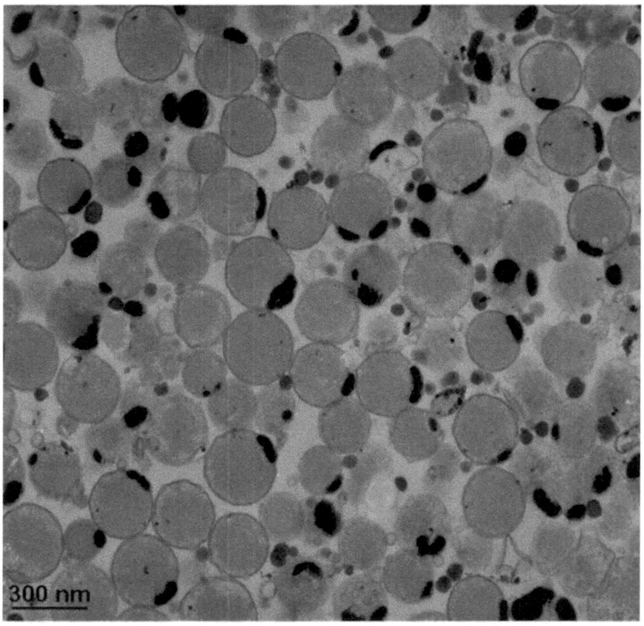

Fig. 22.5. Isolated hydrogenosomes. Hydrogenosomes isolated from the protist *T. foetus* by Percoll-sucrose density centrifugation. The hydrogenosomes are seen as spherical organelles containing a peripheral vesicle, which is a distinct sub-compartment from the hydrogenosomal matrix. Note the calcium deposit in the hydrogenosome peripheral vesicles, seen as *black areas*. Bar, 200 nm.

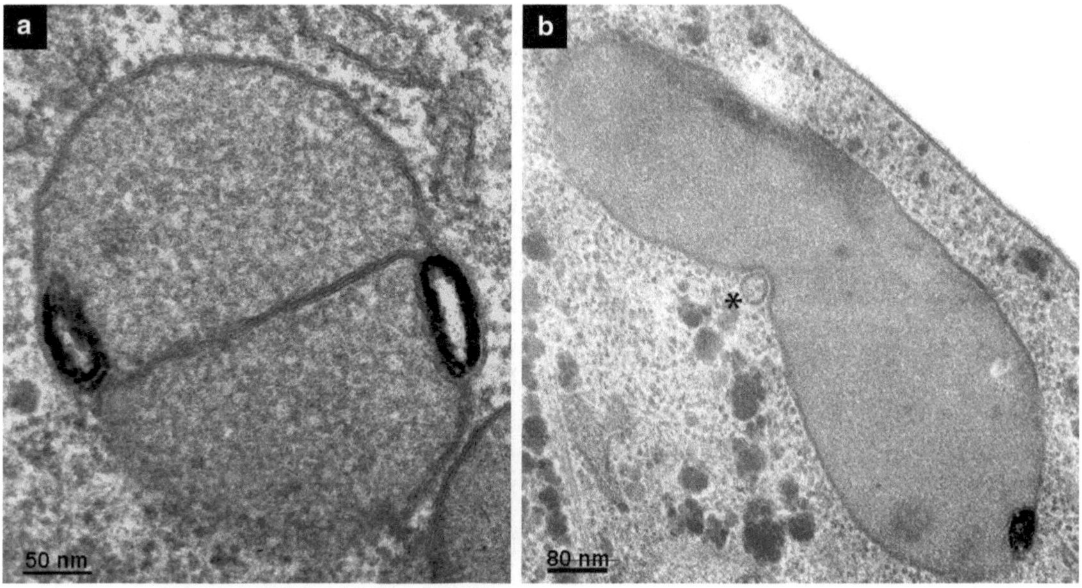

Fig. 22.6. Dividing hydrogenosomes. Views of the process of partition (**a**) and segmentation (**b**) of two dividing hydrogenosomes. (**a**) Thin section of a hydrogenosome dividing via the partition process. The hydrogenosome becomes larger and an invagination of the inner hydrogenosomal membrane is observed, gradually dividing the hydrogenosomal matrix in two compartments. Note the two peripheral vesicles with calcium deposits on opposite sides of the dividing organelle. Bar, 50 nm; (**b**) The organelle in process of segmentation, where it is elongated showing a constriction in the central region. Note that a membranous profile seems to 'strangle' the organelle (*asterisk*). Bar, 80 nm.

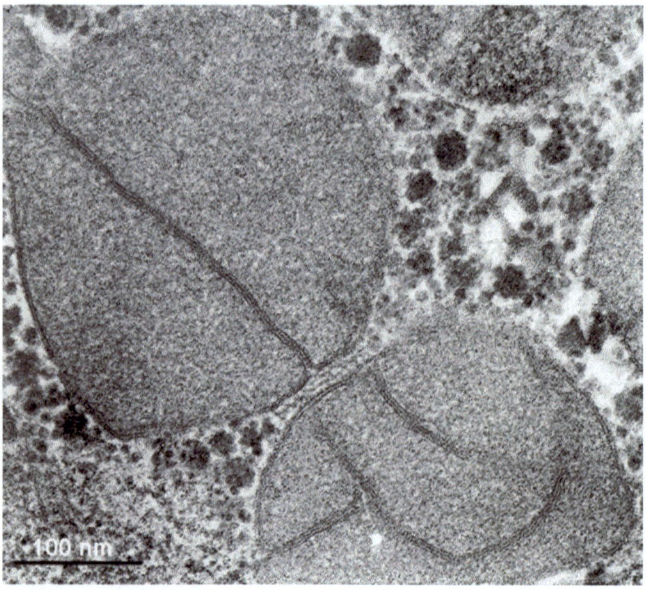

Fig. 22.7. Routine preparation of hydrogenosomes of the anaerobic fungus *Neocallimastix frontalis*. Note that all hydrogenosomes are enveloped by a double membrane and on occasion some cristae are observed in the organellar matrix. Bar, 100 nm.

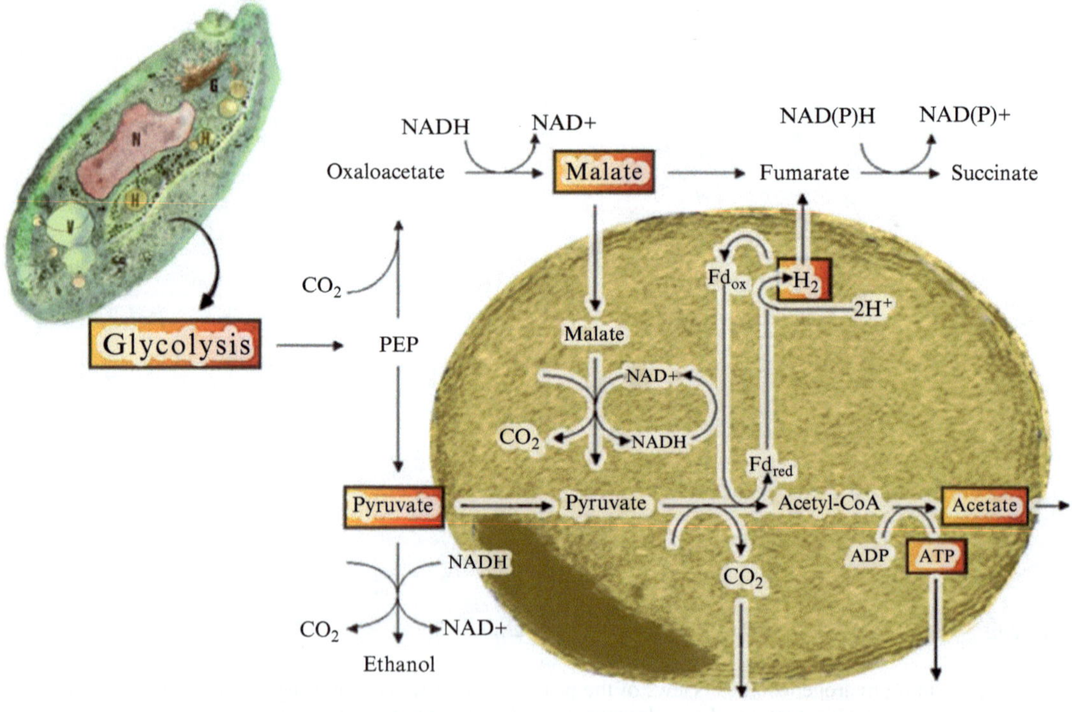

Fig. 22.8. Metabolic pyruvate metabolism of *Trichomonas hydrogenosomes*. The process starts in the cytosol, where glycolysis takes place. The glycolysis give rise to intermediate products, such as pyruvate and malate, which enter in the hydrogenosome. Pyruvate is oxidatively decarboxylated and the electrons are transferred primarily to protons resulting in H_2 formation. Oxidative pyruvate decarboxylation in *T. vaginalis*, in contrast to other eukaryotic organisms, is catalyzed by a different enzyme, pyruvate:ferredoxin oxidoreductase, an enzyme found in several bacteria and in a number of microorganisms.

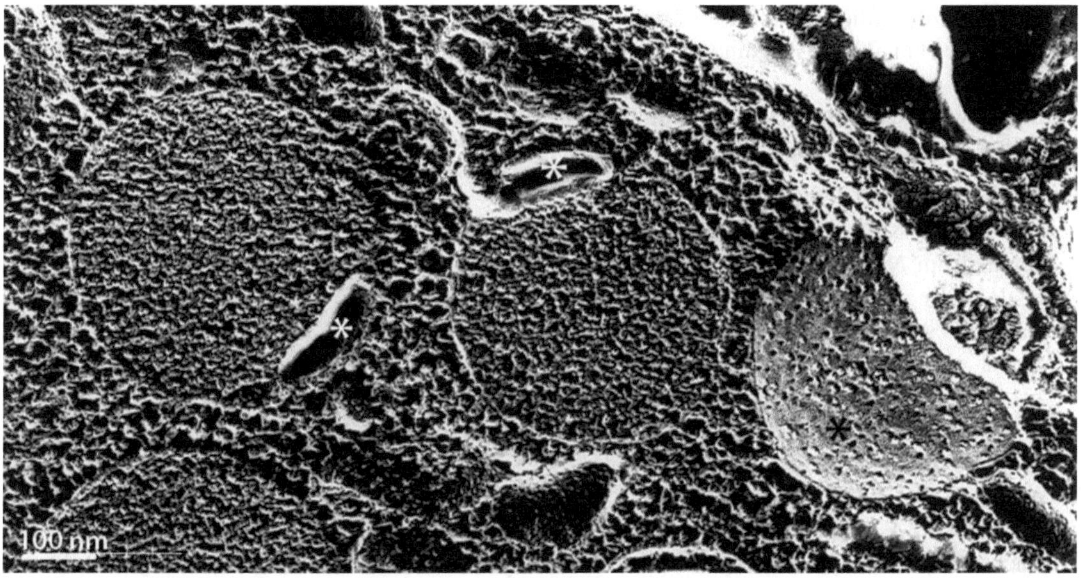

Fig. 22.9. Freeze-etching after rapid cryo-preservation by slam-freezing and rotatory shadowing exhibits different structural features of hydrogenosomes. Notice the peripheral vesicle is a distinct compartment (*white asterisks*) and the hydrogenosomes shows the matrix. A hydrogenosome (*black asterisk*) shows the organellar envelope where intramembranous particles are present. Bar, 100 nm.

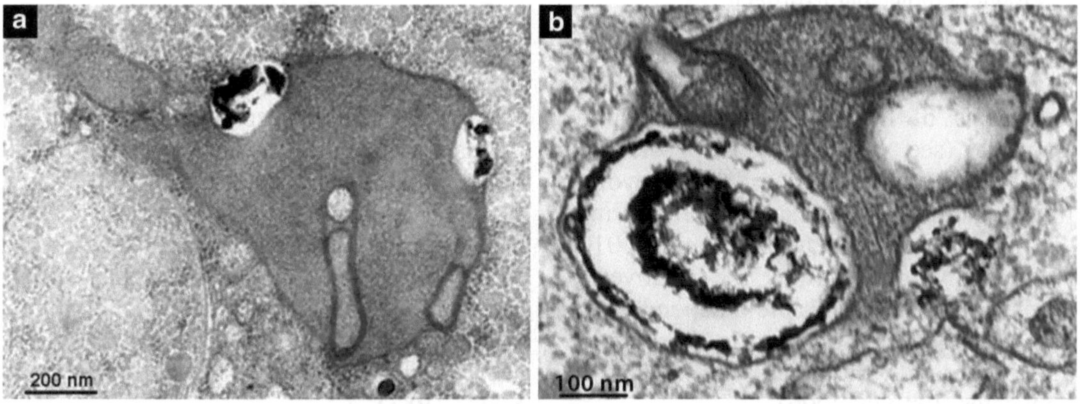

Fig. 22.10. Abnormal hydrogenosomes. They are seen when cells are submitted to stress conditions, such as incubation with drugs (*e.g.* metronidazole, colchicines or cytochalasins). The hydrogenosomes exhibit very abnormal shapes and sizes, which can reach 2 μm. Note that the organelles are not spherical as in the routine preparations and can have internal membranes and abnormally enlarged peripheral vesicles. Bars, (**a**) 200 nm; (**b**) 100 nm.

- Contain mitochondrial-type 70 kDa heat shock protein and Cpn60 (Bui et al. 1996)
- Contain the NADH dehydrogenase module of mitochondrial complex I (Hrdý et al. 2004)
- Hydrogenosomal proteins are synthesized on free polyribosomes (Lahti and Johnson 1991)
- Both are able to store calcium (Fig. 22.5) (Benchimol et al. 1982)
- Are able to utilize oxygen as a terminal electron acceptor (Cerkasov et al. 1978)
- Have a relationship with the endoplasmic reticulum
- Contain frataxin, a conserved mitochondrial protein (Dolezal et al. 2007)

- Presence of cardiolipin is still controversial (Andrade et al. 2006; Guschina et al. 2009).

The hydrogenosome differs from mitochondria in these aspects:

- Lacks a genome (Clemens and Johnson 2000), with the possible exception of hydrogenosomes from *Nyctotherus ovalis* (Akhmanova et al. 1998)
- Lacks a respiratory chain
- Lacks cytochromes and members of complexes I–IV, with the exception of NADH dehydrogenase 51 kDa (Ndh51) and 24 kDa (Ndh24) subunits (Dyall et al. 2004; Hrdý et al. 2004)
- Lacks the F_0–F_1 ATPase,
- Lacks the tricarboxylic acid cycle
- Lacks oxidative phosphorylation (Hrdý et al. 2008)
- Lacks sensitivity to metabolic inhibitors such as rotenone and cyanide (Cerkasov et al. 1978)
- Contains hydrogenase
- Produces molecular hydrogen

Mitochondrial-type proteins were found in hydrogenosomes and similarities in the biogenesis of hydrogenosomes and mitochondria (see Chap. 24), supported the hypothesis that these organelles evolved from a single α-proteobacterial endosymbiont (reviewed in Shiflett and Johnson 2010).

Thus, hydrogenosomes are relatives of mitochondria with multiple independent origins in the diverse eukaryotic lineages like the parabasalids, the ciliates, or the chytridiomycetes (Embley 2006).

III. The Hydrogenosome Morphology

In trichomonads (Figs. 22.1 and 22.2) (Benchimol et al. 1996a), fungi (Fig. 22.7) (Benchimol et al. 1997) and free-living ciliates (Finlay and Fenchel 1989) hydrogenosomes are spherical or slightly elongated organelles (Figs. 22.4, 22.5, 22.6 and 22.7). Even though the organelle morphology may vary, hydrogenosomes in trichomonads possess a peripheral vesicle (Figs. 22.4b, 22.5 and 22.9) that is a distinct compartment within the organelle (Díaz and De Souza 1997).

A. Hydrogenosome Size

Hydrogenosome sizes vary according to the species or if the cell is submitted to stress conditions (Fig. 22.10) (Benchimol 1999, 2001; Madeiro and Benchimol 2004). In routine and normal conditions, hydrogenosomes of trichomonads (Figs. 22.1, 22.2, 22.4 and 22.5) possess an average diameter of 300 nm, but they may reach 2 μm under stress conditions (Fig. 22.10), such as drug treatment. In this situation, drugs such as β-lapachone, hydroxyurea, metronidazole, cytochalasins and other reagents such as fibronectin provoked profound morphological changes in this organelle. Thus, bizarre and giant forms, as well as alterations in the organellar matrix and in the peripheral vesicle occur (Fig. 22.10).

B. The Hydrogenosome Envelope

Hydrogenosomes are enveloped by two very thin and very closely apposed unit membranes in all species studied (Figs. 22.4 and 22.7) (Benchimol and De Souza 1983; Benchimol et al. 1997; van der Giezen et al. 1997). As a general rule no space is observed between the two membranes. Each membrane has a thickness of 6 nm and may present a certain undulation (Benchimol and De Souza 1983).

Invaginations of the hydrogenosome membrane were occasionally observed in some species, delineating inner compartments (Fig. 22.7) (Benchimol et al. 1996a). Some of the compartments present the same morphology and electron density as the hydrogenosomal matrix while others had a lower density and presented tubular structures (Benchimol et al. 1996a).

Ciliate hydrogenosomes in *Metopus contortus* and *Cyclidium porcatum* present internal membranes and look like mitochondria

(Fenchel and Finlay 1995). In addition, these hydrogenosomes are also calcium stores and displayed a membrane potential, which are similar to features of mitochondria (Biagini et al. 1997).

C. The Peripheral Vesicle

In most trichomonads a special compartment is found at the periphery of hydrogenosomes (Figs. 22.4b, 22.5 and 22.9). While *T. foetus* hydrogenosomes present one or two peripheral vesicles (Fig. 22.5), *T. vaginalis* hydrogenosomes exhibit several vesicles at the organelle periphery (Fig. 22.4b). The hydrogenosome peripheral vesicle varies in size and electron density. Morphometric analysis showed that it represents 8.6 % of the whole organelle in *T. foetus*. The peripheral vesicle is completely surrounded by two closely apposed unit membranes. The inner portion of this compartment is very distinct from the hydrogenosomal matrix. In some preparations it appears empty, whereas in others preparations it appears filled with an deposits of calcium salts, for example, when the cells are fixed in a solution containing calcium (Fig. 22.5) (Benchimol and De Souza 1983; Benchimol et al. 1996a), or when the cells are submitted to freeze-fracture and deep-etching (Benchimol 2000).

D. The Matrix of the Hydrogenosome

The hydrogenosome matrix is homogeneous, with a granular appearance and differs from the cytoplasmic matrix (Figs. 22.1, 22.3, 22.4 and 22.9). Occasional calcium deposits are seen as electron-dense spots in the matrix of some hydrogenosomes. This deposition occurs when cells are incubated in the presence of calcium ions in the fixative, or processed cytochemically for the localization of Ca^{2+}, such as when the pyroantimonate technique is applied (Benchimol et al. 1982).

A dense amorphous or crystalline core, also known as nucleoid has been described in some hydrogenosomes (Honigberg and Brugerolle 1990). However, observations indicated that this core is not a usual structure, appearing only when the cells are incubated in the presence of drugs or when good fixation is not achieved (Benchimol et al. 1996a).

IV. Hydrogenosomes Biogenesis

Being devoid of a genome, all *T. vaginalis* hydrogenosomal proteins are nuclearly encoded, synthesized in the cytosol and subsequently targeted and translocated into the organelle. Many hydrogenosomal matrix proteins contain conserved N-terminal presequences that are similar to sequences known to target proteins to the mitochondrial matrix (Bradley et al. 1997). In the case of the ferredoxin protein, the presequence has been shown to be necessary for targeting and translocation of the protein into hydrogenosomes in vitro (Bradley et al. 1997).

Hydrogenosomes, as almost all other organelles, grow by proliferation of preexisting organelles (Fig. 22.6). Studies performed concerning the membrane machineries required for biogenesis of the organelle and metabolite exchange have used a combination of mass spectrometry, immunofluorescence microscopy, in vitro import assays and reverse genetics. These studies identified components of the outer membrane (TOM) and inner membrane (TIM) protein translocases, which are some of the membrane proteins of the hydrogenosome (Rada et al. 2011). Hydrogenosomes were shown to possess a pathway that allows the assembly of C-tail-anchored proteins into the outer membrane of hydrogenosomes. In addition. candidates for metabolite exchange across the outer hydrogenosomal membrane were identified, including multiple isoforms of the β-barrel proteins, Hmp35 and Hmp36. Furthermore, five homologs of the ATP/ADP carrier Hmp31, a inner membrane MCF-type metabolite carriers, were identified (Rada et al. 2011).

The targeting of matrix proteins is dependent on N-terminal cleave-able presequences (Bradley et al. 1997; Smíd et al. 2008) or internal targeting signals. The presequences are removed in the hydrogenosomal matrix by a dimeric hydrogenosomal processing peptidase that shares a common origin with the mitochondrial processing peptidase, MPP (Mentel et al. 2008).

Hydrogenosomes, like mitochondria, may divide by three distinct processes: (1) partition (Fig. 22.6a) (2) segmentation (Fig. 22.6b), and (3) heart-form. In the partition process, rounded hydrogenosomes, in a bulky form, are further separated by a membranous internal transversal septum (Fig. 22.6a) separating the organelle matrix into two compartments (Benchimol et al. 1996b). During segmentation, a necklace of intramembranous particles delimiting the outer hydrogenosomal membrane in the region of organelle division was observed by freeze-etching (Benchimol and Engelke 2003). In the segmentation process, the hydrogenosome grows, becoming elongated with the appearance of a constriction in the central portion (Benchimol et al. 1996b). Microfibrillar membranous structures (Fig. 22.6b) appear to help the furrowing process, ending with the total fission of the organelle.

The most common form of division observed in the hydrogenosomes of trichomonas is the segmentation process, whereas the partition is the most usual division process observed in the hydrogenosomes of the fungus *Neocallimastix frontalis* (Benchimol, unpublished observations). All three forms of division can be found during any phase of the cell cycle, and the organellar division is not synchronized (Benchimol and Engelke 2003). Interestingly, these three modes of division and cell phase-independent timing were also described previously in mitochondria (Tandler and Hoppel 1973).

All hydrogenosomal proteins are encoded in the nucleus and imported into hydrogenosomes. Hydrogenosomal proteins in *T. vaginalis* were shown to be synthesized on free polyribosomes, released in the cytoplasm and subsequently translated into the organelle (Lahti and Johnson 1991). In addition, mitochondrial-type proteins were found in hydrogenosomes and similarities in the biogenesis of hydrogenosomes and mitochondria, supported the hypothesis that these organelles evolved from a single α-proteobacterial endosymbiont (reviewed in Shiflett and Johnson 2010).

It has been demonstrated (Rada et al. 2011) that hydrogenosomes and mitochondria share common core membrane components required for protein import and metabolite exchange, but they also revealed remarkable differences that reflect the functional adaptation of hydrogenosomes to anaerobic conditions and the peculiar evolutionary origin.

V. Hydrogenosome Metabolism

The hydrogenosomes are considered an anaerobic form of mitochondria (see Chaps. 12 and 21), which produce hydrogen and ATP, although the majority of mitochondrial pathways and the organellar genome were lost during the mitochondrion-to-hydrogenosome transition. Under anaerobic conditions, isolated hydrogenosomes produce equimolar amounts of CO_2, acetate, molecular hydrogen and ATP from pyruvate in a process of substrate-level phosphorylation of ADP to ATP but not via oxidative phosphorylation as in mitochondria (Hrdý et al. 2008).

The trichomonad metabolism starts in the cytosol of the cell housing the hydrogenosome, where glycolysis takes place; various intermediates of the glycolysis give rise to products, such as the pyruvate, which corresponds to the classical glycolysis pathway observed in other eukaryotic cells. Malate can also be produced and like pyruvate, enters into the hydrogenosome (Fig. 22.8). Within the hydrogenosome, pyruvate is oxidatively decarboxylated to acetyl coenzyme A and CO_2. During pyruvate decarboxylation electrons are transferred primarily to protons with H_2 formation and in this reaction CO_2 is liberated and the acyl moiety is transferred to coenzyme A (CoA) to form acetyl–CoA. This latter product is converted to acetate and the energy of the thioesther bond is conserved in two successive steps, resulting in substrate level phosphorylation. Oxidative pyruvate decarboxylation in *T. vaginalis*, differently from the other eukaryotic cells, is catalyzed by pyruvate:ferredoxin oxidoreductase, an enzyme found in several bacteria and in a number of microorganisms. Electrons released from pyruvate are transferred to ferredoxin, a low molecular weight electron carrier protein. The ferredoxin represents the key electron transport component of the hydrogenosome and once reduced by pyruvate:ferredoxin oxidoreductase, it is reoxidized with protons as terminal electrons acceptors through the action of a hydrogenase and, thus producing molecular hydrogen. In addition to protons being reduced to molecular hydrogen by the action of hydrogenase, acetate, ATP and CO_2 are also liberated (Fig. 22.8). According to Müller (1990) no electron transport occurs across the hydrogenosomes membranes.

To mediate reduction of reactive oxygen species (ROS), *T. vaginalis* contains a bacterial-type thioredoxin reduction system within the hydrogenosome (Coombs et al. 2004). Enzymes responsible for iron–sulphur (Fe–S) cluster assembly and frataxin, which is a conserved mitochondrial protein involved in iron metabolism that is typically found in mitochondria, have been localised to the *T. vaginalis* hydrogenosome (Carlton et al. 2007; Dolezal et al. 2007). The organism, however, lacks peroxisomes and the glutathione reducing pathway normally present in mitochondria.

VI. Hydrogenosomes Under Drug Treatments

Hydrogenosomes are the target of the 5-nitroimidazole drugs used in trichomoniasis treatment. When trichomonads were treated with drugs, such as metronidazole, colchicines, cytochalasins, among others, important alterations are observed in hydrogenosome morphology: giant organelles, invaginations of the hydrogenosome membrane delineating inner compartments, abnormal sizes and shapes and distinct electron density in the hydrogenosomal matrix (Fig. 22.10).

Metronidazole and related 5-nitroimidazoles are the only drugs available for the treatment of human urogenital trichomoniasis caused by the protozoan parasite The hydrogenosome constitutes the main site of the initial effect of this drug. In the 1950s it was discovered that a strain of *Streptomyces* was able to produce a nitroimidazole derivative named metronidazole, exhibiting an anti-trichomonal activity. It is well-known that oxidative decarboxylation of pyruvate, which is coupled to ATP synthesis and which is also linked to ferredoxin-mediated electron transport, occurs in the hydrogenosome. This pathway is responsible for metabolic activation of 5-nitroimidazole drugs. Metronidazole is administered in an inactive form and is reduced to its cytotoxic radical anion within the hydrogenosome. Electron transport components in the organelle, pyruvate:ferredoxin oxidoreductase and ferredoxin, donate a single electron to the drug, converting it to a cytotoxic free radical. Metronidazole inhibits growth and kills trichomonads. Electrons required for metronidazole activation are released from pyruvate by the activity of pyruvate:ferredoxin oxidoreductase and transferred to the drug by a low-redox-potential carrier, the ferredoxin. A novel pathway involved in the drug activation within the hydrogenosome was described (Hrdý et al. 2005). These authors showed that trichomonads acquire high-level metronidazole resistance only after both pyruvate- and malate-dependent pathways of metronidazole activation are eliminated from the hydrogenosomes. The source of electrons is malate, another major hydrogenosomal substrate, which is oxidatively decarboxylated to pyruvate and CO_2 by the NAD-dependent malic enzyme. The electrons released during this reaction are transferred from NADH to ferredoxin by NADH dehydrogenase, which is homologous to the catalytic module of mitochondrial complex I. However, in contrast to the mitochondrial complex I module that receives electrons from NADH, the hydrogenosome complex donates electrons to ferredoxin.

A. Other Drugs Interfering in Hydrogenosome Metabolism

Putrescine analogue such as the 1,4-diamino-2-butanone (DAB) (Reis et al. 1999), β-lapachone, a natural naphthoquinone present in the bark of the South American Lapacho tree causes an intense effect on culture growth and in the hydrogenosomes in trichomonads (Benchimol, unpublished). Morphological alterations were found in hydrogenosomes when trichomonads were treated with other substances, such as fibronectin (Benchimol 2001), hydroxyurea (Ribeiro et al. 2002), Concanavalin A, EDTA (Granger et al. 2000), hydrogen peroxide (Mariante et al. 2003); cytochalasin B (Madeiro and Benchimol 2004). The hydrogenosomes modified their morphology presenting a high diversity in size and shape (see Fig. 22.10), suggesting a mechanism that seems to compensate for the stress provoked by the drugs.

B. Effect of Zinc on Hydrogenosomes

The hydrogenosome constitutes the main site of the initial effect of zinc, where the hydrogenosomal vesicle increases in electron density and size. Electron spectroscopy imaging and the electron energy loss spectrum showed the presence of zinc, calcium and oxygen in the electron-dense areas of the hydrogenosome (Benchimol et al. 1993). Zinc has been described in prostatic fluid and it has been demonstrated that the normal concentration found in men is lethal for most isolates of *T. vaginalis*.

VII. Iron and Hydrogenosomes

A. Iron-Deprivation and Hydrogenosomal Proteins

Iron is essential for the hydrogenosome function and is involved in the expression of hydrogenosomal proteins, thus indicating the presence of iron-dependent control of gene transcription in *T. foetus* (Vanácová et al. 2001). The effect of iron availability on hydrogenosomal behavior, metabolic activity, and expression of hydrogenosomal enzymes has been examined.

Protein analysis of purified hydrogenosomes from iron-restricted cells showed decreased levels of proteins corresponding to PFOR, malic enzyme, hydrogenase and ferredoxin. The impaired function of hydrogenosomes under iron-restricted conditions was compensated by the activation of the cytosolic pathway, mediating conversion of pyruvate to ethanol via acetaldehyde; this metabolic switch was fully reversible. Thus, when trichomonads are iron-deprived the cells decrease growth and after some days, the culture exhibits several characteristics of cell death (Benchimol, unpublished).

VIII. The Proteome of Hydrogenosomes

Hydrogenosome proteomics is very important to increase our understanding of hydrogenosome functions and evolution. Ideally, the complete genome sequence of an organism must be known, techniques for cell fractionation and hydrogenosome purification must be established, and axenic culture methods must be available, to facilitate a powerful proteomic analyses by 2D electrophoresis and mass spectrometry. One of these hydrogenosomes-harboring organisms that fulfills these prerequisites is *T. vaginalis*, which has a published genome sequence (Carlton et al. 2007).

The proteome of purified hydrogenosomes was analyzed using multiple fractionation techniques by Schneider et al. (2011) and 569 proteins were identified in the *T. vaginalis* hydrogenosome uncovering new members of known hydrogenosomal pathways and revealing new metabolic pathways present in this unique organelle. Interestingly, the Schneider group showed that the number of identified proteins is substantially lower than the 1,000–1,500 proteins reported for fungal and animal mitochondrial proteomes. Of the 569 proteins in the hydrogenosomes proteome, many appeared to be associated with the external surface of hydrogenosomes, including large numbers of GTPases and ribosomal proteins. The same authors also demonstrated that ~30 % of the hydrogenosome proteome are formed of proteins with important functions such as amino acid and energy metabolism, Fe–S cluster assembly, flavin-mediated catalysis, oxygen stress response, membrane translocation, chaperonin functions, proteolytic processing and ATP hydrolysis.

Analyses of the hydrogenosome proteome revealed pathways common to and distinct from mitochondria, as well as glycolytic proteins were also found to be associated with the hydrogenosome proteome (Schneider et al. 2011). However, approximately 18 % of the hydrogenosomal proteome was found to be composed of hypothetical proteins of unknown function. In addition, homologues of several proteins known to be involved in protein import and the biogenesis of yeast mitochondria are present in the *T. vaginalis* hydrogenosome. These include mitochondrial-like chaperones Hsp70, Hsp60 and Hsp10, a processing peptidase and putative members of the translocation machinery (Pam18 and Tim17/22/23 orthologs) (Carlton et al. 2007; Shiflett and Johnson 2010).

Acknowledgements

This work was supported by CNPq (Conselho Nacional de Desenvolvimento Científico e Tecnológico), PRONEX (Programa de Núcleo de Excelência), FAPERJ (Fundação Carlos Chagas Filho de Amparo à Pesquisa do Estado do Rio de Janeiro), and AUSU (Associação Universitária Santa Úrsula).

References

Akhmanova A, Voncken F, van Alen T, van Hoek A, Boxma B, Vogels G, Veenhuis M, Hackstein JH (1998) A hydrogenosome with a genome. Nature 396:527–528

Andrade RI, Einicker-Lamas M, Bernardo RR, Previatto LM, Mohana-Borges R, Morgado-Diaz J, Benchimol M (2006) Cardiolipin in hydrogenosomes: evidence of symbiotic origin. Eukaryot Cell 5:784–787

Benchimol M (1999) Hydrogenosome autophagy in *Tritrichomonas foetus*: an ultrastructural and cytochemical study. Biol Cell 91:165–174

Benchimol M (2000) Ultrastructural characterization of the isolated hydrogenosome in *Tritrichomonas foetus*. Tiss Cell 32:1–9

Benchimol M (2001) Hydrogenosome morphological variation induced by fibronectin and other drugs in *Tritrichomonas foetus* and *Trichomonas vaginalis*. Parasitol Res 87:215–222

Benchimol M, De Souza W (1983) Fine structure and cytochemistry of the hydrogenosome of *Tritrichomonas foetus*. J Protozool 30:422–425

Benchimol M, Engelke F (2003) Hydrogenosome behavior during the cell cycle in *Tritrichomonas foetus*. Biol Cell 95:283–293

Benchimol M, Elias CA, De Souza W (1982) Ultrastructural localization of calcium in the plasma membrane and in the hydrogenosome of *Tritrichomonas foetus*. Exp Parasitol 54:277–284

Benchimol M, Almeida JC, Lins U, Gonçalves NR, de Souza W (1993) Electron microscopic study of the effect of zinc on *Tritrichomonas foetus*. Antimicrob Agents Chemother 37:2722–2726

Benchimol M, Almeida JCA, De Souza W (1996a) Further studies on the organization of the hydrogenosome in *Tritrichomonas foetus*. Tiss Cell 28:287–299

Benchimol M, Johnson PJ, De Souza W (1996b) Morphogenesis of the hydrogenosome: an ultrastructural study. Biol Cell 87:197–205

Benchimol M, Durand R, Almeida J (1997) A double membrane surrounds the hydrogenosomes of the anaerobic fungus Neocallimastix frontalis. FEMS Microbiol 154:277–282

Biagini GA, Hayes AJ, Suller MTE, Winters C, Finlay BJ, Lloyd D (1997) Hydrogenosomes of *Metopus contortus* physiologically resemble mitochondria. Microbiology 143:1623–1629

Bradley PJ, Lahti CJ, Plumper E, Johnson PJ (1997) Targeting and translocation of proteins into the hydrogenosome of the protist *Trichomonas*: similarities with mitochondrial protein import. EMBO J 16:3484–3493

Bui ET, Bradley PJ, Johnson PJ (1996) A common evolutionary origin for mitochondria and hydrogenosomes. Proc Natl Acad Sci U S A 93:9651–9656

Carlton JM, Hirt RP, Silva JC, Delcher AL, Schatz M, Zhao Q, Wortman JR, Bidwell SL, Alsmark UC, Besteiro S, Sicheritz-Ponten T, Noel CJ, Dacks JB, Foster PG, Simillion C, Van de Peer Y, Miranda-Saavedra D, Barton GJ, Westrop GD, Müller S, Dessi D, Fiori PL, Ren Q, Paulsen I, Zhang H, Bastida-Corcuera FD, Simoes-Barbosa A, Brown MT, Hayes RD, Mukherjee M, Okumura CY, Schneider

R, Smith AJ, Vanacova S, Villalvazo M, Haas BJ, Pertea M, Feldblyum TV, Utterback TR, Shu CL, Osoegawa K, de Jong PJ, Hrdy I, Horvathova L, Zubacova Z, Dolezal P, Malik SB, Logsdon JM Jr, Henze K, Gupta A, Wang CC, Dunne RL, Upcroft JA, Upcroft P, White O, Salzberg SL, Tang P, Chiu CH, Lee YS, Embley TM, Coombs GH, Mottram JC, Tachezy J, Fraser-Liggett CM, Johnson PJ (2007) Draft genome sequence of the sexually transmitted pathogen *Trichomonas vaginalis*. Science 315:207–212

Cerkasov J, Cerkasovová A, Kulda J, Vilhelmová D (1978) Respiration of hydrogenosomes of *Tritrichomonas foetus*. I. ADP-dependent oxidation of malate and pyruvate. J Biol Chem 253:1207–1214

Clemens DL, Johnson PJ (2000) Failure to detect DNA in hydrogenosomes of *Trichomonas vaginalis* by nick translation and immunomicroscopy. Mol Biochem Parasitol 106:307–313

Coombs GH, Westrop GD, Suchan P, Puzova G, Hirt RP, Embley TM, Mottram JC, Müller S (2004) The amitochondriate eukaryote *Trichomonas vaginalis* contains a divergent thioredoxin-linked peroxiredoxin antioxidant system. J Biol Chem 279:5249–5256

Díaz JAM, De Souza W (1997) Purification and biochemical characterization of the hydrogenosomes of the flagellate protist *Tritrichomonas foetus*. Eur J Cell Biol 74:85–91

Dolezal P, Dancis A, Lesuisse E, Sutak R, Hrdý I, Embley TM, Tachezy J (2007) Frataxina conserved mitochondrial protein, in the hydrogenosome of Trichomonas vaginalis. Eukaryot Cell 6:1431–1438

Dyall SD, Koehler CM, Delgadillo-Correa MG, Bradley PJ, Plümper E, Leuernberger D, Turck CW, Johnson PJ (2000) Presence of a member of the mitochondrial carrier family in hydrogenosomes: conservation of membrane targeting pathways between hydrogenosomes and mitochondria. Mol Cell Biol 20:2488–2497

Dyall SD, Yan W, Delgadillo-Correa MG, Lunceford A, Loo JA, Clarke CF, Johnson PJ (2004) Non-mitochondrial complex I proteins in a hydrogenosomal oxidoreductase complex. Nature 431:1103–1107

Embley TM (2006) Multiple secondary origins of the anaerobic lifestyle in eukaryotes. Philos Trans R Soc Lond B Biol Sci 361:1055–1067

Embley TM, Hirt RP (1998) Early branching eukaryotes? Curr Opin Genet Dev 8:624–629

Fenchel T, Finlay BJ (eds) (1995) Ecology and Evolution in Anoxic Worlds. Oxford University Press, Oxford

Finlay BJ, Fenchel T (1989) Hydrogenosomes in some anaerobic protozoa resemble mitochondria. FEMS Microbiol Lett 65:311–314

Granger BL, Warwood SJ, Benchimol M, De Souza W (2000) Transient invagination of flagella by *Tritrichomonas foetus*. Parasitol Res 86:699–709

Guschina AI, Harris KM, Maskrey B, Goldelberg B, Lloyd D, Harwood JL (2009) The microaerophilic flagellate, *Trichomonas vaginalis*, contains unusual acyl lipids but no detectable cardiolipin. J Eukaryot Microbiol 56:52–57

Honigberg MB, Brugerolle G (1990) Structure. In: Honigberg BM (ed) Trichomonads Parasitic in Humans. Springer, New York, pp 5–35

Hrdý I, Hirt RP, Dolezal P, Bardonova L, Foster PG, Tachezy J, Embley TM (2004) *Trichomonas hydrogenosomes* contain the NADH dehydrogenase module of mitochondrial complex I. Nature 432:618–622

Hrdý I, Cammack R, Stopka P, Kulda J, Tachezy J (2005) Alternative pathway of metronidazole activation in *Trichomonas vaginalis* hydrogenosomes. Antimicrob Agents Chemother 49:5033–5036

Hrdý I, Tachezy J, Müller M (2008) Metabolism of trichomonad hydrogenosomes. In: Tachezy J (ed) Hydrogenosomes and mitosomes: mitochondria of anaerobic eukaryotes. Springer, Berlin, pp 113–145

Johnson PJ, Lahti CJ, Bradley PJ (1993) Biogenesis of the hydrogenosome: an unusual organelle in the anaerobic protist *Trichomonas vaginalis*. J Parasitol 79:664–670

Lahti CJ, Johnson PJ (1991) *Trichomonas vaginalis* hydrogenosomal proteins are synthesized on free polyribosomes and may undergo processing upon maturation. Mol Biochem Parasitol 46:307–310

Lahti CJ, d'Oliveira CE, Johnson PJ (1992) Beta-succinyl-coenzyme A synthetase from *Trichomonas vaginalis* is a soluble hydrogenosomal protein with an amino-terminal sequence that resembles mitochondrial presequences. J Bacteriol 174: 6822–6830

Lindmark DG, Müller M (1973) Hydrogenosome, a cytoplasmic organelle of the anaerobic flagellate, *Tritrichomonas foetus*, and its role in pyruvate metabolism. J Biol Chem 248:7724–7728

Madeiro RF, Benchimol M (2004) The effect of drugs in cell structure of *Tritrichimonas foetus*. Parasitol Res 92:159–170

Mariante RM, Guimarães CA, Linden R, Benchimol M (2003) Hydrogen peroxide induces caspase activation and programmed cell death in the amitochondrial *Tritrichomonas foetus*. Histochem Cell Biol 120:129–141

Mentel M, Zimorski V, Haferkamp P, Martin W, Henze K (2008) Protein import into hydrogenosomes of *Trichomonas vaginalis* involves both N-terminal and internal targeting signals: a case study of thioredoxin reductases. Eukaryot Cell 7:1750–1757

Müller M (1990) Biochemistry. In: Honigberg BM (ed) Trichomonads parasitic in humans. Springer, New York, pp 36–83

Müller M (1993) The hydrogenosome. J Gen Microbiol 139:2879–2889

Rada P, Dolez¡al P, Jedelsky PL, Bursac D, Perry AJ, Sedinova M, Smískova K, Novotny M, Beltrán NC, Hrdý I, Lithgow T, Tachezy J (2011) The core components of organelle biogenesis and membrane transport in the hydrogenosomes of *Trichomonas vaginalis*. PLoS ONE 6(9):e24428. doi:10.1371

Reis IA, Martinez MP, Yarlett N, Johnson PJ, Silva-Filho FC, Vannier-Santos MA (1999) Inhibition of polyamine synthesis arrests trichomonad growth and induces destruction of hydrogenosomes. Antimicrob Agents Chemother 43:1919–1923

Ribeiro KC, Vetö Arnholdt AC, Benchimol M (2002) *Tritrichomonas foetus*: induced division synchrony by hydroxyurea. Parasitol Res 88:627–631

Schneider RE, Brown MT, Shiflett AM, Dyall SD, Hayes RD, Xie Y, Loo JA, Johnson PJ (2011) The *Trichomonas vaginalis* hydrogenosome proteome is highly reduced relative to mitochondria, yet complex compared with mitosomes. Int J Parasitol 41:1421–1434

Shiflett AM, Johnson PJ (2010) Mitochondrion-related organelles in eukaryotic protists. Annu Rev Microbiol 64:409–429

Smíd O, Matusková A, Harris SR, Kucera T, Novotný M, Horváthová L, Hrdý I, Kutejová E, Hirt RP, Embley TM, Janata J, Tachezy J (2008) Reductive evolution of the mitochondrial processing peptidases of the unicellular parasites *Trichomonas vaginalis* and *Giardia intestinalis*. PLoS Pathog 4:e1000243

Tandler B, Hoppel L (1973) Division of giant mitochondria during recovery from cuprizone intoxication. J Cell Biol 56:266–272

van der Giezen M, Sjollema KA, Artz RR, Alkema W, Prins RA (1997) Hydrogenosomes in the anaerobic fungus *Neocallimastix frontalis* have a double membrane but lack an associated organelle genome. FEBS Lett 408:147–150

Vanácová S, Rasoloson D, Rázga J, Hrdý I, Kulda J, Tachezy J (2001) Iron-induced changes in pyruvate metabolism of *Tritrichomonas foetus* and involvement of iron in expression of hydrogenosomal proteins. Microbiology 147:53–62

Chapter 23

Biogenesis of Chloroplasts

Simon Geir Møller*
Department of Biological Sciences, St. John's University, New York, NY 11439, USA

Jodi Maple** and Daniela Gargano**
Faculty of Science and Technology, Centre for Organelle Research, University of Stavanger, N-4036 Stavanger, Norway

Summary

Chloroplasts belong to a diverse family of plant organelles called plastids that perform essential functions, including important steps in many biosynthetic pathways. All plastids differentiate from proplastids through a complex process, in which numerous events must be coordinated and integrated into the overall developmental pathway of the cell. Due to the overwhelming importance of chloroplasts as sites of oxygenic photosynthesis the differentiation of chloroplasts from proplastids has been most studied. Chloroplast biogenesis begins with the perception of light, which triggers the coordinated expression of genetic information contained in both the nuclear and plastid genomes. Subsequently the chloroplast protein import machinery plays a major role in organelle biogenesis, mediating the import of nuclear-encoded proteins into the organelle. This process is challenged by the complex organization of the chloroplast sub-compartments. The conversion of sunlight into chemical energy by the

*Author for correspondence, e-mail: mollers@stjohns.edu

**Author contributed equally with all other contributors.

M.F. Hohmann-Marriott (ed.), The Structural Basis of Biological Energy Generation, Advances in Photosynthesis and Respiration 39, pp. 435–449, DOI 10.1007/978-94-017-8742-0_23,

photosynthetic machinery requires thylakoid membranes, a specialized membrane system found in chloroplasts, and this process involving a complex cascade of biochemical and structural events. Here we will address the major molecular events following the initiation of chloroplast biogenesis, culminating in the formation of the mature chloroplast and the segregation of plastids to daughter cells during cell division.

I. Introduction

Chloroplasts are remnants of a cyanobacteria-like ancestor that was engulfed by a eukaryotic host cell approximately 1.2 billion years ago (McFadden 2001). During evolution chloroplasts developed specialized functions in different tissues and cell types, giving rise to a diverse group of other non-photosynthetic, interconvertible plastids (Fig. 23.1). Chloroplasts are the most prominent form of plastids, essential not only for photosynthesis, but also for nitrogen fixation, the synthesis of amino acids, fatty acids, purine and pyrimidine bases, isoprenoids and tetrapyrroles and for the synthesis of the lipid components for their own membrane structures (Tetlow et al. 2005).

Plastids are fully integrated into the life cycle of photosynthetic eukaryotes and have retained a semi-autonomous character, a minimal genetic machinery and genes for a small number of proteins. As a result of horizontal gene transfer, the coding capacity of the present-day plastid is reduced to only about 130 genes and the majority of plastid proteins are encoded in the nucleus (Sugita and Sugiura 1996). The transfer of genes from the endosymbiont to the nucleus of the host cell provided an opportunity for increased control of the plastid by the host cell, both at the level of cell-type specific expression of plastid genes directed by the nucleus and the control of the expression and import of nuclear encoded proteins into the plastids.

All plastids, including chloroplasts, develop from proplastids, small (0.5–1 μm in diameter) undifferentiated organelles present in meristematic cells of roots and shoots (Cran and Possingham 1972). As plant cells develop, plastids can undergo different patterns of differentiation depending on the committed developmental pathway within the cell. However, the differentiation pathways are dynamic and in response to cellular and environmental signals, plastid differentiation can be reversible where mature plastid types can redifferentiate from one vplastid type to another (Fig. 23.1). The majority of plastid research has focused on the chloroplast which will be the focus of this chapter.

Upon germination a seedling must establish an independent energy source before those stores in the seed are depleted. Attainment of this is dependent on the formation of photosynthetically active chloroplasts. Whilst proplastids are simple organelles, mature chloroplasts are large (5–10 μm in diameter), structurally complex organelles, which house the functional components required for both the light-driven and light-independent enzymatic reactions of photosynthesis. All plastids are bound by a double membrane called the envelope and chloroplasts have in addition a third internal membrane system, the thylakoid membrane, where the light-dependent reactions of photosynthesis take place. Consequently, the molecular mechanisms involved in the development of a proplastid to a mature chloroplast consist of a series of complex events

Abbreviations: ARC – Accumulation and replication of chloroplast; ATP – Adenosine-5'-triphosphate; Cry – Cryptochromes; GTP – Guanosine-5'-triphosphate; LHCII – Light harvesting complex II; NADPH – Nicotinamide adenine dinucleotide phosphate; NEP – Nuclear encoded polymerase; PDV – Plastid division proteins; PEP – Plastid encoded polymerase; PIFs – phytochrome-interacting factor; PSI – Photosystem I; PSII – Photosystem II; TIC – Translocon at the inner envelope membrane of chloroplasts; TOC – Translocon at the outer envelope membrane of chloroplasts; VIPP – Vesicle-inducing protein in plastids

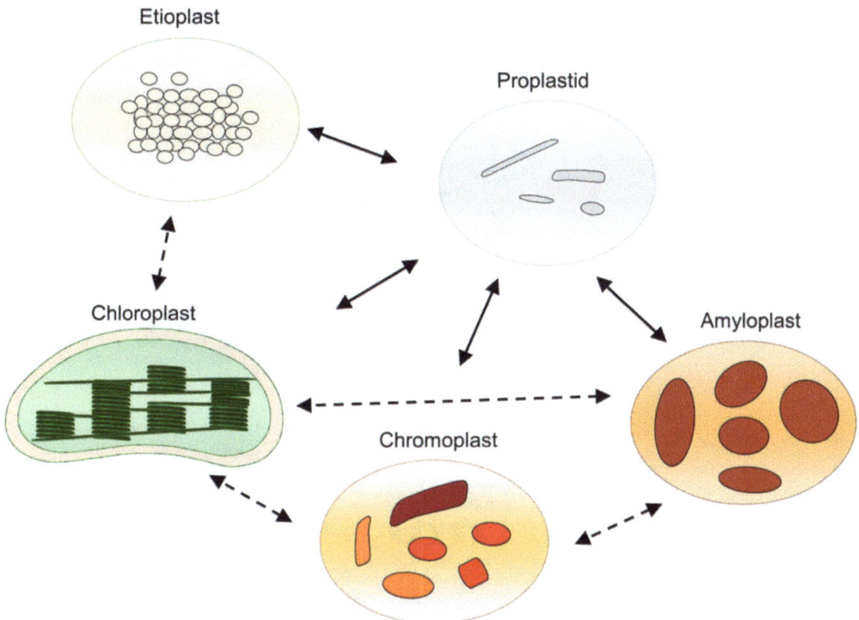

Fig. 23.1. The main forms of plastids in higher plants. All plastids develop from proplastids, undifferentiated organelles present in meristematic cells of roots and shoots. Plastids can undergo different patterns of differentiation depending on the developmental stage of the plant and environmental conditions. Plastids can store chlorophyll (chloroplast), carotenoids (chromoplasts) and starch (amyloplasts). *Solid lines* indicate conversions that occur under normal conditions; *dashed lines* specify conversions that are possible under a limited set of conditions or that are unusual.

and involve extensive structural and functional changes of the organelle.

Here we discuss progress in understanding the molecular mechanisms behind chloroplast biogenesis, including the initiation of photomorphogenesis, gene-expression, protein import and assembly of the thylakoid membrane system (Fig. 23.2). As well as considering molecular aspects of plastid differentiation, we will also address the replication of chloroplasts and fine tuning of the differentiation process.

II. Proplastid to Chloroplast Differentiation

The fundamental control point activating the proplastid-to-chloroplast differentiation programme is the perception of light. When light intensity is below a certain threshold level hypocotyl elongation is promoted, cotyledon opening is inhibited and the development of proplastids in leaves is arrested at an intermediate stage resulting in the formation of etioplasts (Von Arnim and Deng 1996) (Fig. 23.1). Once light has triggered the transition from dark growth (skotomorphogenesis) to light growth (photomorphogenesis) several molecular events are initiated including gene expression, protein import, and thylakoid membranes assembly (Fig. 23.2) ultimately culminating in the formation of mature chloroplasts.

A. Light Perception and the Initiation of Photomorphogenesis

Perception of light is mediated by a suite of photoreceptors, which detect specific wavelengths. Of primary importance in chloroplast biogenesis are the phytochromes (primarily PhyA and PhyB), which perceive red and far-red light, and the cryptochromes

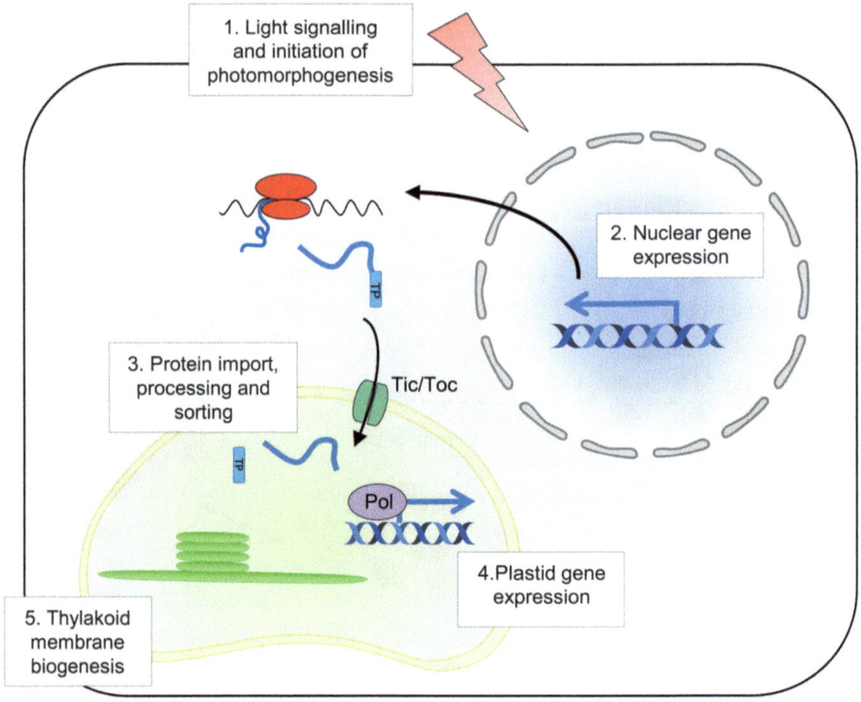

Fig. 23.2. Early events in chloroplast biogenesis. (1) The differentiation of proplastids to chloroplasts is triggered by the perception of light. (2) This in turn leads to the coordinated expression of nuclear genes. (3) Many nucleus-encoded proteins are translated in the cytoplasm and targeted for import into chloroplasts by a transit peptide (TP), which directs protein translocation through the Toc and Tic complex. After import many proteins are processed to remove the TP and proteins not destined for the stroma are directed to their proper location. (4) Plastid gene expression is regulated throughout differentiation by the activities of two classes of polymerase (pol); the nuclear encoded polymerase and plastid encoded polymerase. (5) The thylakoid network is assembled to generate a fully photosynthetically active chloroplast.

(Cry1 and Cry2), which respond to blue and UVA light. Extensive research efforts have identified a complex and interconnecting signalling network downstream of these photoreceptors (Reviewed in Chory 2010) and key events in triggering photomorphogenesis will be summarized only briefly.

Downstream of the photoreceptors are two main light signalling branches, the COP1-HY5 pathway and the PIF pathway (Fig. 23.3). COP1, along with the COP10, DET1, DDB1 (CDD) complex, is a central repressor of photomorphogenesis, repressing light signalling in darkness (Osterlund et al. 2000; Schroeder et al. 2002; Yanagawa et al. 2004). COP1 is an E3 ubiquitin ligase which tags light response activators, such as the bZIP transcription factor HY5, for targeting to the 26S proteasome for subsequent degradation (Osterlund et al. 2000). In darkness, inactive Cry dimers are localized to the nucleus where they bind COP1. Blue light leads to the photoactivation of CRY1 and triggers CRY1 relocation to the cytoplasm. Consequently, in the light COP1 is mostly absent from the nucleus and HY5 is no longer ubiquitinated and degraded, ultimately leading to an increase in the expression of photomorphogenesis genes (Saijo et al. 2003) (Fig. 23.3).

In parallel to this, phytochrome signalling acts through a family of basic helix–loop–helix (bHLH) transcription factors called PIFs (for phytochrome-interacting factor). Several PIFs have been identified in *Arabidopsis* and function mainly as repressors of photomorphogenesis (Castillon et al. 2007) (Fig. 23.3). For example, PIF3

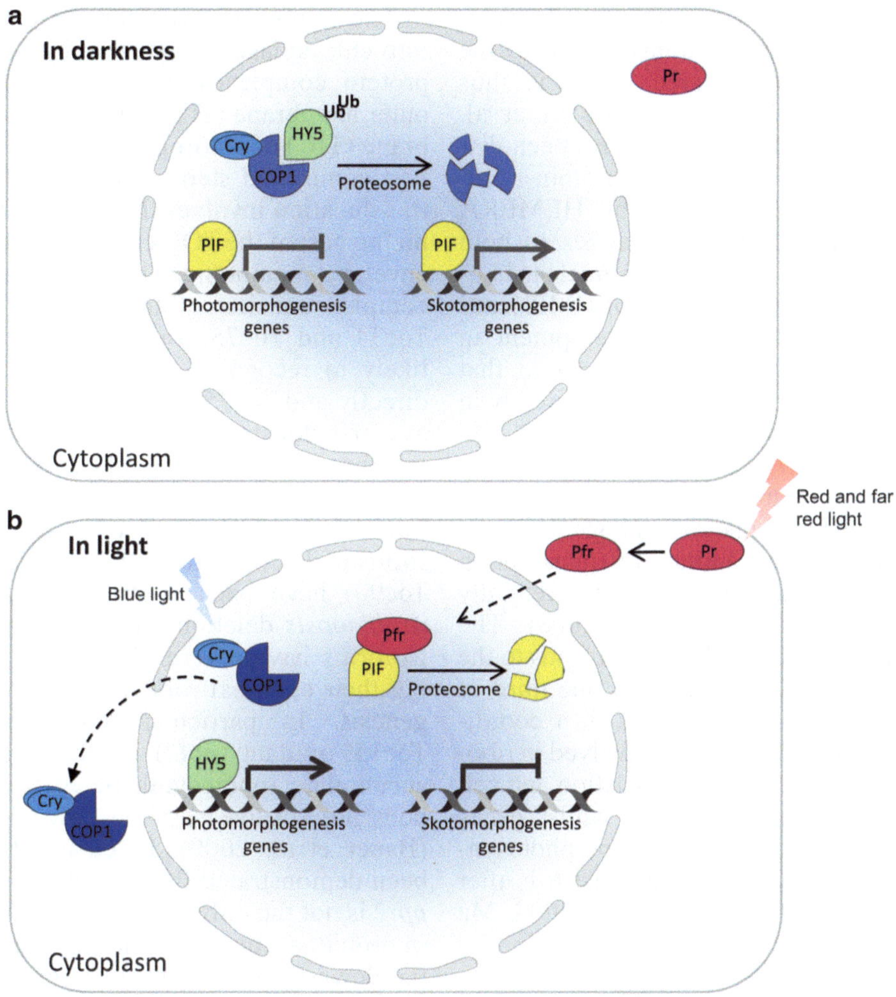

Fig. 23.3. A simplified model of the initiation of light signaling during photomorphogenesis. (**a**) In darkness, inactive CRY1 dimers are bound to COP1 in the nucleus. COP1 is an E3 ligase and promotes the ubiquitiation (Ub) of photomorphogenesis-promoting transcription factors, such as HY5. Polyubiquitinated HY5 is subsequently degraded by the 26S proteasome. Phytochrome dimers are in their inactive (Pr) form and localize to the cytoplasm. Skotomorphogenesis promoting transcription factors, such as PIF3, are bound to target promoters and also inhibit the transcription of photomorphogenesis-related genes. (**b**) In light, *blue light* triggers the activation of CRY1 dimers, leading to their translocation to the cytoplasm. This leads to accumulation of HY5 in the nucleus, which in turn activates transcription of photomorphogenesis genes. In parallel, red and far red light cause the conversion of phytochrome dimers from their inactive Pr form to their active Pfr form. Active Pfr dimers translocate to the nucleus where they bind to PIF3, ultimately leading to its degradation. This de-represses the expression of photomorphogenesis related genes. *Dashed lines* indicate the translocation of photoreceptors in response to light.

blocks the transcription of genes involved in chlorophyll biosynthesis and photosystem I (PSI) biogenesis (Ni et al. 1998). In the dark, PIFs are active and regulate gene expression to promote skotomorphogenic growth. In the light phytochrome dimers are converted from their inactive Pr form to their active Pfr form and translocate from the cytoplasm to the nucleus, where they localize to foci called phytochrome nuclear bodies (Quail 2002). Here the phytochromes bind to PIFs, ultimately leading to their phosphorylation and subsequent degradation. The block in the transcription of the photomorphogensis-related genes is thus released (Al-Sady et al. 2006; Ni et al. 1999) (Fig. 23.3). Recently it has been discovered that PIF1 and PIF3 protein accumulation is also regulated by HEMERA (*HMR*; pTAC12), which localizes to both the periphery of phytochrome nuclear bodies and to chloroplasts. Loss of HMR leads to a block in chloroplast development in response to light and it is proposed that HMR influences gene expression directly in both the nucleus and chloroplasts (Chen et al. 2010; Pfalz et al. 2006).

Shortly after light exposure, even when light-stimulated photomorphogenesis is barely observable, families of transcription factors are rapidly and differentially expressed (Lopez-Juez et al. 2008). The switch from dark to light growth leads to the differential regulation of up to one third of all nuclear *Arabidopsis* genes in a coordinated manor: Initially genes involved in ribosome biogenesis, protein translation and cell proliferation are upregulated whilst genes involved in plastid biogenesis and photosynthesis are largely induced at least 6 h after light exposure (Lopez-Juez et al. 2008; Ma et al. 2001).

B. Chloroplast Protein Import

Approximately 3,000 different nuclear-encoded proteins must be imported into chloroplasts during chloroplast biogenesis to fulfil the variety of cellular processes involving chloroplasts (Martin et al. 2002). Most chloroplast proteins are synthesized as precursor proteins (preproteins) in the cytosol and are imported post-translationally into the organelle. Preproteins bear an N-terminal transit peptide, responsible for the specificity of targeting which is proteolytically degraded after successful import. Once the preproteins have been guided to the chloroplast by chaperones, they interact with receptors on the membrane surface and are transported through the membranes in a GTP- and ATP-dependent manner.

The major pathway for protein import into chloroplasts is mediated by two multiprotein complexes or translocons of the outer membrane (TOC) and the inner membrane (TIC) of chloroplasts (Fig. 23.4). The first committed step in chloroplast protein translocation involves the preprotein recognizing part of the TOC complex on the outer envelope membrane. The core of the TOC complex consists of the proteins Toc159, Toc34 and Toc75. Toc159 and Toc34 are likely to recognize the precursor proteins directly and Toc75 (a β-barrel protein) acts as a translocation channel across the outer membrane (Hinnah et al. 2002; Kessler et al. 1994). In *Arabidopsis* two isoforms of Toc34 (atToc33 and atToc34) and four isoforms of Toc159 (Toc159, Toc132, Toc120, Toc90) have been identified (Fig. 23.4). *Arabidopsis* deletion mutants of the major isoforms have severe phenotypes suggesting their essential roles in chloroplast biogenesis. In particular, the *Arabidopsis* Toc159 mutant (*ppi2*) is albino, does not accumulate photosynthetic proteins and does not survive past the cotyledon stage (Bauer et al. 2000). Fascinatingly, it has been demonstrated that receptor absence in *ppi2* is not the only reason for the reduction in protein accumulation in plastids but also the downregulation of a suite of photosynthetic genes indicating tight regulation through retrograde signaling (Kakizaki and Inaba 2010). The *AtToc33* knockout mutant (*ppi1*) shows a pale green phenotype and is defective in the import and accumulation of several photosynthetic proteins (Kubis et al. 2003). It has been suggested that Toc complexes with different isoforms are able to regulate protein import of different classes of preproteins important for chloroplast differentiation (Ivanova et al. 2004).

Once the preproteins have crossed the intermembrane space the preproteins encounter the Tic translocon. Tic110, Tic40 and Hsp93 represent a minimal functional Tic unit (Fig. 23.4), where regulatory subunits are dynamically associated. Tic110 is most abundant mediating precursor binding, chaperone recruitment on the stromal

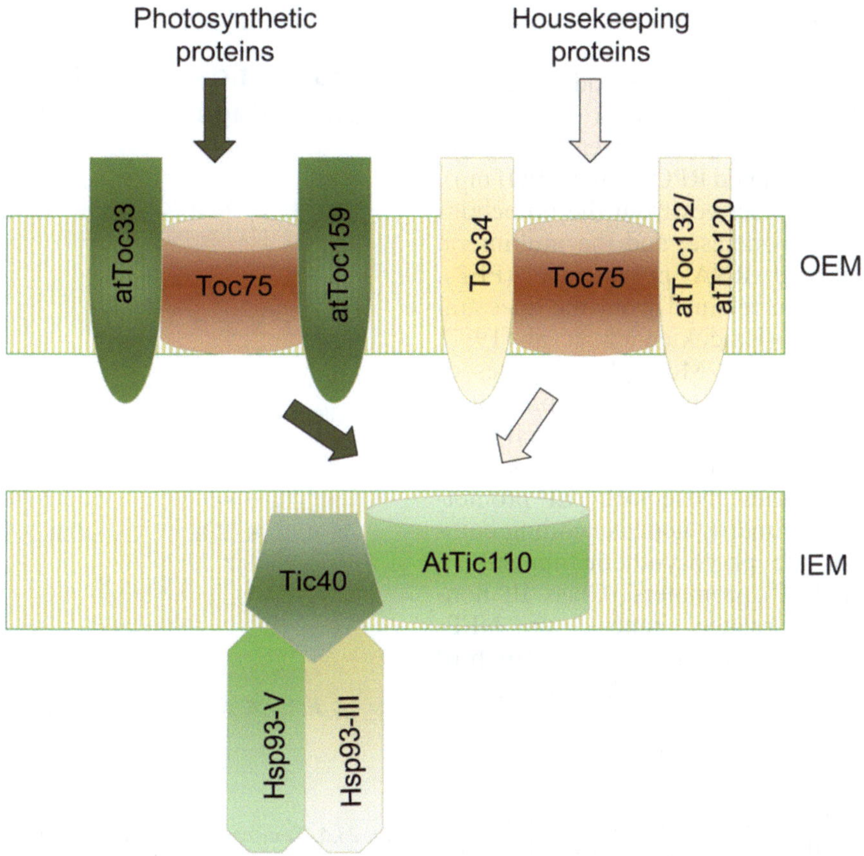

Fig. 23.4. A simplified illustration of the main components of chloroplast import apparatus. Nuclear-encoded chloroplast proteins are first translated on cytosolic ribosomes and then transported to the chloroplast outer envelope membrane (OEM). The TOC complex on the outer envelope membrane recognizes the preproteins. Once the preproteins have crossed the intermembrane space they encounter the Tic translocon on the inner envelope membrane (IEM). Green color indicates proteins demonstrated to be involved in the import of photosynthetic proteins; *yellow* indicates the ones involved in the import of housekeeping proteins.

side of the inner membrane as well as constituting a protein translocation channel (Lubeck et al. 1996). Reduction in *atTic110* expression results in a pale green phenotype, inappropriate plant growth and a marked decrease in thylakoid membranes (Inaba et al. 2003). In *Arabidopsis* there are two nuclear-encoded *hsp93* genes: *hsp93-III* (*clpC2*) and *hsp93-V* (*clpC1*). Whilst the *hsp93-III* mutant displays no phenotype the *hsp93-V* mutant contains chloroplasts with reduced thylakoid membranes, decreased amounts of PSI and photosystem II (PSII) proteins and reduced import efficiency (Constan et al. 2004). Similarly, null mutants of *AtTic40* display a pale green phenotype, slow growth and have less grana stacks in the thylakoids (Chou et al. 2003).

Stromal proteins may also be targeted to the thylakoid lumen which requires a second transit peptide involving either the Sec machinery-dependent or the twin-arginine translocation (Tat) protein translocation pathways (Robinson et al. 2001). Added to this is the fact that not all plastid proteins are imported through the conventional TOC and TIC complexes. Indeed research has shown that proteins can also be imported into chloroplasts through alternative pathways (Miras et al. 2007).

C. Plastid Gene Expression

The expression of plastid-encoded genes is mediated by two types of RNA polymerases of different origin, the nuclear-encoded polymerase (NEP; named RPOTp and RPOTmp), which probably derives from the mitochondrial RNA polymerase, and the multimeric plastid-encoded RNA polymerase (PEP), which has been retained from the ancestral endosymbiont (Hajdukiewicz et al. 1997; Shiina et al. 2005). Most plastid genes or operons are preceded by multiple promoters allowing transcription by NEP as well as by PEP. Plastid genes can generally be ascribed to three categories: those with PEP promoters (Class I; primarily genes encoding components of the photosynthetic apparatus), NEP- and PEP- promoters (Class II; non-photosynthesis-related genes), or NEP-promoters only (Class III; a small group of genes) (Fig. 23.5) (Hajdukiewicz et al. 1997).

Selective transcription of the groups of genes by the NEP and PEP provides a mechanism of differential gene expression during chloroplast development. For example, genes with NEP promoters are preferentially transcribed early in chloroplast development (Fig. 23.5). These provide mainly housekeeping functions and components of the plastid genetic machinery, including *rrn*, *atpB*, *atpI* and *clpP* and subunits of the PEP. This leads to a dramatic increase in the transcription and translation activity in the chloroplast. Once the NEP function has led to the translation of the PEP chloroplast-encoded subunits, the PEP becomes more important in the expression of photosynthesis-related genes, such as *rbcL*, *psbA* and *psbD* (Fig. 23.5) (Hajdukiewicz et al. 1997), ultimately leading to the synthesis and assembly of the photosynthetic machineries.

Chloroplast maturation is accompanied by the repression of NEP activity. The replacement of NEP activity by PEP activity in chloroplasts defines a switch in differential gene expression and a commitment to photosynthetic development. The molecular mechanism underlying the switch is however questionable; it has been hypothesized that glutamyl-tRNA, a product of PEP transcription, binds to and inhibits the transcriptional activity of NEP, however, the specificity of this interaction is unclear (Bohne et al. 2009; Hanaoka et al. 2005). Alternatively, thylakoid sequestration may down-regulate NEP activity; however, given that the plastid transcription machinery is membrane-associated

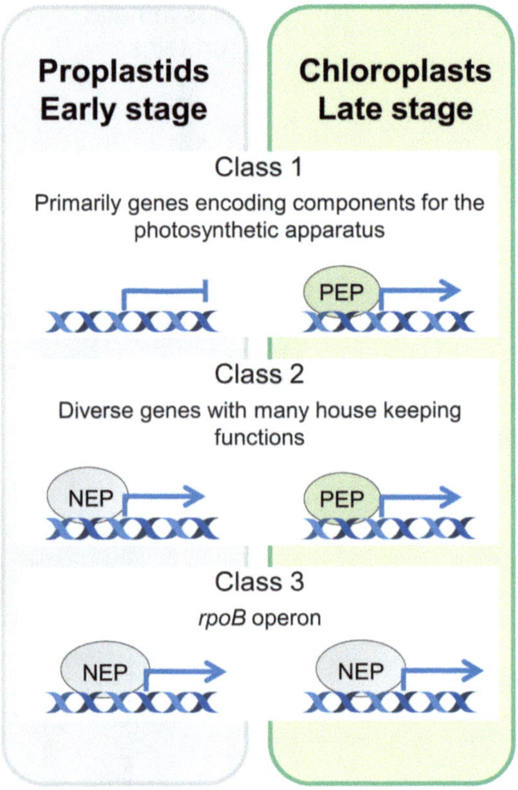

Fig. 23.5. Transcription from NEP- or PEP-dependent promoters in proplastids and chloroplasts. Differential gene expression during chloroplast development is partially based on recognition of distinct classes of promoters by NEP and PEP. In proplastids the NEP is most active and responsible for the transcription of housekeeping genes (Class 2), as well as a small class of genes including the *rpoB* operon, which encodes for subunits of the PEP (Class 3). In the proplastid the photosynthetic genes are not transcribed. During chloroplast biogenesis the subunits of the PEP are transcribed and the activity of the PEP becomes more important as the chloroplast develops. In the mature chloroplast the photosynthetic genes (Class 1) and the housekeeping genes (Class 2) are transcribed by the PEP.

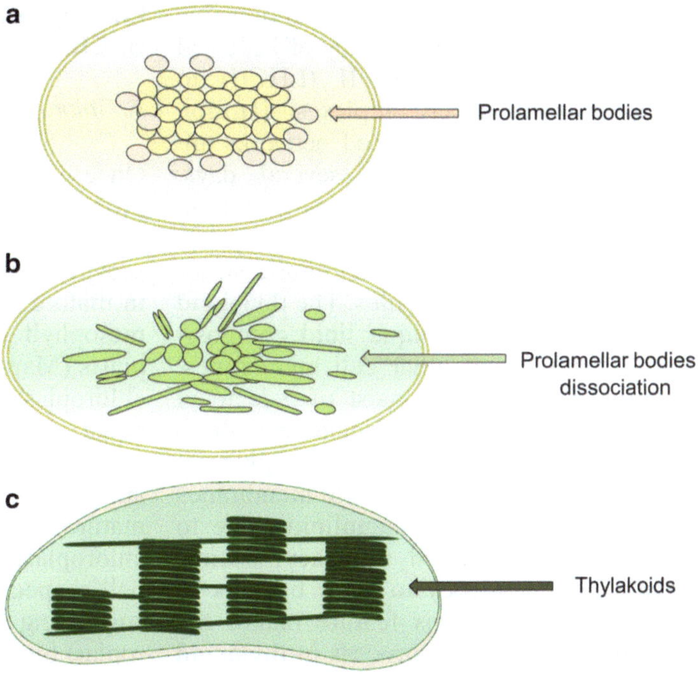

Fig. 23.6. Thylakoid formation in the presence of light. (**a**) Etioplast are characterized by the presence of a semicrystalline structure called prolamellar bodies (PLBs). (**b**) Upon illumination the PLBs dissociate and lamellar structures begin to be assemble. (**c**) Fully developed thylakoid networks are observed after several days depending on the species and light conditions.

more work is required to ascertain whether thylakoid association affects NEP activity (Azevedo et al. 2008).

After the switch to PEP activity in chloroplasts, the transcription of PEP-transcribed genes remains under nuclear control through the action of nuclear encoded sigma factors, which provide promoter specificity to PEP. The *Arabidopsis* genome encodes six sigma factors, which are expressed in a developmental- and light-dependent manner. Several of the sigma factors are plastid-localized and have overlapping functions and the functional role of sigma factors during chloroplast biogenesis is yet to be fully understood (Fujiwara et al. 2000; Isono et al. 1997).

D. Thylakoid Membrane Biogenesis

Thylakoid membranes are found exclusively in organisms performing photosynthesis. In plants the thylakoid membranes are arranged into highly interconnected lamellar networks by the presence of cylindrical stacks called grana, which are linked to each other by unstacked membrane regions called stroma lamellae (Adam et al. 2010).

Thylakoid ontogeny in plant is highly dependent on the development of the chloroplasts from undifferentiated proplastids. Proplastids contain very few internal membranes, often found as vesicles containing little or no photosyntethic complexes. In the absence of light proplastids differentiate into etioplasts, which contain prolamellar bodies and perforated lamellae (Fig. 23.6). The prolamellar body is a paracrystalline structure consisting of lipids, NADPH, the precursor of chlorophyll α (protochlorophyllide) and the enzyme that calalyzes its photoreduction to chlorophyllide (Vothknecht and Westhoff 2001).

In the presence of light, proplastids develop into mature chloroplast (Fig. 23.6). Under normal conditions, membrane formation and protein complex assembly occur in a coordinated manner. First, prolamellar bodies start to disassemble and the formation of long, unconnected lamellae is observed (Sperling et al. 1998). How this transition occurs is still unclear. Next, lamellar regions begin to overlap and the assembly and incorporation of photosynthetic complexes into the membrane leads to the formation of the grana. The establishment of granal domains is likely driven by the tendency of PSII and light harvesting complex II (LHCII) to assemble into super-complexes (Horton et al. 1991). Fully developed networks are completed within one to several days, depending on the plant species and light conditions.

The formation of any membrane structure requires lipid molecules. The thylakoid membranes have a unique lipid composition different from other cell membranes. The thylakoid lipids consist of galactosyl diglycerides and both monogalactosyl diacylglycerol and digalactosyl diacylglycerol are exclusively found in plastid membranes (Block et al. 1983). Galactolipids, due to two highly unsaturated fatty acyl chains, have a poor tendency to form bilayers in aqueous solution. This feature makes the lipid bilayer assembly process in thylakoid membranes rather complex. The formation of bilayer structures is an energetic process that requires LHCII and membrane proteins that have to be combined with lipids. It has indeed been shown that increasing the protein to lipid ratio induces the formation of ordered lamellar structures (Simidjiev et al. 2000).

How the lipid trafficking is carried out is controversially discussed. The vesicular transport mechanism theory is supported by the observation of membrane vesicles in leaves exposed to normal and low temperatures (Morre et al. 1991). Analysis of *Arabidopsis* mutants led to the identification of VIPP1 (vesicle-inducing protein in plastids 1), a protein conserved from cyanobacteria to plants (Kroll et al. 2001; Westphal et al. 2001), that was found to be associated with both the inner envelope and the thylakoid membranes. At low temperatures a the *vipp1* mutant does not accumulate vesicles, suggesting that VIPP1 may be involved in the budding of vesicles from the plastid envelope membrane (Kroll et al. 2001). Alternatively, lipids may be shuttled to the thylakoid membranes through direct connections with the inner envelope membrane (Shimoni et al. 2005).

III. Regulation and Maintenance of Chloroplast Populations

A. Chloroplast Division

Once chloroplasts have differentiated from proplastids, chloroplast must divide in order to match cell division and expansion and to provide the full complement of chloroplast in mature mesophyll cells. *Arabidopsis* leaf mesophyll cells contain over 100 chloroplasts (Marrison et al. 1999). Proplastid and chloroplast division is regulated and coordinated with mesophyll cell division and expansion, which is supposed to assure the size and photosynthetic competence of the mature chloroplast. During leaf development chloroplasts become larger, whereas dumbbell-shaped plastids became less common, suggesting that division occurs early in chloroplast biogenesis (Pyke 1999).

Chloroplasts divide by binary fission whereby constriction of the inner and outer membranes occurs at the division site (Fig. 23.7). This process is driven by the coordinated action of two distinct machineries located on the inner and outer chloroplast envelopes. The division machinery involves proteins derived from the cytokinetic machinery of cyanobacteria and proteins originating from the eukaryotic host cell. FtsZ is thought to be one of the first proteins of the inner machinery to arrive at the division site where it acts as a scaffold for other proteins (Osteryoung et al. 1998). It is believed that polymerization of the FtsZ proteins into a ring structure (Z-ring) is the initiating event in plastid division. The *Arabidopsis* plastid division component ARC6 is required for the stability and the maintenance of the Z-ring whilst its paralogue (PARC6) appears to destabilize the Z-ring (Glynn et al. 2009; Vitha et al. 2003). Moreover, the Min proteins regulate the exact placement of the Z-ring in order to ensure the generation of two equally sized daughter plastids (Colletti et al. 2000; Maple et al. 2002). Once the stromal ring is formed and fixed to the inner membrane the plastid division proteins (PDV) PDV1 and

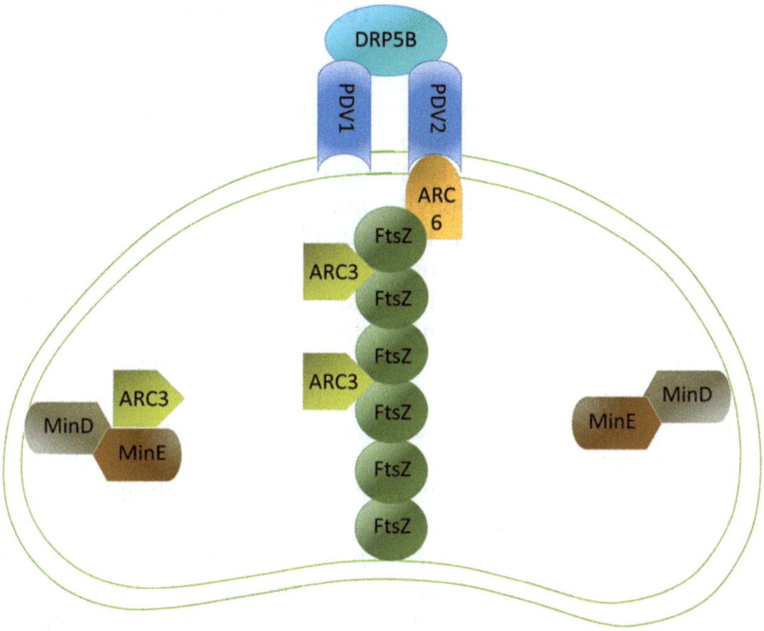

Fig. 23.7. Model of the plastid division machinery. Chloroplasts divide by binary fission whereby constriction of the inner and outer membranes occurs at the division site. Initially the FtsZ proteins (AtFtsS1-1, AtFtsZ2-1 and AtFtsZ2-2) form a Z-ring at the centre of chloroplasts. ARC6 and ARC3 are recruited to the Z-ring through specific interactions with AtFtsZ2-1 and AtFtsZ1-1, respectively. ARC6 is required for the stability and the maintenance of the Z-ring while ARC3 and Min proteins regulate the exact placement of the Z-ring. PDV1 and PDV2 localize to ring-like structures on the cytosolic surface of the outer envelope membrane and recruit DRP5B to the division site to constitute the cytosolic division machinery. Coordination and signalling between the two division machineries may occur through a direct interaction between ARC6 and PDV2.

PDV2, in the outer membrane, recruit DRP5B (cytosolic dynamin-like protein) around the chloroplast in alignment with the inner ring (Gao et al. 2003; Miyagishima et al. 2006). It has been shown that when the interaction between PDV2 and ARC6 is disrupted PDV2 is unable to recognize the chloroplast division site and subsequently unable to recruit DRP5B to the correct place (Glynn et al. 2008). This demonstrates that the two division machineries are coordinated with each other. Recently, it has been found that the PDV levels determine the rate of chloroplast division (Okazaki et al. 2009). Since the level of PDV2 has been shown to decrease during leaf development, whilst the levels of FtsZ and DRB5 remain similar, it suggests that PDV proteins may play a regulatory role in development. Furthermore, application of exogenous cytokinin leads to an increase in PDV2 levels, accompanied by an increase in chloroplast number in cotelydons, supporting a role of PDV proteins during development.

The rate of chloroplast development is an important determinant of the photosynthetic capacity of leaves. The control of chloroplast division and maintenance of chloroplast numbers is of fundamental importance. *Arabidopsis* accumulation and replication of chloroplast mutants (*arc* mutants) show severely decreased numbers of chloroplasts and have a significantly reduced photosynthetic capacity and altered thylakoid architecture. This in turn leads to a different light harvesting capacity that seems also to affect their ability to adapt to changes in growth conditions and light intensity (Austin II and Webber 2005).

B. Fine Tuning of the Differentiation Pathway

Whilst all green tissues contain chloroplasts, different cells within the tissue can harbor different chloroplast complements, which can further vary in terms of size and levels of membrane development and chlorophyll accumulation. This demonstrates that the process of chloroplast biogenesis is controlled in a cell-specific manner. The mechanisms in place to fine tune the process are not well understood. However, a number of mutants have been identified with defects in both chloroplast biogenesis and the cell cycle, which represent potential candidates to link plastid differentiation, plastid division and cell division. For example, CRUMPLED LEAF (CRL) is targeted to the plastid envelope and *crl* mutants shows defects in not only plastid division and expansion, but also in overall plant development (Asano et al. 2004; Chen et al. 2009). Additionally, deficiencies in the two *CDT1* genes (*AtCDT1a* and *AtCDT1b*) *lead to cell cycle abnormalities and large*, unevenly divided chloroplasts (Raynaud et al. 2005). *AtCDT1a* and *AtCDT1b* are key components needed to initiate nuclear DNA replication and are also targeted to plastids where they interact with ARC6, suggesting that they play a major role in regulating events in both the nucleus and plastid (Raynaud et al. 2005).

In addition to cell-specific variations in chloroplast biogenesis, chloroplasts must also sense and adapt to environmental fluctuations that the cell and tissue responds to. For example, within a leaf, chloroplasts experience different levels of light and in order to adapt to these fluctuations in light levels chloroplast acclimatize by adjusting the proportion of light energy used to drive photosynthesis. Recent work has highlighted the GLK transcription factors, which regulate a large suite of genes involved in light-harvesting and thylakoid protein complex formation in a cell-autonomous fashion, thereby enabling the photosynthetic capacity of each cell in a leaf to be regulated independently (Waters et al. 2008).

IV. Conclusions

Chloroplast biogenesis is a complex and highly regulated process and our understanding of the molecular mechanisms involved during proplastid to chloroplast differentiation has improved dramatically in the last decade. Although the molecular framework of chloroplast biogenesis is starting to be unraveled, it remains a major challenge to dissect the precise control mechanisms, which integrate environmental, cellular and temporal factors that affect chloroplast biogenesis and development. It is also intriguing to consider that differentiation from proplastids to chloroplasts is probably a default developmental pathway, with non-photosynthetic forms of plastids arising after the evolution of chloroplasts. It will be fascinating to discover, what mechanisms are in place to prevent chloroplast biogenesis in certain tissues and instead promote the vastly different pathways of plastid differentiation, which lead to the development non-photosynthetic plastids. Recent years have also seen an increase in our understanding of the components essential for chloroplast division, but the mechanisms in place to control the number of plastids in a cell-specific manner remains unclear. Recent identification of the role of both the PDV proteins and hormones will add new impetus to this field of research.

Acknowledgments

Plastid division research in our laboratory is funded by the Norwegian Research Council.

References

Adam Z, Charuvi D, Tsabari O, Knopf RR, Reich Z (2010) Biogenesis of thylakoid networks in angiosperms: knowns and unknowns. Plant Mol Biol. doi:10.1007/s11103-010-9693-5

Al-Sady B, Ni W, Kircher S, Schafer E, Quail PH (2006) Photoactivated phytochrome induces rapid

PIF3 phosphorylation prior to proteasome-mediated degradation. Mol Cell 23:439–446

Asano T, Yoshioka Y, Kurei S, Sakamoto W, Machida Y (2004) A mutation of the CRUMPLED LEAF gene that encodes a protein localized in the outer envelope membrane of plastids affects the pattern of cell division, cell differentiation, and plastid division in *Arabidopsis*. Plant J 38:448–459

Austin J II, Webber AN (2005) Photosynthesis in *Arabidopsis thaliana* mutants with reduced chloroplast number. Photosynth Res 85:373–384

Azevedo J, Courtois F, Hakimi MA, Demarsy E, Lagrange T, Alcaraz JP, Jaiswal P, Marechal-Drouard L, Lerbs-Mache S (2008) Intraplastidial trafficking of a phage-type RNA polymerase is mediated by a thylakoid RING-H2 protein. Proc Natl Acad Sci U S A 105:9123–9128

Bauer J, Chen K, Hiltbunner A, Wehrli E, Eugster M, Schnell D, Kessler F (2000) The major protein import receptor of plastids is essential for chloroplast biogenesis. Nature 403:203–207

Block MA, Dorne AJ, Joyard J, Douce R (1983) Preparation and characterization of membrane fractions enriched in outer and inner envelope membranes from spinach chloroplasts II. Biochemical characterization. J Biol Chem 258: 13281–13286

Bohne AV, Weihe A, Borner T (2009) Transfer RNAs inhibit Arabidopsis phage-type RNA polymerases. Endocytobiosis Cell Res 19:63–69, 7p

Castillon A, Shen H, Huq E (2007) Phytochrome interacting factors: central players in phytochrome-mediated light signaling networks. Trends Plant Sci 12:514–521

Chen Y, Asano T, Fujiwara MT, Yoshida S, Machida Y, Yoshioka Y (2009) Plant cells without detectable plastids are generated in the crumpled leaf mutant of *Arabidopsis thaliana*. Plant Cell Physiol 50:956–969

Chen M, Galvao RM, Li M, Burger B, Bugea J, Bolado J, Chory J (2010) *Arabidopsis* HEMERA/pTAC12 initiates photomorphogenesis by phytochromes. Cell 141:1230–1240

Chory J (2010) Light signal transduction: an infinite spectrum of possibilities. Plant J 61:982–991

Chou ML, Fitzpatrick LM, Tu SL, Budziszewski G, Potter-Lewis S, Akita M, Levin JZ, Keegstra K, Li HM (2003) Tic40, a membrane-anchored co-chaperone homolog in the chloroplast protein translocon. EMBO J 22:2970–2980

Colletti KS, Tattersall EA, Pyke KA, Froelich JE, Stokes KD, Osteryoung KW (2000) A homologue of the bacterial cell division site-determining factor MinD mediates placement of the chloroplast division apparatus. Curr Biol 10:507–516

Constan D, Froehlich JE, Rangarajan S, Keegstra K (2004) A stromal Hsp100 protein is required for normal chloroplast development and function in *Arabidopsis*. Plant Physiol 136:3605–3615

Cran DG, Possingham JV (1972) Variation of plastid types in spinach. Protoplasma 74:345–356

Fujiwara M, Nagashima A, Kanamaru K, Tanaka K, Takahashi H (2000) Three new nuclear genes, sigD, sigE and sigF, encoding putative plastid RNA polymerase sigma factors in *Arabidopsis thaliana*. FEBS Lett 481:47–52

Gao H, Kadirjan-Kalbach D, Froehlich JE, Osteryoung KW (2003) ARC5, a cytosolic dynamin-like protein from plants, is part of the chloroplast division machinery. Proc Natl Acad Sci U S A 100:4328–4333

Glynn JM, Froehlich JE, Osteryoung KW (2008) *Arabidopsis* ARC6 coordinates the division machineries of the inner and outer chloroplast membranes through interaction with PDV2 in the intermembrane space. Plant Cell 20:2460–2470

Glynn JM, Yang Y, Vitha S, Schmitz AJ, Hemmes M, Miyagishima SY, Osteryoung KW (2009) PARC6, a novel chloroplast division factor, influences FtsZ assembly and is required for recruitment of PDV1 during chloroplast division in *Arabidopsis*. Plant J 59:700–711

Hajdukiewicz PT, Allison LA, Maliga P (1997) The two RNA polymerases encoded by the nuclear and the plastid compartments transcribe distinct groups of genes in tobacco plastids. EMBO J 16:4041–4048

Hanaoka M, Kanamaru K, Fujiwara M, Takahashi H, Tanaka K (2005) Glutamyl-tRNA mediates a switch in RNA polymerase use during chloroplast biogenesis. EMBO Rep 6:545–550

Hinnah SC, Wagner R, Sveshnikova N, Harrer R, Soll J (2002) The chloroplast protein import channel Toc75: pore properties and interaction with transit peptides. Biophys J 83:899–911

Horton P, Ruban AV, Rees D, Pascal AA, Noctor G, Young AJ (1991) Control of the light-harvesting function of chloroplast membranes by aggregation of the LHCII chlorophyll-protein complex. FEBS Lett 292:1–4

Inaba T, Li M, Alvarez-Huerta M, Kessler F, Schnell DJ (2003) atTic110 functions as a scaffold for coordinating the stromal events of protein import into chloroplasts. J Biol Chem 278:38617–38627

Isono K, Shimizu M, Yoshimoto K, Niwa Y, Satoh K, Yokota A, Kobayashi H (1997) Leaf-specifically expressed genes for polypeptides destined for chloroplasts with domains of sigma70 factors of bacterial RNA polymerases in *Arabidopsis thaliana*. Proc Natl Acad Sci U S A 94:14948–14953

Ivanova Y, Smith MD, Chen K, Schnell DJ (2004) Members of the Toc159 import receptor family represent distinct pathways for protein targeting to plastids. Mol Biol Cell 15:3379–3392

Kakizaki T, Inaba T (2010) New insights into the retrograde signaling pathway between the plastids and the nucleus. Plant Signal Behav 5:196–199

Kessler F, Blobel G, Patel HA, Schnell DJ (1994) Identification of two GTP-binding proteins in the chloroplast protein import machinery. Science 266:1035–1039

Kroll D, Meierhoff K, Bechtold N, Kinoshita M, Westphal S, Vothknecht UC, Soll J, Westhoff P (2001) VIPP1, a nuclear gene of Arabidopsis thaliana essential for thylakoid membrane formation. Proc Natl Acad Sci U S A 98:4238–4242

Kubis S, Baldwin A, Patel R, Razzaq A, Dupree P, Lilley K, Kurth J, Leister D, Jarvis P (2003) The Arabidopsis pp i1 mutant is specifically defective in the expression, chloroplast import, and accumulation of photosynthetic proteins. Plant Cell 15: 1859–1871

Lopez-Juez E, Dillon E, Magyar Z, Khan S, Hazeldine S, de Jager SM, Murray JA, Beemster GT, Bogre L, Shanahan H (2008) Distinct light-initiated gene expression and cell cycle programs in the shoot apex and cotyledons of *Arabidopsis*. Plant Cell 20:947–968

Lubeck J, Soll J, Akita M, Nielsen E, Keegstra K (1996) Topology of IEP110, a component of the chloroplastic protein import machinery present in the inner envelope membrane. EMBO J 15:4230–4238

Ma L, Li J, Qu L, Hager J, Chen Z, Zhao H, Deng XW (2001) Light control of *Arabidopsis* development entails coordinated regulation of genome expression and cellular pathways. Plant Cell 13:2589–2607

Maple J, Chua NH, Moller SG (2002) The topological specificity factor AtMinE1 is essential for correct plastid division site placement in *Arabidopsis*. Plant J 31:269–277

Marrison JL, Rutherford SM, Robertson EJ, Lister C, Dean C, Leech RM (1999) The distinctive roles of five different ARC genes in the chloroplast division process in *Arabidopsis*. Plant J 18:651–662

Martin W, Rujan T, Richly E, Hansen A, Cornelsen S, Lins T, Leister D, Stoebe B, Hasegawa M, Penny D (2002) Evolutionary analysis of *Arabidopsis*, cyanobacterial, and chloroplast genomes reveals plastid phylogeny and thousands of cyanobacterial genes in the nucleus. Proc Natl Acad Sci U S A 99:12246–12251

McFadden GI (2001) Chloroplast origin and integration. Plant Physiol 125:50–53

Miras S, Salvi D, Piette L, Seigneurin-Berny D, Grunwald D, Reinbothe C, Joyard J, Reinbothe S, Rolland N (2007) Toc159- and Toc75-independent import of a transit sequence-less precursor into the inner envelope of chloroplasts. J Biol Chem 282:29482–29492

Miyagishima SY, Froehlich JE, Osteryoung KW (2006) PDV1 and PDV2 mediate recruitment of the dynamin-related protein ARC5 to the plastid division site. Plant Cell 18:2517–2530

Morre DJ, Sellden G, Sundqvist C, Sandelius AS (1991) Stromal low temperature compartment derived from the inner membrane of the chloroplast envelope. Plant Physiol 97:1558–1564

Ni M, Tepperman JM, Quail PH (1998) PIF3, a phytochrome-interacting factor necessary for normal photoinduced signal transduction, is a novel basic helix-loop-helix protein. Cell 95: 657–667

Ni M, Tepperman JM, Quail PH (1999) Binding of phytochrome B to its nuclear signalling partner PIF3 is reversibly induced by light. Nature 400:781–784

Okazaki K, Kabeya Y, Suzuki K, Mori T, Ichikawa T, Matsui M, Nakanishi H, Miyagishima SY (2009) The PLASTID DIVISION1 and 2 components of the chloroplast division machinery determine the rate of chloroplast division in land plant cell differentiation. Plant Cell 21:1769–1780

Osterlund MT, Wei N, Deng XW (2000) The roles of photoreceptor systems and the COP1-targeted destabilization of HY5 in light control of *Arabidopsis* seedling development. Plant Physiol 124:1520–1524

Osteryoung KW, Stokes KD, Rutherford SM, Percival AL, Lee WY (1998) Chloroplast division in higher plants requires members of two functionally divergent gene families with homology to bacterial ftsZ. Plant Cell 10:1991–2004

Pfalz J, Liere K, Kandlbinder A, Dietz KJ, Oelmuller R (2006) pTAC2, -6, and -12 are components of the transcriptionally active plastid chromosome that are required for plastid gene expression. Plant Cell 18:176–197

Pyke KA (1999) Plastid division and development. Plant Cell 11:549–556

Quail PH (2002) Phytochrome photosensory signalling networks. Nat Rev Mol Cell Biol 3:85–93

Raynaud C, Perennes C, Reuzeau C, Catrice O, Brown S, Bergounioux C (2005) Cell and plastid division are coordinated through the prereplication factor AtCDT1. Proc Natl Acad Sci U S A 102: 8216–8221

Robinson C, Thompson SJ, Woolhead C (2001) Multiple pathways used for the targeting of thylakoid proteins in chloroplasts. Traffic 2:245–251
Saijo Y, Sullivan JA, Wang H, Yang J, Shen Y, Rubio V, Ma L, Hoecker U, Deng XW (2003) The COP1-SPA1 interaction defines a critical step in phytochrome A-mediated regulation of HY5 activity. Genes Dev 17:2642–2647
Schroeder DF, Gahrtz M, Maxwell BB, Cook RK, Kan JM, Alonso JM, Ecker JR, Chory J (2002) De-etiolated 1 and damaged DNA binding protein 1 interact to regulate *Arabidopsis* photomorphogenesis. Curr Biol 12:1462–1472
Shiina T, Tsunoyama Y, Nakahira Y, Khan MS (2005) Plastid RNA polymerases, promoters, and transcription regulators in higher plants. Int Rev Cytol 244:1–68
Shimoni E, Rav-Hon O, Ohad I, Brumfeld V, Reich Z (2005) Three-dimensional organization of higher-plant chloroplast thylakoid membranes revealed by electron tomography. Plant Cell 17:2580–2586
Simidjiev I, Stoylova S, Amenitsch H, Javorfi T, Mustardy L, Laggner P, Holzenburg A, Garab G (2000) Self-assembly of large, ordered lamellae from non-bilayer lipids and integral membrane proteins in vitro. Proc Natl Acad Sci U S A 97:1473–1476
Sperling U, Franck F, van Cleve B, Frick G, Apel K, Armstrong GA (1998) Etioplast differentiation in *Arabidopsis*: both PORA and PORB restore the prolamellar body and photoactive protochlorophyllide-F655 to the cop1 photomorphogenic mutant. Plant Cell 10:283–296
Sugita M, Sugiura M (1996) Regulation of gene expression in chloroplasts of higher plants. Plant Mol Biol 32:315–326
Tetlow IJ, Rawsthorne S, Rines C, Emes MJ (2005) Plastid metabolic pathways. In: Moller SG (ed) Plastids. Blackwell, Oxford, pp 60–109
Vitha S, Froehlich JE, Koksharova O, Pyke KA, van Erp H, Osteryoung KW (2003) ARC6 is a J-domain plastid division protein and an evolutionary descendant of the cyanobacterial cell division protein Ftn2. Plant Cell 15:1918–1933
von Arnim A, Deng XW (1996) Light control of seedling development. Annu Rev Plant Physiol Plant Mol Biol 47:215–243
Vothknecht UC, Westhoff P (2001) Biogenesis and origin of thylakoid membranes. Biochim Biophys Acta 1541:91–101
Waters MT, Moylan EC, Langdale JA (2008) GLK transcription factors regulate chloroplast development in a cell-autonomous manner. Plant J 56: 432–444
Westphal S, Heins L, Soll U, Vothknecht UC (2001) Vipp1 deletion mutant of Synechocystis: a connection between bacterial phage shock and thylakoid biogenesis. Proc Natl Acad Sci U S A 98: 4243–4248
Yanagawa Y, Sullivan JA, Komatsu S, Gusmaroli G, Suzuki G, Yin J, Ishibashi T, Saijo Y, Rubio V, Kimura S, Wang J, Deng XW (2004) *Arabidopsis* COP10 forms a complex with DDB1 and DET1 in vivo and enhances the activity of ubiquitin conjugating enzymes. Genes Dev 18:2172–2181

Chapter 24

Mitochondrial Biogenesis and Quality Control

Jason A. Mears*
Department of Pharmacology and Center for Mitochondrial Disease,
Case Western Reserve University School of Medicine,
Cleveland, OH 44106, USA

Summary

Mitochondrial biogenesis is a complex process that facilitates an increase in mitochondrial content to meet cellular energy demands. The large majority of proteins needed for this task are encoded in the nucleus, while the mitochondrial genome encodes for just 13 gene products. Nevertheless, mitochondrial-encoded proteins are essential for organelle function. Therefore, a delicate coordination of cytoplasmic and mitochondrial protein synthesis is needed.

*Author for correspondence, e-mail: jam348@case.edu

M.F. Hohmann-Marriott (ed.), The Structural Basis of Biological Energy Generation, Advances in Photosynthesis and Respiration 39, pp. 451–476, DOI 10.1007/978-94-017-8742-0_24,

Furthermore, protein products from the nucleus and mitochondria must be targeted to specific sub-compartments within the organelle, where they assemble into multi-subunit complexes essential for cellular respiration. Mitochondrial DNA replication and lipid synthesis are also coordinately regulated to generate a proportional increase in the mitochondrial mass. As mitochondrial content increases, mitochondrial fission is needed to distribute the increased bioenergetic capacity throughout the cell. Mitochondrial division also serves to isolate damaged regions of the organelle that can be targeted for mitophagy. Mitochondrial fusion promotes mixing of DNA, protein, and lipid content within the mitochondrial network and counteracts fragmentation, thereby preventing additional damage and associated cell death. Several physiological factors stimulate transcriptional elements that govern mitochondrial proliferation. Often, these signals enhance mitochondrial biogenesis, but quality control measures are needed to prevent the propagation of deleterious mitochondria.

I. Introduction

Mitochondria are essential eukaryotic organelles that act as the major supplier of cellular ATP through oxidative phosphorylation (see Chap. 12). In addition, mitochondria serve critical roles in calcium maintenance, apoptosis, and the biosynthesis of various cellular molecules, including lipids, cholesterol, nucleotides, heme, and steroids. Therefore, tight regulation of mitochondrial mass and mitochondrial function is vital. Accordingly, mitochondrial abundance, morphology, and function are finely tuned to meet cell-specific energetic, metabolic, and signaling demands. This regulation is dependent on physiological cues, which can differ significantly between cell types.

Cells from different tissues contain differing numbers of mitochondria, ranging from a just a few to more than a thousand. Given this diversity, regulation of mitochondrial growth is coupled to different physiological or environmental signals that regulate the activation of specific transcription factors (Attardi and Schatz 1988; Moyes and Hood 2003). In the heart, mitochondria are highly abundant, and produce more than 90 % of the cell's energy (Barth et al. 1992). Energy demands in neuronal cells also lead to an increased mitochondrial mass relative to cell volume. Genetic, metabolic, and dietary factors that influence mitochondrial proliferation impact health and disease in these tissues.

Mitochondrial biogenesis regulates mitochondrial content and maintains a healthy organelle network. Increasing mitochondrial mass offers a unique challenge to cells, because mitochondrial growth requires coordinated expression of proteins encoded by nuclear and mitochondrial genomes. The mitochondrial genome encodes only 13 proteins in mammals (Fig. 24.1a), while the nuclear genome encodes for hundreds of proteins required for mitochondrial functions, including oxidative phosphorylation, calcium homeostasis, and apoptosis.

Mitochondria also exhibit characteristics indicative of their bacterial origin. Mitochondrial DNAs (mtDNAs) have features derived from their bacterial ancestors and must replicate during proliferation. Furthermore, mitochondria have retained their own transcription and translation machineries, which are critical for generating key components in respiratory chain complexes (see Chaps. 12 and 21). Due to its endosymbiotic origin, the mitochondrial interior/cytoplasm (matrix) is separated from the cell cytoplasm by a double membrane (Fig. 24.1b). While the outer membrane of mitochondria is quite permeable, the inner

Abbreviations: AMP – Adenosine monophosphate; ATP – Adenosine triphosphate; CL – Cardiolipin; GMP – Guanosine monophosphate; H_2O_2 – Hydrogen peroxide; PI – Phosphatidylinositol; PS – Phosphatidylserine; PG – Phosphatidylglycerol; PA – Phosphatidic acid; ROS – Reactive oxygen species

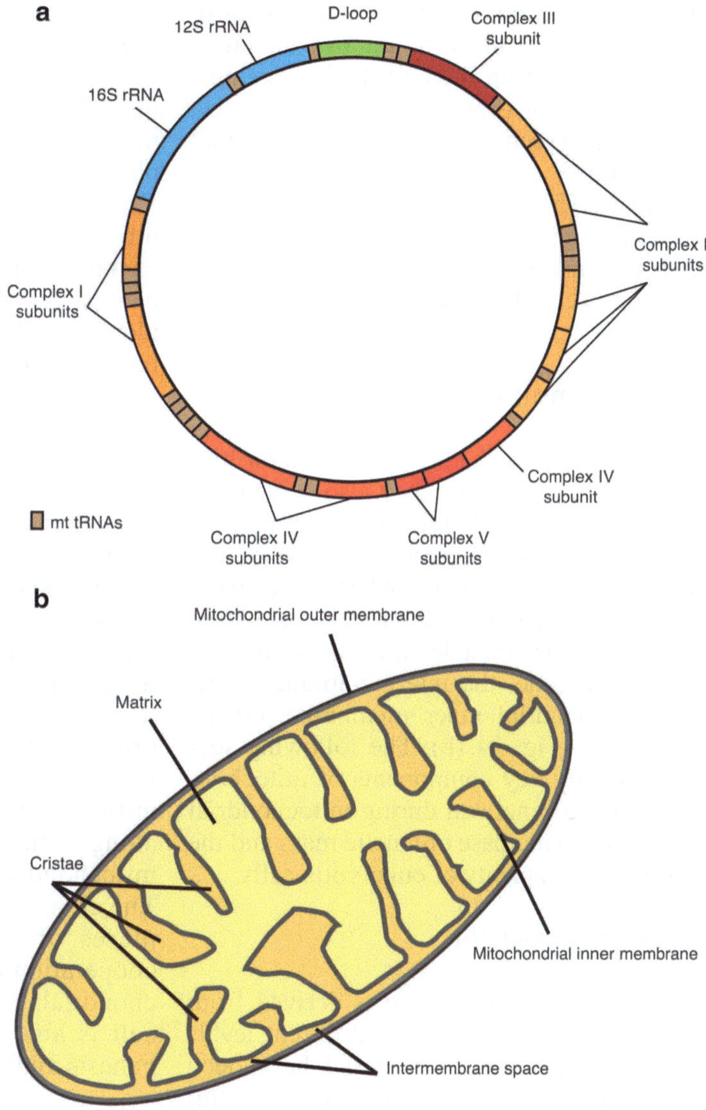

Fig. 24.1. Mitochondrial components. (**a**) Mitochondrial DNA organization. (**b**) Schematic representation of a mitochondrion. Major features and sub-compartments are highlighted.

membrane is protein-rich and impermeable to most molecules. Thus, specific transport systems have evolved to facilitate transport of specific compounds and polypeptides into and out of mitochondria. The innermost membrane system is highly invaginated, resulting in internal compartments called cristae, (Fig. 24.1b) which add to the complexity of the mitochondrial structure. This complexity is utilized to regulate the surface area of the inner mitochondrial membrane, and cristae structure is intimately linked to the formation of respiratory complexes and their subsequent function (Davies et al. 2011, 2012; De los Rios Castillo et al. 2011).

When cells sense a deficiency in respiratory function, mitochondrial biogenesis ensues to meet energy demands. Some mitochondrial signals are transmitted to the nucleus to induce mitochondrial gene expression leading to production of a larger mitochondrial network. This process faces the caveat that a deficiency in respiratory function may be due to mitochondrial damage and/or dysfunction. Therefore, quality control measures are in place to remove damaged mitochondria from the cell. If damage persists too long or is too severe to be repaired, mitochondria produce signals that are essential for the onset of the mitochondrial-induced apoptosis. This process involves the release of cytochrome *c* into the cytoplasm, which leads to activation of cytosolic caspases and, eventually, cell death (Kroemer et al. 1998).

II. Mitochondrial Components

Mitochondria are characterized by their bioenergetic functions, but they also serve roles in cell signaling and molecular biosynthesis. The membrane organization of mitochondria creates sub-compartments where specialized functions are maintained in relative isolation. These regions include the mitochondrial outer membrane, the intermembrane space, the mitochondrial inner membrane, cristae and matrix (Fig. 24.1b). The following sections identify key components of mitochondria that are generated during mitochondrial biogenesis to increase organelle mass and the bioenergetic capacity of eukaryotic cells.

A. Mitochondrial DNA

The mitochondrial genome is derived from an early endosymbiont, and different levels of the original genetic material have been preserved in present-day organisms. After incorporation of symbiotic bacteria, mitochondrial evolution has involved the progressive loss of genes to the nucleus. Due to evolutionary pressures, the size and packaging of mtDNA is remarkably different between distinct organisms, ranging from the 570 kilobase (kb), circular molecules in corn plants (Lonsdale et al. 1984) to the linear, 46 kb molecules in Tetrahymena (Borst et al. 1980), and to the ~16 kb, circular molecules found in animals (Attardi 1985). The mtDNA from yeast and mammals are the most extensively studied, and despite differences in genome size, there are significant similarities.

In yeast, mtDNA is a circular, ~75 kb molecule that contains coding sequence for two ribosomal RNAs (rRNAs), a set of transfer RNAs (tRNAs), and conserved inner membrane proteins essential for oxidative phosphorylation (Borst and Grivell 1978; Tzagoloff et al. 1979). Similarly, human mtDNA is a circular, 16.5 kb molecule that encodes for two rRNAs and 22 tRNAs needed for mitochondrial protein synthesis along with 13 essential inner membrane proteins that co-assemble with nuclear-encoded proteins to form complex I, complex III (see Chap. 8), complex IV (see Chap. 10), and complex V (see Chap. 6), of the respiratory chain (Fig. 24.1a). Transfer of electrons through these complexes is essential for generating the electrochemical gradient coupled to synthesis of ATP.

Even with their relative independence, mitochondria are absolutely dependent on the genome of the host organism in which they reside. The vast majority of mitochondrial proteins are encoded in the nucleus. In fact, all of the proteins that mediate mitochondrial protein and RNA import along with all of the enzymes within the mitochondrial matrix are nuclear-encoded. The co-assembly of the multi-subunit complexes of the respiratory chain requires cooperation between the nuclear and mitochondrial genomes. As a result, the host cell is able to control mitochondrial biogenesis and function.

B. Mitochondrial Lipid Content

When compared to mtDNA, the role of lipids in mitochondrial protein function has not been studied as extensively. Phosphatidylcholine (PC) and phosphatidylethanolamine (PE) are the major phospholipid constituents of mitochondrial membranes. Phosphatidylinositol (PI), phosphatidylserine (PS), phosphatidylglycerol (PG), phosphatidic acid (PA) and cardiolipin (CL) are less abundant. However, CL is enriched in the mitochondrial inner membrane (Zambrano et al. 1975). PC, PI, PS and PA are synthesized in the endoplasmic reticulum (ER) and need to be imported.

Conversely, PE, PG and CL are synthesized within the mitochondrion (Daum and Vance 1997). Eukaryotic cells with defects in the mitochondrial synthesis of PE, PG, and CL have been isolated and exhibit abnormal mitochondrial morphology or stability. Still, it is not clear how phospholipid biosynthesis within the ER is regulated and contributes to mitochondrial biogenesis.

Ongoing research is focused on elucidating functional interactions between mitochondria and the endoplasmic reticulum (ER) as well as mitochondrial components involved in lipid biogenesis and mitochondrial morphology. Mitochondrial membrane remodeling is regulated by interactions between mitochondria and the ER, as components of the mitochondrial fusion machinery (see Sect. III.B) are involved in forming ER-mitochondria contact sites (de Brito and Scorrano 2008). These sites are proposed to be involved in calcium signaling and lipid transfer between the adjacent membranes. In yeast, translocator and maintenance protein 41 (Tam41) and prohibitins regulate the synthesis and assembly of mitochondrial inner membrane phospholipids and proteins (Kutik et al. 2008; Osman et al. 2009a). Prohibitins also regulate mitochondrial protein degradation, stability of the mitochondrial genome, and cristae morphology (Osman et al. 2009b).

Recent studies have shown that particular phospholipids serve a vital role in the organization and function of inner membrane proteins. CL is critical for the organization and function of many protein complexes, including components of the respiratory chain (Eble et al. 1990). Interestingly, the amount of CL in the mitochondrial inner membrane is depleted in response to various stimuli, including hormone levels (Paradies and Ruggiero 1990), chronic muscle contraction (Takahashi and Hood 1993), and ischemia (Lesnefsky et al. 2001). During aging, cytochrome *c* oxidase is depleted, while addition of exogenous CL can reverse this trend (Paradies et al. 1997).

C. Mitochondrial Proteins

Since mtDNA encodes only 13 proteins, nuclear genes are required for mitochondrial biogenesis. A cellular machinery had to evolve to mediate the transport of nuclear-encoded proteins to mitochondria. Moreover, these events must be coordinated with translation and assembly of mitochondrial-encoded proteins to generate additional mitochondrial machineries that maintain organelle function. In mammals, the mitochondrial proteome includes ~1,000 proteins, and about 50 % of mitochondrial proteins are expressed in a tissue-specific manner (Mootha et al. 2003). Therefore, the mitochondrial proteome likely includes a large number of proteins that are dedicated to specialized functions in addition to factors that are essential for mitochondrial biogenesis (i.e. proteins involved in energy metabolism).

There are three major classes of mitochondrial proteins encoded in the nucleus. The first class includes proteins that directly or indirectly support catalysis required for mitochondrial energy generation and synthesis of biological macromolecules. As mentioned before, the large majority of proteins within the respiratory chain complexes are nuclear encoded. Additionally, a large majority of mitochondrial proteins are synthesized in the cytosol and transported into specialized mitochondrial compartments, including the mitochondrial inner membrane and matrix, where they perform fatty acid oxidation, contribute to the tricarboxylic acid cycle (TCA cycle), build heme, and synthesize certain amino acids. A second class of nuclear genes encodes proteins needed for the replication and expression of the mitochondrial genome. These protein factors include nucleic acid polymerases, RNA processing enzymes, transcription and replication factors, translation factors, and ribosomal proteins (see Sect. IV.A–C.).

Lastly, a third class includes proteins needed for import and assembly of proteins targeted to mitochondria. Interestingly, the mitochondrial inner membrane is extremely protein-rich (60–70 % by weight). In a recent study, 850 different proteins were identified in *Saccharomyces cerevisiae* mitochondria (Reinders et al. 2006), and roughly 30 % of these proteins encode for integral membrane proteins. Since the majority of these proteins must be imported from the cytosol, nuclear-encoded mitochondrial proteins contain specific signal sequences directing them to the mitochondrial inner membrane (Chacinska et al. 2009). Receptor proteins within large macromolecular complexes on the mitochondrial outer membrane recognize these signal sequences and facilitate translocation through the mitochondrial membranes. The complexity of the factors involved in mitochondrial protein integration will be discussed in a later section of this review (see Sect. IV. D.)

III. Mitochondrial Dynamics

Mitochondria are in constant movement within cells. To facilitate mitochondrial biogenesis and distribution of the mtDNA and proteome, mitochondria undergo constant cycles of fission and fusion (Fig. 24.2). In addition to mitochondrial proliferation and distribution within the cell, mitochondrial dynamics also affect organelle morphology, cell bioenergetics, and cell injury or death (Chan 2006a, b). Mitochondrial fission increases the number of mitochondria within the cell in concert with biogenesis and also provides a mechanism for isolating defective organelles targeted for degradation (Youle and Narendra 2011). Conversely, mitochondrial fusion is required for organelle mixing to distribute mitochondrial components through the organelle network. These dynamic events lead to alterations in mitochondrial size, number, and mass, and these changes are regulated by various physiologic stimuli and differentiation states.

A. Mitochondrial Fission

Drp1 (Dnm1 in yeast) is a ~80 kDa GTPase protein that regulates mitochondrial division in eukaryotic cells (Smirnova et al. 1998, 2001) (Fig. 24.2a). Drp1 is mostly cytosolic and forms distinct punctae on mitochondria that correspond to sites of ensuing fission. Mitochondrial fission is inhibited in cells expressing a dominant-negative GTPase mutation of Drp1 (Frank et al. 2001) or in cells containing siRNA that target Drp1 (Gandre-Babbe and van der Bliek 2008). Conversely, over-expression of functional Drp1 results in excessive fission leading to small fragmented mitochondria.

In vitro, Drp1 and Dnm1 forms ordered, oligomeric structures in the presence of a lipid template or non-hydrolysable GTP

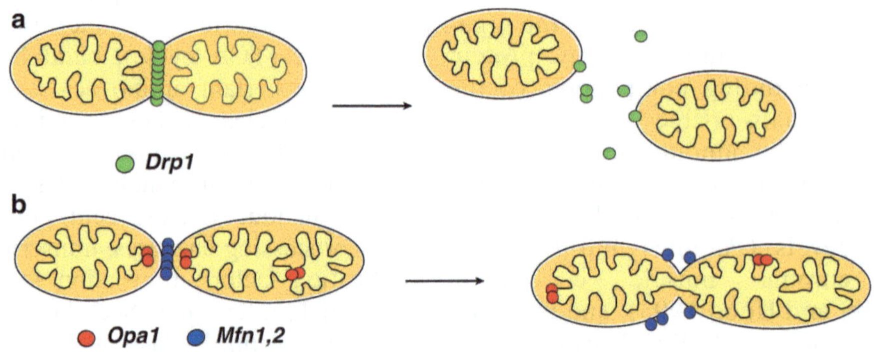

Fig. 24.2. Mammalian mitochondrial dynamics. (**a**) Mitochondrial fission is mediated by dynamin-related protein 1 (Drp1). (**b**) Mitochondrial fusion is regulated by mitofusins (Mfn1,2) on the outer mitochondrial membrane and Optic Atrophy 1 (Opa1) on the inner membrane.

(Yoon et al. 2001; Ingerman et al. 2005; Mears et al. 2011). The self-assembly properties of Drp1 can be disrupted by mutations within functional domains, and these alterations influence GTPase activity (Chang et al. 2010).

Additional partner proteins interact with Drp1 and Dnm1 in the mitochondrial fission complex, and distinct partners have been identified in mammals and yeast. Mitochondrial fission factor (Mff), mitochondrial elongation factor-1/mitochondrial dynamics proteins of 49 and 51 kDa (MIEF1/MiD49/51), and mitochondrial fission 1 protein (Fis1) have all been identified as Drp1 partners in mammals, while mitochondrial division protein 1 (Mdv1), Caf4 and Fis1 have been shown to interact with Dnm1 in yeast. In mammals, Mff, MIEF1/MiD49/51, and Fis1 are all anchored in the mitochondrial outer membrane by a transmembrane segment (Yoon et al. 2003; Gandre-Babbe and van der Bliek 2008; Palmer et al. 2011; Zhao et al. 2011). The cytoslic domains of these proteins are proposed to be involved in interactions with Drp1. It is still unclear what the distinct roles of Mff, MIEF1/MiD49/51, and Fis1 are in mediating mitochondrial fission and how each affect Drp1 function. The yeast mitochondrial fission complex contains Fis1, which recruits Dnm1 to the mitochondrial outer membrane (Mozdy et al. 2000; Zhang and Chan 2007). Additional adaptor proteins, Caf4 and Mdv1, are not conserved in mammalian cells, even though they are essential for mitochondrial fission in yeast (Tieu and Nunnari 2000; Griffin et al. 2005).

Additional proteins are known to co-localize with Drp1 in vivo and may affect Drp1 self-assembly and activity. The translocase of IMM 8 homologue A (Timm8a/DDP) is released from the intermembrane space of mitochondria and binds to cytosolic Drp1 to promote its translocation to mitochondrial fission sites (Arnoult et al. 2005). Endophilin B1 (Endo B1) contains a Bin-Amphiphysin-Rvs (BAR) domain, which promotes mitochondrial membrane curvature. Endo B1 also localizes to the cytosol and interacts with Bax, a pro-apoptotic factor (Cuddeback et al. 2001; Pierrat et al. 2001). At the mitochondrial surface, this interaction promotes Bax insertion in the outer membrane during the apoptotic cascade. Drp1 also can enhance Bax oligomerization (Montessuit et al. 2010), which suggests that these proteins work in concert to facilitate outer mitochondrial membrane permeabilization and subsequent cell death.

Post-translational modifications of Drp1 affect mitochondrial dynamics, and these have been extensively characterized. S-Nitrosylation of Drp1 enhances GTPase activity and oligomer formation, leading to excessive mitochondrial fragmentation in diseased neurons (Cho et al. 2009). Phosphorylation and dephosphoylation at distinct sites also contribute to changes in mitochondrial morphology in response to specific stimuli (Chang and Blackstone 2007; Cribbs and Strack 2007). Sumoylation and ubiquitylation of Drp1 also affect the amount of mitochondrial division that occurs within the cell (Wasiak et al. 2007; Park et al. 2010), which can have serious consequences on mitochondrial biogenesis and subsequent function.

B. Mitochondrial Fusion

The integral membrane proteins mitofusin 1 and mitofusin 2 (Mfn1,2, Fzo1 in yeast) and optic atrophy 1 (Opa1, Mgm1 in yeast) control mitochondrial fusion (Fig. 24.2b). Mfn1,2 reside in the mitochondrial outer membrane, while Opa1 sits at the mitochondrial inner membrane. In yeast, an additional adaptor protein, Ugo1, tethers Fzo1 and Mgm1, but no comparable factor has been discovered in mammals.

At an early stage of mitochondrial fusion, Mfn1,2 are thought to form homo- or hetero-oligomers that tether adjacent mitochondria (Benard and Karbowski 2009). Subsequent membrane fusion likely requires the energy from GTP hydrolysis. Knockout of Mfn1,2 results in the formation of small fragmented mitochondria, highlighting their conserved role in mediating mitochondrial fusion

(Chen et al. 2003, 2007). When compared with the loss of Mfn 2 in mammals, Mfn1 depletion leads to a more severe mitochondrial fragmentation phenotype with a concurrent increase in cell death (Bach et al. 2003; Chen et al. 2005).

Cells lacking Opa1 also exhibit increased mitochondrial fragmentation (Sesaki et al. 2003). Opa1 together with scaffolding proteins such as prohibitins, are thought to be involved in the maintenance of cristae morphology, possibly by assembling into higher-order structures (Merkwirth and Langer 2008). More recently, Opa1 (Ban et al. 2010) and Mgm1 (Meglei and McQuibban 2009) have been shown to oligomerize and associate with negatively charged lipids, like CL. Interestingly, Opa1 localizes to the mitochondrial inner membrane and the intermembrane space due to differential proteolytic cleavage. The electrochemical gradient across the inner membrane has been shown to influence cleavage and, thereby, affects cristae morphology and release of cytochrome *c* leading to cell death. Conversely, Mfn1,2-deficient cells exhibit extensive mitochondrial fragmentation, but apoptosis is not enhanced (Chen et al. 2003; Chen and Chan 2005). Thus, factors that lead to mitochondrial fission may contribute to apoptosis, but they are not required for cell death.

IV. Biogenesis: Increasing Mitochondrial Mass

The mechanisms involved in controlling mitochondrial proliferation are starting to be elucidated. With constant mitochondrial turnover, an increase in mitochondrial mass means that the synthesis of both mitochondrial and nuclear-encoded proteins are added to pre-existing organelles in such a way that functional consistency is preserved. Mitochondrial fission then works to generate a well-distributed mitochondrial network throughout the cell. In the end, mitochondrial biogenesis requires the coordinated actions of roughly 1,000 genes and 20 % of cellular proteins, and a complex regulatory network of transcription factors.

A. Replication of the Mitochondrial Genome

Unlike nuclear DNA, which only replicates once during cell division, mtDNA is continuously recycled (Bogenhagen and Clayton 1977). Replication of mtDNA is performed by a number of nuclear-encoded proteins (Fig. 24.3a), including the mitochondrial DNA polymerase γ (Pol γ, encoded by the POLG gene) (Kaguni 2004) and the accessory p55 protein (encoded by the POLG2 gene). A 195 kDa, hetero-trimeric replication complex, composed of the catalytic Pol γ subunit and two p55 subunits, synthesizes new mtDNA (Fig. 24.3a). The Pol γ subunit has DNA polymerase, $3' \rightarrow 5'$ exonuclease, and 5′-deoxyribose phosphate lyase activities (Longley et al. 1998a, b). The p55 subunits act as DNA binding factors that increase the affinity of the replication complex for template DNA, which enhances polymerase activity (Lim et al. 1999).

Additional components contribute to mtDNA replication. The mitochondrial single-stranded binding protein (mtSSB, Fig. 24.3a) stabilizes single-stranded regions of DNA at replication forks and promotes Pol γ activity (Mikhailov and Bogenhagen 1996). Twinkle (Fig. 24.3a) is a $5' \rightarrow 3'$ helicase that unwinds double-stranded mtDNA at the replication fork, facilitating mtDNA synthesis (Korhonen et al. 2003). Twinkle may also regulate mtDNA copy number within cells (Tyynismaa et al. 2004). There is also a strong correlation between the levels of transcription factors and mtDNA (Larsson et al. 1998), which may indicate that these factors bind to mtDNA to protect against oxidative damage (Ghivizzani et al. 1994). Similarly, the expression of the mtSSB protein is up-regulated when mtDNA content increases, which may help prevent mtDNA damage and regulate mitochondrial biogenesis.

The overall amount of mtDNA varies by cell type (Veltri et al. 1990) and also depends

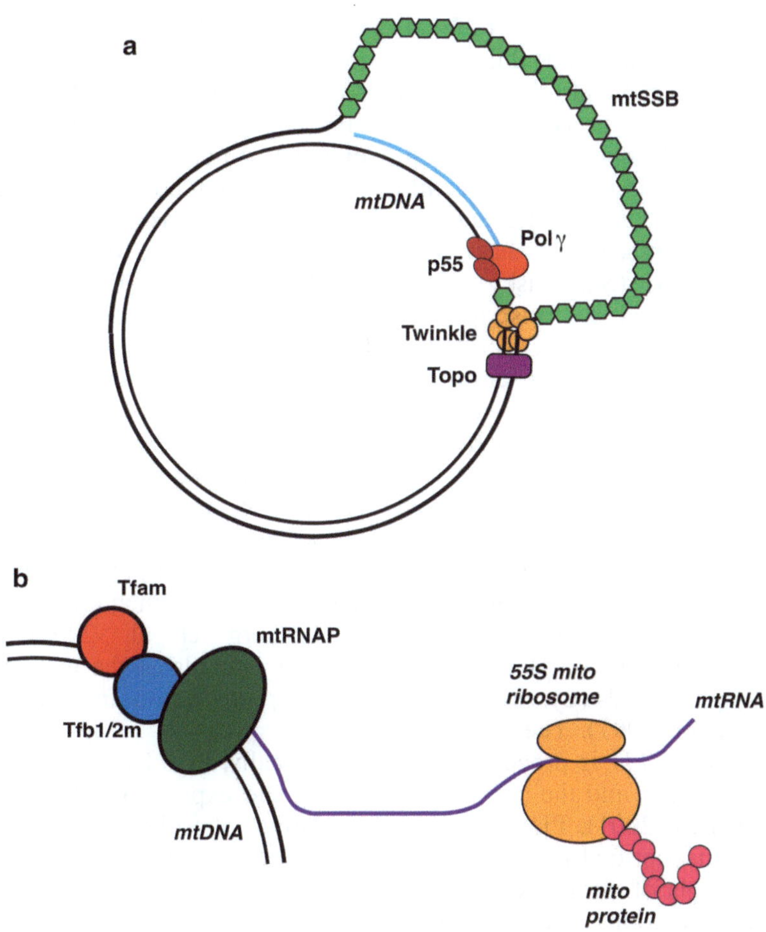

Fig. 24.3. Mitochondrial replication, transcription, and translation. (**a**) Factors involved in mitochondrial replication are highlighted. Mitochondrial DNA (mtDNA), mitochondrial single-stranded binding protein (mtSSB). (**b**) A schematic representation of mitochondrial transcription and translation is shown. Mitochondrial RNA polymerase (RNAPmt) transcribes mitochondrial mRNA (POLRMT) in concert with transcriptional factors (Tfam and Tfb1/2). Mitochondrial protein synthesis is then mediated by the *55S* mitochondrial ribosome.

on local energy requirements (Marcuello et al. 2005). After the mitochondrion has enough mtDNA, membrane area, and membrane proteins, it can undergo fission to form two separate organelles. Evidence suggests that mitochondria can also undergo fusion and exchange genetic material between one another (Chen et al. 2007).

Factors regulating the expression of proteins in the mtDNA replication machinery are not well known. Nuclear respiratory factor-1 (NRF-1) is one transcription factor that has been identified to play such a role. NRF-1 expression is related to cellular ATP levels. NRF-1 binds to promoter regions of POLG and POLG2. NRF-1 also regulates the expression of many mitochondrial proteins and transcription factors, which may help to coordinate mtDNA replication with mitochondrial proliferation (Scarpulla 2006).

B. Transcription of Mitochondrial Genes

The initiation of transcription in mitochondria requires three types of proteins (Fig. 24.3a): the mitochondrial RNA polymerase (mtRNAP), mitochondrial transcription factor A (Tfam), and mitochondrial transcription factors B1 and B2 (Tfb1m, Tfb2m). The proteins mtRNAP, Tfam, and Tfb1m or Tfb2m assemble at mitochondrial promoters and facilitate transcription. The molecular details of the complex involved in mitochondrial transcription were recently determined (Ringel et al. 2011), and the basal transcription machinery has been reconstituted in vitro (Smidansky et al. 2011). Many genes lack a transcriptional initiation site because they are transcribed as part of larger, concatenated transcriptional units, which require additional processing to generate mature RNAs (Rorbach and Minczuk 2012).

In yeast, mtDNA transcription is initiated at ~20 transcriptional units throughout the genome (Poyton and McEwen 1996). In vertebrates, transcription is initiated bi-directionally within the d-loop of mtDNA (Fig. 24.1a) at two promoters, PH and PL for heavy (H) and light strands (L), respectively (Shadel and Clayton 1997). The polygenic transcript directed by PH is processed into 14 tRNAs, 12 mRNAs, and the two rRNAs (Garesse and Vallejo 2001). PL directs the synthesis of a transcript that is processed to one mRNA and eight of the 22 tRNAs present in the mitochondrion. Both promoters (PH and PL) share a critical upstream enhancer that serves as the recognition site for Tfam, which stimulates transcription through specific binding of these enhancer elements. Tfam has been shown to bend and unwind DNA, which promotes transcription upon binding mtDNA (Fisher et al. 1992). In addition to specific promoter recognition, Tfam binds non-specific DNA with high affinity. This property along with its abundance in mitochondria suggests that it plays a role in the stabilization and maintenance of mtDNA through its binding to non-promoter sites (Parisi et al. 1993).

In addition to Tfam, a highly interconnected network of transcription factors regulates nuclear genes encoding mitochondrial proteins during mitochondrial biogenesis. This regulation also encompasses proteins that control replication and transcription of the mitochondrial genome. The molecular regulators of this inter-genomic coordination include nuclear-encoded regulatory proteins, such as NRF-1 and NRF-2, Tfam, Tfb1m and Tfb2m, and the cyclic-AMP response element binding protein (CREB). These factors also regulate expression of genes encoding proteins in the respiratory chain complexes (Herzig et al. 2000; Scarpulla 2002). Peroxisomal proliferator-activated receptor-γ (PPARγ) co-activator 1 α (PGC1-α) and PGC1-β act upstream to modulate these nuclear regulatory proteins.

The PGC-1 family of proteins governs the mitochondrial genome through regulation of nuclear-encoded genes that regulate mtDNA replication, transcription, and translation. Genes regulated by PGC-1 proteins include POLG, TWINKLE, TFAM, TFB1M, TFB2M, and POLRMT, which encodes the mtRNAP (Falkenberg et al. 2007). Additionally, PGC1-α stimulates the expression of other transcription factors involved in the coordinated expression of mitochondrial genes. Specifically, PGC1-α helps to directly coordinate the expression of genes involved in oxidative phosphorylation and has been identified as a major regulator of mitochondrial biogenesis (Knutti and Kralli 2001; Puigserver and Spiegelman 2003). An increase in the mRNA and protein levels of PGC1-α accompany mitochondrial proliferation during adaptive thermogenesis (Lowell and Spiegelman 2000) and muscle regeneration (Duguez et al. 2002). Additionally, overexpression of PGC1-α leads to mitochondrial proliferation in myoblasts (Wu et al. 1999) and the hearts of the transgenic mice (Lehman et al. 2000). These observations highlight the important role of PGC1-α in regulating mitochondrial biogenesis.

Nuclear-encoded transcription factors were also identified as being required for the expression of the cytochrome *c* gene to form a functional respiratory complex (Scarpulla 1997). The cytochrome *c* promoter contains a recognition site for NRF-1 (Evans and Scarpulla 1989). Additional NRF-1 target genes include separate components of the respiratory chain complexes (Virbasius et al. 1993). The regulatory network of NRF-1

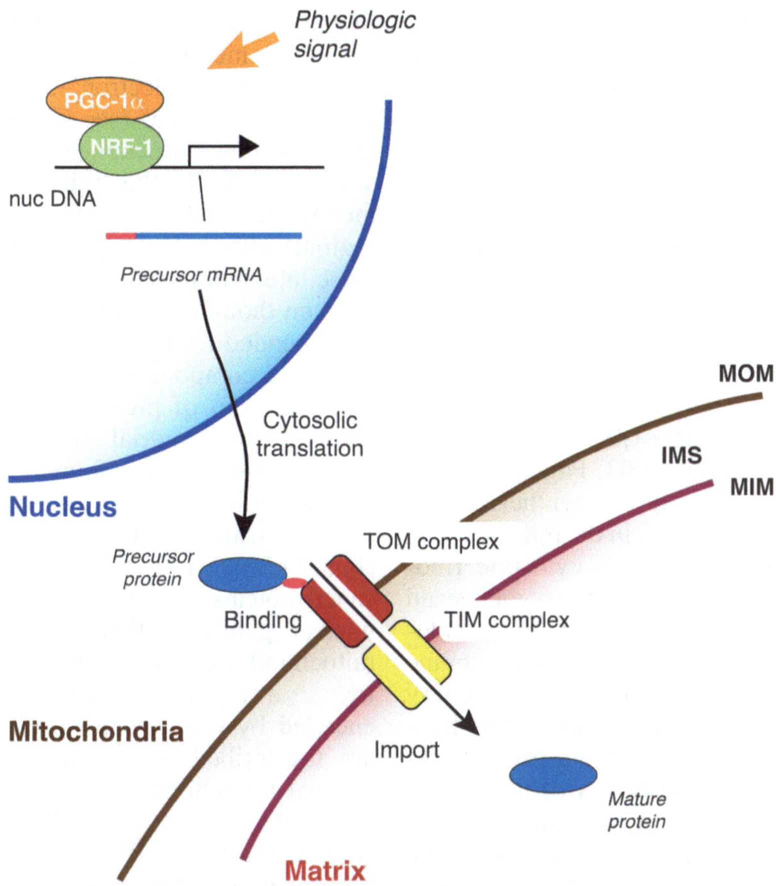

Fig. 24.4. Signals influence mitochondrial biogenesis. Physiologic signals, including calcium and oxidative stress, promote expression of nuclear genes through transcription factors (ex. PGC-1 *α* and NRF-1). Precursor proteins are then synthesized in the cytosol and bind to the mitochondrial protein import machinery (TOM and TIM complexes) to facilitate translocation. Proteolysis removes the target sequence to yield mature protein. Mitochondrial outer membrane (MOM), mitochondrial inner membrane (MIM), intermembrane space (IMS).

extends to nuclear proteins as well. NRF-1 and NRF-2 trigger the expression of nuclear genes coding for proteins involved in assembly of the respiratory chain complexes, subunits of the mtDNA replication and transcription machinery, mitochondrial and cytosolic enzymes of the heme biosynthetic pathway, and components of mitochondrial protein import. NRF-1 activity can also be regulated by phosphorylation and/or interactions with PGC-1 proteins (see Fig. 24.4). In some cases, NRF-1 phosphorylation influences its nuclear translocation, DNA binding, and transcriptional activity (Gugneja and Scarpulla 1997). Interestingly, NRF-1 targets the nuclear TFAM gene, which is consistent with the idea that transcription factors coordinate interactions between the nucleus and mitochondria during mitochondrial biogenesis.

PGC-1 partners are linked to additional transcription factors that regulate mitochondrial biogenesis. PGC1-α co-activates the nuclear receptor PPARγ, a key regulator of nuclear genes coding for enzymes with function in the mitochondrial fatty acid oxidation (Vega et al. 2000). PGC1 proteins were found to co-activate estrogen-related receptors (ERRs), ERRα and ERRβ, during mitochondrial biogenesis (Huss et al. 2002; Schreiber et al. 2003). Transcriptional regulation is linked to mitochondrial dynamics as well. PGC-1β increases expression of Mfn2 by co-activation of ERRα (Liesa et al. 2008). These observations suggest that ERRs and PPARs may drive distinct but overlapping mitochondrial pathways downstream of PGC1 proteins. The ability of these co-regulators to interact with multiple DNA-binding factors enables the integration of signals throughout the broad mitochondrial gene expression system.

C. Translation of Mitochondrial Proteins

Mitochondria contain ~1,000 different proteins (Sickmann et al. 2003; Pagliarini et al. 2008). Of these, ~1 % are synthesized on mitochondrial ribosomes in the matrix, while the rest are synthesized on cytosolic ribosomes. Mitochondrial proteins synthesized in the matrix retain features of their endosymbiotic ancestor and are crucial membrane subunits of the respiratory machinery (Fig. 24.1a). Mitochondria also contain hundreds of proteins derived from prokaryotes that are synthesized in the cytosol, because the genes encoding them were transferred to the nucleus during evolution. As eukaryotic cells evolved, some proteins of non-mitochondrial origin developed functional roles within mitochondria, and these factors are also encoded by nuclear genes and synthesized in the cytosol.

Mitochondrial translation systems more closely resemble the standard prokaryotic translation machinery than their eukaryotic analogue. For example, the range of antibiotics inhibiting mitochondrial translation coincides with antibiotics that inhibit prokaryotic ribosomes (Borst and Grivell 1971). The total number of mitochondrial translation products varies from 8 in yeast (Pel and Grivell 1993), to 13 in mammals (Chomyn et al. 1986), and about 20 in plants (Levings and Brown 1989). Therefore, mitochondrial translation is tailored to a small subset of proteins, and the mitochondrial ribosome has evolved under distinct environmental pressures to translate a shrinking mitochondrial genome. All of the proteins synthesized by the mammalian mitochondrial translational system are localized in the mitochondrial inner membrane where they function as subunits in the electron transfer and ATP synthase complexes (see Chap. 6). There is evidence that mitochondrial ribosomes are associated with the mitochondrial inner membrane and are enriched in cristae regions, although the proteins required for these interactions have not been defined.

Even though the mitochondrial ribosome is descended from a bacterial ribosome, the structural components are noticeably different. The ratio of protein to rRNA mass in animal mitochondrial ribosomes (2:1) is inverted from the ratio found in bacterial and archaeal ribosomes (1:2), which yields a more buoyant 55S mitochondrial ribosome (Fig. 24.3b). All of the mitochondrial ribosomal proteins are encoded by rapidly evolving nuclear genes (O'Brien 2002), while the mitochondrial rRNAs, which are also evolving at a rapid rate (Kurland 1992), are encoded by the mitochondrial genome and are transcribed within the mitochondrion. Animal mitochondrial ribosomes have only two rRNA species, a 12S rRNA in the small subunit and a 16S rRNA in the large subunit (Glotz et al. 1981) in contrast to larger homologous rRNAs found in typical bacterial ribosomes. These smaller mitochondrial rRNAs are not shortened at random positions throughout the primary sequence. Rather, they lack specific regions of secondary structure or entire domains (Sharma et al. 2003; Mears et al. 2006). The small rRNA of yeast mitochondria also lacks the equivalent of a Shine-Dalgarno sequence (Dekker et al. 1993). As in many eukaryotic cytoplasmic mRNAs, the 5′ leader sometimes contains one or more short reading frames, which could mediate translational control of downstream coding sequences.

Animal mitochondria have just 22 tRNAs (Fig. 24.1a), one for each amino acid except for leucine and serine, which have two. The tRNAs found in mammalian mitochondria have several unusual features distinguishing them from their bacterial and eukaryotic equivalents. In a trend similar to that observed in mitochondrial rRNAs, mitochondrial tRNAs are generally shorter than tRNAs from other bacteria and eukaryotes and often lack nucleotides that play important roles in creating the L-shaped tertiary structure of cytoplasmic tRNAs (Zagryadskaya et al. 2004). Chemical and enzymatic probing has led to the idea that these tRNAs fold into the basic cloverleaf structure of canonical tRNAs, but lack a number of conserved tertiary interactions and have a weaker three-dimensional structure (Helm et al. 2000).

Several factors required for the general processes of mitochondrial translation initiation, elongation, termination and recycling have been identified, and a number of detailed studies on these factors have been performed. The cycle of polypeptide chain synthesis in mammalian mitochondria is similar to the process in prokaryotes (Spremulli et al. 2004). Mitochondrial translation involves two initiation factors, mitochondrial initiation factor 2 (mtIF2) and 3 (mtIF3). Although IF1 has been viewed as a universal translational initiation factor, no homologous protein exists in mitochondria. The elongation phase of mitochondrial translation has been conserved during mitochondrial evolution, and all three prokaryotic elongation factors, mtEF-Tu, mtEF-Ts, and mtEF-G, are conserved in mammalian mitochondria. The properties of these elongation factors have been reviewed extensively (Smits et al. 2010). In the current model for mitochondrial translation termination, the mRNA stop codon is recognized by mitochondrial release factor 1a (mtRF-1a). The process through which mtRF1a dissociates from mitochondrial ribosomes is unknown.

D. Protein Import and Sorting

Proteins encoded in the mitochondria and nucleus must work in concert to maintain the mitochondrial proteome. This coordinated expression of mitochondrial proteins has resulted in a complex environment for protein import and folding. At present, four principal pathways of mitochondrial protein import have been characterized: (1) the presequence pathway to the matrix and inner membrane, (2) the carrier protein pathway to the inner membrane, (3) the redox-regulated import pathway to the inter-membrane space, and (4) the β-barrel pathway into the outer membrane. Each pathway requires multisubunit protein complexes that act as chaperones and gating factors. Assembly of these factors is essential to expand metabolic capacity of cells during mitochondrial biogenesis. Several detailed reviews of protein import are available (Eilers et al. 1988; Becker et al. 2009; Chacinska et al. 2009).

Nuclear-encoded proteins targeted to mitochondrial sub-compartments are translated in the cytoplasm and require cytosolic chaperones for trafficking and import (Young et al. 2003). Heat shock protein (HSP)70 and HSP90 serve this functional role maintaining precursor proteins in an unfolded state and preventing protein aggregation during delivery to the translocase of the outer membrane (TOM) complex (Neupert and Herrmann 2007), which serves as a central gate through the outer membrane (Fig. 24.4). In addition to receptor proteins, the TOM complex is comprised of nuclear-encoded proteins that form a protein-conducting channel and regulate the structural integrity of the complex. A sorting and assembly machinery, or SAM complex, exists to integrate membrane proteins into the mitochondrial outer membrane as well. For example, the SAM complex is required for integration and assembly of the TOM complex. The diversity of proteins requires that the SAM complex be able to insert β-barrel proteins and some α-helical proteins (Stojanovski et al. 2007).

Nearly all mitochondrial pre-proteins targeted to the inner membrane and matrix are imported via the TOM complex (Fig. 24.4), and two distinct mechanisms for translocation of precursor proteins through the channel have been demonstrated. First, proteins with a mitochondrial targeting sequence (MTS) are translocated as unfolded polypeptides with the amino-terminal sequence passing first. The MTS continues its journey through a series of binding sites on the translocase of the inner membrane, or TIM complex, to facilitate entry into the matrix or integration into the mitochondrial inner membrane (Fig. 24.4) (Esaki et al. 2004; Chacinska et al. 2005). In the second mechanism, precursor proteins are translocated through the outer membrane in a loop configuration, as opposed to unfolded strands. This topological difference leaves both termini of the protein exposed to the cytosol while a central segment is positioned towards the mitochondrial intermembrane space (Wiedemann et al. 2001).

Often, the MTS signal is cleaved by mitochondrial proteases after targeting (Fig. 24.4). However, numerous non-cleavable targeting/sorting signals have been identified within mature regions of mitochondrial proteins, which reveal the diversity of import pathways. Examples of non-MTS import include all mitochondrial outer membrane proteins, the majority of intermembrane space proteins, several inner membrane proteins with multiple transmembrane helices, and a few matrix proteins.

V. Signals for Biogenesis

The number of mitochondria within a cell varies widely between different organisms, where some cells contain a single mitochondrion, while others contain several thousand. Similarly, different tissues exhibit differing levels of mitochondrial mass. The number of mitochondria within a cell can also be dynamic, and the energetic demands often dictate the fluctuations that occur. Various cell signals regulate mitochondrial proliferation, and the following sections examine these diverse stimuli.

A. Calcium Flux

Links between calcium concentration and mitochondrial gene expression have been observed, and the regulation of transcription factors through calcium signaling can regulate mitochondrial abundance (Williams and Rosenberg 2002). Effector proteins sense changes in the intracellular calcium and ATP levels and alter cellular bioenergetics. This adaptation is critical in muscle tissues, where chronic energy stress stimulates mitochondrial biogenesis.

Regulation of mitochondrial proliferation is mediated by calcium/calmodulin-dependent protein kinase (CAMK) and protein kinase C (PKC). In response to contractile activity in skeletal muscle, these enzymes trigger a cascade of events that lead to an increase in oxidative fibers along with an increase in mitochondrial number (Holloszy and Coyle 1984; Chin et al. 1998). Transgenic mice with a constitutively active, skeletal muscle-specific CAMK exhibit increased mitochondrial biogenesis, which demonstrates the calcium-dependent regulation of mitochondrial gene expression (Wu et al. 2002). Additionally, over-expression of PGC-1α in the skeletal muscle of transgenic mice leads to an induction of genes involved in mitochondrial oxidative metabolism (Lin et al. 2002). Myogenic cell lines were also used to show that CaMK activates PGC-1α transcription (Handschin et al. 2003).

The activity of AMP kinases (AMPKs), which are regulated by changes in the intracellular ATP/AMP ratio, was shown to influence PGC-1α levels and mitochondrial abundance (Zong et al. 2002). NRF-1 is also induced as part of the adaptation of skeletal muscle to exercise training (Murakami et al. 1998). Similar results were observed using cultured myotubes with elevated calcium to mimic exercise-induced mitochondrial biogenesis (Ojuka et al. 2003). Rats treated with a creatine analog to mimic muscle adaptations seen during exercise were found to activate AMPK and NRF-1, leading to increased cytochrome *c* content and mitochondrial density (Bergeron et al. 2001).

B. Energy Demands

Cellular demands for increased energy supply are often met by an increase in respiratory activity, usually accompanied by an increase in mitochondrial mass. Under conditions in which yeast cells are dependent on mitochondria for energy via oxidative metabolism, growth and division of the cell is matched by an increase in mitochondrial mass. When an energy deficit is sensed in mammalian cells, mitochondrial biogenesis is induced by the activity of sirtuins (SIRTs) or AMPK (Lagouge et al. 2006; Jager et al. 2007). AMPK also indirectly activates SIRT1, which suggests a convergence of these two pathways (Canto et al. 2009).

Conversely, a decrease in mitochondrial biogenesis and function are seen during aging, physical inactivity, obesity, and insulin resistance. These changes often parallel decreases in the expression of transcriptional regulators, like PGC-1α, PGC-1β, and NRF-1 (Handschin and Spiegelman 2006; Reznick et al. 2007). In contrast to decreases in expression by inactivity, PGC-1α expression is induced by exercise in mammalian skeletal muscle (Goto et al. 2000; Baar et al. 2002; Terada et al. 2002; Pilegaard et al. 2003). Furthermore, regular exercise is known to increase mitochondrial activity and content in rats (Holloszy 1967) and humans (Gollnick et al. 1972).

C. Cell Cycle Regulation

Cell cycle control of mitochondrial biogenesis remains controversial. The majority of mtDNA replication occurs in the late S and G2 phases of the cell cycle (Bogenhagen and Clayton 1977). Consequently, the replication of mtDNA is not linked with mitochondrial proliferation. As stated earlier, the mtDNA number varies between different tissues, and a defined range is usually preserved (Moraes 2001). The number of mtDNA within a cell can be modulated to meet cellular energy demands, and alterations in the copy number of mtDNA are observed during different stages of cell growth and differentiation (Renis et al. 1989).

Recent studies showed that mitochondrial mass and mitochondrial membrane potential increased from early G1 to G1/S phase of the cell cycle (Lee et al. 2007). Additionally, mtDNA levels increased from G1/S to G2, concomitant with an increase of NRF-1 level. Taken together, these studies suggest that mitochondrial biogenesis and dynamics are closely regulated throughout the cell cycle.

D. Oxidative Stress

Mitochondrial biogenesis is stimulated when mitochondrial function is disrupted by oxidative stress, nitric oxide (NO) or carbon monoxide (Nisoli et al. 2003; Suliman et al. 2007). Increased oxidative stress in cells leads to elevated levels of both NRF-1 and Tfam mRNAs as a means to increase mtDNA levels and respiration activity (Miranda et al. 1999). On the other hand, rodents subjected to chronic hypoxia also exhibit increased mitochondrial biogenesis (Meerson et al. 1973). Furthermore, mitochondria extracted from hypoxic hearts show increased rates of ATP synthesis (Eells et al. 2000).

In mammals, NO induces an increase in mitochondrial biogenesis in a variety of cell types (Nisoli et al. 2003). This NO effect is dependent on cyclic-GMP signaling and linked to PGC-1α activation. Furthermore, focused disruption of the endothelial NO synthase (eNOS) gene in vivo causes a severe reduction in mitochondrial mass. Therefore, the NO signaling pathway may represent a conserved mechanism by which many types of mammalian cells regulate mitochondrial content.

Excess production of reactive oxygen species (ROS) has been shown to elicit additional signaling pathways as a response to oxidative stress. Non-lethal concentrations of hydrogen peroxide (H_2O_2) induce an increase in mitochondrial mass and mtDNA copy number within human cells (Lee et al. 2000). The mRNA levels of PGC-1 and NRF-1 also increase when cells are treated with H_2O_2. Moreover, human cells respond to defective respiratory function by promoting expression of Tfam and NRF-1 through an H_2O_2-dependent signaling pathway

(Suzuki et al. 1998). Oxidative stress induced by injection of lipopolysaccharides (LPS) increased NRF-1 activity, Tfam gene expression, and stimulated mtDNA replication and cell proliferation in rat liver (Suliman et al. 2003). It is also worth noting that induction of PGC1 proteins up-regulates a broad spectrum of ROS detoxification systems (St-Pierre et al. 2003). Taken together, these observations clearly suggest that oxidative stress is one of the factors involved in the increase of mitochondrial abundance in mammalian cells.

VI. Quality Control: Removing Mitochondrial Damage

Collectively, mitophagy and mitochondrial biogenesis work to maintain steady-state mitochondrial numbers needed to meet cellular metabolic demands. Quality control measures ensure that damaged mitochondrial components, or the mitochondrion as a whole, are selectively removed to maintain a functional organelle network (Fig. 24.5). Accumulation of dysfunctional mitochondria triggers a cascade of events resulting in increased mitochondrial division to isolate the damage. However, excessive fragmentation of mitochondria can ultimately lead to cell death.

A. Dysfunctional Mitochondria

Human mtDNA is more susceptible to oxidative damage and acquires mutations at a higher rate than nuclear DNA (Richter et al. 1988). This increase is likely due to the mtDNA environment, which includes high levels of ROS, lack of DNA-packaging complexes (i.e. histones), and limited ability to repair DNA damage. To deal with higher levels of ROS, mammalian cells have developed antioxidant enzymes (Ames et al. 1993). Although these counter measures can dispose of ROS, a small amount may elude these mechanisms and induce damage to cellular constituents including nucleic acids, proteins, and lipids. Therefore, excessive ROS production by mitochondria is harmful to the entire cell, and over-production can eventually culminate in cell death.

Damage to DNA by oxidative stress comprises oxidative damage to bases and sugar phosphates as well as single- or double-strand breaks in DNA. Moreover, adjacent single-strand breaks can be converted to double-strand breaks during replication. Unless the damages are repaired or removed, both single- and double-strand breaks in mtDNA lead to an inefficient, if not useless, templates for replication (Souza-Pinto et al. 1999). Consequently, the accumulation of oxidative damage can inhibit replication and affect mtDNA copy number.

B. Mitochondrial Degradation

Protein turnover and degradation are an important part of mitochondrial biogenesis. Proteins within different mitochondrial sub-compartments exhibit distinct half-lives, and this difference likely reflects separate degradation mechanisms (Hare 1990). Mitochondrial proteases are responsible for protein degradation to remove potentially harmful polypeptides from the cell and to control the concentrations of various enzymes and regulatory proteins (Fig. 24.5) (Baker et al. 2011). Degradation of mitochondrial fragments is mediated by lysosomal proteases via the process of autophagocytosis (Fig. 24.5c), and this releases amino acids that can be used for other purposes (Takeshige et al. 1992).

Misfolded proteins can arise through various sources of damage, including ROS-induced protein damage, mtDNA mutations, and errors in cytosolic and protein translation. Accumulation of dysfunctional proteins is monitored by chaperones and proteases within each mitochondrial sub-compartment. Several ubiquitin ligases localize to the mitochondrial outer membrane to regulate mitochondrial dynamics by ubiquitylating proteins involved in organelle fission and fusion (Neutzner et al. 2008). Studies have also demonstrated a role for ubiquitylated-protein degradation with the p97 protein, which regulates mitochondrial morphology

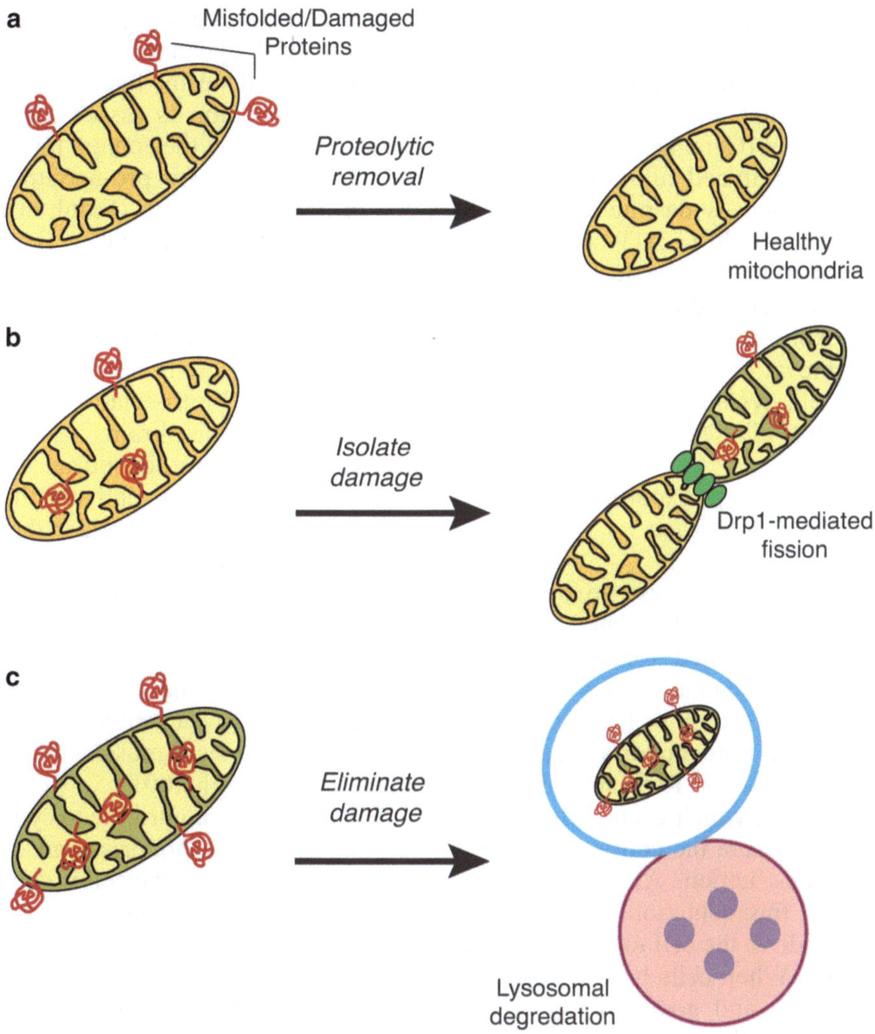

Fig. 24.5. Quality control in mitochondrial biogenesis. (**a**) Misfolded and/or damaged proteins can be removed via ubiquitin-mediated proteolysis to yield healthy mitochondria. (**b**) Alternatively, dysfunctional proteins or damaged compartments within mitochondria can be isolated through mitochondrial fission. (**c**) Damaged mitochondria are targeted for mitophagy to selectively retain organelles with functional respiration.

and degradation of cell death factors (Jarosch et al. 2002).

Mitophagy is intimately linked with mitochondrial dynamics (Fig. 24.5c), which are believed to be important in maintaining cell homeostasis. Parkin, a ubiquitin ligase, accumulates only on mitochondria that have lost their electrochemical gradient. In this scenario, mitofusins are ubiquitylated by parkin, and subsequent proteasome-dependent degradation prevents the fusion of defective organelles. Thus, dysfunctional organelles within the mitochondrial network are isolated and subsequently degraded by mitophagy (Gegg et al. 2010; Tanaka et al. 2010). Similarly, it has been shown that mitochondria preferentially undergo fusion only when they possess an intact electrochemical gradient, and mitochondria without a suitable membrane potential are more likely to remain isolated and eventually are targeted for autophagy (Twig et al. 2008).

Drp1-induced mitochondrial fragmentation is propagated through dephosphorylation of Drp1 by calcineurin, which promotes Drp1 translocation for the mitochondrial outer membrane (Cereghetti et al. 2008). Drp1 knockdown in cell cultures was shown to suppress mitophagy despite observations that mtDNA was lost, ATP levels were depleted, and mitochondria lost respiratory function (Parone et al. 2006). Concomitantly, Drp1 over-expression leads to excessive fission and removal of mitochondria from the eukaryotic cell, eventually leading to apoptosis (Arnoult et al. 2005).

C. Cell Death: Damage Exceeds Quality Control Capabilities

During oxidative stress response, mitochondrial proliferation is stimulated to meet energy demands needed for cell survival, including repair of damage and synthesis of new proteins. On the other hand, proliferation of defective mitochondria leads to an excessive production of ROS and additional oxidative damage, which eventually reaches a tipping point where the cell is irreversibly targeted for destruction. At this threshold, mitochondria fragmentation and loss of cytochrome *c* drive the cell to an irreversible apoptotic fate. When cells have a high antioxidant capacity and good quality control measures to limit the amount of oxidation, mitochondrial numbers within the cell increase. In this case, mitochondrial biogenesis will lead to increased energy levels within the cell.

The exact mechanism by which apoptosis occurs is not well understood, but it is clear that mitochondria play a key role in regulating cell death. The release of cytochrome *c* from the mitochondria to the cytosol is associated with an increase in the activity of caspases, a family of proteases that play a key role in the process of apoptosis (Shimizu et al. 1999; Goldstein et al. 2000). Proteins in the B-cell lymphoma 2 (Bcl-2) family play distinct and opposing roles in mediating cell death by either promoting (pro-apoptotic) or preventing (anti-apoptotic) cytochrome *c* release.

Recent evidence suggests that abnormal mitochondrial dynamics may contribute to cell injury and death. Increased mitochondrial fragmentation is associated with permeabilization of the outer mitochondrial membrane to release pro-apoptotic factors that promote cell death (Frank et al. 2001; Estaquier and Arnoult 2007). Furthermore, both rotenone and 6-hydroxydopamine have been shown to induce Drp1-dependent mitochondrial fragmentation and oxidative stress (Barsoum et al. 2006; Gomez-Lazaro et al. 2008). S-nitrosylation of Drp1 in Huntington's disease brains has been identified to promote mitochondrial fragmentation and subsequent cell death. Similar observations have been made in the brains of Alzheimer's disease patients (Cho et al. 2009). In general, affected tissues of patients with neuromuscular diseases exhibit augmented oxidative stress, as well as an increase in mitochondria and mtDNA molecules. Similarly, excessive exercise has been shown to induce apoptosis in muscle tissue (Phaneuf and Leeuwenburgh 2001).

Several factors contribute to cytochrome *c* release from mitochondria, and once cytochrome *c* is released, the cell is committed to death. Disruption of mitochondrial membrane potential triggers a series of events leading to the opening of a permeability pore. Formation of this mitochondrial outer membrane pore, coupled with loss of cristae structure, leads to the release of cytochrome c and pro-apoptotic molecules from the intermembrane space of mitochondria to the cytosol (Kroemer et al. 1998). Additionally, increased ROS may induce the peroxidation of CL in the mitochondrial inner membrane, which triggers dissociation of cytochrome *c* from CL (Imai et al. 2003). During apoptosis, the apoptotic caspase cascade is activated and nuclear DNA is fragmented. Alternatively, necrosis can occur due to the collapse of the respiratory function, resulting in over-production of ROS and insufficient ATP supplies. The loss of bioenergetic capacity along with the continued accumulation of oxidative stress culminates in cell death.

VII. Conclusion

Within different organisms and between distinct tissues, mitochondria have evolved under select genetic pressures to meet the energy demands of different eukaryotic cells. Despite these differences, the process of mitochondrial biogenesis retains many conserved features. Given the complexity of the process, an immensely interwoven network of transcription factors is used to control the expression of genes that mediate mitochondrial proliferation. Almost all mitochondrial proteins are nuclear encoded, but the select few proteins encoded by mitochondria are supremely important for functional oxidative phosphorylation. Therefore, cytoplasmic and mitochondrial protein synthesis must be synchronized to accumulate equivalent levels of mitochondrial components (DNA, proteins, and lipid) while increasing overall mitochondrial mass. This process involves significant regulation at the levels of protein transcription, translation, folding and import into designated mitochondrial sub-compartments.

Mitochondrial fission and fusion are intimately linked with mitochondrial biogenesis. A disperse mitochondrial network provides energy to remote regions within the cell, and fission prevents perinuclear clustering of mitochondria. Combining content from isolated mitochondria is mediated by fusion, which has the potential to rescue defective organelles that have accumulated damage due to oxidative stress. However, extensively damaged mitochondria are targeted for removal through isolation by Drp1-induced fragmentation and eventual engulfment during mitophagy to recycle molecules needed for continued mitochondrial regeneration. This continuous cycle of building and tearing down subunits within the mitochondrial network works to selectively maintain a healthy pool of organelles optimized for energy production. Overall, this process lacks apparent efficiency. However, the benefit of the observed complexity is that mitochondria can dynamically adapt to host cell energy demands and maintain cell homeostasis.

Acknowledgements

JAM is supported by the American Heart Association (Grant ID: 12SDG9130039).

References

Ames BN, Shigenaga MK, Hagen TM (1993) Oxidants, antioxidants, and the degenerative diseases of aging. Proc Natl Acad Sci USA 90:7915–7922

Arnoult D, Rismanchi N, Grodet A, Roberts RG, Seeburg DP, Estaquier J, Sheng M, Blackstone C (2005) Bax/Bak-dependent release of DDP/TIMM8a promotes Drp1-mediated mitochondrial fission and mitoptosis during programmed cell death. Curr Biol 15(23):2112–2118

Attardi G (1985) Animal mitochondrial DNA: an extreme example of genetic economy. Int Rev Cytol 93:93–145

Attardi G, Schatz G (1988) Biogenesis of mitochondria. Annu Rev Cell Biol 4:289–333

Baar K, Wende AR, Jones TE, Marison M, Nolte LA, Chen M, Kelly DP, Holloszy JO (2002) Adaptations of skeletal muscle to exercise: rapid increase in the transcriptional coactivator PGC-1. FASEB J 16:1879–1886

Bach D, Pich S, Soriano FX, Vega N, Baumgartner B, Oriola J, Daugaard JR, Lloberas J, Camps M, Zierath JR, Rabasa-Lhoret R, Wallberg-Henriksson H, Laville M, Palacin M, Vidal H, Rivera F, Brand M, Zorzano A (2003) Mitofusin-2 determines mitochondrial network architecture and mitochondrial metabolism. A novel regulatory mechanism altered in obesity. J Biol Chem 278:17190–17197

Baker MJ, Tatsuta T, Langer T (2011) Quality control of mitochondrial proteostasis. Cold Spring Harb Perspect Biol 3:a007559. doi:10.1101/cshperspect.a007559

Ban T, Heymann JA, Song Z, Hinshaw JE, Chan DC (2010) OPA1 disease alleles causing dominant optic atrophy have defects in cardiolipin-stimulated GTP hydrolysis and membrane tubulation. Hum Mol Genet 19:2113–2122

Barsoum MJ, Yuan H, Gerencser AA, Liot G, Kushnareva Y, Graber S, Kovacs I, Lee WD, Waggoner J, Cui J, White AD, Bossy B, Martinou JC, Youle RJ, Lipton SA, Ellisman MH, Perkins GA, Bossy-Wetzel E (2006) Nitric oxide-induced mitochondrial fission is regulated by dynamin-related GTPases in neurons. EMBO J 25:3900–3911

Barth E, Stammler G, Speiser B, Schaper J (1992) Ultrastructural quantitation of mitochondria and myofilaments in cardiac muscle from 10 different animal species including man. J Mol Cell Cardiol 24:669–681

Becker T, Gebert M, Pfanner N, van der Laan M (2009) Biogenesis of mitochondrial membrane proteins. Curr Opin Cell Biol 21:484–493

Benard G, Karbowski M (2009) Mitochondrial fusion and division: regulation and role in cell viability. Semin Cell Dev Biol 20:365–374

Bergeron R, Ren JM, Cadman KS, Moore IK, Perret P, Pypaert M, Young LH, Semenkovich CF, Shulman GI (2001) Chronic activation of AMP kinase results in NRF-1 activation and mitochondrial biogenesis. Am J Physiol Endocrinol Metab 281:E1340–E1346

Bogenhagen D, Clayton DA (1977) Mouse L cell mitochondrial DNA molecules are selected randomly for replication throughout the cell cycle. Cell 11:719–727

Borst P, Fase-Fowler F, Hoeijmakers JH, Frasch AC (1980) Variations in maxi-circle and mini-circle sequences in kinetoplast DNAs from different Trypanosoma brucei strains. Biochim Biophys Acta 610:197–210

Borst P, Grivell LA (1971) Mitochondrial ribosomes. FEBS Lett 13:73–88

Borst P, Grivell LA (1978) The mitochondrial genome of yeast. Cell 15:705–723

Canto C, Gerhart-Hines Z, Feige JN, Lagouge M, Noriega L, Milne JC, Elliott PJ, Puigserver P, Auwerx J (2009) AMPK regulates energy expenditure by modulating NAD+ metabolism and SIRT1 activity. Nature 458:1056–1060

Cereghetti GM, Stangherlin A, Martins de Brito O, Chang CR, Blackstone C, Bernardi P, Scorrano L (2008) Dephosphorylation by calcineurin regulates translocation of Drp1 to mitochondria. Proc Natl Acad Sci USA 105:15803–15808

Chacinska A, Lind M, Frazier AE, Dudek J, Meisinger C, Geissler A, Sickmann A, Meyer HE, Truscott KN, Guiard B, Pfanner N, Rehling P (2005) Mitochondrial presequence translocase: switching between TOM tethering and motor recruitment involves Tim21 and Tim17. Cell 120:817–829

Chacinska A, Koehler CM, Milenkovic D, Lithgow T, Pfanner N (2009) Importing mitochondrial proteins: machineries and mechanisms. Cell 138:628–644

Chan DC (2006a) Mitochondrial fusion and fission in mammals. Annu Rev Cell Dev Biol 22:79–99

Chan DC (2006b) Mitochondria: dynamic organelles in disease, aging, and development. Cell 125:1241–1252

Chang CR, Blackstone C (2007) Cyclic AMP-dependent protein kinase phosphorylation of Drp1 regulates its GTPase activity and mitochondrial morphology. J Biol Chem 282(30):21583–21587

Chang CR, Manlandro CM, Arnoult D, Stadler J, Posey AE, Hill RB, Blackstone C (2010) A lethal de novo mutation in the middle domain of the dynamin-related GTPase Drp1 impairs higher order assembly and mitochondrial division. J Biol Chem 285:32494–32503

Chen H, Chan DC (2005) Emerging functions of mammalian mitochondrial fusion and fission. Hum Mol Genet 14(Spec No. 2):R283–R289

Chen H, Detmer SA, Ewald AJ, Griffin EE, Fraser SE, Chan DC (2003) Mitofusins Mfn1 and Mfn2 coordinately regulate mitochondrial fusion and are essential for embryonic development. J Cell Biol 160:189–200

Chen H, Chomyn A, Chan DC (2005) Disruption of fusion results in mitochondrial heterogeneity and dysfunction. J Biol Chem 280:26185–26192

Chen H, McCaffery JM, Chan DC (2007) Mitochondrial fusion protects against neurodegeneration in the cerebellum. Cell 130:548–562

Chin ER, Olson EN, Richardson JA, Yang Q, Humphries C, Shelton JM, Wu H, Zhu W, Bassel-Duby R, Williams RS (1998) A calcineurin-dependent transcriptional pathway controls skeletal muscle fiber type. Genes Dev 12:2499–2509

Cho DH, Nakamura T, Fang J, Cieplak P, Godzik A, Gu Z, Lipton SA (2009) S-nitrosylation of Drp1 mediates beta-amyloid-related mitochondrial fission and neuronal injury. Science 324:102–105

Chomyn A, Cleeter MW, Ragan CI, Riley M, Doolittle RF, Attardi G (1986) URF6, last unidentified reading frame of human mtDNA, codes for an NADH dehydrogenase subunit. Science 234:614–618

Cribbs JT, Strack S (2007) Reversible phosphorylation of Drp1 by cyclic AMP-dependent protein kinase and calcineurin regulates mitochondrial fission and cell death. EMBO Rep 8:939–944

Cuddeback SM, Yamaguchi H, Komatsu K, Miyashita T, Yamada M, Wu C, Singh S, Wang HG (2001) Molecular cloning and characterization of Bif-1. A novel Src homology 3 domain-containing protein that associates with Bax. J Biol Chem 276:20559–20565

Daum G, Vance JE (1997) Import of lipids into mitochondria. Prog Lipid Res 36(2–3):103–130

Davies KM, Strauss M, Daum B, Kief JH, Osiewacz HD, Rycovska A, Zickermann V, Kuhlbrandt W (2011) Macromolecular organization of ATP synthase and complex I in whole mitochondria. Proc Natl Acad Sci USA 108:14121–14126

Davies KM, Anselmi C, Wittig I, Faraldo-Gomez JD, Kuhlbrandt W (2012) Structure of the yeast F1Fo-ATP synthase dimer and its role in shaping the mitochondrial cristae. Proc Natl Acad Sci USA 109:13602–13607

de Brito OM, Scorrano L (2008) Mitofusin 2 tethers endoplasmic reticulum to mitochondria. Nature 456:605–610 de los Rios Castillo D, Zarco-Zavala M, Olvera-Sanchez S, Pardo JP, Juarez O, Martinez F, Mendoza-Hernandez G, Garcia-Trejo JJ, Flores-Herrera O (2011) Atypical cristae morphology of human syncytiotrophoblast mitochondria: role for complex V. J Biol Chem 286:23911–23919

Dekker PJ, Papadopoulou B, Grivell LA (1993) In-vitro translation of mitochondrial mRNAs by yeast mitochondrial ribosomes is hampered by the lack of start-codon recognition. Curr Genet 23:22–27

Duguez S, Feasson L, Denis C, Freyssenet D (2002) Mitochondrial biogenesis during skeletal muscle regeneration. Am J Physiol Endocrinol Metab 282:E802–E809

Eble KS, Coleman WB, Hantgan RR, Cunningham CC (1990) Tightly associated cardiolipin in the bovine heart mitochondrial ATP synthase as analyzed by 31P nuclear magnetic resonance spectroscopy. J Biol Chem 265:19434–19440

Eells JT, Henry MM, Gross GJ, Baker JE (2000) Increased mitochondrial K(ATP) channel activity during chronic myocardial hypoxia: is cardioprotection mediated by improved bioenergetics? Circ Res 87:915–921

Eilers M, Verner K, Hwang S, Schatz G (1988) Import of proteins into mitochondria. Philos Trans R Soc Lond B Biol Sci 319:121–126

Esaki M, Shimizu H, Ono T, Yamamoto H, Kanamori T, Nishikawa S, Endo T (2004) Mitochondrial protein import. Requirement of presequence elements and tom components for precursor binding to the TOM complex. J Biol Chem 279:45701–45707

Estaquier J, Arnoult D (2007) Inhibiting Drp1-mediated mitochondrial fission selectively prevents the release of cytochrome c during apoptosis. Cell Death Differ 14:1086–1094

Evans MJ, Scarpulla RC (1989) Interaction of nuclear factors with multiple sites in the somatic cytochrome c promoter. Characterization of upstream NRF-1, ATF, and intron Sp1 recognition sequences. J Biol Chem 264:14361–14368

Falkenberg M, Larsson NG, Gustafsson CM (2007) DNA replication and transcription in mammalian mitochondria. Annu Rev Biochem 76:679–699

Fisher RP, Lisowsky T, Parisi MA, Clayton DA (1992) DNA wrapping and bending by a mitochondrial high mobility group-like transcriptional activator protein. J Biol Chem 267:3358–3367

Frank S, Gaume B, Bergmann-Leitner ES, Leitner WW, Robert EG, Catez F, Smith CL, Youle RJ (2001) The role of dynamin-related protein 1, a mediator of mitochondrial fission, in apoptosis. Dev Cell 1:515–525

Gandre-Babbe S, van der Bliek AM (2008) The novel tail-anchored membrane protein Mff controls mitochondrial and peroxisomal fission in mammalian cells. Mol Biol Cell 19:2402–2412

Garesse R, Vallejo CG (2001) Animal mitochondrial biogenesis and function: a regulatory cross-talk between two genomes. Gene 263:1–16

Gegg ME, Cooper JM, Chau KY, Rojo M, Schapira AH, Taanman JW (2010) Mitofusin 1 and mitofusin 2 are ubiquitinated in a PINK1/parkin-dependent manner upon induction of mitophagy. Hum Mol Genet 19:4861–4870

Ghivizzani SC, Madsen CS, Nelen MR, Ammini CV, Hauswirth WW (1994) In organello footprint analysis of human mitochondrial DNA: human mitochondrial transcription factor A interactions at the origin of replication. Mol Cell Biol 14(12):7717–7730

Glotz C, Zwieb C, Brimacombe R, Edwards K, Kossel H (1981) Secondary structure of the large subunit ribosomal RNA from *Escherichia coli*, *Zea mays* chloroplast, and human and mouse mitochondrial ribosomes. Nucleic Acids Res 9:3287–3306

Goldstein JC, Waterhouse NJ, Juin P, Evan GI, Green DR (2000) The coordinate release of cytochrome c during apoptosis is rapid, complete and kinetically invariant. Nat Cell Biol 2(3):156–162

Gollnick PD, Piehl K, Saubert CW 4th, Armstrong RB, Saltin B (1972) Diet, exercise, and glycogen changes in human muscle fibers. J Appl Physiol 33:421–425

Gomez-Lazaro M, Bonekamp NA, Galindo MF, Jordan J, Schrader M (2008) 6-Hydroxydopamine (6-OHDA) induces Drp1-dependent mitochondrial fragmentation in SH-SY5Y cells. Free Radic Biol Med 44:1960–1969

Goto M, Terada S, Kato M, Katoh M, Yokozeki T, Tabata I, Shimokawa T (2000) cDNA cloning and mRNA analysis of PGC-1 in epitrochlearis muscle in swimming-exercised rats. Biochem Biophys Res Commun 274:350–354

Griffin EE, Graumann J, Chan DC (2005) The WD40 protein Caf4p is a component of the mitochondrial fission machinery and recruits Dnm1p to mitochondria. J Cell Biol 170:237–248

Gugneja S, Scarpulla RC (1997) Serine phosphorylation within a concise amino-terminal domain in nuclear respiratory factor 1 enhances DNA binding. J Biol Chem 272:18732–18739

Handschin C, Rhee J, Lin J, Tarr PT, Spiegelman BM (2003) An autoregulatory loop controls peroxisome proliferator-activated receptor gamma coactivator 1alpha expression in muscle. Proc Natl Acad Sci USA 100:7111–7116

Handschin C, Spiegelman BM (2006) Peroxisome proliferator-activated receptor gamma coactivator 1 coactivators, energy homeostasis, and metabolism. Endocr Rev 27:728–735

Hare JF (1990) Compartmentation and turnover of the low density lipoprotein receptor in skin fibroblasts. J Biol Chem 265:21758–21763

Helm M, Brule H, Friede D, Giege R, Putz D, Florentz C (2000) Search for characteristic structural features of mammalian mitochondrial tRNAs. RNA 6:1356–1379

Herzig RP, Scacco S, Scarpulla RC (2000) Sequential serum-dependent activation of CREB and NRF-1 leads to enhanced mitochondrial respiration through the induction of cytochrome c. J Biol Chem 275:13134–13141

Holloszy JO (1967) Biochemical adaptations in muscle. Effects of exercise on mitochondrial oxygen uptake and respiratory enzyme activity in skeletal muscle. J Biol Chem 242:2278–2282

Holloszy JO, Coyle EF (1984) Adaptations of skeletal muscle to endurance exercise and their metabolic consequences. J Appl Physiol 56:831–838

Huss JM, Kopp RP, Kelly DP (2002) Peroxisome proliferator-activated receptor coactivator-1alpha (PGC-1alpha) coactivates the cardiac-enriched nuclear receptors estrogen-related receptor-alpha and -gamma. Identification of novel leucine-rich interaction motif within PGC-1alpha. J Biol Chem 277:40265–40274

Imai H, Koumura T, Nakajima R, Nomura K, Nakagawa Y (2003) Protection from inactivation of the adenine nucleotide translocator during hypoglycaemia-induced apoptosis by mitochondrial phospholipid hydroperoxide glutathione peroxidase. Biochem J 371:799–809

Ingerman E, Perkins EM, Marino M, Mears JA, McCaffery JM, Hinshaw JE, Nunnari J (2005) Dnm1 forms spirals that are structurally tailored to fit mitochondria. J Cell Biol 170:1021–1027

Jager S, Handschin C, St-Pierre J, Spiegelman BM (2007) AMP-activated protein kinase (AMPK) action in skeletal muscle via direct phosphorylation of PGC-1alpha. Proc Natl Acad Sci USA 104:12017–12022

Jarosch E, Taxis C, Volkwein C, Bordallo J, Finley D, Wolf DH, Sommer T (2002) Protein dislocation from the ER requires polyubiquitination and the AAA-ATPase Cdc48. Nat Cell Biol 4:134–139

Kaguni LS (2004) DNA polymerase gamma, the mitochondrial replicase. Annu Rev Biochem 73:293–320

Knutti D, Kralli A (2001) PGC-1, a versatile coactivator. Trends Endocrinol Metab 12:360–365

Korhonen JA, Gaspari M, Falkenberg M (2003) TWINKLE Has 5′→3′ DNA helicase activity and is specifically stimulated by mitochondrial single-stranded DNA-binding protein. J Biol Chem 278:48627–48632

Kroemer G, Dallaporta B, Resche-Rigon M (1998) The mitochondrial death/life regulator in apoptosis and necrosis. Annu Rev Physiol 60:619–642

Kurland CG (1992) Evolution of mitochondrial genomes and the genetic code. Bioessays 14:709–714

Kutik S, Rissler M, Guan XL, Guiard B, Shui G, Gebert N, Heacock PN, Rehling P, Dowhan W, Wenk MR, Pfanner N, Wiedemann N (2008) The translocator maintenance protein Tam41 is required for mitochondrial cardiolipin biosynthesis. J Cell Biol 183:1213–1221

Lagouge M, Argmann C, Gerhart-Hines Z, Meziane H, Lerin C, Daussin F, Messadeq N, Milne J, Lambert P, Elliott P, Geny B, Laakso M, Puigserver P, Auwerx J (2006) Resveratrol improves mitochondrial function and protects against metabolic disease by activating SIRT1 and PGC-1alpha. Cell 127:1109–1122

Larsson NG, Wang J, Wilhelmsson H, Oldfors A, Rustin P, Lewandoski M, Barsh GS, Clayton DA (1998) Mitochondrial transcription factor A is necessary for mtDNA maintenance and embryogenesis in mice. Nat Genet 18:231–236

Lee HC, Yin PH, Lu CY, Chi CW, Wei YH (2000) Increase of mitochondria and mitochondrial DNA in response to oxidative stress in human cells. Biochem J 348:425–432

Lee S, Kim S, Sun X, Lee JH, Cho H (2007) Cell cycle-dependent mitochondrial biogenesis and dynamics in mammalian cells. Biochem Biophys Res Commun 357(1):111–117

Lehman JJ, Barger PM, Kovacs A, Saffitz JE, Medeiros D, Kelly DP (2000) Peroxisome proliferator-activated receptor gamma coactivator-1 promotes cardiac mitochondrial biogenesis. J Clin Invest 106:847–856

Lesnefsky EJ, Slabe TJ, Stoll MS, Minkler PE, Hoppel CL (2001) Myocardial ischemia selectively depletes cardiolipin in rabbit heart subsarcolemmal mitochondria. Am J Physiol Heart Circ Physiol 280:H2770–H2778

Levings CS 3rd, Brown GG (1989) Molecular biology of plant mitochondria. Cell 56:171–179

Liesa M, Borda-d'Agua B, Medina-Gomez G, Lelliott CJ, Paz JC, Rojo M, Palacin M, Vidal-Puig A, Zorzano A (2008) Mitochondrial fusion is increased by the nuclear coactivator PGC-1beta. PLoS One 3:e3613

Lim SE, Longley MJ, Copeland WC (1999) The mitochondrial p55 accessory subunit of human DNA polymerase gamma enhances DNA binding, promotes processive DNA synthesis, and confers N-ethylmaleimide resistance. J Biol Chem 274:38197–38203

Lin J, Wu H, Tarr PT, Zhang CY, Wu Z, Boss O, Michael LF, Puigserver P, Isotani E, Olson EN, Lowell BB, Bassel-Duby R, Spiegelman BM (2002) Transcriptional co-activator PGC-1 alpha drives the formation of slow-twitch muscle fibres. Nature 418:797–801

Longley MJ, Prasad R, Srivastava DK, Wilson SH, Copeland WC (1998a) Identification of 5′-deoxyribose phosphate lyase activity in human DNA polymerase gamma and its role in mitochondrial base excision repair in vitro. Proc Natl Acad Sci USA 95:12244–12248

Longley MJ, Ropp PA, Lim SE, Copeland WC (1998b) Characterization of the native and recombinant catalytic subunit of human DNA polymerase gamma: identification of residues critical for exonuclease activity and dideoxynucleotide sensitivity. Biochemistry 37:10529–10539

Lonsdale DM, Hodge TP, Fauron CM (1984) The physical map and organisation of the mitochondrial genome from the fertile cytoplasm of maize. Nucleic Acids Res 12:9249–9261

Lowell BB, Spiegelman BM (2000) Towards a molecular understanding of adaptive thermogenesis. Nature 404:652–660

Marcuello A, Gonzalez-Alonso J, Calbet JA, Damsgaard R, Lopez-Perez MJ, Diez-Sanchez C (2005) Skeletal muscle mitochondrial DNA content in exercising humans. J Appl Physiol 99:1372–1377

Mears JA, Sharma MR, Gutell RR, McCook AS, Richardson PE, Caulfield TR, Agrawal RK, Harvey SC (2006) A structural model for the large subunit of the mammalian mitochondrial ribosome. J Mol Biol 358:193–212

Mears JA, Lackner LL, Fang S, Ingerman E, Nunnari J, Hinshaw JE (2011) Conformational changes in Dnm1 support a contractile mechanism for mitochondrial fission. Nat Struct Mol Biol 18:20–26

Meerson FZ, Gomzakov OA, Shimkovich MV (1973) Adaptation to high altitude hypoxia as a factor preventing development of myocardial ischemic necrosis. Am J Cardiol 31:30–34

Meglei G, McQuibban GA (2009) The dynamin-related protein Mgm1p assembles into oligomers and hydrolyzes GTP to function in mitochondrial membrane fusion. Biochemistry 48:1774–1784

Merkwirth C, Langer T (2008) Mitofusin 2 builds a bridge between ER and mitochondria. Cell 135: 1165–1167

Mikhailov VS, Bogenhagen DF (1996) Effects of Xenopus laevis mitochondrial single-stranded DNA-binding protein on primer-template binding and 3′→5′ exonuclease activity of DNA polymerase gamma. J Biol Chem 271:18939–18946

Miranda S, Foncea R, Guerrero J, Leighton F (1999) Oxidative stress and upregulation of mitochondrial biogenesis genes in mitochondrial DNA-depleted HeLa cells. Biochem Biophys Res Commun 258:44–49

Montessuit S, Somasekharan SP, Terrones O, Lucken-Ardjomande S, Herzig S, Schwarzenbacher R, Manstein DJ, Bossy-Wetzel E, Basanez G, Meda P, Martinou JC (2010) Membrane remodeling induced by the dynamin-related protein Drp1 stimulates Bax oligomerization. Cell 142:889–901

Mootha VK, Bunkenborg J, Olsen JV, Hjerrild M, Wisniewski JR, Stahl E, Bolouri MS, Ray HN, Sihag S, Kamal M, Patterson N, Lander ES, Mann M (2003) Integrated analysis of protein composition, tissue diversity, and gene regulation in mouse mitochondria. Cell 115:629–640

Moraes CT (2001) What regulates mitochondrial DNA copy number in animal cells? Trends Genet 17:199–205

Moyes CD, Hood DA (2003) Origins and consequences of mitochondrial variation in vertebrate muscle. Annu Rev Physiol 65:177–201

Mozdy AD, McCaffery JM, Shaw JM (2000) Dnm1p GTPase-mediated mitochondrial fission is a multistep process requiring the novel integral membrane component Fis1p. J Cell Biol 151:367–380

Murakami T, Shimomura Y, Yoshimura A, Sokabe M, Fujitsuka N (1998) Induction of nuclear respiratory factor-1 expression by an acute bout of exercise in rat muscle. Biochim Biophys Acta 1381:113–122

Neupert W, Herrmann JM (2007) Translocation of proteins into mitochondria. Annu Rev Biochem 76:723–749

Neutzner A, Benard G, Youle RJ, Karbowski M (2008) Role of the ubiquitin conjugation system in the maintenance of mitochondrial homeostasis. Ann N Y Acad Sci 1147:242–253

Nisoli E, Clementi E, Paolucci C, Cozzi V, Tonello C, Sciorati C, Bracale R, Valerio A, Francolini M, Moncada S, Carruba MO (2003) Mitochondrial biogenesis in mammals: the role of endogenous nitric oxide. Science 299:896–899

O'Brien TW (2002) Evolution of a protein-rich mitochondrial ribosome: implications for human genetic disease. Gene 286:73–79

Ojuka EO, Jones TE, Han DH, Chen M, Holloszy JO (2003) Raising Ca2+ in L6 myotubes mimics effects of exercise on mitochondrial biogenesis in muscle. FASEB J 17:675–681

Osman C, Haag M, Potting C, Rodenfels J, Dip PV, Wieland FT, Brugger B, Westermann B, Langer T (2009a) The genetic interactome of prohibitins: coordinated control of cardiolipin and phosphatidylethanolamine by conserved regulators in mitochondria. J Cell Biol 184:583–596

Osman C, Merkwirth C, Langer T (2009b) Prohibitins and the functional compartmentalization of mitochondrial membranes. J Cell Sci 122:3823–3830

Pagliarini DJ, Calvo SE, Chang B, Sheth SA, Vafai SB, Ong SE, Walford GA, Sugiana C, Boneh A, Chen WK, Hill DE, Vidal M, Evans JG, Thorburn DR, Carr SA, Mootha VK (2008) A mitochondrial protein compendium elucidates complex I disease biology. Cell 134:112–123

Palmer CS, Osellame LD, Laine D, Koutsopoulos OS, Frazier AE, Ryan MT (2011) MiD49 and MiD51, new components of the mitochondrial fission machinery. EMBO Rep 12:565–573

Paradies G, Ruggiero FM (1990) Stimulation of phosphate transport in rat-liver mitochondria by thyroid hormones. Biochim Biophys Acta 1019:133–136

Paradies G, Ruggiero FM, Petrosillo G, Quagliariello E (1997) Age-dependent decline in the cytochrome c oxidase activity in rat heart mitochondria: role of cardiolipin. FEBS Lett 406:136–138

Parisi MA, Xu B, Clayton DA (1993) A human mitochondrial transcriptional activator can functionally replace a yeast mitochondrial HMG-box protein both in vivo and in vitro. Mol Cell Biol 13:1951–1961

Park YY, Lee S, Karbowski M, Neutzner A, Youle RJ, Cho H (2010) Loss of MARCH5 mitochondrial E3 ubiquitin ligase induces cellular senescence through dynamin-related protein 1 and mitofusin 1. J Cell Sci 123:619–626

Parone PA, James DI, Da Cruz S, Mattenberger Y, Donze O, Barja F, Martinou JC (2006) Inhibiting the mitochondrial fission machinery does not prevent Bax/Bak-dependent apoptosis. Mol Cell Biol 26(20):7397–7408

Pel HJ, Grivell LA (1993) The biology of yeast mitochondrial introns. Mol Biol Rep 18:1–13

Phaneuf S, Leeuwenburgh C (2001) Apoptosis and exercise. Med Sci Sports Exerc 33:393–396

Pierrat B, Simonen M, Cueto M, Mestan J, Ferrigno P, Heim J (2001) SH3GLB, a new endophilin-related protein family featuring an SH3 domain. Genomics 71:222–234

Pilegaard H, Saltin B, Neufer PD (2003) Exercise induces transient transcriptional activation of the PGC-1alpha gene in human skeletal muscle. J Physiol 546:851–858

Poyton RO, McEwen JE (1996) Crosstalk between nuclear and mitochondrial genomes. Annu Rev Biochem 65:563–607

Puigserver P, Spiegelman BM (2003) Peroxisome proliferator-activated receptor-gamma coactivator 1 alpha (PGC-1 alpha): transcriptional coactivator and metabolic regulator. Endocr Rev 24:78–90

Reinders J, Zahedi RP, Pfanner N, Meisinger C, Sickmann A (2006) Toward the complete yeast mitochondrial proteome: multidimensional separation techniques for mitochondrial proteomics. J Proteome Res 5:1543–1554

Renis M, Cantatore P, Loguercio Polosa P, Fracasso F, Gadaleta MN (1989) Content of mitochondrial DNA and of three mitochondrial RNAs in developing and adult rat cerebellum. J Neurochem 52:750–754

Reznick RM, Zong H, Li J, Morino K, Moore IK, Yu HJ, Liu ZX, Dong J, Mustard KJ, Hawley SA, Befroy D, Pypaert M, Hardie DG, Young LH, Shulman GI (2007) Aging-associated reductions in AMP-activated protein kinase activity and mitochondrial biogenesis. Cell Metab 5:151–156

Richter C, Park JW, Ames BN (1988) Normal oxidative damage to mitochondrial and nuclear DNA is extensive. Proc Natl Acad Sci USA 85:6465–6467

Ringel R, Sologub M, Morozov YI, Litonin D, Cramer P, Temiakov D (2011) Structure of human mitochondrial RNA polymerase. Nature 478:269–273

Rorbach J, Minczuk M (2012) The post-transcriptional life of mammalian mitochondrial RNA. Biochem J 444:357–373

Scarpulla RC (1997) Nuclear control of respiratory chain expression in mammalian cells. J Bioenerg Biomembr 29(2):109–119

Scarpulla RC (2002) Nuclear activators and coactivators in mammalian mitochondrial biogenesis. Biochim Biophys Acta 1576(1–2):1–14

Scarpulla RC (2006) Nuclear control of respiratory gene expression in mammalian cells. J Cell Biochem 97:673–683

Schreiber SN, Knutti D, Brogli K, Uhlmann T, Kralli A (2003) The transcriptional coactivator PGC-1 regulates the expression and activity of the orphan nuclear receptor estrogen-related receptor alpha (ERRalpha). J Biol Chem 278:9013–9018

Sesaki H, Southard SM, Yaffe MP, Jensen RE (2003) Mgm1p, a dynamin-related GTPase, is essential for fusion of the mitochondrial outer membrane. Mol Biol Cell 14:2342–2356

Shadel GS, Clayton DA (1997) Mitochondrial DNA maintenance in vertebrates. Annu Rev Biochem 66:409–435

Sharma MR, Koc EC, Datta PP, Booth TM, Spremulli LL, Agrawal RK (2003) Structure of the mammalian mitochondrial ribosome reveals an expanded functional role for its component proteins. Cell 115:97–108

Shimizu S, Narita M, Tsujimoto Y (1999) Bcl-2 family proteins regulate the release of apoptogenic cytochrome *c* by the mitochondrial channel VDAC. Nature 399:483–487

Sickmann A, Reinders J, Wagner Y, Joppich C, Zahedi R, Meyer HE, Schonfisch B, Perschil I, Chacinska A, Guiard B, Rehling P, Pfanner N, Meisinger C (2003) The proteome of *Saccharomyces cerevisiae* mitochondria. Proc Natl Acad Sci USA 100:13207–13212

Smidansky ED, Arnold JJ, Reynolds SL, Cameron CE (2011) Human mitochondrial RNA polymerase: evaluation of the single-nucleotide-addition cycle on synthetic RNA/DNA scaffolds. Biochemistry 50:5016–5032

Smirnova E, Shurland DL, Ryazantsev SN, van der Bliek AM (1998) A human dynamin-related protein controls the distribution of mitochondria. J Cell Biol 143:351–358

Smirnova E, Griparic L, Shurland DL, van der Bliek AM (2001) Dynamin-related protein Drp1 is required for mitochondrial division in mammalian cells. Mol Biol Cell 12(8):2245–2256

Smits P, Smeitink J, van den Heuvel L (2010) Mitochondrial translation and beyond: processes implicated in combined oxidative phosphorylation deficiencies. J Biomed Biotechnol 2010:737385

Souza-Pinto NC, Croteau DL, Hudson EK, Hansford RG, Bohr VA (1999) Age-associated increase in 8-oxo-deoxyguanosine glycosylase/AP lyase activity in rat mitochondria. Nucleic Acids Res 27:1935–1942

Spremulli LL, Coursey A, Navratil T, Hunter SE (2004) Initiation and elongation factors in mammalian mitochondrial protein biosynthesis. Prog Nucleic Acid Res Mol Biol 77:211–261

St-Pierre J, Lin J, Krauss S, Tarr PT, Yang R, Newgard CB, Spiegelman BM (2003) Bioenergetic analysis of peroxisome proliferator-activated receptor gamma coactivators 1alpha and 1beta (PGC-1alpha and PGC-1beta) in muscle cells. J Biol Chem 278:26597–26603

Stojanovski D, Guiard B, Kozjak-Pavlovic V, Pfanner N, Meisinger C (2007) Alternative function for the mitochondrial SAM complex in biogenesis of alpha-helical TOM proteins. J Cell Biol 179:881–893

Suliman HB, Carraway MS, Piantadosi CA (2003) Postlipopolysaccharide oxidative damage of mitochondrial DNA. Am J Respir Crit Care Med 167:570–579

Suliman HB, Carraway MS, Tatro LG, Piantadosi CA (2007) A new activating role for CO in cardiac mitochondrial biogenesis. J Cell Sci 120:299–308

Suzuki H, Kumagai T, Goto A, Sugiura T (1998) Increase in intracellular hydrogen peroxide and upregulation of a nuclear respiratory gene evoked by impairment of mitochondrial electron transfer in human cells. Biochem Biophys Res Commun 249:542–545

Takahashi M, Hood DA (1993) Chronic stimulation-induced changes in mitochondria and performance in rat skeletal muscle. J Appl Physiol 74(2): 934–941

Takeshige K, Baba M, Tsuboi S, Noda T, Ohsumi Y (1992) Autophagy in yeast demonstrated with proteinase-deficient mutants and conditions for its induction. J Cell Biol 119:301–311

Tanaka A, Cleland MM, Xu S, Narendra DP, Suen DF, Karbowski M, Youle RJ (2010) Proteasome and p97 mediate mitophagy and degradation of mitofusins induced by Parkin. J Cell Biol 191:1367–1380

Terada S, Goto M, Kato M, Kawanaka K, Shimokawa T, Tabata I (2002) Effects of low-intensity prolonged exercise on PGC-1 mRNA expression in rat epitrochlearis muscle. Biochem Biophys Res Commun 296:350–354

Tieu Q, Nunnari J (2000) Mdv1p is a WD repeat protein that interacts with the dynamin-related GTPase, Dnm1p, to trigger mitochondrial division. J Cell Biol 151:353–366

Twig G, Elorza A, Molina AJ, Mohamed H, Wikstrom JD, Walzer G, Stiles L, Haigh SE, Katz S, Las G, Alroy J, Wu M, Py BF, Yuan J, Deeney JT, Corkey BE, Shirihai OS (2008) Fission and selective fusion govern mitochondrial segregation and elimination by autophagy. EMBO J 27:433–446

Tyynismaa H, Sembongi H, Bokori-Brown M, Granycome C, Ashley N, Poulton J, Jalanko A, Spelbrink JN, Holt IJ, Suomalainen A (2004) Twinkle helicase is essential for mtDNA maintenance and regulates mtDNA copy number. Hum Mol Genet 13:3219–3227

Tzagoloff A, Macino G, Sebald W (1979) Mitochondrial genes and translation products. Annu Rev Biochem 48:419–441

Vega RB, Huss JM, Kelly DP (2000) The coactivator PGC-1 cooperates with peroxisome proliferator-activated receptor alpha in transcriptional control of nuclear genes encoding mitochondrial fatty acid oxidation enzymes. Mol Cell Biol 20:1868–1876

Veltri KL, Espiritu M, Singh G (1990) Distinct genomic copy number in mitochondria of different mammalian organs. J Cell Physiol 143:160–164

Virbasius JV, Virbasius CA, Scarpulla RC (1993) Identity of GABP with NRF-2, a multisubunit activator of cytochrome oxidase expression, reveals a cellular role for an ETS domain activator of viral promoters. Genes Dev 7:380–392

Wasiak S, Zunino R, McBride HM (2007) Bax/Bak promote sumoylation of DRP1 and its stable association with mitochondria during apoptotic cell death. J Cell Biol 177:439–450

Wiedemann N, Pfanner N, Ryan MT (2001) The three modules of ADP/ATP carrier cooperate in receptor recruitment and translocation into mitochondria. EMBO J 20:951–960

Williams RS, Rosenberg P (2002) Calcium-dependent gene regulation in myocyte hypertrophy and remodeling. Cold Spring Harb Symp Quant Biol 67:339–344

Wu H, Kanatous SB, Thurmond FA, Gallardo T, Isotani E, Bassel-Duby R, Williams RS (2002) Regulation of mitochondrial biogenesis in skeletal muscle by CaMK. Science 296:349–352

Wu Z, Puigserver P, Andersson U, Zhang C, Adelmant G, Mootha V, Troy A, Cinti S, Lowell B, Scarpulla RC, Spiegelman BM (1999) Mechanisms controlling mitochondrial biogenesis and respiration through the thermogenic coactivator PGC-1. Cell 98:115–124

Yoon Y, Pitts KR, McNiven MA (2001) Mammalian dynamin-like protein DLP1 tubulates membranes. Mol Biol Cell 12:2894–2905

Yoon Y, Krueger EW, Oswald BJ, McNiven MA (2003) The mitochondrial protein hFis1 regulates mitochondrial fission in mammalian cells through an interaction with the dynamin-like protein DLP1. Mol Cell Biol 23:5409–5420

Youle RJ, Narendra DP (2011) Mechanisms of mitophagy. Nat Rev Mol Cell Biol 12:9–14

Young JC, Hoogenraad NJ, Hartl FU (2003) Molecular chaperones Hsp90 and Hsp70 deliver preproteins to the mitochondrial import receptor Tom70. Cell 112:41–50

Zagryadskaya EI, Kotlova N, Steinberg SV (2004) Key elements in maintenance of the tRNA L-shape. J Mol Biol 340:435–444

Zambrano F, Fleischer S, Fleischer B (1975) Lipid composition of the Golgi apparatus of rat kidney and liver in comparison with other subcellular organelles. Biochim Biophys Acta 380:357–369

Zhang Y, Chan DC (2007) Structural basis for recruitment of mitochondrial fission complexes by Fis1. Proc Natl Acad Sci USA 104:18526–18530

Zhao J, Liu T, Jin S, Wang X, Qu M, Uhlen P, Tomilin N, Shupliakov O, Lendahl U, Nister M (2011) Human MIEF1 recruits Drp1 to mitochondrial outer membranes and promotes mitochondrial fusion rather than fission. EMBO J 30:2762–2778

Zong H, Ren JM, Young LH, Pypaert M, Mu J, Birnbaum MJ, Shulman GI (2002) AMP kinase is required for mitochondrial biogenesis in skeletal muscle in response to chronic energy deprivation. Proc Natl Acad Sci USA 99:15983–15987

Subject Index

M.F. Hohmann-Marriott (ed.), The Structural Basis of Biological Energy Generation, Advances in Photosynthesis and Respiration 39, pp. 477–483, DOI 10.1007/978-94-017-8742-0,

MIX
Papier aus verantwortungsvollen Quellen
Paper from responsible sources
FSC® C105338

If you have any concerns about our products,
you can contact us on
ProductSafety@springernature.com

In case Publisher is established outside the EU,
the EU authorized representative is:
Springer Nature Customer Service Center GmbH
Europaplatz 3, 69115 Heidelberg, Germany

Printed by Libri Plureos GmbH
in Hamburg, Germany